ANNUAL REVIEW OF
PLANT PHYSIOLOGY
AND PLANT
MOLECULAR BIOLOGY

ANNUAL REVIEW OF PLANT PHYSIOLOGY AND PLANT MOLECULAR BIOLOGY

VOLUME 48, 1997

RUSSELL L. JONES, *Editor*
University of California, Berkeley

CHRISTOPHER R. SOMERVILLE, *Associate Editor*
Carnegie Institution of Washington, Stanford, California

VIRGINIA WALBOT, *Associate Editor*
Stanford University

http://annurev.org   science@annurev.org   415-493-4400
ANNUAL REVIEWS INC.   4139 EL CAMINO WAY   P.O. BOX 10139   PALO ALTO, CALIFORNIA 94303-0139

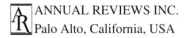 ANNUAL REVIEWS INC.
Palo Alto, California, USA

International Standard Serial Number: 1040-2519
International Standard Book Number: 0-8243-0648-1
Library of Congress Catalog Card Number: A-51-1660

Typesetting by Ruth McCue Saavedra and the Annual Reviews Inc. Editorial Staff

PRINTED AND BOUND IN THE UNITED STATES OF AMERICA

ANNUAL REVIEWS INC. is a nonprofit scientific publisher established to promote the advancement of the sciences. Beginning in 1932 with the *Annual Review of Biochemistry*, the Company has pursued as its principal function the publication of high-quality, reasonably priced *Annual Review* volumes. The volumes are organized by Editors and Editorial Committees who invite qualified authors to contribute critical articles reviewing significant developments within each major discipline. The Editor-in-Chief invites those interested in serving as future Editorial Committee members to communicate directly with him. Annual Reviews Inc. is administered by a Board of Directors, whose members serve without compensation.

For the convenience of readers, a detachable order form/envelope is bound into the back of this volume.

Annual Review of Plant Physiology and Plant Molecular Biology
Volume 48 (1997)

CONTENTS

SOME RELATED ARTICLES IN OTHER *ANNUAL REVIEWS*

From the *Annual Review of Biochemistry*, Volume 66 (1997)

From Chemistry to Biochemistry to Catalysis to Movement, WP Jencks

Mechanistic Aspects of Enzymatic Catalysis: Lessons from Comparison of RNA and Protein Enzymes, GJ Narlikar and D Herschlag

Replication Protein A: A Heterotrimeric, Single-Stranded DNA-Binding Protein Required for Eukaryotic DNA Metabolism, MS Wold

Basic Mechanisms of Transcript Elongation and Its Regulation, SM Uptain, CM Kane, and MJ Chamberlin

Molecular Basis for Membrane Phospholipid Diversity: Why Are There So Many Lipids?, W Dowhan

Molybdenum-Cofactor–Containing Enzymes: Structure and Mechanism, C Kisker, H Schindelin, and DC Rees

Dynamic O-Linked Glycosylation of Nuclear and Cytoskeletal Proteins, GW Hart

Regulation of Eukaryotic Phosphatidylinositol-Specific Phospholipase C and Phospholipase D, WD Singer, HA Brown, and PC Sternweis

Clathrin-Coated Vesicle Formation and Protein Sorting: An Integrated Process, SL Schmid

Protein Folding: The Endgame, M Levitt, M Gerstein, E Huang, S Subbiah, and J Tsai

Regulation of Phosphoenolpyruvate Carboxykinase (GTP) Gene Expression, RW Hanson and L Reshef

G-Protein Mechanisms: Insights from Structural Analysis, SR Sprang

Ribosomes and Translation, R Green and HF Noller

The ATP Synthase—A Splendid Molecular Machine, PD Boyer

Subtractive Cloning: Past, Present, and Future, CG Sagerström, BI Sun, and HL Sive

Force Effects on Biochemical Kinetics, SM Khan and MP Sheetz

Transcriptional Regulation by Cyclic AMP, M Montminy

Protein Import into Mitochondria, W Neupert

From the *Annual Review of Biophysics and Biomolecular Structure*, Volume 25 (1996)

Bridging the Protein Sequence-Structure Gap by Structure Predictions, B Rost and C Sander

The Sugar Kinase/Heat-Shock Protein 70/Actin Superfamily: Implications of Conserved Structure for Mechanism, JH Hurley

Lipoxygenases: Structural Principles and Spectroscopy, BJ Gaffney

The Dynamics of Water-Protein Interactions, RG Bryant

Antibodies as Tools to Study the Structure of Membrane Proteins: The Case of the Nicotinic Acetylcholine Receptor, BM Conti-Fine, S Lei, and KE McLane

Engineering the Gramicidin Channel, RE Koeppe II and OS Andersen
Computational Studies of Protein Folding, RA Friesner and JR Gunn
Protein Function in the Crystal, A Mozzarelli and GL Rossi
Visualizing Protein—Nucleic Acid Interactions on a Large Scale With the
  Scanning Force Microscope, C Bustamante and C Rivetti
Use of ^{19}F NMR to Probe Protein Structure and Conformational Changes,
  MA Danielson and JJ Falke
Circular Oligonucleotides: New Concepts in Oligonucleotide Design, ET Kool
Modeling DNA in Aqueous Solutions: Theoretical and Computer Simulation
  Studies on the Ion Atmosphere of DNA, B Jayaram and DL Beveridge

From the *Annual Review of Cell and Developmental Biology*,
Volume 12 (1996)

Import and Routing of Nucleus-Encoded Chloroplast Proteins, K Cline and
  R Henry
Signal-Mediated Sorting of Membrane Proteins Between the Endoplasmic
  Reticulum and the Golgi Apparatus, RD Teasdale and MR Jackson
Actin: General Principles From Studies In Yeast, KR Ayscough and DG Drubin
Cross-Talk Between Bacterial Pathogens and Their Host Cells, JE Galán and
  JB Bliska
Acquisition of Identity in the Developing Leaf, AW Sylvester, L Smith, and
  M Freeling
Mitotic Chromosome Condensation, D Koshland and A Strunnikov
Peroxisome Proliferator-Activated Receptors: A Nuclear Receptor Signaling
  Pathway In Lipid Physiology, T Lemberger, B Desvergne, and W Wahli
Transport Vesicle Docking: Snares and Associates, SR Pfeffer
Signaling by Extracellular Nucleotides, AJ Brake and D Julius
Structure-Function Analysis of the Motor Domain of Myosin, KM Ruppel and
  JA Spudich
Endocytosis and Molecular Sorting, Ira Mellman

From the *Annual Review of Phytopathology*, Volume 35 (1997)

Concepts of Active Defense Mechanisms, S Hutcheson
The Impact of Ti Plasmid-Derived Gene Vectors on the Study of the Mechanism of
  Action of Phytohormones, J St. Schell, R Walden, and C Koncz
Plant Viruses and Vaccines, JE Johnson
hrp Genes and other Avr Genes, P Lindgren
Molecular Strategies for Developing Nematode-Resistant Plants, M Conkling and
  C Opperman
Appressorium Morphogenesis, RA Dean
Systemic Acquired Resistance, J-P Metraux
Engineering Disease Resistance, LS Melchers and PJM van den Elzen
Use of Beneficial Bacteria in Plant-Disease Control, L Thomashow

From the *Annual Review of Genetics*, Volume 30 (1996)

N. E. Tolbert

Annu. Rev. Plant Physiol. Plant Mol. Biol. 1997. 48:1–25

THE C_2 OXIDATIVE PHOTOSYNTHETIC CARBON CYCLE

N. E. Tolbert

Department of Biochemistry, Michigan State University, East Lansing, Michigan 48824–1319

KEY WORDS: C_2 and C_3 photosynthetic carbon cycles, C_4 plants, O_2 and CO_2 compensation point, photorespiration, ratio atmospheric CO_2 and O_2, Rubisco

ABSTRACT

The C_2 oxidative photosynthetic carbon cycle plus the C_3 reductive photosynthetic carbon cycle coexist. Both are initiated by Rubisco, use about equal amounts of energy, must regenerate RuBP, and result in exchanges of CO_2 and O_2 to establish rates of net photosynthesis, CO_2 and O_2 compensation points, and the ratio of CO_2 and O_2 in the atmosphere. These concepts evolved from research on O_2 inhibition, glycolate metabolism, leaf peroxisomes, photorespiration, $^{18}O_2/^{16}O_2$ exchange, CO_2 concentrating processes, and a requirement for the oxygenase activity of Rubisco. Nearly 80 years of research on these topics are unified under the one process of photosynthetic carbon metabolism and its self-regulation.

CONTENTS

1

1040-2519/97/0601-0001$08.00

Introduction

I chose the subject of the C_2 cycle for my prefatory chapter because this part of photosynthetic carbon metabolism needs to be better understood. It represents a part of my research that occurred in bits and pieces over many years. This synthesis is hindsight. For a long time I did not foresee conclusions to be drawn from research on glycolate metabolism. Now the term "C_2 oxidative photosynthetic carbon cycle" parallels the nomenclature for the "C_3 reductive photosynthetic carbon cycle" (Figure 1). The term "photorespiratory carbon cycle" should not be used for the C_2 cycle, because that implies photorespiration is a separate process with separate CO_2 and O_2 pools. Rather, there is one process of photosynthetic carbon metabolism that is the sum of the C_2 plus C_3 cycles. Less informative nomenclature based on names of investigators, such as Calvin/Benson or Hatch/Slack cycles, is slowly disappearing. Topics related to the C_2 cycle started at about the dates shown in Table 1, which also lists some events of my life. Metabolic pathways and enzymatic properties to support the C_2 cycle are in the references. Our recent report of an O_2 compensation point (Γ) during photosynthesis (10) has extended our viewpoint, but because it has not yet been extensively debated, those parts of this chapter addressing this are more speculative. A more extensive review of the O_2 Γ may be found in a manuscript in preparation with Erwin Beck. For 50 years, investigators have studied glycolate and P-glycolate synthesis and its function, regulation, and inhibition. I am extending this with speculation on regulation of global CO_2 and O_2 concentrations. It will not be easy to overcome 150 years of dogma that photosynthesis is only O_2 evolution and CO_2 fixation (it is O_2 and CO_2 exchange), and the physiological impact that Rubisco is also an oxygenase. Most difficult of all is to understand that at the recent low atmospheric CO_2 and high O_2 about half of photosynthetic energy has been used by the C_2 cycle.

Photosynthesis has been divided into two parts: light reactions for ATP synthesis plus NADP reduction with O_2 evolution and photosynthetic carbon metabolism to use the ATP and NADPH. There are two parts to carbon metabolism as initiated by the dual activities of ribulose bisphosphate carboxylase/oxygenase (Rubisco). It catalyzes carboxylation of ribulose bisphosphate (RuBP) with $^{14}CO_2$ to form two molecules of carboxyl labeled 3-P-glycerate of the C_3 reductive photosynthetic carbon cycle. Rubisco is also an oxygenase that catalyzes oxidation of RuBP by addition (fixation) of an $^{18}O_2$ to RuBP to form one carboxyl [$^{18}O_2$]P-glycolate and one 3-P-glycerate of the C_2 oxidative photosynthetic carbon cycle. The C_2 and C_3 cycles coexist and together are photosynthetic carbon metabolism (Figure 1). The ratio of CO_2 and O_2 concentrations in the chloroplast determines the flow of carbon between the two cycles.

At low global CO_2 equilibrium level of <0.03% CO_2 and ~21% O_2, CO_2 fixation by the C$_3$ cycle in C$_3$ plants is about three times more than CO_2 production by the C$_2$ cycle. Thus we have net CO_2 fixation and O_2 evolution. However, the C$_2$ cycle uses about three times more energy per CO_2 turnover. Thus, in air with high light intensity nearly equal amounts of photosynthetic energy are used by the C$_3$ cycle for net photosynthesis as are consumed by the C$_2$ cycle (2, 9). A lower limit of CO_2 and an upper limit of O_2 are compensation points (Γ), beyond which net photosynthesis is zero and plants senesce from excess photooxidation of storage carbohydrates by oxygenase activity of Rubisco and the C$_2$ cycle. The O_2 Γ with 220 ppm CO_2 at 20°C is about 23%

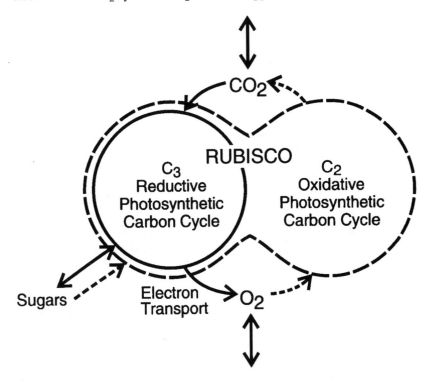

Figure 1 The C$_2$ and C$_3$ cycles of photosynthetic carbon metabolism. This scheme (9, 10) emphasizes that the two cycles coexist, that both represent about equal parts of the process, that carboxylase and oxygenase activities of Rubisco initiate both, and that there is just one CO_2 and one O_2 pool. Although the C$_2$ cycle has been called photorespiration, it is not a separate process. Rather, the two cycles create a necessary balance during photosynthesis for net CO_2 and O_2 exchange.

Table 1 Glycolate oxidase, glycolate synthesis, glycolate pathway, and other related events of my life

1920	O Warburg published that O_2 inhibited photosynthesis. That year I was one year old in Twin Falls, Idaho
1930–1950	Photosynthetic investigators considered O_2 inhibition of electron transport
1940–1941	Ruben's search for the first product of $^{11}CO_2$ fixation. My BS, Department of Chemistry, University of California, Berkeley
1941–1943	University of California, Davis, Department of Viticulture and Enology, a diversion for me to use the carbon products of photosynthesis
1943–1946	World War II, Captain, US Air Force Intelligence
1946–1949	Glycolate oxidase, PhD thesis with RH Burris, University Wisconsin
1950	In Calvin's group with Benson and Bassham. $^{14}CO_2$ fixation by the C_3 cycle and $[^{14}C]$ glycolate, glycine, and serine found to be major early products
1950–1952	US Atomic Energy Commission grants for plant science. Glycolic oxidase light activation at USDA, Beltsville
1952–1958	Glycolate metabolism and glycolate excretion by algae. Biology Division at Oak Ridge National Laboratory, Organizer of Gatlinburg Photosynthesis Conferences
1958–1962	Professor of Biochemistry Michigan State University. Glycolate pathway; P-glycolate phosphatase. Concept of photorespiration developed by many investigators
1960–1968	The CO_2 compensation point. C_4 cycle and C_4 plants. Construction of biochemistry building with RG Hansen and WA Wood
1968–1975	P-glycolate synthesis by oxygenase activity of Rubisco. Glycolate metabolism in peroxisomes from leaves or liver
1975	The C_2 oxidative photosynthetic carbon cycle. Algal CO_2 concentrating processes. Effect of increasing atmospheric CO_2 on plants
1989	Professor Emeritus—too soon
1996	The oxygen compensation point. Glycolate metabolism in chloroplasts. Regulation of the ratio of atmospheric CO_2 and O_2 by Rubisco

O_2 for C_3 plants (10). The ratio of CO_2 and O_2 concentration in the chloroplast limits net photosynthesis, as well as sets the CO_2 and O_2 atmospheric ratio on a short-term basis. Long-term, i.e. centuries, geological global carbon and O_2 cycles provide sources of CO_2 and sinks for O_2, a process that slowly establishes atmospheric equilibrium levels. Besides the ratio of CO_2 and O_2 concentrations, many other plant growth factors also determine rate of net photosynthesis. These are not integrated into this summary, but among them are temperature, light intensity, water, nutrient, stomata openings, canopy coverage, processes for concentrating CO_2 and exclusion of O_2, membrane diffusion, and geological storage reservoirs of CO_2 and O_2. Changes in any of these alter total photosynthesis, but the short-term CO$_2$ and O$_2$ equilibria are based on kinetic properties of the immense amount of Rubisco.

From the time of the discoveries of O$_2$ inhibition of CO$_2$ fixation, glycolate oxidase, photorespiration, oxygenase activity of Rubisco, and leaf peroxisomes (Table 1), a continuing question has been their purpose. These phenomena occur at 21% O$_2$ and increase at higher O$_2$ levels, as if they were the result of O$_2$ toxicity. Many of us sought ways to reduce this O$_2$ inhibition to increase crop yield. So far there has been no evidence that research targeted at modifying Rubisco has been agronomically successful. This may be because O$_2$ inhibition of CO$_2$ fixation is now an essential, environmental part of photosynthesis. In addition, incomplete data have limited our understanding, and even now we are just publishing about the O$_2$ Γ (10) and a second glycolate oxidizing system in chloroplasts (1). Earlier, G Lorimer & J Andrews pointed out that oxygenase activity of Rubisco was probably unavoidable because there are no binding sites for CO$_2$ or O$_2$ substrates, and I am adding that this dual specificity is an essential part of photosynthetic carbon metabolism. The O$_2$ Γ is a consequence of oxygenase activity of Rubisco and is an upper limit on the permissible atmospheric O$_2$ level for plant growth, just as the CO$_2$ Γ limits lowering the atmospheric CO$_2$.

My Start in Low CO$_2$

The story about O$_2$ regulation of photosynthesis began in 1920, a year after I was born, when Otto Warburg published that air (21% O$_2$) inhibited CO$_2$ fixation by algae. I grew up on a farm at Twin Falls, Idaho, where recreation was fishing and hunting in canyons along the Snake River. From then till well past my graduate school, photosynthesizers discussed how O$_2$ might oxidize chloroplast electron transport components. Rabinowitch wrote that if O$_2$ inhibition effected respiration, all their hypotheses were off. It did, because O$_2$ inhibition of CO$_2$ fixation is the C$_2$ cycle, which is glycolate synthesis and metabolism.

The small rural grade school that I attended, at first by a horse-drawn covered wagon as a school bus, taught only the three basics and baseball, but my mother supplemented the basics with nightly drills around kerosene lamps. I read Zane Grey, but I never learned to spell well or master languages, even later after suffering through Latin, French, and German. I must have been a disappointment to my mother, who had a Master's degree in English from the University of Chicago. I am not an extrovert, but I became a daydreamer from sitting all day on a farm cultivator. I do best when talking about science.

On the farm, my father trained me to do irrigation and stack hay. It was somehow assumed that we would go to college, and I, being the oldest, led all four of us to major in science. In the first two years at the University of Southern Idaho in Pocatello, I remember best Sunday hikes up the canyons into the mountains. After two years, my father suggested that if I returned home, he would buy a sheep ranch for us near Ketchum or Sun Valley. That temptation has often been remembered, but I went instead to the University of California, Berkeley, where I was a B+ student with that inferiority complex. I deviated once from chemistry, when I took an intensive course in biochemistry for premed majors and pulled my only C. Thus, I was challenged. At this time, the first radioactive, $^{11}CO_2$ photosynthetic experiments were done by Ruben (until he was killed by phosgene gas) while I was in a nearby lab. Melvin Calvin, another assistant professor in the Chemistry Department, replaced him. At graduation in 1941, I was told to find employment at Shell Oil Research down by the Bay, near Berkeley. Upon going there, the aromas from their labs and mud flats made me long for a sheep ranch in Idaho. Returning from that interview, I stopped at the student employment office on campus and learned that the Head of Viticulture and Enology at UC Davis was there looking for an organic chemist. Within half hour I had a job, conceptually nearer the farm, to analyze brandy for $150 a month. My first publications were in the *Journal of Food Research,* on pH and tannins. Maynard Amerine, Professor of Enology and American's foremost wine connoisseur, invited AD Webb and me to live with him in Davis while he taught us wine tasting. Until then I had been a teetotaler, but suddenly I was a wine and brandy expert. My parents may have been in shock, but they said nothing.

At that time (about 1940), the atmospheric CO_2 level was around 300 ppm. It had been first measured in 1890 at about 290 ppm and concern was expressed even then that it might be rising. In graduate school in the late 1940s we used 303 ppm as the CO_2 level. In 1957, when the CO_2 level was 315 ppm, NSF began funding measurements of the CO_2 concentrations atop Mauna Loa volcano in Hawaii. Atmospheric CO_2 is now about 380 ppm. I use CO_2 levels in Arctic ice cores to indicate an atmospheric CO_2 equilibrium for the past 165,000 years that oscillated around 235 ± 45 ppm with 21% O_2. This seems to have been an equilibrium whose highs of 280 ppm CO_2 corresponded with interglacial periods and whose lows around 190 ppm occurred during glacial periods. I hope this chapter will promote discussion about whether competition between atmospheric CO_2 and O_2 levels for Rubisco catalysts during photosynthesis limited CO_2 level on the downside and maintained global O_2 level at 21%. Now we are in the midst of a mammoth atmospheric experiment in which the CO_2 concentration has been raised about 50% to 380 ppm. At this current CO_2 level, the permissible O_2 level from an equilibrium based on the specificity of Rubisco could rise over a long period of time to around 28% O_2 (10).

In the Chemistry Department at Berkeley my senior thesis for M Randall, coauthor of the textbook on physical chemistry, listed the literature citations in Beilstein and Chem Abstracts on light-activation of silver halides. In 1942, as my draft number rose toward the top, I applied to the Air Force for officer training as a photographic expert on the basis of that compilation. After boot camp in Florida and classes at Yale University, assignments were New England photo lab at Bradley Field, Connecticut; photo intelligence school at Orlando, Florida; briefing B-24 crews at Muroc Air Force Base in Southern California; air force intelligence in the Philippines (purple heart); and Japan. This captain had traveled a long way from that sheep ranch in Idaho. Following the war, I settled down as a graduate student in biochemistry at the University of Wisconsin in Madison with RH Burris as PhD advisor, mentor, and long-time friend. He started me on polyphenol oxidase, which seemed to be responsible for browning of potato tubers. O_2 and CO_2 gas exchange measurements were run with a Warburg apparatus with 18 manometers. Today, Warburg analyses have fallen into complete oblivion, because they are deemed too complicated. Instead, O_2 changes are measured one at a time in an O_2 electrode, and CO_2 concentrations are assayed by an IR analyzer. For uptake of $^{14}CO_2$ and metabolism of ^{14}C-labeled substrates, Warburg flasks and Thunberg tubes have some distinct advantages. After I started graduate school, Burris, one of the authors of *Manometric Techniques*, invented the circular Warburg, and I was his first graduate student to use it. It was a real advancement to sit in one place while rotating and reading the rack of oscillating manometers.

I have been helped greatly by my brother, Bert, and sister, Marian. Bert and I shared the same room at home and at college and took the same courses. He became an expert on synthesis and use of ^{14}C-labeled compounds in biological research. He made 1- and 2-[^{14}C]glycolates, which I used to determine the glycolate pathway. M Calvin felt that glycolate was labeled from $^{14}CO_2$ via the sugar phosphates of the C$_3$ cycle (he was right), and so glycolate was relegated to the discard heap. Thanks! This left me the other half of photosynthetic carbon, the C$_2$ cycle.

The C2 Cycle

In the 1950s, glycolate and photorespiration topics were not mainstream, and we had to struggle to be heard. At the Brookhaven Photosynthesis Symposium of 1963, Bessel Kok put his arm around me after my talk outlining reactions of the glycolate pathway (4) and said, "Ed, why keep working on that glycolate? You know it is not an early product during photosynthesis. There are many other more important things to do." Fortunately, I did not follow his advice.

My traditional biochemical research approach has been first to find and characterize enzymes and second to place them in metabolic pathways by studies with isotopically labeled substrates. A third approach, starting in the 1960s, was to use mutants to confirm or predict biochemical pathways. C_2 cycle mutants in Arabidopsis were developed by CR Somerville. It was gratifying that reactions of the C_2 pathway, as determined earlier, were confirmed with several mutants. We can now speculate that failure to isolate a mutant of peroxisomal glycolate oxidase is consistent with alternate glycolate metabolism in the chloroplast (1). The techniques of molecular biology became available as we worked on the algal CO_2 pump. For a complete research program, all four approaches should be used.

In 1948, Carl Clagett finished his PhD thesis with RH Burris on glycolate oxidase, and I continued my PhD research on this new oxidase, function unknown. Glycolate (CH_2OH-COOH) is oxidized with O_2 uptake to glyoxylate (CHO-COOH) and H_2O_2. Glyoxylate was trapped as a hydrazone or oxidized to CO_2 and HCOOH depending on the amount of catalase remaining to remove H_2O_2. From the start we thought that a glycolate/glyoxylate system might be a terminal oxidase to waste energy, which was then a popular concept. For that, I Zelitch and S Ochoa in 1957 characterized a glyoxylate reductase, which turned out to be the same as a glycerate dehydrogenase isolated in 1956 by B Vennesland. From our studies in the 1960s and 1970s this enzyme became known as hydroxypyruvate reductase, because it had a low K_m of 10 μM for hydroxypyruvate (CH_2OH-CO-COOH), but a high K_m of about 500 μM for glyoxylate, which was rapidly transaminated to glycine instead. However, possible peroxidation of glyoxylate and hydroxypyruvate by H_2O_2 has been considered numerous times by others. Conceptually, the C_2 cycle is a highly regulated, complex, terminal oxidase system involving extensive metabolic reactions to use excess photosynthetic energy, but it is not two, unregulated, coupled reactions of an oxidase and a reductase. For 20 years these enzymes were considered to be soluble or cytosolic, and it was not until 1968 that we realized that their cellular location was in the fragile, single-membrane-bound peroxisomes (5, 6).

The glycolate pathway part of the C_2 cycle is channeled irreversibly toward glycine and serine formation (2–7, 11). Enzymatic reactions by Rubisco and P-glycolate phosphatase in the chloroplast, glycolate oxidase and glyoxylate:glutamate aminotransferase in peroxisomes and glycine decarboxylase in mitochondria are physiologically irreversible, consecutive reactions that start carbon flow around the C$_2$ cycle. The glycolate pathway only occurs in light because ATP and NADPH are needed to regenerate RuBP for the C$_2$ and C$_3$ cycles. When continuing around the C$_2$ cycle, serine hydroxymethyltransferase in mitochondria converts two glycines to one serine, NH$_3$, NADH, and a CO$_2$. An aminotransferase in peroxisomes converts the serine into hydroxypyruvate, which is reduced to glycerate. I consider these reactions as a reversible, glycerate pathway that can occur in light or darkness. ^{14}C metabolic studies with WA Wood showed that 3-carbon serine or glycerate were condensed to a 6-carbon hexose and ended up in sucrose. Because glycerate kinase is in chloroplasts, we assumed that the 3-P-glycerate entered that pool shared with the C$_3$ cycle. If you think the C$_2$ cycle is complex, so does everyone else, and at about this point many elect to disregard further the complete C$_2$ cycle. To consider the whole cycle, one needs to add: (*a*) regeneration of RuBP from 3-P-glycerate or from refixing CO$_2$ or from storage carbohydrate via the pentose phosphate pathway (a part of the C$_3$ cycle); (*b*) refixation of NH$_3$ in the chloroplast; (*c*) an aspartate:malate shuttle for balancing reducing equivalents among cell compartments; and (*d*) a glycolate/H$^+$ and/or glycolate/glycerate exchange, the triose phosphate shuttle, and glutamate transport for movement across chloroplast membranes.

Step-wise details of the glycolate pathway part of the C$_2$ cycle were elaborated in the 1950s and 1960s with B Rabson, E Jimenez, P Kearney, F Snyder, K Richardson, J Hess, G Orth, W Bruin, WT Chang, and A Baker. Reactions in the leaf peroxisomes and other organelles were first investigated in the 1970s with A Oeser, C Scharrenberger, T Kisaki, W Chang, D Rehfeld, R Yamasaki, R Gee, P Gruber, S Federick, Y Nakamura, J Hanks, and B Ho. Generally, each person concentrated on one phenomenon or step of the C$_2$ cycle by isolating and characterizing enzymes involved. For example, the properties and physiological role of P-glycolate phosphatase and P-glycerate phosphatase were investigated in the 1960s and 1970s with P Kearney, K Richardson, Y Yu, D Anderson, D Randall, J Christeller, and D Husic.

Rubisco: Phosphoglycolate and Glycolate Biosynthesis

Regulation of one enzyme for one reaction can be complicated, but regulation at a metabolic branch point by an enzyme with dual activities, as Rubisco, is indeed complex. Rubisco has a binding site for RuBP and for an activating CO$_2$ but has no known binding site for its other substrates, CO$_2$ and O$_2$. Thus catalysis is very slow and dependent on CO$_2$ and O$_2$ concentrations in the chloroplast. Our emphasis has been on understanding glycolate biosynthesis and metabolism, which developed into the C_2 cycle and then later into regulation of photosynthetic carbon metabolism relative to the CO_2 to O_2 ratio. More recent topics have dealt with increasing the CO_2 to O_2 ratio from the atmosphere by C_4 plants and the algal CO_2 and HCO_3^- pumps.

In my lab, the 1970s was a decade of productive research on Rubisco and on peroxisomes with funds from NSF and NIH. Properties and regulation of Rubisco were studied with G Lorimer, J Andrews, J Pierce, C Peach, F Ryan, S McCurry, N Hall, R Gee, R Houtz, RM Mulligan, K Cook, and others (2, 3, 5–7, 11). The oxygenase activity of Rubisco in vivo has been underestimated by [^{14}C] glycolate formation during $^{14}CO_2$ fixation or by O_2 uptake. Because of O_2 competition, the carboxylase $K_m(CO_2)$ increases from ~12 μM CO_2 in nitrogen to 26–42 μM CO_2 in air (21% O_2). V_{max} with high CO_2 and air does not change, but high enough CO_2 concentrations (perhaps 1000–1500 ppm) for V_{max} are not available in vivo except in C_4 plants or algae with activated DIC pumps. Thus, in vivo activity of the carboxylase in a C_3 plant is decreased by the O_2 level in air. Likewise, CO_2 inhibits the oxygenase activity of activated Rubisco, but exclusion of CO_2 from oxygenase assays is hard or impossible to do. Activation of both activities of Rubisco requires a high activating CO_2 concentration of 27 μM CO_2, yet this much CO_2 inhibits the oxygenase. If CO_2 is removed to measure the oxygenase, the enzyme inactivates. DB Jordan & WL Ogren devised an isotopic (^{14}C and ^{32}P) procedure for simultaneous chromatographic measurement of 3-PGA and P-glycolate, and from them calculated the K_m and V_{max} and a substrate specificity factor to compare Rubisco from various sources. The factor is approximately 80 for Rubisco from C_3 and C_4 plants, 60 for unicellular green algae, and from ~48 to 9 for photosynthetic bacteria. CO_2 activation of Rubisco is not competitive with O_2, because the activating CO_2 forms a HCO_3^- lysine complex to structure the active site. There is little conjecture on Rubisco inactivation in vivo at low CO_2. With increasing CO_2 concentration, the oxygenase activity of Rubisco will be more fully activated, and the storage carbohydrates can be photo-oxidized by the C_2 cycle at high O_2 levels. On the other hand, at low CO_2 concentration of the CO_2 Γ of a C_3 plant, both activities of Rubisco may be low, because the enzyme may not be fully CO_2 activated. Fixation of HCO_3^- by phosphoenolpyruvate carboxylase (PEPC) is not O_2 sensitive and seems to account for much of the dissolved inorganic carbon (DIC) fixation even by C_3 plants, because the CO_2 to O_2 ratio decreases at either the CO_2 or O_2 Γs.

Voluminous literature over the past 50 years on photorespiration and photosynthetic carbon metabolism contains proposed systems for glycolate synthesis and metabolism and energy balances. Only the oxygenase of Rubisco accounts for all known facts about glycolate synthesis and photorespiration.

An early concern was rapid formation of uniformly [^{14}C] glycolate during ^{14}CO$_2$ fixation in light. This led to proposals for an alternate CO$_2$ fixation pathway involving condensation of two CO$_2$. Because amounts of glycolate were not generally measured, it was not emphasized that S.A. of glycolate was much less (<25%) than S.A. of ^{14}CO$_2$. In addition, [^{14}C] glycolate was labeled with ^{14}CO$_2$ slower than intermediates of the C$_3$ cycle. Later we showed that glycolate is formed photosynthetically in light, even in absence of ^{14}CO$_2$, and of course, the S.A. had dropped to zero. Thus, glycolate synthesis is a photosynthetic, light-dependent photo-oxidation of carbohydrates, whether RuBP comes from newly fixed ^{14}CO$_2$ in the C$_3$ cycle or from storage reserves. Carbons 1 and 2 of RuBP are uniformly labeled during ^{14}CO$_2$ fixation and form uniformly labeled [^{14}C]P-glycolate when RuBP is oxidized by the oxygenase. Calvin's group labeled the first major ^{14}C product on their paper chromatograms as PGA, which can stand for and contained 3-P-glyceric acid and 2-P-glycolic acid. My contribution to that era in Calvin's lab was made the first week I got there in 1950. These chemists were trying to acid hydrolyze ^{14}C-labeled phosphate esters from their chromatograms, but this produced too much salt for rechromatography. As a biochemist, I suggested hydrolysis by a pinch of phosphatase, Polidase S.

A second major requirement for glycolate synthesis during photosynthesis is that one O$_2$ is incorporated into its carboxyl group, which, in turn, leads to carboxyl ^{18}O-labeled glycine and serine. P-glycolate synthesis by Rubisco is enhanced with higher oxygen or with decreasing CO$_2$ concentrations. K_m(O$_2$) for the oxygenase activity of Rubisco has been reported between 260 and 400 μM, or about the O$_2$ level in aerated water. CO$_2$ to O$_2$ ratio becomes the governing factor for the amount of glycolate synthesis relative to net CO$_2$ fixation.

Papers in the 1960s also proposed glycolate synthesis by peroxidation of carbohydrates formed during photosynthesis. Specifically, transketolase in the C$_3$ cycle transfers the top two carbon atoms of sedoheptulose-7-P or xylulose-5-P to its cofactor, thiamine pyrophosphate, to form a C$_2$ complex, which can be nonenzymatically oxidized by H$_2$O$_2$ to form glycolate. This reaction, however, would not incorporate ^{18}O$_2$ into the carboxy group nor form P-glycolate. In 1965, an NIH sabbatical fellowship took me to Freiburg, Germany, with H Holzer, an expert on transketolase in yeast. Angelica Oser assisted me in countless Warburg measurements of O$_2$ uptake by isolated spinach chloroplasts. We found much O$_2$ uptake without a time lag, upon adding P-glycolate or glycolate. She continued on this puzzle at Michigan until we heard C de Duve and H Beever lecture on peroxisomes and realized that our chloroplast preparations were contaminated with peroxisomes. The transketolase hypothesis was dropped, and we became experts on leaf peroxisomes. My associates for 30 years afterward know that I never ceased to wonder how chloroplasts could oxidize P-glycolate as fast as glycolate without a time lag to hydrolyze it to glycolate. Never ignore data. The explanation may be in our recent paper (1) on an alternate system in chloroplasts, which oxidizes either P-glycolate or glycolate.

O_2 and CO_2 Compensation Points

By decreasing the CO_2 concentration while the O_2 level remained constant at 21%, J Decker in the 1950s found that at about 50 ppm CO_2, net CO_2 uptake by C_3 plants had declined to zero. This has been called the CO_2 Γ in 21% O_2. At lower CO_2 levels in light there was CO_2 evolution, which became greater than dark respiration, and the C_3 plants senesced. The CO_2 Γ increases with increasing O_2 concentrations (10). With a given level of O_2, there is a lowest concentration of CO_2 at which net photosynthetic CO_2 exchange is zero, and with even lower CO_2 or higher O_2, net CO_2 evolution occurs. In a simplification, the CO_2 Γ has been considered as the CO_2 concentration with a given level of O_2, when CO_2 evolution from the C_2 cycle equals the rate of CO_2 fixation by the C_3 cycle. Historically, the CO_2 Γ became well known in the 1960s from an unsuccessful search for C_4-like varieties of C_3 wheat and soybean plants in closed, survival chambers with C_4 corn plants. With funds from Union Carbide (D Manning), W Smith and HS Ku with us likewise surveyed unsuccessfully hundreds of compounds with C_3 plants for altering the CO_2 Γ to see if they differentially effected the dual activities of Rubisco (notebooks of unpublished data). After that failure we became involved with processes to alter the CO_2 to O_2 ratio and thus the C_2 and C_3 cycles.

The immense O_2 level in air does not significantly vary, but small changes in the very low CO_2 level would alter the CO_2 to O_2 ratio and should change Rubisco reactions. Since O_2 diffuses extremely rapidly out of the leaf, O_2 exclusion does not appear to exist except for the thick-walled bundle sheath cells of C_4 plants that separate Rubisco inside from O_2 generated in the chloroplasts of the mesophyll cells. Four processes are known to change the CO_2 concentration around Rubisco. These are the C_4 cycle in C_4 and CAM plants, algal DIC concentrating processes, altering atmospheric CO_2 concentrations, and carbonic anhydrase in the chloroplast to accelerate HCO_3^- conversion to CO_2. The algal CO_2 or HCO_3^- pumps appear to directly import DIC, and although less is known about them, they should be simpler than the C_4 cycle. The algal DIC pumps use photosynthetic energy and are suppressed by high CO_2. Our group in this area has included J Moroney, D Husic, DW

Husic, A Goyal, Y Shiraiwa, J Thielman, S Dietrick, R Togasaki, and B Wilson.

The CO_2 Γ of C_4 plants is ~2 ppm CO_2. By concentrating CO_2 and partially excluding O_2 from bundle sheath cells of C_4 plants, the oxygenase activity of Rubisco is suppressed. Furthermore, C_4 plants initially fix HCO_3^- by PEPC, which has a much lower $K_m(HCO_3^-)$ than the $K_m(CO_2)$ for Rubisco. In addition, at pH > 8, there is at least a 100-fold higher concentration of HCO_3^- than CO_2. Important, but unrecognized earlier, is that HCO_3^- fixation by PEPC is not inhibited by O_2! No wonder C_4 plants are more CO_2 efficient at low CO_2 or high O_2. Part of photosynthetic energy used by the C_2 cycle in C_3 plants is used for the C_4 cycle. An ecologically exciting part of this C_4 story (9) is that as long as atmospheric CO_2 level remains less than ~380 ppm CO_2, the C_4 plant has been more CO_2 efficient than C_3 plants. However, above ~380 ppm CO_2, the C_4 plant is nearly light saturated and does not further increase its rate of CO_2 fixation with more CO_2, because it must use energy for its CO_2 pump to get CO_2 to Rubisco in the bundle sheath cells. At ~380 ppm CO_2, C_3 and C_4 plants are about equal in photosynthetic efficiency for CO_2 fixation. As CO_2 continues to rise, C_3 plants will increase their CO_2 fixation rate with the excess energy that they no longer use on the C_2 cycle, and become more efficient than C_4 plants. This change has occurred in a geological moment of one century, as CO_2 level has increased from 280 to 380 ppm. From now on with further increases in CO_2, the C_3 plant should be photosynthetically superior to a C_4 plant.

With all the work over 35 years on C_4 plants and CO_2 Γ and realization that Rubisco and the C_2 cycle involved O_2 uptake, you can ask what the reasons were that no one found an O_2 Γ to match a CO_2 Γ. After all, Rubisco is both a carboxylase and an oxygenase. First, we have all considered photosynthesis as only CO_2 fixation. To do an experiment to measure the O_2 Γ, the CO_2 concentration must be held constant, and small changes in the high O_2 concentration must be measured over long periods. This is technically difficult and requires an O_2 leak-proof chamber. Second, when the CO_2 level is held constant at 220 or 350 ppm, rates of CO_2 fixation are reduced between 5% O_2 up to 100% O_2, but are never completely inhibited. To repeat: because of O_2 inhibition, rate of net CO_2 fixation and O_2 evolution decreases with increasing O_2, but not to a zero rate of CO_2 fixation. Investigators of $^{18}O_2/^{16}O_2$ exchange likewise either used low O_2 or found increasing O_2 exchange with increasing O_2 or decreasing CO_2, but never zero CO_2 fixation. Thus, it was implied that there was no photosynthetic compensation point at high O_2. Wrong! Although there is no CO_2 Γ with higher O_2, there is an O_2 Γ based on zero net O_2 change (10), which also limits net photosynthesis and plant growth just as much as the CO_2

Γ at low CO_2. The O_2 Γ may be thought of as that O_2 concentration when the rate of O_2 evolution has decreased to the rate of O_2 uptake in a given CO_2 concentration.

I was honored by a Senior Scientist award from the German Humbolt Foundation. First, I spent six months with H Senger at the University of Marburg examining with J Thielman the DIC pump in *Scenedesmus*. Then at the University of Bayreuth, with Erwin Beck, we pondered why there was no O_2 Γ for plants. A photosynthetic chamber was built with controls for light, temperature, and CO_2 and O_2 concentrations. While holding CO_2 constant, the rate of O_2 change was measured at different O_2 concentrations from 2 to 90% with young tobacco or spinach plants. With increasing O_2, net rate of CO_2 fixation and of O_2 evolution decreased as expected from O_2 inhibition. With 220 ppm CO_2, the O_2 evolution had decreased to zero at 23% O_2 or with 350 ppm CO_2 at 27% O_2 (10). These are the O_2 Γs, when net rate of O_2 exchange at a fixed CO_2 level and temperature is zero. Nevertheless, at the O_2 Γ, the rate of CO_2 fixation had only fallen ~60% because of O_2 inhibition, although O_2 evolution had fallen 100%. With O_2 levels over the O_2 Γ, net O_2 uptake occurred (no O_2 evolution), and at 30 to 40%, O_2 rates of O_2 uptake increased to rates nearly equal to CO_2 uptake at low O_2. At first, one might say "impossible if photosynthesis is occurring." This were as if at low O_2 and/or sufficient CO_2, CO_2 fixation was the acceptor of the photosynthetic energy, but at high O_2 over the O_2 Γ, O_2 was the sole acceptor of photosynthetic energy. Above the O_2 Γ, plants slowly senesced. Our first conclusion is that an O_2 Γ exists and limits plant growth not far above present 21% O_2 with past low levels of CO_2. At a low CO_2 to high O_2 ratio, O_2 fixation by Rubisco functions as the alternate electron acceptor during photosynthesis rather than CO_2. GC-MS analysis of the products in tobacco leaves after several hours in high O_2 over the O_2 Γ found a big increase in malate. So E Beck and I proposed that increasing O_2 inhibition of the carboxylase of Rubisco reached a point where net photosynthesis, when measured as O_2 evolution, reached zero. Continued fixation of CO_2 could be due to O_2-insensitive, HCO_3^- fixation by the abundant PEPC in C_3 plants. The products, oxaloacetate and malate, are not reduced by the C_3 cycle to carbohydrate with accompanying O_2 evolution. For O_2 evolution by photosynthetic electron transport, NADP reduction must occur and be used for CO_2 reduction by the C_3 cycle. In the absence of enough CO_2 in air or the presence of too much O_2, net O_2 exchange in the light became O_2 uptake, and the plants senesced. The concept of an O_2 Γ opens up a new field dealing with photosynthetic limitation of the atmospheric O_2 levels, just as the CO_2 Γ limits CO_2 removal. Our research on this is being done after my retirement and without a grant. Unhappiness is at the lack of understanding, earlier, of all the physiological aspects of Rubisco and, now, at the lack of financial support for plant physiology research.

Photosynthetic Quotient and Quantum Efficiency

Earlier studies on photosynthesis had decreed a photosynthetic quotient of one CO_2 fixed and one O_2 evolved from photolysis of water. As long as the CO_2 level was high, over ~1500 ppm, and/or the O_2 level low, the oxygenase activity of Rubisco would be suppressed. As the O_2 level increases toward 21% with low CO_2 or surpasses the O_2 Γ, a higher percentage of DIC fixation is for HCO_3^- by O_2-insensitive PEPC. Because the complete C_2 cycle results in no net CO_2 or O_2 exchange as long as the O_2 level is below the O_2 Γ (7, 9), the C_2 cycle is normally a Hill process and only reduces the amount, but it does not change the ratio of one CO_2 uptake to one O_2 evolved. However, the quantum efficiency decreases from O_2 inhibition because of consumption of energy by the C_2 cycle. At O_2 levels above the O_2 Γ, which blocks the C_3 cycle, there is continued CO_2 fixation into malate and net O_2 uptake, so the ratio of CO_2 evolution to O_2 uptake becomes meaningless. Past avoidance of these problems is illustrated in a quote by my colleague, N Good, in his prefatory chapter in the *Annual Review of Plant Physiology* (1986, pp. 5–6):

> We succeeded by using isotopically labeled oxygen as the substrate for respiration, since the oxygen produced from water by photosynthesis is, of course, unlabeled. The problem was of considerable importance in those days because the great Otto Warburg had decreed that photosynthesis requires an immense uptake of oxygen, reconsuming three quarters of the photosynthetically produced oxygen. We succeeded in showing that Warburg was wrong....I have sometimes been asked how we came to miss the phenomenon of photorespiration since our instrument was ideally suited for detecting it. The simple answer is that we did not miss photorespiration at all. We observed it and adjusted conditions to eliminate it, using low concentrations of oxygen and high concentrations of carbon dioxide. We chose these conditions specifically to avoid a phenomenon that was already well known but irrelevant to our concerns. Photorespiration has been described many years before by Warburg, who called it "photocombustion."

Warburg was right for air with 21% O_2 and low CO_2. He worked with algae in small closed Warburg flasks in light, which led to rapid accumulation of high O_2, high pH, a low CO_2/O_2 ratio, and domination by the oxygenase activity of Rubisco for P-glycolate synthesis. I prize a letter in 1958 from Warburg congratulating us on discovering glycolate excretion by algae, because "its less reduced state would explain low quantum efficiency of 4." He then published that up to 80% of CO_2 fixed by *Chlorella* during photosynthesis was glycolate. At the time, I had no explanation for this, but I now can speculate that the glycolate would have come from oxidation of storage carbohydrate by Rubisco and did not represent newly fixed and reduced CO_2. Thus, Warburg calculated a high quantum efficiency on much glycolate formation during low CO_2 fixation or O_2 evolution. These two examples are from past dogma by top scientists in photosynthesis. They did not know about the oxygenase activity of Rubsico nor consider photosynthesis to be both "CO_2 fixation and evolution" as well as "O_2 evolution and O_2 fixation."

Booby Traps with Algae

Much research in photosynthesis has been conveniently done with algae. In order to use small amounts of $^{14}CO_2$ because of expense and radioactivity, cultures were used in small closed flasks. Rapid growth on 2 to 5% CO_2 suppressed the algal CO_2 pump so they were like a C_3 plant. When first put on low CO_2, O_2 accumulated as the CO_2 level dropped, and the CO_2 to O_2 ratio favored the C_2 cycle with glycolate and malate formation. This trapped Warburg and every one else. After I and P Zill published that algae specifically excrete glycolate, T Fogg, other marine biologists, and we visualized a photosynthetic glycolate symbiosis with flora in the ocean. However, in lakes mere traces of glycolate were much lower than expected, as the algae are adapted to low, free CO_2 and do not excrete much glycolate.

Unicellular green algae do not have peroxisomes or glycolate oxidase but rather appear to metabolize glycolate at the slow rate of ~1 μmole • mg Chl^{-1} • hr^{-1} by a mitochondrial glycolate dehydrogenase (work done with E Nelson, S Frederick, DW Husic, A Goyal, P Kehlenbeck, N Selph). However, when glycolate metabolism is blocked by an aminotransferase inhibitor (aminooxyacetate) for glyoxylate conversion to glycine, algae excrete at a rate of up to 15 μmole glycolate and some glyoxylate. This high excretion rate, which varies with the CO_2 to O_2 ratio, has been used to indicate the rate of glycolate production. For 20 years we have procrastinated about these data. Either we could not assay the mitochondrial glycolate dehydrogenase or there was another enzyme for oxidizing glycolate. Now we think the latter may be the case (1).

Leaf type peroxisomes are present in the multicellular charophyceae line of algae, which is the only line of algae that evolved into higher plants. In many other unicellular algae, these peroxisomal enzymes appear to be in their mitochondria. The small multicellular *Mougeotia* in the Charophyceae, as well as *Chara* and *Nitella,* have peroxisomes with enzymes for the C_2 cycle (8). Is there some special reason why leaf-type peroxisomes had to be present before higher plants could evolve? Rubisco is in all algae along with much P-glycolate phosphatase, as studied with D Husic and D Randall. The $K_m(O_2)$ for Rubisco oxygenase and for peroxisomal glycolate oxidase is about 260 μM, which is obtained by bubbling water with air, and both activities increase with higher O_2 (8). Thus, leaf type peroxisomes in green algae should not have functioned before the atmospheric O_2 increased (nor plants evolve?).

Before the above was known about algae, I proposed in 1970 a comparison of photorespiration in plants with that in marine algae. Up to that time, most work on glycolate had been done with plants or freshwater algae. NSF funded a three-month study of photorespiration by 20 plant physiologists and algologists aboard the *Alpha Helix* research vessel at Lizard Island on the Great Barrier Reef, before there were many tourists. This was a scientific trip that proved I had been right in not returning to the farm in Idaho. Our papers occupied an issue of *Australian Journal of Plant Physiology,* but we did not find much evidence of photorespiration or glycolate excretion. This resulted in turning my attention to CO_2-concentrating processes in algae. I need to emphasize how much such field research taught us bench scientists, and how it rewarded us for a lifetime of hard research.

Other topics that were discussed on the Barrier Reef were coral deposits and symbiosis between algae and the polyps for fixed nitrogen, fixed CO_2 (glycerol), and phosphate. $CaCO_3$ deposits represent immense reservoirs of CO_2 that have formed during algal photosynthesis, just as fossil fuels are reduced carbon deposits. Many algae produce an external carbonic anhydrase to accelerate the conversion of HCO_3^- to CO_2 plus OH^-. As the CO_2 is fixed, the OH^- titrates a second HCO_3^- to $CO_3^=$, which forms insoluble salts (8). In an unbuffered culture of green algae in the light with $NaHCO_3$, the pH quickly rises to 8.5 to 9, and CO_2 fixation stops. When the pH is high, the CO_2 concentration is low and the algae turn to O_2 fixation with the excretion of glycolate, which acid (Pk$_a$ of 3.8) might neutralize some of the HCO_3^- and $CO_3^=$. Carbonate deposits during photosynthesis are a sink for atmospheric CO_2, but these topics have not been pursued vigorously, and funds for work like our Great Barrier Reef trip are sadly not now in vogue.

What If? Is Photorespiration the Best Term?

I tell this story to emphasize that Rubisco is both an oxygenase and a carboxylase and that O_2 and CO_2 fixation coexist. Remember that compounds of the C_2 cycle are not labeled by $^{14}CO_2$ as fast as in the C_3 cycle, and that $^{18}O_2$ labels the C_2 cycle but not the C_3 cycle. In 1940, Ruben used radioactive $^{11}CO_2$ and Calvin and colleagues from 1946 on used $^{14}CO_2$ to find the first products of photosynthetic CO_2 fixation. About 20 years later, G Lorimer, J Andrews, and I, using $^{18}O_2$ with GC-MS analyses, examined the first products of $^{18}O_2$ fixation (P-glycolate, glycolate, glycine, and serine) as expected from the C_2 cycle. Now just suppose that Calvin's group could have used a radioactive isotope of oxygen or that my major professor, RH Burris, who had a home-made mass spectrometer at that time for ^{15}N analyses, had used $^{18}O_2$. The C_2 cycle might have been discovered first and be the photosynthetic carbon cycle. This story is not a criticism, Dr. Burris, just a lament. One reason the reverse did not happen was that during the preceding 150 years, photosynthesis was conceptually CO_2 fixation, and no one conceived of O_2 fixation. After all, photosynthesis evolves O_2, and measurements of net O_2 uptake in the light never occurred until now (10). Terms such as photodecomposition by O Warburg and photorespiration by J Decker, G Krotkov, W Jackson and R Volk, M Gibbs, I Zelitch, W Ogren, B Osmond, myself, and others were not considered as photosynthesis. $^{18}O_2/^{16}O_2$ exchange and O_2 as an energy acceptor was considered photorespiration and as a wasteful, protective mechanism when CO_2 was low. Photosynthesis was by dogma only O_2 evolution but not O_2 fixation. We must recognize that during photosynthesis, CO_2 and O_2 are both fixed and evolved (Figure 1) to arrive at net gas exchanges. Old dogma obscured CO_2 and O_2 exchange by the dual reactions of Rubisco. ^{14}C-labeling and analyses of the components of the C_2 cycle greatly underestimated their amount (especially from unlabeled storage carbohydrate). The term photorespiration implies distinct processes for O_2 uptake and CO_2 evolution that are masked by refixation of CO_2 and O_2 evolution. Calculations that were made of the energy consumed by photorespiration were based on the glycolate pathway, but not on the complete C_2 cycle. Total energy consumed should have included RuBP oxidation by the glycolate pathway followed by CO_2 refixation and resynthesis of RuBP. Both C_2 and C_3 cycles require regeneration of RuBP by reactions of the C_3 cycle for continued cycling (Figure 1). The energy used for resynthesis of RuBP is a major cost of either cycle. In a complete turn of the C_2 cycle, there is as much O_2 uptake as O_2 evolution and as much CO_2 evolution as CO_2 refixation. This is the case normally when there is sufficient CO_2 for net CO_2 fixation and O_2 evolution. In a closed box when the CO_2 concentration is less than the CO_2 Γ, there is net CO_2 evolution, or when O_2 is higher than the O_2 Γ, there is net O_2 uptake in the light from oxidation of the carbohydrate reserve. Then the C_3 cycle is blocked, and O_2 fixation is an alternate energy consumer to CO_2 assimilation.

If the term photorespiration has been confusing to others, it has also misled investigators. I have used photorespiration as a crutch to refer to several related processes (1, 5, 8, 9, 11). It has been used for glycolate biosynthesis, glycolate pathway, peroxisomal respiration, C_2 oxidative photosynthetic carbon cycle, O_2 inhibition of photosynthesis, $^{18}O_2/^{16}O_2$ exchange, and the dark CO_2 burst.

The two processes of photosynthesis and photorespiration have been considered so different that we used two abbreviations for them, PS and PR, and two pools of CO_2 and two of O_2 that change in opposite manner. These facts miss the purpose and function of the C$_2$ cycle and the reason Rubisco is an oxygenase. Let us develop the realization that the oxygenase activity of Rubisco and the subsequent C$_2$ cycle are essential and equal parts of photosynthetic carbon metabolism as the CO_2 decreases or the O_2 increases. The combination of the C$_2$ and C$_3$ cycles provides a check and balance on net photosynthesis and atmospheric concentrations of CO_2 and O_2. Photosynthetic carbon metabolism has its own Hill process or alternate electron acceptor in O_2 fixation. Photosynthesis is self-limiting by balancing the atmospheric CO_2 to O_2 ratio within limits set by the specificity of Rubisco. At lower CO_2 or higher O_2 than the Γs, plants cannot grow. We all should be proud to be a part of the plant science community as these concepts evolve. This group includes plant physiologists and photosynthesizers who have investigated CO_2 fixation and the C$_3$ cycle, and O_2 fixation and the C$_2$ cycle. Although photosynthetic carbon metabolism, as a section or symposium topic, has disappeared from national biochemistry meetings, it flourishes at plant physiology meetings, International Photosynthetic Congresses, the Gordon Conferences, and other special symposia.

The question I am raising is whether the term "photorespiration" is appropriate to represent the oxygenase activity of Rubisco, because the oxygenase is as much a part of photosynthesis as is the carboxylase activity. It would be more informative to use "O_2 exchange or CO_2 exchange" during photosynthesis. In this chapter, I have also used the term "O_2 fixation" to parallel "CO_2 fixation." This could be confusing, because normally there is no measurable net O_2 fixation even though O_2 exchange as well as CO_2 exchange is occurring. Emphasis is needed in photosynthetic carbon metabolism that Rubisco is both a carboxylase for CO_2 fixation with O_2 evolution, as well as an oxygenase for O_2 fixation with CO_2 evolution. Both cycles consume photosynthetic energy. We need to comprehend that the two Rubisco reactions together establish a ratio of CO_2 to O_2 in the atmosphere that is based upon its specificity. The CO_2 to O_2 ratio only recently reached an atmospheric equilibrium, with the geological pools. Maintenance of this ratio is a primary result of photosynthetic CO_2 and O_2 exchange. We should not imply that the oxygenase activity of Rubisco is wasteful, unavoidable, undesirable, or a separate process. Instead, for the global CO_2 and O_2 balance, both the C$_2$ and C$_3$ cycles coexist and are essential. Understanding this will be difficult for every one—ecologists, geologists, and legislators. However, first scientists like us must agree on these interpretations.

Regulation of Atmospheric CO_2 and O_2 Levels

Plant scientists are not geologists, yet our research is important to environmental regulation. Recognizing this, I asked JCG Walker, a leading geologist, to speak at my retirement symposium on the role of photosynthesis in regulating atmospheric CO_2 and O_2 levels. His paper (in Reference 12) is a superb summary of the global carbon and oxygen cycles. He used photosynthesis to supply the atmosphere with O_2 and to remove CO_2, but he concluded that otherwise photosynthesis has absolutely nothing to do with regulating the CO_2 and O_2 concentration in air. This conclusion indicates a need to understand one another's data. We need to consider the great mass of plants and algae and great differences in the kinetics of enzymatic reactions (seconds), such as Rubisco, versus geological times for nonphotosynthetic changes in the storage pools. Do rapid photosynthetic changes in low CO_2 concentrations alter the atmosphere CO_2 and O_2 ratio on a short time basis, while the nonbiological changes may take centuries? The Mauna Loa data on CO_2 show a short-term decrease in summer of about 15 ppm in atmospheric CO_2, even in the presence of CO_2 replacement from CO_2 pools in the global carbon cycle. This rapid photosynthetic CO_2 depletion must be replaced, otherwise the CO_2 level would quickly decrease to where plants would have little net photosynthesis in 21% O_2. Lowering of CO_2 much below past equilibria levels of 235 ± 45 ppm may not be possible for C_3 plants, because oxygenase of Rubisco would dominate in 21% O_2. The O_2 Γ with lower CO_2 levels, rather than the extremely low CO_2 Γ from PEPC, may be the short-term limiting factor in 21% O_2 for plant growth. Fortunately, now that CO_2 is rising, the oxygenase activity is decreasing relative to the carboxylase. I propose that the short-term CO_2 to O_2 ratio in the atmosphere over decades is set by the O_2 and CO_2 specificity of Rubisco, whereas the large storage reserves of carbon and oxygen more slowly alter these gases over many centuries. In the distant past, as long as there were low O_2 and/or high atmospheric CO_2 to lower the oxygenase of Rubisco, the CO_2 to O_2 equilibrium ratio from its specificity was not reached. Now our grand environmental experiment with increasing CO_2 may have started from a low CO_2 to O_2 equilibrium and is being conducted in the presence of high O_2 and rising temperatures. Considerable thought will be needed to evaluate whether or how fast an increase might occur in the O_2 level toward a higher O_2 Γ.

Photosynthetic investigators have justified their research as the basic process for plant growth for food, fuel, and fiber and O_2 for life. We have also expressed concern about increasing atmospheric CO_2 with an increase in the greenhouse effect. Another dimension with which we have not dealt is regulation of atmospheric O_2 concentration or ratio of CO_2 and O_2. The CO_2 level of the past 165,000 years, as found in the ice cores, has shifted between 190 ppm to 280 ppm. The activity of Rubisco and distribution of C_3 and C_4 plants should have varied within this CO_2 to O_2 range because of changes only in the CO_2 level, as the O_2 concentration probably did not fluctuate very fast. 300 mya, atmospheric O_2 level was about 35%, much higher than today. To reach 35% O_2 as the O_2 Γ, C_3 plants must have ~700 ppm CO_2 (10). As the CO_2 decreased, the O_2 level must have also decreased to 21%. Decreasing CO_2 would have affected the CO_2 and O_2 competition for activities of Rubisco unless the O_2 also decreased to maintain a near constant ratio. C_3 plants cannot live in 300 ppm CO_2 and 35% O_2. This research area becomes a new topic for photosynthetic carbon metabolism, ecology, and geology to understand how the CO_2 and O_2 balance in the atmosphere is maintained by Rubisco.

As I drafted this prefatory chapter, I realized what a small part of this story (even work done by my associates) could be told in this limited length. It will take a lot of work to assemble all the previous papers from the past 80 years on O_2 and CO_2 effects on photosynthesis and plants and put them into perspective. In addition, the data are not all in, and different investigators will certainly continue to disagree on some points. Younger investigators can be excused for not understanding our past turmoil.

Eclectic Research

There are constant temptations to spread one's research program too broadly. Such dilution has been dignified by claims of being a generalist in plant physiology. Our annual national meetings, textbooks, and Annual Reviews promote diversification. Many younger plant scientists do not have a focused plan in research but follow general areas and techniques. The most focused postdoc I had was W Outlaw, who would only work on guard-cell metabolism, although I knew nothing about it. I used much of my time and energy on many other topics besides those discussed in this prefatory chapter on the C_2 cycle. Those other associates were equally competent and appreciated. For the record, I have selected the reason and results of six of these other areas.

1. Research on phosphorylcholine in the 1950s and 1960s was started with H Wiebe under the general Atomic Energy Commission (AEC) theme of using radio isotopes for studies of metabolism. Only one ^{32}P-labeled phosphate ester had a unique paper chromatographic RF of ~0.9 with 80% phenol-water (it was a Zwitter ion at the pH of phenol) but only ~0.15 with butanol-propionic acid (like other ionized phosphate esters). AA Benson's lab joined us in identifying it, and then B Martin, K Tanaka, and A Golke

studied its metabolism over 10 years. It is an immense, rapidly labeled pool (~20% of the total P) in plants moving in the xylem sap to the leaves, after formation in the roots. It is the only phosphate ester in plants that is transported. Much more needs to be done on its general role in regulating phosphate and lipid metabolism.

2. When searching for the importance of phosphorylcholine, I substituted a chlorine for the phosphate by synthesizing chlorocholine chloride, which was trivialized to CCC or Cycocel. CCC turned out to be an effective plant growth regulator for thick, short-stem wheat, rice, chrysanthemums, and poinsettias (work with S Wittwer). CCC was considered an antigibberellin after H Kende and colleagues found that it blocked GA synthesis, but alas, CCC had no apparent effect on phosphorylcholine metabolism. My bad guess led to a successful commercial product and I joined the Plant Growth Regulatory Society. You never know.

2. During my brief postdoc stay in Calvin's lab in 1950, P Pearson, Chief of the Biology Branch at the AEC, came to the lab looking for an assistant in plant physiology. Because most of Calvin's people were chemists with strong convictions about staying in California, it took but a moment for Calvin to decide that I should go to Washington, DC. That was a complete diversion from studying glycolate metabolism. At first I learned a lot from travel to projects in the USA, AEC labs, and Einewetok. Within two years, I wised up to the fact that giving out dollars was not going to get me much further in science. That was the first and last administrative job I ever took. I adopted the attitude that the perfect scientific environment is to be blessed with supportive, competent, leave-you-alone administrators. One such person, A Hollander, Director of Biology Division at AEC Oak Ridge National Lab, let me return to full-time research in plant physiology, meaning glycolate and sedoheptulose metabolism with P Zill, A Krall, and others. R Rabson, a postdoc with us, did valuable ^{14}C tracer work on the reactions of the glycolate pathway. He later proved me wrong about Washington, DC, by building a highly significant career supervising Department of Energy grants for research on plants.

4. When I was administrating plant physiology grants, S Hendricks, at USDA, Beltsville, and R Withrow, at Smithsonian, told me I was too young and rather stupid to have gotten out of research so early. Hendricks found me space and a technician in his lab, and I moonlighted studies on activation of glycolate oxidase in etiolated leaves with his spectrometer for whole plants. The action spectra was the same as photosynthesis—no big surprise even then. However, I was grateful for being kept scientifically alive. These two were experts on light action spectra for phytochrome and seed germination,

and their interest rubbed off on me, as indicated by latter publications with postdocs on seed germination.

5. After publications beginning in 1968 about leaf peroxisomes, I fancied myself a peroxisome biochemist. Michigan State University had established a medical school, and even though I avoided teaching medical students, NIH provided us funds for research on liver peroxisomes (6). Just as we all incorrectly thought that leaf peroxisomal metabolism was just a wasteful process, we also wondered whether peroxisomal respiration in animals, in contrast to mitochondrial respiration, might be controlling body weight. That grabbing hypothesis has not been supported by enough research to be accepted. Peroxisomes from liver, kidney, and elsewhere in the body have glycolate oxidase, and they oxidize long-chain fatty acids as do peroxisomes in germinating fatty seeds. How glycolate is formed in animals is not proven nor is its metabolic function understood. This research deviation from plants probably decreased our productivity in photosynthetic carbon metabolism, but striking similarities between leaf and liver peroxisomes augmented our peroxisomal studies with M Markwell, J Krahling, E McGroarty, R Donaldson, M Redinbaugh, S Furuta, J Uhlig, P Murphy, S Haas, S Vandor, and B Hsieh. Our lab and those of H Beevers and E Newcomb became places for research on plant peroxisomes. One of the most striking comparisons between plant and animal peroxisomes is how acetate from peroxisomal, long-chain fatty acid β oxidation is utilized. Animals form carnitine fatty acyl carrier complexes to move acetate and medium-length fatty acids from peroxisomes to other parts of the cell (work done with L Bieber's group). In the mitochondria, these carnitine derivatives are transferred to CoA and metabolized by the TCA cycle. Plant peroxisomes do not use carnitine as a carrier for fatty acids but rather completely degrade the fatty acid in their peroxisomes to acetyl CoA and condense it with glyoxylate to form malate. When this research on the glyoxylate cycle was initiated by H Beevers, he called peroxisomes by the term glyoxysomes. I followed de Duve's nomenclature for peroxisomes, which he originally discovered. In plants, this difference in nomenclature is not resolved and is confusing to others, so we have sometimes used "leaf peroxisomes" for the C$_2$ cycle. However, some peroxisomes are in all tissues of plants and animals.

6. I initially got interested in glycerol synthesis from the Great Barrier Reef expedition. Glycerol is an end product, just as is sucrose, of photosynthetic CO$_2$ fixation by symbiotic algae, zooxanthellae, which excrete it along with some glycolate to feed coral polyps. The triose phosphate shuttle between chloroplasts and cytosol transports the photosynthate, and there is just one

reduction step needed to form glycerol-P from dihydroxyacetone-P (DHAP). I was challenged by remarks from my contemporaries that there was no glycerol-P dehydrogenase in leaves. The secret, which R Gee and G Santora discovered for us, is that the plant enzyme is severely inhibited by trace amounts of any fats, lipids, and detergents in homogenates. A Goyal, RU Byerrum, T Kirsch, and D Gerber joined our project. There are major forms in the cytoplasm, in chloroplast, and in algae. They are very active as DHAP reductases for glycerol-P synthesis at pH 7.5, but nearly inactive as a glycerol-P dehydrogenase even at pH 9.5, so we call them DHAP reductases. A glycerol-P/glycerol metabolic cycle exists in algae for glycerol accumulation, and a longer-term goal would be to put this algal system in plants for stress protection.

Reviews, References

With the proliferation of research, the number of references to a given topic becomes immense. A review seldom can include all references that are pertinent. An excuse to cite mainly your own is to say the number is restricted by the editor. Some particularly important paper(s) of your competitors need to be cited to show modesty and to appease in a grant application. The older literature is generally not cited because those authors are no longer around to review your paper or argue. In fact, by ignoring the older literature you can claim more originality (but also ignorance). Thus, older literature is never adequately covered, and your own work is never given enough credit.

I had thought that after retirement I would keep busy by writing reviews, but that was false thinking. In the first place, there is no inspiration to write reviews if you cannot experimentally test out the new ideas that float to the top. I am grateful to E Beck in Bayreuth, Germany, for doing our experiments on measuring the O_2 Γ, because otherwise I would be less motivated to write about nearly 80 years of work on O_2 inhibition of photosynthesis. Another issue is that it is horrifying how fast one becomes obsolete in research if you stop working on it. Keeping up requires attending meetings, scanning journals, and discussing research every day with active laboratory investigators. If one stops these activities for a year or even a few months, you are completely out, including the writing of reviews. Good reviews cannot be done in the absence of research. If you are a Distinguished Emeritus Professor, you might get away with memoirs without references about once, as in a prefatory chapter.

So then, what do you do with your office files? There is nothing so dear, yet so cold, as old lab notebooks, nothing so nostalgic as one's reprint and card file—but only to you. I have a collection since 1945 of most of the important reprints on photosynthetic carbon metabolism, peroxisomes, and photorespiration, but it is of little value to others. However, I will hang on to it just as dearly as to life. With this tirade about references, I follow the editor's instructions to maybe cite a few, and, of course, they are all my own.

ACKNOWLEDGMENTS

This research occurred over a long period of time that was robbed from my family. I am most grateful to them for tolerating my single-minded devotion to it, which remains my only work and hobby. In the first draft of this chapter, I tried to mention names and results of all graduate students, postdocs, and visitors, who did the work. The review was far too long. For each of these associates our paths ran together for a valuable time, and I am forever grateful.

Life hereafter continues in our genes; memory rests in our writings.

Literature Cited

1. Goyal A, Tolbert NE. 1996. Association of glycolate oxidation with photosynthetic electron transport in plant and algal chloroplasts. *Proc. Natl. Acad. Sci. USA* 93: 3319–24

2. Husic DW, Husic HD, Tolbert NE. 1987. The oxidative photosynthetic carbon cycle. *Crit. Rev. Plant Sci.* 5:45–100

3. Paech C, McCurry SD, Pierce J, Tolbert NE. 1978. Active site of ribulose-1,5-bisphosphate carboxylase/oxygenase. In *Photosynthetic Carbon Assimilation,* ed. HW Siegleman, G Hind, pp. 227–43. New York: Plenum

4. Tolbert NE. 1963. Glycolate pathway. In *Photosynthesis Mechanisms in Green Plants,* Publ. 1147, pp. 648–62. Washington, DC: Natl. Acad. Sci./Nat. Res. Counc.

5. Tolbert NE. 1980. Photorespiration. In *The Biochemistry of Plants,* ed. P Stumpf & E Conn, Vol. 2: *Metabolism and Respiration,* ed. DD Davies, pp. 488–525. New York: Academic

6. Tolbert NE. 1981. Metabolic pathways in peroxisomes and glyoxysomes. *Annu. Rev. Biochem.* 50:133–57

7. Tolbert NE. 1983. The oxidative photosynthetic carbon cycle. In *Current Topics in Plant Biochemistry and Physiology,* ed. DD

Randall, DG Blevin, R Larson, 1:63–77. Columbia: Univ. Missouri

8. Tolbert NE. 1992. The role of peroxisomes and photorespiration in regulating atmospheric CO$_2$. In *Phylogenetic Changes in Peroxisomes of Algae. Phylogeny of Plant Peroxisomes,* ed. H Stabenau, pp. 428–42. Oldenburg, Ger: Univ. Oldenburg Press

9. Tolbert NE. 1994. The role of photosynthesis and photorespiration in regulating atmospheric CO$_2$ and O$_2$. See Ref. 12, pp. 8–33

10. Tolbert NE, Benker C, Beck E. 1995. The oxygen and carbon dioxide compensation points of C3 plants: possible role in regulating atmospheric oxygen. *Proc. Natl. Acad. Sci. USA* 92:11230–33

11. Tolbert NE, Husic HD, Husic DW, Moroney JV, Wilson BJ. 1985. Relationship of glycolate excretion to the DIC pool in microalgae. In *Inorganic Carbon Uptake by Aquatic Photosynthetic Organisms,* ed. WJ Lucas, JA Berry, pp. 211–23. Rockville, MD: Am. Soc. Plant Physiol.

12. Tolbert NE, Preiss J. 1994. *Regulation of Atmospheric CO$_2$ and O$_2$ by Photosynthetic Carbon Metabolism.* New York: Oxford Univ. Press. 272 pp.

Annu. Rev. Plant Physiol. Plant Mol. Biol. 1997. 48:27–50

TRANSPORT OF PROTEINS AND NUCLEIC ACIDS THROUGH PLASMODESMATA

Soumitra Ghoshroy, Robert Lartey, Jinsong Sheng, and Vitaly Citovsky

Department of Biochemistry and Cell Biology, State University of New York, Stony Brook, New York 11794-5215

KEY WORDS: cell-to-cell movement of plant viruses, intercellular communication, protein transport, nucleic acid transport

---

ABSTRACT

Despite a potentially key role in cell-to-cell communication, plant intercellular connections—the plasmodesmata—have long been a biological "black box." Little is known about their protein composition, regulatory mechanisms, or transport pathways. However, recent studies have shed some light on plasmodesmal function. These connections have been shown to actively traffic proteins and protein–nucleic acid complexes between plant cells. This review describes these transport processes—specifically, cell-to-cell movement of plant viruses as well as endogenous cellular proteins—and discusses their possible mechanism(s). For comparison and to provide a broader perspective on the plasmodesmal transport process, the current model for nuclear import, the only other known example of transport of large proteins and protein–nucleic acid complexes through a membrane pore, is summarized. Finally, the function of plasmodesmata as communication boundaries within plant tissue is discussed.

CONTENTS

INTRODUCTION

Plasmodesmata are cytoplasmic bridges that span the walls separating plant cells. Bernhardi suggested the existence of such channels, as well as the possibility of intercellular communication that they would permit, almost two centuries ago (6). Three quarters of a century passed before Tangl first observed plasmodesmata (104), and it was not until the turn of the century that Strasburger gave them their present name (102). Although the existence of such cell wall bridges has long been known, the isolation of plasmodesmata and the elucidation of their constituent proteins and regulatory mechanisms have remained elusive to this day. In recent studies, plasmodesmata have been implicated not only in transport of small molecules such as water, ions, and photoassimilates, but in intercellular traffic of proteins, nucleic acids, and protein–nucleic acid complexes. This ability to transport large molecules and multimolecular complexes has been described for only one other biological channel, the nuclear pore.

   In this review, we summarize the structure of plasmodesmata, discuss their function in intercellular transport of macromolecules, and describe possible similarities between the mechanisms for nuclear import and plasmodesmal transport. In addition, the role of plasmodesmata as conduits for viral infection is discussed.

STRUCTURE OF PLASMODESMATA

Plasmodesmata are large membrane-lined pore structures that interconnect adjacent plant cells. The current structural model of a simple plasmodesma derives from the electron microscopy experiments that used high-pressure freezing to preserve fine details of tobacco leaf plasmodesmata (28). According to this study, the endoplasmic reticulum (ER) passes through the plasmodesma and is both surrounded by and filled with regularly spaced globular particles approximately 3 nm in diameter. This membrane-protein complex is called the desmotubule or appressed ER (Figure 1). The particles associated with the outer leaflet of the appressed ER appear to be connected to those on the inner leaflet by proteinaceous filaments. Further, globular particles are embedded in the inner leaflet of the plasma membrane, greatly restricting the

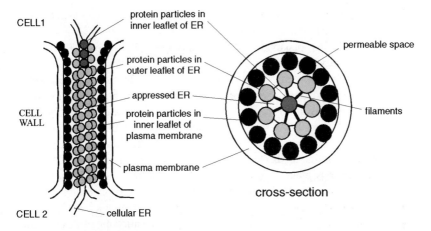

CELL1

protein particles in
inner leaflet of ER

permeable space

protein particles in
outer leaflet of ER

appressed ER

CELL
WALL

protein particles in
inner leaflet of
plasma membrane

filaments

plasma membrane

cross-section

CELL 2   cellular ER

longitudinal section

Figure 1   Structure of a simple plasmodesma. See text for details. Adapted from References 11, 17, and 28.

open space (28). In addition to the ER, plasmodesmata may associate with cytoskeletal elements. The presence of actin in plasmodesmata has been proposed (114), and recent results suggested that myosin, which interacts with actin and produces force through hydrolysis of ATP, also colocalizes with plasmodesmata (88).

Plasmodesmata follow a complex developmental pathway. There are two types of plasmodesmata, which may be defined morphologically as simple and branched, or developmentally as primary and secondary (25, 63). Primary or simple plasmodesmata are found predominantly in young tissue and consist of a simple, single channel, as pictured in Figure 1 (25). The creation of primary plasmodesmata is a function of cell plate formation during cytokinesis (37, 62, 64). The ER is positioned across and perpendicular to the cell plate (52, 86). It then becomes appressed by the developing cell plate and, together with the plasma membrane, provides cytoplasmic continuity between cells. Thus, the primary plasmodesma is essentially an incomplete separation between two daughter cells. Secondary or branched plasmodesmata are found in older tissues and show a higher degree of variability, often with many channels leading into a larger central cavity (25). Because the number of plasmodesmata in a given area is constant in both older and younger tissue, it is believed that secondary plasmodesmata are derived in most cases from preexisting primary plasmodesmata (26, 63). The secondary plasmodesmata differ functionally in several ways from simple plasmodesmata, most notably in response to viral infection (25) (see below).

Transport through plasmodesmata appears to be very complex. Ordinarily, only small microchannels are available for passive transport between the globular particles. This unassisted transport through plasmodesmata, studied using microinjected dyes and fluorescently labeled dextrans, appears to be limited to molecules up to 1.5–2.0 nm in diameter, equivalent to a molecular mass of 0.75–1.0 kDa (105, 116). Several factors decrease plasmodesmal permeability. These include divalent cations (Ca^{2+}, Mn^{2+}, and Sr^{2+}), phorbol esters, aromatic amino acids, and phosphoinositides (5, 33, 62, 107). Only one known endogenous plant protein, KN1, encoded by the maize *knotted-1* homeobox gene, increases plasmodesmal permeability (61; see also below). However, many plant viruses have evolved the ability to increase the plasmodesmal size exclusion limit during local and systemic spread of infection. The mechanism by which this increase in plasmodesmal permeability occurs is unknown. For example, it is possible that a conformational change in the aforementioned filaments enlarges the permeable space of plasmodesmal channels by pulling the globular particles into the appressed ER (Figure 2) (11).

Our present knowledge of plasmodesmal structure and composition is purely descriptive and has been derived mainly from electron microscopy studies. No functional plasmodesmal proteins have been definitely identified. However, many initial reports on cloning plasmodesmata genes or purifying plasmodesmata-associated proteins (67, 68) either have been disputed or require further substantiation (73). This is in contrast to the animal counterparts of plasmodesmata—gap junctions—which have been long since purified and characterized, and whose encoding genes have been cloned (57). The main reason for such difference is technical. Unlike gap junctions, plasmodesmata are firmly embedded in the plant cell wall and are thus recalcitrant to purification. One way to circumvent this difficulty is to use plant viruses, which specifically interact with plasmodesmata during infection, as a molecular tool to identify and characterize the plasmodesmal transport pathway and its protein components.

PLASMODESMAL TRANSPORT OF PROTEINS AND NUCLEIC ACIDS

Viral Cell-to-Cell Movement Proteins

Katherine Esau first postulated that viruses moved throughout the plant via plasmodesmata (34). Since then, viral spread through plant intercellular con-

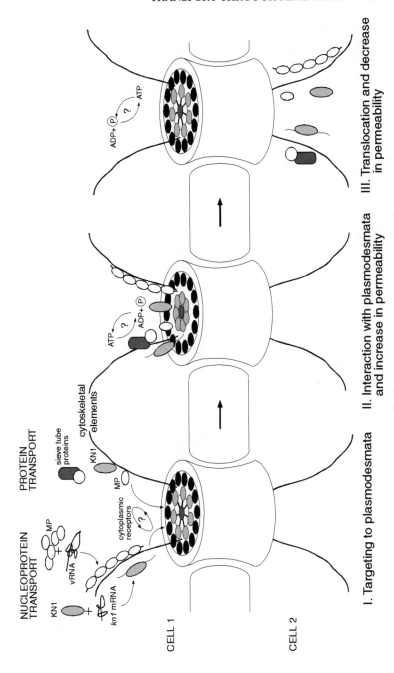

Figure 2 A model for plasmodesmal transport of proteins and protein–nucleic acid complexes. For explanation, see text.

nection has been shown to occur in two major steps: local and systemic (Figure 3). Following initial infection, usually by mechanical or insect-mediated inoculation, many plant viruses spread from cell to cell through plasmodesmata until they reach the vascular system; the viruses are then transported systemically through the vasculature. Presumably, viral spread through the vascular tissue is a passive process, occurring with the flow of photoassimilates (reviewed in 59). In contrast, the cell-to-cell movement is an active function, requiring specific interaction between the virus and plasmodesmata. This interaction is mediated by virus-encoded nonstructural movement proteins that act to increase plasmodesmal permeability and transport viral nucleic acids through the enlarged plasmodesmal channels (reviewed in 11, 63, 64).

MOVEMENT PROTEIN–NUCLEIC ACID COMPLEXES   The best studied cell-to-cell movement protein is the 30-kDa protein (P30) of tobacco mosaic virus (TMV) (22). To date, P30 has been suggested to possess three biological activities. It is thought to bind TMV RNA, forming an extended P30-RNA complex that can penetrate the plasmodesmal channel (12, 16). It also may interact with the

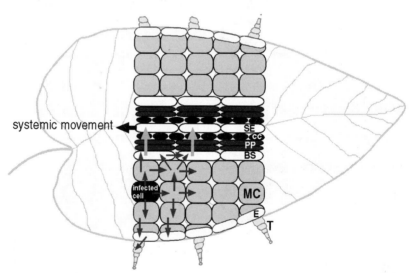

Figure 3 Cellular routes for local and systemic movement of plant viruses. T, trichome; E, epidermal cells; MC, mesophyll cells; BS, bundle sheath cells; PP, phloem parenchyma cells; CC, companion cells; SE, sieve elements. Arrows indicate viral movement. Viral spread between trichome, epidermis, and mesophyll cells represents local, cell-to-cell movement. Plasmodesmata between bundle sheath and phloem parenchyma cells are thought to mediate transition from local to systemic movement, which then proceeds through the sieve elements to other plant organs (25, 63).

cytoskeletal elements to facilitate transport of the P30-TMV RNA complexes from the cell cytoplasm to plasmodesmata (51, 66). Finally, P30 functions to increase the size exclusion limit of plasmodesmata (116). On the basis of these activities, a model for the P30-mediated intercellular transport of TMV RNA has been proposed (Figure 2) (11, 18, 117). In the initially infected cell, P30 is produced by translation of the invading genomic TMV RNA. This protein then associates with a certain portion of the viral RNA molecules, sequestering them from replication and mediating their transport into the neighboring uninfected host cells. In vitro studies showed that P30 binds both RNA and single-stranded DNA (ssDNA), but not double-stranded DNA (dsDNA), cooperatively and without sequence specificity (12, 16). This sequence-nonspecific binding explains the observations that coinfection with TMV can allow cell-to-cell movement of plant viruses that normally do not spread through plasmodesmata (reviewed in 3, 10). Electron microscopic experiments revealed that P30 binding unfolds the nucleic acid molecule, creating an extended protein-RNA complex of 2.0–2.5 nm in diameter (16) (Figure 2). Because the free-folded TMV RNA has been estimated to be 10 nm in diameter (40), association with P30 likely shapes it in a thinner, transferable form capable of being transported through plasmodesmal channels.

Following the demonstration that TMV P30 binds single-stranded nucleic acids (12, 16), cell-to-cell movement proteins from many other plant viruses were also found to exhibit similar activity. For example, P1 of cauliflower mosaic virus (CaMV) was shown to associate with both ssDNA and RNA in long, thin complexes that closely resemble P30-ssDNA and P30-RNA complexes (13, 106). P1 binding affinity to RNA was higher than that toward ssDNA, which suggests that P1-CaMV RNA complexes may be involved in the cell-to-cell spread of this virus (13). This TMV-like mechanism for cell-to-cell movement may coexist with the better-characterized spread of CaMV in the form of a whole viral particle through virus-modified plasmodesmata (56). In addition to TMV and CaMV, the movement proteins of alfalfa mosaic virus (AMV), red clover necrotic mosaic dianthovirus (RCNMV), and several other viruses have been shown to bind single-stranded nucleic acids (81, 83, 97). Similar to P30, these movement proteins bound nucleic acids cooperatively (13, 81, 97) [though the movement protein of RCNMV did not appear to significantly extend the bound RNA molecules (38)]. Thus, transport through plasmodesmata via formation of movement protein–nucleic acid intermediates may represent a common mechanism for cell-to-cell spread of many plant viruses.

Although practically all examined movement proteins most likely function to transport viral nucleic acids, the degree of transport selectivity varies. For example, the P30 protein of TMV can bind and, by implication, transport any single-stranded nucleic acid (12, 16). In contrast, the RCNMV movement protein is capable of trafficking only ssRNA but not ssDNA or dsDNA (38), whereas the bean dwarf mosaic geminivirus (BDMV) BL1 movement protein facilitates transport of dsDNA but not ssDNA or ssRNA molecules (79). In the latter case, the transport of dsDNA seems incompatible with biochemical and genetic evidence obtained with another bipartite geminivirus, squash leaf curl virus (SqLCV), for which the transported form has been proposed to be viral genomic ssDNA (83). Because BL1 only weakly binds nucleic acids (83), it interacts with the second geminivirus movement protein—BR1 (95)—which directly associates with the transported nucleic acid molecule (83). BR1 most likely binds the viral nucleic acid to transport it out of the host cell nucleus where geminiviruses, such as SqLCV, replicate (44). BR1-ssDNA complexes (or BR1-dsDNA complexes, in the case of BDMV) then associate with BL1, which mediates the plasmodesmal transport (see also below).

MOVEMENT PROTEIN-CYTOSKELETON INTERACTION   Because TMV RNA translation and, therefore, the production of P30 occurs in the host cell cytoplasm (82), P30-TMV RNA cell-to-cell transport complexes most likely are formed in the cytoplasmic compartment. How, then, do these complexes arrive at plasmodesmata before the cell-to-cell movement? Recent data suggest that P30 interacts with microtubules and, to a lesser extent, with actin (51, 66). This interaction was inferred from colocalization of the wild-type P30, transiently expressed in tobacco protoplasts, primarily with microtubules and, occasionally, with actin filaments (66). P30 association with tubulin and actin was also demonstrated using in vitro binding assays (66). Furthermore, P30 tagged by translational fusion with the jellyfish green fluorescent protein (GFP) formed a filamentous network following transient expression in plant protoplasts. These filamentous arrays of P30-GFP were best detected 18 to 20 h following expression; after 48 to 72 h, most P30-GFP appeared as aggregates along the periphery of the protoplasts (66). Finally, P30-GFP was introduced into the TMV genome, and the resulting modified virus retained infectivity (51). The fluorescent P30-GFP expressed in tobacco protoplasts and leaf tissue following infection formed an intracellular network that coaligned with cellular microtubules (however, no association of P30 with F-actin was detected in these experiments) (51). Although movement proteins of other plant viruses have not yet been tested for interaction with cytoskeleton, it is likely that such interaction will be found for many viral species.

Taken together, these observations suggest that P30 may interact with the cytoskeletal elements in the host cell cytoplasm and use them as tracks to migrate to the cell periphery and, ultimately, to plasmodesmata (Figure 2). Because P30 is most likely associated with the viral RNA during infection, it is the P30-TMV RNA complex that may interact with the host cell microtubules and microfilaments during transport to plasmodesmata (Figure 2). Alternatively, interaction with the cytoskeleton may anchor P30 or P30-TMV RNA complexes in the cytoplasm, gradually releasing them in response to as yet unidentified signals for the plasmodesmal transport to occur. In this case, association with microtubules and microfilaments would function as a regulatory mechanism for plasmodesmal transport rather than the targeting apparatus. Similar regulation by cytoplasmic anchoring has been described for nuclear import of the NF-κB transcription factor (reviewed in 29) (see also below).

INCREASE IN PLASMODESMAL PERMEABILITY   Once the P30-TMV RNA cell-to-cell transport complex reaches the plasmodesmal channel, it must traverse it to enter the neighboring host cell. Although the estimated 2.0–2.5 nm diameter of this nucleoprotein complex (16) is relatively small, it still is incompatible with the 1.5-nm diameter of the intact plasmodesmal channel (116). To allow movement, therefore, P30 increases plasmodesmal permeability (Figure 2). Originally, the ability of P30 to increase plasmodesmal size exclusion limit was detected by injection of fluorescently labeled dextrans into leaf mesophyll of transgenic tobacco expressing P30 (116). Unlike the wild-type tobacco mesophyll plasmodesmata that can traffic dextrans of up to 0.75–1.0 kDa, the P30 transgenic plants exhibited plasmodesmal size exclusion limit of almost 10 kDa (116). These transgenic plants are responding to P30 steady state expression, accumulation, and activity. Thus, it remained unclear whether the increase in plasmodesmal size exclusion limit was due to activation of an endogenous plasmodesmal transport pathway or induction of a specific host response to this viral protein. Experiments using wild-type tobacco plants and in vitro–produced P30 suggested an answer to this question. Direct microinjection of purified P30 into the wild-type tobacco mesophyll resulted in a relatively fast (3–5 min) size exclusion limit increase up to 20 kDa, which suggests that P30 functions via an existing plasmodesmal transport machinery (111). Importantly, the increased size exclusion limit of 10–20 kDa (111, 116) corresponds to the 5–9 nm–diameter of the dilated channel, potentially allowing unrestricted traffic of 2.0–2.5 nm–wide P30-TMV RNA complexes (16).

P30 microinjection experiments provided another important clue to its function. It was noted that large fluorescent dextrans not only moved into the cells adjacent to the microinjected cell but traveled as far as 20 to 50 cells away from the site of injection (111). The observations suggested that P30 itself must have moved through plasmodesmata to induce the increase in size exclusion limit in the distant mesophyll cells, providing the first evidence that these plasmodesmata can traffic protein molecules. An alternative and unlikely possibility that P30 induced an intracellular signaling pathway without its own cell-to-cell movement was ruled out later by immunolocalization experiments that showed that microinjected P30 does move between plant cells (112).

Similarly to single-stranded nucleic acid binding, the increase in plasmodesmal size exclusion limit is most likely a property of many viral movement proteins. Presently, the cell-to-cell movement proteins of RCNMV, AMV, cucumber mosaic virus (CMV), tobacco rattle virus (TRV), potato virus X, and the BL1 movement protein of BDMV were shown to enable transport of large fluorescent dextrans between plant cells (2, 23, 38, 79, 85, 109). RCNMV, CMV, and BDMV movement proteins were also shown to move from cell to cell themselves (27, 38, 79, 112). Collectively, these experiments have established the use of large fluorescent dextrans as an assay for interaction between plasmodesmata and viral movement proteins. Recent observations, however, questioned the biological relevance of this approach. P30 was coinjected with fluorescent dextrans into the tobacco trichomes that are arranged in a linear file of cells and, consequently, allow better visualization of intercellular transport (112). Surprisingly, no size exclusion limit increase could be detected, although trichome cells support viral infection and cell-to-cell movement. However, when P30 was fused translationally to a reporter β-glucuronidase (GUS) enzyme, the microinjected GUS-P30 protein efficiently moved between trichome cells (112). This result strongly suggests that P30 is a *cis*-acting mediator of plasmodesmal transport. That is, P30 must be physically associated with the transported molecule such as viral RNA or the reporter GUS enzyme. This idea is consistent with most known examples of protein transport and targeting. For example, the increase in the nuclear pore size exclusion limit during nuclear import does not allow passage of protein molecules that are not directly associated with the nuclear localization signal (NLS) sequence (43). Cell-to-cell transport of large fluorescent dextrans coinjected with P30 into the leaf mesophyll, then, may be only an "afterglow" of the P30 biological activity when mesophyll (but not trichome) cell plasmodesmata remain open following microinjection or overexpression of the movement protein. The true plasmodesmal transport, then, likely requires direct interaction between the movement protein and the transported molecule.

Phloem Proteins

Viral movement proteins are by far the best studied example of protein and protein–nucleic acid transport through plasmodesmata. However, in host-pathogen interactions, the invading microorganism generally does not invent novel metabolic pathways; instead, it insinuates into existing cellular processes and adapts them for use in its life cycle. Thus, the plasmodesmal transport of viruses probably reflects the ongoing traffic of macromolecules between plant cells. Although it is logical to assume that such transport may have an important function in plant growth and development, there is surprisingly little evidence to support this idea. The first indication of plasmodesmal trafficking of endogenous macromolecules derives from the studies on the phloem. This specialized plant tissue, which functions to transport photoassimilates from the source leaves to the rest of the plant, is composed of sieve element cells interconnected by large (0.35–1.00 μm) pores derived from primary plasmodesmata (63). During differentiation, the sieve elements lose their nuclei, tonoplasts, microtubules, microfilaments, and ribosomes. Only the plasma membrane, a parietal endoplasmic reticulum, plastids, and mitochondria remain in the functional sieve element cell (35). Yet these mature sieve elements function for many months or even years (90). Thus, the maintenance of these cells must involve transport of all necessary proteins from the neighboring companion cells (Figure 3), which are connected to sieve elements by specialized deltoid-shaped plasmodesmata (63).

Newly synthesized proteins have been found in rice, castor bean, and wheat sieve element cells (36, 74, 94). The rice phloem sap contains more than 100 different proteins, ranging from 10 to 70 kDa (74), whereas the soluble proteins from the wheat sieve elements are enriched in 34–36 kDa species (36). A recent study attempted to identify the transported proteins. Using immunodetection techniques, ubiquitin and several chaperonins, such as HSP70 and GroEL homologs, were detected in the castor bean sieve element exudates (96). The ubiquitin-to–total protein ratio in the exudate was significantly higher than that in the surrounding plant tissue (96). Accumulation of ubiquitin in the sieve elements suggests that the intercellular transport of this protein is an active process rather than a simple diffusion down the concentration gradient. However, because ubiquitin is a relatively small molecule (8.5 kDa), facilitated diffusion through the companion cell–sieve element plasmodesmata cannot be excluded. Finally, this study did not detect ribulose-1,5-biphosphate carboxylase-oxygenase (Rubisco) in the phloem sap (96). Although Rubisco resides in the chloroplast, its small subunit is encoded by the nuclear genome and produced in the cell cytoplasm (48). The lack of its transport into the sieve element, therefore, supports the idea that protein traffic through the companion cell–sieve element plasmodesmata may be specific and that not all companion cell proteins can cross this cellular boundary.

Protein transport from companion cells into the sieve elements may occur by a mechanism similar to that of the cell-to-cell transport of plant viruses. Unlike plant virus movement proteins, however, the phloem sap proteins most likely do not mediate transport of nucleic acids such as messenger RNA (mRNA) molecules. Because sieve elements lack ribosomes, transporting mRNA into these cells will have little biological sense. Although the pumpkin phloem lectin PP2 protein was detected both in the companion cells and in sieve elements (101), its mRNA was localized exclusively in the companion cells (7).

Other Cellular Proteins

Until very recently, viral movement proteins and nucleic acids and phloem exudate proteins were the only macromolecules known to traverse plasmodesmata. One approach to identifying additional proteins that move from cell to cell is to compare the cellular distribution of a protein to that of its mRNA. The presence of the tested protein but not its mRNA in a certain type of cell would suggest intercellular transport. Using this criterion, a protein encoded by the maize *knotted1* (*kn1*) homeobox gene (110) was identified as a candidate for intercellular movement. Immunolocalization studies have detected the KN1 protein both in the outer (epidermal) layer and in the adjacent interior layer of the maize shoot apical meristem (61). In contrast, the *kn1* mRNA was found only within the interior cell layer (61). The possibility of KN1 intercellular transport was then directly tested by microinjection of the fluorescently labeled KN1 protein into tobacco and maize leaf mesophyll. Technical difficulties precluded microinjection into the apical meristem, the natural functional site of KN1. Microinjected KN1 moved very rapidly (within 1–2 s) between leaf mesophyll cells both in maize and in tobacco (61). It is puzzling that practically no nuclear import of KN1 was detected in these experiments (61) (since KN1 is a transcription factor, it must function exclusively in the cell nucleus). While it is possible that KN1 nuclear targeting in the leaf mesophyll is not efficient, ectopic expression of KN1 in tobacco leaves has been shown to alter cell fate, suggesting functional nuclear import of this protein (100).

Microinjection of unlabeled KN1 induced cell-to-cell movement of 20–39-kDa fluorescent dextrans as well as a 20-kDa cytosolic protein (soybean trypsin inhibitor), which indicates that KN1 dilates plasmodesmata similarly to the movement proteins of plant viruses (61). Like the viral movement proteins that transport viral nucleic acids through plasmodesmata, KN1 was shown to traffic RNA molecules. Unlike viral movement proteins, however, KN1 transported only its own mRNA (61). If both KN1 and *kn1* mRNA move between cells, it is unclear why only the protein is found in the outer layer of the shoot apical meristem in vivo (see above). Furthermore, neither KN1 nor *kn1* mRNA are detected in the expanding leaves which flank the shoot apical meristem (61). If microinjected KN1 can move through leaf plasmodesmata (61), there must be a mechanism to prevent such movement in vivo.

Cell-to-cell transport of KN1 supports the idea that plasmodesmata may function as conduits for developmental signals and morphogens. Preliminary data suggest that protein products of *deficiens* (DEF) and *globosa* (GLO) genes of *Antirrhinum,* which are involved in floral organ specification, traffic intercellularly and can increase plasmodesmal permeability (70). Based on the ability of DEF and GLO to move from cell to cell, it has been proposed that most floral regulators traverse plasmodesmata and function by creating an intercellular signal gradient leading to specific pattern formation (70).

POSSIBLE MECHANISMS FOR PLASMODESMAL TRANSPORT

The mechanism by which macromolecules are transported through plasmodesmata is unknown. However, it may be possible to predict the key features of this process by comparing it to nuclear transport, the only other known example of traffic of large proteins and protein–nucleic acid complexes through a membrane pore. Before making such predictions, we first summarize the mechanism of the nuclear import process.

Highlights of Nuclear Import

NUCLEAR LOCALIZATION SIGNAL (NLS)   Molecular transport across the nuclear envelope involves a great diversity of proteins and nucleic acids. This transport is bidirectional and occurs exclusively through the nuclear pore complex (NPC). While relatively small molecules (up to 60 kDa) (reviewed in 1, 42, 78) diffuse through the NPC, transport of larger molecules occurs by an active mechanism mediated by specific nuclear localization signal (NLS) sequences contained in the transported molecule (reviewed in 39). Transport of some small endogenous nuclear proteins such as the H1 histone (21 kDa) also occurs by an active process (8).

With few exceptions (e.g. influenza virus nucleoprotein NLS, yeast Gal4 protein NLS), all NLSs can be classified in two general groups: (*a*) the SV40 large T antigen NLS (PKKKRKV) motif and (*b*) the bipartite motif consisting of two basic domains separated by a variable number (but not less than 4) of spacer amino acids and exemplified by the nucleoplasmin NLS (KR-X_{10}KKKL). The first domain of a bipartite NLS usually consists of two adjacent basic residues, whereas the second domain contains three out of five basic amino acids (reviewed in 30). To date, most NLSs found in plant proteins belong to the bipartite type (89).

NLS RECEPTORS NLS receptors, which have been implicated in protein nuclear import in several organisms, belong to a multiprotein family that includes animal karyopherin α (importin 60) and yeast Kap60 (formerly Srp1) (32, 45, 87 and references therein, 91, 113). They are thought to recognize and bind the NLS sequence of the transported protein molecule and direct it to the nuclear pore. Once the receptor-NLS–bearing protein complex is at the nuclear pore, animal karyopherin β (importin 90) or yeast Kap95 proteins mediate binding to the NPC proteins, nucleoporins. This binding is followed by translocation into the nucleus, which requires the GTPase Ran (or its yeast homolog Gsp1) (20, 69, 71) and the Ran-interacting protein p10 (75). During translocation, karyopherin α is thought to enter the nucleus together with the bound NLS-containing protein while karyopherin β remains at the nuclear pore orifice (46).

Thus, nuclear import involves two discrete steps: (a) binding of NLSs to the cytoplasmic receptors, which then direct the transported molecule to the NPC, and (b) transport through the nuclear pore (76, 92). While interaction with NLS receptors and NPC binding is energy independent, the actual translocation requires metabolic energy in the form of ATP and GTP (69, 77).

REGULATION OF NUCLEAR IMPORT Many nuclear proteins function only at a specific developmental stage or in response to certain external stimuli such as hormones. Nuclear entry of these proteins, therefore, is tightly regulated. Signal masking and cytoplasmic anchoring represent the two major mechanisms for regulation of nuclear import (29). Regulation by NLS masking is exemplified by the nuclear import of the glucocorticoid receptor. In the absence of hormone, this protein is bound in the cytoplasm to the HSP90 chaperone protein, which masks the NLS (9, 84). When activated by the binding of a steroid hormone, the receptor molecule is released from HSP90, and its NLS is revealed, directing the receptor-hormone complex into the cell nucleus to activate gene expression (49, 84).

Nuclear import of the NF-κB transcription factor is a paradigm of regulation by cytoplasmic anchors. NF-κB, which usually consists of 50- and 65-kDa subunits (41), is present in the cytoplasm of nonactivated cells as a complex with an inhibitory protein I-κB (4). I-κB contains ankyrin repeats (50) originally found in the erythrocyte membrane anchoring protein ankyrin (65). These repeats are thought to promote association of I-κB, and its cognate

NF-κB, with the cellular matrix (115). Stimulation of cells with a variety of agents (e.g. phorbol esters, interleukin 1) activates cellular kinases that phosphorylate I-κB, resulting in the release of NF-κB and its migration into the cell nucleus (98). Thus, the selectivity of the nuclear pore is itself constant, and the regulation of transport involves conversion of the transported molecule from a nontransferable form to a transferable form (29).

Implications for the Mechanism of Plasmodesmal Transport

Similar to nuclear import, transport through plasmodesmata most likely consists of two major steps: (a) recognition of the transported molecule in the cell cytoplasm and its targeting to the plasmodesmal channel and (b) translocation (Figure 2). Proteins and protein–nucleic acids complexes destined for cell-to-cell transport may be recognized by their putative targeting sequence, the plasmodesmata localization signal (PLS). Although no such signal has been identified, mutational analysis of KN1 by alanine scanning identified only one mutant unable to increase plasmodesmal permeability (61). This mutation affected three basic amino acids of the putative KN1 NLS (61), raising a possibility that basic amino acid residues are involved in plasmodesmal targeting as well. Consistent with this idea, the carboxyl terminal part of the 100–amino acid–long P30 domain required for interaction with plasmodesmata (111) also contains many (9 out of 19) basic residues (93). PLSs may also mediate transport of nucleic acids associated with the PLS-containing protein, such as P30-TMV RNA and KN1-*kn1* mRNA complexes (Figure 2). Similar transport of nucleic acids via protein import pathway has been proposed for the nuclear import of Agrobacterium T-DNA associated with the bacterial VirD2 and VirE2 proteins (15, 17, 19, 47, 54, 119), as well as for the influenza virus genomic RNA-NP nucleoprotein complexes (80).

The putative PLS potentially interacts with specific cytoplasmic receptors (Figure 2). These as yet unidentified receptor proteins may function to transport the PLS-containing protein to the plasmodesmal annulus. Alternatively, the transported protein may be guided to plasmodesmata simply by association with cytoskeletal tracks (see above). In this case, the question of specific targeting, i.e. whether there are microfilament and microtubule arrangements that are specific for or lead only to plasmodesmata, remains to be resolved.

Once at the plasmodesmal channel, the transported protein or protein-PLS receptor complexes must increase plasmodesmal permeability to allow translocation. By analogy with nuclear import, GTPase and/or ATPase activities may be involved. An ATPase activity has been localized to plasmodesmata (24, 118), but further studies are necessary to ascertain the molecular nature of this enzyme. It is also possible that interaction of the transported protein with cytoskeleton actively affects the plasmodesmal annulus, increasing its size exclusion limit for translocation. The actual mechanism of plasmodesmal gating will be elucidated only with purification and characterization of the protein components of this channel.

Like nuclear import, which is tightly controlled, plasmodesmal transport should also be stringently regulated. Interaction of viral movement proteins with plasmodesmata may interfere with normal intercellular communication and, thus, be detrimental to the host plant. It is therefore likely that a mechanism exists to regulate the activity of P30 and, possibly, cellular proteins capable of plasmodesmal transport (see above). Irreversible deposition of P30 in the central cavity of secondary plasmodesmata has been proposed to represent one such mechanism (14). Electron microscopy studies have localized P30 to the secondary but not primary plasmodesmata (25). Conversely, microinjected P30 increases plasmodesmal permeability in young tobacco leaves devoid of secondary plasmodesmata (111). Thus, secondary plasmodesmata may represent a cellular compartment where P30 is inactivated or sequestered following its function. It is also possible that P30 inactivation is mediated by a phosphorylation reaction (Figure 2). A plant cell wall–associated protein kinase has been shown to specifically phosphorylate P30 at its carboxyl-terminal serine and threonine residues (14). This P30 kinase activity was developmentally regulated (14), correlating with the formation of secondary plasmodesmata within tobacco leaf (25). The possibility that the P30 kinase is a functional component of secondary plasmodesmata is being tested.

By analogy to nuclear import, another possible mechanism for plasmodesmal regulation is cytoplasmic anchoring. P30 interaction with cytoskeleton may serve such function by immobilizing P30 in the cell cytoplasm. This interaction may also mask the putative PLS sequence on the transported protein. Regardless of its molecular basis, the regulation of plasmodesmal transport probably determines the various communication domains thought to exist in plants.

CELLULAR DOMAINS SPECIFIED BY LEAF PLASMODESMATA

For decades, the interconnection of plant cells has been thought to form a single continuum, termed symplast (72). Recent data, however, suggest that intercellular communication and transport are modulated to produce specific tissue domains. The best-characterized cellular domains are those in the leaf tissue (25, 62, 63, 112), though other plant organs have been shown to contain similar subdivisions or symplastic domains (108). Individual domains within a plant leaf are thought to be delimited by boundaries with different plasmodesmal permeability. Figure 3 and Table 1 summarize our knowledge about these compartments.

The first cellular boundary in a leaf is specified by plasmodesmata between trichome, epidermal, and mesophyll cells. Plasmodesmal permeability within a five-cell-long tobacco trichome has been found remarkably different from that in the mesophyll and epidermal cell layers. Whereas only small molecules with the molecular mass of up to 0.75–1.0 kDa pass freely within mesophyll and epidermis (105, 116), significantly larger 3-kDa dextrans have been shown to move between trichome cells. No movement of these molecules into the adjacent mesophyll and epidermal tissues was detected (112). Another difference between trichome and mesophyll plasmodesmata was revealed using direct microinjection of the P30 movement protein. Unlike its ability to increase plasmodesmal size exclusion limit and allow diffusion of 10–20-kDa dextrans (111, 116), P30 completely failed to dilate trichome plasmodesmata (112). P30 itself, however, efficiently moved between trichome cells (112). These observations indicate functional differences between plasmodesmata of trichomes and the adjacent epidermal and mesophyll cells.

The second cellular domain as defined by plasmodesmal permeability of its constituent cells is mesophyll tissue. As mentioned above, the apparent plasmodesmal size exclusion limit in mesophyll cells is 0.75–1.0 kDa (105, 116). This threshold can be raised to 10–20 kDa by direct microinjection or stable expression of viral cell-to-cell movement proteins (111, 116). In addition, this tissue has been shown to traffic the KN1 transcription factor that was also found to increase the size exclusion limit of mesophyll plasmodesmata up to 39 kDa (61). As the most convenient substrate for microinjection and electron microscopy, mesophyll plasmodesmata are perhaps the best-studied type of plant intercellular connections.

Another transport boundary is between mesophyll and vascular tissue. In higher plants, phloem and xylem are arranged in bundles delimited from other tissues by a ring of cells termed the bundle sheath. Bundle sheath cells, therefore, are located at the physical border between leaf mesophyll and vasculature. In terms of plasmodesmal function, however, this boundary is not apparent. No significant differences in the size exclusion limit or interaction with viral movement proteins have been detected between bundle sheath and the mesophyll (25). Plasmodesmata between bundle sheath cells and the adjacent phloem parenchyma, however, may represent a significant transport boundary. It has specifically been shown that these intercellular channels interact with the TMV P30 movement protein differently than plasmodesmata in the mesophyll or between mesophyll and bundle sheath (25).

Table 1 Summary of structural and functional characteristics of leaf plasmodesmata[a]

Cellular boundary	Type of PD	PD morphology	SEL (kDa)	MP transport	MP Accumulation	Apparent increase in SEL (dKa)	CP requirement for viral movement	Transport of cellular proteins
Trichome/epidermis–mesophyll	Primary, secondary	Single, branched	≤3.0	Yes	Yes	None	No	?
Mesophyll–mesophyll	Primary, secondary	Single, branched	<1.0	Yes	Yes	<10–39	No	KN1
Mesophyll–bundle sheath	Primary, secondary	Single, branched	<1.0	?	Yes	<10	No	?
Bundle sheath—phloem parenchyma	Primary, secondary	Single, branched	<1.0	?	Yes	None	Yes	?
Phloem parenchyma–companion cell	Primary, secondary	Single, branched	?	?	No	?	Yes(?)	? (No protein traffic from CC into PP)
Companion cell–sieve element	Mainly secondary	Deltoid–shaped	≥3.0	?	No	?	Yes(?)	Sieve tube proteins move from CC to SE

[a]PD, plasmodesmata; SEL, size exclusion limit; MP, viral cell-to-cell movement protein; CP, viral coat protein; CC, companion cells; PP, phloem parenchyma; SE, sieve elements.

Within mesophyll as well as between mesophyll and the bundle sheath, P30 is solely responsible for the increase in plasmodesmal permeability and, consequently, for the cell-to-cell movement of the virus (25, 111, 116). TMV mutants that lack coat protein (CP) move normally from cell to cell in these tissues (21, 31, 103). To move systemically, however, the virus must enter the phloem tissue. This event requires the presence of the viral CP (53). Microinjection and electron microscopy studies demonstrated that P30 accumulates within the central cavity of the secondary plasmodesmata that separate bundle sheath cells from the phloem (25). The accumulation pattern was identical to that observed within mesophyll plasmodesmata. In contrast to its activity in mesophyll cells, however, P30 was unable to dilate the bundle sheath–phloem parenchyma plasmodesmata (25). This observation suggests that plasmodesmata between bundle sheath and phloem parenchyma cells represent a critical boundary for the onset of viral systemic movement. Once this barrier is crossed, the virus enters the host plant vasculature, resulting in systemic infection. It is tempting to speculate that TMV CP mediates viral movement through these boundary plasmodesmata. Furthermore, because P30 also recognizes these connections (25), it is possible that CP and P30 act in concert to affect plasmodesmal permeability and viral transport. It is also likely that this mechanism governs both the viral entry into the phloem and exit from the vascular tissue into the mesophyll of uninfected leaves.

Plasmodesmata separating phloem parenchyma from companion cells differ from those between phloem parenchyma and the bundle sheath. Sieve element–companion cell plasmodesmata have a characteristic deltoid shape (63) and allow diffusion of 3-kDa dextran molecules; unfortunately, the exact size exclusion limit for these channels has not been determined because of technical difficulties (55). They do not accumulate P30 (25) and may not require it for viral transport. Thus, CP alone may be sufficient to allow viral movement through these channels. Moreover, the plasmodesmata at the phloem parenchyma–companion cell boundary do not allow traffic of companion cell endogenous proteins (63). In contrast, these proteins constantly move through plasmodesmata connecting the companion cells and sieve elements (36, 74, 94, 96) (see above).

FUTURE PERSPECTIVES

In recent years, the previously dormant field of plasmodesmal macromolecular traffic has gained much impetus. Solution of plasmodesmal transport mechanisms will profoundly affect our understanding of intercellular signaling and plant-pathogen interaction. Undoubtedly, biochemical and structural characterization of plasmodesmata and microinjection studies will continue to enhance our knowledge of intercellular communication and its role in plant development and morphogenesis. The true progress, however, may come from future applications of the genetic approach to dissect the molecular pathway for plasmodesmal transport. For example, Arabidopsis mutants may be isolated that are resistant to viral systemic and/or cell-to-cell spread, potentially because of specific blockage in transport through plasmodesmata. Presently, several such mutants have been identified, and their characterization is under way (RT Lartey, J Sheng & V Citovsky, unpublished results). In addition, several Arabidopsis ecotypes have been described in which viral systemic spread is restricted (58, 60, 99). Ultimately, isolation of additional mutants will saturate the entire plasmodesmal transport pathway and reveal its functional components.

ACKNOWLEDGMENTS

We thank Gail McLean for critical reading of this manuscript. Our research is supported by grants from National Institutes of Health (Grant No. R01-GM50224), US Department of Agriculture (Grant No. 94-02564), and US-Israel Binational Research and Development Fund (BARD) (Grant No. US-2247-93) to VC.

> Visit the *Annual Reviews home page* at http://www.annurev.org.

Literature Cited

1. Akey CW. 1992. The nuclear pore complex. *Curr. Opin. Struct. Biol.* 2:258–63
2. Angell SM, Davies C, Baulcombe DC. 1996. Cell-to-cell movement of potato virus X is associated with a change in the size exclusion limit of plasmodesmata in trichome cells of *Nicotiana clevelandii. Virology* 216:197–201
3. Atabekov JG, Taliansky ME. 1990. Expression of a plant virus-coded transport function by different viral genomes. *Adv. Virus Res.* 38:201–48
4. Baeuerle PA, Baltimore D. 1988. I-κB: a specific inhibitor of the NF-KB transcription factor. *Science* 242:540–46
5. Baron-Eppel O, Hernandes D, Jiang L-W, Meiners S, Schindler M. 1988. Dynamic continuity of cytoplasmic and membrane compartments between plant cells. *J. Cell Biol.* 106:715–21
6. Bernhardi JJ. 1805. *Beobachtungen Ÿber Pflanzengefüsse und eine neue Artderselben.* Erfurt
7. Bostwick DE, Dannenhoffer JM, Skaggs MI, Lister RM, Larkins BA, et al. 1992. Pumpkin phloem lectin genes are specifically expressed in companion cells. *Plant Cell* 4:1539–48
8. Breeuwer M, Goldfarb DG. 1990. Facilitated nuclear transport of histone H1 and other small nucleophilic proteins. *Cell* 60: 999–1008
9. Cadepond F, Schweizer-Groyer G, Segard-Maurel I, Jibard N, Hollenberg SM, et al. 1991. Heat shock protein 90 is a critical factor in maintaining glucocorticosteroid receptor in a nonfunctional state. *J. Biol. Chem.* 266:5834–41
10. Carr RJ, Kim KS. 1983. Evidence that bean golded mosaic virus invaded nonphloem tissue in double infections with tobacco mosaic virus. *J. Gen. Virol.* 64:2489–92

11. Citovsky V. 1993. Probing plasmodesmal transport with plant viruses. *Plant Physiol.* 102:1071–76

12. Citovsky V, Knorr D, Schuster G, Zambryski P. 1990. The P30 movement protein of tobacco mosaic virus is a single strand nucleic acid binding protein. *Cell* 60: 637–47

13. Citovsky V, Knorr D, Zambryski P. 1991. Gene I, a potential movement locus of CaMV, encodes an RNA binding protein. *Proc. Natl. Acad. Sci. USA* 88:2476–80

14. Citovsky V, McLean BG, Zupan J, Zambryski P. 1993. Phosphorylation of tobacco mosaic virus cell-to-cell movement protein by a developmentally regulated plant cell wall-associated protein kinase. *Genes Dev.* 7:904–10

15. Citovsky V, Warnick D, Zambryski P. 1994. Nuclear import of *Agrobacterium* VirD2 and VirE2 proteins in maize and tobacco. *Proc. Natl. Acad. Sci. USA* 91:3210–14

16. Citovsky V, Wong ML, Shaw A, Prasad BVV, Zambryski P. 1992. Visualization and characterization of tobacco mosaic virus movement protein binding to single-stranded nucleic acids. *Plant Cell* 4: 397–411

17. Citovsky V, Zambryski P. 1993. Transport of nucleic acids through membrane channels: snaking through small holes. *Annu. Rev. Microbiol.* 47:167–97

18. Citovsky V, Zambryski P. 1995. Transport of protein–nucleic acid complexes within and between plant cells. *Membr. Protein Transp.* 1:39–57

19. Citovsky V, Zupan J, Warnick D, Zambryski P. 1992. Nuclear localization of *Agrobacterium* VirE2 protein in plant cells. *Science* 256:1803–5

20. Corbett AH, Koepp DM, Schlenstedt G, Lee MS, Hopper AK, et al. 1995. Rna1p, a Ran/TC4 GTPase activating protein, is required for nuclear import. *J. Cell Biol.* 130: 1017–26

21. Dawson WO, Bubrick P, Grantham GL. 1988. Modifications of the tobacco mosaic virus coat protein gene affecting replication, movement and symptomatology. *Phytopathology* 78:783–89

22. Deom CM, Shaw MJ, Beachy RN. 1987. The 30-kilodalton gene product of tobacco mosaic virus potentiates virus movement. *Science* 327:389–94

23. Derrick PM, Barker H, Oparka KJ. 1992. Increase in plasmodesmatal permeability during cell-to-cell spread of tobacco rattle tobravirus from individually inoculated cells. *Plant Cell* 4:1405–12

24. Didehvar F, Baker DA. 1986. Localization of ATPase in sink tissue of *Ricinus*. *Ann. Bot.* 40:823–28

25. Ding B, Haudenshield JS, Hull RJ, Wolf S, Beachy RN, et al. 1992. Secondary plasmodesmata are specific sites of localization of the tobacco mosaic virus movement protein in transgenic tobacco plants. *Plant Cell* 4:915–28

26. Ding B, Haudenshield JS, Willmitzer L, Lucas WJ. 1993. Correlation between arrested secondary plasmodesmal development and onset of accelerated leaf senescence in yeast acid invertase transgenic tobacco plants. *Plant J.* 4:179–90

27. Ding B, Li Q, Nguyen L, Palukaitis P, Lucas WJ. 1995. Cucumber mosaic virus 3a protein potentiates cell-to-cell trafficking of CMV RNA in tobacco plants. *Virology* 207:345–53

28. Ding B, Turgeon R, Parthasarathy MV. 1992. Substructure of freeze-substituted plasmodesmata. *Protoplasma* 169:28–41

29. Dingwall C. 1991. Transport across the nuclear envelope: enigmas and explanations. *BioEssays* 13:213–18

30. Dingwall C, Laskey RA. 1991. Nuclear targeting sequences: a consensus? *Trends Biochem. Sci.* 16:478–81

31. Dorokhov YL, Alexandrova NM, Miroschnichenko NA, Atabekov JG. 1983. Isolation and analysis of virus-specific ribonucleoprotein of tobacco mosaic virus-infected tobacco. *Virology* 127:237–52

32. Enenkel C, Blobel G, Rexach M. 1995. Identification of a yeast karyopherin heterodimer that targets import substrate to mammalian nuclear pore complexes. *J. Biol. Chem.* 270:16499–502

33. Erwee MG, Goodwin PB. 1984. Characterization of the *Egeria densa* leaf symplast: response to plasmolysis, deplasmolysis and to aromatic amino acids. *Protoplasma* 122:162–68

34. Esau K. 1948. Some anatomical aspects of plant virus disease problems. II. *Bot. Rev.* 14:413–49

35. Evert RF. 1990. Dicotyledons. In *Sieve Elements*, ed. H-D Behnke, RD Sjolund, pp. 103–37. New York: Springer-Verlag

36. Fisher DB, Wu Y, Ku MSB. 1992. Turnover of soluble proteins in the wheat sieve tube. *Plant Physiol.* 100:1433–41

37. Franceschi VR, Ding B, Lucas WJ. 1994. Mechanism of plasmodesmata formation in characean algae in relation to evolution of intercellular communication in higher plants. *Planta* 192:347–58

38. Fujiwara T, Giesman-Cookmeyer D, Ding B, Lommel SA, Lucas WJ. 1993. Cell-to-cell trafficking of macromolecules through

plasmodesmata potentiated by the red clover necrotic virus movement protein. *Plant Cell* 5:1783–94

39. Garcia-Bustos J, Heitman J, Hall MN. 1991. Nuclear protein localization. *Biochim. Biophys. Acta* 1071:83–101

40. Gibbs AJ. 1976. Viruses and plasmodesmata. In *Intercellular Communication in Plants: Studies on Plasmodesmata,* ed. BES Gunning, AW Robards, pp. 149–64. Berlin: Springer-Verlag

41. Gilmore TD. 1990. NF-KB, KBF1, *dorsal* and related matters. *Cell* 62:841–43

42. Goldfarb D, Michaud N. 1991. Pathways for the nuclear transport of proteins and RNAs. *Trends Cell Biol.* 1:20–24

43. Goldfarb DS, Gariepy J, Schoolnik G, Kornberg RD. 1986. Synthetic peptides as nuclear localization signals. *Nature* 322: 641–44

44. Goodman RM. 1981. Geminiviruses. In *Handbook of Plant Virus Infection and Comparative Diagnosis,* ed. E Kurstak, pp. 879–910. New York: Elsevier

45. Gorlich D, Prehn S, Laskey RA, Hartman E. 1994. Isolation of a protein that is essential for the first step of nuclear import. *Cell* 79:767–78

46. Gorlich D, Vogel F, Mills AD, Hartmann E, Laskey RA. 1995. Distinct functions for the two importin subunits in nuclear protein import. *Nature* 377:246–48

47. Guralnick B, Thomsen G, Citovsky V. 1996. Transport of DNA into the nuclei of Xenopus oocytes by a modified VirE2 protein of Agrobacterium. *Plant Cell* 8:363–73

48. Gutteridge S, Gatenby AA. 1995. Rubisco synthesis, assembly, mechanism, and regulation. *Plant Cell* 7:809–19

49. Ham J, Parker MG. 1989. Regulation of gene expression by nuclear hormone receptors. *Curr. Opin. Cell Biol.* 1:503–11

50. Hatada EN, Nieters A, Wulczyn G, Naumann M, Meyer R, et al. 1992. The ankyrin repeat domains of the NF-κB precursor p105 and the protooncogene *bcl-3* act as specific inhibitors of NF-κB DNA binding. *Proc. Natl. Acad. Sci. USA* 89: 2489–93

51. Heinlein M, Epel BL, Beachy RN. 1995. Interaction of tobamovirus movement proteins with the plant cytoskeleton. *Science* 270:1983–85

52. Hepler PK. 1982. Endoplasmic reticulum in the formation of the cell plate and plasmodesmata. *Protoplasma* 111:121–33

53. Hilf ME, Dawson WO. 1993. The tobamovirus capsid protein functions as a host-specific determinant of long-distance movement. *Virology* 193:106–14

54. Howard EA, Zupan JR, Citovsky V, Zambryski P. 1992. The VirD2 protein of A. tumefaciens contains a C-terminal bipartite nuclear localization signal: implications for nuclear uptake of DNA in plant cells. *Cell* 68:109–18

55. Kempers R, Prior DAM, van Bel AJE, Oparka KJ. 1993. Plasmodesmata between sieve elements and companion cells in extracellular phloem of *Cucurbita maxima* stems permit intercellular passage of fluorescent 3 kDa probes. *Plant J.* 4: 567–75

56. Kitajima EW, Lauritis JA. 1969. Plant virions in plasmodesmata. *Virology* 37:681–85

57. Kumar NM, Gilula NB. 1996. The gap junction communication channel. *Cell* 84:381–88

58. Lee S, Stenger DC, Bisaro DM, Davis KR. 1994. Identification of loci in *Arabidopsis* that confer resistance to geminivirus infection. *Plant J.* 6:525–35

59. Leisner SM, Howell SH. 1993. Long-distance movement of viruses in plants. *Trends Microbiol.* 1:314–17

60. Leisner SM, Turgeon R, Howell SH. 1993. Effects of host plant development and genetic determinants on the long-distance movement of cauliflower mosaic virus in *Arabidopsis. Plant Cell* 5:191–202

61. Lucas WJ, Bouche-Pillon S, Jackson DP, Nguyen L, Baker L, et al. 1995. Selective trafficking of KNOTTED1 homeodomain protein and its mRNA through plasmodesmata. *Science* 270:1980–83

62. Lucas WJ, Ding B, van der Schoot C. 1993. Plasmodesmata and the supracellular nature of plants. *New Phytol.* 125:435–76

63. Lucas WJ, Gilbertson RL. 1994. Plasmodesmata in relation to viral movement within leaf tissues. *Annu. Rev. Phytopathol.* 32:387–411

64. Lucas WJ, Wolf S, Deom CM, Kishore GM, Beachy RN. 1990. Plasmodesmata-virus interaction. In *Parallels in Cell-to-Cell Junctions in Plant and Animals,* ed. AW Robards, H Jongsma, WJ Lucas, J Pitts, D Spray, pp. 261–72. Berlin: Springer-Verlag

65. Lux SE, John KM, Bennett V. 1990. Analysis of cDNA for human erythrocyte ankyrin indicates a repeated structure with homology to tissue-differentiation and cell-cycle control proteins. *Nature* 344:36–42

66. McLean BG, Zupan J, Zambryski P. 1995. Tobacco mosaic virus movement protein associates with the cytoskeleton in tobacco cells. *Plant Cell* 7:2101–14

67. Meiners S, Schindler M. 1989. Charac-

terization of a connexin homologue in cultured soybean cells and diverse plant organs. *Planta* 179:148–55

68. Meiners S, Xu A, Schindler M. 1991. Gap junction protein homologue from *Arabidopsis thaliana:* evidence for connexins in plants. *Proc. Natl. Acad. Sci. USA* 88: 4119–22

69. Melchior F, Paschal B, Evans J, Gerace L. 1993. Inhibition of nuclear import by the nohydrolyzable analogues of GTP and identification of the small GTPase Ran/TC4 as an essential transport factor. *J. Cell Biol.* 123:1649–59

70. Mezitt LA, Lucas WJ. 1996. *A role for macromolecular trafficking in floral morphogenesis.* Presented at Int. Workshop Basic Appl. Res. Plasmodesmal Biol., 3rd, pp. 82–86. Zichron-Yakov, Israel

71. Moore MS, Blobel G. 1993. The GTP-binding protein Ran/TC4 is required for protein import into the nucleus. *Nature* 365: 661–63

72. Munch E. 1930. *Die Stoffbewegung in der Pflanze.* Jena: Fischer

73. Mushegian AR, Koonin EV. 1993. The proposed plant connexin is a protein kinase-like protein. *Plant Cell* 5:998–99

74. Nakamura S, Hayashi H, Mori S, Chino M. 1993. Protein phosphorylation in the sieve tubes of rice plants. *Plant Cell Physiol.* 34:927–33

75. Nehrbass U, Blobel G. 1996. Role of the nuclear transport factor p10 in nuclear import. *Science* 272:120–22

76. Newmeyer DD, Forbes DJ. 1988. Nuclear import can be separated into distinct steps *in vitro:* nuclear pore binding and translocation. *Cell* 52:641–53

77. Newmeyer DD, Lucocq JM, Burglin TR, De Robertis EM. 1986. Assembly *in vitro* of nuclei active in nuclear protein transport: ATP is required for nucleoplasmin accumulation. *EMBO J.* 5:501–10

78. Nigg EA, Baeuerle PA, Luhrmann R. 1991. Nuclear import-export: in search of signals and mechanisms. *Cell* 66:15–22

79. Noueiry AO, Lucas WJ, Gilbertson RL. 1994. Two proteins of a plant DNA virus coordinate nuclear and plasmodesmal transport. *Cell* 76:925–32

80. O'Neill RE, Jaskunas R, Blobel G, Palese P, Moroianu J. 1995. Nuclear import of influenza virus RNA can be mediated by viral nucleoprotein and transport factors required for protein import. *J. Biol. Chem.* 270:22701–4

81. Osman TAM, Hayes RJ, Buck KW. 1992. Cooperative binding of the red clover necrotic mosaic virus movement protein to single-stranded nucleic acids. *J. Gen. Virol.* 73:223–27

82. Palikaitis P, Zaitlin M. 1986. Tobacco mosaic virus: infectivity and replication. In *The Rod-Shaped Viruses,* ed. MHV Van Regenmortel, H Fraenkel-Conrat, pp. 105–31. New York: Plenum

83. Pascal E, Sanderfoot AA, Ward BM, Medville R, Turgeon R, et al. 1994. The geminivirus BR1 movement protein binds single-stranded DNA and localizes to the cell nucleus. *Plant Cell* 6:995–1006

84. Picard D, Yamamoto K. 1987. Two signals mediate hormone-dependent nuclear localization of the glucocorticoid receptor. *EMBO J.* 6:3333–40

85. Poirson A, Turner AP, Giovane C, Berna A, Roberts K, et al. 1993. Effect of the alfalfa mosaic virus movement protein expressed in transgenic plants on the permeability of plasmodesmata. *J. Gen. Virol.* 74:2459–61

86. Porter KR, Machado RD. 1960. Studies on the endoplasmic reticulum. IV. Its form and distribution during mitosis in cells of onion root tip. *J. Biophys. Biochem. Cytol.* 7: 167–80

87. Powers MA, Forbes DJ. 1994. Cytosolic factors in nuclear import: what's importin? *Cell* 79:931–34

88. Radford J, White RG. 1996. *Preliminary localization of myosin to plasmodesmata.* Presented at Int. Workshop Basic Appl. Res. Plasmodesmal Biol., 3rd, pp. 37–38 Zichron-Yakov, Israel

89. Raikhel NV. 1992. Nuclear targeting in plants. *Plant Physiol.* 100:1627–32

90. Raven JA. 1991. Long-term functioning of enucleate sieve elements: possible mechanisms of damage avoidance and damage repair. *Plant Cell Environ.* 14:139–46

91. Rexach M, Blobel G. 1995. Protein import into nuclei: association and dissociation reactions involving transport substrate, transport factors, and nucleoporins. *Cell* 83: 683–92

92. Richardson WD, Mills AD, Dilworth SM, Laskey RA, Dingwall C. 1988. Nuclear pore migration involves two steps: rapid binding at the nuclear envelope followed by slower translocation through nuclear pores. *Cell* 52:655–64

93. Saito T, Imai Y, Meshi T, Okada Y. 1988. Interviral homologies of the 30K proteins of tobamoviruses. *Virology* 167:653–56

94. Sakuth T, Schobert C, Pecsvaradi A, Eichholz A, Komor E, et al. 1993. Specific proteins in the sieve-tube exudate of *Ricinus communis* L. seedlings: separation, characterization and *in vivo* labelling. *Planta* 191:207–13

95. Sanderfoot AA, Lazarowitz SG. 1995. Co-operation in viral movement: the geminivirus BL1 movement protein interacts with BR1 and redirects it from the nucleus to the cell periphery. *Plant Cell* 7:1185–94

96. Schobert C, Großmann P, Gottschalk M, Komor E, Pecsvaradi A, et al. 1995. Sieve-tube exudate from *Ricinus communis* L. seedlings contains ubiquitin and chaperones. *Planta* 196:205–10

97. Schoumacher F, Erny C, Berna A, Godefroy-Colburn T, Stussi-Garaud C. 1992. Nucleic acid binding properties of the alfalfa mosaic virus movement protein produced in yeast. *Virology* 188:896–99

98. Shirakawa F, Urizel SB. 1989. *In vitro* activation and nuclear translocation of NF-κB catalysed by cyclic AMP-dependent protein kinase and protein kinase C. *Mol. Cell. Biol.* 9:2424–30

99. Simon AE. 1994. Interactions between *Arabidopsis thaliana* and viruses. In *Arabidopsis*, ed. EM Meyerowitz, CR Somerville, pp. 685–704. Cold Spring Harbor, NY: Cold Spring Harbor Lab.

100. Sinha NR, Hake S. 1994. The *Knotted1* leaf blade is a mosaic of blade, sheath and auricle identities. *Dev. Genet.* 15:401–14

101. Smith LM, Sabnis DD, Johnson RPC. 1987. Immunochemical localization of phloem lectin from *Cucurbita maxima* using peroxidase and colloidal gold labels. *Planta* 170:461–70

102. Strasburger E. 1901. Ueber Plasmaverbindungen pflanzlicher Zellen. *Jahrb. Wiss. Bot.* 36:493–601

103. Takamatsu N, Ishiakwa M, Meshi T, Okada Y. 1987. Expression of bacterial chloramphenicol acetyltransferase gene in tobacco plants infected by TMV-RNA. *EMBO J.* 6:307–11

104. Tangl E. 1879. Ueber offene Communicationen zwischen den Zellen des Endosperms einiger Samen. *Jahrb. Wiss. Bot.* 12:170–90

105. Terry BR, Robards AW. 1987. Hydrodynamic radius alone governs the mobility of molecules through plasmodesmata. *Planta* 171:145–57

106. Thomas CL, Maule AJ. 1995. Identification of the cauliflower mosaic virus movement protein RNA binding domain. *Virology* 206:1145–49

107. Tucker EB. 1988. Inositol biphosphate and inositol triphosphate inhibit cell-to-cell passage of carboxyfluorescein in staminal hairs of *Setcreasea purpurea*. *Planta* 174:358–63

108. van der Schoot C, Dietrich MA, Storms M, Verbeke JA, Lucas WJ. 1995. Establishment of a cell-to-cell communication pathway between separate carpels during gynoecium development. *Planta* 195:450–55

109. Vaquero C, Turner AP, Demangeat G, Sanz A, Serra MT, et al. 1994. The 3a protein from cucumber mosaic virus increases the gating capacity of plasmodesmata in transgenic tobacco plants. *J. Gen. Virol.* 75:3193–97

110. Vollbrecht E, Veit B, Sinha NR, Hake S. 1991. The developmental gene *Knotted-1* is a member of a maize homeobox gene family. *Nature* 350:241–43

111. Waigmann E, Lucas W, Citovsky V, Zambryski P. 1994. Direct functional assay for tobacco mosaic virus cell-to-cell movement protein and identification of a domain involved in increasing plasmodesmal permeability. *Proc. Natl. Acad. Sci. USA* 91:1433–37

112. Waigmann E, Zambryski P. 1995. Tobacco mosaic virus movement protein-mediated transport between trichome cells. *Plant Cell* 7:2069–79

113. Weis K, Mattaj IW, Lamond AI. 1995. Identification of hSRP1-α as a functional receptor for nuclear localization sequences. *Science* 268:1049–53

114. White RG, Badelt K, Overall RL, Vesk M. 1994. Actin associated with plasmodesmata. *Protoplasma* 180:169–84

115. Whiteside ST, Goodbourn S. 1993. Signal transduction and nuclear targeting: regulation of transcription factor activity by subcellular localization. *J. Cell Sci.* 104:949–55

116. Wolf S, Deom CM, Beachy RN, Lucas WJ. 1989. Movement protein of tobacco mosaic virus modifies plasmodesmatal size exclusion limit. *Science* 246:377–79

117. Zambryski P. 1995. Plasmodesmata: plant channels for molecules on the move. *Science* 270:1943–44

118. Zheng G-C, Nie X-V, Wang Y-X, Jian L-C, Sun L-H, et al. 1985. Cytochemical localization of ATPase activity during cytomixis in pollen mother cells of David lily-*Lilium davidii* var. Willmottiae and its relation to the intercellular migrating chromatin substance. *Acta Acad. Sin.* 27:26–32

119. Zupan J, Citovsky V, Zambryski P. 1996. *Agrobacterium* VirE2 protein mediates nuclear uptake of ssDNA in plant cells. *Proc. Natl. Acad. Sci. USA* 93:2392–97

Annu. Rev. Plant Physiol. Plant Mol. Biol. 1997. 48:51–66

AUXIN BIOSYNTHESIS

Bonnie Bartel

Department of Biochemistry and Cell Biology, Rice University, Houston, Texas 77005

KEY WORDS: indole-3-acetic acid, IAA, conjugate hydrolysis, phytohormone

ABSTRACT

Indole-3-acetic acid (IAA) is the most abundant naturally occurring auxin. Plants produce active IAA both by de novo synthesis and by releasing IAA from conjugates. This review emphasizes recent genetic experiments and complementary biochemical analyses that are beginning to unravel the complexities of IAA biosynthesis in plants. Multiple pathways exist for de novo IAA synthesis in plants, and a number of plant enzymes can liberate IAA from conjugates. This multiplicity has contributed to the current situation in which no pathway of IAA biosynthesis in plants has been unequivocally established. Genetic and biochemical experiments have demonstrated both tryptophan-dependent and tryptophan-independent routes of IAA biosynthesis. The recent application of precise and sensitive methods for quantitation of IAA and its metabolites to plant mutants disrupted in various aspects of IAA regulation is beginning to elucidate the multiple pathways that control IAA levels in the plant.

CONTENTS

51

1040-2519/97/0601-0051$08.00

INTRODUCTION

Indole-3-acetic acid (IAA) is the most abundant naturally occurring auxin. Plants produce active IAA both by de novo synthesis and by releasing IAA from conjugates. Because the various pathways implicated in IAA biosynthesis have been reviewed recently (2, 61, 63), this review emphasizes recent genetic experiments and complementary biochemical analyses that are beginning to unravel the complexities of IAA biosynthesis in plants. It is becoming evident that several routes for de novo IAA synthesis exist in plants. This multiplicity has contributed to the current situation in which no pathway of IAA biosynthesis in plants has been unequivocally established. Genetic and biochemical experiments have demonstrated both tryptophan-dependent and tryptophan-independent routes of IAA biosynthesis. Analyses of plant mutants disrupted in various aspects of IAA regulation, including de novo synthesis and conjugate hydrolysis, will elucidate the multiple pathways that control IAA levels in the plant and clarify their relative importance in various environmental conditions and developmental stages.

DE NOVO IAA BIOSYNTHESIS

Analysis of IAA Biosynthesis

Several possible pathways for IAA biosynthesis by plants have been proposed. The pathways responsible for IAA biosynthesis in various plant-associated microbes (reviewed in 17, 59, 64) and suggested to exist in plants are shown in Figure 1. Evidence for the relevance of these pathways in plants includes the presence of the intermediates, interconversion of different compounds, and partial purification of enzymes (reviewed in 61). Although all the pathways shown in Figure 1 share tryptophan as a common precursor, it is clear that plants also use tryptophan-independent pathways in the biosynthesis of IAA.

Recent advances in stable isotope labeling techniques and microscale IAA isolation and analysis have proved invaluable for determining the contribution of potential precursors to the final IAA pool (reviewed in 63) and have been particularly useful for analyzing plant mutants disrupted in various aspects of IAA metabolism. D_2O is a useful reagent for in vivo labeling studies because it is accessible to all compartments, is incorporated early in the shikimate pathway, and does not require prior knowledge of the relevant intermediates (2, 61, 63). The use of in vivo stable isotope labeling to determine the relevance of suggested precursor-product relationships also avoids the problems of compartment mixing that are inherent to interconversions observed in plant ex-

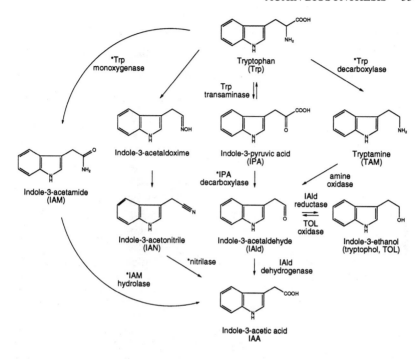

Figure 1 Tryptophan-dependent pathways of IAA biosynthesis. The pathways shown have been implicated in both plant and microbial IAA biosynthesis. Genes encoding enzymes indicated with an asterisk have been cloned from plant-associated bacteria (17, 59, 64). Genes encoding tryptophan decarboxylase (18) and nitrilase (4, 6, 7, 83) have been cloned from plants.

tracts. In isotope-enrichment experiments, the percent incorporation of heavy isotopes must be greater in a precursor than a product. Thus these experiments can indicate whether a particular precursor is a plausible intermediate in a particular pathway, but in isolation cannot define a pathway.

Plants maintain most IAA in conjugated forms (2), which complicates IAA quantitation. These conjugates must be hydrolyzed before analysis in order to measure total IAA in a particular tissue. Free IAA, thought to be the active form of the hormone, is measured in a parallel sample. The contribution of enzymatic hydrolysis of these conjugates to IAA production in vivo is considered at the end of this review.

Little is known about the factors that regulate IAA levels in vivo. To confirm that a particular treatment affects synthesis, effects on turnover and conjugate metabolism have to be considered. For example, this type of detailed analysis has revealed that transgene-mediated cytokinin overproduction down-regulates IAA levels by decreasing synthesis rather than increasing conjugation or turnover (19).

Tryptophan-Dependent IAA Biosynthesis

Although many recent experiments implicate a tryptophan precursor rather than tryptophan itself as the predominant IAA precursor, other experiments continue to demonstrate conversion of tryptophan to IAA. Many studies have demonstrated that plant extracts can convert tryptophan to IAA (reviewed in 61). Cultured carrot cells or excised hypocotyls are able to synthesize IAA from tryptophan (57, 70), maize coleoptile tips incorporate label from tryptophan into IAA (39), and bean seedlings from which the cotyledons are removed (to remove tryptophan supplied from hydrolyzed storage proteins) are able to synthesize IAA from exogenously supplied labeled tryptophan (10). It is possible that the same plant uses different pathways for IAA biosynthesis at different developmental stages (57), in different tissues (39), or under different environmental conditions.

Mutants that Overproduce IAA

Much has been learned about the roles of IAA and other hormones in plant development by expressing bacterial genes involved in hormone metabolism in transgenic plants (28, 38). To fully understand how plants normally synthesize IAA, however, and to assess the relative importance of potentially redundant pathways, it will be necessary to isolate and analyze plant mutants disrupted in each endogenous pathway. Recently, several mutants with altered IAA metabolism have been described, and these are summarized here.

TRYPTOPHAN AUXOTROPHS   Studies of *Zea mays* and *Arabidopsis thaliana* mutants with defects in tryptophan biosynthesis argue against tryptophan as a necessary precursor of IAA. The maize *orange pericarp* (*orp*) mutant is a tryptophan auxotroph because of mutations in both loci encoding tryptophan synthase β (86) (Figure 2). This mutant accumulates 50-fold higher levels of IAA conjugates than wild type, and isotopic enrichment of tryptophan and IAA following labeling with D_2O is inconsistent with tryptophan as a major precursor of IAA in these plants (87). In support of these results, incubation with [^{15}N]anthranilate efficiently labels both tryptophan and IAA in wild-type plants, but only IAA in *orp* mutant seedlings, and labeled tryptophan fails significantly to label IAA in wild-type or *orp* maize (87). In addition, indole rather than tryptophan is an important IAA precursor in maize endosperm (34, 68, 69). In another monocot, *Lemna gibba,* enriching the endogenous tryptophan pool to

98% [^{15}N]Trp does not lead to significant [^{15}N]IAA accumulation or changes in free IAA levels (1).

Similar experiments have been carried out with the *A. thaliana trp2* (44) and *trp3* (66) mutants, which are conditional auxotrophs blocked in the final and penultimate steps in tryptophan biosynthesis, respectively (Figure 2). Although free IAA levels are maintained at wild-type levels in these mutants, there is a 19- to 36-fold overproduction of IAA conjugates and a 6- to 11-fold

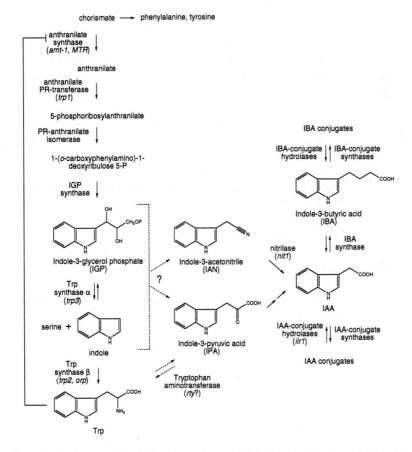

Figure 2   Possible tryptophan-independent pathways of IAA biosynthesis in plants. On the left is the tryptophan biosynthetic pathway, and mutants discussed in the text are indicated in italics. The possible intermediacy of indole-3-acetonitrile (IAN) and indole-3-pyruvate (IPA) in tryptophan-independent IAA biosynthesis is discussed in the text. Hypothetical conversions are indicated with dashed lines.

accumulation of the possible IAA precursor indole-3-acetonitrile (62). Consistent with this biochemical analysis, the *trp2* and *trp3* mutants do not display phenotypes indicative of an overproduction of free IAA (44, 66). Incubation of *trp2* seedlings with [^{15}N]anthranilate results in labeling of 39% of the free IAA pool with ^{15}N, while only 13% of the tryptophan pool is labeled, ruling out tryptophan as a predominant IAA precursor in these plants (62). *trp1-1*, an *A. thaliana* conditional tryptophan auxotroph blocked at the second step of tryptophan biosynthesis (anthranilate phosphoribosyltransferase; see Figure 2) that accumulates anthranilate (45, 72), has similar levels of IAA and IAA conjugates as wild-type plants (62), indicating that the branchpoint for IAA biosynthesis lies after this step. Thus most IAA in the *trp2* and *trp3* mutants, and probably in wild-type *A. thaliana* as well, is not made from tryptophan, but rather from an intermediate in the pathway between anthranilate and tryptophan.

FEEDBACK-INSENSITIVE ANTHRANILATE SYNTHASE   Anthranilate synthase converts chorismate, the common precursor of three aromatic amino acids, to anthranilate (Figure 2). This branchpoint enzyme is subject to allosteric feedback inhibition by the end product tryptophan. Mutants (*amt-1, trp5*) in the *A. thaliana ASA1* gene [encoding the major isoform of the α subunit of anthranilate synthase (60)] that result in feedback-insensitive anthranilate synthase contain elevated levels of free tryptophan because of increased flux through the tryptophan biosynthetic pathway (41, 42, 47). The *amt-1* mutant also overproduces IAA conjugates, free indole-3-butyric acid (IBA; see below), and IBA conjugates to levels about threefold over wild type. Interestingly, this mutant maintains normal free IAA levels (54).

Similarly, mutants of *L. gibba* selected for resistance to the toxic tryptophan analog α-methyltryptophan contain feedback-insensitive anthranilate synthase and accumulate tryptophan (78). As in the *A. thaliana amt-1* mutant, this increased flux through the tryptophan pathway correlates with increased IAA biosynthesis, although in this system the details of the subsequent IAA metabolism differ. Whereas the *A. thaliana* mutant maintains normal free IAA levels by increasing conjugate levels, the *L. gibba* mutant maintains slightly more free IAA and the same conjugated IAA concentration but turns over IAA 10 times faster than the parental line (78).

Thus, increasing levels of IAA (and tryptophan) precursors increases synthesis of IAA, as inferred from increased levels of IAA conjugates or turnover rates. Precursors can be increased by rendering anthranilate synthase feedback insensitive and therefore increasing flux through the pathway (54, 78) or by blocking downstream biosynthetic steps (62, 87), which also leads to anthranilate synthase induction due to tryptophan starvation (3, 66). These results suggest that IAA synthesis is partially limited in wild-type plants by precursor availability. It is important to note that in each of these examples, the plant maintains relatively normal free IAA levels, which indicates that plants can use several mechanisms, including degradation and conjugation, to maintain constant free IAA levels in the face of fluctuations in IAA synthesis, regardless of their nutritional state.

ROOTY Mutants have been found, however, that disrupt this homeostasis. An *A. thaliana* mutant that overproduces both free IAA and IAA conjugates has been described by several laboratories and has been given various names [e.g. *alf1* (12), *sur1* (11), *ivr, rty* (37), *hls3* (46)]. In marked contrast with the *trp2*, *trp3,* and *amt-1* mutants, which overaccumulate conjugated but not free IAA and display relatively normal morphologies, the *rty* mutant has a number of striking phenotypes, which include an increased number of lateral and adventitious roots, the absence of an apical hook when grown in the dark, epinastic cotyledons and leaves, and sterility due to lack of flower development. These effects can be phenocopied by applying synthetic auxins to *A. thaliana* plants (11, 37, 46). Consistent with these phenotypic abnormalities, free IAA levels in mutant tissue are increased two- to sixfold with similar increases in conjugated IAA (11, 37, 46). As both amide- and ester-linked conjugates of IAA accumulate along with free IAA in the *rty* mutant, it is unlikely that this mutant is disrupted in conjugate formation.

The *axr1-3* mutant, which displays reduced sensitivity to auxin (23, 48), partially suppresses the *rty* mutant (37). The *axr1-3* mutant also suppresses the phenotypic consequences of IAA overproduction resulting from expression of the *Agrobacterium tumefaciens* IAA biosynthetic gene, *iaaM* (71). These results support the hypothesis that the *rty* phenotype results from increased IAA production rather than the IAA overproduction being a secondary consequence of the altered morphology.

All 13 of the described mutants with a rooty phenotype [*alf1-1* (12); *sur1-1* to *sur1-7* (11); *rty-1, ivr-1,* and *ivr-2* (37); *hls3-1* (46); and *rty-3* (26)] are allelic and recessive, suggesting that the *RTY* gene product is a unique regulator of auxin homeostasis (37). Recently, the *RTY* gene has been cloned by T-DNA tagging and shown to encode a protein similar to tyrosine aminotransferases (26), which implies that an aminotransferase normally acts to limit free IAA accumulation. The RTY aminotransferase might act to enhance IAA (or IAA conjugate) degradation or to limit IAA synthesis. It is also possible that the RTY protein functions more indirectly, for example in the synthesis of an endogenous inhibitor of IAA synthesis.

In one of the proposed pathways of IAA biosynthesis (Figure 1), an aromatic aminotransferase catalyzes the conversion of tryptophan to indole-3-pyruvate (IPA). If this were a predominant IAA biosynthetic pathway in plants, one would expect an aminotransferase mutant to be deficient in IAA biosynthesis. However, the observation that the *rty* mutant overproduces IAA suggests that the normal function of the *RTY* gene product might be to convert IPA to tryptophan, thereby controlling IPA conversion to IAA (Figure 2). This speculative model requires a tryptophan-independent pathway of IPA biosynthesis. Whereas isotopic enrichment following labeling of tomato shoots with D_2O is consistent with IPA being a precursor of IAA, labeling of tryptophan in the same experiment was insufficient for tryptophan to be the primary IPA precursor in this system (15, 61). Moreover, in the plant-associated bacterium *Azospirillum brasilense,* 90% of the IAA is synthesized via a tryptophan-independent pathway when tryptophan is absent from the media (65), and disrupting the *A. brasilense* gene encoding IPA decarboxylase (16) reduces IAA production by 95% (65). Thus, in both tomato (15, 61) and in *A. brasilense* (17), IPA can be synthesized independently of tryptophan, and IPA remains a candidate intermediate in the tryptophan-independent synthesis of IAA in plants (Figure 2). Because aminotransferases are reversible, this model is not inconsistent with tryptophan-dependent IAA biosynthesis observed in extracts supplied with large amounts of tryptophan.

FASS A second *A. thaliana* mutant that overproduces IAA is the *fass* mutant. This mutant accumulates twofold more free IAA, normal ester-linked IAA, and twofold less amide-linked IAA than wild-type plants, suggesting a possible defect in IAA conjugation (25). The *fass* mutant has extreme defects in organ expansion and was originally isolated as a seedling-lethal mutant (81). The *fass* mutant phenotype does not resemble the *rty* mutant phenotype, despite both mutants having free IAA levels increased to a similar extent. Unlike the *rty* mutant, the *fass* mutant is not closely phenocopied by exogenous auxin application, suggesting that the defects in IAA metabolism observed in *fass* might be secondary to some other defect. Analysis of a morphologically identical mutant indicates that this defect might be an absence of cortical microtubular arrays necessary for correct alignment of cell division planes (82).

A LARGE *L. GIBBA* MUTANT A large frond mutant of *L. gibba, jsR₁,* accumulates free IAA during certain phases of growth (75). Interestingly, this mutant lacks detectable IAA conjugates, which suggests that it might be disrupted in conjugate formation (75). Whether the large frond phenotype cosegregates with the IAA accumulation has not been reported.

Mutants with Possible Defects in IAA Biosynthesis

Auxotrophic mutants have been invaluable in deciphering biosynthetic pathways in both microbes and plants (67). Even without a higher plant IAA auxotroph, the analyses of IAA overproduction mutants summarized above have begun to illuminate the pathways of IAA biosynthesis. While the multiplicity of pathways of IAA synthesis and their potential compartmentalization complicate both biochemical and genetic analyses, a mutant disrupted in only one pathway might still be viable.

NITRILASES   A direct approach to isolate IAA biosynthetic mutants is to screen for mutants resistant to potential IAA precursors that remain sensitive to IAA. Several experiments suggest that indole-3-acetonitrile (IAN) is a candidate IAA precursor. IAN accumulates 6- to 11-fold in the *A. thaliana trp3* and *trp2* mutants that overaccumulate IAA conjugates (62), and four genes (*NIT1–NIT4*) encoding nitrilase enzymes that hydrolyze IAN in vitro have been cloned from *A. thaliana* (4, 6, 7). IAN has auxin-like effects when supplied exogenously to *A. thaliana* seedlings, and an IAN-resistant mutant containing a point mutation in the *NIT1* gene has been isolated (J Normanly, P Grisafi, GR Fink & B Bartel, unpublished data).

Experiments with transgenic plants provide additional evidence regarding the function of the *A. thaliana* nitrilase genes. When the *A. thaliana NIT2* gene is expressed in transgenic tobacco, the resultant plants display increased sensitivity to the auxin effects of exogenously supplied IAN and acquire the ability to hydrolyze IAN to IAA (73). Similarly, overexpression of *NIT2* in *A. thaliana* increases IAN sensitivity (J Normanly, P Grisafi, GR Fink & B Bartel, unpublished data). Biochemical analysis of IAA metabolism in the *nit1* mutant and nitrilase overproducing lines can now be used to test the contribution of the IAN pathway to the biosynthesis of IAA in *A. thaliana*.

The generality of the IAN pathway has been questioned by the limited distribution of the nitrilase enzyme, which in a 1964 survey of 21 plant families was found only in members of the Cruciferae, Gramineae, and Musaceae families (79). Genes quite similar to the *A. thaliana NIT* genes have been isolated from tobacco (83), suggesting that the pathway may be more widespread than previously appreciated. Unlike *A. thaliana,* however, tobacco seedlings fail to hydrolyze exogenously supplied IAN to IAA (73), which indicates either that the tobacco nitrilase genes are not active at the developmental stage assayed, or that they use a different substrate.

ALF3   In the *alf3* mutant, both the lateral and primary root meristems arrest growth and eventually die (12). These phenotypes can be prevented by exogenous application of indole or IAA, but not tryptophan (12). This mutant separates a requirement for IAA in lateral root initiation, which is normal in *alf3*, from a requirement for IAA in root meristem maintenance, which is blocked in the mutant. The ability of indole to substitute for IAA in this rescue is particularly intriguing, because indole, like tryptophan, lacks auxin effects when supplied exogenously to wild-type *A. thaliana* plants. The paradox of a mutant apparently deficient in indole but having no obvious tryptophan deficit suggests compartmentalization or channeling of indole destined for IAA biosynthesis versus indole destined for tryptophan biosynthesis. The IAA metabolites of this mutant await characterization.

IBA Biosynthesis

Indole-3-butyric acid (IBA), a naturally occurring auxin that is found in a variety of species, including maize, peas, and *A. thaliana* (reviewed in 22), has long been used commercially to promote rooting. Like IAA, endogenous IBA is found in free and conjugated forms, and exogenously applied IBA is rapidly conjugated by the plant (22). In both maize (49, 52, 53) and *A. thaliana* (50), IBA is synthesized from IAA via a chain elongation reaction similar to those found in fatty acid biosynthesis (53). The IBA synthase purified from maize uses acetyl CoA and ATP as cofactors (52). The conversion of IBA to IAA has also been reported (21, 22), so it is not yet clear whether IBA is itself an auxin or whether IBA exerts auxin activity via its conversion to IAA.

IAA CONJUGATE HYDROLYSIS

In addition to de novo synthesis, plants produce active IAA by hydrolyzing conjugates of IAA with other molecules. The ability to form IAA conjugates is widely distributed in the plant kingdom from mosses to angiosperms (77), and most of the IAA in plant tissues is in a conjugated form (2, 13). The structures of numerous IAA conjugates have been elucidated, and these include conjugates of the carboxyl group of IAA to sugars, high molecular weight glycans, amino acids, and peptides (2, 13). Growing evidence suggests that different conjugates perform different functions in the plant. Certain conjugates, such as IAA-Asp, can be intermediates in IAA destruction (58, 84, 85). Often IAA conjugates may serve as reservoirs of inactive IAA that can be hydrolyzed to provide the plant with active hormone. In maize, IAA-inositol can be transported from the kernel to the shoot, and conjugate hydrolysis, rather than de novo synthesis, supplies the developing seedling with IAA (20). Seven-day-old maize seedlings incorporate D_2O into tryptophan but not IAA, which indicates that de novo IAA synthesis has not yet begun (35). Thus conjugate hydrolysis must be considered along with de novo synthesis when considering input into the free IAA pool.

Biochemistry of Conjugate Hydrolysis

Studies have suggested a relationship between the biological activity of IAA–amino acid conjugates and their hydrolysis rates. This correlation was inferred from the study of IAA conjugates as slow release forms of auxin in tissue culture (29) and directly demonstrated in a bean internode curvature assay, in which the auxin activity of IAA–amino acid conjugates was directly related to the amount of free IAA released by the tissue (9). In support of this role, an enzyme hydrolyzing IAA-alanine has been partially purified from bean (14) and carrot (43), and a maize activity hydrolyzing IAA-*myo*-inositol has been characterized (27). Enzymes that hydrolyze IAA-glucose have been detected in a number of plants, including corn (40), oats, potatoes, beans (33), and Chinese cabbage (51).

Hydrolysis of IAA–amino acid conjugates has been studied in some detail in extracts of Chinese cabbage. This system contains isozymes that hydrolyze IAA-Ala, IAA-Asp, or IAA-Phe but not IAA conjugates with inositol or the amino acids Gly, Val, or Ile (51). Inoculation with the causative agent of clubroot disease, *Plasmodiophora brassicae,* correlates with dramatic increases in both amide-linked IAA conjugates and a hydrolytic activity specific for IAA-Asp (51). This induction of a specific hydrolase isozyme in response to a specific challenge supports the hypothesis that the various conjugate hydrolases might function to supply free IAA in response to a variety of needs.

Genetics of Conjugate Hydrolysis

Certain exogenously applied IAA conjugates mimic IAA in several bioassays (5, 9, 24, 30, 55, 56, 76), which suggests either that these conjugates are auxins themselves or that the plant or tissue hydrolyzes the conjugates to release free IAA. This observation was exploited in the isolation of the first gene encoding an IAA conjugate hydrolase. The *ilr1* mutant was isolated by screening for *A. thaliana* plants able to elongate roots on inhibitory concentrations of IAA-leucine (5). The gene defective in this mutant was cloned using a map-based approach and shown to encode an IAA–amino acid hydrolase with preference for IAA-Leu and IAA-Phe among the conjugates tested (5). Isolation of the *ILR1* gene allowed the identification of a family in *A. thaliana* that contains at least six genes encoding related IAA conjugate hydrolases. Additional IAA–amino acid conjugate-insensitive mutants may have defects in other members of this gene family. For example, the *iar3* mutant, which was iso-lated on the basis of resistance to the IAA-like effects of IAA-Ala, is mutated in a homolog that encodes an IAA-Ala–specific hydrolase (D Goetz & B Bartel, unpublished data).

Examination of the amino acid sequences of the IAA–amino acid hydro-lases identified to date provide some clues about the localization of conjugate hydrolysis in the plant. Each member of the ILR1-like family has a hydropho-bic leader sequence that is predicted to target the secretory pathway. More-over, one member of the family, ILL2, terminates with the motif -His-Asp-Glu-Leu, which has been shown to dictate retention in the lumen of the ER (8). This is intriguing because the major auxin–binding protein from maize, ABP1, also has an ER retention signal (31, 32) and is largely localized to the ER (80), although its site of action may well be the plasma membrane (36). Some members of the ILR1 family display related carboxyl-terminal sequences (5), while others do not (B Bartel, unpublished data), which suggests an added layer of complexity in the compartmentalization of conjugate hydrolysis. However, the actual subcellular location of these enzymes has not yet been directly determined, and the only report on subcellular localization of IAA conjugates reports a cytosolic location (74).

Analysis of the expression patterns of the *ILR1*-like genes and the substrate specificities of the corresponding hydrolases is likely to determine whether the members of this family perform unique, overlapping, or fully redundant func-tions. The relevance of these enzymes in IAA metabolism in vivo is confirmed by the observation that the *iar1* mutant partially suppresses the *alf1* (an allele of *rty,* see above) mutant (B Bartel, unpublished data), which proliferates lateral and adventitious roots presumably because of the overproduction of free IAA (12).

CONCLUSIONS AND PROSPECTS

The pathways of IAA biosynthesis in plants have for years been elusive targets for plant biologists. It is becoming clear that multiple pathways exist for de novo IAA synthesis in plants and that a number of plant enzymes can liberate IAA from conjugates. Much work remains both in defining the IAA biosyn-thetic pathways and in determining their relative importance in supplying the plant with free IAA. The complexity and redundancy seen in IAA biosynthesis likely reflects the importance to the plant of exact modulation of this essential growth regulator. The recent development of precise and sensitive methods for quantitation of IAA and its metabolites in combination with the isolation of plant mutants disrupted in various aspects of IAA metabolism makes this an exciting time in the auxin field.

ACKNOWLEDGMENTS

I thank Janet Braam, John Celenza, Susan Gibson, Seiichi Matsuda, and Neil Olszewski for critical comments on the manuscript, and Jerry Cohen and other colleagues for reprints and unpublished data. Work in my laboratory is supported by the National Institutes of Health (GM54749-01), the Texas Advanced Technology Program (003604-065), and the Robert A. Welch Foundation (C-1309).

> Visit the *Annual Reviews home page* at http://www.annurev.org.

Literature Cited

1. Baldi BG, Maher BR, Slovin JP, Cohen JD. 1991. Stable isotope labeling, in vivo, of D- and L-tryptophan pools in *Lemna gibba* and the low incorporation of label into indole-3-acetic acid. *Plant Physiol.* 95: 1203–8
2. Bandurski RS, Cohen JD, Slovin JP, Reinecke DM. 1995. Auxin biosynthesis and metabolism. See Ref. 17a, pp. 39–65
3. Barczak AJ, Zhao J, Pruitt KD, Last RL. 1995. 5-Fluoroindole resistance identifies tryptophan synthase beta subunit mutants in *Arabidopsis thaliana. Genetics* 140: 303–13
4. Bartel B, Fink GR. 1994. Differential regulation of an auxin-producing nitrilase gene family in *Arabidopsis thaliana. Proc. Natl. Acad. Sci. USA* 91:6649–53
5. Bartel B, Fink GR. 1995. ILR1, an amidohydrolase that releases active indole-3-acetic acid from conjugates. *Science* 268: 1745–48
6. Bartling D, Seedorf M, Mithöfer A, Weiler EW. 1992. Cloning and expression of an *Arabidopsis* nitrilase which can convert indole-3-acetonitrile to the plant hormone, indole-3-acetic acid. *Eur. J. Biochem.* 205: 417–24
7. Bartling D, Seedorf M, Schmidt RC, Weiler EW. 1994. Molecular characterization of two cloned nitrilases from *Arabidopsis thaliana:* key enzymes in the biosynthesis of the plant hormone indole-3-acetic acid. *Proc. Natl. Acad. Sci. USA* 91: 6021–25

8. Bednarek SY, Raikhel NV. 1992. Intracellular trafficking of secretory proteins. *Plant Mol. Biol.* 20:133–50
9. Bialek K, Meudt WJ, Cohen JD. 1983. Indole-3-acetic acid (IAA) and IAA conjugates applied to bean stem sections. *Plant Physiol.* 73:130–34
10. Bialek K, Michalczuk L, Cohen JD. 1992. Auxin biosynthesis during seed germination in *Phaseolus vulgaris. Plant Physiol.* 100:509–17
11. Boerjan W, Cervera M-T, Delarue M, Beeckman T, Dewitte W, et al. 1995. *superroot*, a recessive mutation in Arabidopsis, confers auxin overproduction. *Plant Cell* 7:1405–19
12. Celenza JL, Grisafi PL, Fink GR. 1995. A pathway for lateral root formation in *Arabidopsis thaliana. Genes Dev.* 9:2131–42
13. Cohen JD, Bandurski RS. 1982. Chemistry and physiology of the bound auxins. *Annu. Rev. Plant Physiol.* 33:403–30
14. Cohen JD, Slovin JP, Bialek KH, Chen KH, Derbyshire MK. 1988. Mass spectrometry, genetics, and biochemistry: understanding the metabolism of indole-3-acetic acid. In *Biomechanisms Regulating Growth and Development*, ed. GL Steffens, TS Rumsey, pp. 229–41. Dordrecht: Kluwer
15. Cooney TP, Nonhebel HM. 1991. Biosynthesis of indole-3-acetic acid in tomato shoots: measurement, mass-spectral identification and incorporation of ^{2}H from ^{2}H$_2$O into indole-3-acetic acid, D- and L-trypto-

phan, indole-3-pyruvate and tryptamine. *Planta* 184:368–76

16. Costacurta A, Keijers V, Vanderleyden J. 1994. Molecular cloning and sequence analysis of an *Azospirillum brasilense* indole-3-pyruvate decarboxylase gene. *Mol. Gen. Genet.* 243:463–72

17. Costacurta A, Vanderleyden J. 1995. Synthesis of phytohormones by plant-associated bacteria. *Crit. Rev. Microbiol.* 21:1–18

17a. Davies PJ, ed. 1995. *Plant Hormones.* Dordrecht: Kluwer

18. De Luca V, Marineau C, Brisson N. 1989. Molecular cloning and analysis of cDNA encoding a plant tryptophan decarboxylase: comparison with animal dopa decarboxylases. *Proc. Natl. Acad. Sci. USA* 86:2582–86

19. Eklöf S, Astot C, Blackwell J, Moritz T, Olsson O, Sandberg G. 1996. Auxin-cytokinin interactions in wild-type and transgenic tobacco. *Plant and Cell Physiol.* In press

20. Epstein E, Cohen JD, Bandurski RS. 1980. Concentration and metabolic turnover of indoles in germinating kernels of *Zea mays* L. *Plant Physiol.* 65:415–21

21. Epstein E, Lavee S. 1984. Conversion of indole-3-butyric acid to indole-3-acetic acid by cuttings of grapevine (*Vitus vinifera*) and olive (*Olea europa*). *Plant Cell Physiol.* 25:697–703

22. Epstein E, Ludwig-Müller J. 1993. Indole-3-butyric acid in plants: occurrence, synthesis, metabolism and transport. *Physiol. Plant.* 88:382–89

23. Estelle M, Somerville C. 1987. Auxin-resistant mutants of *Arabidopsis thaliana* with an altered morphology. *Mol. Gen. Genet.* 206:200–6

24. Feung C-S, Hamilton RH, Mumma RO. 1977. Metabolism of indole-3-acetic acid. IV. Biological properties of amino acid conjugates. *Plant Physiol.* 59:91–93

25. Fisher RH, Barton MK, Cohen JD, Cooke TJ. 1996. Hormonal studies of *fass,* an Arabidopsis mutant that is altered in organ elongation. *Plant Physiol.* 110:1109–21

26. Gopalraj M, Tseng T-S, Olszewski N. 1996. The *ROOTY* gene of Arabidopsis encodes a protein with highest similarity to aminotransferases. *Plant Physiol.* 111S:114

27. Hall PJ, Bandurski RS. 1986. [³H]Indole-3-acetyl-*myo*-inositol hydrolysis by extracts of *Zea mays* L. vegetative tissue. *Plant Physiol.* 80:374–77

28. Hamill JD. 1993. Alterations in auxin and cytokinin metabolism of higher plants due to expression of specific genes from patho-

genic bacteria: a review. *Aust. J. Plant Physiol.* 20:405–23

29. Hangarter RP, Good NE. 1981. Evidence that IAA conjugates are slow-release sources of IAA in plant tissues. *Plant Physiol.* 68:1424–27

30. Hangarter RP, Peterson MD, Good NE. 1980. Biological activities of indoleacetylamino acids and their use as auxins in tissue culture. *Plant Physiol.* 65:761–67

31. Hesse T, Feldwisch J, Balshüsemann D, Bauw G, Puype M, et al. 1989. Molecular cloning and structural analysis of a gene from *Zea mays* (L.) coding for a putative receptor for the plant hormone auxin. *EMBO J.* 8:2453–61

32. Inohara N, Shimomura S, Fukui T, Futai M. 1989. Auxin-binding protein located in the endoplasmic reticulum of maize shoots: molecular cloning and complete structure. *Proc. Natl. Acad. Sci. USA* 86:3564–68

33. Jakubowska A, Kowalczyk S, Leznicki AJ. 1993. Enzymatic hydrolysis of 4-*O* and 6-*O*-indole-3-ylacetyl-β-D-glucose in plant tissues. *J. Plant Physiol.* 142:61–66

34. Jensen PJ, Bandurski RS. 1994. Metabolism and synthesis of indole-3-acetic acid (IAA) in *Zea mays.* *Plant Physiol.* 106:343–51

35. Jensen PJ, Bandurski RS. 1996. Incorporation of deuterium into indole-3-acetic acid and tryptophan in *Zea mays* seedlings grown on 30% deuterium oxide. *J. Plant Physiol.* 147:697–702

36. Jones AM. 1994. Auxin-binding proteins. *Annu. Rev. Plant Physiol. Plant Mol. Biol.* 45:393–420

37. King JJ, Stimart DP, Fisher RH, Bleecker AB. 1995. A mutation altering auxin homeostasis and plant morphology in Arabidopsis. *Plant Cell* 7:2023–37

38. Klee HJ, Lanahan MB. 1995. Transgenic plants in hormone biology. See Ref. 17a, pp. 340–53

39. Koshiba T, Kamiya Y, Iino M. 1995. Biosynthesis of indole-3-acetic acid from L-tryptophan in coleoptile tips of maize (*Zea mays* L.). *Plant Cell Physiol.* 36:1503–10

40. Kowalczyk S, Bandurski RS. 1990. Isomerization of 1-*O*-(indole-3-ylacety-β-D-glucose: enzymatic hydrolysis of 1-*O,* 4-*O,* and 6-*O*-indole-3-ylacetyl-β-D-glucose and the enzymatic synthesis of indole-3-acetyl glycerol by a hormone metabolizing complex. *Plant Physiol.* 94:4–12

41. Kreps JA, Ponappa T, Dong W, Town CD. 1996. Molecular basis of α-methyltryptophan resistance in *amt-1,* a mutant of *Arabidopsis thaliana* with altered tryptophan metabolism. *Plant Physiol.* 110:1159–65

42. Kreps JA, Town CD. 1992. Isolation and characterization of a mutant of *Arabidopsis thaliana* resistant to α-methyltryptophan. *Plant Physiol.* 99:269–75

43. Kuleck GA, Cohen JD. 1993. An indole-3-acetyl-amino acid hydrolytic enzyme from carrot cells. *Plant Physiol. Suppl.* 102:60

44. Last RL, Bissinger PH, Mahoney DJ, Radwanski ER, Fink GR. 1991. Tryptophan mutants in *Arabidopsis:* the consequences of duplicated tryptophan synthase β genes. *Plant Cell* 3:345–58

45. Last RL, Fink GR. 1988. Tryptophan-requiring mutants of the plant *Arabidopsis thaliana. Science* 240:305–10

46. Lehman A, Black R, Ecker JR. 1996. *HOOKLESS1,* an ethylene response gene, is required for differential cell elongation in the Arabidopsis hypocotyl. *Cell* 85:183–94

47. Li J, Last RL. 1996. The *Arabidopsis thaliana trp5* mutant has a feedback-resistant anthranilate synthase and elevated soluble tryptophan. *Plant Physiol.* 110:51–59

48. Lincoln C, Britton JH, Estelle M. 1990. Growth and development of the *axr1* mutants of *Arabidopsis. Plant Cell* 2:1071–80

49. Ludwig-Müller J, Epstein E. 1992. Indole-3-acetic acid is converted to indole-3-butyric acid by seedlings of *Zea mays* L. In *Progress in Plant Growth Regulation,* ed. CM Karssen, LC van Loon, D Vreugdenhil, pp. 188–93. Dordrecht: Kluwer

50. Ludwig-Müller J, Epstein E. 1994. Indole-3-butyric acid in *Arabidopsis thaliana.* III. In vivo biosynthesis. *Plant Growth Regul.* 14:7–14

51. Ludwig-Müller J, Epstein E, Hilgenberg W. 1996. Auxin-conjugate hydrolysis in Chinese cabbage: characterization of an amidohydrolase and its role during infection with clubroot disease. *Physiol. Plant.* 97:627–34

52. Ludwig-Müller J, Hilgenberg W. 1995. Characterization and partial purification of indole-3-butyric acid synthetase from maize (*Zea mays*). *Physiol. Plant.* 94:651–60

53. Ludwig-Müller J, Hilgenberg W, Epstein E. 1995. The in vitro biosynthesis of indole-3-butyric acid in maize. *Phytochemistry* 40:61–68

54. Ludwig-Müller J, Sass S, Sutter EG, Wodner M, Epstein E. 1993. Indole-3-butyric acid in Arabidopsis thaliana. I. Identification and quantification. *Plant Growth Regul.* 13:179–87

55. Magnus V, Hangarter RP, Good NE. 1992. Interaction of free indole-3-acetic acid and its amino acid conjugates in tomato hypocotyl cultures. *J. Plant Growth Regul.* 11:67–75

56. Magnus V, Nigovic B, Hangarter RP, Good NE. 1992. *N*-(Indole-3-ylacetyl)amino acids as sources of auxin in plant tissue culture. *J. Plant Growth Regul.* 11:19–28

57. Michalczuk L, Ribnicky DM, Cooke TJ, Cohen JD. 1992. Regulation of indole-3-acetic acid biosynthetic pathways in carrot cell cultures. *Plant Physiol.* 100:1346–53

58. Monteiro AM, Crozier A, Sandberg G. 1988. The biosynthesis and conjugation of indole-3-acetic acid in germinating seeds of *Dalbergia dolchipetala. Planta* 174:561–68

59. Morris RO. 1995. Genes specifying auxin and cytokinin biosynthesis in prokaryotes. See Ref. 17a, pp. 318–39

60. Niyogi KK, Fink GR. 1992. Two anthranilate synthase genes in Arabidopsis: defense-related regulation of the tryptophan pathway. *Plant Cell* 4:721–33

61. Nonhebel HM, Cooney TP, Simpson R. 1993. The route, control and compartmentation of auxin synthesis. *Aust. J. Plant Physiol.* 20:527–39

62. Normanly J, Cohen JD, Fink GR. 1993. *Arabidopsis thaliana* auxotrophs reveal a tryptophan-independent biosynthetic pathway for indole-3-acetic acid. *Proc. Natl. Acad. Sci. USA* 90:10355–59

63. Normanly J, Slovin JP, Cohen JD. 1995. Rethinking auxin biosynthesis and metabolism. *Plant Physiol.* 107:323–29

64. Patten CL, Glick BR. 1996. Bacterial biosynthesis of indole-3-acetic acid. *Can. J. Microbiol.* 42:207–20

65. Prinsen E, Costacurta A, Michiels K, Vanderleyden J, Van Onckelen H. 1993. *Azospirillum brasilense* indole-3-acetic acid biosynthesis: evidence for a nontryptophan dependent pathway. *Mol. Plant-Microbe Interact.* 6:609–15

66. Radwanski ER, Barczak AJ, Last RL. 1996. Characterization of tryptophan synthase alpha subunit mutants of *Arabidopsis thaliana. Mol. Gen. Genet.* In press

67. Reid JB, Howell SH. 1995. Hormone mutants and plant development. See Ref. 17a, pp. 448–85

68. Rekoslavskaya NI. 1995. Pathways of indoleacetic acid and tryptophan synthesis in developing maize endosperm: studies in vitro. *Russ. J. Plant Physiol.* 42:143–51

69. Rekoslavskaya NI, Bandurski RS. 1994. Indole as a precursor of indole-3-acetic acid in *Zea mays. Phytochemistry* 35:905–9

70. Ribnicky DM, Ilic N, Cohen JD, Cooke TJ. 1996. The effects of exogenous auxins on

endogenous indole-3-acetic acid metabolism: the implications for carrot somatic embryogenesis. *Plant Physiol.* 112:549–58

71. Romano CP, Robson PRH, Smith H, Estelle M, Klee M. 1995. Transgene-mediated auxin overproduction in *Arabidopsis:* hypocotyl elongation phenotype and interactions with the *hy6-1* hypocotyl elongation and *axr1* auxin-resistant mutants. *Plant Mol. Biol.* 27:1071–83

72. Rose AB, Casselman AL, Last RL. 1992. A phosphoribosylanthranilate transferase gene is defective in blue fluorescent *Arabidopsis thaliana* tryptophan mutants. *Plant Physiol.* 100:582–92

73. Schmidt RC, Müller A, Hain R, Bartling D, Weiler EW. 1996. Transgenic tobacco plants expressing the *Arabidopsis thaliana* nitrilase II enzyme. *Plant J.* 9:683–91

74. Sitbon F, Edlund A, Gardeström P, Olsson O, Sandberg G. 1993. Compartmentalization of indole-3-acetic acid metabolism in protoplasts isolated from leaves of wild-type and IAA-overproducing transgenic tobacco plants. *Planta* 191:274–79

75. Slovin JP, Cohen JD. 1988. Levels of indole-3-acetic acid in *Lemna gibba* G-3 and in a large *Lemna* mutant regenerated from tissue culture. *Plant Physiol.* 86:522–26

76. Soskic M, Klaic B, Magnus V, Sabljic A. 1995. Quantitative structure-activity relationships for *N*-(indole-3-ylacetyl)amino acids used as sources of auxin in plant tissue culture. *Plant Growth Regul.* 16: 141–52

77. Sztein AE, Cohen JD, Slovin JP, Cooke TJ. 1995. Auxin metabolism in representative land plants. *Am. J. Bot.* 82:1514–21

78. Tam YY, Slovin JP, Cohen JD. 1995. Selection and characterization of α-methyltryptophan-resistant lines of *Lemna gibba* showing a rapid rate of indole-3-acetic acid turnover. *Plant Physiol.* 107:77–85

79. Thimann KV, Mahadevan S. 1964. Nitrilase. I. Occurrence, preparation, and general properties of the enzyme. *Arch. Biochem. Biophys.* 105:133–41

80. Tian HC, Klämbt D, Jones AM. 1995. Auxin-binding protein 1 does not bind auxin within the endoplasmic reticulum despite this being the predominant subcellular location for this hormone receptor. *J. Biol. Chem.* 270:26962–69

81. Torres-Ruiz RA, Jürgens G. 1994. Mutations in the *FASS* gene uncouple pattern formation and morphogenesis in *Arabidopsis* development. *Development* 120:2967–78

82. Traas J, Bellini C, Nacry P, Kronenberger J, Bouchez D, Caboche M. 1995. Normal differentiation patterns in plants lacking microtubular preprophase bands. *Nature* 375:676–77

83. Tsunoda H, Yamguchi K. 1995. The cDNA sequence of an auxin-producing nitrilase homologue in tobacco (Access. No. D63331). *Plant Physiol.* 109:339

84. Tsurumi S, Wada S. 1986. Dioxindole-3-acetic acid conjugates formation from indole-3-acetylaspartic acid in *Vicia* seedlings. *Plant Cell Physiol* 27:1513–22

85. Tuominen H, Östin A, Sandberg G, Sundberg B. 1994. A novel metabolic pathway for indole-3-acetic acid in apical shoots of *Populus tremula* (L.) X *Populus tremuloides* (Michx.). *Plant Physiol.* 106: 1511–20

86. Wright AD, Moehlenkamp CA, Perrot GH, Neuffer MG, Cone KC. 1992. The maize auxotrophic mutant *orange pericarp* is defective in duplicate genes for tryptophan synthase β. *Plant Cell* 4:711–19

87. Wright AD, Sampson MB, Neuffer MG, Michalczuk L, Slovin JP, Cohen JD. 1991. Indole-3-acetic acid biosynthesis in the mutant maize *orange pericarp,* a tryptophan auxotroph. *Science* 254:998–1000

Annu. Rev. Plant Physiol. Plant Mol. Biol. 1997. 48:67–87
Copyright © 1996 by Annual Reviews Inc. All rights reserved

THE SYNTHESIS OF THE STARCH GRANULE

A.M. Smith, K. Denyer, and C. Martin

John Innes Centre, Colney Lane, Norwich NR4 7UH, United Kingdom

KEY WORDS: ADPglucose pyrophosphorylase, amylopectin, amylose, starch-branching enzyme, starch synthase

ABSTRACT

This review describes and discusses the implications of recent discoveries about how starch polymers are synthesized and organized to form a starch granule. Three issues are highlighted. 1. The role and importance of ADPglucose pyrophosphorylase in the generation of ADPglucose as the substrate for polymer synthesis. 2. The contributions of isoforms of starch-branching enzyme, starch synthase, and debranching enzyme to the synthesis and ordered packing of amylopectin molecules. 3. The requirements for and regulation of the synthesis of amylose.

CONTENTS

This review is dedicated to the memory of our mentors and friends Professor Tom ap Rees (University of Cambridge), who died on October 3, 1996, and Professor Harold Woolhouse (Director of the John Innes Institute, 1980–1991), who died on June 19, 1996.

INTRODUCTION

Starch is both the major component of yield in the world's main crop plants and an important raw material for many industrial processes. The advent of

67

routine genetic manipulation of these plants in the past decade has made it possible, in theory at least, to increase yield and to provide novel raw materials through alteration of the pathway of starch synthesis (60, 88). This prospect has fueled an enormous amount of research, aimed primarily at characterizing the three enzymes directly involved in starch synthesis—ADPglucose pyrophosphorylase, starch synthase, and starch-branching enzyme—and cloning the genes that encode them. Improved understanding of these enzymes (70) has outstripped understanding of the nature and regulation of the processes in which they are involved. If rational manipulation of flux through the pathway and the structure of the starch granule is to be realized, predictive models of the processes are required. In this review we consider recent developments in understanding of three interrelated processes central to starch synthesis: the generation of ADPglucose as the substrate for the synthesis of starch polymers, the synthesis and packing of amylopectin molecules to form the starch granule, and the synthesis of amylose.

THE GENERATION OF ADPGLUCOSE

It is generally accepted that the enzyme ADPglucose pyrophosphorylase (AGPase) is responsible in all plant organs for the synthesis of ADPglucose, the substrate for the synthesis of starch polymers (Figure 1; 1, 67). In spite of its universal importance, however, the nature and location of this enzyme show remarkable variation within and between organs and between species. In this section, we discuss the extent of this variation and its possible significance for the generation of ADPglucose.

The enzyme consists of large and small subunits, which show considerable similarities but can be distinguished by features of their primary amino acid sequences. The sequences of the small subunits are highly conserved between species, whereas those of the large subunits are more divergent (75). Probably most plants contain small multigene families encoding one or both subunits, members of which display different patterns of expression. There are considerable differences between species in the number and expression patterns of genes encoding each subunit. For example, in potato, four different transcripts—three encoding large and one encoding small subunits—have been identified. All of the large subunit genes are expressed in the tuber, but only two of the three are expressed in leaves (50). Expression of each of the genes encoding the subunits is differently regulated by developmental and metabolic signals (61, 62, 65). In bean (*Vicia faba*), one large and two small subunit genes are expressed in the embryo (89). The two small subunit genes show different spatial and temporal patterns of expression within the developing

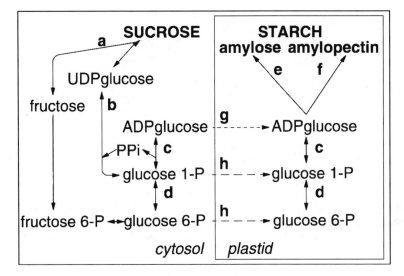

Figure 1   The major metabolites and enzymes involved in the conversion of sucrose to starch in storage organs. Carbon is shown entering the plastid either as a hexose phosphate (73) or as ADPglucose. Enzymes are: a, sucrose synthase; b, UDPglucose pyrophosphorylase; c, ADPglucose pyrophosphorylase; d, phosphoglucomutase; e, starch synthase (GBSSI); f, starch synthase and starch-branching enzyme; g, ADPglucose transporter; h, hexose phosphate transporter. PPi: inorganic pyrophosphate.

embryo, and one is expressed in pods and leaves whereas the other appears to be embryo-specific (89).

In maize there are probably three genes encoding the small and at least two genes encoding the large subunits. Transcripts for one of the small subunit genes (*BRITTLE2*) are detected only in the endosperm, but transcripts from a second gene (*AGP1*) are present in the embryo and also at low levels in the endosperm; expression of a third gene has been detected only in the leaf (33, 71). In barley there are two genes encoding the large subunit—one expressed only in the endosperm and the other primarily in the leaves—but only one encoding the small subunit (82, 86). However, the small subunit gene appears to produce two transcripts, one expressed only in the endosperm and the other primarily in the leaves but also in the endosperm. The two transcripts are identical for 90% of their length but differ at their 5′ ends, apparently because of the use of independent promoters upstream of each of two alternative 5′ exons (82).

Until recently, it was widely believed that AGPase was located exclusively in plastids. Evidence for this included direct localization studies using isolated plastids, and the ability of plastids from several types of organ to take up hexose phosphates and convert them to starch (discussed in 1, 67, 74). Two sorts of observations had suggested that this might not be the case in the developing endosperm of maize and barley. First, the subunit proteins predicted from transcripts abundant in the endosperm of barley lack obvious transit peptides at their N termini (87). Mature subunit proteins from the endosperms of both barley and maize are of the same apparent mass as proteins produced from full-length cDNAs, indicating that cleavage of a transit peptide does not occur in vivo (33, 87). Second, the effect of mutations at the *BRITTLE1* (*BT1*) locus on metabolite levels in developing maize endosperm indicates that ADPglucose is synthesized outside the amyloplast in this organ. The *BT1* gene encodes a protein located in the amyloplast envelope, the amino acid sequence of which is related to those of adenylate transporters (12). Elimination of this protein results in a fivefold decrease in the rate of starch synthesis and a twelvefold increase in the amount of ADPglucose in the endosperm (72). The simplest explanation of this observation is that ADPglucose is synthesized in the cytosol, and that the product of the *BT1* gene is a transporter that allows movement of ADPglucose from the cytosol to the amyloplast.

These observations suggest that AGPase may be cytosolic in endosperm, but do not constitute convincing evidence that this is the case (1). To investigate this possibility, we compared activities of AGPase with those of enzymes known to be confined to the plastid or the cytosol, in preparations of amyloplasts isolated from developing endosperms of maize and barley (19, 81). This revealed that 95% of the activity of AGPase is cytosolic in maize endosperm and that 85% is cytosolic in barley endosperm. Further evidence for both plastidial and cytosolic forms of the enzyme in these organs was provided by antisera to the small subunit of AGPase. These recognized two proteins of slightly different molecular masses in crude homogenates of endosperm, one of which was enriched in amyloplast preparations. The use of a brittle2 mutant of maize confirmed that the cytosolic and plastidial forms of the small subunit are different gene products. The mutant—which lacks the major form of the small subunit, and has a reduced rate of starch synthesis in the endosperm (33)—lacked the small subunit protein apparently located in the cytosol, and all of its AGPase activity was associated with the amyloplast rather than the cytosol. These data indicate that the AGPase transcripts abundant in the endosperm in maize and barley encode the subunits that constitute the major, cytosolic form of the enzyme. In barley, the plastidial form of the enzyme in the endosperm may be composed of the same subunits as the enzyme in leaf chloroplasts (81). In maize, the plastidial form of the enzyme may be encoded by the *AGP1* and *AGP2* genes that are known to be expressed at low levels in the endosperm (19).

At present there is no good evidence for quantitatively important cytosolic forms of AGPase in any organs other than maize and barley endosperm. The phenomenon is not even universal among cereals: Studies of isolated amyloplasts from wheat endosperm concluded that all or most of the AGPase in this organ is plastidial (1). Synthesis of ADPglucose in the cytosol must have profound consequences for cytosolic metabolism, in particular the regulation of pyrophosphate concentration (Figure 1; 45), and the selection pressures responsible for the cytosolic location of AGPase in maize and barley remain obscure.

Variation in the subunit composition of AGPase is expected to confer variation in kinetic properties. Most AGPases are subject to allosteric regulation by 3-phosphoglycerate (3-PGA, an activator) and inorganic phosphate (an inhibitor) (70). Studies of subunits expressed in *Escherichia coli* indicate that small subunits alone can form a catalytically active enzyme, and that the large subunit is responsible for modulation of the sensitivity of the enzyme to effectors (6, 40, 70). Together with the observation that there is a low level of sequence conservation between large subunits (75), these results suggest that expression of different large subunit genes in different organs may lead to variation in the regulatory properties of AGPases between these organs (57, 70).

The extent of variation in the properties of the enzyme between organs is unclear. There are great differences in the sensitivity to effectors of AGPase purified from different plant sources (38, 46, 47, 70), and from different organs of the same plant. For example, the enzymes from bean and barley leaves are reported to be more sensitive to effectors than those from the storage organs (embryo and endosperm respectively) of these species (47, 89). However, some of the reported differences in properties may be artifactual, reflecting different degrees of proteolytic cleavage of the subunits during enzyme purification (70).

The allosteric properties of AGPase are of potential importance in determining its role in control of the rate of starch synthesis. This is illustrated by two cases in which alterations in the allosteric properties of the enzyme, rather than its maximum catalytic activity, apparently bring about a change in the rate of starch synthesis. First, mutations in a gene (*STA1*) encoding one subunit of AGPase in the unicellular alga *Chlamydomonas* lead to a 95% reduction in starch synthesis. They dramatically reduce the sensitivity of AGPase to activation by 3-PGA but have no effect on its activity measured in the absence of activator (5, 85). Second, transposon-induced mutagenesis of the gene encoding the major large subunit of AGPase in the maize endosperm (*SHRUNKEN2*) has allowed the isolation of a line with a significant increase (about 15%) in the amount of starch per seed (34). The total activity of AGPase in the endosperm of this variant (Rev6) is actually reduced by about half, but the activity appears to be less sensitive to inhibition by phosphate.

Overall, we suggest that variation in subunit composition, and consequently in properties, of AGPase within and between plant organs reflects variation in the way in which flux through the pathway of starch synthesis is regulated. The demands for regulation of flux through the pathway in leaves, for example, are very different from those in storage organs generally (74), and the few quantitative studies so far indicate that AGPase plays a much more important role in flux control in leaves than in storage organs (20, 66). We suspect that the demands for regulation and the importance of AGPase in flux control also differ between different types of leaf and storage organ, with different environmental conditions, and through the development of individual organs. AGPase is often referred to as a rate-limiting step in starch synthesis: This is an unhelpful generalization that ignores both the undoubted variation in the importance of the enzyme between organs and the general complexity of the relationship between flux and the properties and activities of enzymes (77).

THE SYNTHESIS AND ORGANIZATION OF AMYLOPECTIN MOLECULES

Starch consists of two sorts of glucose polymer: highly branched amylopectin and relatively unbranched amylose. The latter comprises about 20–30% of most storage starches but is not essential to the formation of a granule. The basic structure of the granule is dictated by the packing of amylopectin molecules in organized arrays (31). Amylopectin consists of chains of $\alpha 1,4$ linked glucose units, branched by $\alpha 1,6$ linkages (Figure 2*a*). There is general agreement that the chains within the granule are radially arranged with their nonreducing ends pointing toward the surface, and are organized into alternating crystalline and amorphous lamellae with a periodicity of 9 nm (43). The lamellae are believed to reflect the arrangement of chains into clusters. Within clusters, chains associate to form double helices that pack together in ordered arrays to give the crystalline lamellae. The amorphous lamellae contain the branch points (Figure 2*b,c*). The organization of chains into clusters is consistent with the observed polymodal distribution of chain lengths within amylopectin molecules. Peaks in the distribution occur at chain lengths of 12–16 glucose units, about 40 glucose units, and about 70 glucose units, and chains of these lengths are proposed to span one, two, and three crystalline lamellae,

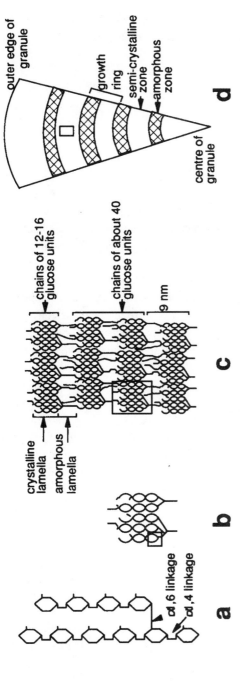

Figure 2   Schematic representation of levels of organization within the starch granule. The boxes within the diagrams in panels *b*, *c*, and *d* represent the area occupied by the structure in the preceding panel. (*a*) Structure of two branches of an amylopectin molecule, showing individual glucose units. (*b*) A single cluster within an amylopectin molecule, showing association of adjacent branches to form double helices. (*c*) Arrangement of clusters to form alternating crystalline and amorphous lamellae. The crystalline lamellae are produced by the packing of double helices in ordered arrays. Chains of 12–16 glucose units span one cluster; chains of about 40 glucose units span two clusters. (*d*) Slice through a granule, showing alternating zones of semicrystalline material, consisting of crystalline and amorphous lamellae, and amorphous material.

respectively (37). It has been proposed from studies of potato starch that the lamellae themselves are organized as interdigitating helices, packed in tetragonal arrays (68).

Regions of alternating crystalline and amorphous lamellae form concentric zones within the granules, on the order of hundreds of nanometers in width. These semicrystalline zones alternate with amorphous zones, in which the organization of the amylopectin chains is not understood. A semicrystalline-amorphous repeat is referred to as a growth ring (Figure 2d; 31, 32).

Within this basic picture there is a large amount of genetically, developmentally, and environmentally induced variation. The chain-length distribution of amylopectin and the packing of double helices within the clusters varies from one organ to another, between species, and between individual cultivars within species (e.g. 37). Although there is variation in the length and packing of the chains that comprise them, the periodicity of the lamellar repeats is highly invariant. A repeat structure of 9 nm has been observed in starches from several different organs (43). The nature and size of growth rings within the granule is very dependent upon growth conditions. Growth rings in the starch of wheat endosperm, for example, are apparently abolished by growth of the plant in a constant environment rather than under a day-night regime (11).

The synthesis of amylopectin and its organization to form a granule are highly integrated and regulated processes. Neither the synthesis of glucans via enzymes of starch synthesis in vitro nor the recrystallization of solubilized amylopectin results in a glucan with the levels of structural organization of a starch granule, and the $\alpha 1,4$, $\alpha 1,6$ linked glucans of bacteria do not form ordered granules. Attempts to explain the synthesis of the starch granule must consider the synthesis of $\alpha 1,4$ linkages, the generation of $\alpha 1,6$ linkages at particular frequencies, the periodic nature of the synthesis that allows the organization of the chains into clusters, the organization of the clusters in superhelices, and the alternation of semicrystalline and amorphous zones in growth rings. The synthesis of $\alpha 1,4$ and $\alpha 1,6$ linkages, via starch synthase and starch-branching enzyme, respectively, has in the past received the most attention. Recently, however, the study of reconstituted systems and of mutant plants and algae with altered starch structure has led to new hypotheses that link the synthesis of amylopectin with its packing to form a granule. We shall consider the contribution of starch synthases, starch-branching enzymes, and other enzymes of starch metabolism to the synthesis of amylopectin and its packing to form a granule, and describe the new hypotheses in this light.

Starch Synthases

It is widely accepted that amylopectin is elaborated at the surface of the starch granule by starch synthase and starch branching enzyme in the soluble fraction of the amyloplast. In most of the organs thus far examined, soluble starch synthase activity can be separated into at least two fractions by anion-exchange chromatography, but in many of these cases the number of different gene products with starch synthase activity and their relative contributions to the soluble activity is not known (70, 74). Where proteins with soluble starch synthase activity have been characterized in detail, a complex picture is emerging. Although distinct classes of isoforms can be defined on the basis of similarity in amino acid sequence, molecular mass, and antigenic properties (56, 57), plant organs vary greatly in the classes they possess and in the relative contribution of the classes to soluble starch synthase activity. This can be illustrated by a comparison of the pea embryo with the potato tuber. Both organs contain starch synthases belonging to a widely distributed class of isoforms referred to as SSII (22, 25, 26). SSII contributes more than 60% of the soluble activity in the embryo, but a maximum of 15% of the soluble activity in the tuber. About 80% of the activity in the tuber is accounted for by SSIII, an isoform very different in amino acid sequence from SSII (56).

For the most part, it is not clear whether individual isoforms of starch synthase make qualitatively different contributions to amylopectin synthesis. Evidence that they do comes from study of mutant lines of pea and *Chlamydomonas*. Mutations at the *RUG5* locus of pea eliminate the activity of SSII, and mapping experiments indicate that the gene encoding SSII lies at the *RUG5* locus. The effect of the mutation upon the starch of the embryo is dramatic: Although content is barely affected through much of embryo development, the starch granule is grossly misshapen, and the distribution of chain lengths in the amylopectin is very different from that of wild-type starch. There are more very short chains (<15 glucose units) and very long chains, and fewer chains of 15–45 glucose units, than in amylopectin of wild-type embryos (16, 53). It appears that SSII is specifically responsible for the elongation of very short chains to create the chains that form the basis of the clusters within amylopectin, and that other isoforms are unable to carry out this function.

Mutations at the *STA3* locus of *Chlamydomonas* give rise to a starch broadly similar in structure to that of rug5 mutants of pea (32). The mutations appear to eliminate specifically the activity of one of two isoforms of starch synthase that contribute the soluble activity. The amylopectin in mutant lines has more very short chains (of 2–7 glucose units) and fewer longer chains (of 8–60 glucose units) than the amylopectin in wild-type lines.

These observations indicate that a specific isoform of starch synthase is responsible for the synthesis of the chains that lie within individual clusters of amylopectin. However, analysis of potato tubers in which activity of either SSII or SSIII has been significantly reduced through expression of antisense RNA has not thus far supported this idea. Reduction of SSII activity had almost no detectable effect on the starch of the tuber (26). Reduction of SSIII—resulting in the loss of almost 80% of the soluble activity of the tuber—caused deep fissuring of the starch granules, but did not have effects like those of the *RUG5* mutation upon the structure of the amylopectin (56; our unpublished data). These results raise the possibility that there are qualitative as well as quantitative differences between organs in the contributions of individual isoforms of starch synthase to the synthesis of amylopectin. At this stage, we cannot assume that it will be possible to construct a widely applicable model that relates particular classes of isoform to specific aspects of amylopectin synthesis.

In addition to the activity in the soluble fraction of the amyloplast, starch synthase activity is also found tightly bound to starch granules. It has been assumed that this granule-bound activity is responsible solely for the synthesis of amylose (74). Much of the granule-bound activity in storage organs is attributable to a class of isoform known as granule-bound starch synthase I (GBSSI). Elimination of this isoform (encoded at the *WAXY* loci of cereals, the *AMF* locus of potato, and the *LAM* locus of pea) by mutations essentially eliminates amylose from the starch, and reduction of its activity through the expression of antisense RNA in potato tubers reduces amylose content (17, 57). However, the following observations show that there are other granule-bound isoforms that could contribute to amylopectin rather than amylose synthesis, and that GBSSI itself can potentially contribute to amylopectin synthesis.

All starches probably contain at least one isoform of starch synthase other than GBSSI (73). In most cases, these are granule-bound fractions of isoforms that also contribute to the soluble activity. For example, pea starch contains SSII, the isoform that contributes much of the soluble activity of the embryo (21). Potato starch contains both SSII and SSIII, the isoforms that together contribute more than 90% of the soluble activity of the tuber (25, 56). These granule-bound starch synthases are probably not involved in amylose synthesis. Mutants of pea, maize, and potato that lack GBSSI contain no detectable amylose yet retain a normal proportion of their other isoforms bound to the granule (17, 26, 38). It is possible that these proteins are inactive in vivo and are simply trapped within the granule by the crystallization of their amylopectin product. Most of the activity of intact, isolated starch granules is contributed by GBSSI, and the activity of other isoforms becomes apparent only when the granules are disrupted (38). However, the fact that different soluble isoforms become buried within the granule to very different extents might indicate that the bound forms play a specific role in starch synthesis. For example, the SSII isoform in pea and potato is largely granule bound, whereas the SSIII isoform in potato is largely soluble (26, 27, 56). The extent to which soluble proteins become bound appears to be an intrinsic property of the protein rather than of the starch: Pea SSII is largely granule bound in pea lines with normal, high-amylose, and amylose-free starch, and when expressed in lines of potatoes with normal and amylose-free starch (26).

Evidence that GBSSI is involved in amylopectin as well as amylose synthesis in vivo comes from two sources. First, the amylopectin of mutants of cereals, potatoes, and *Chlamydomonas* that lack GBSSI differs from that of wild-type lines (16, 28, 54, 80). The amylopectin of waxy mutants of rice, for example, lacks a fraction of very long chains present in the amylopectin of wild-type lines (80). Lines of *Chlamydomonas* with mutations at the *STA2* locus—which lack the granule-bound isoform of starch synthase responsible for amylose synthesis—are also deficient in a fraction of amylopectin with very long chains (16). Second, GBSSI elongates chains within amylopectin in isolated starch granules. Incorporation of $[^{14}C]$glucose from ADP$[^{14}C]$glucose in granules isolated from wild-type pea embryos, potato tubers, and sweet-potato roots is mainly into long chains in the amylopectin fraction of the starch (3, 18). Overall, it seems likely that GBSSI elongates chains within the amylopectin of normal starches, but its precise contribution to amylopectin structure is unclear.

Starch-Branching Enzymes

The nature of the enzyme that catalyzes the formation of the $\alpha1,6$ linkages of amylopectin might be expected to be of great importance in determining amylopectin structure. The fact that plant organs almost invariably contain multiple isoforms of starch-branching enzyme (SBE) (10, 57, 73) raises the possibility that different forms create chains of different lengths or branch points at different frequencies. Multiple forms of SBE could thus give rise to the branching pattern and polymodal distribution of chain lengths that underlie the cluster structure of amylopectin. This idea has been promoted by recent, detailed studies of the nature and properties of isoforms of SBE. All the isoforms of SBE for which information is available fall into one of two classes on the basis of differences in primary sequence (referred to as A and B; see 10), although it is possible that other types of isoform exist in some organs (27, 78). It appears that the SBE activity of plant organs is contributed by both an A and a B isoform. For example, isoforms IIa and I of maize endosperm, III and I of rice endosperm, I and II of pea embryo (10), and II and I of potato tuber (51) fall into classes A and B, respectively. The A (II) and B (I) isoforms of maize endosperm differ both in their substrate affinities and in the length of branches they preferentially create. In vitro, isoform A preferentially branches amylopectin, whereas isoform B preferentially branches amylose. With amylose as a substrate, isoform B preferentially transfers longer chains than isoform A (36, 79). When expressed in a strain of *E. coli* that lacks a glycogen-branching enzyme, both isoforms can branch the linear product of the bacterial glycogen synthase to give a glycogen-like polymer. Consistent with their actions in vitro, the glycogen synthesized by isoform A has more shorter chains (6–9 glucose units) and fewer longer chains (greater than 14 glucose units) than the glycogen synthesized by isoform B (35). It is likely that these differences in properties of the maize isoforms define general differences between the A and B classes.

The differences in properties between the A and B isoforms have led to the suggestion that isoform B participates in vivo in the synthesis of the long and intermediate length chains that will span clusters, whereas isoform A participates in the synthesis of the shorter chains that lie wholly within clusters (70, 79). This idea is potentially testable through study of mutant and transgenic plants in which one isoform is eliminated or severely reduced in activity. Mutations at the *AMYLOSE-EXTENDER* (*AE*) loci of cereals and the *RUGOSUS* (*R*) locus of peas lead specifically to the loss of an A isoform (7, 10, 58, 76). The amylopectin in ae mutant endosperms and r mutant embryos displays an increase in average chain length relative to that of the wild type (e.g. 2, 13, 53). There is, however, no dramatic change in the distribution of chain lengths among chains of up to 50 glucose units, and the structural periodicity of 9 nm within the semicrystalline regions of the granule is not affected by the mutations (44). No mutations affecting the B isoform of SBE have been described, but dramatic reduction of the activity of this isoform in potato tuber through expression of antisense RNA is reported to have only minor effects on the structure of amylopectin (29).

Overall, these observations do not suggest that the A and B isoforms play distinct and essential roles in creating the cluster structure of amylopectin. Their interpretation is, however, fraught with problems. The mutations bring about several changes likely to influence amylopectin structure, in addition to the specific loss of one isoform of SBE. They alter the ratio of total SBE to starch synthase activity and reduce the rate of starch synthesis, and may alter the pattern of change in SBE activity through development (10, 83).

To provide a clearer view of the roles of the two isoforms in determining amylopectin structure, we have studied the effect of the *R* mutation upon amylopectin synthesized during a single photoperiod in pea leaves. The mutation causes a 10-fold decrease in SBE activity in the leaf but has relatively little effect on the rate of starch synthesis. It slightly increases the abundance of longer relative to shorter chains in the amylopectin, but has no effect on the polymodal distribution of chain lengths. The amylopectin of both wild-type and mutant leaves displays a strongly polymodal distribution of chain lengths, with maxima at lengths of 12, 15, and 19–21 glucose units (83, 84). In this organ at least, it appears that isoform A plays no specific role in establishing the basic distribution of chain lengths. We suggest that the polymodal distribution of chain lengths in amylopectin is not primarily a consequence of the fact that plants possess two distinct classes of SBE isoforms. Explanations for the distribution of chain lengths, and hence the organization of chains into clusters, must be sought elsewhere.

The Role of Debranching Enzymes

Mutations at the *SUGARY1* (*SU1*) locus of maize and the *SUGARY* locus of rice dramatically reduce the starch content of the endosperm and cause its partial replacement by a water-soluble, very highly branched glucan known as phytoglycogen. The mutations alter the activity of several enzymes of starch metabolism in the developing endosperm, but in both species they decrease debranching activity, defined by the hydrolysis of pullulan (24, 63, 64, 69). These phenotypes have given rise to the idea that the branching pattern of normal amylopectin is determined by SBE and debranching enzyme (DBE) acting in concert: Phytoglycogen is argued to be the product of SBE acting alone or at reduced activities of DBE (64, 69).

This idea has recently been supported by the discovery of a phytoglycogen-accumulating mutant of *Chlamydomonas* which, like the cereal mutants, is deficient in debranching activity. Mutations at the *STA7* locus cause a loss of starch and its replacement by smaller amounts (8% or less of wild-type starch) of a glucan with a structure similar to that of glycogen. Mutants specifically lack a DBE capable of debranching amylopectin. They appear to have normal activities and complements of isoforms of other enzymes of starch metabolism (59). Following this discovery, Ball and colleagues have proposed a model for the formation and organization of amylopectin clusters at the periphery of the growing starch granule in which DBE is an essential component (4). The sequence of events that they envisage is as follows.

Soluble starch synthase elongates very short chains at the periphery of the granule. Initially these chains are of insufficient length to act as substrates for SBE—which acts preferentially upon chains in double helical conformation (9)—and they remain unbranched. When they reach an appropriate length for branching to occur, branches are created through the action of SBE and starch synthase. DBE removes the outer chains from this unorganized glucan created by SBE and starch synthase, but does not have access to branch points formed close to the organized, double-helical zone. The action of DBE thus leaves a zone of short chains arising from branch points at the top of the double-helical region, and a further round of elongation by soluble starch synthase is initiated.

This model is an important step forward in attempts to explain the integration of the synthesis of amylopectin with its packing to form a granule. There is clearly much work to be done to refine and test it. One of the most immediate challenges lies in the complexity of DBEs in higher plants (55). Multiple isoforms of DBE have been identified in several developing starch-storing organs. In maize endosperm, for example (23), there are isoforms that preferentially debranch pullulan (pullulanase- or limit-dextrinase-like DBEs, or R-enzyme) and isoforms that debranch amylopectin but cannot debranch pullulan (isoamylase-like DBEs). Because of this complexity, the relationship between the *SUGARY* mutations and the biochemically identified debranching enzymes of cereal endosperms is not yet understood. Although the gene at the *SU1* locus of maize encodes a protein similar in primary amino acid sequence to bacterial isoamylases rather than pullulanases (41), the activity depleted in the su1 endosperm has been identified by its ability to debranch pullulan. The *SUGARY* locus of rice lies on a different chromosome from the gene encoding the pullulan-DBE that is reduced in activity in endosperms of sugary mutants (63, 64). The question of whether one or both types of DBE could participate in amylopectin synthesis in the manner proposed by Ball remains open.

THE SYNTHESIS OF AMYLOSE

Amylose molecules appear to exist as single helices within the starch granule, interspersed with amylopectin in amorphous regions (32, 42). Their precise location in relation to the ordered amylopectin matrix remains unclear. As discussed above, amylose synthesis in storage organs is a specific function of the GBSSI class of isoforms of starch synthase. The precise mechanism of synthesis—in particular the reason why the product of GBSSI remains unbranched—remains to be resolved. It has been suggested that location of

GBSSI on the granule renders its products inaccessible to SBE, but the existence of other granule-bound starch synthases incapable of amylose synthesis and of granule-bound SBE casts doubt on this model (21, 73). It is evident that GBSSI must possess specific properties different from those of other isoforms, and detailed comparison of its structure-function relationships with those of other isoforms is likely to yield valuable information. However, the location of GBSSI within the starch granule in vivo may confer on it properties that are not evident in the test tube. Three observations suggest that the synthesis of amylose in vivo is integrated in a complex way with the synthesis of the granule matrix: the nonuniform distribution of amylose within granules with reduced GBSSI activity, the synthesis of amylopectin rather than amylose via GBSSI in isolated starch granules, and the positive correlations observed in some species between the rate of starch synthesis and its amylose content.

Tubers of potatoes in which GBSSI activity has been reduced through expression of antisense RNA have reduced amounts of amylose, and there is a general correlation between GBSSI activity and amylose content (28). The reduction in amylose is not, however, uniform through the starch granule. The core of the granule stains blue with iodine, indicating the presence of amylose, whereas the periphery stains the red color typical of amylose-free starches (48). This effect is not brought about by changes in GBSSI expression during tuber development and must be due to a change in the rate of amylose relative to amylopectin synthesis during the growth of the granule. It has been proposed to reflect changes in the concentration of GBSSI protein at the granule surface as the granule grows (48). It is argued that a minimum concentration is required to allow amylose synthesis to occur and that the concentration falls as the granule surface area increases through development. In wild-type plants, the concentration is always adequate to allow amylose synthesis, but in plants with reduced GBSSI the concentration at some point falls below the critical threshold, amylose synthesis ceases, and subsequent growth of the matrix is entirely through amylopectin synthesis (48). An alternative explanation assumes that synthesis of amylose occurs within the matrix formed by the synthesis of amylopectin rather than at the outer surface of the granule. In wild-type plants, amylose synthesis keeps pace with the synthesis of the amylopectin matrix such that available space within the matrix becomes filled with amylose to within a small distance of the granule surface. In plants with reduced GBSSI, amylose synthesis fails to keep pace with the synthesis of the matrix, and a peripheral zone in which amylose molecules are too small or infrequent to stain with iodine increases in width as the granule grows. These hypotheses raise the important question of the location of amylose synthesis—at the granule surface at the same time as amylopectin synthesis or within the amylopectin matrix. There are some grounds for believing that the latter location is the more likely. The amylose content of potato starch is not increased above 20% by increases in GBSSI activity, leading to the suggestion that the maximum amylose content is determined by the space available for its deposition within the amylopectin matrix (28).

Although GBSSI undoubtedly synthesizes amylose in starch granules in vivo, this is not the case in isolated, washed starch granules. As discussed above, incorporation of glucose from ADPglucose via GBSSI in isolated granules is almost exclusively into a fraction of long chains within amylopectin molecules. It thus appears that amylose synthesis requires some component removed from the granules during isolation. We have shown that this component is likely to be malto-oligosaccharides of low molecular mass. Soluble extracts of pea embryo and potato tuber stimulate amylose synthesis from ADPglucose in granules isolated from these organs (18). This stimulation is reduced by treatment of the extracts with α-glucosidase, an enzyme that converts malto-oligosaccharides to glucose. Addition of pure malto-oligosaccharides, from maltose to malto-heptaose, to isolated starch granules stimulates the overall rate of starch synthesis and promotes amylose synthesis at the expense of amylopectin synthesis. Glucose does not have this effect (18). It is not yet clear how malto-oligosaccharides act on GBSSI within the granule to promote amylose synthesis. Although they undoubtedly act as primers for the synthesis of short glucans (52), they may well also affect the affinity of the enzyme for existing amylose molecules. The malto-oligosaccharides needed for amylose synthesis could be produced in vivo via starch phosphorylase, and by hydrolysis of glucans. They could also be generated by cleavage of α1,6 linkages by debranching enzyme during the synthesis of clusters of amylopectin in the model proposed by Ball and colleagues (described above). Whatever the source of malto-oligosaccharides, these findings suggest that the synthesis of amylose is intimately linked to the synthesis and degradation of other glucans within the amyloplast.

There is a complex relationship between the rate of amylose synthesis and the overall rate of starch synthesis in storage organs, which may be mediated via the availability of ADPglucose to GBSSI. Evidence for this is as follows. First, reduction or elimination of amylose via mutation or expression of antisense RNA for GBSSI does not result in a decrease in starch content (15, 49). Presumably the rate of amylopectin synthesis is increased in the mutant and transformed organs so that starch content is normal. This implies that there is competition for ADPglucose between GBSSI and other isoforms of starch synthase. Second, mutations at the *RB* locus of pea, which reduce the rate of starch synthesis by 40% through their effects on AGPase (39), reduce amylose content from 35% to 25% of the starch (8). The simplest explanation of this phenomenon is that a reduction in ADPglucose levels caused by the mutation affects GBSSI activity more than that of other isoforms, perhaps because GBSSI has an intrinsically lower affinity for ADPglucose, or because it is active inside the granule matrix where ADPglucose availability is dictated by diffusion and the activity of the soluble starch synthases. This type of relationship between ADPglucose concentration and amylose synthesis has also been suggested to explain the effects on starch structure of mutations at the *STA1* and *STA5* loci of *Chlamydomonas,* which reduce the activities of AGPase and phosphoglucomutase, respectively (85). Mutants have 5–10% of the wild-type starch content but completely lack amylose. Explanations of the relationship between rate and amylose content based solely on ADPglucose concentration must, however, be treated with caution in the light of the effects of malto-oligosaccharides on amylose synthesis discussed above. Changes in malto-oligosaccharide concentration resulting from changes in the rate of starch synthesis might also bring about a selective alteration of the rate of amylose synthesis. Whatever the underlying cause, a general relationship between starch structure and the rate of starch synthesis provides a possible explanation for the growth rings within the starch granule. Diurnal variation in the rate of supply of sucrose to the storage organ and thus in the rate of starch synthesis could lead to diurnal variation in the structure of starch.

We suggest from the above observations that four factors are potentially important in determining the rate of amylose synthesis and the amylose content of starch in storage organs. These are the amount of GBSSI protein, the availability of ADPglucose, the availability of malto-oligosaccharides, and the physical space available within the matrix created by the synthesis of amylopectin. The relative importance of these factors in determining the rate of synthesis may differ from one organ to another, but the space available within the matrix could well be the primary determinant of the final amylose content of storage starches in wild-type organs (28). This would provide an explanation of the rather narrow range of variation in amylose content found in such starches.

ACKNOWLEDGMENTS

We are very grateful to Tom ap Rees, Steven Ball, Curt Hannah, and Yasunori Nakamura for providing us with unpublished information used in the preparation of this article, and to Sam Zeeman and Rod Casey for their comments.

Literature Cited

1. ap Rees T. 1995. Where do plants make ADP-Glc? In *Sucrose Metabolism, Biochemistry, Physiology and Molecular Biology: Proc. Int. Symp. Sucrose Metabolism,* ed. HG Pontis, GL Salerno, EJ Echeverria, pp. 143–55. Rockville, MD: Am. Soc. Plant Physiol.

2. Baba T, Arai Y. 1984. Structural characterization of amylopectin and intermediate material in amylomaize starch granules. *Agric. Biol. Chem.* 48:1763–75

3. Baba T, Yoshii M, Kainuma K. 1987. Acceptor molecule of granular-bound starch synthase from sweet-potato roots. *Starch* 39:52–56

4. Ball S, Guan H-P, James M, Myers A, Keeling P, et al. 1996. From glycogen to amylopectin: a model explaining the biogenesis of the plant starch granule. *Cell* 86:349–52

5. Ball S, Marianne T, Dirick L, Fresnoy M, Delrue B, Decq A. 1991. A *Chlamydomonas reinhardtii* low-starch mutant is defective for 3-phosphoglycerate activation and orthophosphate inhibition of ADP-glucose pyrophosphorylase. *Planta* 185: 17–26

6. Ballicora MA, Laughlin MJ, Fu Y, Okita TW, Barry GF, Preiss J. 1995. Adenosine 5′-diphosphate-glucose pyrophosphorylase from potato tuber. Significance of the N terminus of the small subunit for catalytic properties and heat stability. *Plant Physiol.* 109:245–51

7. Bhattacharyya MK, Smith AM, Ellis THN, Hedley C, Martin C. 1990. The wrinkled-seed character of pea described by Mendel is caused by a transposon-like insertion in a gene encoding starch branching enzyme. *Cell* 60:115–22

8. Bogracheva TY, Davydova NI, Genin YV, Hedley CL. 1995. Mutant genes at the *r* and *rb* loci affect the structure and physicochemical properties of pea seed starches. *J. Exp. Bot.* 46:1905–13

9. Borovsky D, Smith EE, Whelan WJ, French D, Kikumoto S. 1979. The mechanism of Q-enzyme action and its influence on the structure of amylopectin. *Arch. Biochem. Biophys.* 198:627–31

10. Burton RA, Bewley JD, Smith AM, Bhattacharyya MK, Tatge H, et al. 1995. Starch branching enzymes belonging to distinct enzyme families are differentially expressed during pea embryo development. *Plant J.* 7:3–15

11. Buttrose MS. 1962. The influence of environment on the shell structure of starch granules. *J. Cell Biol.* 14:159–67

12. Cao H, Sullivan TD, Boyer CD, Shannon JC. 1995. *Bt1,* a structural gene for the major 39–44 kDa amyloplast membrane polypeptides. *Physiol. Plant.* 95:176–86

13. Colonna P, Mercier C. 1984. Macromolecular structure of wrinkled- and smooth-pea starch components. *Carbohydr. Res.* 126:233–47

14. Craig J, Smith A, Wang TL, Lloyd J, Hedley C. 1995. Biochemistry of new wrinkled-seeded mutants of pea. In *Improving Production and Utilisation of Grain Legumes, Proc. 2nd Eur. Conf. Grain Legumes, Copenhagen,* p. 396. Paris: Assoc. Eur. Rech. Protéagineux

15. Creech RG. 1965. Genetic control of carbohydrate synthesis in maize endosperm. *Genetics* 52:1175–86

16. Delrue B, Fontaine T, Routier F, Decq A, Weiruszeski J-M, et al. 1992. Waxy *Chlamydomonas reinhardtii:* monocellular algal mutants defective in amylose biosynthesis and granule-bound starch synthase activity accumulate a structurally modified amylopectin. *J. Bacteriol.* 174:3612–20

17. Denyer K, Barber LM, Burton R, Hedley CL, Hylton CM, et al. 1995. The isolation and characterization of novel low-amylose mutant of *Pisum sativum. Plant Cell Environ.* 18:1019–26

18. Denyer K, Clarke B, Hylton C, Tatge H, Smith AM. 1996. The elongation of amylose and amylopectin chains in isolated starch granules. *Plant J.* In press

19. Denyer K, Dunlap F, Thorbjørnsen T, Keeling P, Smith AM. 1996. The major form of ADP-glucose pyrophosphorylase in maize (*Zea mays* L.) endosperm is extraplastidial. *Plant Physiol.* 12:779–86

20. Denyer K, Foster J, Smith AM. 1995. The contributions of adenosine 5′-diphosphoglucose pyrophosphorylase and starch-branching enzyme to the control of starch synthesis in developing pea embryos. *Planta* 97:57–62

21. Denyer K, Sidebottom C, Hylton CM, Smith AM. 1993. Soluble isoforms of starch synthase and starch-branching enzyme also occur within starch granules in developing pea embryos. *Plant J.* 4: 191–98

22. Denyer K, Smith AM. 1992. The purification and characterisation of the two forms of soluble starch synthase from developing pea embryos. *Planta* 186:609–67

23. Doehlert DC, Knutson CA. 1991. Two classes of starch debranching enzymes from developing maize kernels. *J. Plant Physiol.* 138:566–72

24. Doehlert DC, Kuo TM, Juvik JA, Beers EP, Duke SH. 1993. Characteristics of carbohydrate metabolism in sweetcorn (*sugary-1*) endosperms. *J. Am. Soc. Hortic. Sci.* 188:661–66

25. Edwards A, Marshall J, Denyer K, Sidebottom C, Visser RGF, et al. 1996. Evidence that a 77-kilodalton protein from the starch of pea embryos is an isoform of starch synthase that is both soluble and granule bound. *Plant Physiol.* 112:87–97

26. Edwards A, Marshall J, Sidebottom C, Visser RGF, Smith AM, Martin C. 1995. Biochemical and molecular characterisation of a novel starch synthase from potato tubers. *Plant J.* 8:283–94

27. Fisher DK, Gao M, Kim K-N, Boyer CD, Guiltinan MJ. 1996. Allelic analysis of the maize amylose-extender locus suggests that independent genes encode starch-branching enzymes IIa and IIb. *Plant Physiol.* 110:611–19

28. Flipse E, Keetels CJAM, Jacobsen E, Visser RGF. 1996. The dosage effect of the wildtype GBSS allele is linear for GBSS activity but not for amylose content: absence of amylose has a distinct influence on the physico-chemical properties of starch. *Theor. Appl. Genet.* 92:121–27

29. Flipse E, Suurs L, Keetels CJAM, Kossmann J, Jacobsen E, Visser RGF. 1996. Introduction of sense and antisense cDNA for branching enzyme in the amylose-free potato mutant leads to physico-chemical changes in the starch. *Planta* 198:340–47

30. Fontaine T, D'Hulst C, Maddelein M-L, Routier F, Pépin TM, et al. 1993. Towards an understanding of the biogenesis of the starch granule: evidence that *Chlamydomonas* soluble starch synthase II controls the synthesis of intermediate size glucans of amylopectin. *J. Biol. Chem.* 268: 16223–30

31. French D. 1984. Organization of starch granules. In *Starch: Chemistry and Technology,* ed. RL Whistler, JN BeMiller, EF Paschall, pp. 183–247. Orlando: Academic

32. Gidley MJ. 1992. Structural order in starch granules and its loss during gelatinisation. In *Gums and Stabilisers for the Food Industry 6,* ed. GO Phillips, PA Williams, DJ Wedlock, pp. 87–92. Oxford: IRL

33. Giroux MJ, Hannah LC. 1994. ADP-glucose pyrophosphorylase in *shrunken-2* and *brittle-2* mutants of maize. *Mol. Gen. Genet.* 243:400–8

34. Giroux MJ, Shaw J, Barry G, Cobb BG, Greene T, et al. 1996. A single gene mutation that increases maize seed weight. *Proc. Natl. Acad. Sci. USA* 93:5824–29

35. Guan HP, Kuriki T, Sivak M, Preiss J. 1995. Maize branching enzyme catalyzes synthesis of glycogen-like polysaccharide in *glgB*-deficient *Escherichia coli. Proc. Natl Acad. Sci. USA* 92:964–67

36. Guan HP, Preiss J. 1993. Differentiation of the properties of the branching isozymes from maize (*Zea mays*). *Plant Physiol.* 102: 1269–73

37. Hizukuri S. 1986. Polymodal distribution of the chain lengths of amylopectins, and its significance. *Carbohydr. Res.* 147: 342–47

38. Hylton CM, Denyer K, Keeling PL, Chang M-T, Smith AM. 1996. The effect of *waxy* mutations on the granule-bound starch synthases of barley and maize endosperms. *Planta* 198:230–37

39. Hylton CM, Smith AM. 1992. The *rb* mutation of peas causes structural and regulatory changes in ADP glucose pyrophosphorylase from developing embryos. *Plant Physiol.* 99:1626–34

40. Iglesias AA, Barry GF, Meyer C, Bloksberg L, Nakata PA, et al. 1993. Expression of the potato tuber ADP-glucose pyrophosphorylase in *Escherichia coli. J. Biol. Chem.* 268: 1081–86

41. James MG, Robertson DS, Myers AM. 1995. Characterization of the maize gene *sugary1,* a determinant of starch composition in kernels. *Plant Cell* 7:417–29

42. Jane J, Xu A, Radosavljevic M, Seib PA. 1992. Location of amylose in normal starch granules. I. Susceptibility of amylose and amylopectin to cross-linking reagents. *Cereal Chem.* 69:405–9

43. Jenkins PJ, Cameron RE, Donald AM. 1993. A universal feature in the structure of starch granules from different botanical sources. *Starch* 45:417–20

44. Jenkins PJ, Donald AM. 1995. The influence of amylose on starch granule structure. *Int. J. Biol. Macromol.* 17:315–21

45. Kleczkowski LA. 1994. Glucose activation and metabolism through UDP-glucose pyrophosphorylase in plants. *Phytochemistry* 37:1507–15

46. Kleczkowski LA, Villand P, Lüthi E, Olsen OA, Preiss J. 1993. Insensitivity of barley endosperm ADP-glucose pyrophosphorylase to 3-phosphoglycerate and orthophosphate regulation. *Plant Physiol.* 101: 179–86

47. Kleczkowski LA, Villand P, Preiss J, Olsen OA. 1993. Kinetic mechanism and regula-

tion of ADP-glucose pyrophosphorylase from barley (*Hordeum vulgare*) leaves. *J. Biol. Chem.* 268:6228–33

48. Kuipers AGJ, Jacobsen E, Visser RGF. 1994. Formation and deposition of amylose in the potato tuber starch granule are affected by the reduction of granule-bound starch synthase gene expression. *Plant Cell* 6:43–42

49. Kuipers AGJ, Vreem JTM, Meyer H, Jacobsen E, Feenstra WJ, Visser RGF. 1992. Field evaluation of antisense RNA mediated inhibition of GBSS gene expression in potato. *Euphytica* 59:83–91

50. La Cognata U, Willmitzer L, Müller-Röber B. 1995. Molecular cloning and characterization of novel isoforms of potato ADP-glucose pyrophosphorylase. *Mol. Gen. Genet.* 246:538–48

51. Larsson C-T, Hofvander P, Khoshnoodi J, Ek B, Rask L, Larsson H. 1996. Three isoforms of starch synthase and two isoforms of starch-branching enzyme are present in potato tuber starch. *Plant Sci.* 117:9–16

52. Leloir LF, De Fekete MAR, Cardini CE. 1961. Starch and oligosaccharide synthesis from uridine diphosphate glucose. *J. Biol. Chem.* 236:636–41

53. Lloyd JR. 1995. *Effect and interactions of Rugosus genes on pea (*Pisum sativum*) seeds.* PhD thesis. Univ. East Anglia, Norwich, UK

54. Maddelein M-L, Libessart N, Bellanger F, Delrue B, D'Hulst C, et al. 1994. Towards an understanding of the biogenesis of the starch granule: determination of granule-bound and soluble starch synthase functions in amylopectin synthesis. *J. Biol. Chem.* 269:25150–57

55. Manners DJ. 1985. Starch. In *Biochemistry of Storage Carbohydrates in Green Plants,* ed. PM Dey, RA Dixon, pp. 149–203. London: Academic

56. Marshall J, Sidebottom C, Debet M, Martin C, Smith AM, Edwards A. 1996. Identification of the major starch synthase in the soluble fraction of potato tubers. *Plant Cell.* 8:1121–35

57. Martin C, Smith AM. 1995. Starch biosynthesis. *Plant Cell* 7:971–85

58. Mizuno K, Kawasaki T, Shimada H, Satoh H, Kobayashi E, et al. 1993. Alteration of the structural properties of starch components by the lack of an isoform of starch branching enzyme in rice seeds. *J. Biol. Chem.* 268:19084–91

59. Mouille G, Maddelein M-L, Libessart N, Tagala P, Decq A, et al. 1996. Preamylopectin processing: a mandatory step for starch biosynthesis in plants. *Plant Cell.* 8:1353–66

60. Müller-Röber BT, Kossmann J. 1994. Approaches to influence starch quantity and starch quality in transgenic plants. *Plant Cell Environ.* 17:601–13

61. Müller-Röber BT, Kossmann J, Hannah LC, Willmitzer L, Sonnewald U. 1990. One of two different ADP-glucose pyrophosphorylase genes responds strongly to elevated levels of sucrose. *Mol. Gen. Genet.* 224:136–46

62. Müller-Röber B, La Cognata U, Sonnewald U, Willmitzer L. 1994. A truncated version of an ADP-glucose pyrophosphorylase promoter from potato specifies guard-cell selective expression in transgenic plants. *Plant Cell* 6:601–12

64. Nakamura Y, Umemoto T, Takahata Y, Komae K, Amano E, Satoh H. 1996. Changes in the structure of starch and enzyme activities affected by *sugary* mutations in developing rice endosperm. Possible role of starch debranching enzyme (R-enzyme) in amylopectin biosynthesis. *Physiol. Plant.* 97:491–98

63. Nakamura Y, Umemoto T, Ogata N, Kuboki Y, Yano M, Sasaki T. 1996. Starch debranching enzyme (R-enzyme or pullulanase) from developing rice endosperm: purification, cDNA and chromosomal localization of the gene. *Planta* 199:209–18

65. Nakata PA, Okita T. 1996. *Cis*-elements important for the expression of the ADP-glucose pyrophosphorylase small-subunit are located both upstream and downstream from its structural gene. *Mol. Gen. Genet.* 250:581–92

66. Neuhaus HE, Stitt M. 1990. Control analysis of photosynthate partitioning: Impact of reduced activity of ADP-glucose pyrophosphorylase or plastid phosphoglucomutase on the fluxes to starch and sucrose in *Arabidopsis thaliana* (L.) Heynh. *Planta* 182:445–54

67. Okita TW. 1992. Is there an alternative pathway for starch synthesis? *Plant Physiol.* 100:560–64

68. Oostergetel GT, van Bruggen EFJ. 1993. The crystalline domains in potato starch granules are arranged in a helical fashion. *Carbohydr. Polym.* 21:7–12

69. Pan D, Nelson OE. 1984. A debranching enzyme deficiency in endosperms of the *sugary-1* mutants of maize. *Plant Physiol.* 74:324–28

70. Preiss J, Sivak M. 1996. Starch synthesis in sinks and sources. In *Photoassimilate Distribution in Plants and Crops,* ed. E Zam-

ski, AA Schaffer, pp. 63–94. New York: Dekker

71. Prioul J-L, Jeannette E, Reyss A, Grégory N, Giroux M, et al. 1994. Expression of ADP-glucose pyrophosphorylase in maize (*Zea mays* L.) grain and source leaf during grain filling. *Plant Physiol.* 104:179–87

72. Shannon JC, Pien F-M, Liu K-C. 1996. Nucleotides and nucleotide sugars in developing maize endosperm. *Plant Physiol.* 110:835–43

73. Smith AM, Denyer K, Martin C. 1995. What controls the amount and structure of starch in storage organs? *Plant Physiol.* 107:673–77

74. Smith AM, Martin C. 1993. Starch biosynthesis and the potential for its manipulation. In *Biosynthesis and Manipulation of Plant Products*, ed. D Grierson, pp. 1–54. Glasgow: Blackie

75. Smith-White BJ, Preiss J. 1992. Comparison of proteins of ADP-glucose pyrophosphorylase from diverse sources. *J. Mol. Evol.* 34:449–64

76. Stinard PS, Robertson DS, Schnable PS. 1993. Genetic isolation, cloning, and analysis of *Mutator*-induced, dominant antimorph of the maize *amylose-extender1* locus. *Plant Cell* 5:1555–66

77. Stitt M. 1995. The use of transgenic plants to study the regulation of plant carbohydrate metabolism. *Aust. J. Plant Physiol.* 22:635–46

78. Sun C, Sathish P, Ek B, Deiber A, Jansson C. 1996. Demonstration of in vitro starch branching enzyme activity for a 51/50-kDa polypeptide isolated from developing barley (*Hordeum vulgare*) caryopses. *Physiol. Plant.* 96:474–83

79. Takeda Y, Guan H-P, Preiss J. 1993. Branching of amylose by the branching isoenzymes of maize endosperm. *Carbohydr. Res.* 240:253–63

80. Takeda Y, Hizukuri S. 1987. Structures of rice amylopectins with high and low affini-

ties for iodine. *Carbohydr. Res.* 168:79–88

81. Thorbjørnsen T, Villand P, Denyer K, Olsen O-A, Smith AM. 1996. Distinct isoforms of ADPglucose pyrophosphorylase occur inside and outside the amyloplasts in barley endosperm. *Plant J.* 10:243–50

82. Thorbjørnsen T, Villand P, Kleczkowski L, Olsen O-A. 1996. A single gene encodes two different transcripts for the ADP-glucose pyrophosphorylase small subunit from barley (*Hordeum vulgare*). *Biochem. J.* 131:149–54

83. Tomlinson KL. 1995. *Starch Synthesis in Leaves of Pea (*Pisum sativum L.*).* PhD thesis. Univ. East Anglia, Norwich, UK

84. Tomlinson KL, Lloyd JR, Smith AM. 1996. Importance of isoforms of starch-branching enzyme in determining the structure of starch in pea leaves. *Plant J.* In press

85. Van den Koornhuyse N, Libessart N, Delrue B, Zabawinski C, Decq A, et al. 1996. Control of starch composition and structure through substrate supply in the monocellular alga *Chlamydomonas reinhardtii*. *J. Biol. Chem.* 271:16281–87

86. Villand P, Aalen R, Olsen O-A, Lüthi E, Lönneborg A, Kleczkowski L. 1992. PCR amplification and sequences of cDNA clones for the small and large subunits of ADP-glucose pyrophosphorylase from barley tissues. *Plant Mol. Biol.* 19:381–89

87. Villand P, Kleczkowski L. 1994. Is there an alternative pathway for starch biosynthesis in cereal seeds? *Z. Naturforsch. Teil C* 49: 215–19

88. Wasserman BP, Harn C, Mu-Forster C, Huang R. 1995. Progress towards genetically modified starches. *Cereal Foods World* 40:810–17

89. Weber H, Heim U, Borisjuk L, Wobus U. 1995. Cell-type specific, coordinate expression of two ADP-glucose pyrophosphorylase genes in relation to starch biosynthesis during seed development of *Vicia faba* L. *Planta* 195:352–61

Annu. Rev. Plant Physiol. Plant Mol. Biol. 1997. 48:89–108

CHEMICAL CONTROL OF GENE EXPRESSION

C. Gatz

Pflanzenphysiologisches Institut, Georg-August-Universität Göttingen, Untere Karspüle 2, 37073 Göttingen, Germany

KEY WORDS: elicitors, safeners, wound response, steroids, tetracycline

ABSTRACT

Promoters that respond to otherwise inactive chemicals will enhance the tools available for analyzing gene function in vivo and for altering defined traits of plants at will. Approaches to provide such tools have yielded plant promoters that respond to compounds activating defense genes. In addition, the transfer of regulatory elements from prokaryotes, insects, and mammals has opened new avenues to construct chemically inducible promoters that respond to signals normally not recognized by plants. This review describes results and applications of these two approaches.

CONTENTS

INTRODUCTION

Chemical gene induction systems provide an essential tool for the temporal and quantitative control of transferred genes in vivo. Such systems have appli-

89

cations in many areas of basic and applied biology, including the study of gene function, cell lineage ablation experiments, enhanced synthesis of recombinant proteins, and expression of commercially valuable traits. For plant molecular biology purposes, the ideal regulatable expression system should have the following features: (*a*) expression levels should be very low in the absence of the chemical and (*b*) should increase rapidly to high levels upon application of the inducer; (*c*) the ideal chemical should be nontoxic to the plant and all other organisms in the plant's ecosystem; (*d*) it should not induce pleiotropic effects in treated plants; (*e*) it should be easily applicable in the field and in the greenhouse by spraying, (*f*) or under tissue culture conditions by adding it to the synthetic medium; (*g*) depending on the application, different derivatives of the inducing chemical should be available, one that moves systemically and a second one that stays at the site of application; (*h*) induction should be efficacious at a low use rate, (*i*) and a second compound should be available that abrogates induction; (*j*) a chemically inducible system should also be combinable with tissue-specific expression. This review describes recent approaches to constructing such an ideal expression system, as well as results of experiments obtained for different reasons that can be discussed in the context of inducible gene systems for heterologous genes.

PLANT PROMOTERS

Plant regulatory sequences responsive to chemical treatment are attractive because their use requires only the cloning of the responsive promoter upstream of the coding region of the gene of interest. However, the intrinsic disadvantage of this approach is that native genes controlled by these regulatory sequences are also induced upon addition of the chemical regulator. Thus, it is important to chose an inducer that affects a set of genes that does not interfere with normal growth and development. In this context, promoters responding to the five classical phytohormones or to other plant growth regulators can be excluded. In addition, metabolic signals or nutrients are difficult to handle, because starvation for one component adversely affects the physiological condition of the plant. Four different groups of chemicals that elicit changes in plant physiology are considered potential inducers of gene expression: chemicals inducing genes required for systemic acquired resistance (SAR), elicitors, safeners, and wound signals.

Chemicals Inducing Systemic Acquired Resistance

Upon infection of a resistant plant with a necrotizing pathogen, SAR develops in uninfected tissue and provides protection against a wide range of pathogens.

It is associated with the systemic appearance of at least five families of pathogen-related (PR) proteins (47) as well as additional sets of proteins (44, 85). PR proteins are a heterogeneous group of low molecular mass proteins that are induced in plants not only by pathogen infection but also by exogenously applied chemicals. Most responses are characterized in tobacco. Arabidopsis shows very similar but distinct responses. Chemicals identified thus far to induce PR protein accumulation are salicylic acid (SA) (87), ethylene (62), xylanase (49), polyacrylic acid (28), barium chloride, 2-thiouracil (88), ethephon (81), amino acids and derivatives such as α-amino-butyric acid (5) and DL-amino-n-butanoic acid (12), thiamine (5), 2,6,-dichloroisonicotinic acid (INA) (85), and benzol (1,2,3)thiadiazole-7-cabothioic acid S-methyl ester (BTH) (21, 45).

Cloning of the genes for different PR proteins and other SA inducible proteins allowed the analysis of their regulation at the transcriptional level. mRNAs for PR-1 to PR-5 as well as mRNAs for four additional sets of proteins can be induced by SA, INA, and BTH (21, 86). These mRNAs represent 1% of the total mRNA in induced tissues (85), which makes the corresponding promoters attractive for high inducible expression of transgenes. Ethephon, an ethylene releasing chemical, induces transcription of a subset of SA responsive promoters in tobacco (10) and Arabidopsis (46), but ethylene is only effective in tobacco (62). Transcription of *PR-2* is inducible by thiamine (15), and transcription of *PR-1* is induced by the same set of chemicals that induce accumulation of the protein (52). Thiamine acts upstream of SA, INA downstream, and polyacrylic acid enters the signal transduction chain at different points (52).

To investigate whether these promoters can mediate chemical induction of heterologous genes, promoters for PR-1a (7, 56, 77, 80), a glycine-rich protein (79), and basic and acidic β-glucanases (79) were fused to different reporter genes. The best-studied promoter is the *PR-1a* promoter. In the uninduced state, GUS (β-glucucoridase) activities range from 10 U [U = pmol 4-MU produced min^{-1} (mg protein)$^{-1}$; 80] to 48 U (77), to 4000 U [56; units indicated by the authors in μmol 4-MU (fresh weight)$^{-1}$ have been converted to the U defined above for the sake of comparability].[1] High GUS activities reported by Ohshima et al (56) might be due to the position of the cauliflower

---

[1] When discussing efficiencies of expression systems, comparison of GUS units can be misleading. Data on the widely used cauliflower mosaic virus (CaMV) 35S promoter activity, for instance, vary over a broad range. CaMV 35S–mediated GUS activities were reported as 113,000 U [average of 10 plants (8)], 321 U [one selected plant (40)], 9000 U [average of 15 plants (67)], 500 U [highest expressing plant (13)], and 130,000 U [one selected plant (42)].

mosaic virus (CaMV) 35S promoter close to the *PR-1a* promoter (7). Surprisingly, upon SA induction, GUS activities increased only 5- to 10-fold after 1–3 days of induction (leaf disks floated on 1–2 mM SA, or SA painted onto leaves). Under field conditions, spraying with 50 mM SA led to 10-fold induction after 8 days. By day 20, GUS levels had dropped to background levels (84). The 10-fold induction stands in striking contrast to the 1000- to 10,000-fold increase that is observed at the *PR-1a* mRNA level. Uknes et al (77) found a high level of *uidA* mRNA induction upon SA treatment, indicating that inefficient translation under inducing conditions might be the reason for the poor induction of GUS enzyme activities. The presence of a heat shock element–like motif in the *PR-1a* promoter suggests an evolutionary relationship between heat shock and pathogen response, allowing the speculation that translation of nonstress mRNAs is reduced (7). However, the *PR-1a* promoter has been used to induce *Bacillus thuringiensis* δ-endotoxin expression in transgenic plants (91). Insect feeding damage was inhibited in plants that had been pretreated with INA, but not in untreated isogenic lines. From this experiment, it can be concluded that the window between the uninduced and the induced state is suitable for at least some accplications in the field.

The most preferable inducer for the *PR-1a* promoter is the chemical BTH, which is most recently described (30). While both SA and INA are potent inducers, crop tolerance problems are associated with their use. Only a narrow safety margin separates the rates at which the compounds are efficacious and the rate at which they are strongly phytotoxic (43). In contrast, BTH does not seem to be phytotoxic. After foliar spray of BTH, the *PR-1a* promoter starts to respond after 12 h, reaching its maximum 3 days after application. mRNA levels stay induced over at least 20 days, a longer-lasting response when compared with SA or INA (21). In addition, GUS activities were induced up to 100-fold, using the same constructs that were only 10-fold inducible with SA (21, 77). SA can be used for local induction, because it induces gene expression only in the leaf tissues that have been treated (81), whereas BTH moves systemically through the plant.

The *PR-1a* promoter is also induced in senescing leaves, by UV-B (10, 34), SO_2, ozone (71), and other treatments that result in oxidative stress, implying that its activity might not be completely controllable outside the laboratory. PR-1 proteins are present in sepals of flowers, (50) and a chimeric *PR-1a:uidA* gene was expressed in mesophyll cells of flowering plants (77). Low *PR-1a* mRNA levels were detected in leaves of flowering plants but not in flowers (53), though Côté et al (15) report *PR-1* gene expression during sepal development. *PR-1* genes are also expressed during early seed development (15). In the light of this developmental expression pattern, transgenic plants encoding a potentially lethal gene under *PR-1a* control might have reduced fertility. PR proteins have also been shown to accumulate in undifferentiated callus cultures (2, 3), and *PR-1* mRNAs have been detected in regenerating tobacco shoots because of elevated cytokinin levels (52). However, Uknes et al (77) did not observe any PR-1a induction by 6-benzylaminopurine (BAP), 1-naphthaleneacetic acid (NAA), or kinetin. Thus, it remains to be investigated whether a potentially lethal gene expression cassette can be kept silent under the tissue culture conditions used to regenerate transformants. The promoters for the glycine-rich proteins and the acid and basic glucanases are even less suitable, because they have higher background activities (15, 79).

When using the *PR-1a* promoter in combination with BTH, it has to be considered that PR proteins, chitinases, glucanases, catalases, stress proteins, cyclophilines, glutathione S-transferases (GSTs), alternative oxidases, and manganese superoxide dismutase are induced and reflect a gross change in plant metabolism (44). Despite this, to date the *PR-1a* promoter is the most promising promoter for applications in the field. In combination with BTH, it is very likely to be useful for expressing genes coding for proteins that are harmful for certain pathogens. As discussed in the context of *B. thuringiensis* toxin, continuous expression of these proteins will favor adaptive processes of the pathogen (85). Thus, BTH can not only be used as a general inducer of SAR, but also for the induction of more specific defense proteins.

Elicitors

SAR is a late response within a coordinated resistance strategy that is induced primarily by elicitors (6). Biotic elicitors originate either from the host plant (endogenous elicitors) or from the plant pathogen (exogenous elicitors). Macromolecules such as oligosaccharides, glycoproteins, peptides, and phospholipids from fungi (9) have been identified as biotic elicitors. Abiotic elicitors are heavy metals such as $HgCl_2$ or $AgNO_3$. Genes coding for enzymes of the phenylpropanoid metabolism such as phenylammonia-lyase (PAL), chalcone synthase (CHS), and chalcone isomerase (CHI) are induced first after elicitor treatment, followed by the induction of hydroxyproline-rich glycoproteins (HRGPs), peroxidases, chitinases, pectinases, etc (6). Lamb et al (16) showed that a bean *chs* promoter fused to the *uidA* gene in transgenic tobacco was inducible as much as 18-fold by a glucan elicitor from *Phytophthora megasperma* and $HgCl_2$ (16). The authors discuss the application of these promoters as a convenient and sensitive assay for screening potential elicitors of the defense response rather than as a tool for inducible expression of heterologous genes. Because elicitors act upstream of the SAR response, the response is even more pleiotropic. In addition, induction of the phenyl-propanoid pathway may lead to accumulation of undesirable metabolites that are harmful to normal plant growth (35). For selected applications, it might be favorable that some pathogenesis-specific promoters are only active in the infected tissue. A chimeric potato *prp1-1:barnase* gene has been used to induce local cell death upon infection with *Phytophthora infestans* (74). Reduced glutathione was also shown to act as an activator of elicitor-induced genes in suspension-cultured cells or protoplasts of bean, soybean, and alfalfa (11, 92), but its use in transgenic plants is unexplored.

Safeners

Safeners are a group of structurally diverse chemicals used to increase the plant's tolerance to the toxic effects of an herbicidal compound. Examples of these compounds include naphthalic anhydride and N,N-diallyl-2,2-dichloroacetamide (DDCA), which protect maize (36) and sorghum (57) against thiocarbamate herbicides; cyometrinil, which protects sorghum against metochlor (18); triapenthenol, which protects soybeans against metribuzin (82); and substituted benzenesulfonamides, which improve the tolerance of several cereal crop species to sulfonylurea herbicides (1). The biochemical basis for the action of the safeners lies in their ability to accelerate the metabolic detoxification of herbicidal compounds in treated plants. Safener responsive genes were reported 10 years ago (89). Safener-induced genes typically encode detoxifying enzymes such as GSTs, cytochrome P-450 mixed function oxygenases, and other proteins of unknown function (20). Hershey & Stoner (38) cloned two maize cDNAs (*In2-1* and *In2-2*) from a 2-chlorobenzenesulfonamide (2-CBSU) induced maize cDNA library using a differential screening procedure. The deduced In2-1 polypeptide has 50% similarity with the amino acid sequence of a Type III GST from potato (52a). The In2-2 polypeptide indicated no significant similarity to known sequences. *In2-1* and *In2-2* mRNA levels were undetectable in untreated maize seedlings and were induced in roots and leaves after addition of 200 mg/l 2-CBSU into the hydroponic medium. Maximum levels of both mRNA species were reached in roots after 6 h and in leaves after 12 h and remained constant for 2 days. Upon removal of plants from the inducing medium, mRNA levels declined within 2–3 days. Foliar application of the safener (1 kg/ha) also led to induction of gene expression in both leaves and meristems after 48 h. Both genes are not responsive to wounding, heat shock, and hydroponic treatment with 100 mM NaCl, 100 mM urea, 20 mM proline, or 100 mg/l gibberellic acid. Transcription of *In2-1*, but not of

In2-2, was minimally responsive to treatment with abscisic acid, indolacetic acid, indolebutyric acid, SA, and salicylamide. The level of induction is not as high as that seen for SAR-related genes. It has not yet been published how well the promoter works in combination with heterologous genes, and whether it is useful for species other than maize.

Jepson et al (41) cloned *GST* cDNAs from a library prepared from a safener-induced maize. Transcripts corresponding to *GST-27* were visible in untreated seedling roots, but in aerial parts of the plants, signals were visible only after application of the safener through the roots (20 mg/L). Foliar application of the safener led to an increase of *GST-27* mRNA after 8 h, which returned to background levels after 2–3 days. Wounding, treatment with SA, and ethylene did not affect transcript abundance, but an increase of expression was observed after treatment with chemicals having phytotoxic effects or in senescent leaves. As for *In2-1* and *In2-2*, it has not yet been shown whether the *GST-27* promoter might be suitable for the regulated expression of transgenes. The constitutive expression in roots, however, does not make this promoter a prime candidate for this application.

GST genes have also been cloned from a number of dicot species, such as soybean (78) and tobacco (17, 75) and these complex gene families were recently reviewed (52a). The soybean *GH2/4* promoter has been analyzed in transgenic tobacco plants (78). In the absence of inducers, chimeric *uidA* transgenes have background activity of 100 to 200 U, which is considerably higher than that of the *PR-1a* fusion (77, 80). The construct was inducible (up to 100-fold) by active and inactive auxins, SA, abscisic acid, kinetin, heavy metals, methyl jasmonate (MeJA), and H_2O_2. In particular, the inducibility by auxins makes this promoter unsuitable for transgene regulation, as many transformation protocols require auxin for the regeneration process of transformed cells. *GH2/4* promoter activity was found in root tips in the absence of inducers. As in the case of the SAR genes, the inducing chemical activates a number of stress or defense genes, though a different set is induced. For applications where these pleiotropic effects can be tolerated, the *PR-1a* promoter seems to be superior to the *GH2/4* promoter, because it has a wider window between the induced and the uninduced state, at least at the mRNA level. Moreover, GST mRNAs decline 4–8 h after induction, even if the inducing chemical is still present (75).

Chemicals Mediating Wound Responses

Similar to the defense response, wounding rapidly induces proteins at the wounding site (PAL, CHS, CHI, HGRPs). Analogous to the SAR, a defined set of genes [e.g. for proteinase inhibitors, polyphenyl oxidases, and proteins of yet unknown function like LAP and TD (68)] are induced systemically. Experiments with transgenic tobacco showed that *cis*-acting sequences of a potato *proteinase inhibitor II* gene could confer wound inducibility on other genes (42, 76). Under the control of the *proteinase inhibitor II* promoter, transcription of the *uidA* gene increased dramatically in wounded and systemic leaves of transgenic tobacco and potato (42). A chemical signal that initiates only the expression of systemically induced genes would have the same advantages as BTH, because only a subset of the wound-inducible genes is induced, thus reducing the extent of pleiotropic effects.

One systemically mobile signal is the 18–amino acid peptide systemin (58), which has been proposed to mediate gene induction via a lipid-based signaling pathway called the octadecanoid pathway. In this pathway, linolenic acid, released from membranes in response to the wound signals, is converted to jasmonic acid (JA), which leads to the transcriptional activation of defense genes. However, JA and its methyl ester, methyl-jasmonate (Me-JA), have diverse physiological effects (70, 73).

Bestatin, a compound, which was first isolated from culture filtrates of *Streptomyces olivoreticuli,* activates the signaling pathway downstream of systemin perception, thus activating gene expression without eliciting the accumulation of JA (68). When supplied to young tomato plants through their cut stems, bestatin specifically induces the accumulation of several mRNAs of wound response genes in leaves. These mRNAs code for proteinase inhibitor I, pro-systemin, polyphenol oxidase, and four other proteins of unknown function. mRNAs for PR proteins were not induced. An initial increase in inhibitor I mRNA was observed 2 h after supplying the plants with the inducer, and mRNA levels continued to rise thereafter up to 24 h. Activation of chimeric reporter genes by bestatin has not yet been shown.

HETEROLOGOUS ELEMENTS OF GENE REGULATION

As outlined in the introduction, the ideal inducer should affect expression only of the transgene. This requirement favors the use of well-characterized regulatory elements from evolutionarily distant organisms, such as yeast, *Escherichia coli, Drosophila melanogaster,* or mammal cells, which respond to chemical signals that are usually not encountered by higher plants. On this basis, three different concepts of gene control can be realized: transcriptional promoter-repression and promoter-activation (Figures 1 and 3) and posttranslational control of protein function.

Promoter-Repressing Systems

TETRACYCLINE-INDUCIBLE GENE EXPRESSION   The repression principle is based on sterical interference of a repressor protein with proteins important for transcription (Figure 1). It is a common mechanism in bacteria but it is found infrequently in higher eukaryotes, where protein-protein interactions are the primary mechanism to mediate stimulating or inhibitory effects on the transcription machinery. Two bacterial repressor/operator systems (Lac and Tet) have been employed to control the activity of polymerase II promoters according to the prokaryotic paradigm. The first paper reporting that a bacterial repressor protein can be used to control the activity of a modified CaMV 35S promoter in transient assays appeared in 1988 (27). The Tet repressor (TetR) encoded by the *E. coli* transposon Tn*10* regulates expression of the tetracycline (tc) resistance

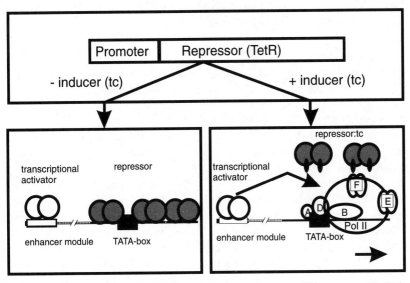

Figure 1   Schematic representation of a tc-inducible repression system. The repressor (TetR) is synthesized under the control of a strong constitutive promoter. The target promoter contains plant enhancer modules upstream of the TATA-box as well as operator sequences in the vicinity of the TATA-box. The DNA is represented as a string of white squares, the operators are indicated in black. In the absence of the inducer tc (*small black symbols*), repressor molecules (*gray circles*) interfere with transcription initiation. In the presence of tc, the repressor dissociates from the DNA, and repression is relieved, allowing assembly of the multifactorial initiation complex, which contains TFIID, TFIIA, TFIIB, TFIIF, TFIIE, other associated factors, as well as polymerase II, and transcription is induced (*arrow*). Tissue specificity of the system can be achieved by choosing appropriate enhancer modules of the target promoter.

gene in *E. coli,* was used in this study. The DNA-binding activity of this protein is abolished by very low amounts of the antibiotic tc (39). Because tc readily enters eukaryotic cells, it is a favorable chemical inducer for laboratory experiments.

Systematic analysis of the effect of repressor-operator-complexes in different positions within the CaMV 35S promoter (22, 37) led to the design of a tightly repressible 35S promoter that contains one *tet* operator directly upstream of the TATA-box and two *tet* operators downstream of the TATA-box (24, 25). Sequence changes in these regions did not reduce expression from the CaMV 35S promoter. Repression depends on high intracellular repressor concentrations [1×10^6 molecules per cell (26)], because the repressor must compete with at least 40 proteins that assemble around the TATA-box to form a competent transcription initiation complex (Figure 1; 66). In tobacco, expression of the tc inducible promoter can be modulated up to 500-fold (25). The induction factor is independent of position effects. Highly expressing plants have background levels of 2000 U that can be induced to 180,000 U. Low-expressing plants, which have GUS activities indistinguishable from background, contain GUS levels ranging between 1000 and 2000 U in the induced stage. If tc is infiltrated into single leaves, full induction at the RNA level is reached within 10 min using tc concentrations as low as 0.1 mg/l (26). Induction at the whole plant level is most efficient in plants growing in hydroponic culture, with fresh tc (1mg/l) added to the nutrient solution every other day (25). For maximal induction under these conditions 10–14 days are needed, and tc must be constantly available in the medium. Foliar spray has so far been ineffective, but plants can also be grown in sand or rockwool, if the setup allows drainage of the tc-containing nutrient solution (14). In tissue culture containers, transpiration is not sufficient for homogenous distribution of the inducer, but expression in roots and in leaves touching the medium is highly induced. When used at a concentration of 1 mg/l in trays with rockwool, tc caused only a marginal reduction in height, chlorophyll content, and assimilation rates of tomato plants, but a 70% reduction of root dry weight (14). Similar observations have been made for tobacco (C Gatz, unpublished observations). Regulation is stringent enough to completely suppress the *rolB* and *rolC* phenotype caused by expression of *Agrobacterium* genes (Figure 2; 19, 65). Because of the short half-life of the antibiotic in the plant, fresh tc must be added to the hydroponic culture every other day. This allows for the genes to also be transiently expressed. For instance, plants with a strong *rolB* phenotype developed normal shoots after omission of the antibiotic. The system has also been used to express a dominant negative mutant of the TGA-family of transcription factors in transgenic tobacco plants (64), allowing the conditional reduction of the amounts of transcription factor complex ASF-1.

Although this system works well in tobacco, applications in other plant species have not been very convincing, except for potato (R Höfgen & L Willmitzer, personal communication). In tomato, high levels of TetR cause reduced shoot dry weight and leaf chlorophyll content, reduced leaf size, and an altered photosynthetic physiology when grown between July and September (14). This phenotype was almost completely reversed by the application of tc. In addition, the phenotype was not visible when plants were grown in November and December. Many laboratories have tried to establish the TetR system in Arabidopsis. Repressor concentrations sufficient for transcriptional control cannot be tolerated in Arabidopsis, a phenomenon that has also been reported for mammalian cells (31).

IPTG-INDUCIBLE GENE EXPRESSION   Efforts to use the Lac repressor-operator system to establish IPTG (isopropyl-β-D-thiogalactopyranoside) inducible tran-

Figure 2 Isogenic cuttings of transgenic tobacco plants encoding the *rolB* gene from *Agrobacterium rhizogenes* under the control of the tc inducible promoter. Only the *left plant* was treated with tc. In the absence of tc (*right plant*), the *rolB* phenotype is completely suppressed, leading to normal growth. In the presence of tc (*left plant*), induction of *rolB* leads to a dramatic reduction in growth. The picture was taken five weeks after the onset of induction. (Figure 2e from Reference 65.)

scription gave promising results first. Lac operators were inserted into the vicinity of the TATA box of the CAB promoter (90). Although the constructs were stably integrated into the genome of tobacco plants, 15-fold induction with IPTG was not done at the whole plant level, but only in protoplasts.

Promoter Activating Systems

STEROID-DEPENDENT GENE EXPRESSION   A different approach to construct a chemically inducible system for higher plants is to use transcriptional activators from other higher eukaryotes. The mammalian glucocorticoid receptor (GR), which activates eukaryotic transcription only in the presence of steroids such as dexamethasone has been used previously to establish a regulatory system in *Schizosaccharomyces pombe* (61). In transiently transformed tobacco cells, transcription of a target promoter containing GR-binding sites upstream of a TATA-box strictly depended on the presence of GR and dexamethasone (69). The difference between the uninduced and induced levels was 150-fold, with the absolute level of expression approximately 1/10 of that seen using a CaMV 35S promoter-driven reporter gene. In stably transformed Arabidopsis plants, however, this arrangement of regulatory elements did not work (48).

It was discovered several years ago that the hormone-binding domain (HBD) of GR and other steroid receptors can be used as a molecular switch to regulate heterologous proteins in *cis* (59). The unliganded HBDs of all vertebrate steroid receptors are known to assemble into a protein complex containing heat shock protein 90 (hsp 90). This complex, which is released upon hormone binding, most likely effects inactivation of protein function by steric hindrance. In the case of GR, nuclear localization is affected. HBD has been shown to regulate protein function both in mammals and *Saccharomyces cerevisiae,* which suggests that the regulatory machinery has been conserved in evolution (51). By fusing HBD to maize transcription factor R, a steroid-inducible transcriptional activator (R-GR) was generated (48). The R-GR fusion protein was transformed into an Arabidopsis mutant (*ttg*) that can be complemented with maize transcription factor R. The R-GR fusion protein allowed steroid-inducible complementation of the phenotype (trichome development and anthocyanin synthesis) and was used for epidermal cell fate mapping experiments. The estrogen receptor was also used, but gave higher constitutive expression (48).

Using a similar approach, a fusion protein consisting of the Arabidopsis transcriptional activator Athb-1 and the activation domain of *Herpes simplex* virion protein 16 (VP16) was fused to HBD of GR (4). Athb-1-VP16-GR displayed the same function as Athb-1-VP16 only in the presense of 10 mM dexamethasone.

Fusing HBDs to transcriptional activators that would recognize binding sites not present in plant promoters might be a sensible way to construct a steroid inducible expression system. Louvion et al (51) constructed a fusion protein consisting of the DNA-binding domain of GAL4, the estrogen receptor hormone binding domain, and the activation domain of VP16. This chimeric activator stimulated reporter genes under the control of a chimeric promoter containing GAL4-binding sites exclusively in the presence of steroid hormone. Whether this three-domain-protein can also confer hormone-dependent gene expression in transgenic plants has not been published. It has been reported, however, that the DNA-binding domain of GAL4 is inefficiently translated in transgenic plants, indicating that this problem must be solved first (63). The construction of such a chimeric transcription factor is not always achievable. Tight regulation appears to be critically dependent on the intramolecular structure of the chimeric protein, especially the relative position between HBD and the domain whose function is to be regulated. In mammalian cells, an E1A-HBD fusion protein has a fully hormone-dependent transcriptional activation function, but both nuclear localization and transformation function remain constitutively active (60).

COPPER-DEPENDENT GENE EXPRESSION   Another eukaryotic ligand-dependent activator is ACE1, a copper-dependent transcriptional activator from yeast. Mett et al (54) found that ACE1 regulates transcription of a suitable target promoter in a copper-dependent manner in transgenic plants. GUS activities could be induced from 40 U to up to 2000 U either by adding copper to the hydroponic solution or by foliar spray. Because copper is likely to affect other processes in plants, it remains to be established how specific this inducer is.

TETRACYCLINE-DEPENDENT GENE EXPRESSION   A third strategy to exploit the activation principle is based on the construction of fusion proteins between transcriptional transactivation domains and bacterial repressor proteins such as the Lac repressor or TetR. The favorable thermodynamic properties of the TetR/tet operator/tc interaction prompted Gossen & Bujard (32) to construct a tc controlled transactivator (tTA) by fusing the VP16 activation domain to the C-terminus of TetR. tTA can regulate gene expression from a target promoter containing 7 tet operators upstream of a minimal promoter over a range of five orders of magnitudes in stably transformed HeLa cells (Figure 3). The same principle was shown to work in transgenic tobacco plants (86), thus establishing a promoter system that can be shut off in the presence of tc. This system has been successfully applied for measuring mRNA decay rates in tobacco BY-2 cells (29). Because of the fast uptake of tc by suspension cultured cells, the target promoter can be shut off very efficiently, which allows the observation of first-order decay of transcripts within 15 min after tc treatment. The tTA-dependent promoter provides an important alternative to using general inhibitors of polymerase II like actinomycin D. Actually, it proved to be essential in the analysis of the effect of the 3' untranslated region of one of the SAUR transcripts on mRNA stability. The destabilizing effect of the sequence was not visible when actinomycin D was used for half-life studies, which indicates that some mRNA decay pathways require ongoing transcription to function. Because the tTA-based system has not been optimized as thoroughly as the tc inducible promoter, its potential has not yet reached its best. Expression levels reach 30% of the levels reached by the inducible system and drop as transgenic plants age (86). The tTA system has also been successfully transferred into Arabidopsis (M Roever, U Treichelt, C Gatz, J Schiemann & R Hehl, manuscript in preparation) and the moss *Physcomytrella patens* (94), where the silencing of gene expression does not seem to occur.

As discussed by Gossen & Bujard (32) and Weinmann et al (86), stringent control of transcription in higher eukaryotes is more likely to be achieved by promoter activation than by repression, most likely because transcriptional activators have free access to their target sites, whereas repressors compete with endogenous transcription factors for binding (Figures 1 and 3). Thus, higher levels of a repressor protein are needed for the same degree of occupancy of target sites. In addition, 50% occupancy of binding sites can be sufficient for transcriptional activation but definitely not sufficient for stringent repression. Nonetheless, a regulation over a range of three orders of magnitude has been achieved in trypanosomes based on the repression principle, using again TetR (93). In this case, a promoter driven by polymerase I was used, and repression was only efficient when the construct was integrated in the nontranscribed spacer of ribosomal DNAs. The tTA system is definitely intriguing because of its low background activity in the presence of tc. However, plants must be permanently cultivated on tc to silence the transgene, a difficult task considering that plants require light and tc is unstable in the light. Furthermore, inactivation of tc is slower than uptake of tc for induction of gene expression.

A promising alternative is the use of a mutant TetR that shows a "reverse phenotype" (33). This reverse repressor binds DNA only in the presence of tc. By fusing this repressor derivative to the VP16 activation domain (rtTA), Gossen et al (33) developed a tc-inducible promoter system that is based on the activation system. Preliminary experiments indicate, however, that rtTA does

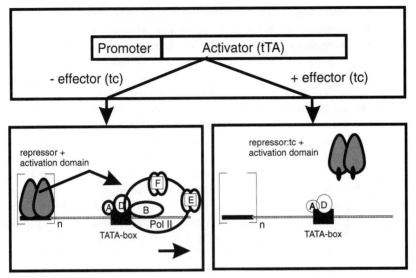

Figure 3 Schematic representation of a tc-dependent activation system. The activator (tTA), consisting of a repressor moiety (TetR) and an activation domain (VP16), is synthesized under the control of a plant promoter. The target promoter consists of multimerized operator sequences upstream of the TATA-box. The DNA is represented as a string of *white squares,* the operators are indicated in *black.* In the absence of the effector tc (*small black symbols*), the transactivator (*pear-shaped symbols*) binds to the operator sequences and stimulates assembly of the transcription complex and subsequent transcription (*arrow*). In the presence of tc, the transactivator dissociates from the DNA, and transcription is not activated. It is unclear whether some factors of the transcription initiation complex are constantly bound to the TATA-box, as indicated in this model. Tissue specificity of the system can be achieved by tissue-specific expression of the transactivator.

not work in transgenic tobacco or transgenic Arabidopsis plants. Although mRNA levels similar to tTa mRNA levels were found, no protein was detectable in Western blot analysis using TetR antibodies (C Gatz, HM Rupp & T Schmülling, personal communication).

Comparison to Promoters Used in Animals

Similar to the situation in plants, tc and steroid derivatives are the most promising regulators in mammalian cells and transgenic mice. Because many experiments are of potential applicability to medical problems, regulated systems are of high economic interest, which is certainly a factor that pushes further refinements. Moreover, these inducible systems might be used one day for a safer application of gene therapy. For plants, however, steroids and tc cannot be used in the field, making independent developments of regulatory systems for use in the field and the laboratory necessary. Inducers such as BTH (30) will be important for field applications, whereas steroids and tc can be used for studying gene function in the laboratory.

Since 1992, over 30 papers have employed a tTA-dependent expression system, either in combination with a reporter gene or with a gene of biological interest (72). These numbers document the potential of the system and far exceed the number of papers that report use of the tc system in plants. However, most of the publications describe its use in cultured cells, where the difference between induced and uninduced expression levels can be several orders of magnitude. The unfavorable kinetics of activation by depletion of tc can be handled easily by washing out the antibiotic. However, cautions have been raised regarding the general efficacy of the system in all cell or tissue types (72). As the use of plant cell cultures is far less widely applicable, reports of the use of the tc dependent regulatory elements are less frequent.

For transgenic mice, only a few applications of the tTA-dependent expression system have been described. The difference in expression levels between induced and uninduced tissues in transgenic mice is only 100 in optimal cases and varies in different tissues (23), supporting the notion that the tTA system works less efficiently in transgenic organisms as compared with homogeneous cells in culture. The tc inducible system, which works in transgenic tobacco plants, yields induction factors of 500, which should be sufficient for a number of experiments (25). However, a drawback of this system is that the tc-inducible promoter does not work in Arabidopsis, which has become the preferred model system in plant research.

In addition to the tc-based regulatory systems, steroids have been shown to specifically induce gene expression in animals. To avoid pleiotropic effects, mammalian steroids are avoided, a problem that is not encountered in plants. RU 486, a progesterone antagonist, has been used specifically to induce gene expression in transgenic mice (83). The dosage of RU 486 required for induction of transcription through a chimeric activator is significantly lower than that required for antagonizing progesterone action. In addition, the Drosophila ecdysone receptor has been shown to mediate ecdysone-specific gene expression when co-expressed with the retinoid X receptor. These newly developed systems are worth being tested in plants, in order to extend the tools available for regulated expression of transgenes.

Literature Cited

1. Amuti K, Sweetser PB. 1987. Herbicidal antidotes. *US Patent No. 4645527*
2. Antoniw JF, Kueh JSH, Walkey DGA, White RF. 1991. The presence of patho-genesis-related proteins in callus of Xanthi-nc tobacco. *Phytophathol. Z.* 101:179–84
3. Antoniw JF, Ooms G, White RF, Wullems GJ, van Vloten-Doting L. 1983. Pathogene-sis related proteins in plants and tissues of *Nicotiana tabacum* transformed by *Agro-bacterium tumefaciens*. *Plant Mol. Biol.* 2 :317–20
4. Aoyama T, Dong C-H, Wu Y, Carabelli M, Sessa G, et al. 1995. Ectopic expression of the *Arabidopsis* transcriptional activator Athb-1 alters leaf cell fate in tobacco. *Plant Cell* 7:1773–85
5. Asselin A, Grenier J, Côté F. 1985. Light-influenced extra-cellular accumulation of b(pathogenesis-related) proteins in *Nico-tiana* green tissue induced by various chemicals or prolonged floating on water. *Can. J. Bot.* 63:1276–83
6. Baron C, Zambryski PC. 1995. The plant response in pathogenesis, symbiosis, and wounding: variations on a common theme? *Annu. Rev. Genet.* 29:107–27
7. Beilmann A, Pfitzner AJP, Goodman HM, Pfitzner UM. 1991. Functional analysis of the pathogenesis-related 1a protein gene minimal promoter region. *Eur. J. Biochem.* 196:415–21
8. Benfey PN, Ren L, Chua N-H. 1989. The CaMV 35S promoter contains at least two domains which can confer different devel-opmental and tissue-specific expression patterns. *EMBO J.* 8:2195–201
9. Benhamou N. 1996. Elicitor-induced plant defence pathways. *Trends Plant Sci.* 1: 233–40
10. Brederode FT, Linthorst HJM, Bol JF. 1991. Differential induction of acquired re-sistance and PR gene expression in tobacco by virus infection, ethephon treatment, UV light and wounding. *Plant Mol. Biol.* 17: 1117–25
11. Choudhary AD, Lamb CJ, Dixon RA. 1990. Stress responses in alfalfa (*Medicago sativa*). VI. Differential responsiveness of chalcone synthase induction to fungal elici-tors or glutathione in electroporated proto-plasts. *Plant Physiol.* 94:1802–7
12. Cohen Y. 1993. Local and systemic control of *Phytophthora infestans* in tomato plants by DL-3-amino-n-butanoic acids. *Phy-topathology* 84:55–59
13. Comai L, Moran P, Maslyar D. 1990. Novel and useful properties of a chimeric plant promoter combining CaMV 35S and MAS elements. *Plant Mol. Biol.* 15:373–81
14. Corlett JE, Myatt SC, Thompson AJ. 1996. Toxicity symptoms caused by high expres-sion of Tet repressor in tomato (*Lycopersi-con esculentum* Mill. L.) are alleviated by tetracycline. *Plant Cell Environ.* 19: 447–54
15. Côté F, Cutt JR, Asselin A, Klessig DF. 1991. Pathogenesis-related acidic β-1,3 glucanase genes of tobacco are regulated by both stress and developmental signals. *Mol. Plant-Microbe Interact.* 4:173–81
16. Doerner PW, Stermer B, Schmid J, Dixon RA, Lamb CJ. 1990. Plant defense gene promoter-reporter gene fusions in trans-genic plants: tools for identification of novel inducers. *BioTechnology* 8:845–48
17. Droog FNJ, Hooykaas PJJ, Libbenga KR, van der Zaal EJ. 1993. Proteins encoded by an auxin-regulated gene family of tobacco share limited but significant homology with glutathione S-transferases and one member indeed shows in vitro GST activity. *Plant Mol. Biol.* 21:965–72
18. Ellis JF, Peek JW, Boehle J, Muller G. 1980. Effectiveness of a new safener for protect-ing sorghum (*Sorghum bicolor*) from me-tolachlor injury. *Weed Sci.* 28:1–5
19. Faiss M, Strnad M, Redig P, Dolezal K, Hanus J, et al. 1996. Chemically induced expression of the *rolC*-encoded β-glucosi-dase in transgenic tobacco plants and analysis of cytokinin metabolism: rolC does not hydrolyze endogenous cytokinin glucosides in planta. *Plant J.* 10:33–46
20. Farago S, Brunold C, Kreuz K. 1994. Her-bicide safeners and glutathione metabo-lism. *Phys. Plant.* 91:537–42
21. Friedrich L, Lawton KA, Ruess W, Masner W, Specker N, et al. 1996. A benzothiadia-zole derivative induces systemic acquired resistance in tobacco. *Plant J.* 10:61–70
22. Frohberg C, Heins L, Gatz C. 1991. Char-acterization of the interaction of plant tran-scription factors using a bacterial repressor protein. *Proc. Natl. Acad. Sci. USA* 88: 10470–74
23. Furth PA, Onge LS, Böger H, Gruss P, Gossen M, et al. 1994. Temporal control of gene expression in transgenic mice by a tetracycline responsive promoter. *Proc. Natl. Acad. Sci. USA* 91:9302–6
24. Gatz C. 1995. Novel inducible/repressible gene expression systems. In *Methods in Cell Biology*, ed. DW Galbraith, DP Bour-

que, HJ Bohnert, 50:411–24. New York: Academic

25. Gatz C, Frohberg C, Wendenburg R. 1992. Stringent repression and homogeneous derepression by tetracycline of a modified CaMV 35S promoter in intact transgenic tobacco plants. *Plant J.* 2:397–404

26. Gatz C, Kaiser A, Wendenburg R. 1991. Regulation of a modified CaMV 35S promoter by the Tn*10*-encoded Tet repressor in transgenic tobacco. *Mol. Gen. Genet.* 227: 229–37

27. Gatz C, Quail PH. 1988. Tn*10*-encoded Tet repressor can regulate an operator-containing plant promoter. *Proc. Natl. Acad. Sci. USA* 85:1394–97

28. Gianinazzi S, Kassanis B. 1974. Virus resistance induced in plants by polyacrylic acid. *J. Gen. Virol.* 23:1–9

29. Gil P, Green PJ. 1995. Multiple regions of the *Arabidopsis* SAUR-AC1 gene control transcript abundance: the 3′ untranslated region fuctions as an mRNA instability determinant. *EMBO J.* 15:1678–86

30. Görlach J, Volrath S, Knauf-Beiter G, Hengy G, Beckhove U, et al. 1996. Benzothiadiazole, a novel class of inducers of systemic acquired resistance, activates gene expression and disease resistance in wheat. *Plant Cell* 8:629–43

31. Gossen M, Bonin AL, Bujard H. 1993. Control of gene activity in higher eukaryotic cells by prokaryotic regulatory elements. *Trends Biochem. Sci.* 18:471–75

32. Gossen M, Bujard H. 1992. Tight control of gene expression in mammalian cells by tetracycline-responsive promoters. *Proc. Natl. Acad. Sci. USA* 89:5547–51

33. Gossen M, Freundlieb S, Bender G, Mueller G, Hillen W, Bujard H. 1995. Transcriptional activation by tetracycline in mammalian cells. *Science* 268:1766–69

34. Green R, Fluhr R. 1995. UV-B–induced PR-1 accumulation is mediated by active oxygen species. *Plant Cell* 7:203–12

35. Hahlbrock K, Scheel D. 1989. Physiology and molecular biology of phenylpropanoid metabolism. *Annu. Rev. Plant Physiol. Plant Mol. Biol.* 40:347–69

36. Hatzios KK. 1983. Herbicide antidotes: development, chemistry, and mode of action. *Adv. Agron.* 36:265–316

37. Heins L, Frohberg C, Gatz C. 1992. The Tn*10* encoded Tet repressor blocks early but not late steps of assembly of the RNA Polymerase II initiation complex in vivo. *Mol. Gen. Genet.* 232:328–31

38. Hershey HP, Stoner TD. 1991. Isolation and characterization of cDNA clones for RNA species induced by substituted ben-zenesulfonamides in corn. *Plant Mol. Biol.* 17:679–90

39. Hillen W, Berens C. 1994. Mechanisms underlying expression of Tn*10* encoded tetracycline resistance. *Annu. Rev. Microbiol.* 48:345–69

40. Jefferson RA, Kavanagh RH, Bevan MW. 1987. GUS fusions: β-glucuronidase as a sensitive and versatile gene fusion marker in higher plants. *EMBO J.* 6:3901–7

41. Jepson I, Lay VJ, Holt DC, Bright SWJ, Greenland AJ. 1994. Cloning and characterization of maize herbicide safener-induced cDNAs encoding subunits of glutathione S-transferase isoforms I, II, IV. *Plant Mol. Biol.* 26:1855–66

42. Keil M, Sanchez-Serrano JJ, Willmitzer L. 1989. Both wound inducible and tuber specific expression are mediated by the promoter of a single member of the potato proteinase inhibitor II gene family. *EMBO J.* 8:1323–30

43. Kessmann H, Staub T, Hofmann C, Maetzke T, Herzog J, et al. 1995. Induction of systemic acquired disease resistance in plants by chemicals. *Annu. Rev. Phyto pathol.* 32:439–59

44. Klessig DF, Malamy J. 1994. The salicylic acid signal in plants. *Plant Mol. Biol.* 126:1439–58

45. Lawton KA, Friedrich L, Hunt M, Weymann K, Delaney T, et al. 1996. Benzothiadiazole induces disease resistance in *Arabidopsis* by activation of the systemic acquired resistance signal transduction pathway. *Plant J.* 10:71–82

46. Lawton KA, Potter SL, Uknes S, Ryals J. 1994. Acquired resistance signal transduction in *Arabidopsis* is ethylene independent. *Plant Cell* 6:581–88

47. Linthorst HJM. 1991. Pathogenesis-related proteins of plants. *Crit. Rev. Plant Sci.* 10: 123–50

48. Lloyd AM, Schena M, Walbot V, Davis RW. 1994. Epidermal cell fate determination in *Arabidopsis:* Patterns defined by a steroid-inducible regulator. *Science* 266:436–39

49. Lotan T, Fluhr R. 1990. Xylanase, a novel elicitor of pathogenesis-related proteins in tobacco, uses a nonethylene pathway for induction. *Plant Physiol.* 93:811–17

50. Lotan T, Ori N, Fluhr R. 1989. Pathogenesis-related proteins are developmentally regulated in tobacco flowers. *Plant Cell* 1:881–87

51. Louvion J-F, Havaux-Copf B, Picard D. 1993. Fusion of GAL4-VP16 to a steroid-binding domain provides a tool for gratuitous induction of galactose-responsive genes in yeast. *Gene* 131:129–34

52. Malamy J, Sanchez-Casas P, Hennig J, Guo A, Klessig DF. 1996. Dissection of the salicylic acid signaling pathway in tobacco. *Mol. Plant-Microbe Interact.* 6:474–82

52a. Marr K. 1996. The functions and regulation of glutathione s-transferases in plants. *Annu. Rev. Plant Physiol. Plant Mol. Biol.* 47:127–58

53. Memelink J, Hoge JHC, Schilperoort RA. 1987. Cytokinin stress changes the developmental regulation of several defense-related genes in tobacco. *EMBO J.* 6: 3579–83

54. Mett VL, Lochhead LP, Reynolds PHS. 1993. Copper controllable gene expression system for whole plants. *Proc. Natl. Acad. Sci. USA* 90:4567–71

55. No D, Yao TP, Evans RM. 1996. Ecdysone-inducible gene expression in mammalian cells and transgenic mice. *Proc. Natl. Acad. Sci. USA* 93:3346–51

56. Ohshima M, Itoh H, Matsuoka M, Murakami T, Ohashi Y. 1990. Analysis of stress-induced salicylic acid-induced expression of the pathogenesis-related 1a-protein gene in transgenic tobacco. *Plant Cell* 2:95–106

57. Parker C. 1983. Herbicide antidotes: a review. *Pestic. Sci.* 14:40–48

58. Pearce G, Strydom D, Johnson S, Ryan CA. 1991. A polypeptide from tomato leaves induces wound-inducible proteinase inhibitor proteins. *Science* 253:895–98

59. Picard D. 1994. Regulation of protein function through expression of chimaeric proteins. *Curr. Top. Biotech.* 5:511–15

60. Picard D, Salser SJ, Yamamoto KR. 1988. A movable and regulable inactivation function within the steroid binding domain of the glucocorticoid receptor. *Cell* 54: 1073–80

61. Picard D, Schena M, Yamamoto KR. 1990. An inducible expression vector for both fission and budding yeast. *Gene* 86:257–61

62. Raz V, Fluhr R. 1993. Ethylene signal is induced via protein phosphorylation events in plants. *Plant Cell* 5:523–30

63. Reichel C, Feltkamp D, Walden R, Steinbiss HH, Schell J, Rosahl S. 1995. Inefficient expression of the DNA-binding domain of GAL4 in transgenic plants. *Plant Cell Rep.* 14:773–76

64. Rieping M, Fritz M, Prat S, Gatz C. 1994. A dominant negative mutant of PG13 suppresses transcription from a cauliflower mosaic virus 35S truncated promoter in transgenic tobacco plants. *Plant Cell* 6: 1087–98

65. Röder FT, Schmülling T, Gatz C. 1994. Efficiency of the tetracycline-dependent

gene expression system: complete suppression and efficient induction of the *rolB* phenotype in transgenic plants. *Mol. Gen. Genet.* 243:32–38

66. Roeder RG. 1991. The complexities of eukaryotic transcription initiation: regulation of preinitiation complex assembly. *Trends Biochem.* 16:402–8

67. Sanger M, Daubert S, Goodman RM. 1990. Characteristics of a strong promoter from Figwort mosaic virus: comparison with the analogous 35S promoter from cauliflower mosaic virus and the regulated mannopine synthase promoter. *Plant Mol. Biol.* 14: 433–43

68. Schaller A, Bergey DR, Ryan CA. 1995. Induction of wound response genes in tomato leaves by bestatin, an inhibitor of aminopeptidase. *Plant Cell* 7:1893–98

69. Schena M, Lloyd AM, Davis RW. 1991. A steroid-inducible gene expression system for plant cells. *Proc. Natl. Acad. Sci. USA* 88:10421–25

70. Sembdner G, Parthier B. 1993. The biochemistry and the physiological and molecular actions of jasmonates. *Annu. Rev. Plant Physiol. Plant Mol. Biol.* 44:569–89

71. Sharma YK, León J, Raskin I, Davis KR. 1996. Ozone-induced responses in *Arabidopsis thaliana:* the role of salicylic acid in the accumulation of defense related transcripts and induced resistance. *Proc. Natl. Acad. Sci. USA* 93:5099–104

72. Shokett PE, Schatz DB. 1996. Diverse strategies for tetracycline-regulated inducible gene expression. *Proc. Natl. Acad. Sci. USA* 93:5173–76

73. Staswick PE. 1992. Jasmonate, genes, and fragrant signals. *Plant Physiol.* 99:804–7

74. Strittmatter G, Janssens J, Opsomer C, Botterman J. 1995. Inhibition of fungal disease development in plants by engineering controlled cell death. *BioTechnology* 13: 1085–89

75. Takahashi Y, Nagata T. 1992. *parB:* an auxin regulated gene encoding glutathione S-transferase. *Proc. Natl. Acad. Sci. USA* 89:56–59

76. Thornburg RW, An G, Cleveland TE, Johnson R, Ryan CA. 1987. Wound-inducible expression of a potato inhibitor II-chloramphenicol acetyltransferase gene fusion in transgenic plants. *Proc. Natl. Acad. Sci. USA* 84:744–48

77. Uknes S, Dincher S, Friedrich L, Negrotto D, Williams S, et al. 1993. Regulation of pathogenesis-related protein-1a gene expression in tobacco. *Plant Cell* 5:159–69

78. Ulmasov T, Ohmiya A, Hagen G, Guilfoyle T. 1995. The soybean *GH2/4* gene that en-

codes a glutathione S-transferase has a promoter that is activated by a wide range of chemical agents. *Plant Physiol.* 108: 919–27

79. Van de Rhee MD, Lemmers R, Bol JF. 1993. Analysis of regulatory elements involved in stress-induced and organ-specific expression of tobacco acidic and basic β-1,3-glucanase genes. *Plant Mol. Biol.* 21: 451–61

80. Van de Rhee MD, Van Kan JAL, González-Jaén MT, Bol JF. 1990. Analysis of regulatory elements involved in the induction of two tobacco genes by salicylate treatment and virus infection. *Plant Cell* 2:357–66

81. Van Loon LC, Antoniw JF. 1982. Comparison of the effects of salicylic acid and ethephon with virus-induced hypersensitivity and acquired resistance in tobacco. *Neth. J. Plant Pathol.* 88:237–56

82. Vavrina CS, Phatak SC. 1988. Efficacy of triapenthenol as a safener against metribuzin injury in soybean (*Glycine max*) cultivars. *J. Plant Growth Regul.* 7:67–75

83. Wang Y, O'Malley BW, Tsai SY, O'Malley BW. 1994. A regulatory system for use in gene transfer. *Proc. Natl. Acad. Sci. USA* 91:8180–84

84. Ward ER, Ryals JA, Miflin BJ. 1993. Chemical regulation of transgene expression in plants. *Plant Mol. Biol.* 22:361–66

85. Ward ER, Uknes SJ, Williams SC, Dincher SS, Wiederhold DL, et al. 1991. Coordinate gene activity in response to agents that induce systemic acquired resistance. *Plant Cell* 3:1085–94

86. Weinmann P, Gossen M, Hillen W, Bujard H, Gatz C. 1994. A chimeric transactivator allows tetracycline-responsive gene expression in whole plants. *Plant J.* 5:559–69

87. White RF. 1979. Acetylsalicylic acid (aspirin) induces resistance to tobacco mosaic virus in tobacco. *Virology* 99:410–12

88. White RF, Dumas E, Shaw P, Antoniw JF. 1986. The chemical induction of PR (b) proteins and resistance to TMV infection in tobacco. *Antivir. Res.* 6:177–85

89. Wiegand RC, Shah DM, Moezer TJ, Harding EI, Collier JD, et al. 1986. Messenger RNA encoding a glutathione S-transferase responsible for herbicide tolerance in maize is induced in response to safener treatment. *Plant Mol. Biol.* 7:235–46

90. Wilde RJ, Shufflebottom E, Cooke S, Jasinska I, Merryweather A, et al. 1992. Control of gene expression in tobacco cells using a bacterial operator-repressor system. *EMBO J.* 11:1251–59

91. Williams S, Friedrich L, Dincher S, Carozzi N, Kessmann H, et al. 1992. Chemical regulation of *Bacillus thuringiensis* d-endotoxin expression in transgenic plants. *BioTechnology* 10:540–43

92. Wingate VPM, Lawton MA, Lamb CJ. 1988. Glutathione causes a massive and selective induction of plant defense genes. *Plant Physiol.* 87:206–10

93. Wirtz E, Clayton C. 1995. Inducible gene expression in trypanosomes mediated by a prokaryotic repressor. *Science* 268: 1179–83

94. Zeidler M, Gatz C, Hartmann E, Hughes J. 1995. Tetracycline regulated reporter gene expression in the moss *Physcomytrella patens*. *Plant Mol. Biol.* 30:199–205

Annu. Rev. Plant Physiol. Plant Mol. Biol. 1997. 48:109–136

REGULATION OF FATTY ACID SYNTHESIS

John B. Ohlrogge

Department of Botany and Plant Pathology, Michigan State University, East Lansing, Michigan 48824

Jan G. Jaworski

Chemistry Department, Miami University, Oxford, Ohio 45056

KEY WORDS: acetyl-CoA carboxylase, oil seeds, triacylglycerol, feedback, metabolic control

ABSTRACT

All plant cells produce fatty acids from acetyl-CoA by a common pathway localized in plastids. Although the biochemistry of this pathway is now well understood, much less is known about how plants control the very different amounts and types of lipids produced in different tissues. Thus, a central challenge for plant lipid research is to provide a molecular understanding of how plants regulate the major differences in lipid metabolism found, for example, in mesophyll, epidermal, or developing seed cells. Acetyl-CoA carboxylase (ACCase) is one control point that regulates rates of fatty acid synthesis. However, the biochemical modulators that act on ACCase and the factors that in turn control these modulators are poorly understood. In addition, little is known about how the expression of genes involved in fatty acid synthesis is controlled. This review evaluates current knowledge of regulation of plant fatty metabolism and attempts to identify the major unanswered questions.

CONTENTS

109

INTRODUCTION

All cells in a plant must produce fatty acids, and this synthesis must be tightly controlled to balance supply and demand for acyl chains. For most plant cells, this means matching the level of fatty acid synthesis to membrane biogenesis and repair. Depending on the stage of development, time of the day, or rate of growth, these needs can be highly variable, and therefore rates of fatty acid biosynthesis must be closely regulated to meet these changes. In some cell types, the demands for fatty acid synthesis are substantially greater. Obvious examples are oil seeds, which during development can accumulate as much as 60% of their weight as triacylglycerol. Another example is epidermal cells, which traffic substantial amounts of fatty acids into surface wax and cuticular lipid biosynthesis. In leek, even though the epidermis is less than 4% of the total fresh weight of the leaf, as much as 15% of the leaf lipid is found in a single wax component, a C31 ketone (52). How do cells regulate fatty acid synthesis to meet these diverse and changeable demands for their essential lipid components? We are only beginning to understand the answer to this question.

In this review, we focus on questions about regulation of fatty acid synthesis. Several reviews in recent years provide excellent overviews of fatty acid synthesis, and we do not duplicate those efforts except where necessary for clarity. An excellent and comprehensive review of plant fatty acid metabolism has recently been published (33), and several other recent reviews covering plant lipid metabolism, molecular biology and biotechnological aspects of plant fatty acids have also appeared (10, 44, 57, 64, 97, 105). Because there is much yet to be learned about regulation of this essential and ubiquitous pathway, we often dwell on what is unknown. This approach is intended to provide the reader with a clearer sense of the major questions of fatty acid metabolism that remain to be answered before a reasonable understanding of this regulation is achieved.

Compartmentalizaton and the Need for Interorganelle Communication

Overall fatty acid synthesis, and consequently its regulation, may be more complicated in plants than in any other organism (Figure 1). Unlike in other organisms, plant fatty acid synthesis is not localized within the cytosol but occurs in an organelle, the plastid. Although a portion of the newly synthesized acyl chains is then used for lipid synthesis within the plastid (the prokaryotic pathway), a major portion is exported into the cytosol for glycerolipid assembly at the endoplasmic reticulum (ER) or other sites (the eukaryotic pathway)

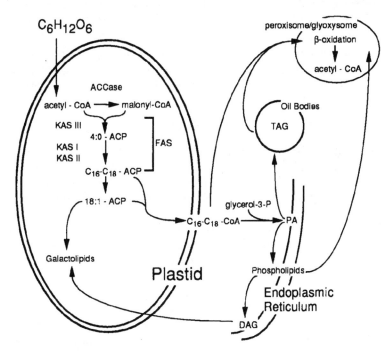

Figure 1 Simplified schematic of overall flow of carbon through fatty acid and lipid metabolism in a generalized plant cell. Because acyl chains are used in every subcellular compartment but are produced almost exclusively in the plastid, interorganellar communication must balance the production and use of these acyl chains.

(82, 100). In addition, some of the extraplastidial glycerolipids return to the plastid, which results in considerable intermixing between the plastid and ER lipid pools. Both the compartmentalization of lipid metabolism and the intermixing of lipid intermediates in these pools present special requirements for the regulation of plant fatty acid synthesis. Foremost is the need for regulatory signals to cross organellar boundaries. Because fatty acids are produced in the plastid, but are principally esterified outside this organelle, a system for communicating between the source and the sinks for fatty acid utilization is essential. The nature of this communication and the signal molecules involved remain an unsolved mystery.

In addition to having regulatory mechanisms that control overall levels of individual lipids, plants must also regulate rates of fatty acid synthesis under circumstances in which there are large shifts in the demand by major pathways of lipid metabolism located both in and out of the plastid. For example, consider the consequence of the Arabidopsis mutant *act1*, which has a mutation in the plastidial acyl transferase that directs newly synthesized fatty acid into thylakoid glycerolipids (49). This mutant has the remarkable ability to compensate for loss of the prokaryotic pathway by diverting nearly all the fatty acids into the phospholipids of the eukaryotic pathway. It then funnels unsaturated diacylglycerol from the ER phospholipids back into the plastidial lipids, with only minor change in the overall composition of either of these membranes.

OVERVIEW OF FATTY ACID SYNTHESIS: THE ENZYME SYSTEMS

The simplest description of the plastidial pathway of fatty acid biosynthesis consists of two enzyme systems: acetyl-CoA carboxylase (ACCase) and fatty acid synthase (FAS). ACCase catalyzes the formation of malonyl-CoA from acetyl-CoA, and FAS transfers the malonyl moiety to acyl carrier protein (ACP) and catalyzes the extension of the growing acyl chain with malonyl-ACP. In nature, ACCase occurs in two structurally distinct forms: a multifunctional homodimeric protein with subunits >200 kDa, and a multisubunit ACCase consisting of four easily dissociated proteins. Ideas about the structure of plant ACCases have undergone considerable evolution in the past few years. Until 1992, most researchers had concluded that plant ACCase was a large (>200 kDa) multifunctional protein similar to that of animal and yeast. This type of enzyme had been purified from several dicot and monocot species, and partial cDNA clones were available. However, in 1993, Sasaki and co-workers demonstrated that the chloroplast genome of pea encodes a subunit of an

ACCase with structure related to the β subunit of the carboxyltransferase found in the multisubunit ACCase of *Escherichia coli* (88). A flurry of interest in this topic has resulted in rapid extension of these initial studies. It has now been clarified that dicots and most monocots have both forms of ACCase, a >200-kDa homodimeric ACCase (probably localized in the cytosol) and a heteromeric ACCase with at least four subunits in the plastid (2, 48, 80). It is the heteromeric plastid form of the ACCase that provides malonyl-CoA for fatty acid synthesis.

In addition to the β-carboxyltransferase subunit characterized by Sasaki, clones are now available for the biotin carboxylase (94), biotin carboxyl carrier protein (BCCP) (17), and α-subunit of the carboxyltransferase (95). The four subunits are assembled into a complex by gel filtration with a size of 650–700 kDa. However, the subunits easily dissociate such that ACCase activity is lost, which accounts for the failure to identify the multisubunit form of ACCase for many years. Because the β–carboxyltransferase (β-CT) subunit is plastome encoded, whereas the other three subunits are nuclear encoded, assembly of a complete complex requires coordination of cytosolic and plastid production of the subunits. Little is know about this coordination and assembly. However, several-fold overexpression and antisense of the biotin carboxylase subunit does not alter the expression of BCCP, which suggests that a strict stoichiometric production of subunits may not be essential (92).

The structure of ACCase in Gramineae species is different in that these species lack the heteromeric form of ACCase and instead have two types of the homodimeric enzyme (89). An herbicide-sensitive form is localized in plastids, and a resistant form is extraplastidial. Because both Gramineae and dicot plastid FAS are regulated by light and dependent on ACCase activity, it will be of considerable interest to discover whether the two structurally very different forms of ACCase are subject to the same or different modes of regulation.

The structure of FAS has many analogies to ACCase structure. In nature, both multifunctional and multisubunit forms of the FAS are found. In addition, as in the case of ACCase, the plastidial FAS found in plants is very similar to the *E. coli* FAS and is the easily dissociated multisubunit form of the enzyme. It is now well established in both plants and bacteria that the initial FAS reaction is catalyzed by 3-ketoacyl-ACP III (KAS III), which results in the condensation of acetyl-CoA and malonyl-ACP (37, 106). Subsequent condensations are catalyzed by KAS I and KAS II. Before a subsequent cycle of fatty acid synthesis begins, the 3-ketoacyl-ACP intermediate is reduced to the saturated acyl-ACP in the remaining FAS reactions, catalyzed sequentially by the 3-ketoacyl-ACP reductase, 3-hydroxyacyl-ACP dehydrase, and the enoyl-ACP reductase.

In addition to ACCase and FAS, discussion of the regulation of fatty acid synthesis must also consider those reactions that precede and follow these two enzyme systems. It is not fully understood which reactions are responsible for providing acetyl-CoA to ACCase, but extensive experiments with leaf tissue indicate that acetyl-CoA synthetase can rapidly convert acetate to acetyl-CoA, and therefore free acetate may be an important carbon source (83–85). Other possible sources of acetyl-CoA include synthesis from pyruvate by a plastidial or mitochondrial pyruvate dehydrogenase (14, 51) or citrate lyase (61). Because acetyl-CoA is a central metabolite required for synthesis of isoprenoids, amino acids, and many other structures, it is likely that more than one pathway provides acetyl-CoA for fatty acid synthesis, and the source may depend on tissue and developmental stage (40).

The final products of FAS are usually 16:0- and 18:0-ACP, and the final fatty acid composition of a plant cell is in large part determined by activities of several enzymes that use these acyl-ACPs at the termination phase of fatty acid synthesis. The relative activities of these enzymes therefore regulate the products of fatty acid synthesis. Stearoyl-ACP desaturase modifies the final product of FAS by insertion of a *cis* double bond at the 9 position of the C18:0-ACP. Reactions of fatty acid synthesis are terminated by hydrolysis or transfer of the acyl chain from the ACP. Hydrolysis is catalyzed by acyl-ACP thioes-

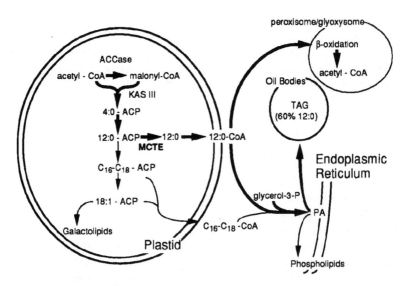

Figure 2 Overexpression of the California Bay medium-chain thioesterase (MCTE) results in activation of β-oxidation as well as increased expression of the enzymes of fatty acid biosynthesis.

terases, of which there are two main types: one thioesterase relatively specific for 18:1-ACP and a second more specific for saturated acyl-ACPs (24, 39). Fatty acids that have been released from ACPs by thioesterases leave the plastid and enter into the eukaryotic lipid pathway (82), where they are primarily esterified to glycerolipids on the ER. In some plants, thioesterases with specificity for shorter chain length acyl-ACPs are capable of prematurely terminating fatty acid synthesis, thereby regulating the chain length of the products incorporated into lipids. For example, thioesterases specific for medium-chain fatty acids are responsible for high levels of C10 and C12 fatty acids in triacylglycerols of California Bay, *Cuphea,* and other species (19, 23, 71). Acyl transferases in the plastid, in contrast to thioesterases, terminate fatty acid synthesis by transesterifying acyl moieties from ACP to glycerol, and they are an essential part of the prokaryotic lipid pathway leading to plastid glycerolipid assembly (82, 100).

BIOCHEMICAL CONTROLS: WHICH ENZYMES REGULATE FATTY ACID SYNTHESIS?

Control in metabolic pathways is not a dictatorship, nor is it an Athenian democracy; rather, it is a halfway house in which (to quote Orwell) "all pigs are equal but some are more equal than others" (101).

A number of approaches have been used by biochemists to identify where regulation occurs in a pathway. One approach depends on examination of the in vitro properties of enzymes of a pathway. Enzymes that have low activity relative to other members of the pathway are frequently considered as potentially rate limiting. Other properties, such as changing activities during developmental regulation of the pathway, or influence (activation or inhibition) on the enzyme by intermediates of the pathway can provide additional evidence toward identification of control points.

Analysis of Rate-Limiting Steps

ACCase is frequently considered the first committed step in the fatty acid biosynthetic pathway. In animals (30) and in yeast (111), there is evidence for ACCase as a major regulatory enzyme in fatty acid production. Therefore, for some time this enzyme was also proposed to be rate determining for plant fatty acid biosynthesis. Several lines of evidence supported this suggestion: Acetate or pyruvate were incorporated into acetyl-CoA in the dark by isolated chloroplasts, but malonyl-CoA and fatty acids were formed only in the light (58). Thus, the light-dependent step of fatty acid synthesis appeared to be at the

ACCase reaction. Eastwell & Stumpf (26) found that chloroplast and wheat germ ACCase were inhibited by ADP and suggested this may account for light-dark regulation of the enzyme. Nikolau & Hawke (63) characterized the pH, Mg, ATP, and ADP dependence of maize ACCase activity and concluded that changes in these parameters between dark and light conditions could account for increased ACCase activity upon illumination of chloroplasts. Finally, ACCase activity and protein levels are coincident with increases and decreases in oil biosynthesis in developing seeds (e.g. 80, 107; but see 40a).

However, in vitro approaches are limited because they show only that the enzyme has in vitro properties consistent with control, not that it actually does control in vivo metabolism. There are numerous examples of enzymes whose biochemical properties imply control but which have subsequently been found to have little role in controlling flux in vivo (96). Thus, there is no obligatory link between enzyme characteristics observed in vitro and regulatory properties in vivo.

The sites of metabolic control of a pathway can be more reliably identified by examination of in vivo properties of enzymes. Although it is often technically difficult, examining the concentrations of the substrates and products of each enzymatic step in a pathway provides information on which reactions are at equilibrium and which are displaced from equilibrium. This information is important because an essential feature of almost all regulatory enzymes is that the reaction that they catalyze is displaced far from its thermodynamic equilibrium. This is a requirement for regulation because if an enzyme has sufficient activity in vivo to bring its reaction to or near equilibrium, then changes in the activity of the enzyme will lead to no net change in flux through the pathway (81). On the basis of this criterion, it is possible to classify enzymes within a pathway as either nonregulatory or potentially regulatory based simply on an examination of the pools of substrates and products for each reaction. Further evidence of actual control can be obtained by examining changes in pool sizes when flux through the pathway is altered. It can be shown both theoretically and experimentally that when flux through the pathway is reduced at the regulatory reaction, the substrate pool for the regulatory step will increase and the product pool will usually decrease. Thus, an important experimental procedure for identifying regulatory steps is to examine changes in the pool sizes of intermediates when flux through the pathway is altered.

How can these principles be applied to evaluating control of plant fatty acid biosynthesis? Most of the substrates and intermediates of plant FAS are attached to acyl carrier protein (ACP). Analysis of acyl-ACPs is aided because the chain length of fatty acids attached to ACP alters the mobility of the protein in native or urea PAGE gels. Because of these alterations in mobility, most of the acyl-ACP intermediates of fatty acid synthesis can be resolved, and when transferred to nitrocellulose, antibodies to ACP can provide sensitive detection at nanogram levels. Although the acyl-ACP intermediates have a half life in vivo of only a few seconds (37, 99), by rapidly freezing tissues in liquid nitrogen it has been possible to determine the relative concentrations of free, nonacylated ACPs and of the individual acyl-ACPs. Analysis of acyl-ACP pools has been used to study regulation of FAS in spinach leaf and seed (72) in chloroplasts (74) in developing castor seeds (73) and in tobacco suspension cultures (93).

The initial examination of the composition of the acyl-ACP pools provided information about the potential regulatory reactions in plant fatty acid biosynthesis. The various saturated acyl-ACP intermediates between 4:0 and 14:0 occur in approximately equal concentrations. Because the 3-ketoacyl-ACPs, enoyl-ACPs, or 3-hydroxyacyl-ACPs, which are substrates for the two reductases and dehydrase reactions, were not detected, it is likely that these reactions are close to equilibrium and that the in vivo activities of these enzymes are in excess. Thus it is not likely that these enzymes are regulatory. In contrast, the concentration of acetyl-ACP was considerably above that of malonyl-ACP. This result suggests that the acetyl-CoA carboxylase reaction, which has an equilibrium constant slightly favoring malonyl-CoA formation, is significantly displaced from equilibrium and therefore potentially regulatory (37, 72, 74). The condensing enzymes can also be considered displaced from equilibrium because of the concentration of malonyl-ACP and the saturated acyl-ACPs.

Is Acetyl-CoA Carboxylase Rate-Limiting?

To obtain more information on sites of regulation, the changes in pool sizes when flux through the fatty acid biosynthetic pathway changes were examined. The rate of spinach leaf fatty acid biosynthesis in the dark is approximately one sixth the rate observed in the light (9). In the light, the predominant form of ACP was the free, nonacylated form, whereas acetyl-ACP represented about 5–6% of the total ACP (72). In the dark, the level of acetyl-ACP increased substantially with a corresponding decrease in free ACP, such that acetyl-ACP was now the predominant form of ACP. In similar experiments, when chloroplasts are shifted to the dark, malonyl-ACP and malonyl-CoA disappear within a few seconds, and acetyl-ACP levels increase over a period of several minutes. The rapid decrease in malonyl-ACP and malonyl-CoA when fatty acid synthesis slows, together with the increase in acetyl-ACP and lack of change in other intermediate acyl-ACP pools all lead to the conclusion that ACCase activity is the major determinant of light/dark control over FAS rates in leaves.

The above experiments on acyl-ACP and acyl-CoA pools have been carried out with dicot plants. Gramineae species such as maize and wheat have a substantially different (homodimeric) structure of ACCase (described above). Is this homomeric enzyme also involved in regulating leaf fatty acid synthesis? This question was addressed by a different approach toward evaluating metabolic control. Page et al (69) took advantage of the susceptibility of maize and barley plastid ACCase to the herbicides fluaxifop and sethoxidim. When chloroplasts or leaves were incubated with herbicides and radiolabeled acetate, a flux control coefficient of 0.5 to 0.6 was calculated for acetate incorporation into lipids. Flux control coefficients of this magnitude indicate strong control by the ACCase reaction over fatty acid synthesis (18, 41). Thus, in a wide variety of species and tissues, both in vivo and in vitro experiments point to ACCase as a major regulatory point for plant fatty acid synthesis.

Other Potential Rate-Limiting Steps

Although the best evidence available so far implicates ACCase activity as a primary determinant of fatty acid synthesis rates, the concept of a single rate-determining reaction is clearly an over-simplification. Control over flux is frequently shared by more than one enzyme in a pathway, and furthermore, the relative contributions of enzymes to control are variable. For example, under low irradience, only ADP-glucose pyrophosphorylase exerted strong control over CO_2 incorporation into starch, whereas under high irradience, phospho-glucoisomerase and phosphoglucomutase also exerted control (101). A similar situation may apply to fatty acid synthesis. For example, chloroplasts incubated in the light have high rates of fatty acid biosynthesis, but these rates can be further stimulated by addition of Triton-X-100. Under these conditions, malonyl-CoA and malonyl-ACP levels increase fivefold or greater, but the increase in fatty acid synthesis is only 30–80% (74). One possible implication of this observation is that under high flux conditions, biochemical regulation of ACCase may be most effective in decreasing fatty acid synthesis rates, and a large increase in ACCase activity may yield a relatively small increase in the overall flux through the pathway. These results further suggest that under high flux conditions through ACCase, the condensing enzymes or other factors may begin to limit FAS rates. All the saturated acyl-ACP intermediates of fatty acid synthesis from 4:0 to 14:0 are detected in approximately equal quantities in extracts from leaf or seed. This suggests that activities of all the KAS isoforms are approximately equal. Furthermore, it suggests that if the rate of any one of the KAS isoforms was to increase, the effect would be small, because the other KAS isoforms would be limiting. Evidence to support this has come from the increased expression of spinach KAS III in tobacco under the control of the CaMV 35S promoter (104). KAS III activity was increased 20- to 40-fold, with no effect on either the quantity or composition of the fatty acids. Analysis of the ACP and acyl-ACP intermediates revealed that the major form of ACP in the transgenic leaves was 4:0-ACP, which suggested that KAS I was now the limiting condensing enzyme in this tissue, and prevented any significant change in the overall flux through the pathway.

FEEDBACK REGULATION

Most biochemical pathways are controlled in part by a feedback mechanism which fine-tunes the flux of metabolites through the pathway. Whenever the product of a pathway builds up in the cell to levels in excess of needs, the end product inhibits the activity of the pathway. In most cases this inhibition occurs at a regulatory enzyme which is often the first committed step of the pathway. When the activity of the regulatory enzyme is reduced, all subsequent reactions are also slowed as their substrates become depleted by mass-action. Because enzyme activity can be rapidly changed by allosteric modulators, feedback inhibition of regulatory enzymes provides almost instantaneous control of the flux through the pathway.

Evidence from leaves, isolated chloroplasts, and suspension culture cells strongly implicates ACCase as a major point of flux control. If ACCase is the valve that determines flow of carbon toward fatty acids, the key question now becomes "How is this enzyme's activity regulated?" In animals and yeast, it has long been considered that fatty acid synthesis is partly controlled by feedback on ACCase by long-chain acyl-CoAs. Since acyl-CoAs are one end product of the FAS pathway, they were tested in vitro and found to strongly inhibit ACCase at submicromolar concentrations (29). Although this inhibition seems logical, it has been called into question by the discovery that acyl-CoA–binding proteins exist at high concentrations in the cytosol of animals (76), yeast (46), and plants (28, 34). Because these proteins have extremely high affinity for acyl-CoAs, the concentration of free acyl-CoA in the cytoplasm may be only nanomolar, a level unlikely to inhibit ACCase. These studies further emphasize the difficulty in extrapolating in vitro enzyme studies to in vivo conditions.

The in vitro analyses of plant ACCase have indicated that pH, Mg, and ATP/ADP can explain much of the light-dark modifications of ACCase activity observed in leaves (63). However, other modifications in the enzyme's activity are apparently involved. For example, light activation of fatty acid synthesis in chloroplasts does not require ATP synthesis by light (58) and has been reported to be stimulated by photosystem (PS) I but not PS II activity (70). In addition, Sauer & Heise (90) reported that ACCase activity was higher after rapid lysis and assay of light-incubated chloroplasts than after similar treatment of dark-incubated chloroplasts. The initial ACCase activity from chloroplasts incubated in the light is three- to fourfold higher than from either chloroplasts incubated 5 min in dark or from chloroplasts incubated in the light followed by 2 min in dark (K Nakahira & JB Ohlrogge, unpublished information). After lysis, ACCase activity from the dark-incubated chloroplasts increased until it reached light levels within 3–4 min. Thus, ACCase activity appears to be transiently inactivated or inhibited in dark-incubated chloroplasts. Because the lysis of chloroplasts into assay buffer would result in over a 100-fold dilution of stromal contents, any differences in the ACCase activity can not be attributed to contributions by the stroma contents (such as ATP) during the assay. At present there is not an explanation for this transient inactivation. Animal ACCase is known to be inactivated by phosphorylation (43). However, addition of protein kinase and phosphatase inhibitors to chloroplast lysates did not substantially alter the patterns.

In an effort to examine under in vivo conditions whether feedback inhibition of fatty acid synthesis occurs in plants, Shintani & Ohlrogge (93) added exogenous lipids to tobacco suspension cultures. The incorporation of ^{14}C-acetate into fatty acids (but not sterols) was rapidly decreased by the addition of lipids to the cultures. Thus, the cells apparently have a mechanism that senses the supply of fatty acids and responds by decreasing the de novo production of more fatty acid. Furthermore, examination of the pools of acyl-ACP intermediates indicated increases in acetyl-ACP, decreases in long-chain acyl-ACPs, and no change in medium-chain acyl-ACPs. These responses are identical to those observed when fatty acid synthesis is decreased by shifting leaves or chloroplasts to the dark and are expected if ACCase activity is decreased. Therefore, ACCase regulation can also account for the decrease of fatty acid production in response to exogenous lipids. Immunoblot analysis indicated that ACCase protein levels did not change during the feedback inhibition, indicating that biochemical controls are primarily responsible for the reduced ACCase and fatty acid synthesis activity.

What Is the Feedback System?

The first end products of plastid fatty acid synthesis are the long-chain acyl-ACPs, and therefore these seem logical candidates for feedback regulators of fatty acid synthesis. However, when fatty acid synthesis slows in the dark or in response to exogenous lipids, the long-chain acyl-ACP pools drop significantly. Therefore, these molecules have the opposite concentration response expected from a feedback inhibitor. Furthermore, in vitro assays have failed to reveal substantial inhibition of ACCase by acyl-ACPs (80).

Acyl-ACPs may play a role in the regulation of KAS III, however. When KAS III in seed homogenates of *Cuphea* was challenged with acyl-ACPs, as little as 0.5 µM 10:0-ACP was able to cause 50% inhibition (12). Similar results were obtained using 1 µM 10:0-ACP with homogenates of *Brassica* seed or spinach leaf. In each case, maximum inhibition was observed with the 10:0-ACP, compared with longer-chain acyl-ACPs, and the source of the ACP was *E. coli*. We have incubated a purified preparation of spinach KAS III with 10:0-ACP (ACP source: spinach) at concentrations up to 0.6 µM and failed to observe any inhibition (B Hinneberg-Wolf & J Jaworski, unpublished information). This suggests that inhibition of KAS III activity observed in the plant homogenates was indirect. A corresponding study carried out using purified preparations of *E. coli* KAS III and 100-µM 16:0- or 18:1-ACP resulted in a 50% inhibition (77). Because these levels far exceed the intracellular levels observed in *E. coli,* the physiological significance of this inhibition remains to be demonstrated.

Several other potential feedback inhibitors such as acyl-CoA, free fatty acids, and glycerolipids also fail to strongly inhibit the plant ACCase at physiological concentrations (80). Because FAS occurs inside the plastid but the major utilization of the products of fatty acid synthesis is at the ER membranes, it is likely that feedback regulation must allow communication across the plastid envelope. At this time we do not have any clear indications of what molecules are involved in feedback regulation of plastid fatty acid synthesis, and their discovery remains a major challenge for plant biochemists.

Control of Substrate and Cofactor Supply

Another way to control flux through a pathway is to regulate delivery of substrates and cofactors to the pathway. In animals (30) and some oleaginous yeast (7), there is evidence that the accumulation of acetyl-CoA via the ATP:citrate lyase reaction is a major determinant of fatty acid synthesis rates. Although conclusive evidence is not yet available for plants, indications are that acetyl-CoA supply does not usually limit plastid fatty acid production. If acetyl-CoA levels were limiting rates of fatty acid production, a decrease in acetyl-CoA level would be expected during high rates of fatty acid synthesis. However, almost all CoA found in chloroplasts is in the form of acetyl-CoA, and despite very large differences in rates of fatty acid synthesis in light or dark, acetyl-CoA levels in chloroplasts remain almost unchanged (74). In developing seeds, levels of acetyl-ACP are higher than in light-grown leaves, and the level does not change substantially throughout seed development, despite major changes in rates of FAS (73). Again, these results suggest that acetyl-CoA concentrations in seed plastids are high and do not change during development.

Although we tentatively conclude that carbon supply does not limit fatty acid production in leaves and most seeds in normal plants, there are examples of transgenic plants that may give some indication of what is required for carbon limitation to occur. The targeting of the enzymes of the polyhydroxy-butyrate (PHB) pathway into the chloroplasts of *Arabidopsis thaliana* resulted in an accumulation of PHB of up to 14% dry weight of the plant (59). Synthesis of this large carbon sink also requires the plastidial acetyl-CoA pool, and yet there was no detectable deleterious effect on fatty acid biosynthesis. Thus, it was concluded that a mechanism must exist that allows the plastid to synthesize the required acetyl-CoA in response to additional metabolic demand for PHB production. In another example, over-expression of the *E. coli* ADP:glucose pyrophosphorylase in developing *Brassica napus* seeds leads to a large increase in starch content of seeds and a 50% decrease in oil content (6). One interpretation of this result is that in this case, diverting carbon to starch storage was sufficient to "starve" the fatty acid pathway for available carbon precursors.

As in the case of substrates, consideration should be given to cofactors or energy as potentially limiting under certain conditions. Fatty acid synthesis is an energy-demanding pathway that requires at least 7 ATP and 14 NAD(P)H to assemble an 18-carbon fatty acid. However, energy is usually not a limitation in overall plant growth. In chloroplasts, ATP and NADPH can be derived from photophosphorylation and electron transport. In the dark or in nongreen tissues, glycolysis and the oxidative pentose phosphate pathway can provide the ATP and reductant (1, 21).

WHAT DETERMINES HOW MUCH OIL IS PRODUCED BY A SEED?

The oil content of seeds of different plant species varies from under 4% of dry weight (e.g. *Triticum sativum*) to over 60% (e.g. *Ricinus communis*). Furthermore, whereas leaves, roots, and other vegetative tissues usually contain less than 10% lipid by dry weight, most of which is polar membrane lipids, seed lipids usually contain over 95% neutral storage lipids in the form of triacylglycerol (TAG). An understanding of how plants regulate fatty acid metabolism to achieve these major differences in lipid content and composition is not yet available.

The only enzyme unique to TAG biosynthesis is diacylglycerol acyltransferase (DAGAT). Several lines of evidence suggest that tissue-specific expression of DAGAT is not sufficient to explain either the high proportion of TAG in seeds or the high level of oil in some seeds. 1. DAGAT activity is found not only in seeds but also in leaves (15, 55). 2. Certain types of stress such as ozone cause leaves to produce high proportions of TAG (87). Even detaching leaves and floating them in buffer plus 0.5 M sorbitol is sufficient to increase the accumulation of neutral lipids several fold (11). 3. Addition of fatty acids to the surface of spinach leaves results in a substantial proportion being incorporated into TAG (86). Clearly, leaves as well as seeds have the capacity for TAG synthesis, and therefore specific expression of DAGAT seems insufficient to explain abundant TAG synthesis in seeds. To understand the high proportion of TAG in seeds, it may be useful to consider whether oil synthesis is controlled by the supply (source) of fatty acids or fatty acid precursors or by the demand (sink) for fatty acids.

Control by Fatty Acid Supply (Source)

Control of oil production in seeds may reflect a response to the high rate of fatty acid synthesis in this tissue. If DAGAT is present at similar levels in most tissues, then high levels of TAG synthesis may occur in seeds because high levels of fatty acid are produced. Data that support this concept are available for *Cuphea* and *Chlamydomonas*. Addition of excess exogenous fatty acid and glycerol to developing *Cuphea* cotyledons rapidly resulted in rates of TAG accumulation several-fold higher than observed on the plant (5). In *Chlamydomonas*, addition of exogenous lipids (PC liposomes) to cultures resulted in up to 10-fold increases in TAG accumulation (32). Thus, it appears to be the fatty acid substrates, not the utilization enzymes that limit TAG synthesis in *Cuphea* and *Chlamydomonas*, and the capacity for TAG synthesis is greater than actually used. Further examples supporting this theory come from comparisons of oleaginous (oil-accumulating) versus nonoleaginous yeast. Extensive studies by Ratledge and coworkers have led to the conclusion that differences between these two types of species are controlled by the *production* rather than utilization of fatty acids (7). In particular, the production of acetyl-CoA through the action of ATP:citrate lyase is considered to control the flux of carbon into storage lipids.

Control by Demand (Sink)

Evidence that utilization of fatty acids can increase rates of fatty acid synthesis is available from experiments with *E. coli* (38, 65, 108). When a plant 12:0-ACP thioesterase is overexpressed in *E. coli* cells, fatty acid synthesis is increased substantially. Such cultures accumulate at least 10-fold more total fatty acid (most in the form of free lauric acid) than control cultures. In these experiments, the removal of the products of fatty acid synthesis to a metabolically inert end product (free fatty acid) appeared to release feedback inhibition on acetyl-CoA carboxylase, resulting in higher malonyl-CoA and fatty acid production (65). In a completely different type of experiment, massive overexpression of cloned membrane proteins resulted in no change of the membrane's phospholipid to protein ratio, but rather more fatty acid was produced to accommodate the excess proteins (110, 112). Therefore, the rate of fatty acid synthesis can apparently be regulated by the demand for fatty acids needed for membrane synthesis. Extrapolating this concept to plant seeds: Perhaps fatty acid synthesis increases in response to the high demand for acyl chains brought on by the depletion (or increase) of a key metabolic intermediate. A hypothetical example to illustrate one possibility might be that high expression of DAGAT (or some other acyl transferase) leads to a depletion of acyl-CoAs. This could in turn lead to release of feedback inhibition (probably indirectly) of ACCase by acyl-CoA and thus provide a mechanism whereby removal of acyl-CoA by the acyltransferases could stimulate fatty acid production.

Additional recent evidence suggests that both of the mechanisms above may be involved in determining seed oil content. When a transit peptide is added to the cytosolic ACCase of Arabidopsis and this chimeric gene is expressed in *B. napus* under control of the napin promoter, the oil content of the seeds is increased approximately 5% (79). Thus, increasing the supply of fatty acids at the first step in the pathway leads to increased oil content. Even larger increases (>20%) in Arabidopsis and *B. napus* seed oil content were achieved when a yeast lysophosphatidic acid acyltransferase (LPAAT) was expressed in developing seeds (114). Because overexpression of coconut (60) or *Limnanthes* (50) LPAAT did not alter oil content, the ability of the yeast enzyme to increase oil may involve its lack of regulation in the plant host.

Analysis of developing *B. napus* seeds expressing high levels of the California Bay 12:0-ACP thioesterase (MCTE) provided intriguing and surprising results that emphasize how much we need to learn before we have a good understanding of what determines the amount of oil in a seed (see Figure 2). Voelker et al (109) found that increased expression of the MCTE in seeds resulted in linear increases of lauric acid (12:0) in TAG up to ~35%. However, for quantities beyond 35%, the correlation with MCTE began to reach a plateau. To achieve 60 mol% 12:0, a further 10-fold increase in MCTE expression was required. At high levels of MCTE, expression of fatty acid β-oxidation as well as enzymes of the glyoxylate cycle were induced (68). Thus, these seeds appear to produce more 12:0 than can be metabolized to TAG by the *Brassica* acyltransferases or other enzymes whose normal substrates are C16 and C18 fatty acids, and the excess 12:0 signals the induction of the catabolic pathway. Thus, a portion of the fatty acid synthesis in these seeds is involved in a futile cycle of synthesis and breakdown of 12:0. Surprisingly, despite induction of fatty acid β-oxidation, the total amount of TAG in these MCTE-expressing seeds is not substantially reduced. How is oil content maintained if a significant amount of the lauric acid produced by FAS is being broken down in a futile cycle? Analysis of the fatty acid biosynthetic enzymes revealed that the levels of ACP and several of the fatty acid biosynthetic enzymes (acetyl-CoA carboxylase, 18:0-ACP desaturase, KAS III, etc) had increased two- to threefold at midstage development of high MCTE-expressing seeds. These results have several important implications. First, a coordinate induction of the enzymes of the fatty acid pathway had occurred, presumably to compensate for the lauric acid lost through β-oxidation or the shortage of long-chain fatty acids in these seeds. This suggests that the enzymes for the entire FAS pathway may be subject to a system of global regulation perhaps similar to lipid biosynthesis genes of yeast (16, 91). Second, these results indicate that although *B. napus* seeds are relatively high in oil content (ca 40%) the expression of the FAS enzymes is not at a maximum and can be induced a further two- to threefold over the levels found at midstage of seed development. Finally, the results suggest that these seeds might be preprogrammed to produce a particular amount of oil, and the levels of expression the fatty acid pathway may adjust to meet the prescribed demand for TAG synthesis.

What determines the level of TAG in these cells and what are the signals that result in increased expression of the FAS pathway? One major difficulty with interpreting the above data occurs if we attempt to fit it to the minimal model comprised of the currently understood roles of the pathway enzymes. However, recent results clearly indicate that there are still major gaps in our understanding of lipid and TAG biosynthesis. Analysis of the pathway for petroselinic acid ($18:1\Delta^6$) from 16:0 in coriander indicated that this plant has evolved a specialized 3-ketoacyl-ACP synthase (KAS), not found in other plants, that has high activity with $16:1\Delta^4$-ACP (13). Similarly, a KAS from *Cuphea wrightii* that has homology to plant KAS II was coexpressed with the *C. wrightii* MCTE in Arabidopsis, and this resulted in dramatic increase in the levels of 10:0 and 12:0 in the seeds compared with seeds expressing the MCTE alone. Both studies suggest that there are additional specialized enzymes for seed lipid metabolism that were not predicted on the basis of previous biochemical understanding. At this time there is no clear idea how these enzymes interact or what additional specialized functions they may have in oil synthesis. Presumably other enzymes or regulatory interactions are yet to be discovered that have a role in determining the quantity as well as quality of the oil.

WHAT HAVE WE LEARNED FROM TRANSGENIC PLANTS AND MUTANTS?

Fatty Acid Composition of Seeds Can Be More Radically Altered Than Other Tissues

The composition of fatty acids produced in plants is primarily determined by thioesterases, condensing enzymes, and desaturases. Manipulation of the thioesterases and desaturases in transgenic plants has been highly successful in producing major modifications of the chain length and level of unsaturation of plant seed oils (for reviews, see 44, 57, 66, 105). An additional new insight into regulation of fatty acid metabolism was recently obtained from expression of enzymes producing unusual fatty acids in other nonseed tissues. For example, transgenic expression of the 12:0-ACP thioesterase in *B. napus* under control of the constitutive 35S promoter resulted in lauric acid production in the seeds but not in leaves or other tissues (27). However, chloroplasts isolated from the *B. napus* leaves produced up to 35% lauric acid. These results indicate that some tissues expressing MCTE produce lauric acid but that it is subsequently degraded. In support of this hypothesis, the activity of isocitrate lyase of the glyoxylate cycle was induced in leaves expressing the California Bay thioesterase. A similar mechanism may apply to the production of hydroxy fatty acids which accumulate in seeds but not leaves of Arabidopsis transformed with the castor oleate hydroxylase under control of the 35S promoter (8). Together, these results imply that nonseed tissues may have general mechanisms to degrade unusual or excess fatty acids and thereby prevent their incorporation into membranes.

Manipulation of Oil Quantity

Plant breeding and mutation studies have demonstrated that the amount of oil in a seed can be varied over a wide range. A classic example is the selection for high and low oil maize that over a period of almost 100 years resulted in lines ranging from 0.5% to 20% lipid (25). Arabidopsis mutants with both increased seed oil (36) and reduced triacylglycerol (42) have been reported. In the latter example, not only was oil content reduced but 18:3 levels were doubled, 18:1 and 20:1 levels were reduced, and several enzymes of lipid metabolism had altered activity. These pleiotropic effects of a single gene mutation illustrate the complexities and inter-relationships of lipid metabolism which are difficult to explain based on our current models of seed oil biosynthesis.

Many Enzymes of Fatty Acid Synthesis Are Present in Excess

Although there has been much success in manipulating chain length and unsaturation of plant seed oils in transgenic plants, directed alterations in oil quantity are just beginning to be achieved. A number of the core enzymes of fatty acid synthesis have been overexpressed or underexpressed in transgenic soybean or *B. napus* seeds (45). Overexpression of ACP, KAS III, KAS I, KAS II, oleoyl-ACP thiosterase (FatA), or saturate-preferring acyl-ACP thioesterase (FatB) individually has not resulted in increased seed oil content. It is not known whether increased expression of combinations of these components might be more effective. As discussed above, increased ACCase and a yeast acyltransferase have been reported to increase oil in *B. napus* seeds.

Complete suppression of any of the core enzymes of FAS would be expected to reduce fatty acid synthesis and seed oil content. In support of this, cosuppression of FatA or KAS I in soybean resulted in reduced embryo oil content (45), and antisense of enoyl-reductase in *B. napus* gave shrunken seeds (113). Antisense of the tobacco biotin carboxylase under control of the 35S promoter was found to result in stunted plants with a 26% reduction in leaf fatty acid content (92). However, these effects were only observed when the reduction in BC level was 80% or greater. At 50% reductions in BC, no phenotype could be detected. Similar results were obtained with antisense of stearoyl-ACP desaturases (47). Many other experiments on antisense of enzymes of plant carbohydrate metabolism have also found that phenotypes (if observed at all) only occur when enzyme level is reduced 80% or more (102). Furthermore, the impact of an enzyme's reduction is usually dependent on growth conditions. These results suggest that under most conditions, many of the enzymes of plant metabolism are present in functional excess. Additional evidence that enzymes of lipid metabolism are expressed in excess comes from crosses of mutants. For several mutants of Arabidopsis fatty acid desaturation (*fad2, fad5,* and *fad6*), crosses with wild-type give a near wild-type phenotype rather than a fatty acid composition intermediate between parents. Thus, for several desaturases and other enzymes, gene dosage is not the primary determinant of the enzymes activity, and more complex controls must operate in vivo. In the case of highly regulated enzymes such as ACCase, a variety of mechanisms, such as increased enzyme activation, may compensate for reductions in expression brought on by antisense or mutation.

CONTROL OF GENE EXPRESSION

Multigene Families

A puzzling aspect of the molecular biology of plant fatty acid synthesis is the role of multiple genes. Even within Arabidopsis with its small genome, there is considerable range in the number of genes encoding the different proteins of plant fatty acid synthesis. Acyl carrier protein and the 18:0-ACP desaturases are each encoded by at least five genes, whereas many other enzymes are encoded by a single gene. The expression of the ACP genes has been studied in some detail, and both constitutive and tissue-specific patterns of expression have been observed (3, 35). In several other plant species, with genomes larger than Arabidopsis, tissue-specific desaturases and other enzymes are known to control seed fatty acid composition (67). Although it might be considered advantageous to have multiple genes to allow fine tuning of expression in different tissues, it is clear that this is not essential for many of the genes of fatty acid metabolism as several of the proteins in the pathway are encoded by a single gene. In the case of the glycerolipid desaturases, one gene encodes a plastid isozyme, whereas a second gene encodes a presumably ER localized isozyme. However, two genes encode the 18:2 desaturase of plastids, one of which is temperature regulated (10).

Promoter Analysis

Currently, the promoter of acyl carrier protein genes has been examined in the greatest detail. de Silva et al (22) fused 1.4 kB of a *B. napus* ACP gene (ACP05) to β-glucuronidase (GUS) and determined expression levels in transgenic tobacco. GUS activity increased during seed development, concurrent with lipid synthesis and at its maximum was 100-fold higher than in leaves. Surprisingly, although ACP is not an abundant protein, the activity of the ACP/GUS construct was comparable to that obtained from the strong 35S promoter. Several constructs of a promoter from another Arabidopsis ACP gene, *Acl1.2,* have been fused to GUS and examined after transformation into tobacco (4). Fluorometric analysis indicated strongest expression in developing seeds. However the promoter was active at lower levels in all organs (approx. 50 fold lower in leaves). Histochemical analysis indicated highest *Acll.2* expression in the apical/meristematic regions of vegetative tissues. During initial flower development, *Acll.2* promoter activity was detected in all cell types, but as the flower matured, GUS staining was lost in the sepals, epidermis of the style, and most cells of the anther. Intense staining remained in the ovary, stigma, stylar transmitting tissue, and tapetal and pollen of the anther. Thus, the expression pattern of this particular gene appears complex. Similar analysis of other promoters is needed to determine whether common signals are responsible for such patterns.

Six deletions of the *Acll.2* promoter revealed distinct regions of the promoter involved in vegetative and reproductive development. Expression of *Acll.2* in young leaves dropped to a basal level when an 85-bp domain from −320 to −236 was removed, but expression in seeds was not altered by this deletion. A protein factor was detected in leaves and roots, but not seeds, which binds to the −320 to −236 domain. Seed expression was reduced ~100-fold when the −235 to −55 region was removed. This same region was also essential for high expression in the flower tissues described above.

2.2 kB of a *B. napus* stearoyl-ACP desaturase promoter was fused to GUS (98). Expression was approximately 2.5-fold higher in developing seeds than in young leaves and thus did not show the dramatic differences reported for the ACP promoters above. However, similar to the ACP promoters, strong activity was also observed in tissues undergoing rapid development, including immature flowers, tapetum, and pollen grains.

Recently the enoyl-ACP reductase promoter of Arabidopsis has undergone similar GUS fusion and deletion analysis (20). Unlike ACP and stearoyl-ACP desaturase, there appears to be only a single gene in Arabidopsis encoding the enoyl-ACP reductase. High expression of enoyl-ACP reductase promoter-GUS fusions was again observed in youngest leaf tissues, with vascular tissue and shoot or root apical meristems highest. As each new leaf matured, GUS activity faded. Three domains of the promoter were identified. Seed expression was unchanged by deletion to −47 bp of the transcription start site, indicating that all elements needed for high level seed expression are present in this relatively small region. Removal of an intron in the 5′ untranslated region resulted in increased expression in roots, suggesting the presence of negative regulatory elements in this region.

Although the studies described above suggest that quantitative differences occur between relative expression levels of ACP, stearoyl-ACP desaturase, and enoyl-ACP reductase promoters in different tissues, all of the analyses of FAS promoters so far have reached similar conclusions about highest expression in apical meristems, developing seeds, and flowers. Such a pattern is not surprising, because these tissues are the most rapidly growing or are producing lipids in high amounts for storage. Furthermore, expression of the mRNA for most components of FAS probably is under coordinant control. For example, in situ hybridization of the mRNA for biotin carboxylase, BCCP, and carboxyltransferase subunits of Arabidopsis ACCase indicates close coordination of these three subunits, which is coincident with oil deposition (62). It should also be emphasized that promoters of lipid biosynthetic proteins are likely not unusual in these expression patterns. Highest expression in rapidly growing cells and coordinant regulation would be expected for promoters involved in most primary biosynthetic pathways.

What Controls Promoter Activity of FAS Genes?

A major challenge for the future is to discover how the level of expression of genes of lipid synthesis is controlled. The initial studies of promoters reviewed above have identified domains involved in control of gene expression levels. Efforts are under way to identify transcription factors that may bind to these elements. In addition, genetic approaches to search for mutants with altered regulation of fatty acid metabolism may allow identification of additional controls. By analogy to the study of other organisms, the control of plant FAS genes is likely to involve a complex array of cis and trans acting factors. For example, promoters of animal fatty acid biosynthetic genes are controlled by hormones such as insulin (56), by dietary fatty acids (78), by glucose levels (75), and by differentiation [particularly to adipocytes (54)]. Repression of transcription by negative regulatory elements has been suggested for both yeast (16) and animal lipid biosynthetic genes (103), and in Saccharomyces, common DNA sequences have been identified in the promoters of many genes involved in lipid metabolism (16, 91). As with genes of the glyoxylate and many other pathways, transcription of plant FAS genes is likely dependent on both developmental and metabolic signals (31). Because all indications are that the enzymes of fatty acid synthesis are coordinately regulated, it seems probable that global transcriptional signals may control expression of many or all genes of the pathway [perhaps similar to the R and C locus products that control transcription of the anthocyanin biosynthetic pathway (53)]. Identification of such global controls may provide the most effective means toward manipulating the amount of fatty acid produced in transgenic plants. If metabolic control over oil synthesis is shared among several enzymes, there will be limits to how much the flux through this pathway can be manipulated in transgenic plants using overexpression of one or a few genes. Thus, more complete control may come from identifying transcription factors that can increase expression of the entire pathway.

SUMMARY AND PERSPECTIVES

Fatty acid synthesis is a primary metabolic pathway essential for the function of every plant cell. Its products serve as the central core of membranes in every plant cell, and in specialized cells, fatty acids or fatty acid derivatives act as signal or hormone molecules, as carbon and energy storage, and as a surface layer protecting the plant from environmental and biological stress. Despite these very diverse functions, essentially all fatty acids in a cell are produced from a single set of enzymes localized in the plastid. Understanding how cells regulate the production of these fatty acids and direct them toward their different functions is thus central to understanding a large range of fundamental questions in plant biology. In addition, much interest has recently developed in genetic engineering of the fatty acid biosynthetic pathway to produce new or improved vegetable oils and industrial chemicals. Therefore, knowledge of how cells control the amount of fatty acid they produce may be essential for optimal commercial production of fatty acids.

Our understanding of regulation of fatty acid metabolism is much less developed than that of carbohydrate or amino acid biosynthetic pathways. We now have convincing evidence that ACCase is one enzyme that is involved in regulating fatty acid synthesis rates, and there are indications for other control points. However, this is only the beginning. ACCase might be considered one "slave" enzyme that controls flux into fatty acids but whose activity is dependent upon higher level master control systems in the cell. But what molecules regulate ACCase by feedback or other mechanisms and what metabolic signals or mechanisms control those molecules? How is the global regulation of dozens of genes for lipid synthesis accomplished? As discussed in this review, we have only fragmentary information about the nature of these controls. Thus, understanding regulation of fatty acid synthesis is a rich and relatively unexplored field with much work left to be done.

Literature Cited

1. Agrawal P, Canvin D. 1971. The pentose phosphate pathway in relation to fat synthesis in the developing castor oil seed. *Plant Physiol.* 47:672–75
2. Alban C, Baldet P, Douce R. 1994. Localization and characterization of two structur-

ally different forms of acetyl-CoA carboxylase in young pea leaves, of which one is sensitive to aryloxyphenoxypropionate herbicides. *Biochem. J.* 300:557–65
3. Baerson SR, Lamppa GK. 1993. Developmental regulation of an acyl carrier protein

gene promoter in vegetative and reproductive tissues. *Plant Mol. Biol.* 22:255–67

4. Baerson SR, Vander Heiden MG, Lamppa GK. 1994. Identification of domains in an *Arabidopsis* acyl carrier protein gene promoter required for maximal organ-specific expression. *Plant Mol. Biol.* 26:1947–59

5. Bafor M, Jonsson L, Stobart AK, Stymne S. 1990. Regulation of triacylglycerol biosynthesis in embryos and microsomal preparations from the developing seeds of *Cuphea lanceolata. Biochem. J.* 272:31–38

6. Boddupalli SS, Stark DM, Barry GF, Kishore GM. 1995. *Effect of overexpressing ADPGlc pyrophosphorylase in the oil biosynthesis in canola.* Presented at Biochem. Mol. Biol. Plant Fatty Acids Glycerolipids Symp., South Lake Tahoe, Calif.

7. Botham PA, Ratledge C. 1979. A biochemical explanation for lipid accumulation in *Candida* 107 and other oleaginous micro-organisms. *J. Gen. Microbiol.* 114: 361–75

8. Broun P, Hawker N, Somerville C. 1996. Expression of castor and *lesquerella fendleri* oleate-12 hydroxylases in transgenic plants: effects on lipid metabolism and inferences on structure-function relationships in fatty acid hydroxylases. Presented at Int. Symp. Plant Lipids, 12th, Toronto

9. Browse J, Roughan PG, Slack CR. 1981. Light control of fatty acid synthesis and diurnal fluctuations of fatty acid composition in leaves. *Biochem. J.* 196:347–54

10. Browse J, Somerville CR. 1994. Glycerolipids. In *Arabidopsis,* ed. EM Meyerwitz, CR Somerville, pp. 881–912. Plainview, NY: Cold Spring Harbor

11. Browse J, Somerville CR, Slack CR. 1988. Changes in lipid composition during protoplast isolation. *Plant Sci.* 56:15–20

12. Bruck FM, Brummel H, Schuch R, Spener F. 1996. In vitro evidence for feed-back regulation of β-ketoacyl-acyl carrier protein synthase III in medium-chain fatty acid biosynthesis. *Planta* 198:271–78

13. Cahoon EB, Ohlrogge JB. 1994. Metabolic evidence for the involvement of a Δ^4-palmitoyl-acyl carrier protein desaturase in petroselinic acid synthesis in coriander endosperm and transgenic tobacco cells. *Plant Physiol.* 104:827–37

14. Camp PJ, Randall DD. 1985. Purification and characterization of the pea chloroplast pyruvate dehydrogenase complex. *Plant Physiol.* 77:571–77

15. Cao Y-Z, Huang AHC. 1986. Diacylglycerol acyltransferase in maturing oil seeds of maize and other species. *Plant Physiol.* 82:813–20

16. Chirala SS. 1992. Coordinated regulation and inositol-mediated and fatty acid-mediated repression of fatty acid synthase genes in *Saccharomyces cerevisiae. Proc. Natl. Acad. Sci. USA* 89:10232–36

17. Choi J-K, Yu F, Wurtele ES, Nikolau BJ. 1995. Molecular cloning and characterization of the cDNA coding for the biotin-containing subunit of the chloroplastic acetyl-coenzyme A carboxylase. *Plant Physiol.* 109:619–25

18. Crabtree B, Newsholme EA. 1987. A systematic approach to describing and analysing metabolic control systems. *Trends Biol. Sci.* 12:4–8

19. Davies HM. 1993. Medium chain acyl-ACP hydrolysis activities of developing oilseeds. *Phytochemistry* 33:1353–56

20. de Boer G-J, Fawcett T, Slabas A, Nijkamp J, Stuitje A. 1996. Analysis of the *Arabidopsis* enoyl-ACP reductase promoter in transgenic tobacco. Presented at Int. Symp. Plant Lipids, 12th, Toronto

21. Dennis D, Miernyk J. 1982. Compartmentation of nonphotosynthetic carbohydrate metabolism. *Annu. Rev. Plant Physiol.* 33: 27–50

22. de Silva J, Robinson SJ, Safford R. 1992. The isolation and functional characterisation of a *B. napus* acyl carrier protein-5′ flanking region involved in the regulation of seed storage lipid synthesis. *Plant Mol. Biol.* 18:1163–72

23. Dörmann P, Spener F, Ohlrogge JB. 1993. Characterization of two acyl-ACP thioesterases from developing *Cuphea* seeds specific for medium chain acyl-ACP and oleoyl-ACP. *Planta* 189:425–32

24. Dörmann P, Voelker TA, Ohlrogge JB. 1995. Cloning and expression in *Escherichia coli* of a novel thioesterase from *Arabidopsis thaliana* specific for long chain acyl-acyl carrier proteins. *Arch. Biochem. Biophys.* 316:612–18

25. Dudley J, Lambert RJ, de la Roche IA. 1977. Genetic analysis of crosses among corn strains divergently selected for percent oil and protein. *Crop Sci.* 17:111–17

26. Eastwell KC, Stumpf PK. 1983. Regulation of plant acetyl-CoA carboxylase by adenylate nucleotides. *Plant Physiol.* 72:50–55

27. Eccleston VS, Cranmer AM, Voelker TA, Ohlrogge JB. 1996. Medium-chain fatty acid biosynthesis and utilization in *Brassica napus* plants expressing lauroyl-acyl carrier protein thioesterase. *Planta* 198: 46–53

28. Engeseth NJ, Pacovsky RS, Newman T,

Ohlrogge JB. 1996. Characterization of an acyl-CoA binding protein from *Arabidopsis thaliana*. *Arch. Biochem. Biophys.* 331: 55–62

29. Goodridge AG. 1972. Regulation of the activity of acetyl coenzyme A carboxylase by palmitoyl coenzyme A and citrate. *J. Biol. Chem.* 247:6946–52

30. Goodridge AG. 1985. Fatty acid synthesis in eucaryotes. In *Biochemistry of Lipids and Membranes,* ed. DE Vance, JE Vance, pp. 143–80. Menlo Park, CA: Benjamin/Cummings

31. Graham IA, Denby KJ, Leaver CJ. 1994. Carbon catabolite repression regulates glyoxylate cycle gene expression in cucumber. *Plant Cell* 6:761–72

32. Grenier G, Guyon D, Roche O, Dubertret G, Tremolieres A. 1991. Modification of the membrane fatty acid composition of *Chlamydomonas reinhardtii* cultured in the presence of liposomes. *Plant Physiol. Biochem.* 29:429–40

33. Harwood J. 1996. Recent advances in the biosynthesis of plant fatty-acids. *Biochim. Biophys. Acta* 301:7–56

34. Hills MJ, Dann R, Lydiate D, Sharpe A. 1994. Molecular cloning of a cDNA from *Brassica napus* L. for a homologue of acyl-CoA-binding protein. *Plant Mol. Biol.* 25: 917–20

35. Hlousek-Radojcic A, Post-Beittenmiller D, Ohlrogge JB. 1992. Expression of constitutive and tissue-specific acyl carrier protein isoforms in *Arabidopsis. Plant Physiol.* 98: 206–14

36. James DW Jr, Dooner HK. 1990. Isolation of EMS-induced mutants in *Arabidopsis* altered in seed fatty acid composition. *Theor. Appl. Genet.* 80:241–45

37. Jaworski JG, Post-Beittenmiller D, Ohlrogge JB. 1993. Acetyl-acyl carrier protein is not a major intermediate in fatty acid biosynthesis in spinach. *Eur. J. Biochem.* 213:981–87

38. Jiang P, Cronan JE. 1994. Inhibition of fatty acid synthesis in *Escherichia coli* in the absence of phospholipid synthesis and release of inhibition by thioesterase action. *J. Bacteriol.* 176:2814–21

39. Jones A, Davies HM, Voelker TA. 1995. Palmitoyl-acyl carrier protein (ACP) thioesterase and evolutionary origin of plant acyl-ACP thioesterases. *Plant Cell* 7:359–71

40. Kang F, Rawsthorne S. 1996. Metabolism of glucose-6-phosphate and utilization of multiple metabolites for fatty acid synthesis by plastids from developing oilseed rape embryos. *Planta* 199(2):321–27

40a. Kang F, Ridout C, Morgan CJ, Rawsthorne S. 1994. The activity of acetyl-CoA carboxylase is not correlated with the rate of lipid-synthesis during development of oilseed rape (*Brassica napus* L.) embryos. *Planta* 193:320–25

41. Kascer H, Porteous JW. 1987. Control of metabolism: what do we have to measure? *Trends Biol. Sci.* 12:5–10

42. Katavic V, Reed DW, Taylor DC, Giblin EM, Barton DL, et al. 1995. Alteration of seed fatty acid biosynthesis by an ethyl methanesulfonate-induced mutation in *Arabidopsis thaliana* affecting diacylglycerol acyltransferase activity. *Plant Physiol.* 108:399–409

43. Kim K-H, Lopez-Casillas F, Bai BH, Luo X, Pape ME. 1989. Role of reversible phosphorylation of acetyl-CoA carboxylase in long-chain fatty acid synthesis. *FASEB J.* 3:2250–56

44. Kinney AJ. 1994. Genetic modification of the storage lipids of plants. *Curr. Opin. Biotech.* 5:144–51

45. Kinney AJ, Hitz WD. 1995. *Improved soybean oils by genetic engineering.* Presented at Biochem. Mol. Biol. Plant Fatty Acids Glycerolipids Symp., South Lake Tahoe, Calif.

46. Knudsen J, Faergeman NJ, Skott H, Hummel R, Borsting M, et al. 1994. Yeast acyl-CoA-binding protein: acyl-CoA–binding affinity and effect on intracellular acyl-CoA pool size. *Biochem. J.* 302:479–85

47. Knutzon DS, Thompson GA, Radke SE, Johnson WB, Knauf VC, Kridl JC. 1992. Modification of *Brassica* seed oil by antisense expression of a stearoyl-acyl carrier protein desaturase gene. *Proc. Natl. Acad. Sci. USA* 89:2624–28

48. Konishi T, Shinohara K, Yamada K, Sasaki Y. 1996. Acetyl-CoA carboxylase in higher plants: most plants other than *Gramineae* have both the prokaryotic and the eukaryotic forms of this enzyme. *Plant Cell Physiol.* 37:117–22

49. Kunst L, Browse J, Somerville C. 1988. Altered regulation of lipid biosynthesis in a mutant of *Arabidopsis* deficient in chloroplast glycerol-3-phosphate acyltransferase activity. *Proc. Natl. Acad. Sci. USA* 85: 4143–47

50. Lassner MW, Levering CK, Davies HM, Knutzon DS. 1995. Lysophosphatidic acid acyltransferase from meadowfoam mediates insertion of erucic acid at the sn-2 position of triacylglycerol in transgenic rapeseed oil. *Plant Physiol.* 109:1389–94

51. Lernmark U, Gardestrom P. 1994. Distribu-

tion of pyruvate dehydrogenase complex activities between chloroplasts and mitochondria from leaves of different species. *Plant Physiol.* 106:1633–38

52. Liu DH, Post-Beittenmiller D. 1995. Discovery of an epidermal stearoyl-acyl carrier protein thioesterase: its potential role in wax biosynthesis. *J. Biol. Chem.* 270: 16962–69

53. Lloyd AM, Walbot V, Davis RW. 1992. *Arabidopsis* and *Nicotiana* anthocyanin production activated by maize regulators R and C1. *Science* 258:1773–75

54. MacDougald OA, Lane MD. 1995. Transcriptional regulation of gene expression during adipocyte differentiation. *Annu. Rev. Biochem.* 64:345–73

55. Martin BA, Wilson RF. 1983. Properties of diacylglycerol acyltransferase from spinach leaves. *Lipids* 18:1–6

56. Moustaid N, Beyers RS, Sul SH. 1994. Identification of an insulin response element in the fatty acid synthase promotor. *J. Biol. Chem.* 269:5629–34

57. Murphy DJ. 1994. Biogenesis, function and biotechnology of plant storage lipids. *Prog. Lipid Res.* 33:71–85

58. Nakamura Y, Yamada M. 1979. The light-dependent step of de novo synthesis of long chain fatty acids in spinach chloroplasts. *Plant Sci. Lett.* 14:291–95

59. Nawrath C, Poirier Y, Somerville C. 1994. Targeting of the polyhydroxybutyrate biosynthetic pathway to the plastids of *Arabidopsis thaliana* results in high levels of polymer accumulation. *Proc. Natl. Acad. Sci. USA* 91:12760–64

60. Knutzon DS, Lardizabal KD, Nelson JS, Bleibaum J, Metz J. 1995. *Molecular cloning of a medium-chain-preferring lysophosphatidic acid acyltransferase from immature coconut endosperm.* Presented at Biochem. Mol. Biol. Plant Fatty Acids Glycerolipids Symp., South Lake Tahoe, Calif.

61. Nelson D, Rinne R. 1975. Citrate cleavage enzyme from developing soybean cotyledons. *Plant Physiol.* 55:69–72

62. Nikolau B, Choi J-K, Guan X, Ke J, McKean AL, et al. 1996. *Molecular biology of biotin-containing enzymes required in lipid metabolism.* Presented at Int. Symp. Plant Lipids, 12th, Toronto

63. Nikolau BJ, Hawke JC. 1984. Purification and characterization of maize leaf acetyl-CoA carboxylase. *Arch. Biochem. Biophys.* 228:86–96

64. Ohlrogge J, Browse J. 1995. Lipid biosynthesis. *Plant Cell* 7:957–70

65. Ohlrogge J, Savage L, Jaworski J, Voelker T, Post-Beittenmiller D. 1995. Alteration of acyl-ACP pools and acetyl-CoA carboxylase expression in *E. coli* by a plant medium chain acyl-ACP thioesterase. *Arch. Biochem. Biophys.* 317:185–90

66. Ohlrogge JB. 1994. Design of new plant products: engineering of fatty acid metabolism. *Plant Physiol.* 104:821–26

67. Ohlrogge JB, Browse J, Somerville CR. 1991. The genetics of plant lipids. *Biochim. Biophys. Acta* 1082:1–26

68. Ohlrogge JB, Eccleston VS. 1996. *Coordinate induction of pathways for both fatty acid biosynthesis and fatty acid oxidation in* Brassica napus *seeds expressing lauroyl-ACP thioesterase.* Presented at Int. Symp. Plant Lipids, 12th, Toronto

69. Page RA, Okada S, Harwood JL. 1994. Acetyl-CoA carboxylase exerts strong flux control over lipid synthesis in plants. *Biochim. Biophys. Acta* 1210:369–72

70. Picaud A, Creach A, Tremolieres A. 1991. Studies on the stimulation by light of fatty acid synthesis in *Chlamydomonas reinhardtii* whole cells. *Plant Physiol. Biochem.* 29:441–48

71. Pollard MR, Anderson L, Fan C, Hawkins DJ, Davies HM. 1991. A specific acyl-ACP thioesterase implicated in medium-chain fatty acid production in immature cotyledons of *Umbellularia californica*. *Arch. Biochem. Biophys.* 284:306–12

72. Post-Beittenmiller D, Jaworski JG, Ohlrogge JB. 1991. *In vivo* pools of free and acylated acyl carrier proteins in spinach: evidence for sites of regulation of fatty acid biosynthesis. *J. Biol. Chem.* 266: 1858–65

73. Post-Beittenmiller D, Jaworski JG, Ohlrogge JB. 1993. Probing regulation of lipid biosynthesis in oilseeds by the analysis of the *in vivo* acyl-ACP pools during seed development. In *Seed Oils for the Future,* ed. SL MacKenzie, DC Taylor, pp. 44–51. Champaign, IL: Am. Oil Chem. Soc.

74. Post-Beittenmiller D, Roughan G, Ohlrogge JB. 1992. Regulation of plant fatty acid biosynthesis: analysis of acyl-CoA and acyl-acyl carrier protein substrate pools in spinach and pea chloroplasts. *Plant Physiol.* 100:923–30

75. Pripbuus C, Perdereau D, Foufelle F, Maury J, Ferre P, Girard J. 1995. Induction of fatty-acid-synthase gene expression by glucose in primary culture of rat hepatocytes: dependency upon glucokinase activity. *Eur. J. Biochem.* 230:309–15

76. Rasmussen JT, Rosendal J, Knudsen J. 1993. Interaction of acyl-CoA binding protein (ACBP) on processes for which acyl-

CoA is a substrate, product or inhibitor. *Biochem. J.* 292:907–13

77. Rock CO, Heath RJ. 1996. *Biochemical mechanism for feedback regulation of type II fatty acid synthesis in* E. coli. Presented at *Noble Found. Symp. Biochem. Metab. Aspects 3-Ketoacyl Synthases*, Humacao, Puerto Rico

78. Roder K, Klein H, Kranz H, Beck KF, Schweizer M. 1994. The tripartite DNA element responsible for diet-induced rat fatty acid synthase (FAS) regulation. *Gene* 144:189–95

79. Roesler K, Shintani D, Savage L, Boddupalli S, Ohlrogge J. 1996. *Targeting of the Arabidopsis 250 kDa acetyl-CoA carboxylase to plastids of developing* Brassica napus *seeds*. Presented at Int. Symp. Plant Lipids, 12th, Toronto

80. Roesler KR, Savage LJ, Shintani DK, Shorrosh BS, Ohlrogge JB. 1996. Co-purification, co-immunoprecipitation, and coordinate expression of acetyl-coenzyme A carboxylase activity, biotin carboxylase, and biotin carboxyl carrier protein of higher plants. *Planta* 198:517–25

81. Rolleston FS. 1972. A theoretical background to the use of measured concentrations of intermediates in study of the control of intermediary metabolism. *Curr. Top. Cell. Regul.* 5:47–75

82. Roughan PG, Slack CR. 1982. Cellular organization of glycerolipid metabolism. *Annu. Rev. Plant Physiol.* 33:97–132

83. Roughan PG. 1995. Acetate concentrations in leaves are sufficient to drive in vivo fatty acid synthesis at maximum rates. *Plant Sci.* 107:49–55

84. Roughan PG, Holland R, Slack CR. 1979. On the control of long-chain fatty acid synthesis in isolated intact spinach *(Spinacia oleracea)* chloroplasts. *Biochem. J.* 184: 193–202

85. Roughan PG, Holland R, Slack CR, Mudd JB. 1978. Acetate is the preferred substrate for long-chain fatty acid synthesis in isolated spinach chloroplasts. *Biochem. J.* 184:565–69

86. Roughan PG, Thompson GA Jr, Cho SH. 1987. Metabolism of exogenous long-chain fatty acids by spinach leaves. *Arch. Biochem. Biophys.* 259:481–96

87. Sakaki T, Saito K, Kawaguchi A, Kondo N, Yamada M. 1990. Conversion of monogalactosyldiacylglycerols to triacylglycerols in ozone-fumigated spinach leaves. *Plant Physiol.* 94:766–72

88. Sasaki Y, Hakamada K, Suama Y, Nagano Y, Furusawa I, Matsuno R. 1993. Chloroplast-encoded protein as a subunit of ace-tyl-CoA carboxylase in pea plant. *J. Biol. Chem.* 268:25118–23

89. Sasaki Y, Konishi T, Nagano Y. 1995. The compartmentation of acetyl-coenzyme A carboxylase in plants. *Plant Physiol.* 108: 445–49

90. Sauer A, Heise K-P. 1984. Regulation of acetyl-coenzyme A carboxylase and acetyl-coenzyme A synthetase in spinach chloroplasts. *Z. Naturforsch. Teil C* 39:268–75

91. Schuller HJ, Hahn A, Troster F, Schutz A, Schweizer E. 1992. Coordinate genetic control of yeast fatty acid synthase genes FAS1 and FAS2 by an upstream activation site common to genes involved in membrane lipid biosynthesis. *EMBO J.* 11: 107–14

92. Shintani D, Shorrosh B, Roesler K, Savage L, Kolattukudy PE, Ohlrogge J. 1996. *Alterations of tobacco leaf fatty acid metabolism using antisense-expression and reverse genetic approaches*. Presented at Int. Symp. Plant Lipids, 12th, Toronto

93. Shintani DK, Ohlrogge JB. 1995. Feedback inhibition of fatty acid synthesis in tobacco suspension cells. *Plant J.* 7:577–87

94. Shorrosh BS, Roesler KR, Shintani D, van de Loo FJ, Ohlrogge JB. 1995. Structural analysis, plastid localization, and expression of the biotin carboxylase subunit of acetyl-CoA carboxylase from tobacco. *Plant Physiol.* 108:805–12

95. Shorrosh BS, Savage LJ, Soll J, Ohlrogge JB. 1996. The pea chloroplast membrane-associated protein, IEP96, is a subunit of acetyl-CoA carboxylase. *Plant J.* 10: 261–68

96. Shulman RG, Bloch G, Rothman DL. 1995. In vivo regulation of muscle glycogen synthase and the control of glycogen synthesis. *Proc. Natl. Acad. Sci. USA* 92:8535–42

97. Slabas AR, Fawcett T. 1992. The biochemistry and molecular biology of plant lipid biosynthesis. *Plant Mol. Biol.* 19:169–91

98. Slocombe SP, Piffanelli P, Fairbairn D, Bowra S, Hatzopoulos P, et al. 1994. Temporal and tissue-specific regulation of a *Brassica napus* stearoyl-acyl carrier protein desaturase gene. *Plant Physiol.* 104: 1167–76

99. Soll J, Roughan G. 1982. Acyl-acyl carrier protein pool sizes during steady-state fatty acid synthesis by isolated spinach chloroplasts. *FEBS Lett.* 146:189–92

100. Somerville C, Browse J. 1991. Plant lipids: metabolism, mutants, and membranes. *Science* 252:80–87

101. Stitt M. 1994. Flux control at the level of the pathway: Studies with mutants, transgenic plants having a decreased activity of

enzymes involved in photosynthesis partitioning. In *Flux Control in Biological Systems from Enzymes to Populations and Ecosystems,* ed. ED Schulze, pp. 13–36. Ger: Academic

102. Stitt M, Sonnewald U. 1995. Regulation of metabolism in transgenic plants. *Annu. Rev. Plant Physiol. Mol. Biol.* 46:341–68

103. Swick AG, Lane MD. 1992. Identification of a transcriptional repressor down-regulated during preadipocyte differentiation. *Proc. Natl. Acad. Sci. USA* 39:7895–99

104. Tai H, Jaworski JG. 1995. *Expression of cDNA encoding 3-ketoacyl-acyl carrier protein synthase III (KAS III) in tobacco and* Arabidopsis. Presented at Biochem. Mol. Biol. Plant Fatty Acids Glycerolipids Symp., South Lake Tahoe, Calif.

105. Topfer R, Martini N, Schell J. 1995. Modification of plant lipid synthesis. *Science* 268:681–86

106. Tsay JT, Oh W, Larson TJ, Jackowski S, Rock CO. 1992. Isolation and characterization of the β-ketoacyl-acyl carrier protein synthase-III gene (*fab* H) from *Escherichia coli* K-12. *J. Biol. Chem.* 267: 6807–14

107. Turnham E, Northcote DH. 1983. Changes in the activity of acetyl-CoA carboxylase during rape-seed formation. *Biochem. J.* 212:223–29

108. Voelker TA, Davies HM. 1994. Alteration of the specificity and regulation of fatty acid synthesis of *Escherichia coli* by expression of a plant medium-chain acyl-acyl carrier protein. *J. Bacteriol.* 176:7320–27

109. Voelker TA, Hayes TR, Cranmer AM, Turner JC, Davies HM. 1996. Genetic engineering of a quantitative trait - metabolic and genetic parameters influencing the accumulation of laurate in rapeseed. *Plant J.* 9:229–41

110. von Meyenburg K, Jorgensen B, Deurs BV. 1984. Physiological and morphological effects of overproduction of membrane-bound ATP synthase in *Escherchia coli* K-12. *EMBO J.* 3:1791–97

111. Wakil S, Stoops J, Joshi V. 1983. Fatty acid synthesis and its regulation. *Annu. Rev. Biochem.* 52:537–79

112. Weiner J, Lemire B, Elmes M, Bradley R, Scraba D. 1984. Overproduction of fumarate reductase in *Escherichia coli* induces a novel intracellular lipid-protein organelle. *J. Bacteriol.* 158:590–96

113. White AJ, Elborough KM, Jones H, Slabas AR. 1996. *Transgenic modification of acetyl CoA carboxylase and βketo reductase levels in* Brassica napus: *functional and regulatory analysis.* Presented at Int. Symp. Plant Lipids, 12th, Toronto

114. Zou J-T, Katavic V, Giblin E, Barton D, MacKenzie S, et al. 1996. Modification of seed oil content and acyl composition in the *Brassicaceae* utilizing a yeast sn-2 acyltransferase (*SLC1*-1) gene. Presented at Int. Symp. Plant Lipids, 12th, Toronto

Annu. Rev. Plant Physiol. Plant Mol. Biol. 1997. 48:137–63

MOLECULAR GENETIC ANALYSIS OF TRICHOME DEVELOPMENT IN ARABIDOPSIS

M. David Marks

Department of Genetics and Cell Biology and Department of Plant Biology, University of Minnesota, St. Paul, Minnesota 55108

KEY WORDS: cell differentiation, cell fate, lateral inhibition, transcription factor

ABSTRACT

Two basic questions in developmental biology are: How does a cell know when it should or should not differentiate, and once a cell is committed to differentiate, how is that process controlled? The first process regulates the arrangement or pattern of the various cell types, whereas the second makes cells functionally distinct. Together, these two processes define plant morphogenesis. Trichome development in Arabidopsis provides an excellent model to analyze these questions. First, trichome development in Arabidopsis is a relatively simple process. A single epidermal cell differentiates into a unicellular trichome. Second, this differentiation occurs in a nonrandom pattern on the plant surface. Finally, the process is amenable to genetic analysis because many mutations that affect trichome differentiation do not alter other aspects of plant development. Thus far, more than 20 genes affecting trichome development have been identified. This review examines the current state of our understanding of these genes.

CONTENTS

137

1040-2519/97/0601-0137$08.00

INTRODUCTION

Most higher eukaryotic organisms begin life as a single-celled zygote. During organismal development this cell divides, and the resulting progeny differentiate and acquire special functions. In many plants, the first division of a zygote results in two cells with different fates. One cell will form the suspensor, and the other cell will produce the embryo proper. Much later, divisions of protodermal cells in leaf primordia generate daughter cells that differentiate into vastly different cell types. Thus, from the first to the terminal cell divisions of plant development, control of cell fate is important.

The development of plant leaf hairs, trichomes, provides an excellent system to study the control of cell fate (24, 49, 61–63, 76). First, trichomes develop on the epidermal surface, and all phases of trichome development can be observed. Second, development of trichomes is a relatively simple process. A single epidermal cell differentiates into a single-celled trichome. Third, the development of trichomes can be genetically dissected because normal plant growth and development do not require the presence of trichomes. Finally, understanding trichome development may have practical implications because there is a correlation between the presence of trichomes and resistance to herbivory by certain insect pests (1, 27, 37, 39, 44–46, 53, 68). The genetics of trichome formation has been studied in other plant species (4, 19, 28, 40, 41, 52, 81), but this review focuses on Arabidopsis.

TRICHOME DEVELOPMENT IN ARABIDOPSIS

Trichomes are normally present on the leaves, stems, and sepals of Arabidopsis (Figure 1*A, B, C*). They are normally absent from the roots, hypocotyl, cotyledons, petals, stamens, and carpels. The morphology of trichomes varies from unbranched spikes, which are most commonly found on the stems and sepals, to structures containing two to five branches, which are found on the leaves. Most trichome mutations affect all of the trichomes on a plant. This

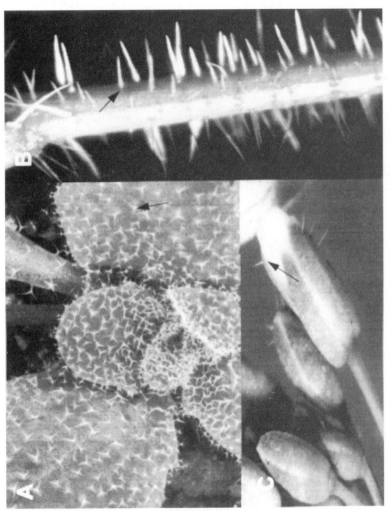

Figure 1   Trichomes on Arabidopsis seedlings. (*A*) Leaf trichomes. (*B*) Stem trichomes. (*C*) Sepal trichomes. Arrows denote trichomes.

suggests that while different trichomes may have different morphologies, their development is controlled by the same genes.

Trichome development proceeds as a wave across the epidermal surface on the leaf (31, 60). The first trichome initiates on the tip of the adaxial surface of the first primordium after it achieves a length of approximately 100 μm (47). As trichomes mature at the leaf tip, new trichomes emerge progressively toward the base (Figure 2A). In addition, new trichomes initiate in between developing trichomes that have been separated from one another by dividing

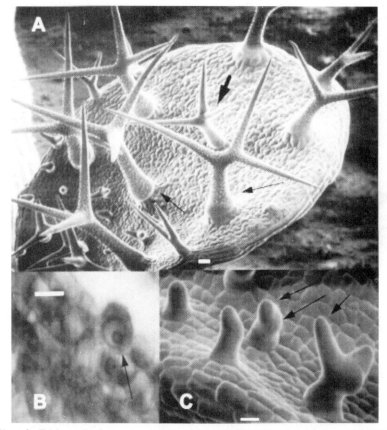

Figure 2 Trichome development on Arabidopsis leaves. (A) Scanning electron micrograph (SEM) of a young leaf with mature and developing trichomes. Thick arrow denotes developing trichome; thin arrow denotes mature trichome. (B) Section through emerging trichome. Arrow denotes enlarged nucleus. (C) SEM of developing trichomes. Arrows denote expanding branches. White bars indicate approximately 12 μm.

epidermal cells. Trichome initiation is found only in regions where epidermal cell division is occurring.

The first detectable step in the commitment to the trichome cell fate is a cessation of cell division; however, nuclear DNA synthesis continues, and the committed cell undergoes at least two rounds of endoreduplication, reaching at least 8N (Figure 2B) (31, 69). Cells surrounding a committed cell continue to divide normally. After the committed cell radially expands to a diameter that is approximately twofold greater than the surrounding cells, it begins to expand preferentially on its outer surface to form a stalk (Figure 2C) (60, 64). As the stalk forms, diffuse growth throughout the cell results in continued radial expansion. The nucleus migrates into the aerial portion of the stalk shortly before secondary protuberances (31), which subsequently expand into the branches, emerge from the aerial tip (Figure 2C). During branch formation, the nucleus undergoes another round of endoreduplication and migrates to the base of the last branch that forms (31). Expansion ceases when a trichome reaches a height of 200–300 μm and a base diameter of approximately 50 μm. During trichome maturation the cell wall thickens to approximately 5 μm, and the trichome surface becomes covered with papillae. In addition, the epidermal cells around the base of a trichome acquire a distinct rectangular shape. It appears as though the trichome base often pushes under the surrounding epidermal cells to create a socket. Thus, the surrounding cells are sometimes referred to as socket cells (31).

Detailed cellular analysis of Arabidopsis trichome development has yet to be completed. Thus, little is known about the role of the endomembrane system or cytoskeleton in trichome morphogenesis.

Trichome development has been used as a marker for leaf heteroblasty (85). Differences in trichome shape and position can be observed in a comparison of the initial and later leaves. The first two to three leaves have adaxial trichomes but lack trichomes on their abaxial surfaces. Later rosette leaves have an increasing number of abaxial trichomes. This progression continues on the bract-like leaves on the stem. The first bracts have trichomes on both adaxial and abaxial surfaces; however, later bracts have diminished numbers of adaxial trichomes, while maintaining their abaxial trichomes.

Trichomes are evenly distributed across the leaf surfaces, but contiguous trichomes are rarely observed. This type of arrangement has been described as an isotropic pattern (25). A statistical analysis of trichome spacing has shown that it is nonrandom (47), that is, it is statistically significant that no trichomes are contiguous. The parameter R was set as a ratio between the measured average distances between nearest neighbor trichomes on the leaf surface and the average nearest neighbor distance expected for a random pattern with the same density (5). A random pattern would have a value of R = 1, whereas a maximum spacing arrangement (i.e. all trichomes equal distance from one another) would result in R = 2.15. A value of R = 1.40, which represents a significant deviation from a random distribution, was observed [P < 0.01 (47)]. This indicates that there is a minimum distance between trichomes.

To study the development of the trichome pattern, the frequency and spacing of initiating trichomes were statistically analyzed (47). In a sample of 2120 epidermal cells on young leaf primordia, it was found that the fraction of cells that commit to the trichome pathway was 0.041. With this population size, if trichome initiation was stochastic, then 16 neighboring trichomes should have been observed. Because none was observed, the probability that trichome initiation was a random event in this trichome sample was less than 10^{-8}.

Because plant cells do not migrate, there are two main ways in which a nonrandom pattern can be generated. First, it is possible a trichome and the cells that surround it are derived from the same cell lineage, and only one cell in the group becomes a trichome; after this cell differentiates continued epidermal cell divisions would always separate developing trichomes. A second patterning mechanism requires cell-to-cell communication. A sectorial analysis was used to analyze the mechanism controlling trichome spacing (47). Plants containing a GUS reporter gene that had been inactivated by a maize Ac transposon were used. The GUS coding sequence was under the control of the CaMV 35S RNA promoter; Ac transposition during early plant development would result in a large clonal sector of GUS positive cells (51). To test the hypothesis that trichome spacing is controlled by cell lineage, trichomes that developed along the border of GUS positive sectors were analyzed. No evidence for a cell lineage associated with trichomes was found. Thus, apparently trichome spacing is controlled by a mechanism involving cell-to-cell communication.

TRICHOME MUTANTS

Mutations affecting trichome initiation, spacing, density, and shape have been recovered. Some of the mutations affect nontrichome developmental processes. Trichome mutants were first used as convenient genetic markers. The glabrous1 (gl1) mutant, which lacks trichomes on most surfaces, was used in early gene mapping studies (67). In 1978 distorted1 (dis1) and distorted2 (dis2) mutants, which have defects in trichome cell expansion, were used to map genes to chromosome 1 (13). In 1982 trichome mutants were used to calculate mutation frequencies generated using several different mutagens (43). In 1988, a review by Haughn & Somerville (24) first documented the possible use of trichome mutants as a model to address questions concerning cell fate and differentiation. In 1994 Hülskamp et al described many new trichome mutants that were recovered from a saturation screen (31). Several other recent reports also describe the characterization of new trichome mutants.

Mutations Affecting Early Trichome Development

The recessive *gl1* and *transparent testa glabra* (*ttg*) mutations have the most dramatic affect on trichome formation (42, 43). Strong loss-of-function mutations in either gene results in a complete loss of trichome formation on most aerial surfaces (Figure 3*A*). The *gl1* mutation only appears to affect trichome development; however, the *ttg* mutation has several developmental consequences. *ttg* plants lack anthocyanin pigments, which causes *ttg* seedlings to lack red pigments and seeds to be yellow instead of reddish brown. *ttg* mutant seeds also lack the polysaccharide mucilage that accumulates in the outer layer of the testa. Aside from the lack of mucilage and normal pigmentation, the *ttg* seed coat develops normally. Finally, *ttg* mutants produce ectopic root hairs (18). The root epidermis of Arabidopsis normally contains two types of cell files. In one file all cells are root hairs, and the cells are slightly less elongated than the cells in the other file type, which contains only nonhair cells. In a *ttg* root, the cell files that normally produce root hairs are unaltered. However, the cells in files that are normally hairless assume the less elongated shape of hair cells and most, but not all, of the cells in these files form hairs (18).

The loss of trichome initiation is not complete in either mutant. Both *gl1* and *ttg* mutants have a few trichomes on the margin of the rosette and cauline leaves (Figure 3*B*). In addition, *ttg* plants often have trichomes near the leaf margin of the adaxial surface (Figure 3*C*). Apparently the marginal trichomes are controlled by genes other than *GL1* or *TTG* (74). Mutations that result in a loss of the marginal trichomes have not been described. Other mutations that affect trichome morphology also affect the morphology of the margin trichomes (MD Marks, unpublished data). Therefore, many genes aside from *GL1* and *TTG* are active in both margin and nonmargin trichomes.

Weak alleles of both *gl1* and *ttg* have been identified. *gl1-2* plants exhibit a partial loss of trichomes, with a marked reduction of trichomes in the midvein region (12). Trichome differentiation is altered because many of the leaf trichomes only form rudimentary spikes. In addition, the trichome spacing pattern is altered in that side-by-side rudimentary trichomes are not uncommon.

Several weak alleles of *ttg* have been identified (48). *ttg-10* plants exhibit clusters of normal and rudimentary trichomes along the leaf margin. Interestingly, the *ttg-10* mutation does not affect all the developmental processes that

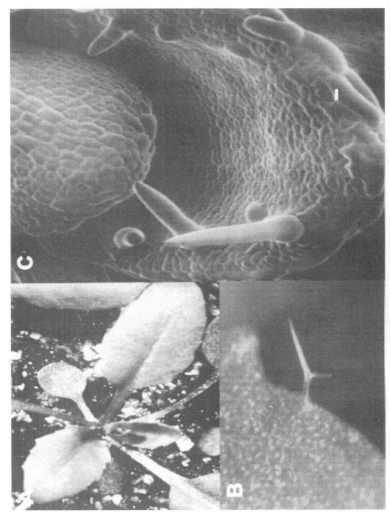

Figure 3 Loss of trichomes on glabrous mutant class. (*A*) *gl1* mutant seedling. (*B*) Trichome on leaf margin of *ttg* mutant seedling. (*C*) SEM of developing leaf on *ttg* seedling. White bar indicates approximately 12 μm.

normally are altered in *ttg* mutants. *ttg-10* plants lack testa pigmentation, but they appear to exhibit normal seedling pigmentation, and their seeds develop mucilage.

Mosaic analysis has been used to determine that the *GL1* gene appears to act cell autonomously. Rédei exposed seeds from a *GL1/gl1* heterozygous plant to X rays (80). He subsequently identified glabrous sectors on the resulting plants. More recently, heterozygous seeds were treated with EMS, and the resulting plants produced glabrous sectors that were not seen on wild-type plants (31). The glabrous patches are thought to result from *GL1* marker loss, uncovering the *gl1* mutant allele in heterozygous plants. The presence of sectors indicates that cells outside the sector cannot provide a diffusable substance or signal to overcome the effect of the *gl1* mutation.

It was noted that the leaves of Landsberg *erecta* (Ler) plants contained fewer trichomes than the leaves of Columbia (Col) plants [Ler ~10 vs Col ~30 on the first leaf (63)]. Using the Lister & Dean recombinant inbred mapping lines (55), which were generated with Ler and Col parents, it was possible to use quantitative trait analysis to map the loci responsible for the differences in trichome density (47). One major locus accounts for 73% of the variance and maps in the region between the markers *erecta* and m220 on chromosome 2 (LOD score: 27.54). This locus has been named *REDUCED TRICHOME NUMBER (RTN)*. The RTN^{Col} allele exhibits incomplete dominance over the RTN^{Ler} allele, because F_1 hybrids have an intermediate number of trichomes on first leaves.

Apparently *RTN* controls the persistence of trichome initiation (47). A comparison of trichome initiation on plants carrying the Col or the Ler alleles revealed that trichome initiation begins on the first leaf primordia after a length of ~100 μm is reached. As the leaf primordia increased from 100 μm in length to 300 μm, equal numbers of trichomes were initiated in each background. However, Ler leaf primordia longer than 300 μm rarely initiated new trichomes, whereas Col leaf primordia over 800 μm long continued to initiate new trichomes.

There are several modes by which *RTN* could influence the timing of trichome initiation (47). It is possible that *RTN* increases the area of lateral inhibition around initiating trichomes. This seems unlikely because the distance between the early initiating trichomes on both Col and Ler leaves are roughly the same. More likely is the possibility that *RTN* is involved in controlling either the length of time during which epidermal cells are competent to respond to a trichome-inductive signal or the length of time they are able to produce the inductive signal. For example, if *TTG* and/or *GL1* expression results in an inductive signal for trichome initiation, then *RTN* could be involved in the transduction of the signal or in positively or negatively regulating the timing of *TTG* and/or *GL1* expression. The incomplete dominance of the Col allele over the Ler allele suggests that the level of *RTN* expression is important. It also is possible that several closely linked genes influencing trichome density are located in the *RTN* region.

Recessive mutations in the *GLABRA2* (*GL2*) gene disrupt normal trichome morphogenesis (43). Two classes of trichomes develop on *gl2* mutants. One class of trichomes has a rudimentary spike that projects upward (Figure 4*A*). Toward the margin of *gl2* leaves, trichomes are more normal in appearance; however, these trichomes are less branched than normal. Mosaic analyses suggest that *GL2* acts cell autonomously (31). Mutations in *GL2* also affect other developmental processes. Like *ttg, gl2* mutants lack seed coat mucilage and produce ectopic hairs in the root (43, 66). However, there is a morphological difference in the root hair phenotype between the two mutants. Normally, hair-bearing cells are shorter than nonhair cells. In *ttg* mutants, all the root epidermal cells assume the morphology of the hair cells and many produce hairs. In contrast, while all the epidermal cells in a *gl2* mutant can produce hairs, the files of cells that normally would not become hairs maintain a more expanded shape.

Recessive mutations in the *GLABRA3* (*GL3*) gene have two effects on trichome development (43). On early leaves there is a decrease in trichome initiation (Figure 3*B*). As in *gl2* mutants and weak *gl1* and *ttg* mutants, the loss of trichome initiation is most striking in the midvein region of the leaves. However, on later leaves trichome initiation is more uniform. The trichomes that do develop tend to be less branched and are more slender than normal. In addition, it has been observed that trichomes in *gl3* mutants undergo fewer rounds of endoreduplication than wild-type trichomes (31).

Mutations in either the *TRYPTYCHON* (*TRY*) or *KAKTUS* (*KAK*) genes result in larger than wild-type trichomes with increased branch formation (31). Trichomes on these mutants also exhibit an increase in endoreduplication. It has been reported that *try* mutants have a greater number of clustered trichomes than are found on wild-type plants (31). Thus, *TRY* also may be involved in controlling the proposed lateral inhibition pathway.

Reduced Branching Mutants

Several recessive mutations appear to reduce trichome branching without affecting trichome initiation. Hülskamp et al divided branching into a primary phase that generates the first branch and a secondary phase that generates subsequent branches (31). The *stichel* (*sti*) mutation eliminates both of these

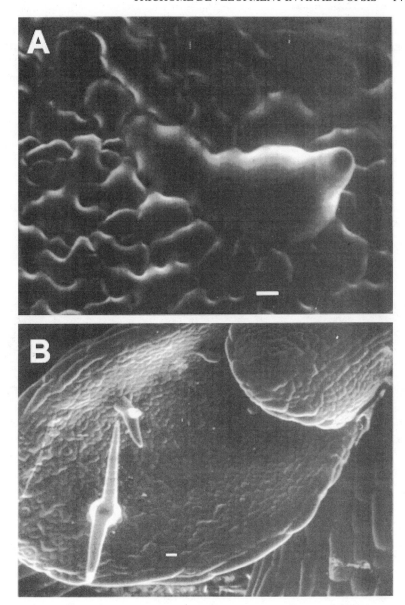

Figure 4   Trichomes on *gl2* and *gl3* mutants. (*A*) SEM of mature trichome on *gl2* mutant leaf. (*B*) SEM of developing trichomes on young *gl3* mutant leaf.

phases (31). Leaf trichomes on *sti* mutants are long, unbranched structures similar to the trichomes found on the stem (31). The trichomes on *stachel* (*sta*) mutants appear to have skipped primary but not secondary branching (31). These trichomes generally have two branches on top of a long stalk. In contrast, trichomes on both the *zwichel* (*zwi*) and *angustifolia* (*an*) mutants appear

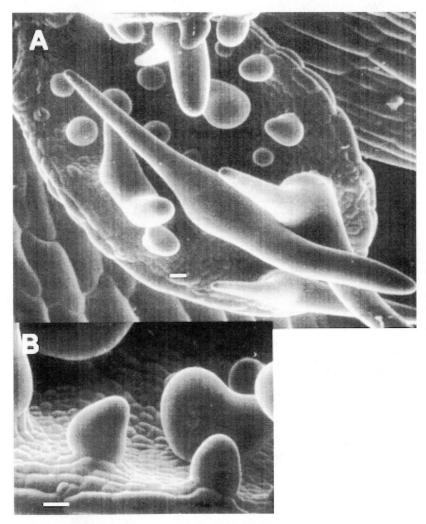

Figure 5 SEM of *zwi* mutant trichomes. (*A*) SEM of young leaf on *zwi* mutant. (*B*) SEM of emerging trichomes on *zwi* leaf. White bars indicate approximately 12 μm.

to undergo only primary branching (31, 43). Trichomes on both of these mutants generally have two branches that originate from a region close to the trichome base (Figure 5). A mosaic analysis suggests that *AN* acts cell autonomously (31). In *an* plants the leaves are narrow and the stems and siliques are twisted.

Trichome Expansion

General trichome expansion is affected by mutations in eight different genes. These recessive mutations include *dis1* and *dis2* as well as *gnarled* (*grl*), *klunker* (*klk*), *spirrig* (*spi*), *wurm* (*wrm*), *crooked* (*crk*), and *alien* (*ali*) (13, 31). Trichomes on these mutants exhibit irregular expansion (illustrated by the developing trichomes on the *dis2* mutant shown in Figure 6). Because these mutants do not display other obvious phenotypes, the genes defined by these mutations may be involved in expansion processes that are unique to trichomes. Alternately, these genes may encode products important for general cell expansion, but may be members of gene families that are expressed in various cell types. As might be expected for this class, mosaic analysis indicates that *DIS2* appears to act cell autonomously (31). In contrast, the same analysis indicates that *DIS1* may act through a noncell-autonomous mechanism because EMS-treated plants lacked *dis1* trichome sectors. As a control, the heterozygous *dis1* plants were also heterozygous for the *an* mutation, and *an* trichome sectors were identified.

The *singed* (*sne*) mutation also results in a general alteration of trichome cell expansion (62). The trichomes appear to develop normally, but the trichome stalk and branches are slightly twisted. In addition, this mutation also causes a shortening of the root hairs.

Maturation Mutants

Several recessive mutations appear to alter the final stages of trichome development. The *under developed trichome* (*udt*) mutation results in trichomes that are slightly more slender than wild-type and that produce underdeveloped papillae toward the tips of the branches (24). Three other mutations, *chablis* (*cha*), *chardonnay* (*cdo*), and *retsina* (*rts*), result in trichomes that lack the rough papillate surface of wild-type mature trichomes (31).

Potikha & Delmer used an elegant screen to isolate a mutant deficient in secondary cellulose deposition in trichome cell walls (77). Trichomes exhibit strong birefringence under polarized light, a characteristic of cell walls containing large amounts of highly ordered cellulose microfibrils. They identified the recessive *trichome birefringence* (*tbr*) mutant that lacks birefringent tri-

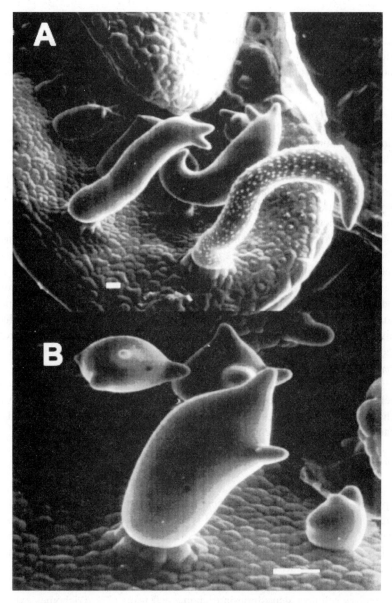

Figure 6   SEM of developing trichomes on *dis2* mutant seedling. (*A*) SEM of *dis2* leaf with mature and developing trichomes. (*B*) SEM of emerging trichomes on *dis2* leaf. White bars indicate approximately 12 μm.

chomes. Mature *tbr* trichomes were essentially wild-type in shape, but had a smooth surface instead of the wild-type rough papillate surface. Quantification of cellulose in trichomes isolated from wild-type and *tbr* plants indicated that *tbr* trichomes have 18% of the cellulose normally found in wild-type trichomes. The *tbr* mutation also may result in a reduction of cellulose in the xylem.

Genetic Interactions

Classical genetic analyses indicate that the *gl1* and *ttg* mutations are epistatic to all other known trichome mutations. Plants doubly mutant for *gl1* or *ttg* and other trichome mutants are glabrous. The genetic relationship between *gl2*, *gl3*, and *try* has also been analyzed (MD Marks, unpublished data; 31). The *gl3/try* double mutant has *gl3*-like trichomes, but retains the clustering phenotype of *try*. This suggests that *GL3* functions downstream of *TRY* for branching, and that the wild-type function of *TRY* may be to inhibit branch formation by directly or indirectly inhibiting *GL3* activity. It also has been found that the *try* mutation suppresses the *gl2* phenotype; this suggests that *TRY* acts downstream of *GL2*. In this case, the function of *GL2* could be to inhibit *TRY* activity, resulting in a promotion of branch formation. The lack of aerial expansion of *gl2* mutant trichomes could result from unrestricted wild-type *TRY* activity. The phenotype of the *gl2/try* and *try/gl3* double mutants would indicate that these three genes act in a simple linear pathway:

GL2—⊣ TRY —⊣ GL3.

The phenotype of the *gl2/gl3* double mutant suggests that this is not the case. With a simple linear relationship, one would predict that a *gl2/gl3* double mutant phenotype would more closely resemble that of *gl3*. Instead, the *gl2/gl3* phenotype is more extreme than either single mutant; even less trichome outgrowth is observed than on *gl2* plants. While this result does not rule out the possibility that a linear pathway exists, it does suggest that *GL2* and *GL3* have some separate functions acting through parallel pathways.

MOLECULAR STUDIES

GL1

GL1 was one of the first Arabidopsis genes isolated by T-DNA tagging. The *gl1-43* allele was identified in a population of plants derived from the 43rd transformant generated by the *Agrobacterium*-mediated seed transformation procedure (15). Unlike strong loss of *GL1* function mutations with no trichomes on either the stems or leaves, *gl1-43* lacked trichomes only on the stems (16). An 8-kb fragment that had the ability to molecularly complement the *gl1* mutation was identified (26). Sequencing that the *GL1* gene encodes a member of the *myb* class of transcription factors (74).

Myb genes have been found in all higher eukaryotic organisms and yeast (3, 36, 83, 84, 86). They contain one to three *myb* domains, each of which appear to form a helix-turn-helix structure (17, 22, 59). The *myb* domains participate in DNA binding and are located at the amino terminus. The *myb* family was first identified as an oncogene associated with the avian myeloblastoma virus (83). The cellular *c-myb* gene was subsequently identified in vertebrates and was found to play an important role in controlling the maturation of white blood cells (54). In animals, *myb* gene families are composed of only a few members.

The first isolated plant gene encoding a transcription factor was the *myb* gene, *C1* (8, 75). The maize *C1* gene regulates anthocyanin synthesis in the aleurone (6). Subsequently it was found that plants contain large *myb* gene families (33, 65, 74). Members of the family typically have two *myb* repeats toward their amino termini and a divergent amino acid sequence toward the carboxy-terminal (49). Most members share very little sequence identity in their carboxy-terminals. A few members have only one *myb* domain (2).

Several plant *myb* genes other than *GL1* and *C1* have a known function. The maize *Pl* gene controls seedling anthocyanin synthesis (7). The maize *P* gene controls steps in the synthesis of the red phlobaphene pigment in ear tissue (23). The *MIXTA* gene of *Antirrhinum* participates in controlling the shape of epidermal petal cells (72).

The *GL1* gene encodes a protein with two myb repeats and a carboxy-terminal domain of approximately 120 amino acids that is not significantly similar to other sequences in the data bases (49, 74). The carboxy-terminal region does contain several clusters of acidic amino acids that may function as transcriptional activators (74). In the weak allele *gl1-2* the molecular lesion is a small deletion that results in the loss of the terminal 27 amino acids (12). The missing region contains one of the acidic clusters. It has been found that the *gl1-1* allele contains a deletion removing the complete *GL1* coding region as well as flanking promoter elements, and the only phenotype is a loss of trichomes (74). Thus, it is likely that the function of *GL1* is restricted to controlling trichome development .

In situ hybridization to localize *GL1* mRNA showed that *GL1* is expressed in fields of cells from which trichomes are initiating (50). Cells committed to the trichome pathway have more *GL1* mRNA activity than the surrounding cells, which might indicate that autocatalytic up-regulation of *GL1* expression occurs once a cell commits to the trichome cell fate. To examine the DNA elements responsible for regulating *GL1* transcription, a GUS construct containing promoter sequences from the 5′ noncoding region of *GL1* was analyzed (74). This construct directed GUS expression to the stipules, which are found at the base of the leaves. Later studies indicated that this expression pattern appears to inconsequential and is not required for the initiation of leaf trichomes (MD Marks, unpublished data). It is possible that the stipule enhancer element is required for the expression of a gene located near *GL1*.

The molecular lesion in the *gl1-43* allele suggested that 3′ noncoding sequences could contain important regulatory sequences. This T-DNA–induced allele contains an insertion that is over 1000 bases downstream of the transcribed 3′ noncoding sequences (74). At first, it appeared that this region only was important for controlling *GL1* expression in the stem because the phenotype of the *gl1-43* mutant is a loss of stem trichomes but not leaf trichomes (64). However, it was found that fusing 5′ and 3′ *GL1* noncoding sequences to the GUS construct resulted in a reconstruction of the pattern of expression found by in situ hybridization (50). It also was found that the 3′ sequences were essential for *GL1* function, because removal of these sequences from a DNA fragment containing the *GL1* gene resulted in a loss of molecular complementation of both leaf and stem trichomes on *gl1* mutants (50). Since a sequence downstream of the insertion in *gl1-43* was required for *GL1* function, the reason the insertion results only in a loss of stem trichomes is not known.

The level of *GL1* expression is critical in controlling trichome initiation. Plants containing a construct with the *GL1* gene under the control of the *CaMV 35S* RNA promoter (35SGL1) accumulate greater than normal levels of *GL1* mRNA (48) and GL1 protein (D Szymanski & MD Marks, unpublished data). In either wild-type or *gl1* mutants, this overexpression of *GL1* results in both a reduction of leaf trichomes and in an induction of ectopic trichomes on organs that are normally glabrous (48). The induction of ectopic trichomes is significant, because it indicates that expression of *GL1* can cause trichome initiation. Since *TTG* may be involved in causing the reduction in leaf trichomes on 35SGL1 plants, a discussion of this phenomenon will follow a review of the possible molecular identity of *TTG*.

TTG

The *TTG* gene has not been isolated. However, an ongoing chromosome walk has narrowed down the location of the gene to a small region on chromosome 5 (A Walker, personal communication). *TTG* may be a homolog of the maize *R* gene (57). In maize, production of the anthocyanin pigment in the aleurone is controlled by the regulatory genes *C1* and *R* (6). *R* encodes a protein with helix-loop-helix DNA binding and protein dimerization motifs characteristic of members of the *L-myc* family of transcription factors (58, 78). Lloyd et al (57) demonstrated that Arabidopsis plants expressing *R* under the control of the *CaMV 35S* promoter (35SR) synthesized greater than normal quantities of anthocyanin and had extra trichomes. The first two leaves had a two- to five-fold increase in trichomes over nontransformed controls. *R* expression also resulted in ectopic trichomes on the petals, stamens, and pistils. *ttg* plants transformed with *R* produced both trichomes and anthocyanin. It was shown subsequently that *R* induced the synthesis of seed coat mucilage in *ttg* plants (MD Marks, unpublished data) and affected the expression of root hairs (18). While *ttg* mutants normally produce root hairs in all files of the epidermis, 35SR *ttg* plants produced few hairs, as would be predicted if *TTG* inhibits hair formation in the nonhair files of the root (18). The correction of the diverse defects in *ttg* plants by *R* suggests that *TTG* may encode an *R* homolog. It also is possible that *TTG* regulates an *R* homolog (57).

A search for Arabidopsis clones with homology to *R* has not resulted in the isolation of *TTG*. Furthermore, several expressed tags (cDNA clones) with sequence similarity to *R* have been identified, but they do not map to *TTG* (A Walker, personal communication). Attempts to use the *R* homolog in *Antirrhinum delila* to complement *ttg* mutants have failed (21, 71). Nevertheless, the *R* gene has proved a valuable tool to study *TTG* function and the control of trichome initiation.

Lloyd et al (56) used a novel approach to induce *R* activity in *ttg* plants. They attached the steroid-binding domain of the rat glucocorticoid receptor to the carboxy-terminal of the *R* gene (RGR). Fusion of the binding domain to other transcription factors previously had been shown to impose hormone dependence on the activity of the fused factor (11, 29). *ttg* plants containing RGR failed to produce trichomes. RGR *ttg* plants grown on agar medium containing as little as 1 nM of dexamethasone produced a few trichomes (56). Plants grown on medium containing 10 nM dex had wild-type numbers of trichomes on their first leaf pair. Nontransformed plants or plants carrying unmodified *R* did not respond to the dex treatment.

RGR plants were used to study trichome development (56). A plant was grown for eight days on soil and then immersed in one μm dex. Sequential scanning electron microscopy (SEM) images were obtained by making casts of molds produced with a dental impression polymer. Twenty-four hours after dex treatment, the surface of the youngest leaf was covered with emerging trichomes. To define the spatial and temporal pattern of trichome initiation, plants were either germinated on dex and then removed from dex or they were started on medium without dex and then transferred to dex-containing medium. Leaves of plants removed from dex had trichomes only at the tips and were glabrous at the base. On plants transferred to dex, the first leaves that formed had trichomes only at the base, whereas the later leaves were covered with trichomes. These results mirror the general impression that is formed by examining wild-type leaves (Figure 1A): Trichome formation proceeds as a wave from the tip (oldest tissue) of the leaf to the base (youngest tissue). The results of Lloyd et al (56) extend the analysis to indicate that the pattern of trichome initiation is the result of the timing of the competence of epidermal cells to respond to *TTG*.

Interactions Between GL1 and TTG

Plants constitutively expressing either *GL1* or *R* were used to test hypotheses about the roles of *GL1* and *TTG* in trichome development (48). To determine whether constitutively expressed *GL1* could bypass the need for *TTG,* crosses were made between 35SGL1 and *ttg* plants. Because glabrous 35SGL1 *ttg* plants were identified in the F2 generation, it was concluded that constitutive *GL1* cannot bypass *TTG*. When 35SR plants were crossed to a *gl1* mutant, glabrous 35SR *gl1* plants were found in the F2. Thus, it is likely that constitutive *TTG* cannot bypass the requirement for *GL1*. To determine the effect of constitutively expressing both *GL1* and *R,* 35SGL1 and 35SR plants were crossed. The constitutive expression of both genes in the same plant had a dramatic impact on trichome initiation. The F_1 plants exhibited abundant trichomes on the hypocotyls and on both the adaxial and abaxial surfaces of the cotyledons. The first and later leaves were densely covered with trichomes on both surfaces, and leaf expansion was severely limited. The results suggest that *TTG* and *GL1* may interact to promote trichome initiation (48). In support of direct physical interaction, it has been shown that antibodies to GL1 can co-precipitate GL1 and R proteins (Marks & Jilk, unpublished date).

   The results also have implications for interpreting the phenotype of 35SGL1 plants (48). These plants make fewer than normal leaf trichomes. However, this phenotype is suppressed in 35SGL1/35SR plants. Two models have been proposed to explain the reduction in leaf trichomes. 1. Excess free GL1 protein (not bound to target DNA sequences) in the nucleus titrates TTG from complexes with GL1 bound to target genes. If transcriptional activation requires the interaction of TTG and GL1 protein at the target promoter, then the titration of TTG by free GL1 could lower the expression of target genes and, in turn, result in fewer trichomes. Excess TTG protein (in the form of R) could prevent the titration. 2. Excess GL1 activates genes that participate in a lateral inhibition of trichome initiation, resulting in fewer trichomes. The two models are not mutually exclusive; it is possible that the lateral inhibition is controlled in part by the ratio of a TTG-GL1 complex vs free GL1.

The hypothesis that *GL1* and *TTG* interact to control lateral inhibition is supported by the interaction between weak alleles of *GL1* and *TTG* (JA Larkin & MD Marks, unpublished data). As described above, it has been found that the plants doubly heterozygous for both weak *ttg* and *gl1* mutant alleles have greater than normal numbers of clustered trichomes; that is, lateral inhibition appears to be reduced. Larkin et al (48) found that plants with one or two copies of the 35SGL1 (35SGL1/-) construct and heterozygous for *TTG* (*TTG/ttg*) have a greater number of leaf trichomes than plants that have one or two copies of 35SGL1 in a homozygous *TTG* background. Approximately 30% of the trichomes on 35SGL1/-*TTG/ttg* were present in clusters. This suggests that the levels of *GL1* and *TTG* expression are important in controlling lateral inhibition.

GL2

The *GL2* gene, like *GL1*, was isolated by T-DNA tagging (82). In a screen of 10,000 transformed lines, seven independent *gl2* mutants were isolated but only one of these, *gl2-2*, had a T-DNA insert linked to the *GL2* locus. A wild-type *GL2* fragment was defined by complementation of transgenic *gl2* (82). *GL2* encodes a 744 aa protein that has an amino-terminal acidic domain followed by the homeodomain (82). Directly downstream of the homeobox is a motif that could encode an antipathic helix that could promote protein dimerization. The amino acid sequence downstream of the putative helix sequence does not show significant similarity to any protein in the databases. A comparison of plant homeodomain genes indicates that *GL2* is most similar to a class that has a leucine zipper domain on the carboxyl side of the homeodomain (38).

In situ hybridization analysis indicates that *GL2 mRNA* is expressed strongly in developing trichomes (82). Immunolocalization of the GL2 protein and the analysis of plants containing a *GL2* promoter GUS reporter gene construct (GL2GUS) indicate a more complex pattern of expression (MD Marks & D Szymanski, unpublished data). Approximately 2000 bases of 5' noncoding sequence were fused to the coding sequence of GUS to create GL2GUS. The staining pattern of GL2GUS plants indicates that *GL2*, like *GL1*, is expressed in fields of epidermal cells before trichome initiation. The level of *GL2* expression increases in cells committing to the trichome cell fate. *GL2* also is expressed in mesophyll cells; GL2 proteins appear in both the cytoplasm and nucleus of nontrichome cells. In developing trichomes GL2 protein is primarily localized to the nucleus.

By genetic analysis, *GL2* functions downstream of *GL1*, but *GL1* does not control the nontrichome expression pattern of *GL2*. Furthermore, the non-trichome expression pattern is observed in *ttg* mutants. Consequently, genes other than *GL1* or *TTG* control the early expression pattern of *GL2*. GL2GUS is strongly expressed in the malformed trichomes present on *gl2* mutants. This result suggests that GL2 protein does not regulate its own expression in developing trichomes.

The analysis of *GL2* promoter deletions has identified a region that is important for controlling *GL2* expression in trichomes. The entire *GL2* coding sequence and approximately 1.5 kb of 5′ noncoding sequence could molecularly complement *gl2*. Removal of approximately 125 bases from the 5′ end of this fragment yielded partial complementation (82). This region contains a myb consensus binding site. Thus, although the GL1 protein does not regulate the early expression pattern of *GL2*, it or another myb protein could influence the expression of *GL2* in developing trichomes by binding to the *myb* binding site.

ZWI

Wild-type leaf trichomes normally have a stalk and three to four branches, whereas *zwi* mutant trichomes have a shortened stalk and only two branches. Three independent *zwi* mutants were isolated from Feldmann's T-DNA lines. Two of these had inserts that co-segregated with the mutant phenotype (73). Plasmid rescue was used to isolate plant DNA flanking the insertions (14). Sequence analysis of the region flanking one of the inserts revealed that the *ZWI* encodes a large kenesin-like protein (73). These gene fragments were used to isolate an intact gene and flanking DNA. A 27-kb fragment could molecularly complement the *zwi* mutant.

Kinesins are microtubule motor proteins characterized by a conserved head domain that comprises the motor domain and a nonconserved tail region, which is thought to participate in binding cargo (20). *ZWI* encodes a kinesin with the motor domain located toward the carboxy terminus (73). In addition, the 5′ portion of *ZWI* encodes a region with similarity to a class IV myosin found in *Acanthamoeba* (30). The function of class IV myosins and the role of the region with similarity are unknown.

While the characterization of *ZWI* was being completed, the sequence became available for a kinesin-like protein from Arabidopsis, isolated because of its ability to bind calmodulin (79). Sequence comparison between *ZWI* and the kinesin-like calmodulin-binding protein (*KCBP*) indicate only a few nucleotide differences, and Southern hybridization analysis indicated that only a single copy of a gene with high similarity to either *ZWI* or *KCBP* exists (73, 79). Thus, *ZWI* and *KCBP* are the same gene. The calmodulin-binding domain is composed of a 21-amino acid sequence located very close to the carboxy terminus of the protein (79). Binding studies indicate ZWI binds to calmodulin with high affinity in the presence of Ca^{2+} (1 μm). These studies strongly suggest that Ca^{2+} and calmodulin are involved in regulating *ZWI* activity.

Northern hybridization analysis showed that *ZWI* is expressed in flowers, leaves, roots, and cultured callus tissues (73, 79). The only discernible phenotype of the *zwi* mutant is the abnormal branching and cell expansion pattern of trichomes. It is possible that only partial loss of function alleles of *zwi* have been identified. However, one of the T-DNA insertions disrupts the region encoding the ATP-binding and microtubule-binding domains (73). The insertion produces an in-frame stop codon that would most likely result in a ZWI protein lacking motor function. There are at least three explanations that can account for the lack of phenotype, aside from altered trichome formation, in plants homozygous for this insertion. 1. It is possible that *ZWI* expression is required only in trichomes. 2. Another gene can function in place of *ZWI* in the nontrichome cells, but not in trichomes. 3. Only the amino terminal of ZWI is required in nontrichome cells.

The ZWI promoter sequences have not yet been identified, but it appears that like *GL1*, sequences in the 3' OTR of *ZWI* are required for its regulation (50, 73). The insertion in one of the tagged mutants is located two kb downstream of the polyadenylation site. This insertion may disrupt a trichome-specific transcriptional enhancer.

Other Genes

Two other genes have been cloned that appear to have important functions in controlling trichome initiation. The first is the *CONSTITUTIVE PHOTOMOR-PHOGENIC (COP1)* gene (9). *COP1* was shown to encode a protein with a zinc finger domain in the amino terminal, followed by a coiled-coiled zipper-like domain, and ends with WD-40 repeats (10). All these domains could participate in protein-protein interactions. Apparently *COP1* functions to repress photomorphic genes in roots and in shoots of dark-grown plants (87). *COP1* may act at the level of transcription because COP1 protein is found in the nucleus in dark-grown plants. In light-grown shoots, COP1 protein is localized to the cytoplasm. Miséra et al characterized plants carrying a lethal allele of *COP1* called *fusca1* (*fus1*) (70). Plants homozygous for *fus1* die as young seedlings. Furthermore, these seedlings accumulate high amounts of anthocyanin. To study the function of *COP1* in the adult seedling, heterozygous fus1 seeds were treated with EMS to induce *fus1/fus1* sectors in the resulting seedlings. Mesophyll sectors had underexpanded cells that accumulated anthocyanin (70). Epidermal sectors, in contrast, exhibited ectopic trichomes. A model has been proposed in which *COP1* is an upstream inhibitor of *TTG* (70).

Recently another *ttg*-like mutant, *ttg2*, has been isolated (34). This mutant has a reduced number of trichomes that are less branched than normal. In addition, like *ttg1*, *ttg2* mutant seeds lack seed coat mucilage and are less pigmented. The gene mutation maps to the bottom of chromosome 2. The mutant was isolated from a population of plants carrying the maize *Ac* transposon that mobilized the endogenous *tag* element into the gene (34). Isolation and characterization of the gene has shown that it encodes a product with similarities to the SPF1 protein of sweet potato that binds SP8 sequences in the promoter of beta-amylase and sporamin genes (32).

CONCLUSIONS AND FUTURE DIRECTIONS

Many mutations define genes that affect discrete aspects of trichome development. Genetic analyses indicate that some of these genes act in linear pathways, whereas others possibly work through unrelated parallel pathways (31, 49). Molecular analyses of several cloned genes led to proposals for specific functions for *GL1*, *GL2*, *TTG* (if *TTG* is an *R* homolog), *TTG2*, and *ZWI*. The first four genes encode transcription factors. For these transcription factors, the lay question is how they are regulated and with which factors do they interact. How do these interactions influence their function? What genes do they regulate? The answers to these questions should identify genes acting both upstream and downstream of these factors. The characterization of many of the regulated genes will extend the analysis of trichome development to the cellular level. However, understanding how the expression of various structural proteins and enzymes, which are likely targets of the regulatory genes, leads to trichome differentiation will be challenging.

A genetic analysis will be essential. Often the sequences of the genes genetically identified will be informative. For example, *ZWI* encodes a gene with a microtubule motor domain. This indicates that the cytoskeleton plays a role in trichome morphogenesis, and *ZWI* now becomes a reagent to probe this role. Some "trichome genes" will turn out to be genes previously characterized in a different context. Again *ZWI* provides an example; *KCBP(=ZWI)* was isolated because of its calmodulin-binding domain. The analysis of *ZWI* can define how Ca^{2+} and calmodulin regulate a developmental program. In the near future, many genetically identified trichome genes will be isolated and characterized. The analysis of these genes likely will result in the characterization of many different aspects of trichome cell type determination and differentiation.

Visit the *Annual Reviews home page* at http://www.annurev.org

Literature Cited

1. Ågren J, Schemske D. 1994. Evolution of trichome number in a naturalized population of *Brassica rapa*. *Am. Nat.* 143:1–13
2. Baranowskij N, Frohberg C, Prat S, Willmitzer L. 1994. A novel DNA binding protein with homology to Myb oncoproteins containing only one repeat can function as a transcriptional activator. *EMBO J.* 13:5283–392
3. Biesalski HK, Doepner G, Tzimas G, Gamulin V, Schroder HC, et al. 1992. Modulation of myb gene expression in sponges by retinoic acid. *Oncogene* 7:1765–74
4. Bowley SR, Lackle SM. 1989. Genetics of nonglandular stem trichomes in Red Cover (*Trifolium pratens* L.). *J. Hered.* 80:472–74
5. Clark PJ, Evans FC. 1954. Distance to nearest neighbor as a measure of spatial relationships in populations. *Ecology* 35:445–53
6. Coe EH, Neuffer MG, Hoisington DA. 1988. The genetics of corn. In *Corn and Corn Improvement. Agron. Monogr. No. 18*, ed. GF Sprague, JW Dudley, pp. 81–236. Madison, WI: Am. Soc. Agron.
7. Cone KC, Cocciolone SM, Moehlenkamp CA, Weber T, Drummond BJ, et al. 1993. Role of the regulatory gene pl in the photocontrol of maize anthocyanin pigmentation. *Plant Cell* 5:1807–16
8. Cone KC, Burr FA, Burr B. 1986. Molecular analysis of the maize anthocyanin regulatory locus *C1*. *Proc. Natl. Acad. Sci. USA* 83:9631–35
9. Deng X-W, Caspar T, Quail P. 1991. *cop1*: a regulatory locus involved in light-controlled development and gene expression in *Arabidopsis*. *Genes Dev.* 5:1172–82
10. Deng X-W, Matsui M, Wei N, Wagner D, Chu A, et al. 1992. COP1, an Arabidopsis regulatory gene, encodes a protein with both a zinc-binding motif and Gβ homologous domain. *Cell* 71:791–801
11. Eilers M, Picard D, Yamamoto K, Bishop J. 1989. Chimaeras of myc oncoprotein and steroid receptors cause hormone-dependent transformation of cells. *Nature* 340:66–68
12. Esch JJ, Oppenheimer DG, Marks MD. 1994. Characterization of a weak allele of the *GL1* gene of *Arabidopsis thaliana*. *Plant Mol. Biol.* 24:203–7
13. Feenstra WJ. 1978. Contiguity of linkage groups I and IV as revealed by linkage relationship of two newly isolated markers *dis-1* and *dis-2*. *Arab. Inf. Serv.* 15:35–38
14. Feldmann K. 1992. T-DNA insertion mutagenesis in Arabidopsis: seed infection/transformation. In *Methods in Arabidopsis Research,* ed. C Koncz, N-H Chua, J Schell, pp. 274–89. Singapore: World Sci.
15. Feldmann KA, Marks MD. 1987. *Agrobacterium*-mediated transformation of germinating seeds of *Arabidopsis thaliana*: a nontissue culture approach. *Mol. Gen. Genet.* 208:1–9
16. Feldmann KA, Marks MD, Christianson ML, Quatrano RS. 1989. A dwarf mutant of *Arabidopsis* generated by T-DNA insertion mutagenesis. *Science* 243:1351–54
17. Frampton J, Leutz A, Gibson T, Graf T. 1989. DNA-binding domain ancestry. *Nature* 342:134
18. Galway ME, Masucci JD, Lloyd AM, Walbot V, Davis RW, Schiefelbein JW. 1994. The *TTG* gene is required to specify epidermal cell fate and cell patterning in the *Arabidopsis* root. *Dev. Biol.* 166:740–54
19. Goffreda JC, Szymkowiak EJ, Sussex IM, Mutschler MA. 1990. Chimeric tomato plants show that aphid resistance and triacylglucose production are epidermal autonomous characters. *Plant Cell* 2:643–49
20. Goldstein LSB. 1993. With apologies to Scheherazade: tails of 1001 kinesin motors. *Annu. Rev. Genet.* 27:319–51

21. Goodrich J, Carpenter R, Coen E. 1992. A common gene regulates pigmentation pattern in diverse plant species. *Cell* 68: 955–64

22. Graf T. 1992. Myb: a transcriptional activator linking proliferation and differentiation in hematopoietic cells. *Curr. Opin. Genet. Dev.* 2:249–55

23. Grotewold E, Drummond B, Bowen B, Peterson T. 1994. The myb-homologous P gene controls phlobaphene pigmentation in maize floral organs by directly activating a flavonoid biosynthetic gene subset. *Cell* 76:543–53

24. Haughn GW, Somerville CR. 1988. Genetic control of morphogenesis in *Arabidopsis. Dev. Genet.* 9:73–89

25. Held LI. 1991. Bristle patterning in Drosophia. *BioEssays* 13:633–40

26. Herman PL, Marks MD. 1989. Trichome development in *Arabidopsis thaliana.* II. Isolation and complementation of the *GLABROUS1* gene. *Plant Cell* 1:1051–55

27. Holt J, Birch N. 1984. Taxonomy, evolution and domestication of *Vicia* in relation to aphid resistance. *Ann. Appl. Biol.* 105: 547–56

28. Hombergen E-J, Bachman K. 1995. RAPD mapping of three QTLs determing trichome formation in Microseris hybrid H27 (Asteraceae: Lactuceae). *Theor. Appl. Genet.* 90:853–58

29. Hope TJ, Huang XJ, McDonald D, Parslow T. 1990. Steroid-receptor fusion of the human immunodeficiency virus type 1 Rev transactivator: mapping cryptic functions of the arginine-rich motif. *Proc. Natl. Acad. Sci. USA* 87:7787–91

30. Horowitz J, Hammer JI. 1990. A new *Acanthamoeba* myosin heavy chain. *J. Biol. Chem.* 265:20646–52

31. Hülskamp M, Miséra S, Jürgens G. 1994. Genetic dissection of trichome cell development in Arabidopsis. *Cell* 76:555–66

32. Ishiguro S, Nakamura K. 1994. Characterization of a cDNA encoding a novel DNA-binding protein, SPF1, that recognizes SP8 sequences in the 5' upstream regions of genes coding for sporamin and beta-amylase from sweet potato. *Mol. Gen. Genet.* 244:563–71

33. Jackson D, Culianez-Macia F, Prescott AG, Roberts K, Martin C. 1991. Expression patterns of *myb* genes from *Antirrhinum* flowers. *Plant Cell* 3:115–25

34. Johnson C, Symth D. 1996. A gene from Arabidopsis that regulates trichome development, seed pigmentation and mucilage production. *Proc. Aust. Soc. Biochem. Mol. Biol.* 28:In press

35. Johnson HB. 1975. Plant pubescence: an ecological perspective. *Bot. Rev.* 41:233–58

36. Katzen AL, Kornberg TB, Bishop JM. 1985. Isolation of the proto-oncogene c-myb from D. melanogaster. *Cell* 41:449–456

37. Kennedy G, Sorenson C. 1985. Role of glandular trichomes in the resistance of *Lycopersion hirsutum* f. *glabratum* to Colorado potato beetle (Coleoptera: Chrysomelidae). *J. Econ. Entomol.* 78:547–51

38. Kerstetter R, Vollbrecht E, Lowe B, Veit B, Yamaguchi J, Hake S. 1994. Sequence analysis and expression patterns divide the maize knotted1-like homeobox genes into two classes. *Plant Cell* 6:1877–87

39. Khan ZR, Ward JT, Norris DM. 1986. Role of trichomes in soybean resistance to cabbage looper, Trichoplusia ni. *Entomol. Exp. Appl.* 42:109–17

40. Kloth RH. 1995. Interaction of two loci that affect trichome density in upland cotton. *J. Hered.* 86:78–80

41. Kloth RH. 1993. New evidence relating the pilose allele and micronaire reading in cotton. *Crop Sci.* 33:683–87

42. Koornneef M. 1981. The complex syndrome of *ttg* mutants. *Arab. Inf. Serv.* 18: 45–51

43. Koornneef M, Dellaert SWM, van der Veen JH. 1982. EMS- and radiation-induced mutation frequencies at individual loci in *Arabidopsis thaliana* (L) Heynh. *Mutat. Res.* 93:109–23

44. Lamb R. 1980. Hairs protect pods of mustard (Brassica hirta 'gisilba') from flea beetle feeding damage. *Can. J. Plant. Sci.* 60: 1439–40

45. Lamb R. 1982. Economics of insecticidal control of flea beetles (Coleoptera: Chrysomelidae) attacking rape in Canada. *Can. Entomol.* 114:827–40

46. Lamb R. 1984. Effects of flea beetles, Phyllotreta ssp. (Chrysomelidae: Coleoptera), on the survival, growth, seed yield and quality of canola, rape and yellow mustard. *Can. Entomol.* 116:269–80

47. Larkin JC, Young N, Prigge M, Marks M. 1996. The control of trichome spacing and number in *Arabidopsis. Development* 122: 997–1005

48. Larkin JC, Oppenheimer DG, Lloyd A, Parozzi ET, Marks MD. 1994. The roles of *GLABROUS1* and *TRANSPARENT TESTA GLABRA genes* in Arabidopsis trichome development. *Plant Cell* 6:1065–76

49. Larkin JC, Oppenheimer DG, Marks MD. 1994. The *GL1* gene and the trichome de-

velopmental pathway in *Arabidopsis thaliana*. In *Results and Problems in Cell Differentiation 20: Plant Promoters and Transcription Factors*, ed. L Nover, pp. 259–75. Berlin: Springer-Verlag

50. Larkin JC, Oppenheimer DG, Pollock S, Marks MD. 1993. Arabidopsis *GLABROUS1* gene requires downstream sequences for function. *Plant Cell* 5:1739–48

51. Lawson EJR, Scofield SR, Sjodin C, Jones JDG, Dean C. 1994. Modification of the 5' untranslated leader region of the maize *Activator* element leads to increased activity in Arabidopsis. *Mol. Gen. Genet.* 245: 608–15

52. Lee JA. 1985. Revision of the genetics of the hairiness-smoothness system of *Gossypium*. *J. Hered.* 76:123–26

53. Levin DA. 1973. The role of trichomes in plant defense. *Q. Rev. Biol.* 48:3–15

54. Lipsick JS, Baluda MA. 1986. The myb oncogene. In *Gene Amplification and Analysis*, Vol. 4, *Oncogenes*, ed. TS Papas, GF Vande Woude, pp. 73–98. New York: Elsevier

55. Lister C, Dean C. 1993. Recombinant inbred lines for mapping RFLP and phenotypic markers in *Arabidopsis thaliana*. *Plant J.* 4:745–50

56. Lloyd AM, Schena M, Walbot V, Davis RW. 1994. Epidermal cell fate determination in Arabidopsis: patterns defined by a steroid-inducible regulator. *Science* 266:436–39

57. Lloyd AM, Walbot V, Davis RW. 1992. Anthocyanin production in dicots activated by maize anthocyanin-specific regulators, *R* and *C1*. *Science* 258:1773–75

58. Ludwig SR, Habera LF, Dellaporta SL, Wessler SR. 1989. *Lc*, a member of the maize *R* gene family responsible for tissue-specific anthocyanin production, encodes a protein similar to transcription factors and contains the Myc homology region. *Proc. Natl. Acad. Sci. USA* 86:7092–96

59. Lüscher B, Eisenman RN. 1990. New light on Myc and Myb. Part II. Myb. *Genes Dev.* 4:2235–41

60. Marks MD. 1994. The making of a plant hair. *Curr. Biol.* 4:621–23

61. Marks MD, Esch J, Herman P, Sivakumaran S, Oppenheimer D. 1991. A model for cell-type determination and differentiation in plants. In *Molecular Biology of Plant Development*, ed. G Jenkins, W Schuch, pp. 77–87. Cambridge: Co. Biol.

62. Marks MD, Esch JJ. 1992. Trichome formation in *Arabidopsis* as a genetic model for studying cell expansion. *Curr. Top. Plant Biochem. Physiol.* 11:131–42

63. Marks MD, Esch JJ. 1994. Morphology and development of mutant and wild type trichomes on the leaves of *Arabidopsis thaliana*. In *Arabidopsis: An Atlas of Morphology and Development*, ed. J Bowman, pp. 56–73. New York: Springer-Verlag

64. Marks MD, Feldmann KA. 1989. Trichome development in *Arabidopsis thaliana*. I. T-DNA tagging of the *GLABROUS1* gene. *Plant Cell* 1:1043–50

65. Marocco A, Wissenbach M, Becker D, Paz-Ares J, Saedler H, Salamini F. 1989. Multiple genes are transcribed in *Hordeum vulgare* and *Zea mays* that carry the DNA binding domain of the myb oncoproteins. *Mol. Gen. Genet.* 210:183–87

66. Masucci J, Rerie W, Foreman D, Zhang M, Galway M, et al. 1996. The homeobox gene *GLABRA2* is required for position-dependent cell differentiation in the root epidermis of Arabidopsis thaliana. *Development* 122: 1253–60

67. McKelvie A. 1965. Preliminary data on linkage groups in Arabidopsis. *Arab. Inf. Serv.* 1S:79–84

68. Meisner J, Mitchell B. 1983. Phagodeterrency induced by two cruciferous plants in adults of the flea beetle *Phyllotreta striolata* (Coleoptera: Chrysomelidae). *Can. Entomol.* 115:1209–14

69. Melaragno J, Mehrota B, Coleman A. 1993. Relationship between endoploidy and cell size in epidemal tissue of Arabidopsis. *Plant Cell* 5:1661–68

70. Miséra S, Müller A, Weiland-Heidecker U, Jürgens G. 1994. The FUSCA genes of Arabidopsis: negative regulators of light responses. *Mol. Gen. Genet.* 244:242–52

71. Mooney M, Desnos T, Harrison K, Jones J, Carpenter R, Coen E. 1995. Altered regulation of tomato and tobacco pigmentation genes caused by the *delila* gene of *Antirrhinum*. *Plant J.* 7:333–39

72. Noda K, Glover B, Linstead P, Martin C. 1994. Flower colour intensity depends on specialized cell shape controlled by a Myb-related transcription factor. *Nature* 369: 661–64

73. Oppenheimer DG, Esch J, Marks MD. 1992. Molecular genetics of *Arabidopsis* trichome development. In *Control of Plant Gene Expression*, ed. DPS Verma, pp. 275–86. Boca Raton, FL: CRC

74. Oppenheimer DG, Herman PL, Esch J, Sivakumaran S, Marks MD. 1991. A *myb*-related gene required for leaf trichome differentiation in Arabidopsis is expressed in stipules. *Cell* 67:483–93

75. Oppenheimer DG, Pollock M, Vacik J, Ericson B, Feldmann K, Marks M. 1996.

Essential role of a novel kinesin-like protein in Arabidopsis trichome morphogenesis. Submitted

76. Paz-Ares J, Ghosal D, Wienand U, Peterson PA, Saedler H. 1987. The regulatory *c1* locus of *Zea mays* encodes a protein with homology to *myb* proto-oncogene products and with structural similarities to transcriptional activators. *EMBO J.* 6:3553–58

77. Potikha T, Delmer D. 1995. A mutant of Arabidopsis thaliana displaying altered patterns of cellulose deposition. *Plant J.* 7:453–60

78. Purugganan MD, Wessler S. 1994. Molecular evolution of the plant *R* regulatory gene family. *Genetics* 138:849–54

79. Reddy A, Safadi F, Narasimholu S, Golovkin M, Hu X. 1996. A novel plant calmodulin-binding protein with a kinesin heavy chain motor domain. *J. Biol. Chem.* 271:7052–60

80. Rédei GP. 1967. Genetic estimate of cellular autarky. *Experientia* 23:584

81. Reeves AF Jr. 1977. Tomato trichomes and mutations affecting their development. *Am. J. Bot.* 64:186–89

82. Rerie WG, Feldmann KA, Marks MD. 1994. The *GLABRA2* gene encodesa homeo domain protein required for normal trichome development in *Arabidopsis*. *Genes Dev.* 8:1388–99

83. Roussel M, Saule S, Lagrou C, Rommens C, Beug H, et al. 1979. Three new types of viral oncogene of cellular origin specific for haematopoietic cell transformation. *Nature* 281:452–55

84. Stober-Grasser U, Brydolf B, Bin X, Grasser F, Firtel RA, Lipsick JS. 1992. *Dictyostelium* MYB: evolution of a DNA-binding domain. *Oncogene* 7:589–96

85. Telfer A, Poethig A. 1994. Leaf development in *Arabidopsis*. In *Arabidopsis,* ed. E Meyerowitz, C Somerville. Plainview, NY: Cold Spring Harbor

86. Tice-Baldwin K, Fink GR, Arndt KT. 1989. BAS1 has an Myb motif and activates HIS4 transcription only in combination with BAS2. *Science* 246:931–35

87. von Arnim A, Deng X-W. 1994. Light inactivation of Arabidopsis photomorphogenic repressor COP1 involves a cell-specific regulation of its nucleocytoplasmic partitioning. *Cell* 79:1035–45

Annu. Rev. Plant Physiol. Plant Mol. Biol. 1997. 48:165–190
Copyright © 1997 by Annual Reviews Inc. All rights reserved

FLUORESCENCE MICROSCOPY OF LIVING PLANT CELLS

Simon Gilroy

Biology Department, The Pennsylvania State University, 208 Mueller Laboratory, University Park, Pennsylvania 16802

KEY WORDS: confocal microscopy, ratio analysis, aequorin, green fluorescent protein, protein biosensors

ABSTRACT

Since its inception, light microscopy has shown the elegance and subtlety with which function is expressed in the form of the cells, tissues, and organs of the plant. Recently, light microscopy has seen a resurgence in use fueled by advances in microscope design and computer-based image analysis. The structural resolution afforded by static, fixed samples is being increasingly supplemented by approaches using fluorescent analogs and selective fluorescent indicators, which visualize the dynamic processes in living, functioning cells. This review describes some of these approaches and discusses how they are taking us a step closer to viewing the intricate complexity with which plants organize and regulate their functions down to the subcellular level.

CONTENTS

1040-2519/97/0601-0165$08.00

INTRODUCTION

Light microscopy is undergoing a renaissance. The continued refinement of the light microscope, development of new optical techniques, major advances in optical probe chemistry, and dramatic recent advances in computer technology have been combined into a suite of advanced microscope techniques. These approaches are giving us an unprecedented view of the complex temporal and spatial interplay of structural elements, ions, and metabolites that allow the cell to function. For example, the view of the functional organization of the cell changed radically with the ability to fix and visualize the microtubule and actin arrays that make up the cytoskeleton. Our ideas then underwent another conceptual shift upon seeing the structurally intricate, but static "skeleton," of fixed samples as a constantly changing, dynamic structure in the living, functioning cell (e.g. 143). Similarly, it took the perfection of fluorescence techniques enabling the visualization of spatial and temporal dynamics of Ca^{2+} in living cells to reveal the ubiquity and subtleness of this ion as an intracellular regulator (132).

The continued development of fluorescent techniques has substantially advanced our understanding of cell function. Approaches such as fluorescence recovery after photobleaching, photoactivation of fluorescence, fluorescence anisotropy, fluorescence lifetime imaging, and confocal and ratio imaging all promise to reveal cellular function with increasing quantitative and spatial resolution (49). Unfortunately, it is impossible to examine all these approaches in a single review. Instead, this review highlights some of the exciting, fluorescence tools being developed and presents an overview of how they are being applied to living, functioning plant cells. These techniques are allowing cell biologists to observe and manipulate the dynamics of structural elements, ions, and metabolites against the ultimate functional backdrop, the cell itself.

MICROSCOPES AND IMAGING TECHNOLOGY

Conventional Fluorescence Imaging

Conventional epifluorescence microscopy coupled to image analysis has become a powerful tool and has yielded fundamental insights into the functioning of the plant cell. The capabilities of this approach are well illustrated by work aimed at defining how plasmodesmatal function is regulated by viruses. Viral movement proteins operate by enlarging the size exclusion limit of plasmodesmata, allowing intercellular transfer of viral nucleic acid (93, 135). This function was largely revealed by monitoring the effect of these proteins on the cell-to-cell transfer of fluorescently labeled dextrans of different sizes

(93; see the chapter on "Transport of Proteins and Nucleic Acids Through Plasmodesmata" in this volume). In an exciting extension of this approach it has been found that the homeobox transcription factor *Knotted-1* (136) can also increase the plasmodesmatal size exclusion limit and move intercellularly (92). Thus, the ability to image and analyze the intercellular transport of fluorescent dyes may well have revealed a class of transcriptional regulators that can move between cells, and has led to the coordinated developmental programs that characterize many aspects of plant development.

Although powerful, such conventional fluorescence imaging can suffer from limitations in spatial resolution, quantitative accuracy, and image degradation due to the out-of-focus blur associated with conventional microscope optics. Fortunately, approaches have been developed to overcome these limitations and dramatically increase the capabilities of quantitative and spatial resolution of the fluorescence microscope.

Fluorescence Ratio Analysis

Ratio analysis corrects quantitative fluorescence imaging from artifacts in signal strength associated with spatial and temporal variations in sample path length, accessible volume, and local indicator concentration (129, 134). The approach has been most widely used with plants to monitor ion levels, principally cytoplasmic levels of Ca^{2+} and pH, because of their fundamental role in signal transduction and ion homeostasis (19). However, the ratio approach also provides an important control in quantitative analysis of structural dynamics such as actin polymerization (49).

Dyes whose fluorescent properties are dependent on the local free concentration of ions, including Ca^{2+}, H^+, Mg^{2+}, K^+, Na^+, Cl^-, and Zn^{2+}, have been synthesized (65), and such indicators are under continuous development. These indicators may show a quantitative change in intensity (single wavelength dyes) or a shift in emission or excitation fluorescence spectrum (ratiometric dyes) upon binding the ion of interest. Generally, for ratiometric dyes, a wavelength of fluorescence that increases with ion concentration is monitored and compared with the intensity at a wavelength that is insensitive to, or decreases with, increasing ion levels. The ratio of these intensities is independent of changes such as alterations in dye concentration, cell thickness, accessible cytoplasmic volume, photobleaching, and dye leakage, which make single wavelength dyes inaccurate (129, 134).

The requirement for using ratio analysis is most dramatically displayed when there is a complex, dynamic, cytoplasmic structure. For example, the root hair shown in Figure 1 has been loaded with the ratiometric $[Ca^{2+}]$ indicating dye Indo-1. When $[Ca^{2+}]$ is calculated from a single wavelength that increases in intensity as $[Ca^{2+}]$ increases, the highest concentration appears along the center of the hair (Figure 1B). This is due to the complex structure of the root hair cytoplasm, where an accumulation of organelles and vesicles may partly exclude the dye from the extreme tip. This problem is coupled to an increased signal from the thicker cytoplasm in the center of the tubular hair, leading to an apparently elevated $[Ca^{2+}]$ along the middle of the hair. The highly localized $[Ca^{2+}]$ gradient at the tip is revealed only when ratio analysis is applied to the data (Figure 1C). This artifact is dramatically demonstrated in the published data on $[Ca^{2+}]$ in pollen tubes. Steep gradients in $[Ca^{2+}]$ are not seen at the tube tip when non-ratioable dyes are used, even when an enhanced spatial view is obtained by using the optical sectioning capabilities of the confocal microscope (e.g. 50, 94). However, ratio imaging has revealed the highly localized tip focused gradient in $[Ca^{2+}]$ associated with the bursts of tube

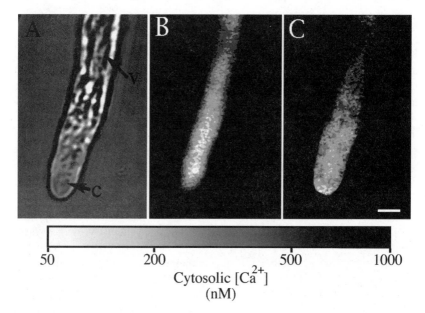

Figure 1 Cytosolic calcium levels in root hairs of *Arabidopsis thaliana*. Roots were acid loaded with Indo-1 and cytosolic-free $[Ca^{2+}]$ determined using confocal imaging. (*A*) Bright field image. (*B*) $[Ca^{2+}]$ calculated from wavelengths (405–430 nm) that increase in intensity with increasing $[Ca^{2+}]$. (*C*) Ratio image of $[Ca^{2+}]$ in the same root hair showing tip-focused gradient. The ratio image was calculated from (intensity at 400–430 nm)/(intensity at 460–480 nm) using a Zeiss LSM 410 UV confocal microscope. v, vacuole; c, cytoplasmically rich tip. Scale bar 10 μm. C Wymer, TN Bibikova & S Gilroy, unpublished data.

growth (e.g. 95, 102, 110, 111, 116). The use of calcium-sensitive dyes indicates a gradient in cytoplasmic-free calcium ($[Ca^{2+}]_c$) may be a fundamental feature of tip growing plant cells, where it may provide a driving force for elongation (e.g. algal rhizoids: 10, 18; pollen tubes: 94, 95, 102, 110, 111, 116; root hairs: 71, and C Wymer, TN Bibikova & S Gilroy, unpublished data).

Cytosolic, organelle, and even apoplastic ion levels are amenable to ratio analysis. However, organelle measurements require finding an indicator that is taken up by the organelle in question, and being able to spatially distinguish the organelle signal from that of the cytoplasm (53). Sequestration of fluorescent Ca^{2+} and pH indicators by organelles has been noted in many plant and animal cell types (52, 117, 118). For example, the pH indicators BCECF and carboxyfluorescein are rapidly loaded into the vacuoles of barley aleurone and root cells where they provide an in vivo method to monitor vacuolar pH (15, 37). Likewise, ER accumulation of Ca^{2+} sensors has been used to estimate Ca^{2+} levels in this organelle (21). However, accumulation of indicators in organelles will compromise cytosolic measurements. Conjugating indicators to dextrans generally (e.g. 10, 102), although not always (84, 117), prevents sequestration problems, and is the method of choice for introducing cytosolic probes into the cell.

When a ratiometric indicator is unavailable, pseudo-ratio analysis has been applied. In this approach, the signals from two unlinked indicators, one sensitive and one insensitive to the molecule under analysis, are used to make the ratio image. The major assumptions for pseudo ratioing are that the two fluorochromes used redistribute identically and photobleach at the same rate. Depending on the indicators chosen, this may not be a valid assumption. Despite this caveat, the pseudo-ratio technique can be extremely informative. For example, a ratioable pH indicator for the acidic apoplastic environment is currently unavailable. However, the "single wavelength" pH indicator Cl-NERF can respond to pH in the range 4 to 6. In an attempt to use this indicator to image apoplastic pH, the cell wall space of corn roots was infiltrated with Cl-NERF and the pH-insensitive dye Texas-red 3000 (131). Ratio analysis revealed that after 45 min of gravistimulation a local acidification (from pH 4.9 to 4.4) occurred in the elongation zones cells of the upper side of the root. Such acidification is predicted by the acid growth model for the tropic response of roots.

Simultaneous Ion Imaging and Electrophysiology

An exciting use of this ion imaging technology is in the simultaneous application of ratio analysis and electrophysiology (30, 103). For example, Schroeder & Hagiwara (124) monitored the effects of abscisic acid (ABA) on ion channel activities (using patch clamping) and cytosolic calcium levels (using ratio photometry) in guard cell protoplasts. They found that there was a direct temporal correlation between the opening of calcium-permeable channels and the increase of cytosolic calcium induced by ABA. Figure 2 shows data from a similar approach using simultaneous confocal ratio imaging of cytoplasmic $[Ca^{2+}]$ and patch clamp analysis of K^+ channel activities in guard cell protoplasts of *Vicia fava*. "Hotspots" of increased cytosolic $[Ca^{2+}]$ are associated with the closing of the K^+ channels. Ratio analysis of cytoplasmic $[Ca^{2+}]$ in intact guard cells also shows increases, hotspots, and oscillations in $[Ca^{2+}]$ associated with stomatal closure (Figure 2D) (3, 55, 57, 77, 96–98, 127, 140), reinforcing the idea of an involvement of cytosolic Ca^{2+} in the guard cell response. Such data have been instrumental in developing the model for Ca^{2+}-dependent signal transduction in ABA action in guard cells (5).

Ratio analysis of changes in guard cell $[Ca^{2+}]$ also provided some of the first indications of just how subtle information encoding by plant cells can be. Guard cells can show stimulus-evoked oscillations in $[Ca^{2+}]$ (98, 127) as do root hairs responding to nodulation factors (46). Stimulus-related information may well be encoded in the frequency, amplitude, and spatial localization of these changes (132).

Fluorescence Resonance Energy Transfer

While ratio analysis provides an approach to increase the quantitative resolution of the fluorescence microscope, a very powerful technology to increase the resolution of spatial interactions between cytoplasmic structures has also emerged in the technique of fluorescence resonance energy transfer (FRET) (70, 133). In FRET analysis, two interacting cell components are labeled with different fluorochromes such that when one, the donor, is excited with its appropriate wavelength of light some part of its emission energy is transferred to the second fluorochrome, the acceptor, which then fluoresces. The efficiency of the energy transfer (E) falls with the inverse of the sixth power of the distance (R) according to the equation $E = [1 + (R/R_o)^6]^{-1}$. R_0 is the characteristic distance at which FRET is 50% efficient and depends on the quantum yield of the donor fluorochrome, the absorbency characteristics of the acceptor, and the overlap in emission and excitation of both (70, 133). In practice, the fluorochromes must be within a few nanometers to show significant FRET. The technique provides the capability to image molecular interactions with superresolution, well beyond the point-to-point resolution of conventional fluorescence microscopy. For example, the interaction between plant EF1 alpha, microtubules, and CaM has recently been demonstrated in vitro (43) and has highlighted the potential role of EF1 alpha as a multifunctional

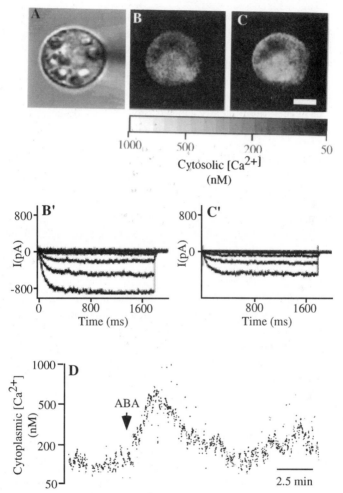

Figure 2 Simultaneous ratio analysis of $[Ca^{2+}]$ and patch clamping of guard cell protoplast of *Vicia fava*. (*A*) Bright field image. (*B, C*) Confocal ratio images of a protoplast as $[Ca^{2+}]$ increases. (*B', C'*) whole cell recordings for the protoplast, at times shown in (*B, C*). Protoplasts were patch clamped in whole cell configuration with the $[Ca^{2+}]$ indicator Indo-1 introduced into the cytoplasm via the patch pipette. Confocal ratio imaging was performed using a Zeiss LSM 410 UV confocal microscope using an 80/20 primary mirror, 460 nm secondary dichroic mirror, and dual detectors (400–430 nm and 460–480 nm). Pipette contained 70 µM Indo-1, 80 mM K glutamate, 20 mM KCl, 2 mM MgCl₂, 10 mM HEPES pH 8.0. Bath solution contained 100 mM KCl, 1 mM CaCl₂, 1 mM MgCl₂, 5 mM MES, 5 mM HEPES pH 5.6. (LA Romano, S Gilroy & SM Assmann, unpublished data.) (*D*) Fluorescence ratio photometry of an intact guard cell microinjected with Indo-1 according to (55). 5 µM ABA was added at the indicated time. (S Gilroy, MD Fricker, ND Read & AJ Trewavas, unpublished data.)

protein in regulating microtubule dynamics (34, 43, 44), actin dynamics (33, 104), protein synthesis (100), phosphatidylinositol signaling (143a), and myriad other cellular functions (34, 43, 44). How closely do these in vitro studies reflect the interactions in the cell? FRET analysis has revealed a very close association between rhodamine-labeled anti–EF1-alpha and fluorescein-labeled anti-tubulin in fixed cells (45). Critically, this approach is applicable to living cells where fluorescently labeled proteins can be coinjected and the dynamics of in vivo interactions analyzed with nanometer resolution.

Three-Dimensional Fluorescence Microscopy

FRET does not represent the only approach available to cell biologists to increase the spatial resolution of the fluorescence microscope. Cells and tissues are three-dimensional objects, and the spatial component of their organization often underlies their function. A prime example of such a three-dimensional link between structure and function is seen in the cortical microtubule array. It is the structural position of the cortical microtubules that defines their role in, for example, the regulation of deposition of cellulose microfibrils in the cell wall (35, 143). Conventional fluorescence microscopy cannot easily capture the three-dimensionality of the cell, and fine details of cellular structure are often obscured by blurred light from other out-of-focus parts of the cell. The two approaches to obtain blur-free "optical sectioning" of cells that have proven highly useful are computational approaches to calculate and remove out-of-focus information and use of the confocal microscope.

COMPUTATIONAL APPROACHES   One approach to solving the problem of out-of-focus blur is to calculate the contribution of this information in each image of the specimen and then subtract it, leaving only the in-focus image. A range of such computational approaches are available (28, 122 and references therein) and can be applied to images of almost any fluorochrome because data are collected using a conventional fluorescence microscope. The newer algorithms also allow dynamic processes to be observed by reducing the requirement for the time-consuming collection of a large, continuous stack of images for the deblurring calculation (28). However, the successful application of these computational methods usually requires powerful computational facilities coupled to a very thorough knowledge of the optical characteristics and aberrations present in the microscope and specimen. Although highly successful in resolving three-dimensional data (e.g. 28, 91, 122), the computational approach has received much less attention by plant biologists than a purely optical way to section specimens, the confocal microscope.

CONFOCAL MICROSCOPY   With the confocal microscope, an object is viewed through a pinhole (or slit) set at a focal point in the optical path. Light from the focal plane of the specimen passes through the pinhole to a photomultiplier detector, whereas light from regions outside the focal plane is largely blocked. Thus, the microscope collects images that are essentially only from the focal plane of the microscope objective.

In the confocal laser scanning microscope (CLSM), illumination is provided by the well-defined wavelength and intense collimated beam of light offered by a laser. The illuminating beam is focused to a spot on the specimen and the resultant in-focus fluorescence emission passes through the confocal aperture to photomultiplier detectors. A point-by-point map of the fluorescence intensity in the focal plane of the specimen is constructed by moving the laser across the sample. By taking sequential optical sections at known z-axis positions a full three-dimensional view of the cell can be calculated. The optical sectioning relies upon the physical principles of the light path and so, unlike the computational methods of deblurring, no calculations or assumptions about the spread of fluorescence from the sample are made. However, by applying deconvolution algorithms to CLSM sections, three-dimensional resolution may be even further enhanced (91, 109). Crucially, no physical sectioning is necessary with either the CLSM or computational approaches to obtaining three-dimensional data. Thus the sample can be a living cell.

Due to their defined wavelengths, the use of lasers as light sources can potentially limit the range of fluorochromes usable by CLSM. However, advances in laser technology, the development of multiple laser confocal microscopes, with the attendant optical corrections for lasers of very different spectral qualities (54, 141), and laser-compatible fluorochrome design are increasingly making wavelength restrictions less significant for confocal applications. CLSM also exhibits relatively poor sensitivity because the confocal aperture excludes much of the fluorescent light. Acquisition of high signal to noise images can be slow, often requiring seconds per image. Slit and disc scanning confocal microscopes are variations on the CLSM that improve image acquisition speed, but inferior image quality and detector efficiency (109, 123) have limited their use with plant specimens.

CLSM is providing an unparalleled view of the three-dimensionality and dynamics of the plant cell. Figures 3A–E show a series of optical sections through an epidermal cell of pea in which the microtubule array has been visualized using rhodamine-labeled tubulin. Each plane within the cell has a unique pattern of microtubules that may well parallel localized differences in cell function. It is known that the microtubules in the outer wall of pea epidermal cell behave very differently from those on the lateral wall (144),

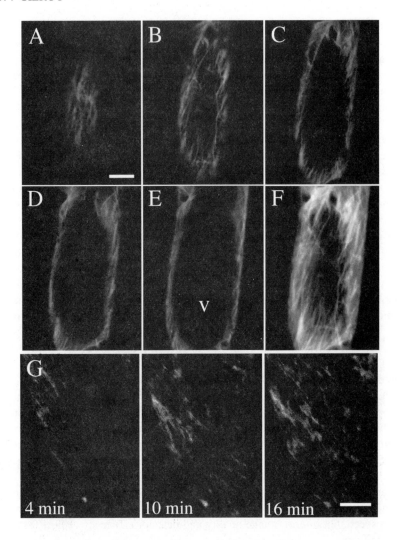

Figure 3 Confocal images of the microtubule array in an epidermal cell of the pea stem. An epidermal cell was microinjected with rhodamine-labeled porcine tubulin, which incorporated into the endogenous microtubule arrays. (*A-E*) Sequential optical sections at 0.5 μm intervals taken with the BioRad MRC1000 confocal microscope. (*F*) Projection of (*A–E*). (*G*) Time course of incorporation of rhodamine-tubulin into the microtubule array. Single optical sections were imaged at the indicated time after microinjection with rhodamine-labeled tubulin. Images were collected using a Bio-Rad MRC-600 confocal microscope and a Kr-Ar laser (95). Note highly variable microtubule distribution both with time (*G*) and with three-dimensional position of the optical section (*A–F*). Scale bar 10 μm. (C Wymer & C Lloyd, unpublished data.)

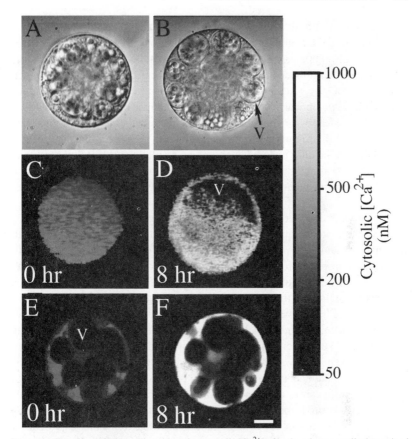

Figure 4   The effect of gibberellic acid on the cytosolic $[Ca^{2+}]$ of barley aleurone cells determined by ratio imaging and confocal ratio imaging. (*A*) Freshly isolated barley aleurone protoplast. (*B*) Aleurone protoplast after 18 h GA treatment showing vacuolation characteristic of the hormonal response in these cells. (*C, D*) Cytosolic $[Ca^{2+}]$ in aleurone protoplasts treated for 0 h or 8 h with 5 μM GA determined by conventional ratio imaging (60). (*E, F*) Equivalent measurements made using confocal ratio imaging. Note the improved spatial resolution, such as visualization of vacuoles, in the confocal images. Confocal ratio imaging was performed as in Figure 1. v, vacuole; scale bar 10 μm.

possibly reflecting a unique role for this face of the cell in environmental sensing. Such an appreciation of the three-dimensional intricacies that underlie cellular regulation is possible because of the confocal technique.

Just as the confocal provides an enhanced three-dimensional view of the cell by combining stacks of optical sections, a single section also provides an improved, blur-free view of cellular organization and control. For example, changes in $[Ca^{2+}]$ in the peripheral cytoplasm of aleurone cells are thought to participate in the hormonal regulation of secretion in these cells (20, 22, 23, 56). Figure 4 compares ratio images of the cytosolic Ca^{2+} levels in a barley aleurone protoplast loaded with the Ca^{2+} indicator Indo-1, taken using conventional fluorescence ratio imaging and confocal ratio imaging. The confocal image is a single optical section through the midplane of the cell. The confocal image clearly resolves cytoplasmic features not seen in the conventional ratio image, e.g. in this figure the vacuoles exclude the indicator and so appear as black holes in the ratio image. The improved resolution afforded by combining confocal analysis and ratio imaging (51, 123) is an example of how combining these new imaging technologies generates extremely powerful tools (49, 51).

TWO-PHOTON MICROSCOPY   A new approach to generating confocal images, the two-photon technique, promises to increase our ability to image into three-dimensional samples even further. Single cells or cells in a monolayer are still the optimal specimen for CLSM. The deeper a confocal section is taken in a tissue, the greater the image degrades (Figure 5) (54). Presently, taking optical sections through more than or two or three cell layers into a tissue represents a significant challenge to the confocal operator. The amount of signal degradation is highly specimen and wavelength dependent (42, 54). Significant signal attenuation occurs in cucumber roots illuminated at 442 nm within 40 μm, whereas 633 nm light was much more penetrating (54). Taylor et al (131) have seen significant attenuation of 488- and 568-nm excitation in as little as 15 μm in corn roots. In contrast, 343-nm illumination can penetrate to >80 μm in the relatively transparent roots of *Arabidopsis thaliana* (Figure 5). Even with the small size and transparency of the Arabidopsis root, we cannot capture a full three-dimensional view of this organ by conventional CLSM. However, two-photon confocal microscopy promises to greatly enhance our ability to view cellular dynamics deep in tissues (39–41).

The two-photon approach uses a mode-locked laser driven by a high-power argon laser to deliver subpicosecond bursts of illumination at a frequency of about 100 MHz. The wavelength of the excitation beam is twice that normally required to excite the fluorochrome in question. Two photons of this longer wavelength laser light are simultaneously absorbed by the fluorochrome and deliver an equivalent excitation energy to one photon of the normal, shorter excitation light. Long-wavelength light is much more penetrating, leading to a capacity for deeper optical sections into tissues with fewer aberrations and image degradation. The confocal nature of the two-photon system arises not from a pinhole or slit on the detector but because the probability of two photons being absorbed simultaneously by a fluorochrome depends quadrati-

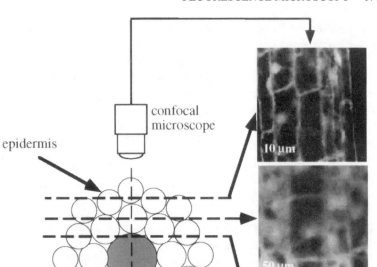

Figure 5   Attenuation of confocal images in root tissues. Roots of *Arabidopsis thaliana* were acid loaded with Indo-1 and optical sections taken within the intact root at the plane indicated in the root cross section, using the Zeiss LSM 410-UV confocal microscope. Note attenuation of signal as sections are taken deeper into the root.

cally on the absorbtion cross section of the molecule (39–41). Therefore, two-photon excitation occurs in a small volume right at the focal plane where the laser is focused, and drops off very rapidly away from this point. Because excitation is limited to this small volume at the focal plane, problems of out-of-focus illumination and phototoxicity are reduced. Two-photon technology is still being developed but promises a clearer three-dimensional view of plant tissues than has previously been possible (64).

FLUORESCENT INDICATORS

In parallel to these developments in imaging technology, fluorescent reporter chemistry has also advanced. The production of new fluorochromes and indicators is providing a significant driving force for the current advances in cell biology and microscopy.

Fluorescent labels with a wide range of excitation and emission wavelengths and increased stability and brightness are under continuous development (65). These fluorochromes are available as a range of reactive derivatizing reagents making selective labeling of cell components increasingly routine (65). In addition, environmentally sensitive fluorophores have received intensive study, leading to the success in the development of fluorescent indicators for a range of ions (discussed above and in 65, 117, 118), and hydrophilic and hydrophobic environments (65).

Fluorescent Analog Cytochemistry

While organic chemists have been improving fluorochrome design, an increased understanding of how protein structure and function are interrelated has led to dramatic advances in the use of fluorescently tagged proteins as probes for cell function. In fluorescent analog cytochemistry, a fluorescently labeled protein is introduced into the cell, incorporates into the pool of normal cellular proteins, and functions in the normal cellular processes. A classic example of this approach is the microinjection of cells with fluorescently labeled tubulin (e.g. 60, 74, 90, 144, 145). Figure 3G shows a time series of confocal images of a pea epidermal cell after microinjection with rhodamine-labeled tubulin. The labeled tubulin incorporates into the microtubule array within a few minutes and then allows the direct visualization of the highly dynamic nature of the cortical cytoskeleton (90, 143). This fluorescent analog approach has also been coupled to high-resolution, time-lapse DIC video microscopy to directly demonstrate the dynamic regulation of preprophase microtubule function and mitotic progression by calcium levels (145) and cyclin-related kinases (74) in the dividing cells of *Tradescantia* stamen hairs.

Protein Biosensors

Fluorescent analog cytochemistry was the first step toward constructing "protein biosensors" that use natural protein specificity to selectively view the metabolic and regulatory activities in the cell (59, 62, 130). Protein biosensors are proteins that have been tagged with an environmentally sensitive fluorophore. The change in activity of the protein, be it a structural rearrangement, ligand binding, or catalytic activity, is translated into a measurable change in fluorescence. By introducing these sensors into the cell, usually by microinjection, the goal of visualizing the wealth of dynamic interactions that make up cellular regulation and activity has become a reality. Proteins are ideal candidates for biosensors because, as natural components of the cell, they are inherently biocompatible. This idea is the basis of the fluorescent protein biosensor (59, 62, 130).

The trick is to target an environmentally sensitive fluorescent label to the correct site in the protein, such that its fluorescence can be affected by the activity of the protein. Achieving selective labeling may require protein engineering. For example, in constructing a probe for myosin light chain kinase, targeting the fluorochrome required engineering a cysteine into the myosin light chain at the required site next to the phosphorylated residue. This cysteine was then selectively labeled with a fluorochrome, acrylodan, that undergoes a change in fluorescence due to the nearby phosphorylation event (113). Thus, with a good understanding of how the protein of interest operates, rational design of the biosensor is possible (59, 62, 130). Notable examples of such "designer probes" include mero-Cam, a fluorescently labeled calmodulin (CaM) that reports CaM activity (61, 63), substrates that allow the imaging of myosin kinase activity (113, 114) and a sensor for cAMP levels based on fluorescently labeled protein kinase A (1). FRET forms the basis of these latter two biosensors.

The elegance of the protein biosensor approach is that it capitalizes on the inherent specificity of the tagged proteins to produce a highly selective probe. As our understanding of the proteins involved in cellular regulation increases, and as protein labeling technology becomes more routine, the range of selective protein biosensors and thus the cellular process we can observe can only continue to expand.

TRANSGENIC REPORTER TECHNOLOGY

In a remarkable example of serendipity, studies of the molecular basis of the bioluminescence of jelly fish have yielded two proteins, aequorin and GFP, that hold great promise in replacing many of the laborious and technically difficult fluorescent dye techniques with the straightforward approaches of molecular biology and plant transformation.

Aequorin

Although not strictly a fluorescence technique, no review of current approaches to visualizing cellular processes in living, functioning plants would be complete without including aequorin-based imaging of cellular Ca^{2+} dynamics. Many jellyfish show green luminescence upon mechanical stimulation. This light comes from a Ca^{2+}-activated photoprotein called aequorin. The aequorin apoprotein combines with a cofactor, coelentrazine, to yield a protein that upon binding Ca^{2+} emits a photon of blue light (475 nm). Aequorin has been used as an indicator of cytoplasmic $[Ca^{2+}]$, being biochemically isolated and then loaded into cells by a variety of techniques ranging from mechanical, chemical, and electrical permeabilization of the plasma membrane to microinjection (e.g. 25, 58, 142; reviewed in 27). The luminescence of the protein is then monitored and converted to changes in $[Ca^{2+}]$. Apoaequorin has been cloned, and the coding sequence is available for transformation of organisms such as *Escherichia coli* (85, 139), Arabidopsis (78), tobacco (78, 86), and a variety of animal cells (e.g. 8, 16, 17, 24, 36, 80). Amazingly, when the transformed cells or organisms are incubated with coelentrazine, active aequorin forms and $[Ca^{2+}]$ can be measured. Thus aequorin technology allows the noninvasive visualization of signaling events in living, functioning tissues, organs, and even whole plants.

Transgenic aequorin was first expressed in the cytosol (85, 86, 139), but recently forms targeted to the nucleus (8, 16, 17), mitochondria (121), chloroplast (78), endoplasmic reticulum (24, 80, 81), tonoplast (83) and plasma membrane (36) have been constructed. Aequorins of altered affinities for Ca^{2+}, and ratioable aequorins are also available (82, 88, 118). Signal levels from plants expressing aequorin are generally too low to allow cellular imaging; however, the use of these organelle-targeted aequorins effectively provides subcellular resolution for the technique.

Changes in the aequorin signal from transgenic plants have been detected by whole plant luminometry (82, 86–89) and highly sensitive cooled CCD cameras (26, 88) in response to myriad stimuli such as touch, oxidative stress, and even circadian rhythms (reviewed in 82). Figure 6 shows luminescence imaging from a tobacco leaf expressing aequorin. At zero time the leaf was subjected to cold stress, and a transient wave of elevated $[Ca^{2+}]$ was seen to follow this stimulus. The aequorin technique continues to reveal the complexity of the whole organ and whole plant Ca^{2+}-signaling network that may well coordinate plant responses to diverse environmental stimuli. At present, the aequorin technique is limited to monitoring $[Ca^{2+}]$. The next challenge is to expand this elegant technology to monitor other ions and compounds.

Green Fluorescent Protein

Green fluorescent proteins are a unique class of proteins also involved in the bioluminescence of many jellyfish. The GFP from *Aequoria victoria* is a 27-kDa protein (238 amino acid) that fluoresces green upon excitation with blue light (29, 115). In the jellyfish, energy for this fluorescence is supplied by both radiative and nonradiative energy transfer from the aequorin photoprotein. GFP remains the only example of a protein where the fluorochrome is

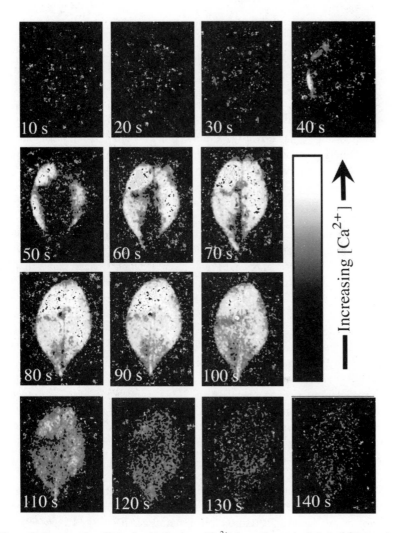

Figure 6 Imaging the effect of cold shock on [Ca^{2+}] in a tobacco leaf. A leaf from a plant transformed with the aequorin gene was removed and luminescence monitored at the indicated times after cold shock, using a cooled CCD camera. Scale bar 10 mm. (MR Knight, unpublished data.)

directly encoded in the amino acid sequence (115). GFP undergoes a cyclization of amino acids 65–67 within its primary structure to produce a cyclic serine-dehydrotyrosine-glycine derivative that forms the chromophore (32, 115). Native GFP from *A. victoria* absorbs optimally at 395 nm (with a weaker absorbance at 470 nm) and emits at 509 nm. The intrinsic fluorescent properties of this protein were revealed when the coding sequence for GFP was isolated and expressed in *E. coli,* yielding a fluorescent product identical to the native Aequoria protein (29, 68, 76).

The first report of the use of transgenic GFP was as a reporter for gene expression in *Caenorhabditis elegans* (29). The nematode was stably transformed with a GFP construct expression under control of the *mec-7* β-tubulin promoter. The GFP selectively accumulated in the cytoplasm of the nematodes touch receptor neurons and was readily apparent using fluorescence micro scopy. GFPs have now been targeted to organelles such as mitochondria (120), chloroplasts (31), nuclei (31, 119), and the secretory apparatus (79), where they become fluorescent and have the potential to allow direct visualization of organelle dynamics in situ.

In plants, GFP has been successfully used as a reporter for gene expression in monocots and dicots in both transient and stable expression systems (9, 31, 64, 69, 73, 99, 106, 107, 126). GFP was induced when tobacco plants that had been stably transformed with an Arabidopsis drought-inducible promoter (RD29A)-GFP or CaB-promoter GFP constructs were subjected to water stress or light stimuli respectively (31). Similarly, when maize or Arabidopsis protoplasts were transfected with an Arabidopsis heat shock promoter-GFP construct, temperature shock led to an accumulation of GFP (73, 126).

Using GFP as an in situ fluorescent tag for proteins has also proved a realizable goal. The entire coding region must be fused to the gene of interest because deletion of even a few of the 238 amino acids of GFP renders it nonfluorescent. When GFP was fused to either the N or C terminus of the *Drosophila exuperantia* gene (137), the fusion proteins were functionally active and highly fluorescent. Other successful GFP fusions include GFP-tubulin (128), the microtubule binding protein tau (14), cyclin (112), viral coat proteins (107), and TMV movement protein (69, 99).

Mutants of GFP have been constructed with shifted emission and excitation spectra (red and blue fluorescent proteins, RFPs, and BFPs), enhanced emission intensities, and rapid fluorophore generation times (38, 47, 66–68). The expression of several genes or the co-localization of several protein fusions could now be simultaneously assayed in a single cell transformed with multiple, spectrally distinct forms of GFP as reporters.

Multiple wavelength forms of GFP are also suited for FRET analysis. The 50% efficiency distance for resonance energy transfer from BFP to GFP (from GFP form Y66H/Y145F to either S65C to S65T) has been calculated as 4.03 nm (68). FRET has been demonstrated between blue and green fluorescent proteins linked by a 25–amino acid poly-histidine linker (68). The FRET was lost when this linker was proteolytically cleaved. However, the FRET did not occur when unlinked GFP and BFP were co-expressed in the restricted volume of the mitochondria (119). Thus, the predicted requirement for very close molecular association of donor and acceptor for FRET to occur is seen in vivo. These results suggest an extremely powerful approach, whereby gene fusions of BFP or GFP could be constructed to reveal interactions between the fusion proteins.

The future for GFP research and usage is exciting. *E. coli* (29, 67, 76), *Caenorhabditis* (29), Drosophila (14, 137), *Dictyostelium* (72), mammalian cells (75, 112, 120), zebra fish (4), plants (9, 31, 64, 69, 73, 99, 106, 107, 126), and yeast (6, 128) can express GFP and remain viable. The work of Haselhof et al (64) has shown that in *Arabidopsis thaliana,* lack of expression is due to an aberrant splice site. By removing this splice site and favoring Arabidopsis codon bias (low AU content) a fluorescent GFP can be stably expressed in this plant. The previous successes in citrus cells (106) using transient expression suggests the splicing events may be species specific or not as critical in transient expression experiments. Similarly, in the successful expression of GFP-fusions using RNA viruses, such as potato virus-x (9, 107), nuclear splicing is circumvented as viral replication is confined to the cytosol. The implication is that through engineering the coding sequence or intelligent use of expression systems, GFP will be usable with almost any plant system.

CAGED PROBES

Using the fluorescence and luminescence technologies outlined above, we can now literally see the dynamics of the structural, biochemical, and regulatory activities of the plant cell. However, some recent advances in optical technology have allowed us to begin to manipulate and control the cell in very regulated ways. By combining these approaches, a suite of approaches has been developed that has begun to allow us to visualize and dissect cellular regulation in the ultimate test-tube setting, the cell itself.

One such approach to manipulating the cytoplasm at the single-cell level is through the use of caged compounds. Caged compounds are molecules whose biological activities have been chemically masked by a photolytic "caging" group. Illuminating the caged compound, with UV light, causes the cage to open and the biologically active molecule to be released (2). The timing, amplitude, and localization of illumination, and therefore of caged probe release, are highly controllable.

The use of caged compounds has greatly expanded over the past few years and the development of commercially available caging "kits" (Molecular

Probes, Eugene, OR) will undoubtedly continue to advance this field. Many caged compounds are related to signal transduction research, including caged neurotransmitters, hormones, nucleotides, ions, and peptides. However, there are now caged fluorochromes, structural proteins, and even enzymes (2, 65). The ability to manipulate and control cell chemistry on demand with light is proving a powerful approach to dissect the role of putative regulators of plant cell function. Caged phytohormones have been synthesized (138) and used to determine an intracellular site for an ABA receptor in guard cells (3). The use of caged Ca^{2+} and caged inositol-1,4,5-trisphosphate has revealed the regulatory role of Ca^{2+} in guiding the tip growth of pollen tubes (94) and perhaps in self-incompatability (50), and in the hormonal regulation of stomatal function (3, 11, 57, 98). The power of the approach is shown in how the use of caged Ca^{2+} has provided a critical link in our understanding of the signal transduction network associated with phytochrome action. Phytochrome has been shown to elicit an increase in cytosolic $[Ca^{2+}]$ (125), and microinjection studies have revealed roles for Ca^{2+} and calmodulin in phytochrome signal transduction (12, 13, 101, 105). Phytochrome has also been shown to elicit rapid changes in protein phosphorylation (48, 108), but the relationship between the Ca^{2+} regulatory system and subsequent protein kinase activities remained obscure. The critical experiment that established a link between these events was when caged calcium and caged IP_3 were electroporated into wheat protoplasts. Upon cage photolysis the increase in Ca^{2+} was shown to elicit the same protein phosphorylation events induced by red light (48).

PERSPECTIVES

Fluorescence microscopy can only continue its technical advances whether it be by designing the next generation of confocal microscopes, the development of new fluorescent ion indicating dyes and protein biosensors, or simply through an increase in the accessibility of video microscopes and the raw computing power for image analysis.

Molecular biology will undoubtedly have significant impact upon how the next generation of cellular probes are designed. The initial success in engineering GFP for altered wavelengths of fluorescence and for compatibility in plants hints at the tremendous potential this protein holds for fluorochrome design. For GFP, the pressing challenge is to reduce the size of the core protein required for the fluorophore. If possible, this would reduce current need to attach the entire 238 amino acids of GFP to each target protein to be labeled, and reduce the possibility of perturbing the function of the fusion protein. GFP and aequorin represent a significant step toward visualizing protein dynamics, protein targeting, and gene expression in an almost noninvasive fashion.

However, a significant goal of all those using these new imaging technologies will continue to be the development of biological systems where response parameters can be viewed in a single living, functioning cell. Where such systems have been developed (e.g. the stomatal guard cell, phytochrome-responsive epidermal cells of tomato, barley aleurone cell, or pollen tubes) application of fluorescence imaging techniques have now truly begun to reveal the amazing subtleties and intricacies of organization and regulation that underlie plant cell function.

ACKNOWLEDGMENTS

The author gratefully acknowledges Marc Knight, Carol Wymer, and Clive Lloyd for access to their unpublished data and Deb Fisher and Sian Ritchie for critical reading of the manuscript. This work was supported by grants from the National Science Foundation (NSF-9513991), United States Department of Agriculture (94-37304-0955), and United States Department of Energy (93ER79239).

> Visit the *Annual Reviews home page* at http://www.annurev.org

Literature Cited

1. Adams SR, Bacskai BJ, Taylor SS, Tsien RY. 1991. Fluorescence ratio imaging of cyclic AMP in single cells. *Nature* 349: 694–97

2. Adams SR, Tsien RY. 1993. Controlling cell chemistry with caged compounds. *Annu. Rev. Physiol.* 55:755–84

3. Allan AC, Fricker MD, Ward JL, Beale MH, Trewavas AJ. 1994. Two transduction pathways mediate rapid effects of abscisic acid in *Commelina communis* guard cells. *Plant Cell* 6:1319–28

4. Amsterdam A, Lin S, Hopkins N. 1995. The *Aequorea victoria* green fluorescent protein can be used as a reporter in live zebra fish embryos. *Dev. Biol.* 171:123–29

5. Assmann SM. 1993. Signal transduction in guard cells. *Annu. Rev. Cell Biol.* 9:345–75

6. Atkins D, Izant JG. 1995. Expression and analysis of the green fluorescent protein gene in the fission yeast *Schizosaccharomyces pombe. Curr. Genet.* 28: 585–88

7. Badlet K, White RG, Overall RL, Vesk M. 1994. Ultrastructure specializations of the cell wall sleeve around the plasmodesmata. *Am. J. Bot.* 81:1422–27

8. Badminton MN, Kendall JM, Salanewby G, Campbell AK. 1995. Nucleoplasmin-targeted aequorin provides evidence for a nuclear calcium barrier. *Exp. Cell Res.* 216: 236–43

9. Baulcombe DC, Chapman S, Santa Cruz S. 1995. Jellyfish green fluorescent protein as a reporter for virus infections. *Plant J.* 7: 1045–53

10. Berger F, Brownlee C. 1993. Ratio confocal imaging of free cytoplasmic calcium gradients in polarizing and polarized *Fucus* zygotes. *Zygote* 1:9–15

11. Blatt MR, Thiel G, Trentham DR. 1990. Reversible inactivation of K+ channels of *Vicia* stomatal guard cells following the photolysis of caged inositol 1,4,5-trisphosphate. *Nature* 346:766–69

12. Bowler C, Chua NH. 1994. Emerging themes of plant signal transduction. *Plant Cell* 6:1529–41

13. Bowler C, Yamagata H, Neuhaus G, Chua NH. 1994. Phytochrome signal trans-

duction pathways are regulated by reciprocal control mechanisms. *Genes Dev.* 8:2188– 202

14. Brand A. 1995. GFP in Drosophila. *Trends Genet. Sci.* 11:324–25

15. Brauer D, Otto J, Tu S. 1995. Selective accumulation of the fluorescent pH indicator BCECF in vacuoles of maize root hair cells. *J. Plant Physiol.* 145:57–61

16. Brini M, Marsault R, Bastianutto C, Pozzan T, Rizzuto R. 1994. Nuclear targeting of aequorin–a new approach for measuring nuclear Ca^{2+} concentration in intact cells. *Cell Calcium* 16:259–68

17. Brini M, Murgia M, Pasti L, Picard D, Pozzan T, Rizzuto R. 1993. Nuclear Ca^{2+} concentration measured with specifically targeted recombinant aequorin. *EMBO J.* 12:4813–19

18. Brownlee C, Pulsford AL. 1988. Visualization of the cytoplasmic Ca2+ gradient in *Fucus serratus* rhizoids: correlation with cell ultrastructure and polarity. *J. Cell Sci.* 91:249–56

19. Bush DS. 1995. Calcium regulation in plant cells and its role in signaling. *Annu. Rev. Plant Physiol. Plant Mol. Biol.* 46:95–122

20. Bush DS. 1996. Effects of gibberellic acid and environmental factors on cytosolic calcium in wheat aleurone cells. *Planta* 199:89–99

21. Bush DS, Biswas AK, Jones RL. 1989. Gibberellic-acid-stimulated Ca^{2+} accumulation in endoplasmic reticulum of barley aleurone: Ca^{2+} transport and steady state levels. *Planta* 178:411–20

22. Bush DS, Jones RL. 1987. Measurement of cytoplasmic calcium in aleurone protoplasts using Indo-1 and Fura-2. *Cell Calcium* 8:455–72

23. Bush DS, Jones RL. 1988. Cytoplasmic calcium and α-amylase secretion from barley aleurone protoplasts. *Eur. J. Cell Biol.* 46:466–69

24. Button D, Eidsath A. 1996. Aequorin targetted to the endoplasmic reticulum reveals heterogeneity in luminal Ca^{2+} concentration and reports agonist- or IP3-induced release of Ca^{2+}. *Mol. Biol. Cell* 7:419–34

25. Callaham DA, Hepler PK. 1991. Measurement of free calcium in plant cells. In *Cellular Calcium, a Practical Approach,* ed. JG McCormack, PH Cobbold, pp. 383–410. Oxford: IRL

26. Campbell AK, Trewavas AJ, Knight MR. 1996. Calcium imaging shows differential sensitivity to cooling and communication in luminous transgenic plants. *Cell Calcium* 19:211–18

27. Campbell PH. 1983. *Intracellular Calcium: Its Universal Role as a Regulator.* Chichester: Wiley

28. Carrington WA, Lynch RM, Moore EDW, Isenberg G, Fogarty KE, Fredrick FS. 1995. Superresolution three-dimensional images of fluorescence in cells with minimal light exposure. *Science* 268:1483–87

29. Chalfie M, Tu Y, Euskirchen G, Ward WW, Prasher DC. 1994. Green fluorescent protein as a marker for gene expression. *Science* 263:802–5

30. Child D, Lischka FW. 1994. Amiloride-insensitive cation conductance in *Xenopus laevis* olfactory neurons: a combined patch clamp and calcium imaging analysis. *Biophys. J.* 66:299–304

31. Chiu W-L, Niwa Y, Zeng W, Hirano T, Kobayashi H, Sheen J. 1996. Engineered GFP as a vital reporter in plants. *Curr. Biol.* 6:325–30

32. Cody CW, Prasher DC, Westler WW, Prendergast FG, Ward WW. 1993. Chemical structure of the hexapeptide chromophore of the aequroea green-fluorescent protein. *Biochemistry* 32:1212–18

33. Collings DA, Wastaneys GO, Miyazaki M, Williamson RE. 1994. Elongation factor-1 alpha is a component of the subcortical actin bundles of characean algae. *Cell Biol. Int.* 18:1019–24

34. Condeelis J. 1995. Elongation factor-1 alpha, translation and the cytoskeleton. *Trends Biochem. Sci.* 20:169–70

35. Cyr RJ. 1994. Microtubules in plant morphogenesis–role of the cortical array. *Annu. Rev. Cell Biol.* 10:153–80

36. Daguzan C, Nicolas MT, Mazars C, Leclerc C, Moreau M. 1995. Expression of membrane targeted aequorin in *Xenopus laevis* oocytes. *Int. J. Dev. Biol.* 39:653–57

37. Davies TGE, Steele SH, Walker DJ, Leigh R. 1996. An analysis of vacuole development in oat aleurone protoplasts. *Planta* 198:356–64

38. Delagrave S, Hawtin RE, Silva CM, Yang MM, Youvan DC. 1995. Red-shifted excitation mutants of the green fluorescent protein. *Bio-Technology* 13:151–54

39. Denk W. 1994. Two-photon scanning photochemical microscopy: mapping ligand gated ion channel distributions. *Proc. Natl. Acad. Sci. USA* 91:6629–33

40. Denk W, Delaney KR, Gelperin A, Kleinfeld D, Strowbridge B, et al. 1994. Anatomical and functional imaging of neurons using 2-photon laser scanning microscopy. *J. Neurosci. Methods* 54:151–62

41. Denk W, Strickler JH, Webb WW. 1990.

Two-photon laser scanning fluorescence microscopy. *Science* 248:73–76

42. Dodt HU. 1993. Infrared-interference video microscopy of living brain slices. *Adv. Exp. Med. Biol.* 333:245–49

43. Durso NA, Cyr RJ. 1994. A calmodulin-sensitive interaction between microtubules and a higher plant homolog of elongation factor-1 alpha. *Plant Cell* 6:893–905

44. Durso NA, Cyr RJ. 1994. Beyond translation: elongation factor 1-alpha and the cytoskeleton. *Protoplasma* 180:99–105

45. Durso NA, Leslie JD, Cyr RJ. 1996. In situ immunocytochemical evidence that a homolog of protein translation elongation factor EF1-alpha is associated with microtubules in carrot cells. *Protoplasma* 190:141–50

46. Ehrhardt DW, Wais R, Long SR. 1996. Calcium spiking in plant root hairs responding to Rhizobium nodulation signals. *Cell* 85:673–81

47. Ehrig T, Okane DJ, Prendergast FG. 1995. Green fluorescent protein mutants with altered fluorescence spectra. *FEBS Lett.* 367:163–66

48. Fallon KM, Shacklock PS, Trewavas AJ. 1993. Detection in vivo of very rapid red light-induced calcium-sensitive protein phosphorylation in etiolated wheat (*Triticum aestivum*) leaf protoplasts. *Plant Physiol.* 101:1039–45

49. Farkas DL, Baxter G, DeBiasio RL, Gough A, Nederlof MA, et al. 1993. Multimode microscopy and the dynamics of molecules, cells, and tissues. *Annu. Rev. Physiol.* 55:785–817

50. Franklin-Tong VE, Ride JP, Read ND, Trewavas AJ, Franklin HH. 1993. The self-incompatibility response in *Papaver rhoeas* is mediated by cytosolic free calcium. *Plant J.* 4:163–77

51. Fricker MD, Blatt MR, White NS. 1993. Confocal ratio imaging of pH in plant cells. In *Biotechnology Applications of Microinjection, Microscopic Imaging and Fluorescence*, ed. PH Bach, CH Reynolds, JM Clark, PL Poole, J Mottley, pp. 153–63. New York: Plenum

52. Fricker MD, Tester M, Gilroy SG. 1993. Fluorescent and luminescent techniques to probe ion activities in living plant cells. See Ref. 95a, pp. 361–78

53. Fricker MD, Tlalka M, Ermantraut J, Obermeyer G, Dewey M, et al. 1994. Confocal fluorescence ratio imaging of ion activities in plant cells. *Scanning Microsc. Suppl.* 8:391–405

54. Fricker MD, White NS. 1992. Wavelength considerations in confocal microscopy of botanical specimens. *J. Microsc.* 166:29–42

55. Gilroy S, Fricker MD, Read ND, Trewavas AJ. 1991. Role of calcium in signal transduction of Commelina guard cells. *Plant Cell* 3:333–44

56. Gilroy S, Jones RL. 1992. Gibberellic acid and abscisic acid coordinately regulate cytoplasmic calcium and secretory activity in barley aleurone protoplasts. *Proc. Natl. Acad. Sci. USA* 89:3591–95

57. Gilroy S, Read ND, Trewavas AJ. 1990. Elevation of cytosolic calcium by caged Ca^{2+} and IP_3 initiates stomatal closure. *Nature* 346:769–71

58. Gilroy S, Hughes WA, Trewavas AJ. 1989. A comparison between Quin-2 and aequorin as indicators of cytoplasmic calcium levels in higher plant protoplasts. *Plant Physiol.* 90:482–91

59. Giuliano KA, Post PL, Hahn KM, Taylor DL. 1995. Fluorescent protein biosensors: measurement of molecular dynamics in living cells. *Annu. Rev. Biophys. Biomed. Struct.* 24:405–34

60. Giuliano KA, Taylor DL. 1995. Measurement and manipulation of cytoskeletal dynamics in living cells. *Curr. Opin. Cell Biol.* 7:4–12

61. Gough AH, Taylor DL. 1993. Fluorescence anisotropy maps calmodulin binding during cellular contraction and locomotion. *J. Cell Biol.* 121:1095–107

62. Hahn K, Kolega J, Montibeller J, DeBiasio R, Post P, et al. 1993. See Ref. 95a, pp. 349–59

63. Hahn KM, DeBiasio R, Taylor DL. 1992. Patterns of elevated free calcium and calmodulin activation in living cells. *Nature* 359:736–38

64. Haseloff J, Amos B. 1995. GFP in plants. *Trends Genet. Sci.* 11:28–329

65. Haugland RP, ed. 1992. *Handbook of Fluorescent Probes and Research Chemicals*. Eugene, OR: Molecular Probes

66. Heim R, Cubitt AB, Tsien RY. 1995. Improved green fluorescent protein. *Nature* 373:663–64

67. Heim R, Prasher DC, Tsien RY. 1994. Wavelength mutations and postranslational autoxidation of green fluorescent protein. *Proc. Natl. Acad. Sci. USA* 91:12501–4

68. Heim R, Tsien RY. 1996. Engineering green fluorescent protein for improved brightness, longer wavelengths and fluorescence resonance energy transfer. *Curr. Biol.* 6:178–82

69. Heinlein M, Epel BL, Padgett HS, Beachy RN. 1995. Interaction of tobacco mosaic

virus movement proteins with the plant cytoskeleton. *Science* 270:1983–85
70. Herman B. 1989. Resonance energy transfer microscopy. *Methods Cell Biol.* 30: 219–43
71. Herrmann A, Felle HH. 1995. Tip growth in root hair cells of *Sinapis alba* L.: significance of internal and external Ca^{2+} and pH. *New Phytol.* 129:523–33
72. Hodkinson S. 1995. GFP in *Dictyostelium*. *Trends Genet. Sci.* 11:327–28
73. Hu W, Cheng C-L. 1995. Expression of aequoria green fluorescent protein in plant cells. *FEBS Lett.* 369:331–34
74. Hush J, Wu LP, John PLC, Hepler LH, Hepler PK. 1996. Plant mitosis promoting factor disassembles the microtubule preprophase band and accelerates prophase progression in Tradescantia. *Cell Biol. Int.* 20:275–87
75. Ikawa M, Kominami K, Yoshimura Y, Tanaka K, Nishimune Y, Okabe M. 1995. Green fluorescent protein as a marker in transgenic mice. *Dev. Growth Differ.* 37: 455–59
76. Inouye S, Tsuji FI. 1994. Green fluorescent protein: expression of the gene and fluorescent characterization of the recombinant protein. *FEBS Lett.* 341:277–80
77. Irving HR, Gehring CA, Parish RW. 1992. Changes in cytosolic pH and calcium precede stomatal movements. *Proc. Natl. Acad. Sci. USA* 89:1790–94
78. Johnson CH, Knight MR, Kondo T, Masson P, Sedbrook J, et al. 1995. Circadian oscillations of cytosolic and chloroplast calcium in plants. *Science* 269:1863–65
79. Kaether C, Gerdes H. 1995. Visualization of protein transport along the secretory pathway using green fluorescent protein. *FEBS Lett.* 369:267–71
80. Kendall JM, Badminton MN, Dormer RL, Campbell AK. 1994. Changes in free calcium in the endoplasmic reticulum of living cells detected using aequorin. *Anal. Biochem.* 221:173–81
81. Kendall JM, Dormer RL, Campbell AK. 1992. Targeting aequorin to the endoplasmic reticulum of living cells. *Biochem. Biophys. Res. Commun.* 189:1008–16
82. Knight H, Knight MR. 1995. Recombinant aequorin methods for intracellular calcium measurement in plants. *Methods Cell Biol.* 49:201–16
83. Knight H, Trewavas AJ, Knight MR. 1996. Cold calcium signaling in Arabidopsis involves two cellular pools and a change in calcium signature after acclimation. *Plant Cell* 8:489–503
84. Knight H, Trewavas AJ, Read ND. 1993.

Confocal microscopy of living fungal hyphae microinjected with Ca^{2+}-sensitive fluorescent dyes. *Mycol. Res.* 97:1505–15
85. Knight MR, Campbell AK, Smith SM, Trewavas AJ. 1991. Recombinant aequorin as a probe for cytosolic free Ca^{2+} in *Escherichia coli. FEBS Lett.* 282:405–8
86. Knight MR, Campbell AK, Smith SM, Trewavas AJ. 1991. Transgenic plant aequorin reports the effects of touch and cold-shock and elicitors on cytoplasmic calcium. *Nature* 352:524–26
87. Knight MR, Knight H, Watkins NJ. 1995. Calcium and generation of plant form. *Philos. Trans. R. Soc. London Ser. B* 350: 83–86
88. Knight MR, Read ND, Campbell AK, Trewavas AJ. 1993. Imaging calcium dynamics in living plants using semi-synthetic recombinant aequorins. *J. Cell Biol.* 121: 83–90
89. Knight MR, Smith SM, Trewavas AJ. 1992. Wind-induced plant motion immediately increases cytosolic calcium. *Proc. Natl. Acad. Sci. USA* 89:4967–71
90. Lloyd CW, Shaw PJ, Warn RM, Yuan M. 1996. Gibberellic acid-induced reorientation of cortical microtubules in living plant cells. *J. Microsc.* 181:140–44
91. Lloyd CW, Venverloo CJ, Goodbody KC, Shaw PJ. 1992. Confocal laser microscopy and 3-dimensional reconstruction of nucleus-associated microtubules in the division plane of vacuolated plant cells. *J. Microsc.* 166:99–109
92. Lucas WJ, Bouche Pillon S, Jackson DP, Nguyen L, Baker L, et al. 1995. Selective trafficking of knotted-1 homeodomain protein and its mRNA through plasmodesmata. *Science* 270:1980–83
93. Lucas WJ, Ding B, Van der Schoot C. 1993. Plasmodesmata and the supracellular nature of plants. *New Phytol.* 125:435–72
94. Malho R, Read ND, Pais MS, Trewavas AJ. 1994. Role of cytosolic calcium in the reorientation of pollen tube growth. *Plant J.* 5:331–41
95. Malhó R, Read ND, Trewavas AJ, Pais MS. 1995. Calcium channel activity during pollen tube growth and reorientation. *Plant Cell* 7:1173–84
95a. Mason WT, ed. 1993. *Fluorescent and Luminescent Probes for Biological Activity: A Practical Guide to Technology for Quantitative Real-Time Analysis.* London: Academic
96. McAinsh MR, Brownlee C, Hetherington AM. 1990. Abscisic acid-induced elevation of guard cell Ca^{2+} precedes stomatal closure. *Nature* 343:186–88

97. McAinsh MR, Brownlee C, Hetherington AM. 1992. Visualizing changes in cytosolic free Ca^{2+} during the response of guard cells to ABA. *Plant Cell* 4:1113–22

98. McAinsh MR, Webb AAR, Taylor JE, Hetherington AM. 1995. Stimulus-induced oscillations in guard cell cytosolic free calcium. *Plant Cell* 7:1207–19

99. McLean BG, Zupan J, Zambryski PC. 1995. Tobacco mosaic virus movement protein associates with the cytoskeleton in tobacco cells. *Plant Cell* 7:2101–14

100. Merrick WC. 1992. Mechanism and regulation of eukaryotic protein synthesis. *Microbiol. Rev.* 56:291–315

101. Millar AJ, McGrath RB, Chua N-H. 1994. Phytochrome phototransduction pathways. *Annu. Rev. Genet.* 28:325–49

102. Miller DD, Callaham DA, Gross DJ, Hepler PK. 1992. Free Ca^{2+} gradient in growing pollen tubes of *Lilium. J. Cell Sci.* 101: 7–12

103. Murray RK, Fleischmann BK, Kotlikoff MI. 1993. Receptor-activated Ca influx in human airway smooth muscle: use of Ca imaging and perforated patch-clamp techniques. *Am. J. Physiol.* 264:485–90

104. Nagata T, Kumagai F, Hasezawa S. 1994. The origin and organization of cortical microtubules during transition between M and G1 phases of the cell cycle as observed in highly synchronized cells of tobacco BY-2. *Planta* 193:567–72

105. Neuhaus G, Bowler C, Kern R, Chua N-H. 1993. Calcium/calmodulin-dependent and -independent phytochrome signal pathways. *Cell* 73:937–52

106. Niedz RP, Sussman MR, Satterlee JS. 1995. Green fluorescent protein: an in vivo reporter of plant gene expression. *Plant Cell Rep.* 14:403–6

107. Oparka KJ, Roberts AG, Prior DAM, Chapman S, Baulcombe D, Santa Cruz S. 1995. Imaging the green fluorescent protein in plants: viruses carry the torch. *Protoplasma* 189:133–41

108. Park M-H, Chae Q. 1989. Intracellular protein phosphorylation in oat (*Avena sativa*) protoplasts by phytochrome action. *Biochem. Biophys. Res. Commun.* 162:9–14

109. Pawley JB. 1995. *Handbook of Confocal Microscopy.* New York: Plenum. 2nd ed.

110. Pierson ES, Miller DD, Callaham DA, Shipley AM, Rivers BA, et al. 1994. Pollen tube growth is coupled to the extracellular calcium ion flux and the intracellular calcium gradient: effect of BAPTA-type buffers and hypertonic media. *Plant Cell* 6: 1815–28

111. Pierson ES, Miller DD, Callaham DA, van Aken J, Hackett G, Hepler PK. 1996. Tip-localized calcium entry during pollen tube growth. *Dev. Biol.* 174:160–73

112. Pines J. 1995. GFP in mammalian cells. *Trends Genet. Sci.* 11:326–27

113. Post PL, DeBiasio RL, Taylor DL. 1995. A fluorescent protein biosensor of myosin II regulatory light chain phosphorylation reports a gradient of phosphorylated myosin II in migrating cells. *Mol. Biol. Cell* 6: 1755–68

114. Post PL, Trybus KM, Taylor DL. 1994. A genetically engineered, protein-based optical biosensor of myosin II regulatory light chain phosphorylation. *J. Biol. Chem.* 269: 12880–87

115. Prasher DC, Eckenrode VK, Ward WW, Prendergast FG, Cormier MJ. 1992. Primary structure of the *Aequoria victoria* green-fluorescent protein. *Gene* 111: 229–33

116. Rathore KS, Cork RJ, Robinson KR. 1991. A cytoplasmic gradient of Ca^{2+} is correlated with growth of lily pollen tubes. *Dev. Biol.* 148:612–19

117. Read ND, Allan WTG, Knight H, Knight MR, Malho R, et al. 1992. Imaging and measurement of cytosolic free calcium in plant and fungal cells. *J. Microsc.* 166: 57–86

118. Read ND, Shacklock PS, Knight MR, Trewavas AJ. 1993. Imaging calcium dynamics in living plant cells and tissues. *Cell Biol. Int.* 17:111–25

119. Rizzuto R, Brini M, DeGiorgi F, Rossi R, Heim R, et al. 1996. Double labeling of subcellular structures with organelle-targeted GFP mutants in vivo. *Curr. Biol.* 6: 183–88

120. Rizzuto R, Brini M, Pizzo P, Murgia M, Pozzan T. 1995. Chimeric green fluorescent protein: a new tool for visualizing subcellular organelles in living cells. *Curr. Biol.* 5:635–42

121. Rizzuto R, Simpson AWM, Brini M, Pozzan T. 1993. Rapid changes of mitochondrial Ca^{2+} revealed by specifically targeted recombinant aequorin. *Nature* 358: 325–27

122. Scalettar BA, Swedlow JR, Sedat JW, Agard DA. 1996. Dispersion, aberration and deconvolution in multiwavelength fluorescence images. *J. Microsc.* 182: 50–60

123. Schild D. 1996. Laser scanning microscopy and calcium imaging. *Cell Calcium* 19: 281–96

124. Schroeder JI, Hagiwara S. 1990. Repetitive increases in cytosolic Ca^{2+} of guard cells by abscisic acid activation of nonselective

Ca²⁺ permeable channels. *Proc. Natl. Acad. Sci. USA* 87:9305–9

125. Shacklock PS, Read ND, Trewavas AJ. 1992. Cytosolic free calcium mediates red light-induced photomorphogenesis. *Nature* 358:753–55

126. Sheen J, Hwang S, Niwa Y, Kobayashi H, Galbraith DW. 1995. Green-fluorescent protein as a new vital marker in plant cells. *Plant J.* 8:777–84

127. Staxen I, Montgomery LT, Hetherington AM, McAinsh MR. 1996. Do oscillations in cytoplasmic calcium encode the ABA signal in stomatal guard cells. *Plant Physiol.* 111:700 (Abstr.)

128. Stearns T. 1995. The green revolution. *Curr. Biol.* 5:262–64

129. Tanasugarn L, McNeil P, Reynolds GT, Taylor DL. 1984. Microspectrofluorimetry by digital image processing: measurement of cytoplasmic pH. *J. Cell Biol.* 98:717–24

130. Taylor DL, Wang Y-L. 1980. Molecular cytochemistry: incorporation of fluorescently molecules as probes of the structure and function of living cells. *Nature* 284: 405–10

131. Taylor DP, Slattery J, Leopold AC. 1996. Apoplastic pH in corn root gravitropism: a laser scanning confocal microscopy measurement. *Physiol. Plant* 97:35–38

132. Tsien RW, Tsien RY. 1990. Calcium channels, stores and oscillations. *Annu. Rev. Cell Biol.* 6:715–60

133. Tsien RY, Bacskai BJ, Adams SR. 1993. FRET for studying intracellular signaling. *Trends Cell Biol.* 3:242–45

134. Tsien RY, Poenie M. 1986. Fluorescence ratio imaging: a new window into intracellular ion signaling. *Trends Biol. Sci.* 11: 450–55

135. Vaquero C, Turner AP, Demangeat G, Sanz A, Serra MT, et al. 1994. The 3a protein from cucumber mosaic virus increases the gating capacity of plasmodesmata in transgenic tobacco plants. *J. Gen. Virol.* 75: 3193–97

136. Volbrecht E, Veit B, Sinha N, Hake S. 1991. The developmental gene Knotted-1 is a member of a maize homeobox gene family. *Nature* 350:241–43

137. Wang SX, Hazelrigg T. 1994. Implications for BCD mRNA localization from spatial distribution of EXU protein in Drosophila oogenesis. *Nature* 369:400–3

138. Ward JL, Beale MH. 1995. Caged plant hormones. *Phytochemistry* 38:811–16

139. Watkins NJ, Knight MR, Trewavas AJ, Campbell AK. 1995. Free calcium transients in chemotactic and nonchemotactic strains of *Escherichia coli* determined using recombinant aequorin. *Biochem. J.* 306:865–69

140. Webb AAR, McAinsh MR, Mansfield TA, Hetherington AM. 1996. Carbon dioxide induces increases in guard cell cytosolic free calcium. *Plant J.* 9:297–304

141. White NS, Errington RJ, Fricker MD, Wood JL. 1996. Aberration control in quantitative imaging of botanical specimens by multidimensional fluorescence microscopy. *J. Microsc.* 181:99–116

142. Williamson R, Ashley CC. 1982. Free Ca²⁺ and cytoplasmic streaming in the alga Chara. *Nature* 296:647–51

143. Wymer C, Lloyd C. 1996. Dynamic microtubules: implications for cell wall patterns. *Trends Plant Sci.* 7:222–28

143a. Yang WN, Burkhart W, Cavallius J, Merrick WC, Boss WF. 1993. Purification and characterization of a phosphatidylinositol 4-kinase activator in carrot cells. *J. Biol. Chem.* 268:392–98

144. Yuan M, Warn RM, Shaw PJ, Lloyd CW. 1992. Dynamic microtubules under the radial and outer tangential walls of microinjected pea epidermal cells observed by computer reconstruction. *Plant J.* 7:17–23

145. Zhang DH, Wadsworth P, Hepler PK. 1990. Microtubule dynamics in living dividing plant cells: confocal imaging of microinjected fluorescent brain tubulin. *Proc. Natl. Acad. Sci. USA* 87:8820–24

Annu. Rev. Plant Physiol. Plant Mol. Biol. 1997. 48:191–222

PHLOEM UNLOADING: Sieve Element Unloading and Post-Sieve Element Transport

J. W. Patrick

Department of Biological Sciences, The University of Newcastle, New South Wales 2308, Australia

KEY WORDS: apoplasm, cellular pathway, symplasm, sucrose, transport mechanisms and controls

ABSTRACT

The transport events from the sieve elements to the sites of utilization within the recipient sink cells contribute to phloem unloading. The phenomenon links sink metabolism and/or compartmentation with phloem transport to, and partitioning between, sinks. The nature of the linkage depends upon the cellular pathway and mechanism of unloading. The common unloading pathway is symplasmic, with an apoplasmic step at or beyond the sieve element boundary reserved for specialized situations. Plasmodesmal conductivity exerts the primary control over symplasmic transport that occurs by diffusion with bulk flow anticipated to be of increasing significance as import rate rises. In the case of an apoplasmic step, efflux across the plasma membranes of the vascular cells occurs by simple diffusion, whereas efflux from nonvascular cells of developing seeds is facilitated and, in some cases, energy coupled. Accumulation of sugars from the sink apoplasm universally occurs by a plasma membrane–bound sugar/proton symport mechanism.

CONTENTS

191

INTRODUCTION

With the realization three decades ago that the axial phloem has spare conductive capacity (88), the research focus on phloem transport shifted from translocation to examine the perceived rate-limiting processes of phloem loading and unloading. Phloem loading has attracted considerable attention (152), whereas investigation of phloem unloading has been less intense. A significant impediment to the study of phloem unloading is that the importing vascular bundles are buried deep within sink tissues with which they form intimate anatomical associations. This presents formidable technical problems for experimental access to, and unambiguous investigation of, the transport events responsible for phloem unloading. These transport events are sieve element unloading (112) arranged in series with post-sieve element transport to the sites of metabolism/compartmentation within the recipient sink cells (117). Predictably, such challenges have been met with the development of novel experimental models accompanied by an array of ingenious and elegant experimental approaches. These have resulted in progress toward reaching a mechanistic understanding of phloem unloading (for recent reviews, see 30, 102, 112, 118). However, as illustrated in this review, understanding is far from complete and many aspects of the phenomenon await quantitative resolution before definitive conclusions can be drawn.

Space constraints dictate that consideration is confined to phloem unloading of sucrose, leaving aside other phloem-sap nutrients such as organic nitrogen (146, 173) and minerals (121, 173). However, in broad terms, the principles and concepts of unloading apply equally to all phloem-mobile solutes. Also not considered are the phloem unloading strategies associated with biotrophic relationships (9, 141) and vascular parasites (19, 63).

SIGNIFICANCE OF PHLOEM UNLOADING

Upon export from the lamina, photoassimilates are unloaded along the entire length of the axial phloem path (147). Moreover, most solutes and water imported by terminal growth and storage sink tissues are delivered by translo-

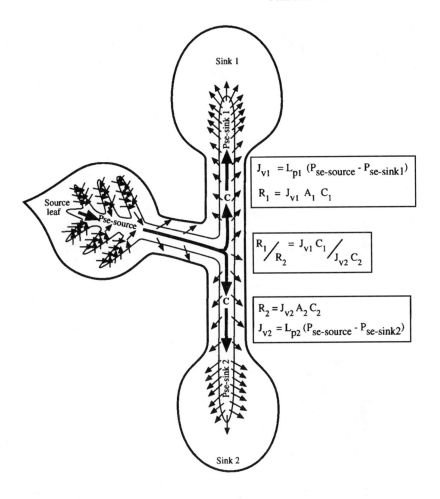

Figure 1   Role of phloem unloading in determining photoassimilate translocation rates to, and partitioning between, competing sinks. The unloading flux from the axial phloem influences the concentration (C) of photoassimilate in the phloem sap and hence, the rate (R) of translocation through the phloem path cross-sectional area (A). At the terminal sinks (Sinks 1 and 2), the unloading flux from the sieve elements sets their turgor pressure ($P_{\text{se-sink}}$), thus influencing the pressure difference ($P_{\text{se-source}} - P_{\text{se-sink}}$) driving the volume flux (J_v) of phloem sap through the phloem path of a given hydraulic conductivity (L_p). Since L_p and A are not limiting translocation (88), then the ratio (partitioning) of the translocation rates to Sinks 1 and 2 is given by $R_1/R_2 = J_{v1}C_1/J_{v2}C_2$. Both J_v and C are influenced by the unloading flux (see above). Arrows indicate direction and flux density of photoassimilate transport in phloem loading of the minor veins in the source leaf (152), translocation through and unloading from the axial path.

cation through the axial phloem arranged in series with phloem unloading (88, 147). Thus, phloem unloading (see Figure 1) is positioned to play a pivotal role in regulating solute translocation to, and partitioning between, sinks (165). Therefore, phloem unloading could serve as a key determinant of crop yield and plant productivity (165).

The above claims can be evaluated using the generally accepted Munch pressure flow hypothesis for phloem translocation (147). The rate of bulk flow is the product of the volume flux (J_v), path cross-sectional area, and the concentration of the transported solute. The latter is influenced by the rate of unloading of each phloem-mobile solute from the axial path (147; see Figure 1). Impacts of phloem unloading in terminal sinks are restricted to effects on J_v. This flux is determined by the difference between the hydrostatic pressure (P) of the sieve elements (se) located at the source and sink, modulated by the hydraulic conductivity (L_p) of the interconnecting axial path (Equation 1).

$$J_v = L_p \left(P_{\text{se-source}} - P_{\text{se-sink}} \right) \qquad 1.$$

Equation 1 shows that the only way phloem unloading in terminal sinks can influence J_v is through altering $P_{\text{se-sink}}$ (see Figure 1). If the water potential of the sieve element sap (ψ_{se}) is in equilibrium with that of the phloem apoplasm (ψ_a), and because $\psi = P - \pi$, it follows that P_{se} will be given by:

$$P_{se} = (P_a - \pi_a) - (-\pi_{se}) \qquad 2.$$

For a range of terminal sink types, P_a is considered to be minimal [e.g. root tips; developing seeds and fruits (118)] and hence P_{se} is determined by:

$$P_{\text{se-sink}} = \pi_{se} - \pi_a \qquad 3.$$

The small volumes of the lumens of the importing sieve elements and of the surrounding sink apoplasm, render π_a and π_{se} sensitive to small changes in the large fluxes of the major osmotic solutes that pass through these pools. Sucrose and potassium plus accompanying anions represent the major osmotic species translocated in the phloem sap (87). Thus, variation in the phloem unloading rates of these solutes will cause shifts in P_{se} (Equation 3) and hence influence J_v of phloem sap translocated to, and partitioned between, sinks (see Equation 1 and Figure 1). The remaining phloem-mobile solutes follow passively in the bulk flow patterns set by the unloading behavior of potassium and sucrose. Unfortunately, information available on the phloem unloading of potassium is too fragmentary to provide a sensible account (see 85, 117, 173, 177, 178). Hence, the remainder of the review focuses on sucrose unloading alone.

The rate of phloem unloading is the product of the unloading flux and the path cross-sectional area through which transport is rate limited. The transport flux across plasma membranes may occur by simple diffusion or be facilitated by membrane porters or channels, whereas diffusion or bulk flow accounts for plasmodesmal transport. The passive transport fluxes are driven by concentration and/or hydrostatic pressure differences sustained by metabolism/compartmentation within the recipient sink cells. Regulation of energy-coupled membrane transport depends upon more sophisticated controls (10). In addition, the conductivity of the post-sieve element pathway modulates the unloading flux (cf Equations 4 and 5). The limiting path cross-sectional area is set by the contiguous cell walls containing the least number of plasmodesmal interconnections or, where unloading includes an apoplasmic step, the plasma membrane surface area supporting solute exchange to or from the sink apoplasm could limit transport. Thus, both physiological and structural properties of the unloading pathway, in concert with sink metabolism/compartmentation, are expected to influence the rate of phloem unloading. This in turn sets the rate of phloem import by the sinks.

CELLULAR PATHWAY OF PHLOEM UNLOADING

Elucidation of the cellular pathway of phloem unloading is central to reaching a mechanistic understanding of the phenomenon as, in large measure, the unloading path determines the key transport events responsible for solute and solvent movement from the sieve element lumens to the recipient sink cells. Definitive conclusions about the nature of the unloading pathways are yet to be drawn. Technical impediments are compounded by the large array of differing sink types. Overall, the unloading path appears to be influenced by sink development and function and hence is a dynamic rather than static property of a particular sink type (112, 118; Figure 2). These issues and concepts are illustrated by reviewing the unloading pathways considered to be operative in key sink types.

Cellular Pathways in Key Sink Types

VEGETATIVE APICES   Photoassimilates are delivered to the subapical region of the root tip through protophloem sieve elements (29). Their plasma membrane surface areas are inadequate to support the observed rates of phloem import of sucrose (32; A Schulz, unpublished data) and water (96). The implied requirement for symplasmic unloading is supported by plasmodesmata linking the sieve elements with the meristematic cells of the root tip (7, 139, 166–168).

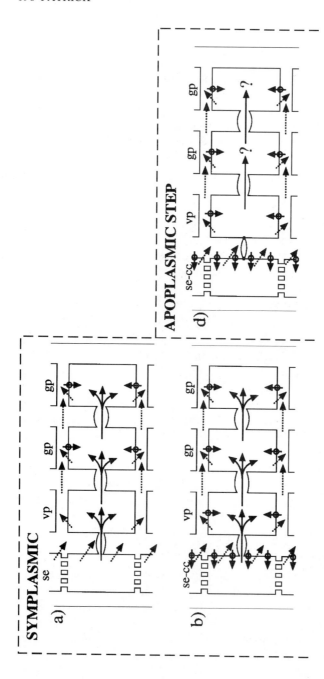

Figure 2 Cellular pathways of phloem unloading through symplasmic routes (*a–c*) and through routes containing an apoplasmic step (*d–f*). Symplasmic unloading from (*a*) the protophloem sieve elements (se) in meristems and (*b*, *c*) from the metaphloem se-cc complexes (se-cc) along the axial path in (*b*) the absence or (*c*) the presence of an apoplasmic barrier (*wavy vertical line*) separating the se-cc complexes from the recipient sink cells. An apoplasmic step (*d–f*) may occur at a number of sites in the post-sieve element pathway. These are: (*d*) from the se-cc complexes along the axial pathway and caused by reversible blockage of the plasmodesmata interconnecting them with the vascular parenchyma (vp); (*e*) from the vascular parenchyma following developmental blockage of the plasmodesmata interconnecting the vascular parenchyma with and within the ground parenchyma (gp); (*continued, next page*)

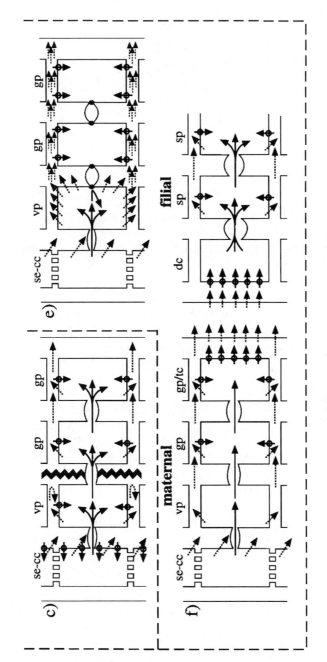

(f) from ground parenchyma cells, specialized for membrane transport and located at the interface between the two genomes of developing seeds or biotrophic relationships. Arrows indicate the direction and flux density of photoassimilate transport through symplasmic routes (*unbroken arrows*), by simple diffusion to and through the apoplasm (*broken arrows*), and by facilitated transport across membranes (*arrows passing through open circles*). Additional abbreviations: tc, transfer cell; dc, dermal cell complex; sp, storage parenchyma.

These form a functional symplasmic route as shown by the movement of the membrane-impermeant dye, 5(6)carboxyfluorescein (CF) imported through the phloem (103, 107). That photoassimilates follow this putative symplasmic route is indicated by such findings as ^{14}C-photoassimilate accumulation is not slowed in the presence of an apoplasmic sucrose trap (23, 32) and is protected from potential extracellular hydrolysis (23, 52). Recently, the capacity of the interconnecting plasmodesmata to support the observed rates of sucrose import has been questioned (7). This opens up the possibility of sym- and apoplasmic paths operating in parallel (7). In the latter case, the absence of sucrose porters from the protophloem (142) would optimize the release of sucrose to the root apoplasm. Here, an extracellular invertase (23, 52) ensures that the leaked sucrose, as well as hexoses (12, 32), are retrieved by a plasma membrane hexose porter (57, 79, 133, 181). However, this apoplasmic component does not appear to be mandatory for sugar movement in root tips. For instance, sugar retrieval only accounts for 20% of the total sugar flux (32), and photoassimilate import is unaltered on blocking membrane transport (32) with the slowly permeable sulfhydryl reagent, parachloromercuribenzenesulfonic acid (PCMBS) (12). Rather, the hexose porters may function as a retrieval mechanism for sugars leaked to the root apoplasm during passage through the symplasmic path (12; cf 81; Figure 2a).

Similar conclusions can be drawn for shoot apices. Here the cells of the apical dome are symplasmically coupled (80), and the vascular supply is linked with the development of leaf primordia (28). For dicotyledenous sink leaves, quantitative structural studies suggest that plasmodesmata provide a uniform conductive pathway from sieve element–companion cell (se-cc) complexes to the mesophyll cells (24). Consistent with symplasmic passage (see Figure 2a) are the findings that imported sucrose is protected from extracellular hydrolysis (51) and that its immediate accumulation is unaffected by PCMBS (136) or anoxia (151). Nevertheless, the low expression of sucrose porters in sink leaves [StSUT (126)], specifically in their differentiating phloem [SUC2 (149)], means that a proportion of the imported sucrose is likely to be leaked to the leaf apoplasm. Thereafter, extracellular hydrolysis (51) would permit retrieval from the apoplasmic sugar pool by the hexose porters expressed in sink leaves (133 and see Figure 2a). In contrast, phloem unloading in developing *Zea mays* leaves may follow an apoplasmic path as the low plasmodesmal frequency forms a significant bottleneck for symplasmic transport (31).

STEM ELONGATION ZONES   Quantitative structural studies indicate that radial transport of photoassimilates from the metaphloem se-cc complexes of elongating stems may follow sym- or apoplasmic routes (177; Figure 2*b* vs 2*d*). Symplasmic transport is considered to be operative in elongating etiolated sunflower hypocotyls (85) and pea epicotyls (135). However, for stems of partially defoliated bean seedlings, sieve element unloading was apoplasmic (89, 90; RM Wood, JW Patrick & CE Offler, unpublished data). These apparent conflicting conclusions may be explained by the fact that both unloading pathways are available, with the operative one depending upon the whole plant source/sink ratio (118; see next section).

MATURE AXIAL PATHWAY   Beyond the root tip, the plasma membrane surface area of the metaphloem se-cc complexes confers the potential for direct exchange to the root apoplasm (A Schulz, unpublished data), while plasmodesmal frequencies indicate symplasmic transport (166–168). The availability of the symplasmic route has been verified by the radial spread of phloem-imported CF into the pith and cortical tissues (118; Figure 2*b*). In contrast with root tips, this symplasmic path is closed off in low source/sink ratio plants at the metaphloem se-cc complex/phloem parenchyma junction (118; but see 103, 107; Figure 2*d*). Modification to a specialist storage function is illustrated by the multicambial tap root of sugar beet that accumulates sucrose to high concentrations (2). A working model of the unloading path in the storage root envisions efflux to the apoplasm from the se-cc complexes, retrieval by the phloem parenchyma, and symplasmic passage to the storage parenchyma cells (40; Figure 2*d*). The proposed model may equally apply to soluble sugar storage in carrot roots (cf 62).

For mature stems, quantitative structural analyses indicate that a potential exists for apo- (58) and symplasmic (58, 134, 157) transport from the metaphloem se-cc complexes to the recipient sink cells (Figures 2*b* and 2*d*). In mature stems of sugarcane, a mandatory symplasmic step is caused by the development of an apoplasmic barrier in the walls of the bundle sheath cells (93; Figure 2*c*). In the absence of such a barrier, the cellular pathway of phloem unloading depends on the prevailing source/sink ratio. This has been detected by the distribution of phloem-imported CF (118) and the differential sensitivity of ^{14}C-photoassimilate accumulation to plasmodesmal rupture or to PCMBS blockage of membrane transport from the stem apoplasm (55, 59, 89, 90, 118). Other observations appear to be consistent with this conclusion. For example, the radial spread of phloem-imported CF indicated that symplasmic unloading paths are operative in stems of soybean plants (53) and mesocotyls of etiolated corn seedlings (27). In contrast, when herbaceous and woody stem segments were held in darkness for several days (i.e. low source/sink ratio), the metaphloem se-cc complexes were symplasmically isolated (153 and references therein). An alternative explanation put forward for these observations is that the interconnecting plasmodesmata are irreversibly blocked once the se-cc complexes are fully differentiated (153 and references therein).

The capacity for rapid and reversible switching between sym- and apoplasmic unloading routes along the axial pathway coincides with the development of neck constrictions in the plasmodesmal canals (80). These ultrastructural changes could confer the capacity for pathway switching by pressure-induced plasmodesmal valving (106). Consistent with this contention is the coincidence of the site of symplasmic blockage (59, 118, 153) with the largest transcellular pressure difference at the se-cc complex/phloem parenchyma boundary (120, 169, 170). Sucrose turnover rates (33) suggest that alterations in the source/sink ratio may exert relatively rapid effects on the osmotic content of the phloem parenchyma. In contrast, the osmotic potential of the phloem sap is maintained by an osmoregulatory mechanism in which sucrose and potassium are interchangeable as the principal osmotica (140). Thus, the pressure difference across the plasmodesmata may be amplified under low source/sink ratio conditions that then could cause their closure (106; Figure 2*b* vs 2*d*).

TERMINAL VEGETATIVE STORAGE SINKS   The potato tuber represents a terminal storage sink that develops from a shoot meristem (108). The symplasmic route of the shoot meristem is retained during tuber development (Figure 2*a*). Thus, plasmodesmata interconnecting the sieve elements to the storage parenchyma cells (101) contribute to post-sieve element transport as indicated by dye coupling (105, 108) and significant inhibition of photoassimilate import following their plasmolytic rupture (104; Figure 2*a*). However, an apoplasmic path may operate in parallel or series as PCMBS slows import (104) and a yeast invertase, expressed within the tuber apoplasm, increases dry matter accumulation (47). The lack of sucrose porter activity in the phloem (179) would enhance exchange to the tuber apoplasm from whence sucrose is retrieved by sucrose porters located in the storage parenchyma cells (109; Figure 2*a*).

TERMINAL REPRODUCTIVE STORAGE SINKS—FLESHY FRUITS   The fruit models most investigated are tomato (61), grape (14), and citrus (48, 66, 67). The early stages of tomato fruit development are characterized by symplasmic unloading routes to the storage parenchyma cells (64, 98; Figure 2*a*). This conclusion is based on the distribution of CF (118, 130) and [^{14}C]glucose (130), protection of phloem-imported sucrose from hydrolysis by an extracellular invertase (17), and the absence of an energy-dependent retrieval system (64, 130). The switch from starch to soluble sugar accumulation, during fruit development (61), is accompanied by changes in the post-sieve element pathway. The symplasmic route to the phloem parenchyma cells is maintained, but the symplasmic route to the storage parenchyma cells is structurally diminished (64, 98). The latter event coincides with a loss in dye coupling at the phloem/storage parenchyma interface (118) and between the storage parenchyma cells (130). The development of this obligatory apoplasmic step (Figure 2e) accords with extracellular hydrolysis of phloem-imported sucrose (17, 18), and the appearance of plasma membrane H^+-ATPase and hexose porter activities (41, 130). The putative hexose/proton symporter accounts for some 70 to 80% of the in vivo hexose accumulation flux by the fruit during the phase of cell expansion (131).

In contrast with tomato fruit, the grape berry and citrus fruit accumulate soluble solutes to high concentrations throughout their development (14, 66). Plasmodesmata interconnect the contiguous cells forming the post-sieve element pathway in both fruit (48, 67; CE Offler, unpublished data). In the case of the grape berry, symplasmic transport could be limited at the phloem/storage parenchyma interface where sufficient plasma membrane surface area is available to support exchange to the fruit apoplasm (CE Offler unpublished data; Figure 2e). Consistent with this conclusion are the high sugar concentrations in the berry apoplasm (8, 70) that are sensitive to changes in phloem import rates (8). An apoplasmic path also may be operative within the stalks of citrus juice vesicles, as ^{14}C-photoassimilate transport is unaffected by disruption of their symplasmic continuity (67; Figure 1e).

TERMINAL REPRODUCTIVE STORAGE SINKS—DEVELOPING SEED   The symplasmic discontinuity between the maternal and filial tissues in seeds (28) necessitates efflux from maternal tissues and subsequent influx by filial tissues. Cereal and grain legume seeds have been the most thoroughly investigated (117, 146). Their unloading pathways share the same common features of extensive symplasmic routes located on either side of apoplasmic exchange localized to the maternal/filial interface (see Figure 2f). During the linear phase of seed fill, import to the filial generation is canalized inward from the maternal vasculature (117, 146). Plasmodesmal cross-sectional areas interconnecting the cell types of the maternal pathway have the potential to support the observed sucrose fluxes (99, 100, 158). Evidence for a functional symplasm has been derived from the movement of CF (119, 159, 162) and ^{14}C-labeled sugars (25, 56, 119). For wheat, lignification and suberization of the pigment strand cell walls (158) prevents inward apoplasmic transport (159, 162), dictating that the nucellar projection transfer cells must be responsible for sucrose efflux to the endosperm cavity (158). The cellular sites of proton-coupled sucrose efflux from coats of *Phaseolus* and *Vicia* seeds (39, 154, 156) have been identified by the histochemical localization of high activities of a plasma-membrane H^+-ATPase to their ground parenchyma and thin-walled parenchyma transfer cells, respectively (164). Similar observations for the filial tissues have led to the conclusion that their cell layers juxtaposed to the sites of efflux from the maternal tissues are responsible for sucrose uptake from the seed apoplasm (6, 83, 84, 160). A symplasmic route for subsequent transfer to the underlying storage cells has been deduced from quantitative structural studies, movement of CF, and the observed slowing of [^{14}C]sucrose transport following plasmolytic rupture of the interconnecting plasmodesmata or by the selective PCMBS-inhibition of uptake by the outermost cell layers (84, 160).

Cellular Pathway—Sink Development and Function Overview

With the caveat that definitive conclusions have yet to be reached for the cellular pathway of phloem unloading in any sink type, a number of circumspect generalizations may be drawn (see Figure 2). First, sieve element unloading invariably includes an apoplasmic component, driven by the large transmembrane sucrose concentration difference (112). The contribution of the apoplasmic route from the sieve elements to the overall unloading flux will depend upon the relative conductances of the apo- and symplasmic paths. Second, symplasmic sieve element unloading and post-sieve element transport is the common principal unloading route (Figures 2a–2c). This undoubtedly reflects the greater transport capacity of the symplasm and its attendant selective advantages (118). An apoplasmic step in the unloading pathway, located at or beyond the sieve element boundary, is reserved for specialized situations (Figures 2d–2f). These include the axial pathway in plants with a lowered source/sink ratio (Figure 2d), sinks that accumulate solutes to high concentrations in the absence of an apoplasmic barrier separating the phloem from the recipient sink cells (Figures 2d and 2e) and unloading pathways that include solute passage between differing genomes such as developing seed (Figure 2f), biotrophic relationships (141), and vascular parasites (19). Third, the apoplasmic step is readily reversible to entrain symplasmic transport (axial pathway) or irreversibly programmed during development of terminal sinks. The above considerations lead to the overall general conclusion that phloem unloading principally occurs symplasmically but, in special situations, includes an apoplasmic step at or beyond the sieve element boundary.

MECHANISMS AND CONTROLS OF PHLOEM UNLOADING

Symplasmic Transport Mechanisms

Symplasmic movement includes intracellular transport (passive transport in meristematic cells possibly supplemented by cytoplasmic streaming in vacuolate cells) arranged in series with intercellular transport via plasmodesmal interconnections. In most circumstances, intracellular transport is unlikely to be rate limiting in meristematic (7) or vacuolate cells (150; but cf 155). The commonly limiting plasmodesmal transport step is considered to occur passively (80). Thus, the symplasmic flux is governed by the transplasmodesmal differences in solute chemical potentials (diffusion) or pressure (bulk flow), modulated by plasmodesmal conductances.

DIFFUSION   For symplasmic transport limited by diffusion through plasmodesmata, transport rate (R_d) can be derived from Fick's first law of diffusion as:

$$R_d = n\,[DA\,(C_{se} - C_s)/l] \qquad\qquad 4.$$

where n is the number of plasmodesmata occupying the path cross-section, D the diffusion coefficient of the diffusing solute within the plasmodesmal canals, A plasmodesma cross-sectional area available for transport, C the solute concentration in the sieve element (se) and sink cytosol (s), and l length of the plasmodesma ultrastructure that most limits transport.

   Although anticipated, unequivocal demonstration of concentration differences between the importing sieve elements and recipient sink cells (Equation 4) are scant because valid estimates of cytoplasmic solute levels are technically difficult to obtain. However, since sugars appear to exchange rapidly and reversibly between the vacuolar and cytoplasmic compartments in a range of sink types (25, 32, 44, 93), measured cellular sugar levels may approximate cytoplasmic concentrations. On these grounds, favorable concentration differences may exist for unloading by diffusion in Zea mays root tips [hexoses (125, but not for sucrose; WK Silk, unpublished data)], in developing wheat seeds [sucrose (44, 45)], and in mature Phaseolus stems [photoassimilates (120)]. For root tips, sucrose is hydrolyzed on reaching the ground tissues (125), and thus subsequent symplasmic diffusion may occur as hexoses. In contrast, diffusion is less likely in sinks that accumulate sucrose to high concentrations (Equation 4). This particularly applies to those situations where transport across the tonoplast into the vacuoles is not energy coupled [e.g. sugarcane stems (93)].

The potential for unloading by diffusion is demonstrated by the slowing of phloem import when the concentration difference (Equation 4) is decreased following exposure of the sink surface to dilute sucrose solutions (34, 36, 118, 138; but cf 161). Model-based assessments of phloem unloading in root tips have concluded that diffusion could be significant [ca 30% (96)] or minimal (7). These disparities reflect differing assumed plasmodesmal conductances (cf 7, 96). Both models require revision now that the sucrose and hexose concentration gradients have been resolved in root tips (125; WK Silk, unpublished data). Other model-based evaluations conclude that diffusion could account for phloem unloading in stems (95) and in developing wheat grains (45). In the case of water exit, the plasma membrane surface area of the sieve elements is sufficient to support exchange of excess water to the sink apoplasm in stems (95), but may not be adequate in seeds (96) or other sinks importing solutes at high rates where pressure differences favor symplasmic unloading by bulk flow (see below).

BULK FLOW   If symplasmic unloading occurs by bulk flow, then the solute transport rate (R_f) is given by the product of the volume flux (J_v), cross-sectional area of the flow path (A) and the concentration (C) of the transported solute. The J_v is set by the product of the hydraulic conductivity (L_p) of the plasmodesmata and the hydrostatic pressure difference between either end of the path summed over the number of plasmodesmata (n). At water equilibrium, the difference between the cell osmotic potential (π) and the water potential of the surrounding apoplasm fluid (ψ_a) determines the cell turgor. Thus, the rate of bulk flow (R_f) is:

$$R_f = nL_p \left[(\pi_{se} - \psi_a) - (\pi_s - \psi_a) \right] AC. \qquad 5.$$

The hydraulic conductivity of capillaries is predicted by Poiseuille's law:

$$L_p = \pi r^4/8, \qquad 6.$$

where r is the plasmodesma radius, η the viscosity of the flowing solution, and L the length of the plasmodesma ultrastructure that rate-limits transport.

For bulk flow to proceed, plasmodesmata must be insensitive to changes in the hydrostatic pressure differences between contiguous cells [e.g. root tips (118)], or if pressure sensitive, the pressure difference must be less than the minimum that induces plasmodesmal closure (106). Furthermore, for roots, the inward-directed transpiration stream restricts unloading by bulk flow to the root tips where outward-directed osmotic differences of 0.7 MPa have been measured between the protophloem sieve elements and the cortical cells (124, 169, 170). Because the water potentials of the phloem and cortical cells are identical (124), the observed osmotic difference corresponds to a hydrostatic pressure difference (see Figure 3a). Centrifugal pressure differences from the phloem of similar magnitude have been observed in elongating hypocotyls of castor bean (86) and deduced in sink leaves (37), mature stems of sugarcane (93), and developing wheat grains (43, 45).

The potential for bulk flow can be tested by manipulating the pressure difference (Equation 5), through exposure of the sink surface to solutions containing nonpermeating osmotica (46, 56, 138, 148, 172, 182). Consistent with unloading by bulk flow, these treatments cause rapid (<10 min) enhancement of photoassimilate import (56, 172), which is linearly dependent upon the external osmotic pressure (138, 173) and is followed by a subsequent decay coincident with turgor recovery (46, 148, 182). The effects can be duplicated using high-molecular-weight polyethylene glycol solutions that sustain the initial pressure difference as their molecular size precludes them from penetrating the sink apoplasm (118, 172; see also 135 but cf 161).

Depending on the assumed value for the plasmodesmal hydraulic conductance, it has been concluded that bulk flow may (96) or may not (7) account for the observed sucrose flux in root tips. However, if the limiting plasmodesmal structure is confined to the plasmodesmal orifice (139) and the canal radius is 2 rather than 1 nm (cf 13), the model of Brete-Harte & Silk (7) predicts pressure gradients of ca 0.15 MPa mm^{-1} (see also Equation 6). Such gradients are in the range estimated for root tips (124). In the case of legume seed coats, bulk flow can account for unloading of photoassimilates if 20% of the plasmodesmata are unoccluded (96). Thus, for those sinks importing photoassimilates at high rates, such as vegetative apices and developing seeds, the above analysis suggests that bulk flow could be responsible for sucrose movement from the sieve elements and possibly through the post-sieve element pathway. As pointed out by Murphy (96), bulk flow may be necessary to dissipate phloem-imported water to prevent a buildup of pressure in the sieve elements.

For growth sinks, the volume of phloem-imported water meets most of the cellular water requirements in root tips (7, 36, 123), elongating stems (135, 178), fruit (60, 71), and developing seeds (111). The relative conductances of the plasma membranes and plasmodesmata favor the latter route (7, 96). Water flow patterns from the phloem in nonexpanding storage sinks are less clear.

Excess phloem-imported water could be dissipated by equilibration with the adjacent xylem transpiration stream along the axial path or by transpiratory loss from the surfaces of terminal sinks such as expanding leaves (144) and certain fruit (4). However, transpiration varies diurnally, and many terminal storage sinks transpire minimal amounts of phloem-imported water. In these

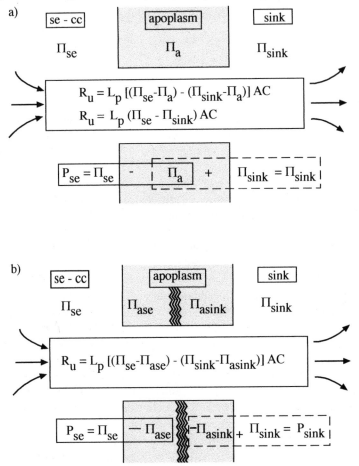

Figure 3 Phloem unloading by bulk flow through a symplasmic route in (*a*) the absence and (*b*) the presence of an apoplasmic barrier (*wavy vertical lines*) separating the se-cc complexes from the recipient sink cells. The rate (*R*) of bulk flow is determined by the pressure difference between the se-cc complexes (P_{se}) and sink cells (P_{sink}), the concentration (*C*) of transported photoassimilate as well as the hydraulic conductivity (L_p) and total cross-sectional area (*A*) of the plasmodesmal canals. The cell turgor pressure is set by the difference in the osmotic pressures of the symplasmic (π_{se}, π_{sink}) and apoplasmic saps (π_a, π_{ase}, π_{asink}).

cases, the water is re-exported in the xylem (69, 72, 111). The cellular site(s) from which phloem-imported water returns to the sink apoplasm is crucial to the question of bulk flow through the sink symplasm. For coats of developing legume seed, Murphy (96) concluded that the observed rates of water exchange to the seed apoplasm require the plasma membrane surface area offered by the ground tissues. Solutes moved in the putative apoplasmic stream of returning water could be retrieved at strategically located apoplasmic barriers to solute but not water movement (e.g. 93, 158) or at xylem discontinuities (42, 73, 97).

Control of Symplasmic Transport

Long-term control of symplasmic phloem unloading could be mediated through developmental shifts in plasmodesmal numbers and in the lengths of their ultrastructure that rate-limit transport [Equations 4 and 5 (80)]. Short-term regulation of symplasmic phloem unloading may be exercised through rapid (s to min) changes in plasmodesmal radii and hence plasmodesmal conductivities [Equations 5 and 6 (13, 139)] or concentration or turgor differences [Equations 4 and 5 (182)].

Regulation of unloading through alteration in the concentration or pressure difference enjoys widespread appeal. It provides a mechanism where metabolic demand of the recipient sink cells directly determines these differences and hence phloem unloading rates (Equations 4 and 5). The linkage is supported by the commonly observed correlation between rates of photoassimilate import and activities of the sucrose hydrolyzing enzymes, invertase, and sucrose synthase (65). More recently, the relationship has been verified in transgenic plants through overexpression of key sink-located enzymes responsible for carbohydrate metabolism (47, 143). The less-than-proportionate response of dry matter import to increases in enzymic activity [e.g. 30% starch increase in transgenic potato tubers overexpressing ADP-glucose pyrophosphorylase by some 10–15-fold (47, 143)] reflects shared control of the sucrose flux at the biochemical and physiological levels (143). At the physiological level, the high phloem sap values greatly attenuate the extent to which compartmentation- or metabolically induced shifts in sink cell concentration/osmolality can influence the concentration or pressure differences (113; and see Equations 4 and 5). This situation applies universally for symplasmic transport by diffusion. However, it only applies for bulk flow in those sinks containing an apoplasmic continuity between the sieve elements and sink cells. Here, the pressure difference is determined by the intracellular osmotic difference (Figure 3a), regulated by the interconversion of imported sugars into nonosmotic compounds within the sink cells. The preceding considerations are consistent with the findings that rates of phloem unloading vary independently of sink-cell sugar levels and turgor pressures in root tips (35, 123, 125; J Pritchard & D Tomos, unpublished data), mature *Phaseolus* stems (118), and developing wheat grains (43). Collectively, these observations suggest that plasmodesmal conductivity exerts a significant regulation over phloem unloading rates (see Equations 4 and 5; Figure 3a).

Optimal plasmodesmal regulation is likely to be mediated at the site of the highest resistance to transport (i.e. the greatest transcellular concentration or osmotic difference). This is located at the se-cc complexes/vascular parenchyma interface of developing wheat grains (43–45), root tips (169, 170), and mature bean stems (120). Recently, Schulz (139) demonstrated that osmotically induced enhancement of unloading in root tips was associated with increased cross-sectional areas of the cytoplasmic annuli in the neck regions of cortical cell plasmodesmata. How the osmotic changes are transmitted to evince such alterations in plasmodesmal ultrastructure is currently unknown. An obvious candidate for a coordinating signal is cell turgor pressure. This is a function of the balance between withdrawal or dilution of osmotically active solutes by sink utilization or growth and their replacement by import from the phloem. Changes in transcellular turgor pressure are known to act directly on plasmodesmal conductivity (106). An indirect action could result from a turgor-induced Ca^{2+} cascade (11) that may lead to an alteration in plasmodesmal conductivity (26). In addition, the model also needs to include feed-forward control by sucrose supply. Such control would include sucrose-responsive gene expression of key enzymes responsible for carbohydrate metabolism and possibly of proteins regulating plasmodesmal conductivity (65).

Potential plasmodesmal control of phloem unloading by bulk flow could be supplemented with effective regulation of the pressure difference (Equation 5) in those sinks where an apoplasmic barrier separates the phloem from the storage tissues [e.g. sugarcane stems (93), suberized endodermis in roots, developing cereal grains (117)]. Here, the apoplasmic barrier permits regulation of the pressure difference by independent changes in π and ψ_a of the sieve elements and sink cells (Equation 5; Figure 3b). The relatively small volumes of the phloem and sink apoplasms mean that incremental shifts in solute partitioning between their apo- and symplasmic compartments have the potential to exert quantitatively profound effects on the phloem/sink pressure difference (113). Currently there is no experimental evidence to support this claim, but there is good evidence that nonphloem cells are capable of turgor regulation (3, 93, 148, 182). This control is of particular significance in sinks, such as sugarcane stems, that accumulate osmotica to high concentrations. Without progressive turgor regulation during the phase of sucrose accumulation (94), the turgor of the storage parenchyma cells would rise. This would cause the pressure difference to decline accordingly and hence dampen the potential for unloading by bulk flow (Equation 5).

The symplasmic concentration of sucrose moved through the post-sieve element pathway (Equation 5) is regulated by vacuolar buffering, metabolism, and retrieval of sugar leaked to the apoplasmic compartment. In the maternal tissues of developing seeds, rapid exchange of sucrose between cytoplasmic and vacuolar compartments (25, 44) serves to buffer supplies to the filial generation (116, 173). Turnover of stored starch may contribute to the maintenance of this sucrose transport pool (128). Hexose porters function to sustain symplasmic transport by retrieval of hexoses leaked to the apoplasmic compartment (cf 81) in growth sinks (93, 133) and developing seeds (JW Patrick, unpublished data).

Apoplasmic Phloem Unloading—Mechanisms and Control

Phloem-imported solutes exchange to the sink apoplasm from the se-cc complexes of the axial path and all cells located along the post-sieve element pathway. However, significant solute exchange to the sink apoplasm occurs at specific cellular sites (see Figure 2). These sites fall into two broad categories. That is, direct exchange from the se-cc complexes of the axial path or exchange at an apoplasmic step following symplasmic transport from the se-cc complexes in terminal sinks (Figure 2). The axial phloem balances unloading from the se-cc complexes into lateral sinks with continued translocation to the terminal sinks that, in contrast, are irreversibly committed to import (153). Thus, the unloading mechanisms and their controls for these two pathway configurations are treated separately.

DIRECT EXCHANGE FROM THE SE-CC COMPLEXES   Although possibly equivocal (see 112), studies with stem segments (1, 175), intact stems (89–91, 112), and isolated vascular strands (54) led to the conclusion that sucrose efflux across the plasma membranes of se-cc complexes to the phloem apoplasm occurs by simple diffusion down a steep concentration difference. In contrast, sucrose retrieval from the apoplasm by the se-cc complexes is carrier mediated (16, 54, 55, 59) and energy coupled in symport with protons returning down their electrochemical gradient generated by a plasma membrane H^+-ATPase (132, 153). The sucrose/proton symporter SUC2 (142, 149) and a phloem-specific isoform of a plasma membrane H^+-ATPase [AHA3 (22)] are expressed throughout the mature phloem path from source to sink. However, they are not expressed in the protophloem located in growth sinks, such as root tips and expanding regions of leaves (142, 149). The sucrose/symporter, PmSUC2 (142), and the AHA3 H^+-ATPase (22) have been immunolocalized to the companion cells in *Plantago major*. This latter observation may explain the absence of the symporter from growth sinks where the sieve elements of the protophloem commonly lack companion cells (28).

Thus, net sucrose efflux from the se-cc complexes is governed by the difference between the rates of passive leakage and sucrose/proton symport retrieval. Regulation of net efflux is dominated by alterations in the activity of the sucrose/proton symporter (153). In this context, there is a growing body of evidence that the symporter is turgor regulated (1, 15, 54, 140). The mechanism of the turgor regulation is yet to be elucidated, but it may be mediated through modulation of H^+-ATPase activity and hence the proton motive force for sucrose uptake (cf 180).

EFFLUX TO THE APOPLASM ALONG THE POST-SIEVE ELEMENT PATHWAY Evidence suggestive of simple or facilitated diffusion, as well as energy-coupled membrane transport, has been reported for photoassimilate efflux from specific cell types located along the post-sieve element pathway (117, 118). For tomato fruit (130) and grape berries (CE Offler, unpublished data), the large transmembrane difference in sucrose concentration across the plasma membranes of the vascular parenchyma cells predict rates of simple diffusion that are comparable to those observed. In tomato fruit, the steep transmembrane concentration differences (129) are maintained through the hydrolysis of the leaked sucrose by an extracellular invertase (18). In contrast with the axial phloem, efflux is optimized by a reduced capacity of the vascular tissues to retrieve sugars leaked to the fruit apoplasm (130, 179).

During the prestorage phase of legume seed development, an extracellular invertase in the seed coats maintains a favorable transmembrane concentration difference to drive sucrose efflux to the seed apoplasm (171). The extracellular invertase activity is retained by tropical cereals but lost from temperate cereals and grain legumes on the commencement of the storage phase (146). Here, sucrose efflux from maternal seed tissues occurs by facilitated membrane transport (117 and references therein; 20, 21, 163). Direct energy coupling is not an obvious requirement, because transmembrane concentration differences may be sufficiently steep to drive facilitated diffusion (20, 45, 163). This assertion is supported by the finding that photoassimilate efflux is insensitive to metabolic inhibitors in seeds of both monocot (122, 161,

163) and dicot species (20, 21; but cf 92, 174, 176). However, this would not appear to be a universal phenomenon, because efflux of sucrose from french and broad bean seed coats is slowed by membrane-permeant metabolic inhibitors (39, 154, 156, 176). In these cases, the responses of sucrose efflux to experimentally imposed alterations in the proton motive force are consistent with the operation of a sucrose/proton antiporter (39, 156).

Apoplasmic osmolality, transduced as a turgor signal (117 and references therein), influences sucrose efflux from the maternal tissues of grain legumes (173) but not cereals (43). Turgor-dependent efflux of photoassimilates is reversible, responds rapidly (min) to changes in solution osmolality, and appears to be localized to the plasma membranes of the seed coat cells responsible for photoassimilate efflux (114; but cf 173). The kinetics of photoassimilate efflux are characterized by an independence at low turgors but, once turgor exceeds a set point, efflux increases proportionately to saturate at elevated turgors (115) (Figure 4b).

RETRIEVAL AS A KEY COMPONENT OF POST-SIEVE ELEMENT TRANSPORT   The chemical form in which the sugars are taken up from the sink apoplasm depends upon whether an extracellular invertase is present. Hexose uptake appears to be central to the carbon economy of developing seeds of tropical grasses (146; but cf 137) and tomato fruit (131). In the case of tomato fruit, the activity of a putative hexose/proton symporter, located on the plasma membranes of the storage parenchyma cells, is considered to play a key role in hexose accumulation (131). Similar circumstances could apply to the prestorage phase of legume seed development (171).

Both sucrose (40, 74, 180) and hexose (40, 180) proton symport activities have been detected in sugar beet tap roots. Comparison of sugar uptake by various protoplast populations suggests that most sugar influx is mediated by the glucose/proton symporter localized to the phloem parenchyma cells (40). Following resynthesis in the cytoplasm (40), sucrose is accumulated in the storage parenchyma vacuoles by a tonoplast sucrose/proton antiporter (49; but see 10). Sugar flux through both the glucose and sucrose symporters is turgor-regulated by the turgor-sensitive activity of a plasma membrane H^+-ATPase (180). As cell turgor rises, sucrose released back to the root apoplasm could be accumulated by other storage parenchyma cells that had not reached their full sucrose content (3).

In the absence of an extracellular invertase, sucrose is taken up intact by the sink cells. Expression of a sucrose porter, SUC1 (50), has been detected in developing ovules of *Plantago,* but its primary function is yet to be determined. Sucrose uptake by grain legume cotyledons, during the storage phase

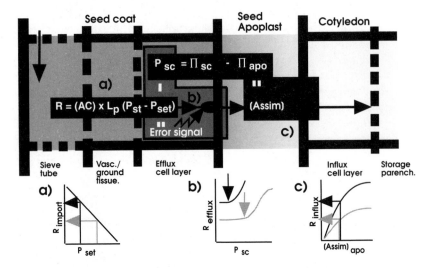

Figure 4 The turgor-homeostat model describing the integration of (*a*) phloem import into a developing legume seed with (*b*) efflux from the seed coat to the seed apoplasm mediated by a turgor-dependent efflux regulated by a turgor homeostat and (*c*) uptake by the cotyledons. Assuming mass flow through the seed coat symplasm, the solute transport rate (R) is the product of total plasmodesmal cross-sectional area (A), concentration (C) of transported solute, hydraulic conductivity of the plasmodesmata (L_p) and the pressure (P) difference between the importing sieve tubes (st) and the parenchyma unloading cells, the pressure of which is governed by the turgor set point (P_{set}). The turgor pressure of the seed coat (P_{sc}) is determined by the osmotic potential difference between the seed coat (π_{sc}) and seed apoplasm (π_{apo}), which is a function of the assimilate concentration [(Assim)]. Deviations (min) of P_{se} from P_{set} (*arrows* in graph *b*) produce error signals that regulate the activity of the efflux mechanism located in the unloading cells (see diagram *b*) to maintain the apoplasmic assimilate concentration (see graph *c*). More long-term (hours) increases of sucrose influx by the cotyledons (*light to dark curves* of graph *c* with associated *arrows* showing the two influxes) are accompanied by adjustments in P_{set} (graph *b* and see *light to dark arrows*) that elicit commensurate increases in the rate of phloem import (graph *a* and note increased rate from *light to dark arrows*). (Reproduced with permission from *Australian Journal of Plant Physiology,* 22:681.)

of seed development, exhibits a biphasic dependence on the external sucrose concentration (68, 75, 82). Sucrose-specific induced transients in the membrane potential difference, cytoplasmic acidification, and extracellular alkalinization demonstrate that the saturable component is mediated by a sucrose/proton symport mechanism (5, 76, 77, 83). The behavior of a proportion of the nonsaturable uptake is also consistent with proton-coupled transport (78). In this context, heterologous expression of a membrane-bound sucrose binding protein (127) in yeast exhibited kinetics comparable with the nonsaturable sucrose porter (110; see also 38). At estimated apoplasmic sucrose concentrations, some 50 to 80% of the total sucrose flux is predicted to be through the sucrose/proton symporter (84, 145). Similarly, circumstantial evidence suggests that some 30 to 100% of the sucrose influx is proton coupled into the endosperm of developing wheat grains (117).

WHOLE PLANT PERSPECTIVE

Phloem Unloading Along the Axial Path

The nature of phloem unloading along the axial path is dynamic and, by responding to the prevailing source/sink balance, regulates photoassimilate supply to the terminal sinks. High source/sink ratio plants accumulate excess photoassimilates along the axial phloem pathway in storage pools (165). Under these conditions, symplasmic exchange to the lateral storage sites largely occurs by diffusion and particularly so where photoassimilates are stored as macromolecules (95). Unloading by bulk flow could be of greater significance in mature axial paths that accumulate osmotically active solutes to high concentrations behind an apoplasmic barrier (sugarcane) and in elongation zones where the imported phloem water is used for volume growth. Concurrent exchange of photoassimilates to, and from the apoplasmic pool, under the control of a turgor-regulated loading mechanism, serves to buffer phloem sap concentrations against short-term shifts in the source/sink balance (90, 91). As the source/sink balance decreases, plasmodesmal conductivity could decline in parallel. Together with turgor-regulated loading, such a response would account for the observed increase in partitioning of translocated photoassimilates to terminal sinks under conditions that reduce the source/sink ratio (123, 165). The final state of plasmodesmal closure results in an exclusive apoplasmic unloading route that optimizes control over partitioning between axial path storage and translocation to the terminal sinks.

Phloem Unloading in Terminal Sinks

How phloem unloading of sucrose and potassium in terminal sinks influences the volume flux of imported phloem sap (Equation 1) depends upon the unloading pathway and mechanism. In contrast with the axial pathway, sieve element unloading universally follows a symplasmic route with the possibility of an apoplasmic step located along the post-sieve element pathway (Figure 2). As the volume flux of imported phloem sap rises, sieve element unloading by bulk flow is anticipated to be of increasing significance.

The influence of unloading by diffusion on the volume flux (Equation 1) is restricted to affecting the turgor pressure of the importing sieve elements by altering their sap osmotic pressure (Equation 3). The magnitude to which sieve element turgor pressure responds to shifts in their sap osmotic pressure is inversely related to the magnitude of the osmotic pressure difference between the sieve element lumen and surrounding apoplasm (Equation 3). This difference is anticipated to be small, because the diminished retrieval capacities of the phloem will contribute to a high apoplasmic osmotic pressure (173). For those sinks where symplasmic unloading may occur by bulk flow, the volume flux is rate-limited by the transport properties of the post-sieve element pathway (Equation 1). That is, the turgor pressure of the recipient sink cells rather than that of the importing sieve elements sets the pressure difference, and the interconnecting plasmodesmata govern the overall hydraulic conductance of the transport path from source to sink.

An apoplasmic step in the post-sieve element pathway is associated with sinks containing different genomes or sinks that accumulate sugars to high concentrations in the absence of an apoplasmic barrier separating the phloem from the storage parenchyma cells (Figures 2*d* and 2*e*). In the latter case, the apoplasmic step permits effective compartmentation of the stored sugars that could otherwise attenuate the concentration or pressure differences driving unloading from the importing sieve elements (Equations 4 and 5). Efflux to the sink apoplasm is a pivotal component of post-sieve element transport linking phloem import with sink demand. In the case of phloem import, efflux influences the concentration or pressure difference driving unloading from the sieve element lumens (Equations 4 and 5). The nature of the linkage with sink demand could be a mass action phenomenon in which the transmembrane concentration differences control efflux to the sink apoplasm by simple or facilitated diffusion. Here, sink demand is signaled through the apoplasmic sucrose level. To prevent dampening phloem pressure and hence import, depletion of apoplasmic sucrose must be offset by the maintenance of apoplasmic osmolality (70). Extracellular sucrose hydrolysis (30) and/or potassium release to the sink apoplasm (3) could serve this osmoregulatory role.

In comparison to the mass action scenario described above, apoplasmic sucrose levels are sustained independently of the apoplasmic sucrose flux in developing seeds (43, 117). The juxtaposition of the cells responsible for photoassimilate exchange to and from the seed apoplasm (Figure 2*f*) provides an optimal environment for the effective transfer of regulatory chemical signals between the two membrane transport events. In this context, for grain legume seeds, a mechanism that accounts for this homeostasis centers around the capacity of the cells responsible for efflux to turgor regulate through turgor-sensitive membrane transport mechanisms (117, 148, 182) (Figure 4). The hydraulic isolation of developing seeds from the vegetative parts of the parent plant ensures that the putative osmotic signals within the seed are not distorted by shifts in whole plant water relations (117). Hence, sink demand for photoassimilates is communicated to the efflux mechanism and ultimately phloem import as a turgor signal to integrate these transport and transfer events (Figure 4). Clearly, other control mechanisms are operative within the developing seed, including regulation of plasmodesmal conductivity (see previous section) that could function in place of (43) or in series with turgor regulation (Figure 4).

FUTURE PROSPECTS

Elucidation of the cellular pathway of phloem unloading in most sinks examined thus far awaits quantitative evaluation. Some progress has been made for unloading paths that include an apoplasmic step. More definitive outcomes will be achieved using transgenic plants. The more common symplasmic unloading route remains unexplored as current technologies are inadequate to quantitatively monitor and manipulate plasmodesmal transport. Once these putative technologies become available, it will be possible to quantitatively resolve the significance of plasmodesmal transport as well as the equally important question of whether symplasmic transport occurs largely by diffusion or mass flow. Elucidation of this latter issue will allow mechanistic definition of how phloem unloading interfaces with phloem translocation and sink demand.

ACKNOWLEDGMENTS

I am especially indebted to my friend and collaborator, Christina Offler, who willingly has given so much to our shared interests in the field. Her comments and advice on earlier drafts, together with those of Vincent Franceschi, were invaluable in crafting the final version of this review. However, I accept full responsibility for errors of science and omission. Gratitude is expressed to Louise Heatherington, who patiently dealt with an ever-changing parade of literature citations, and to numerous colleagues who kindly supplied useful preprint material. Research contributions from the author's laboratory were supported by the Australian Research Council. Finally, the review is dedicated to my mentors, Fred Milthorpe, Phillip Wareing, and Ian Wardlaw, whose inspiration and encouragement have greatly facilitated my modest odyssey into phloem transport.

Literature Cited

1. Aloni B, Wyse R, Griffith S. 1986. Sucrose transport and phloem unloading in stem of *Vicia faba:* possible involvement of a sucrose carrier and osmotic regulation. *Plant Physiol.* 81:482–87

1a. Baker DA, Miburn JA, eds. 1989. *Transport of Photoassimilates.* Harlow: Longman Sci. Tech.

2. Bell CI, Jones J, Milford GFJ, Leigh RA. 1992. The effects of crop nutrition on sugar beet quality. *Asp. Appl. Biol.* 32:19–26

3. Bell CI, Leigh RA. 1996. Differential effects of turgor on sucrose and potassium transport at the tonoplast and plasma membrane of sugarbeet storage root tissue. *Plant Cell Environ.* 19:191–200

4. Blanke MM, Whiley AW. 1995. Bioenergetics, respiration cost and water relations of developing avocado fruit. *J. Plant Physiol.* 145:87–92

5. Bonnemain J-L, Bourquin S, Renault S, Offler CE, Fisher DG. 1991. Transfer cells: structure and physiology. See Ref. 5a, pp. 74–83

5a. Bonnemain J-L, Delrot S, Lucas WJ, Dainty J, eds. 1991. *Recent Advances in Phloem Transport and Assimilate Compartmentation.* Nantes Cedex: Ouest Ed.

6. Bouche-Pillon S, Fleurat-Lessard P, Serrano R, Bonnemain J-L. 1994. Asymmetric distribution of the plasma-membrane H+-ATPase in embryos of *Vicia faba* L. with special reference to transfer cells. *Planta* 193:392–97

7. Bret-Harte MS, Silk WK. 1994. Nonvascular, symplastic diffusion of sucrose cannot satisfy the carbon demands of growth in the primary root tip of *Zea mays. Plant Physiol.* 105:19–33

8. Brown SC, Coombe BG. 1985. Solute accumulation by grape pericarp cells. III. Sugar changes in vivo and the effect of shading. *Biochem. Physiol. Pflanz.* 180: 371–81

9. Brown SM, Oparka KJ, Sprent JI, Walsh KB. 1995. Symplastic transport in soybean root nodules. *Soil Biol. Biochem.* 27: 387–99

10. Bush DR. 1993. Proton-coupled sugar and amino acid transporters in plants. *Annu. Rev. Plant Physiol. Plant Mol. Biol.* 44: 513–42

11. Bush DS. 1995. Calcium regulation in plant cells and its role in signaling. *Annu. Rev. Plant Physiol. Plant Mol. Biol.* 46:95–122

12. Chapleo S, Hall JL. 1989. Sugar unloading in roots of *Ricinus communis* L. III. The extravascular pathway of sugar transport. *New Phytol.* 111:391–96

13. Cleland RE, Fujiwara T, Lucas WJ. 1994. Plasmodesmatal-mediated cell-to-cell transport in wheat roots is modulated by anaerobic stress. *Protoplasma* 178:81–85

14. Coombe BG. 1992. Research on development and ripening of the grapeberry. *Am. J. Enol. Vitic.* 43:101–10

14a. Cronshaw J, Lucas JW, Giaquinta RT, eds. 1986. *Phloem Transport.* New York: Liss

15. Daie J. 1987. Interaction of cell turgor and hormones on sucrose uptake in isolated phloem of celery. *Plant Physiol.* 84: 1033–37

16. Daie J. 1987. Sucrose uptake in isolated phloem of celery is a simple saturable transport system. *Planta* 171:474–82

17. Dali N, Michaud D, Yelle S. 1992. Evidence for the involvement of sucrose phosphate synthase in the pathway of sugar accumulation in sucrose-accumulating tomato fruits. *Plant Physiol.* 99:434–39

18. Damon S, Hewitt J, Nieder M, Bennett AB. 1988. Sink metabolism in tomato fruit. II. Phloem unloading and sugar uptake. *Plant Physiol.* 87:731–36

19. Dawson JH, Musselman LJ, Wolswinkel P, Dorr I. 1994. Biology and control of *Cuscuta. Rev. Weed Sci.* 6:265–317

20. de Jong A, Koerselman-Kooij JW, Schuurmans JAMJ, Borstlap AC. 1996. Characterisation of the uptake of sucrose and glucose by isolated seed coat halves of developing pea seeds: evidence that a sugar facilitator with diffusional kinetics is involved in seed coat unloading. *Planta* 199:486–92

21. de Jong A, Wolswinkel P. 1995. Differences in release of endogenous sugars and amino acids from attached and detached seed coats of developing pea seeds. *Physiol. Plant.* 94:1–6

22. DeWitt ND, Sussman MR. 1995. Immunocytological localisation of an epitope-tagged plasma membrane proton pump

(H+-ATPase) in phloem. *Plant Cell* 7: 2053–67

23. Dick PS, apRees T. 1975. The pathway of sugar transport in roots of *Pisum sativum*. *J. Exp. Bot.* 26:305–14

24. Ding B, Parthasarathy MV, Niklas K, Turgeon R. 1988. A morphometric analysis of the phloem-unloading pathway in developing tobacco leaves. *Planta* 176:307–18

25. Ellis EC, Turgeon R, Spanswick RM. 1992. Quantitative analysis of photosynthate unloading in developing seeds of *Phaseolus vulgaris*. II. Pathway and turgor sensitivity. *Plant Physiol.* 99:644–51

26. Epel BL. 1994. Plasmodesmata: composition, structure and trafficking. *Plant Mol. Biol.* 26:1343–57

27. Epel BL, Bandurski RS. 1990. Tissue to tissue symplastic communication in the shoots of etiolated corn seedlings. *Physiol. Plant.* 79:604–9

28. Esau K. 1965. *Plant Anatomy*. New York: Wiley. 767 pp.

29. Eschrich W. 1983. Phloem unloading in aerial roots of *Monstera deliciosa*. *Planta* 157:540–47

30. Eschrich W. 1989. Phloem unloading of photoassimilates. See Ref. 1a, pp. 206–63

31. Evert RF, Russin WA. 1993. Structurally, phloem unloading in the maize leaf cannot be symplastic. *Am. J. Bot.* 80:1310–17

32. Farrar JF. 1985. Fluxes of carbon in roots of barley plants. *New Phytol.* 99:57–69

33. Farrar JF. 1992. The whole plant: carbon partitioning during development. See Ref. 121a, pp. 163–80

34. Farrar JF, Minchin PEH. 1991. Carbon partitioning in split root systems of barley: relation to metabolism. *J. Exp. Bot.* 42: 1261–71

35. Farrar JF, Minchin PEH, Thorpe MR. 1994. Carbon import into barley roots: stimulation by galactose. *J. Exp. Bot.* 45: 17–23

36. Farrar JF, Minchin PEH, Thorpe MR. 1995. Carbon import into barley roots: effects of sugars and relation to cell expansion. *J. Exp. Bot.* 46:1859–67

37. Fellows RJ, Geiger DR. 1974. Structural and physiological changes in sugar beet leaves during sink to source conversion. *Plant Physiol.* 54:877–85

38. Fieuw S, Franceschi VR, Hitz WD. 1992. Antibody against the sucrose binding protein specifically inhibits sucrose-proton symport by transfer cell protoplasts of developing *Vicia faba* L. seed. *Plant Physiol.* 99S:40

39. Fieuw S, Patrick JW. 1993. Mechanism of photosynthate efflux from *Vicia faba* L.

seed coats. I. Tissue studies. *J. Exp. Bot.* 44:63–74

40. Fieuw S, Willenbrink J. 1990. Sugar transport and sugar-metabolising enzymes in sugar beet storage roots (*Beta vulgaris* ssp. *altissima*). *J. Plant Physiol.* 137:216–23

41. Fieuw S, Willenbrink J. 1991. Isolation of protoplasts from tomato fruit (*Lycopersicon esculentum*): first uptake studies. *Plant Sci.* 76:9–17

42. Findlay N, Oliver KJ, Nii N, Coombe BG. 1987. Solute accumulation by grape pericarp cells. IV. Perfusion of pericarp apoplast via the pedicel and evidence for xylem malfunction in ripening berries. *J. Exp. Bot.* 38:668–79

43. Fisher DB. 1996. Phloem unloading in developing wheat grains. In *Sucrose Metabolism, Biochemistry, Physiology and Molecular Biology,* ed. HG Pontis, GL Salerno, EJ Echeverria, pp. 205–15. Rockville, MD: Am. Soc. Plant. Physiol.

44. Fisher DB, Wang N. 1993. A kinetic and microautoradiographic analysis of [^{14}C] sucrose import by developing wheat grains. *Plant Physiol.* 101:391–98

45. Fisher DB, Wang N. 1995. Sucrose concentration gradients along the post-phloem transport pathway in the maternal tissues of developing wheat grains. *Plant Physiol.* 109:587–92

46. Frensch J, Hsiao TC. 1994. Transient responses of cell turgor and growth of maize roots as affected by changes in water potential. *Plant Physiol.* 104:247–54

47. Frommer WB, Sonnewald U. 1995. Molecular analysis of carbon partitioning in solanaceous species. *J. Exp. Bot.* 46: 587–607

48. Garcia-Luis A, Didehvar F, Guardiola JL, Baker DA. 1991. The transport of sugars in developing fruits of *Satsuma mandarin*. *Ann. Bot.* 68:349–59

49. Getz HP, Klein M. 1995. Characteristics of sucrose transport and sucrose-induced H$^+$ transport on the tonoplast of red beet (*Beta vulgaris* L.) storage tissue. *Plant Physiol.* 107:459–69

50. Ghartz M, Schmeilzer E, Stolz J, Sauer N. 1996. Expression of the *PmSUC1* sucrose carrier gene from *Plantago major* L. is induced during seed development. *Plant J.* 9:93–100

51. Giaquinta RT. 1977. Sucrose hydrolysis in relation to phloem translocation in *Beta vulgaris*. *Plant Physiol.* 60:339–43

52. Giaquinta RT, Lin W, Sadler NL, Franceschi VR. 1983. Pathway of phloem unloading of sucrose in corn roots. *Plant Physiol.* 72:362–67

53. Grignon N, Touraine B, Durand M. 1989. 6(5)carboxyfluorescein as a tracer of phloem sap translocation. *Am. J. Bot.* 76: 871–77

54. Grimm E, Bernhardt G, Rothe K, Jacob F. 1990. Mechanism of sucrose retrieval along the phloem path: a kinetic approach. *Planta* 182:480–85

55. Grimm E, Jahnke S, Rothe K. 1996. Assimilate movement in the petiole of *Cyclamen* and *Primula* is independent from the lateral resorption. *J. Exp. Bot.* In press

56. Grusak MA, Minchin PEH. 1988. Seed coat unloading in *Pisum sativum:* osmotic effects in attached versus excised empty ovules. *J. Exp. Bot.* 39:543–59

57. Harrison MJ. 1996. A sugar transporter from *Medicago truncatula:* altered expression pattern in roots during vesicular-arbuscular (VA) mycorrhizal associations. *Plant J.* 9:491–503

58. Hayes PM, Offler CE, Patrick JW. 1985. Cellular structures, plasma membrane surface areas and plasmodesmatal frequencies of the stem of *Phaseolus vulgaris* L. in relation to radial photosynthate transfer. *Ann. Bot.* 56:125–38

59. Hayes PM, Patrick JW, Offler CE. 1987. The cellular pathway of radial transfer of photosynthates in stems of *Phaseolus vulgaris* L.: effect of cellular plasmolysis and p-chloromercuribenzenesulfonic acid. *Ann. Bot.* 59:635–42

60. Ho LC, Grange RI, Picken AJ. 1987. An analysis of the accumulation of water and dry matter in tomato fruit. *Plant Cell Environ.* 10:157–62

61. Ho LC, Hewitt JD. 1985. Fruit development. In *The Tomato Crop: A Scientific Basis for Improvement,* ed. JC Atherton, J Rudich, pp. 201–39. London: Chapman & Hall

62. Hole CC, Dearman J. 1994. Sucrose uptake by the phloem parenchyma of carrot storage root. *J. Exp. Bot.* 45:7–15

63. Jeschke WD, Baumel P, Rath N, Czygan F-C, Proksch P. 1994. Modelling of the flows and partitioning of carbon and nitrogen in the holoparasite *Cuscuta reflexa* Roxb. and its host *Lupinus albus* L. II. Flows between host and parasite and within the parasitized host. *J. Exp. Bot.* 45: 801–12

64. Johnson C, Hall JL, Ho LC. 1988. Pathways of uptake and accumulation of sugars in tomato fruit. *Ann. Bot.* 59:595–603

65. Koch KE. 1996. Carbohydrate-modulated gene expression in plants. *Annu. Rev. Plant Physiol. Plant Mol. Biol.* 47:509–50

66. Koch KE, Avigne WT. 1990. Postphloem,

nonvascular transfer in citrus: kinetics, metabolism, and sugar gradients. *Plant Physiol.* 93:1405–16

67. Koch KE, Lowell CA, Avigne WT. 1986. Assimilate transfer through citrus juice vesicle stalks: a nonvascular portion of the transport path. See Ref. 14a, pp. 247–58

68. Lanfermeijer FCW, Koerselman-Kooij JW, Borstlap AC. 1991. Osmosensitivity of sucrose uptake by immature pea cotyledons disappears during development. *Plant Physiol.* 95:832–38

69. Lang A. 1990. Xylem, phloem and transpiration flows in developing apple fruits. *J. Exp. Bot.* 41:645–51

70. Lang A, During H. 1991. Partitioning control by water potential gradients: evidence for compartmentation breakdown in grape berries. *J. Exp. Bot.* 42:1117–23

71. Lang A, Ryan K. 1995. Vascular development and sap flow in apple pedicels. *Ann. Bot.* 74:381–88

72. Lang A, Thorpe MR. 1989. Xylem, phloem and transpiration flows in a grape: application of a technique for measuring the volume of attached fruits to high resolution using Archimedes' principle. *J. Exp. Bot.* 40:106–78

73. Lee DR. 1988. Vasculature of the abscission zone of tomato fruit: implications for transport. *Can. J. Bot.* 67:1898–902

74. Lemoine R, Daie J, Wyse R. 1988. Evidence for the presence of a sucrose carrier in immature sugar beet tap roots. *Plant Physiol.* 86:575–80

75. Lichtner FT, Spanswick RM. 1981. Sucrose uptake by developing soybean cotyledons. *Planta* 68:693–98

76. Lichtner FT, Spanswick RM. 1981. Electrogenic sucrose transport in developing soybean cotyledons. *Plant Physiol.* 67: 869–74

77. Lin W. 1985. Energetics of sucrose transport into protoplasts from developing soybean cotyledons. *Plant Physiol.* 78:41–45

78. Lin W. 1985. Linear sucrose transport in protoplasts isolated from developing soybean cotyledons. *Plant Physiol.* 78:649–51

79. Lin W, Schmitt MR, Hitz WD, Giaquinta RT. 1984. Sugar transport in isolated corn root protoplasts. *Plant Physiol.* 76:894–97

80. Lucas WJ, Ding B, van der Schoot C. 1993. Plasmodesmata and the supracellular nature of plants. *New Phytol.* 125:435–76

81. Madore MA, Lucas WJ. 1989. Transport of photoassimilates between leaf cells. See Ref. 1a, pp. 49–78

82. McDonald R, Fieuw S, Patrick JW. 1996. Sugar uptake by the dermal transfer cells of developing cotyledons of *Vicia faba* L. I.

Experimental systems and general transport properties. *Planta* 198:54–63

83. McDonald R, Fieuw S, Patrick JW. 1996. Sugar uptake by the dermal transfer cells of developing cotyledons of *Vicia faba* L. II. Mechanism of energy coupling. *Planta* 198:502–9

84. McDonald R, Wang HL, Patrick JW, Offler CE. 1995. Cellular pathway of sucrose transport in developing cotyledons of *Vicia faba* L. and *Phaseolus vulgaris* L. A physiological assessment. *Planta* 196:659–67

85. McNeil D. 1976. The basis of osmotic pressure maintenance during expansion growth in *Helianthus annuus* hypocotyls. *Aust. J. Plant Physiol.* 3:311–24

86. Meshcheryakov A, Steudle E, Komor E. 1992. Gradients of turgor, osmotic pressure, and water potential in the cortex of the hypocotyl of growing *Ricinus* seedlings. *Plant Physiol.* 98:840–52

87. Milburn JA, Baker DA. 1989. Physicochemical aspects of phloem sap. See Ref. 1a, pp. 345–59

88. Milthorpe FL, Moorby J. 1969. Vascular transport and its significance in plant growth. *Annu. Rev. Plant Physiol.* 20: 117–38

89. Minchin PEH, Ryan KG, Thorpe MR. 1984. Further evidence of apoplastic unloading into the stem of bean: identification of the phloem buffering pool. *J. Exp. Bot.* 35:1744–53

90. Minchin PEH, Thorpe MR. 1984. Apoplastic phloem unloading in the stem of bean. *J. Exp. Bot.* 35:538–50

91. Minchin PEH, Thorpe MR. 1987. Measurement of unloading and reloading of photo-assimilate within the stem of bean. *J. Exp. Bot.* 38:211–20

92. Minchin PEH, Thorpe MR. 1990. Transport of photoassimilates in pea ovules. *J. Exp. Bot.* 41:1149–55

93. Moore PH. 1995. Temporal and spatial regulation of sucrose accumulation in the sugarcane stem. *Aust. J. Plant Physiol.* 22: 661–79

94. Moore PH, Cosgrove DJ. 1991. Developmental changes in cell and tissue water relations parameters in storage parenchyma of sugarcane. *Plant Physiol.* 96:794–802

95. Murphy R. 1989. Water flow across the sieve tube boundary: estimating turgor and some implications for phloem loading and unloading. II. Phloem in the stem. *Ann. Bot.* 63:551–59

96. Murphy R. 1989. Water flow across the sieve tube boundary: estimating turgor and some implications for phloem loading and

unloading. IV. Root tips and seed coats. *Ann. Bot.* 63:571–79

97. O'Brien TP, Summut ME, Lee JW, Smart MG. 1985. The vascular system of the wheat spikelet. *Aust. J. Plant Physiol.* 12: 487–511

98. Offler CE, Horder B. 1992. The cellular pathway of short distance transfer of photosynthates in developing tomato fruit. *Plant Physiol.* 99S:41

99. Offler CE, Patrick JW. 1984. Cellular structures, plasma membrane surface areas and plasmodesmatal frequencies of seed coats of *Phaseolus vulgaris* L. in relation to photosynthate transfer. *Aust. J. Plant Physiol.* 11:79–99

100. Offler CE, Patrick JW. 1993. Pathway of photosynthate transfer in the developing seed of *Vicia faba* L.: a structural assessment of the role of transfer cells in unloading from the seed coat. *J. Exp. Bot.* 44: 711–24

101. Oparka KJ. 1986. Phloem unloading in the potato tuber. Pathways and sites of ATPase. *Protoplasma* 131:201–10

102. Oparka KJ. 1990. What is phloem unloading? *Plant Physiol.* 94:393–96

103. Oparka KJ, Duckett CM, Prior DAM, Fisher DB. 1994. Real-time imaging of phloem unloading in the root tip of *Arabidopsis*. *Plant J.* 6:759–66

104. Oparka KJ, Prior DAM. 1987. [14]C sucrose efflux from the perimedulla of growing potato tubers. *Plant Cell Environ.* 10:667–75

105. Oparka KJ, Prior DAM. 1988. Movement of Lucifer yellow CH in potato tuber storage tissues: a comparison of symplastic and apoplastic transport. *Planta* 176:533–40

106. Oparka KJ, Prior DAM. 1992. Direct evidence for pressure-generated closure of plasmodesmata. *Plant J.* 2:741–50

107. Oparka KJ, Prior DAM, Wright KM. 1995. Symplastic communication between primary and developing lateral roots of *Aribidopsis thaliana*. *J. Exp. Bot.* 46:187–99

108. Oparka KJ, Viola R, Wright KM, Prior DAM. 1992. Sugar transport and metabolism in the potato tuber. See Ref. 121a, pp. 91–114

109. Oparka KJ, Wright KM. 1988. Influence of cell turgor on sucrose partitioning in potato tuber storage tissues. *Planta* 175:520–26

110. Overoorde PJ, Frommer WB, Grimes HD. 1996. A soybean sucrose binding protein independently mediates nonsaturable sucrose uptake in yeast. *Plant Cell* 8:271–80

111. Pate JS, Peoples MB, van Bel AJE, Kuo J, Atkins CA. 1985. Diurnal water balance of the cowpea fruit. *Plant Physiol.* 77:148–56

112. Patrick JW. 1990. Sieve element unloading: cellular pathway, mechanism and control. *Physiol. Plant.* 78:298–308

113. Patrick JW. 1991. Control of phloem transport to and short-distance transfer in sink regions. See Ref. 5a, pp. 167–77

114. Patrick JW. 1993. Osmotic regulation of assimilate unloading from seed coats of *Vicia faba:* role of turgor and identification of turgor-dependent fluxes. *Physiol. Plant.* 89:87–96

115. Patrick JW. 1994. Turgor-dependent unloading of photosynthates from coats of developing seed of *Phaseolus vulgaris L.* and *Vicia faba L.:* turgor homeostasis and set points. *Physiol. Plant.* 90:367–77

116. Patrick JW. 1994. Turgor-dependent unloading of assimilates from coats of developing legume seed: assessment of the significance of the phenomenon in the whole plant. *Physiol. Plant.* 90:645–54

117. Patrick JW, Offler CE. 1995. Post-sieve element transport of sucrose in developing seeds. *Aust. J. Plant Physiol.* 22:681–702

118. Patrick JW, Offler CE. 1996. Post-sieve element transport of photoassimilates in sink regions. *J. Exp. Bot.* 47:1167–77

119. Patrick JW, Offler CE, Wang X-D. 1995. Cellular pathway of photosynthate transport in coats of developing seed of *Vicia faba* L. and *Phaseolus vulgaris* L. I. Extent of transport through the coat symplast. *J. Exp. Bot.* 46:35–47

120. Patrick JW, Turvey PM. 1981. The pathway of radial transfer of photosynthate in decapitated stems of *Phaseolus vulgaris* L. *Ann. Bot.* 47:611–21

121. Pearson JN, Jenner CF, Rengel Z, Graham RD. 1996. Differential transport of Zn, Mn and sucrose along the longitudinal axis of developing wheat grains. *Physiol. Plant.* 97:332–38

121a. Pollock CJ, Farrar JF, Gordon AJ, eds. 1992. *Carbon Partitioning Within and Between Organisms.* Oxford: Bios Sci.

122. Porter GA, Knievael DP, Shannon JC. 1987. Assimilate unloading from maize (*Zea mays* L.) pedicel tissue. II. Effects of chemical agents on sugar, amino acid, and ^{14}C assimilate unloading. *Plant Physiol.* 85:558–65

123. Pritchard J. 1994. The control of cell expansion in roots. *New Phytol.* 127:3–26

124. Pritchard J. 1996. Aphid stylectomy reveals an osmotic step between phloem and cortical cells in barley roots. *J. Exp. Bot.* 47:1519–24

125. Pritchard J, Fricke W, Tomos D. 1996. Turgor-regulation during extension growth and osmotic stress of maize roots: an exam-ple of single-cell mapping. *Plant Soil.* In press

126. Riesmeier JW, Hirner B, Frommer WB. 1993. Potato sucrose transporter expression in minor veins indicates a role in phloem loading. *Plant Cell* 5:1591–98

127. Ripp KG, Viitanen PV, Hitz WD, Franceschi VR. 1988. Identification of a membrane protein associated with sucrose transport into cells of developing soybean cotyledons. *Plant Physiol.* 88:1435–45

128. Rochat C, Wuilleme S, Boutin J-P, Hedley CL. 1995. A mutation at the rb gene, lowering ADPGPPase activity, affects storage product metabolism of pea seed coats. *J. Exp. Bot.* 46:415–21

129. Ruan Y-L, Brady CJ, Patrick JW. 1996. Chemical composition of apoplast sap from intact developing tomato fruit. *Aust. J. Plant Physiol.* 23:9–13

130. Ruan Y-L, Patrick JW. 1995. The cellular pathway of postphloem sugar transport in developing tomato fruit. *Planta* 196: 434–44

131. Ruan Y-L, Patrick JW, Brady CJ. 1996. Protoplast hexose carrier activity is a determinate of genotypic difference in hexose storage in tomato fruit. *Plant Cell Environ.* In press

132. Salmon S, Lemoine R, Jamai A, Bouche-Pillon S. 1995. Study of sucrose and mannitol transport in plasma-membrane vesicles from phloem and nonphloem tissue of celery (*Apium graveolens* L.) petioles. *Planta* 197:76–84

133. Sauer N, Baier K, Gahrtz M, Stadler R, Stolz J, Truernit E. 1994. Sugar transport across the plasma membranes of higher plants. *Plant Mol. Biol.* 26:1671–81

134. Sauter JJ, Kloth S. 1986. Plasmodesmatal frequency and radial translocation rates in ray cells of poplar (*Populus x canadensis* Moench 'robusta'). *Planta* 168:377–80

135. Schmalstig JG, Cosgrove DJ. 1990. Coupling of solute transport and cell expansion in pea stems. *Plant Physiol.* 94:1625–34

136. Schmalstig JG, Geiger DR. 1985. Phloem unloading in developing leaves of sugar beet. I. Evidence for pathway through the symplast. *Plant Physiol.* 79:237–41

137. Schmalstig JG, Hitz WD. 1987. Transport and metabolism of a sucrose analog (1'-fluorosucrose) into *Zea mays* L. endosperm without invertase hydrolysis. *Plant Physiol.* 55:898–901

138. Schulz A. 1994. Phloem transport and differential unloading in pea seedlings after source and sink manipulations. *Planta* 192: 239–49

139. Schulz A. 1995. Plasmodesmatal widening

accompanies the short-term increase in symplastic phloem unloading in pea root tips under osmotic stress. *Protoplasma* 188:22–37

140. Smith JAC, Milburn JA. 1980. Osmoregulation and the control of phloem-sap composition in *Ricinus communis* L. *Planta* 148:28–34

141. Smith SE, Smith FA. 1990. Structure and function of the interfaces in biotrophic symbioses as they relate to nutrient transport. *New Phytol.* 114:1–38

142. Stadler R, Brandner J, Schulz A, Gahrtz M, Sauer N. 1995. Phloem loading by the PmSUC2 sucrose carrier from *Plantago major* occurs into companion cells. *Plant Cell* 7:1545–54

143. Stitt M, Sonnewald U. 1995. Regulation of metabolism in transgenic plants. *Annu. Rev. Plant Physiol. Plant Mol. Biol.* 46:341–68

144. Swanson CA, Hoddinott J, Sij JW. 1976. The effect of selected sink leaf parameters on translocation rates. See Ref. 165a, pp. 347–56

145. Thorne JH. 1982. Characterisation of the active sucrose transport system of immature soybean embryos. *Plant Physiol.* 70: 953–58

146. Thorne JH. 1985. Phloem unloading of C and N assimilates in developing seeds. *Annu. Rev. Plant Physiol.* 36:317–43

147. Thorpe MR, Minchin PEH. 1996. Mechanisms of long- and short-distance transport from sources to sinks. In *Photoassimilate Distribution in Plants and Crops: Source-Sink Relationships,* ed. E Zamski, AA Schaffer, pp. 261–82. New York: Dekker

148. Thorpe MR, Minchin PEH, Williams JHH, Farrar JF, Tomos AD. 1993. Carbon import into developing ovules of *Pisum sativum:* the role of the water relations of the seed coat. *J. Exp. Bot.* 44:937–47

149. Truernit E, Sauer N. 1995. The promoter of the *Arabidopsis thalliana* SUC2 sucrose-H⁺ symporter gene directs expression of β-glucuronidase to the phloem: evidence for phloem loading by SUC2. *Planta* 196: 564–70

150. Tucker EB. 1993. Azide treatment enhances cell-to-cell diffusion in staminal hairs of *Setcreasea purpurea. Protoplasma* 174:45–49

151. Turgeon R. 1987. Phloem unloading in tobacco sink leaves: insensitivity to anoxia indicates a symplastic pathway. *Planta* 171:73–81

152. van Bel AJE. 1993. Strategies of phloem loading. *Annu. Rev. Plant Physiol. Plant Mol. Biol.* 44:253–82

153. van Bel AJE. 1996. Interaction between

sieve element and companion cell and the consequences for photoassimilate distribution: two structural hardware frames with associated physiological software packages in dicotyledons. *J. Exp. Bot.* 47: 1129–40

154. van Bel AJE, Patrick JW. 1984. No direct linkage between proton pumping and photosynthate unloading from seed coats of *Phaseolus vulgaris* L. *Plant Growth Regul.* 2:319–26

155. Walker NA. 1976. The effect of flow in solute transport in plants. See Ref. 165a, pp. 43–50

156. Walker NA, Patrick JW, Zhang W, Fieuw S. 1995. Mechanism of photosynthate efflux from seed coats of *Phaseolus vulgaris:* a chemiosmotic analysis. *J. Exp. Bot.* 46: 539–49

157. Walsh KB, Sky RD, Brown SM. 1996. Pathway of sucrose unloading from the phloem in sugarcane stalk. In *Sugarcane: Research Towards Efficient and Sustainable Production,* ed. JR Wilson, DM Hogarth, JA Campbell, AL Garside, pp. 105–7. Brisbane: CSIRO Div. Trop. Crops Pastures

158. Wang HL, Offler CE, Patrick JW. 1995. Cellular pathway of photosynthate transfer in the developing wheat grain. II. A structural analysis and histochemical studies of the transfer pathway from the crease phloem to the endosperm cavity. *Plant Cell Environ.* 18:373–88

159. Wang HL, Offler CE, Patrick JW, Ugalde TD. 1994. The cellular pathway of photosynthate transfer in the developing wheat grain. I. Delineation of a potential transfer pathway using fluorescent dyes. *Plant Cell Environ.* 17:257–66

160. Wang HL, Patrick JW, Offler CE, Wang X-D. 1995. The cellular pathway of photosynthate transfer in the developing wheat grain. III. A structural analysis and physiological studies of the pathway from the endosperm cavity to the starchy endosperm. *Plant Cell Environ.* 18:389–407

161. Wang N, Fisher DB. 1994. Monitoring phloem unloading and post-phloem transport by microperfusion of attached wheat grains. *Plant Physiol.* 104:7–17

162. Wang N, Fisher DB. 1994. The use of fluorescent tracers to characterise the post-phloem transport pathway of attached wheat grains. *Plant Physiol.* 104:7–17

163. Wang N, Fisher DB. 1995. Sucrose release into the endosperm cavity of wheat grains apparently occurs by facilitated diffusion across the nucellar cell membranes. *Plant Physiol.* 109:579–85

164. Wang X-D, Harrington G, Patrick JW, Offler CE, Fieuw S. 1995. Cellular pathway of photosynthate transport in coats of developing seed of *Vicia faba* L. and *Phaseolus vulgaris* L. II. Principal cellular site(s) of efflux. *J. Exp. Bot.* 46:49–63

165. Wardlaw IF. 1990. The control of carbon partitioning in plants. *New Phytol.* 27: 341–81

165a. Wardlaw IF, Passioura JB, eds. 1976. *Transport and Transfer Processes in Plants.* New York: Academic

166. Warmbrodt RD. 1985. Studies of the root of *Hordeum vulgare* L.: ultrastructure of the seminal root with special reference to the phloem. *Am. J. Bot.* 72:414–32

167. Warmbrodt RD. 1985. Studies on the root of *Zea mays* L.: structure of the adventitious root with respect to phloem unloading. *Bot. Gaz.* 146:169–80

168. Warmbrodt RD. 1986. Structural aspects of the primary tissues of the *Cucurbita pepo* L. root with special reference to the phloem. *New Phytol.* 102:175–92

169. Warmbrodt RD. 1986. Solute concentrations in the phloem and associated vascular and ground tissues of the root of *Hordeum vulgare* L. See Ref. 14a, pp. 435–44

170. Warmbrodt RD. 1987. Solute concentrations in the phloem and apex of the root of *Zea mays. Am. J. Bot* 74:394–402

171. Weber H, Borisjuk L, Heim U, Buchner P, Wobus U. 1995. Seed coat–associated invertases of Faba bean control both unloading and storage functions: cloning of cDNAs and cell type–specific expression. *Plant Cell* 7:1835–46

172. Williams JHH, Minchin PEH, Farrar JF. 1991. Carbon partitioning in split root systems of barley: the effect of osmotica. *J. Exp. Bot.* 42:453–61

173. Wolswinkel P. 1992. Transport of nutrients into developing seeds: a review of physiological mechanisms. *Seed Sci. Res.* 2: 59–73

174. Wolswinkel P, Ammerlaan A. 1983. Phloem unloading in developing seeds of *Vicia faba* L. *Planta* 158:205–15

175. Wolswinkel P, Ammerlaan A. 1983. Sucrose and hexose release by excised stem segments of *Vicia faba* L.: the sucrose-specific stimulating influence of *Cuscuta* on sugar release and the activity of acid invertase. *J. Exp. Bot.* 34:1516–27

176. Wolswinkel P, Ammerlaan A, Kuyvenhoven H. 1983. Effect of KCN and p-chloromercuribenzenesulfonic acid on the release of sucrose and 2-amino(1-^{14}C) isobutyric acid by the seed coat of *Pisum sativum. Physiol. Plant.* 59:375–86

177. Wood RM, Offler CE, Patrick JW. 1996. The cellular pathway of short-distance transfer of photosynthate and potassium in the elongating stem of *Phaseolus vulgaris* L.: a structural assessment. *Ann. Bot.* In press

178. Wood RM, Patrick JW, Offler CE. 1994. The cellular pathway of short-distance transfer of photosynthates and potassium in the elongating stem of *Phaseolus vulgaris* L.: stem anatomy, solute transport and pool sizes. *Ann. Bot.* 73:151–61

179. Wright KM, Oparka KJ. 1991. Sugar uptake and metabolism in sink and source potato tubers. See Ref. 5a, pp. 258–64

180. Wyse RE, Zamski E, Tomos AD. 1986. Turgor regulation of sucrose transport in sugar beet taproot tissue. *Plant Physiol.* 81:478–81

181. Xia J, Saglio H. 1988. Characterization of the hexose transport system in maize root tips. *Plant Physiol.* 88:1015–20

182. Zhang W, Atwell BJ, Patrick JW, Walker NA. 1996. Turgor-dependent efflux of assimilates from coats of developing seed of *Phaseolus vulgaris* L.: water relations of the efflux cells. *Planta* 199:25–33

Annu. Rev. Plant Physiol. Plant Mol. Biol. 1997. 48:223–250

OXYGEN DEFICIENCY AND ROOT METABOLISM: Injury and Acclimation Under Hypoxia and Anoxia

Malcolm C. Drew

Department of Horticultural Sciences, Texas A&M University, College Station, Texas 77843-2133

KEY WORDS: anaerobic, metabolism, fermentation, aerenchyma, programmed cell death

ABSTRACT

Oxygen deficiency in the rooting zone occurs with poor drainage after rain or irrigation, causing depressed growth and yield of dryland species, in contrast with native wetland vegetation that tolerates such conditions. This review examines how roots are injured by O_2 deficiency and how metabolism changes during acclimation to low concentrations of O_2.

In the root apical meristem, cell survival is important for the future development; metabolic changes under anoxia help maintain cell survival by generating ATP anaerobically and minimizing the cytoplasmic acidosis associated with cell death. Behind the apex, where cells are fully expanded, ethylene-dependent death and lysis occurs under hypoxia to form continuous, gas-filled channels (aerenchyma) conveying O_2 from the leaves. This selective sacrifice of cells may resemble programmed cell death and is distinct from cell death caused by anoxia. Evidence concerning alternative possible mechanisms of anoxia tolerance and avoidance is presented.

CONTENTS

223

1040-2519/97/0601-0223$08.00

INTRODUCTION

Although all higher plants require access to free water, excess water in the root environment of land plants can be injurious or even lethal because it blocks the transfer of O_2 and other gases between the soil and the atmosphere. With transient flooding, or irrigation followed by slow drainage, or in natural wetlands, plant roots can become O_2 deficient because of slow transfer of dissolved O_2 in the water-filled pore space of the soil (reviewed in 46, 47). When soil is warm and respiration by microorganisms is stimulated, depletion of O_2 can be complete in less than 24 h, and roots experience a transition from a fully aerobic to anaerobic environment (51, 66).

The adverse effects of excess soil water (transient flooding or waterlogging) on the establishment and yield of many agricultural crops is well documented (for reviews, see 46, 47). Yet wetland species apparently thrive under such conditions of O_2 shortage (90). Is there a fundamental difference in the biochemistry of "flood-tolerant" and "flood-intolerant" species? Perhaps such a difference can be exploited to help develop crops, through molecular biology and plant breeding, that could tolerate longer periods of O_2-shortage.

THE OXYGEN STATUS OF CELLS AND TISSUES

Research advances have come about through the design of O_2-sensitive microelectrodes that are small relative to root diameter, and consume negligible amounts of O_2 (11, 79, 113). These allow direct measurement of O_2 partial pressures in the rhizosphere and within tissues. More rigorous control over the composition of gases used to sparge the rooting zone may have contributed to improved uniformity of results (122), as has more widespread recognition that appreciable fluxes of gases occur within tissues between shoots and roots. Thus, in experiments in which roots are sparged with O_2-free N_2, but the shoots are exposed to ambient air, O_2 concentrations in the solution can be 5–10% [v/v in the equilibrium gas phase (51)]. This implies that such a root system is extremely heterogeneous, with normoxic, hypoxic, and in the apical zones where O_2 consumption rates are high, possibly anoxic cells. Such heterogeneity can be minimized by confining shoots and roots to the same, semiclosed environment so that contamination from ambient O_2 in air is avoided during sparging (6, 85).

Root apical zones, with rapid rates of O_2 consumption and few intercellular spaces to conduct O_2 by gaseous diffusion, are liable to hypoxia even in air when temperatures (and therefore respiration rates) are high (12). At 25°C the critical oxygen pressure (COP)—the O_2 partial pressure or concentration below which O_2 consumption rates begin to be inhibited by the restricted supply of O_2—is about 30% in well-stirred solution and about 10% in a water-vapor saturated gaseous environment (137, 150). The difference in COP values occurs because of the absence of a liquid boundary layer in the gaseous environment. The COP coincided with the O_2 concentration at which ATP/ADP ratios began to decline (157), which indicated a constraint on oxidative phosphorylation. The concentration of dissolved O_2 in water at 20°C in equilibrium with air is 277 μM, so that estimates (105) of the K_m for cytochrome oxidase and the alternate oxidase in mitochondria isolated from soybean roots, 0.14 and 1.7 μM respectively, emphasize that at the mitochondrial level, very low $[O_2]$ develop, despite the high values of the COP. This extreme disparity in $[O_2]$ is explained by a high resistance of cells to the radial diffusion of dissolved O_2.

When excised mature roots were made hypoxic by sparging with about 10% O_2, then separated into cortex/epidermis and stele, typical products of anaerobic metabolism (alanine, ethanol) as well as elevated activities of ADH and PDC (for abbreviations, see Figure 1) were found in the stele but not the cortex (150). Clearly, hypoxia must be associated with extreme radial gradients of $[O_2]$, and this conclusion has a number of important implications. One is that ethanol produced by the anoxic stele could be metabolized by the aerobic cortex (150). Another relates to the Crafts-Broyer hypothesis, which postulated that even in well-aerated environments the final step in radial transfer of ions to the xylem involved a passive leakage, because of stelar anoxia (45). The presence of elevated concentrations of alanine in the stele even when roots were in solution sparged with air, together with the high COP, supports the Crafts-Broyer hypothesis, at least under conditions of rapid respiration. A further implication is that the anoxic core of cells, in which ATP is being generated more slowly by glycolysis and fermentation, receives ATP from the sheath of normoxic cells. Experimental support for this notion comes from study of the radial symplastic movement of a fluorescent ATP analog, TNP-ADP (31).

Direct measurements of $[O_2]$ with microelectrodes demonstrate that hypoxia is a common occurrence in metabolically active root tissues (11, 14, 79).

With maize roots exposed to air, steep gradients of $[O_2]$ were also found (113), with a drop from 20.6% at the epidermal surface to about 10% in the center (apical zone, root in vermiculite) or to as low as 4–5% in the stele (mature zone, root in stirred nutrient solution).

Wetland species have roots with lignified and suberized secondary cell walls that develop within a few millimeters of the root tip (30). These structures help conserve O_2 by preventing its radial loss to the surrounding medium (which in an anaerobic soil at low redox potential is a powerful sink for O_2) so that more O_2 is directed toward the meristematic zone. In the tip zone, loss to the medium does occur, but the effect is to detoxify reduced components and maintain an oxidized rhizosphere. Oxidation of Fe^{2+} to Fe^{3+} at the root surface and in cell walls can give rise to depositions of epidermal/hypodermal ferric hydroxide (iron plaque) (36) with a capacity to scavenge metals. The presence of iron plaque on roots of rice grown in nutrient solution did not appear to interfere with growth, and in the presence of elevated Cu or Ni, these metals became concentrated on the plaque, and toxicity was avoided (68).

METABOLISM UNDER ANOXIA

The ability of roots to grow and to function in the delivery of inorganic nutrients, water, and phytohormones to the shoots and other sinks is essential to plant survival (13, 45–47). Other situations where O_2 supply is critical to plants include seed germination, which sometimes involves anaerobic metabolism (23, 24, 50), survival of submersed coleoptiles and rhizomes (13, 90), oxygenation of root nodules (41), gas exchange in bulky fruits and storage organs under natural conditions and with low O_2 storage (89, 98), and ice encasement of leaves (5, 70).

Under field conditions, flooding and the ensuing O_2 shortage lead to low soil redox potential, and accumulation of reduced substances including NO^-_2, Mn^{2+}, Fe^{2+}, and H_2S, as well as intermediates in microbial carbon metabolism such as acetic and butyric acid (47). Under some conditions, these solutes accumulate to phytotoxic levels and contribute to plant injury. However, characteristic signs of flooding damage occur in plants in solution culture shortly after its deoxygenation (47), suggesting that O_2 shortage is the primary factor in flooded soil to which plants respond, and that deoxygenated solutions mimic the deprivation of O_2 to which roots are exposed in badly aerated soil. But in using such simplified systems, resistance mechanisms that function in a natural environment may be overlooked. For example, in its natural habitat the rhizomatous marshland monocot *Acorus calamus,* which is highly tolerant of anoxia, is exposed to high concentrations of S^-. This enhances accumulation of glutathione, which may serve both to detoxify S^- and act as an antioxidant during postanoxic periods when reactive O_2^- species are potentially damaging (13).

A complication in the plant literature is the use of different terminologies to describe O_2^--depleted cells. In this review, tissues or cells are defined as hypoxic when the O_2 partial pressure limits the production of ATP by mitochondria. Anoxia occurs when the production of ATP by oxidative phosphorylation is negligible relative to that generated by glycolysis and fermentation (119). Normoxia occurs when the O_2 supply does not limit oxidative phosphorylation. The terms O_2 shortage or O_2 deficiency here indicate more general situations or those where the precise degree of O_2 depletion was unknown.

Mechanism of Cellular Injury

All plant cells are able to survive periods of anoxia of an hour or more and sometimes much longer, without cell death. Normally, the ATP content is sufficient for only 1–2 min in cells that are metabolically very active (126, 127). Anaerobic metabolism must therefore contribute to cell survival in the short term by allowing ATP regeneration. At the subcellular level, elongation and swelling of mitochondria are detectable within a few minutes of anoxia (3, 4), but changes in the fine structure are almost as quickly restored upon reintroduction of O_2, with an equally rapid rise in energy metabolism to normoxic levels (4). Only after 15-h anoxia was there irreversible damage to mitochondrial structure, energy metabolism, and cell viability (4). What then leads to the eventual demise of cells that are not resistant to long-term anoxia? Injury and death of roots have been attributed to the accumulation of toxic end products of anaerobic metabolism, to lowering of energy metabolism, or to lack of substrates for respiration. High concentrations of sugars including fructans in roots during anoxia (1, 25, 158) make starvation of respirable substrates seem unlikely, but transport of these sugars to the apical zone is not assured under anoxia. Root tissues are liable to lose reserves of sucrose quickly, because of inhibition of phloem transport to anoxic roots (157), where specifically the unloading step is affected (135). With excised roots, exogenous supplies of glucose boost fermentation rates and energy metabolism (80, 138, 158) and sometimes enhance survival of intact roots (157, 160) or maintain normal mitochondrial structure (153), so that adequate provision of readily respired sugars may sometimes improve survival.

For more than a decade, research has focused more on the fact that cell death under anoxia is closely associated with acidification of the cytoplasm. Noninvasive ^{31}P- and ^{13}C-NMR observations of the root-tips of maize have provided insight into the changes in cytoplasmic pH that accompany the sudden transition from normoxia to anoxia (126, 127, 129). The pH of the cytoplasm (7.3–7.4 units) showed an early decrease that was attributed to an initial production of lactic acid. After about 20 min, the pH remained steady at 6.8, corresponding with a diversion of fermentation to ethanol rather than lactic acid (or other acids such as malic) (Figure 1). This switch in fermentation pathway can be accounted for by the ability of pH <7 to inhibit LDH and activate PDC. Eventually, a second phase of cytoplasmic acidosis occurred that corresponded with a loss of protons from the vacuole (initially at pH 5.8) and coincided with death of cells. Proton-translocating ATPases in the tonoplast normally maintain this steep gradient, but with a decline in energy status their activity is presumed to be restricted, with passive leakage of protons to the cytoplasm. Cytoplasmic acidosis is thus viewed as a determinant of cell death in plant cells, as it is in anoxic animal cells (77, 78). Mutant maize roots, lacking a functional *Alcohol dehydrogenase 1* gene (*Adh1*) and with a diminished ability to carry out glycolysis and fermentation, showed a more rapid cytoplasmic pH drop and succumbed more quickly to anoxia than did wild-type roots. A rapid drop in cytoplasmic pH with anoxia has been confirmed in maize root cells by others using ^{31}P-NMR (52, 140) and in root hairs of *Medicago sativa* using pH-sensitive microelectrodes (53). Working with 50-mm-long seminal root segments, so that the predominant response was that of vacuolated cells, anoxia caused an abrupt drop in cytoplasmic pH from 7.35 to 7.15 that remained stable for at least 6 h (52). Other NMR studies with rice shoots (103) found that anoxia caused an initial drop in cytoplasmic pH from 7.4 to 7.0, followed by an alkalinization of both cytoplasm and vacuole. The metabolic reactions that were consuming protons were not examined. In anoxia-intolerant shoots of wheat, the pH declined continuously from 7.4 to 6.6 without a subsequent rise in pH (103). Regulation of cytoplasmic pH is therefore central to survival.

Although much research has centered around the maize root tip model, it is doubtful whether the pattern of response to anoxia can be generalized to all root tissues. Others working with maize root tips found that cytoplasmic pH decreased more rapidly than the production of lactic acid would suggest but paralleled the decrease in concentration of nucleoside triphosphates (140). This places much more importance on concurrent energy metabolism and the availability of ATP to energize tonoplast H^+ pumps. The initial acidification of the cytoplasm during anoxia cannot be attributed to lactic acid production alone (103, 120) but must include protons released at numerous other metabolic steps, especially in hydrolysis of ATP. Cytoplasmic acidosis in cells of rice and wheat shoots could not be accounted for by lactate production (103). An

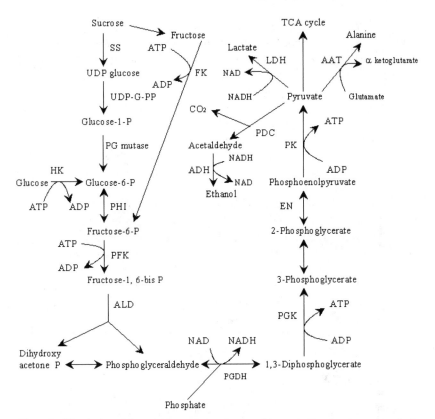

Figure 1 Glycolysis and fermentation pathways in anaerobic respiration showing end products, ATP synthesis, and NAD regeneration. SS, sucrose synthase; UDP-G-PP, UDP glucose pyrophosphorylase; PG mutase, phosphoglucomutase; GK, glucokinase; HK, hexokinase; PHI, phosphohexose isomerase; FK, fructokinase; PFK, phosphofructokinase; ALD, aldolase; PGD, phosphoglyceraldehyde dehydrogenase; PGK, phosphoglycerate kinase; EN, enolase; PK, pyruvate kinase; AAT, alanine aminotransferase; LDH, lactate dehydrogenase; PDC, pyruvate decarboxylase; ADH, alcohol dehydrogenase.

additional consideration is that lactic acid production may be less harmful to viability than supposed. In *Limonium* species, those with greatest flooding tolerance maintained a continuous fermentation to lactic acid, at a rate that matched ethanolic fermentation. Much of the lactic acid appeared in the external medium (123). Experiments aimed at exacerbating the injury to anoxic roots, by overexpression of LDH activity in transgenic tomato did not, however, give the expected result: The fermentative path to lactic acid was insensitive to LDH activity, and rates of lactic acid production by wild-type and transgenic roots were similar (124). A sustained production of lactic acid and export to the external medium has been measured for maize seminal roots distal to the tip, and in nodal roots produced by more mature plants (99). These more mature root zones, in which cells are fully vacuolated, are much more resistant to anoxia than are the metabolically highly active tip zones. Lactic acid fermentation is not necessarily injurious in older root zones, provided lactate and H^+ can be transported to the outer solution. It may be significant that LDH is anoxically inducible in maize and barley roots (66, 81).

The critical role of pH in determining the switch to ethanolic fermentation in anoxic maize root tips was demonstrated in experiments using a permeant weak base (methylamine) to counter cytoplasmic acidosis. Ethanol production rates were then strongly inhibited as cytoplasmic pH returned to a level comparable with that in normoxic cells (55).

Interest in the possibility of ethanol as a cause of cell death during anoxia has diminished with the recognition that exceptionally high concentrations would have to accumulate to interfere in metabolism. Being soluble in the lipid bilayer, ethanol can readily diffuse out of tissues to the surrounding solution where it is diluted or metabolized by microorganisms. Evidence of cell injury following aerobic conversion of ethanol to acetaldehyde has been obtained with carrot cells in suspension culture (117). Whether binding of acetaldehyde to proteins was a causative factor in cell injury was inconclusive.

Resistance to Anoxia

Because plant cells can survive anoxia for at least several hours, the distinction between anoxia-tolerant and anoxia-intolerant is only relative. However, it is clear that species of wetland origin are specialized metabolically so that they can remain viable much longer under conditions where oxidative phosphorylation is suppressed by lack of O_2. Studies have demonstrated the ability to tolerate weeks or months of anoxia in rhizomes and emerging leaves of *Scheonoplectus lacustris, Scirpus maritimus, Typha angustifolia* (13, 20), and in shoots and tubers of an aquatic monocot, *Potamogeton pectinatus* (146) and an amphibious monocot *Acorus calamus* (27). Germination and early growth of embryos and coleoptiles of rice (22, 122) and of some barnyard grass species, such as *Echinochloa phyllopogon* (90) also occurs under anoxia, although further development is O_2 dependent. Other plant cells with a prolonged tolerance to anoxia include beetroot storage tissues (168) and rice cells in suspension culture (107).

Even among these species, medium- to long-term tolerance of anoxia is strictly organ specific: Roots of rice or of *Echinochloa phyllopogon* are as sensitive to O_2 shortage as are roots of maize or barley. The generalized biochemical basis for anoxia tolerance must involve maintenance of glycolysis for generation of ATP, regeneration of NAD from NADH so that glycolysis does not stall, and metabolic end products that are innocuous or are readily transported to the external solution. NAD is required only for the conversion of 3-phosphoglyceraldehyde to 1,3-diphosphoglycerate in glycolysis. This reaction is catalyzed by phosphoglyceraldehyde dehydrogenase; NAD is regenerated equally effectively in ethanolic fermentation or in conversion of pyruvate to lactic acid (Figure 1). Likewise, a net ATP production of 2 mols in substrate-linked phosphorylations in glycolysis is irrespective of the end product. If the starting point for glycolysis is starch, and starch phosphorylases are involved, an additional mol of ATP becomes available per mol of hexose respired. The characteristics of anoxia-tolerant organs of wetland species include a sustained, predominately ethanolic fermentation (13, 22, 27, 49), leakage of ethanol to the external medium (22, 49, 108) or the transpiration stream (60, 100), adequate reserves of carbohydrates to maintain glycolysis and energy metabolism (13, 22, 72, 73, 131, 143), and in some cases high activity of starch phosphorylases (73, 143). In addition, in *Echinochloa phyllopogon* ATP is generated anoxically by continued functioning of part of the TCA cycle, with operation of the pentose phosphate pathway and lipid biosynthesis converting NADH to NAD (57).

Avoidance of cytoplasmic acidosis is critical, and succinate synthesis has been suggested as a means of consuming H^+ during anoxia. There is greater synthesis of succinate relative to lactate in anoxia-tolerant leaf tissues of rice and barnyard grass, compared with anoxia-intolerant wheat and maize leaves (102). However, succinate synthesis was not found in rice roots, and in maize roots it was a minor metabolite under anoxia (52, 129). Protons are also consumed during decarboxylation of glutamate to GABA. Accumulation of GABA in response to anoxia has been noted in several studies [shoots (102), mature roots of maize (52)]. Lowered cytosolic pH levels, induced by hypoxia or addition of a permeant weak acid stimulated GABA synthesis in isolated mesophyll cells (35) and appeared to consume an appreciable fraction of the imposed acid load. However, in 2-mm root tips of maize, the greatest accumulation of GABA occurred after several hours of anoxia, when cell metabolism in general was declining and H^+-producing reactions had ceased (129). In addition, although an ammonium pretreatment increased glutamine levels and thereby enhanced the capacity to make GABA (and alanine) under anoxia, it failed to modify cytoplasmic acidosis. Furthermore, low pH alone was not sufficient to account for activation of glutamate decarboxylase because GABA synthesis began long after the initial fall in cytoplasmic pH. Perhaps a rise in cytosolic [Ca^{2+}], triggered by the low cytoplasmic pH that develops under anoxia, activates glutamate decarboxylase (35).

Alanine is a prominent end product of anaerobic fermentation in roots (10, 52, 65, 121, 149, 150), and alanine amino transferase is an anaerobically inducible enzyme in barley roots (64) but not in maize (159). The functional significance of alanine synthesis under anoxia is unclear. In maize root tips (129), greater synthesis of alanine in ammonium pretreated roots, as noted above, did not modify cytosolic pH, which suggests that it is not associated with a significant production or consumption of protons. Decarboxylation of malate to pyruvate in these same cells between 15–35 min of anoxia coincided with a small rise in cytoplasmic pH, which may indicate a role for malic acid in the transient relief of cytoplasmic acidosis (129). The significance of the production of alanine, succinate, and GABA during anoxia remains in doubt without information on absolute rates of synthesis in roots, relative to the synthesis of ethanol and lactate. A detailed study using labeled glucose has examined quantitatively the C fluxes in intermediary metabolism in normoxic maize roots (44), and comparable information is needed on stressed roots.

Under anoxia, it is possible that pyrophosphate (PPi) might substitute (at least briefly) for ATP as an energy source. Vacuolar H^+-pyrophosphates (V-PPase) might then replace the function of vacuolar H^+-ATPases in pumping H^+ into the vacuole. With rice seedlings, V-PPase showed a 75-fold increase in activity after 6 days of anoxia (28). Preliminary experiments also suggested an increased PPi-dependent H^+ pumping in tonoplast vesicles isolated from anoxically treated plants (28). In rice seedlings, anoxia induced an increase in activity of PPi: Fru-6-P 1 phosphotransferase, which substitutes for ATP-dependent phosphofructokinase in glycolysis (104). When uncouplers or Pi starvation were used to collapse the ATP/ADP and UTP/UDP ratios, PPi levels were relatively insensitive (38), indicating that the energy status of the PPi pool and the nucleotide systems can vary independently. A declining cytosolic pH would also tend to favor PPi as an energy source because it raises the free energy of hydrolysis of PPi while lowering that of ATP (39). However, in root tips of pea (37), anoxia or a respiratory uncoupler failed to modify PPi levels, despite changes in glycolytic rate. This suggested that PPi: Fru-6-P 1 phosphotransferase did not play an important role in relation to glycolysis. Root cells do not have appreciable reserves of pyrophosphate [ca 10 μmol kg^{-1} FW in pea roots (37)], so that the long-term significance of anaerobic metabolic pathways that function with PPi is dubious without a means for generating high-energy phosphate bonds. In anoxic cells, in which the ATP/ADP ratio is low, mitochondrial glucokinases (GK) are strongly inhibited, while cytosolic

GK can transiently generate ATP from glucose 6-phosphate (61). Whether this is important in relation to anoxia tolerance remains to be explored. Availability of NO^-_3 for dissimilatory nitrate reduction (nitrate respiration) has been suggested as a means for production of ATP at the level of complex 1 and regeneration of mitochondrial and cytoplasmic NAD. Higher concentrations of NTP were found in maize root tips in the presence of 25 mM $Ca(NO_3)_2$ (125), and in mature maize roots, 0.2–1.0 mM $NaNO_3$ also enhanced NTP levels, but without concomitant effect on cytoplasmic pH (52). However, with root tips of maize (136), the presence of nitrate failed to modify levels of ATP or any of the adenylates or to enhance viability, and rates of nitrate reduction were negligible relative to NADH generation.

Hypoxic Acclimation to Anoxia

Under natural conditions, roots are unlikely ever to be exposed suddenly to anaerobic conditions. A gradual transition from normoxia to hypoxia to anoxia provides an opportunity for acclimation before conditions become lethal. Under laboratory conditions this transition can be mimicked by making roots hypoxic before anoxic; that is, by exposing them first to subambient $[O_2]$. When seedlings of maize or wheat are subjected to a hypoxic pretreatment (HPT), viability during a subsequent period of anoxia is greatly extended (85, 86, 136, 158, 163), with maintenance of normal subcellular structure (32). The response resembles superficially that of heat shock, in that exposure to a sublethal high temperature induces acclimation to even higher, previously lethal temperatures, but the phenomena probably have little resemblance at the biochemical level.

Understanding the mechanism of acclimation to anoxia may hold the key to identifying genes and gene products that confer anoxia tolerance. Research efforts have focused on two hypotheses or their combination: that acclimation is dependent on improved energy metabolism, or that acclimation involves an improved control of cytoplasmic acidosis. Following HPT, excised maize root tips with exogenous glucose as respiratory substrate maintained high rates of anaerobic respiration for hours and transported ethanol to the external solution. In contrast, in roots of nonacclimated seedlings (NHPT), transferred directly from normoxia to anoxia (anoxic shock), anaerobic respiration collapsed after 1.5 h (80). Improved anoxia tolerance of maize and wheat root tips was also associated with greater concentrations of ATP and total adenylates (85, 136), and with sustained levels of ADH and PDC activities (158; JR Johnson, BG Cobb & MC Drew, unpublished data). Hypoxia increased transcript levels for aldolase, enolase, PDC, ADH1, and ADH2, and during anoxia these mRNA levels declined only gradually (6–8). In anoxically shocked roots (NHPT), increases in transcript levels were transient, usually reaching a peak at 6 h and thereafter quickly disappearing. It appears that hypoxia leads to acclimation in part by inducing early expression of "anaerobic genes," encoding enzymes of glycolysis and ethanolic fermentation. This might serve to both maintain a higher energy metabolism under anoxia and direct metabolism toward ethanolic fermentation with less initial cytoplasmic acidosis, and hence improved viability.

Other studies have identified induction of a lactate transport system from the cytosol to the external solution as an important aspect of hypoxic acclimation. In HPT root tips, there was less cytoplasmic acidosis, and a greater transport of lactate to the external solution relative to NHPT roots suggesting that synthesis of a lactate transport mechanism was induced under hypoxia (163, 165). HPT tomato roots also show an enhanced transport of lactate to the external solution (124). Whether lactate was involved in a lactate-H^+ cotransport mechanism is uncertain at present (164). Although H^+- ATPases functioned under anoxia, and were stimulated by falling cytoplasmic pH, net H^+ extrusion appeared to contribute little to cytoplasmic pH regulation and did not account for the greater ability of HPT root tips to limit cytoplasmic acidosis.

The importance of the greater energy metabolism in HPT root tips has been questioned by experiments in which ATP levels or energy charge were lowered by treatment of acclimated roots during anoxia with fluoride or mannose. ATP decreased to levels at or below those of NHPT root tips, but HPT roots maintained higher cytoplasmic pH values, and viability (166). Higher concentrations of F^- that depressed the rates of ethanolic fermentation to less than half that of HPT controls did, however, lead to less viability, suggesting that a minimum level of ATP was necessary for cell maintenance. If lactic acid production is the initial source of cytoplasmic acidosis, an early shortage of ATP may have minor influence on pH regulation. In the long term, when leakage of H^+ from acidic vacuoles is the main contribution to cytoplasmic acidosis (126, 127), it is difficult to imagine that ATP levels would not have a function in energizing vacuolar H^+- ATPases and cytoplasmic pH regulation.

What explains the collapse of glycolysis and fermentation (80, 165) in NHPT root tips soon after the onset of anoxia? Extracts from HPT and NHPT root tips of maize were assayed for enzyme activities at pH of 7.5 and 6.5, equivalent to those found in the cytoplasm during normoxia and hypoxia, respectively. In excised or intact root tips, sucrose was the principal substrate of glycolysis, until the pool was exhausted. Glucose and fructose were then utilized, consistent with the view that in intact anoxic roots, transport of sucrose is severely limited. Of 10 enzymes of glycolysis and fermentation tested, only GK showed a marked reduction in activity at the lower pH (25). GK is inhibited at low ATP/ADP ratios, and the absolute concentration of ATP in NHPT root tips was calculated to be close to the K_m for GK. In addition, only the activity of GK was near that of the in vivo rate of glycolysis (estimated from the rate of accumulation of ethanol and lactate); the activity of most other enzymes that were assayed were in excess. Incidentally, induction of high levels of GK activity occurs in anoxia-tolerant seedlings of *Echinochloa phyllopogon* but not in anoxia-sensitive *E. crus-pavonis* (56).

In conclusion, hypoxic acclimation to anoxia in root tips seems to be associated with avoidance of lactate accumulation and cytoplasmic acidosis, combined with a sustained level of glycolysis and ethanolic fermentation to generate ATP. Synthesis of enzymes that contribute to the above metabolic pathway, particularly those that regulate it such as glucokinase, may be critical.

Postanoxic Injury

Because no plant organs are known to survive indefinitely under anaerobic conditions, an ability to deal with the consequences of reexposure to O_2 is important. Specifically, metabolism of the reactive oxygen species generated during aerobic metabolism is an important corollary to anoxia tolerance. One protective system involves superoxide dismutase (SOD), converting superoxide radicals to hydrogen peroxide, which is reduced to water by peroxidases or catalases. In anoxia-tolerant rhizomes of *Iris pseudacorus,* SOD activity increased some 13-fold during 28 days of anoxia but failed to increase in *Iris germanica* and *Glyceria maxima,* which tolerate anoxia for much shorter periods (13, 109). This suggests that under anoxia, enzymes are induced that have a role in protection of cell metabolism from a subsequent exposure to O_2. This "anticipation" of the reintroduction of O_2 in rhizomes is explicable in terms of the life cycle, because these organs can overwinter in anaerobic mud and produce leaves that emerge above the water surface in spring, whereupon the rhizome must switch from anaerobic to aerobic metabolism. After only 1–2 h of anoxia, there was greater injury to soybean seedling root tips on exposure to air than for seedlings made anoxic for 5 h before re-aeration (152). Injury correlated with depression of the SOD mRNA level and enzyme activity after 1 h of anoxia, with a subsequent rise in SOD with longer periods of anoxia. Short-term postanoxic injury was avoided if tissue was incubated in ascorbate (152), an antioxidant and free radical scavenger. In roots of intact wheat seedlings made hypoxic or anoxic by sparging the roots with N_2, there was an increase in the pool of the antioxidant, reduced glutathione (GSH). A rapid increase in oxidized glutathione (GSSG) occurred after only 20 min of reoxygenation (2), but this disappeared by 16 h, and there was restoration of a high ratio of GSH/GSSG. Whether the glutathione system was effective in protecting wheat roots is difficult to gauge. Despite the presence of GSH, which presumably served to limit damage, there was a 50% increase in products of lipid peroxidation after 2 h of exposure to oxygen, and the significance in relation to root injury remains to be demonstrated.

With the reintroduction of O_2, ethanol accumulated within roots and their surroundings can be metabolized to acetaldehyde, catalyzed by ADH, and then converted to acetate by acetaldehyde dehydrogenase, and then to acetyl CoA (34). Considerable metabolism of ethanol occurs in submerged roots of intact rice seedlings at night (156) when lack of O_2 transfer from shoots to roots causes the roots to become hypoxic. The ability of roots to tolerate diurnal surges of ethanol metabolism, and resume growth when they are made normoxic following illumination of the leaves and O_2 transfer (156), suggests that acetaldehyde does not accumulate to harmful levels. Whether aerobic metabolism of ethanol is important to root survival or to the C economy of the plant is unknown.

Gene Expression and Protein Synthesis Under Anoxia

Although the apical zone of maize seminal roots is sensitive to anoxic shock, more mature zones distal to the apex including lateral premordia remain viable for several days. Changes in gene expression and protein synthesis in young whole maize roots under anoxia have been intensively studied (reviewed in 19, 134). Because the majority of the root is composed of mature, vacuolate cells of greater anoxia tolerance, the response is not identical to that of the apical zone. Anoxia quickly inhibits protein synthesis in general, with dissociation of polysomes (18). Despite that, a select group of proteins—the anaerobic proteins (ANPs)—continue to be made. Of the 20 ANPs initially identified in maize seedling roots, most are enzymes of glycolysis and fermentation, and the corresponding cDNAs have been cloned (122, 134). An exception is a cDNA with homology to xyloglucan endotransglycosylase that was isolated from anoxic maize roots and may function in cell-wall loosening in aerenchyma formation (134). It is not yet clear whether the response of seedling maize roots provides a generalized biochemical model, because few other species have received detailed investigation. In soybean roots, which are less tolerant of anoxia than maize, only four major proteins were identified, of which one was ADH (132). In addition, the anaerobic response in maize depends on seedling age or development. In root tips there were markedly greater initial levels of *Adh1* activity, and inducibility of ADH1 transcripts and enzyme activity in response to anoxia in seedlings at 3 days, compared with 5 days postimbibition (7).

The basis for the selective synthesis of the ANPs is not well understood: The glycolytic enzymes are required under aerobic conditions as well, and their mRNAs must exist in the anoxic cell alongside thousands of other aerobic messages that remain translatable in an in vitro system (134). It is possible that cytoplasmic acidosis may contribute to the translational control at the level of elongation and termination of protein synthesis (161). Certain proteins in an in vitro translation system were selectively enhanced at low pH. Anoxia-induced phosphorylation of an initiation factor (eIF-4A) could also contribute to depressing protein synthesis (162). Transcripts of *Adh1* are more effective at loading ribosomes under anoxia than are transcripts of genes that are only aerobically expressed, and elongation is greater on *Adh1* transcripts (54). In a transient expression system, mRNA constructs comprising 5' and 3' UTR sequences of *Adh1* fused to the coding region of a reporter gene (GUS) were electroporated into maize protoplasts (17). Enhanced translation under hypoxia (up to 57-fold) depended on the ADH1 5'-UTR as well as on the 3'-UTR; transcript stability was unaffected by hypoxia.

Oxygen deficiency strongly stimulates the transcription of genes encoding ANPs (130, 134)—although not all are translated (29)—suggesting that maintenance of glycolysis and fermentation is afforded a high priority. However, the significance to cell survival is questionable, because we lack evidence that enzyme activity is limiting. ADH levels can be shown to be greatly in excess in anoxic or hypoxic roots relative to the rate of fermentation, and ADH levels above a threshold do not correlate with ability to survive anoxia (7, 86, 128). Of the enzymes of glycolysis and fermentation, only the activities of GC and PDC in root extracts are close to the in vivo rates of fermentation (25, 111, 158). Induction of enolase in anoxia-tolerant *Echinochloa* seedlings may serve some other function, perhaps in stabilizing or protecting cellular structure during anoxia (58). Another possibility is that an excess of glycolytic and fermentative enzymes may be necessary to drive a rapid transition from a low to a high glycolytic flux without change of metabolite concentration (122). In the case of ADH, high levels of activity may be of greater significance during the subsequent aerobic conversion of ethanol to acetaldehyde (150, 156).

Among the enzymes of glycolysis and fermentation, PDC is interesting because its activity increases markedly with lower cytosolic pH (111), which serves to divert metabolism away from lactate and malate fermentation, and because its transcription is anaerobically (8, 134) and hypoxically (8) inducible. The activity of PDC in vitro is close to the in vivo fermentation rate, and it may have an important role in regulating ethanol fermentation. Constitutive overexpression of PDC has been studied in transgenic tobacco (26) transformed with a construct comprising a 35S promoter and a bacteria PDC. This construct yields high rates of ethanolic fermentation under anoxia, but no details on the response to anoxia of roots expressing high levels of PDC activity have been reported.

Parallel induction of mRNA levels of some of the ANPs has been reported (8, 19, 134) suggesting that O_2 shortage causes coordinate induction of a select number of genes. Of four genes (*Adh1, Pdc, Ald, En*) examined in maize root tips and in more mature root zones, with either hypoxia or anoxia, transcript levels increased under conditions of O_2 shortage in at least some tissues. However, the steady state levels of the four mRNAs differed in the rapidity and magnitude of their increase, and the time to reach peak levels, with no discernible common pattern (8). Likewise, three *Pdc* genes were hypoxically induced in maize seedlings but with different levels of mRNA production and kinetics of induction (118). In roots and other organs of *Acorus calamus, Pdc, Adh,* and *Ald* did not appear to be coordinately induced, with peak message levels occurring at different times after the onset of anoxia (27). A differential pattern of accumulation of transcripts of genes associated with glycolysis and fermentation was also found in rice seedlings (roots and shoots analyzed together) in response to submergence stress (151). If there is coordinate regulation of transcription of these ANPs, posttranscriptional processing (62) and message stability are likely to vary for different mRNA species to modify gene expression.

Several of the genes encoding ANPs (*Adh1, Adh2, Sh1, Ald*) in maize, pea, and Arabidopsis share a consensus motif (ARE) in the promoter regions (42, 94, 114, 154); the anaerobic response element may function in induction of gene expression under anoxia, and thus provide a basis for coordinate induction. Nuclear proteins have been identified that bind to the ARE of maize *Adh1* promoter (114) or with sequences outside the ARE (116). However, occurrence of ARE-like sequences is insufficient to account for anoxic induction. Similar sequences appear in the promoters of genes that are not anoxically inducible (42, 94, 118). Furthermore, the promoters of *Adh1* and *Adh2*, both inducible by anoxia, showed very different patterns of binding of nuclear proteins, as revealed by DMS footprinting (116). A cDNA encoding a *trans*-acting DNA protein (43) that binds to a G-box element in the maize *Adh1* promoter has been isolated. The deduced amino acid sequence of this G-box binding factor (GBF) indicates a basic-leucine zipper protein, although the authors point out that evidence concerning its precise mode of regulation of transcription in response to anoxia is inconclusive. Analyses of the *Adh* promoter and the role of DNA-binding proteins have been extended to Arabidopsis (40, 97) and the role of G box elements in relation to ADH induction by ABA, independently of O_2 shortage. A possible link between anoxic gene expression and Ca^{2+} signaling (see next section) may emerge from the fact that the GBFs interact with 14-3-3 proteins that can phosphorylate and can bind Ca^{2+} (43).

SIGNALING OF HYPOXIA AND ANOXIA

Until recently, little attention has been given to the question of how plant cells sense changes in O_2 concentration. Such a sensing mechanism can be inferred from observations of rapid changes in gene expression and levels of enzyme activity with anoxia, or with hypoxia or with re-establishment of normoxia. Are these changes signaled simply by low levels of ATP, or adenylates, or by cytoplasmic acidosis in anoxic cells? Or is there a sensing mechanism for O_2 status, such as the redox of an electron carrier, e.g. heme, closely linked to the cytochrome electron transport chain? Aerobic fermentation was induced in tobacco leaves in the presence of respiratory inhibitors that lowered energy metabolism, but ADH was not induced under these conditions. This observation has been interpreted as indicating that an O_2 sensing mechanism other than energy metabolism is required for ADH induction (26). Similar considerations apply to the sensing of O_2 in animal cells. In brine shrimp embryos, mitochondrial protein synthesis during anoxia is strongly inhibited, but this cannot be accounted for by blockage of the electron transport chain or change in redox state, or adenylate concentrations. A molecular oxygen sensor that serves to rapidly depress protein synthesis has been proposed (93).

Sensing mechanisms that respond to decreasing O_2 concentration have been thoroughly characterized in *Escherichia coli* (33, 71), *Rhizobium meliloti* (95, 110), and *Saccharomyces cerevisiae* (169). In the above bacterial examples, two component sensor-response regulatory systems have been defined both genetically and biochemically. In *R. meliloti* within alfalfa root nodules, an O_2-binding, membrane hemeoprotein constitutes the sensor. This protein responds by autophosphorylating as the initial step in the mechanism by which O_2 concentration induces expression of bacterial nitrogen fixation genes (95, 110). In *S. cerevisiae,* heme is also important in activation and repression of genes in response to O_2 concentration (133, 169). Identification of genes encoding hemoglobin in nonlegume plants (9) has raised the possibility that heme may also contribute to sensing O_2 concentration in higher plants. Preliminary work with maize roots found induction of hemoglobin gene expression with flooding (112). Root hypoxia, more than anoxia, induces some of the

ANPs (6), supporting the notion that O_2 levels are sensed within roots, but as there may be an anoxic core and normoxic sheath, the evidence is ambiguous. All the induction may be confined to the anoxic cells; the normoxic ones could function as sources of ATP for them via the symplasm (31). Stronger support comes from observation of *Adh1* mRNA induction by hypoxia (5–10% O_2) in maize cells in suspension culture or in protoplasts (17, 115) where steep gradients of O_2 concentration across cells or small groups of cells, causing anoxia, would seem unlikely.

Transient changes in the concentration of cytosolic Ca^{2+} are part of a signal transduction pathway triggered by anoxia in maize roots (145). Ruthenium red (RR) was used to inhibit Ca^{2+} release from subcellular organelles, and anoxic induction of *Adh1* and Sh1 mRNAs as well as ADH enzyme activity were blocked. Just 2 h of anoxia was sufficient to kill seedlings treated with RR, compared with the usual viability of about 70 h. Inclusion of Ca^{2+} eliminated the inhibitory effects of RR. Using a fluorescent probe to quantify cytosolic $[Ca^{2+}]$, there was an increase within 1–2 min of the start of anoxia, and this increase was dependent on intracellular not exogenous Ca^{2+} (144). Under aerobic conditions, increasing cytosolic Ca^{2+} with caffeine induced ADH activity as if the cells were anoxic. The rapidity with which $[Ca^{2+}]$ changes during anoxia suggests that it responds to the drop in cytoplasmic pH or to lowered energy metabolism. By contrast, in roots of Arabidopsis seedlings, anoxia did not appear to cause an increase in cytosolic Ca^{2+}, even though *Adh* transcript levels increased (141), but the system of Ca^{2+} detection (transgenic aquorin luminescence) may not have been sufficiently sensitive to detect luminescence from cells within the root tissue. The signaling pathway between increased $[Ca^{2+}]$ and changes in expression of *Adh* and the other ANPs remains to be discovered.

Hypoxia, as distinct from anoxia, may also be signaled by ethylene. The synthesis of ethylene by roots is strongly promoted by hypoxia, but blocked by anoxia (16, 75, 83) because of the requirement for free O_2 for the conversion of ACC to ethylene. The role of ethylene in signaling cell death during aerenchyma formation also implicates Ca^{2+} as a second messenger, as discussed below.

In animal cells, a variety of O_2-sensing mechanisms are beginning to emerge. These include ATP-sensitive K^+ channels (147), which have also been identified in Arabidopsis (148), and O_2-sensitive K^+ and Ca^{2+} channels (59, 96). The significance of these sensors in relation to low-O_2 responses of higher plants is unknown.

ANOXIA AVOIDANCE BY SELECTIVE CELL DEATH AND AERENCHYMA FORMATION

Improved Oxygen Status of Aerenchymatous Roots

Aerenchyma formation is commonly found in the stems and roots of aquatic and flood-tolerant species, developing by cell separation during development (schizogeny) or by cell death and dissolution (lysigeny). Aerenchyma formation in the roots of such species is usually constitutive, requiring no external stimulus. The ability to form lysigenous aerenchyma is widespread even among dry-land vegetation (87). In roots lacking aerenchyma, the resistance to O_2 movement in gas-filled intercellular spaces is relatively large and provides a path length of only about 80 mm before $[O_2]$ becomes too attenuated to support the metabolism of the apical zone (139). For oxygenation over greater distances, aerenchyma is necessary and effective because it cuts back on the number of O_2-consuming cells, and lowers the resistance to gas diffusion or convection. Aerenchyma in maize nodal roots gave improved oxygenation over distances of 300 mm, as determined by higher levels of ATP and adenylate energy charge (48). Oxygen concentrations as high as 16% were measured in gases in young aerenchymatous wheat roots in solution sparged with N_2 with the shoots in air (51). However, even aerenchymatous roots are limited in their ability to conduct O_2 over long distances to the apical zone: In cereal root tips, lowered energy metabolism (48) and increased production of ethanol, alanine, and lactate suggest that anoxia can develop (63, 150). Induction of ADH activity in aerenchymatous roots of *Spartina alterniflora* by flooding of marsh soil also implies that the roots were not fully normoxic (101). Notwithstanding, the effectiveness of internal transport of O_2 to roots through interconnected gas-filled spaces over appreciable distances is well documented (15, 21, 69, 167).

Cell Death in Lysigenous Aerenchyma Formation

The formation of aerenchyma was long thought to be a consequence of anoxia, but analysis of sunflower stems led to the discovery that cellular disintegration is promoted by ethylene when environmental conditions raised tissue concentrations (for earlier references, see 82). In roots of maize, low concentrations of the gas in air ($0.1–1.0$ $\mu L \cdot liter^{-1}$ air) trigger cell death selectively in the cortex of normoxic roots, resulting in an aerenchymatous structure like that induced by hypoxia. Hypoxic roots contain higher concentrations of ethylene than normoxic ones, have higher concentrations of the ethylene precursor ACC (16), and display a greater activity of ACC synthase (74) and ACC oxidase

(75). Anoxia alone fails to induce cell lysis and aerenchyma formation: ACC oxidase requires molecular O_2 to convert ACC to ethylene, and anoxia also lowers the activity of ACC synthase (74). However, this is not the full explanation because anoxic, viable roots cannot respond when supplied with ethylene exogenously (83).

In rice roots, lysigenous aerenchyma is constitutive, developing in the absence of O_2 shortage (84), which suggests that cell death is solely under developmental control. However, involvement of ethylene cannot be ruled out because wall thickenings in the hypodermal layers must entrap endogenous ethylene. In some rice cultivars, exogenous ethylene enhances aerenchyma formation, and Ag^+, an ethylene action inhibitor, reduces it (88).

Induction of the cell lytic enzymes that accompany cell death and accomplish dissolution has received little attention. In hypoxic maize root tips, the activity of cellulase increases some 16-fold immediately before aerenchyma is detectable (74), an increase that is blocked by addition of AVG (an inhibitor of ACC synthesis) and reversed by simultaneous addition of ethylene, indicating that cellulase is induced by the hormone. Phospholipase D (PLD) mRNA and enzyme activity are stimulated by hypoxia early in aerenchyma formation (CJ He, J Jin, PW Morgan & MC Drew, unpublished data), but it is not yet clear whether PLD is solely involved in membrane degradation or is part of a signal transduction pathway.

The ethylene signal transduction pathway leading to cell death has been investigated in hypoxic and normoxic roots (76) by applying various antagonists that modify specific steps in signal transduction in other biological systems. Both cellulase activity and cell death in the root cortex were blocked in hypoxic roots by antagonists of inositol phospholipids, CaCaM, cytosolic $[Ca^{2+}]$, and protein kinases. It was possible to promote cell death in the cortex of normoxic or hypoxic roots by reagents that activate G proteins, raise cytosolic $[Ca^{2+}]$, or inhibit protein phosphatases. The evidence points to a signal transduction pathway culminating in a rise in cytosolic $[Ca^{2+}]$ as necessary for cell death in hypoxic roots (76). Although $[Ca^{2+}]$ is involved in signaling in anoxic roots (145) as well as in hypoxic ones, the consequences in the different tissues, and under different environmental conditions, appear to have little in common. In anoxic cells, acclimation is associated with prolonging cell survival, and in mature hypoxic ones, cell death.

The signal transduction pathway between hypoxia and increased activity of ACC synthase and ACC oxidase is obscure. None of the antagonists or reagents affected ethylene biosynthesis with the exception of caffeine, which was inhibitory (76). Ethylene response mutants in *Arabidopsis thaliana* have been used to define components of an ethylene perception and signal transduction pathway (92) and could help link pharmacological and genetic approaches.

The death of cells specifically in the root cortex, triggered by ethylene, has features in common with the genetically controlled processes of programmed cell death (PCD) or apoptosis in animal cells (91). Recent studies of cell death in higher plants indicate that PCD is involved in the hypersensitive reaction to pathogens (67), death of maturing xylem cells (106), and root cap cells (155), and it is noteworthy that an increased $[Ca^{2+}]$ is implicated in activation of enzymes associated with PCD in animal cells (e.g. 142). It is not yet clear whether ethylene-dependent cell death in the root cortex takes place by PCD, i.e. is directed by the cell nucleus, or by necrosis, in which an external agent such as a toxin initiates degeneration, usually beginning at the plasma membrane. In rice seminal roots, there was evidence of ethylene induction of aerenchyma, and an acidification of cells (as determined by neutral red staining) in the midcortex, before their lysis (M Kawai, PK Samarajeewa, M Nishiguchi & H Uchimiya, unpublished data). Starting from the midcortex, cell death followed sequentially along radial files of cells, which suggests a passage of molecules from cell to cell that induced cell death. However, such a pattern, in the absence of further information, could correspond to either PCD or necrosis.

CONCLUSION

One aim of research into mechanisms of resistance to O_2 shortage is to identify genes regulating metabolic changes that lead to cellular damage, as well as those that contribute to tolerance and to avoidance. Studies involving overexpression and antisense inhibition have scarcely begun in this field, and the potential opportunities for modifying plant response are immense. It should be possible to modify the biochemical mechanisms that lead to acidosis, to enhance pathways that consume protons, to enhance rates of glycolysis and fermentation and thereby improve energy metabolism. A better understanding of the signal transduction pathways that result in ethylene-dependent cell death may help in devising means for enhancing the speed and extent of aerenchyma formation. Information on plant responses to hypoxia or anoxia is limited at present to a handful of species, and it is likely that numerous mechanisms of tolerance to O_2 deficiency reside genetically in an unexplored wetland vegetation. Research in these areas is relevant to practical agriculture and to plant/environment relations in a broader context. Such studies will also contribute to a fundamental understanding of plant biology.

ACKNOWLEDGMENTS

I thank Drs. B Greg Cobb, Ted Fox, Chuan-Jiu He, Page W Morgan, and Mary E Rumpho for helpful discussions on this topic. This review was prepared while the research in the author's laboratory was supported by the USDA National Research Initiative Competitive Grants Program (#93-37100-8922).

Visit the *Annual Reviews home page* at http://www.annurev.org.

Literature Cited

1. Albrecht G, Kammerer S, Praznik W, Wiedenroth EM. 1993. Fructan content of wheat seedlings (*Triticum aestivum* L.) under hypoxia and following re-aeration. *New Phytol.* 123:471–76

2. Albrecht G, Wiedenroth EM. 1994. Protection against activated oxygen following re-aeration of hypoxically pretreated wheat roots: the response of the glutathione system. *J. Exp. Bot.* 45:449–55

3. Aldrich HC, Fel RJ, Hils MH, Akin DE. 1985. Ultrastructural correlates of anaerobic stress in corn roots. *Tissue Cell* 17: 341–46

4. Andreev VY, Generozova IP, Vartapetian BB. 1991. Energy status and mitochondrial ultrastructure of excised pea root at anoxia and postanoxia. *Plant Physiol. Biochem.* 29:171–76

5. Andrews CJ, Pomeroy K. 1989. Metabolic acclimation to hypoxia in winter cereals: Low temperature flooding increases adenylates and survival in ice encasement. *Plant Physiol.* 91:1063–68

6. Andrews DL, Cobb BG, Johnson JR, Drew MC. 1993. Hypoxic and anoxic induction of alcohol dehydrogenase in roots and shoots of seedlings of *Zea mays*: Adh transcripts and enzyme activity. *Plant Physiol.* 101:407–14

7. Andrews DL, Drew MC, Johnson JR, Cobb BG. 1994. The response of maize seedlings of different age to hypoxic and anoxic stress: changes in induction of *Adh1* mRNA, ADH activity and survival of anoxia. *Plant Physiol.* 105:53–60

8. Andrews DL, MacAlpine DM, Cobb BG, Johnson JR, Drew MC. 1994. Differential induction of mRNAs for the glycolytic and ethanolic fermentative pathways by hypoxia and anoxia in maize seedlings. *Plant Physiol.* 106:1575–82

9. Appleby CA, Bogusz D, Dennis ES, Peacock WJ. 1988. A role for haemoglobin in all plant roots? *Plant Cell Environ.* 11: 359–67

10. ap Rees T, Jenkin LET, Smith AM, Wilson PM. 1987. The metabolism of flood tolerant plants. See Ref. 35a, pp. 227–38

11. Armstrong W. 1994. Polarographic oxygen electrodes and their use in plant aeration studies. *Proc. R. Soc. Edinburgh B* 102: 511–27

12. Armstrong W, Beckett PM. 1985. Root aeration in unsaturated soil: a multishelled mathematical model of oxygen diffusion and distribution with and without sectoral wet-soil blocking of the diffusion path. *New Phytol.* 100:293–311

13. Armstrong W, Brändle R, Jackson MB. 1994. Mechanisms of flood tolerance in plants. *Acta Bot. Néerl.* 43:307–58

14. Armstrong W, Strange ME, Cringle S, Beckett PM. 1994. Microelectrode and modelling study of oxygen distribution in roots. *Ann. Bot.* 74:287–99

15. Ashford AE, Allaway WG. 1995. There is a continuum of gas space in young plants of *Avicennia marina. Hydrobiology* 295: 5–11

16. Atwell BJ, Drew MC, Jackson MB. 1988. The influence of oxygen deficiency on ethylene synthesis, 1-aminocyclopropane-1 carboxylic acid levels and aerenchyma formation in roots of *Zea mays* L. *Physiol. Plant.* 72:15–22

17. Bailey-Serres J, Dawe RK. 1996. Both 5′ and 3′ sequences of maize *adh1* mRNA are required for enhanced translation under low-oxygen conditions. *Plant Physiol.* 112: 685–95

18. Bailey-Serres J, Freeling M. 1990. Hypoxic stress-induced changes in ribosomes of maize seedlings. *Plant Physiol.* 94: 1237–43

19. Bailey-Serres J, Kloeckener-Gruissem B,

Freeling M. 1988. Genetic and molecular approaches to the study of the anaerobic response and tissue specific gene expression in maize. *Plant Cell Environ.* 11: 351–57

20. Barclay HM, Crawford RMM. 1982. Plant growth and survival under strict anaerobiosis. *J. Exp. Bot.* 22:541–49

21. Bendix M, Tornberj T, Brix H. 1994. Internal gas transport in *Typha latifolia* L. and *Typha angustifolia* L. I. Humidity-induced pressurization and convective through flow. *Aquat. Bot.* 49:75–89

22. Bertani A, Brambilla I, Menegus F. 1980. Effect of anaerobiosis on rice seedlings: growth, metabolic rate and fate of fermentation products. *J. Exp. Bot.* 31:325–31

23. Bewley JD, Black M. 1994. *Seeds: Physiology of Development and Germination.* New York: Plenum. 2nd ed.

24. Botha FC, Potgieter GP, Botha AM. 1992. Respiratory metabolism and gene expression during seed germination. *Plant Growth Regul.* 11:211–24

25. Bouny JM, Saglio PH. 1996. Glycolytic flux and hexokinase activities in anoxic maize root tips acclimated by hypoxic pretreatment. *Plant Physiol.* 111:187–94

26. Bucher M, Brändle R, Kuhlemeier C. 1994. Ethanolic fermentation in transgenic tobacco expressing *Zymomonas mobilis* pyruvate decarboxylase. *EMBO J.* 13: 2755–63

27. Bucher M, Kuhlemeier C. 1993. Long-term anoxia tolerance: multilevel regulation of gene expression in the amphibious plant *Acorus calamus* L. *Plant Physiol.* 103: 441–48

28. Carystinos GD, MacDonald HR, Monroy AF, Dhindsa RS, Poole RJ. 1995. Vacuolar H⁺- translocating pyrophosphatase is induced by anoxia or chilling in seedlings of rice. *Plant Physiol.* 108:641–49

29. Chourey PS, Talierco EW, Kane EJ. 1991. Tissue-specific expression and anaerobically induced posttranscriptional modulation of sucrose synthase genes in *Sorghum bicolor* M. *Plant Physiol.* 96:485–90

30. Clark LH, Harris WM. 1981. Observation on the root anatomy of rice. *Am. J . Bot.* 68:154–61

31. Cleland RE, Fujiwara T, Lucas WJ. 1994. Plasmodesmal-mediated cell-to-cell transport in wheat roots is modulated by anaerobic stress. *Protoplasma* 178:81–85

32. Cobb BG, Drew MC, Andrews DL, Johnson JR, MacAlpine DM, et al. 1995. How maize seeds and seedlings cope with oxygen deficit. *HortScience* 30:1160–64

33. Compan I, Touati D. 1994. Anaerobic activation of arcA transcription in *Escherichia coli:* roles of Fnr and ArcA. *Mol. Microbiol.* 11:955–64

34. Cossins EA. 1978. Ethanol metabolism in plants. See Ref. 81a, pp. 169–202

35. Crawford LA, Brown AW, Breitkreuz KE, Guinel FC. 1994. The synthesis of γ-aminobutyric acid in response to treatments reducing cytosolic pH. *Plant Physiol.* 104: 865–71

35a. Crawford RMM, ed. 1987. *Plant Life in Aquatic and Amphibious Habitats.* Oxford: Blackwell

36. Crowder AA, Macfie SM. 1986. Seasonal deposit of ferric hydroxide plaque on roots of wetland plants. *Can. J. Bot.* 64:2120–24

37. Dancer JE, ap Rees T. 1989. Effects of 2, 4-dinitrophenol and anoxia on the inorganic pyrophosphate content of the spadix of *Arum maculatum* and the root apices of *Pisum sativum. Planta* 178:421–24

38. Dancer JE, Veith R, Feil R, Komer E, Stitt M. 1990. Independent changes of inorganic pyrophosphate and the ATP/ADP or UTP/UDP ratios in plant cell suspension cultures. *Plant Sci.* 66:59–63

39. Davies JM, Poole RJ, Sanders D. 1993. The computed free energy change of hydrolysis of inorganic pyrophosphate and ATP: apparent significance for inorganic-pyrophosphate-driven reactions of intermediary metabolism. *Biochem. Biophys. Acta* 1141: 29–36

40. de Bruxelles GL, Peacock WJ, Dennis ES, Dolferus R. 1996. Abscisic acid induces the alcohol dehydrogenase gene in Arabidopsis. *Plant Physiol.* 111:381–91

41. Denison F. 1992. Mathematical modeling of oxygen diffusion and respiration in legume root nodules. *Plant Physiol.* 98: 901–7

42. Dennis ES, Gerlach WL, Walker JC, Lavin M, Peacock WJ. 1988. Anaerobic regulated aldolase gene of maize: a chimaeric origin? *J. Mol. Biol.* 202:759–67

43. de Vetten NC, Ferl RJ. 1995. Characterization of a maize G-box binding factor that is induced by hypoxia. *Plant J.* 7: 589–601

44. Dieuaide-Noubhani M, Raffard G, Canioni P, Pradet A, Raymond P. 1995. Quantification of compartmented metabolic fluxes in maize root tips using isotope distribution from ¹³C-or ¹⁴C-labeled glucose. *J. Biol. Chem.* 270:13147–59

45. Drew MC. 1988. Effects of flooding and oxygen deficiency on plant mineral nutrition. In *Advances in Plant Nutrition,* ed. A Lauchli, PB Tinker, 3:115–59. New York: Praeger

46. Drew MC. 1992. Soil aeration and plant root metabolism. *Soil Sci.* 154:259–68

47. Drew MC, Lynch JM. 1980. Soil anaerobiosis, microorganisms and root function. *Annu. Rev. Phytopathol.* 18:37–66

48. Drew MC, Saglio PH, Pradet A. 1985. Higher adenylate energy charge and ATP/ADP ratios in aerenchymatous roots of *Zea mays* in anaerobic media as a consequence of improved internal oxygen transport. *Planta* 165:51–58

49. Duss F, Brändle R. 1982. Flooding tolerance in bulrush (*Schoenoplectus lacustris* L. Palla). V. Synthesis of different fermentation products and transport substances in the rhizome tissue under oxygen deficit. *Flora* 172:217–22

50. Edelstein M, Corbineau F, Kigel J, Nerson H. 1995. Seed coat structure and oxygen availability control low-temperature germination of melon (*Cucumis melo*) seeds. *Physiol. Plant.* 93:451–56

51. Erdmann B, Wiedenroth EM. 1988. Changes in the root system of wheat seedlings following anaerobiosis. III. Oxygen concentration in the roots. *Ann. Bot.* 62: 277–86

52. Fan T W-M, Higashi RM, Lane AN. 1988. An in vivo ^{1}H and ^{31}P NMR investigation of the effect of nitrate on hypoxic metabolism in maize roots. *Arch. Biochem. Biophys.* 266:592–606

53. Felle HH. 1996. Control of cytoplasmic pH under anoxic conditions and its implication for plasma membrane proton transport in *Medicago sativa* root hairs. *J. Exp. Bot.* 47:967–73

54. Fennoy SL, Bailey-Serres J. 1995. Posttranscriptional regulation of gene expression in oxygen-deprived roots of maize. *Plant J.* 7:287–95

55. Fox GG, McCallan NR, Ratcliffe RG. 1994. Manipulating cytoplasmic pH under anoxia: a critical test of the role of pH in the switch from aerobic to anaerobic metabolism. *Planta* 195:324–30

56. Fox TC, Green BJ, Drew MC, Kennedy RA, Rumpho ME. 1996. Anaerobic expression of hexokinase in shoots of *Echinochloa phyllopogan* and *Echinochloa crus-pavonis*. *Plant Physiol.* 111S:254 (Abstr.)

57. Fox TC, Kennedy RA, Rumpho ME. 1994. Energetics of plant growth under anoxia: metabolic adaptations of *Oryza sativa* and *Echinochloa phyllopogon*. *Ann. Bot.* 74: 445–55

58. Fox TC, Mujer CV, Andrews DL, Williams AS, Cobb BG, et al. 1995. Identification and gene expression of anaerobically induced enolase in *Echinochloa phyllopogon* and *Echinochloa crus-pavonis*. *Plant Physiol.* 109:433–43

59. Franco-Obregon A, Ureña J, Lopez-Barneo J. 1995. Oxygen-sensitive calcium channels in vascular smooth muscle and their possible role in hypoxic arterial relaxation. *Proc. Natl. Acad. Sci. USA* 92:4715–19

60. Fulton JM, Erickson AE. 1964. Relation between soil aeration and ethyl alcohol accumulation in xylem exudate of tomatoes. *Soil Sci. Soc. Am. Proc.* 28:610–14

61. Galina A, Reis M, Albuquerque MC, Puyou AG, Puyou MTG. 1995. Different properties of the mitochondrial and cytosolic hexokinases in maize roots. *Biochem. J.* 309:105–12

62. Gallie DR. 1993. Posttranscriptional regulation of gene expression in plants. *Annu. Rev. Plant Physiol. Plant Mol. Biol.* 44: 77–105

63. Gibbs J, de Bruxelle G, Armstrong W, Greenway H. 1995. Evidence for anoxic zones in 2–3 mm tips of aerenchymatous maize roots under low O_2 supply. *Aust. J. Plant Physiol.* 22:723–30

64. Good AG, Muench DG. 1992. Purification and characterization of an anaerobically induced alanine amino transferase from barley roots. *Plant Physiol.* 99:1520–25

65. Good AG, Muench DG. 1993. Long-term anaerobic metabolism in root tissue: metabolic products of pyruvate metabolism. *Plant Physiol.* 101:1163–68

66. Good AG, Paetkau DH. 1992. Identification and characterization of a hypoxically induced maize lactate dehydrogenase gene. *Plant Mol. Biol.* 19:693–97

67. Greenberg JT. 1997. Programmed cell death in plant-pathogen interactions. *Annu. Rev. Plant Physiol. Plant Mol. Biol.* 48: 525–45

68. Greipsson S, Crowder AA. 1992. Amelioration of copper and nickel toxicity by iron plaque on roots of rice (*Oryza sativa*). *Can. J. Bot.* 70:824–30

69. Grosse W, Meyer D. 1992. The effect of pressurized gas transport on nutrient uptake during hypoxia of alder roots. *Bot. Acta* 105:223–26

70. Gudleifsson BE. 1994. Metabolite accumulation during ice encasement of timothy grass (*Phleum pratense* L.) *Proc. R. Soc. Edinburgh B* 102:373–80

71. Guest JR. 1992. Oxygen-regulated gene expression in *Escherichia coli*. *J. Gen. Microbiol.* 138:2253–63

72. Guglielminetti L, Perata P, Alpi A. 1995. Effect of anoxia on carbohydrate metabolism in rice seedlings. *Plant Physiol.* 108: 735–41

73. Hanhijärvi AM, Fagerstedt KV. 1995. Comparision of carbohydrate utilization and energy charge in the yellow flag iris (*Iris pseudacorus*) and garden iris (*Iris germanica*) under anoxia. *Physiol. Plant* 93: 493–97

74. He CJ, Drew MC, Morgan PW. 1994. Induction of enzyme associated with lysigenous aerenchyma formation in roots of *Zea mays* during hypoxia or nitrogen-starvation. *Plant Physiol.* 105:861–65

75. He CJ, Finlayson SA, Drew MC, Jordan WR, Morgan PW. 1996. Ethylene biosynthesis during aerenchyma formation in roots of *Zea mays* subjected to mechanical impedance and hypoxia. *Plant Physiol.* In press

76. He CJ, Morgan PW, Drew MC. 1996. Transduction of an ethylene signal is required for programmed cell death and lysis in the root cortex of maize during aerenchyma formation induced by hypoxia. *Plant Physiol.* 112:463–72

77. Hochachka PW. 1993. *Surviving Hypoxia: Mechanisms of Control and Adaptation.* Boca Raton, FL: CRC Press. 570 pp.

78. Hoffman TL, LaManna JC, Pundik S, Selman WR, Whittingham TS, et al. 1994. Early reversal of acidosis and metabolic recovery following ischemia. *J. Neurosurg.* 81:567–73

79. Hojberg O, Sorensen J. 1993. Microgradients of microbial oxygen consumption in a barley rhizosphere model system. *Appl. Environ. Microbiol.* 59:431–37

80. Hole DJ, Cobb BG, Hole P, Drew MC. 1992. Enhancement of anaerobic respiration in root tips of *Zea mays* following low oxygen (hypoxic) acclimation. *Plant Physiol.* 99:213–18

81. Hondred D, Hanson AD. 1990. Hypoxically inducible barley lactate dehydrogenase: cDNA cloning and molecular analysis. *Proc. Natl. Acad. Sci. USA* 87:7300–4

81a. Hook DD, Crawford RMM, eds. 1978. *Plant Life in Anaerobic Environments.* Ann Arbor, MI: Ann Arbor Press. 564 pp.

82. Jackson MB. 1985. Ethylene and responses of plants to soil waterlogging and submergence. *Annu. Rev. Plant Physiol.* 36:145–74

83. Jackson MB, Fenning TM, Drew MC, Saker LR. 1985. Stimulation of ethylene production and gas-space (aerenchyma) formation in adventitious roots of *Zea mays* L. by small partial pressures of oxygen. *Planta* 165:486–92

84. Jackson MB, Fenning TM, Jenkin W. 1985. Aerenchyma (gas-space) formation in adventitious roots of rice (*Oryza sativa* L.) is not controlled by ethylene or small partial pressures of oxygen. *J. Exp. Bot.* 36: 1566–72

85. Johnson JR, Cobb BG, Drew MC. 1989. Hypoxic induction of anoxia tolerance in roots of *Zea mays. Plant Physiol.* 91: 837–41

86. Johnson JR, Cobb BG, Drew MC. 1994. Hypoxic induction of anoxia tolerance in roots of *Adh1* null *Zea mays. Plant Physiol.* 105:61–67

87. Justin SHF, Armstrong W. 1987. The anatomical characteristics of roots and plant response to soil flooding. *New Phytol.* 106: 465–95

88. Justin SHF, Armstrong W. 1991. Evidence for the involvement of ethene in aerenchyma formation in adventitious roots of rice (*Oryza sativa* L.). *New Phytol.* 118: 49–62

89. Ke D, Yahia E, Hess B, Zhou L, Kader AA. 1995. Regulation of fermentative metabolism in avocado fruit under oxygen and carbon dioxide stresses. *J. Am. Soc. Hortic. Sci.* 120:481–90

90. Kennedy RA, Rumpho ME, Fox TC. 1992. Anaerobic metabolism in plants. *Plant Physiol.* 100:1–6

91. Kerr JFR, Harmon BV. 1991. Definition and incidence of apoptosis: an historical perspective. See Ref. 150a, pp. 5–29

92. Kieber J. 1997. The ethylene response pathway in Arabidopsis. *Annu. Rev. Plant Physiol. Mol. Biol.* 48:277–96

93. Kwast KE, Hand SC. 1996. Acute depression of mitochondrial protein synthesis during anoxia: contributions of oxygen sensing, matrix acidification, and redox state. *J. Biol. Chem.* 271:7313–19

94. Llewellyn DJ, Finnegan EJ, Ellis JG, Dennis ES, Peacock WJ. 1987. Structure and expression of an alcohol dehydrogenase 1 gene from *Pisum sativum* (cv Greenfeast). *J. Mol. Biol.* 195:115–23

95. Lois AF, Ditta GS, Helinski DR. 1993. The oxygen sensor Fix L of *Rhizobium meliloti* is a membrane protein containing four possible trans-membrane segments. *J. Bacteriol.* 175:1103–9

96. Lopez-Barneo J. 1994. Oxygen-sensitive ion channels: How ubiquitous are they? *Trends Neurosci.* 17:133–35

97. Lu G, Paul AL, McCarty DR, Ferl RJ. 1996. Transcription factor veracity: Is GBF3 responsible for ABA-regulated expression of Arabidopsis *Adh*? *Plant Cell* 8:847–57

98. Lushuk JA, Saltveit ME. 1991. Effects of rapid changes in oxygen concentration on

respiration of carrot roots. *Physiol. Plant* 82:559–68

99. MacAlpine DM. 1995. *The expression of lactate dehydrogenase in Zea mays seedlings under hypoxic and anoxic conditions.* MS thesis. Texas A&M Univ., College Station

100. MacDonald RC, Kimmerer TW. 1993. Metabolism of transpired ethanol by eastern cottonwood (*Populus deltoides* Bartr.) *Plant Physiol.* 102:173–79

101. Mendelssohn IA, McKee KL. 1987. Root metabolic response of *Spartina alterniflora* to hypoxia. See Ref. 35a, pp. 239–53

102. Menegus F, Cattaruzza L, Chersi A, Fronza G. 1989. Differences in the anaerobic lactate-succinate production and the changes of cell sap pH for plants with high and low resistance to anoxia. *Plant Physiol.* 90: 29–32

103. Menegus F, Cattaruzza L, Mattana M, Beffagna N, Ragg E. 1991. Responses to anoxia in rice and wheat seedlings: changes in the pH of intracellular compartments, glucose-6-phosphate level, and metabolic rate. *Plant Physiol.* 95:760–67

104. Mertens E, Larondelle Y, Hers H-G. 1990. Induction of pyrophosphate: fructose 6-phosphate 1 phosphotransferase by anoxia in rice seedlings. *Plant Physiol.* 93: 584–87

105. Millar AH, Bergersen FJ, Day DA. 1994. Oxygen affinity of terminal oxidases in soybean mitochondria. *Plant Pysiol. Biochem.* 32:847–52

106. Mittler R, Lam E. 1995. In situ detection of nDNA fragmentation during the differentiation of tracheary elements in higher plants. *Plant Physiol.* 108:489–93

107. Mohanty B, Wilson PM, ap Rees T. 1993. Effects of anoxia on growth and carbohydrate metabolism in suspension cultures of soybean and rice. *Phytochemistry* 34: 75–82

108. Monk LS, Crawford RMM, Brändle R. 1984. Fermentation rates and ethanol accumulation in relation to flooding tolerance in rhizomes of monocotyledonous species. *J. Exp. Bot.* 35:738–45

109. Monk LS, Fagerstedt KV, Crawford RMM. 1989. Oxygen toxicity and superoxide dismutase as an anti-oxidant in physiological stress. *Physiol. Plant.* 76:456–59

110. Monson EK, Weinstein M, Ditta GS, Helinski DR. 1992. The FixL protein of *Rhizobium meliloti* can be separated into a heme-binding oxygen-sensing domain and a functional C-terminal kinase domain. *Proc. Natl. Acad. Sci. USA* 89:4280–84

111. Morrell S, Greenway H, Davies DD. 1990. Regulation of pyruvate decarboxylase in vitro and in vivo. *J. Exp. Bot.* 41:131–39

112. Nie XZ, Silva-Cardenas I, Hill RD. 1996. Function and regulation of hemoglobin in barley aleurone tissue and maize roots. *Plant Physiol.* 111:252 (Abstr.)

113. Ober ES, Sharp RE. 1996. A microsensor for direct measurement of O_2 partial pressure within plant tissues. *J. Exp. Bot.* 47: 447–54

114. Olive MR, Peacock WJ, Dennis ES. 1991. The anaerobic responsive element contains two GC-rich sequences essential for binding a nuclear protein and hypoxic activation of the maize *Adh1* promoter. *Nucleic Acids Res.* 19:7053–60

115. Paul AL, Ferl RJ. 1991. *Adh1* and *Adh2* regulation. *Maydica* 36:129–34

116. Paul AL, Ferl RJ. 1991. In vivo footprinting reveals unique *cis*-elements and different modes of hypoxic induction in maize *Adh1* and *Adh2*. *Plant Cell* 3:159–68

117. Perata P, Vernieri P, Armellini D, Bugnoli M, Tognoni F, Alpi A. 1992. Immunological detection of acetaldehyde-protein adducts in ethanol-treated carrot cells. *Plant Physiol.* 98:913–18

118. Peschke VM, Sachs MM. 1993. Multiple pyruvate decarboxylase genes in maize are induced by hypoxia. *Mol. Gen. Genet.* 240: 206–12

119. Pradet A, Bomsel JL. 1978. Energy metabolism in plants under hypoxia and anoxia. See Ref. 81a, pp. 89–118

120. Ratcliffe RG. 1995. Metabolic aspects of the anoxic response in plant tissue. In *Environment and Plant Metabolism: Flexibility and Acclimation*, ed. N Smirnoff, pp. 111–27. Oxford: Bios Sci.

121. Reggiani R, Cantu CA, Brambilla I, Bertani A. 1988. Accumulation and interconversion of amino acids in rice roots under anoxia. *Plant Cell Physiol.* 30:893–98

122. Ricard B, Couée I, Raymond P, Saglio PH, Saint-Ges V, Pradet A. 1994. Plant metabolism under hypoxia and anoxia. *Plant Physiol. Biochem.* 32:1–10

123. Rivoal J, Hanson AD. 1993. Evidence for a large and sustained glycolytic flux to lactate in anoxic roots of some members of the halophytic genus *Limonium*. *Plant Physiol.* 101:553–60

124. Rivoal J, Hanson AD. 1994. Metabolic control of anaerobic glycolysis: overexpression of lactate dehydrogenase in transgenic tomato roots supports the Davies-Roberts hypothesis and points to a critical role for lactate secretion. *Plant Physiol.* 106: 1179–85

125. Roberts JKM, Andrade FH, Anderson IC.

1985. Further evidence that cytoplasmic acidosis is a determinant of flooding intolerance in plants. *Plant Physiol.* 77:492–94

126. Roberts JKM, Callis J, Jardetzky O, Walbot V, Freeling M. 1984. Cytoplasmic acidosis as a determinant of flooding intolerance in plants. *Proc. Natl. Acad. Sci. USA* 81: 6029–33

127. Roberts JKM, Callis J, Wemmer D, Walbot V, Jardetzky O. 1984. Mechanism of cytoplasmic pH regulation in hypoxic maize root tips and its role in survival under hypoxia. *Proc. Natl. Acad. Sci. USA* 81: 3379–83

128. Roberts JKM, Chang K, Webster C, Callis J, Walbot V. 1989. Dependence of ethanolic fermentation, cytoplasmic pH regulation, and viability on the activity of alcohol dehydrogenase in hypoxic maize root tips. *Plant Physiol.* 89:1275–78

129. Roberts JKM, Hooks MA, Miaullis AP, Edwards S, Webster C. 1992. Contribution of malate and amino acid metabolism to cytoplasmic pH regulation in hypoxic maize root tips studied using nuclear magnetic resonance spectroscopy. *Plant Physiol.* 98:480–87

130. Rowland LJ, Strommer JN. 1986. Anaerobic treatment of maize roots affects transcription of *Adh1* and transcript stability. *Mol. Cell. Biol.* 6:3368–72

131. Rumpho ME, Pradet A, Khalik A, Kennedy RA. 1984. Energy charge and emergence of the coleoptile and radicle at varying oxygen levels in *Echinochloa crus-galli*. *Physiol. Plant* 62:133–38

132. Russell DA, Wong DML, Sachs MM. 1990. The anaerobic response of soybean. *Plant Physiol.* 92:401–7

133. Sabova L, Zeman I, Supek F, Kolarov J. 1993. Transcriptional control of AAC3 gene encoding mitochondrial ADP/ATP translocator in *Saccharomyces cerevisiae* by oxygen, heme and ROX1 factor. *Eur. J. Biochem.* 213:547–53

134. Sachs MM, Subbaiah CC, Saab IN. 1996. Anaerobic gene expression and flooding tolerance in maize. *J. Exp. Bot.* 47:1–15

135. Saglio PH. 1985. Effect of path or sink anoxia on sugar translocation in roots of maize seedlings. *Plant Physiol.* 77:285–90

136. Saglio PH, Drew MC, Pradet A. 1988. Metabolic acclimation to anoxia induced by low (2-4 kPa partial pressure) oxygen pretreatment (hypoxia) in root tips of *Zea mays*. *Plant Physiol.* 86:61–66

137. Saglio PH, Rancillac F, Bruzan F, Pradet A. 1984. Critical oxygen for growth and respiration of excised and intact roots. *Plant Physiol.* 76:151–54

138. Saglio PH, Raymond P, Pradet A. 1980. Metabolic activity and energy charge of excised maize root tips under anoxia: control by soluble sugars. *Plant Physiol.* 66: 1053–57

139. Saglio PH, Raymond P, Pradet A. 1983. Oxygen transport and root respiration of maize seedlings: a quantitative approach using the correlation between ATP/ADP and the respiration rate controlled by oxygen tension. *Plant Physiol.* 72:1035–39

140. Saint-Ges V, Roby C, Bligny R, Pradet A, Douce R. 1991. Kinetic studies of the variations of cytoplasmic pH, nucleotide triphosphates(^{31}P-NMR) and lactate during normoxic and anoxic transitions in maize root tips. *Eur. J. Biochem.* 200: 477–82

141. Sedbrook JC, Kronebusch PJ, Borisy GG, Trewavas AJ, Masson PH. 1996. Transgenic AEQUORIN reveals organ-specific cytosolic Ca^{2+} responses to anoxia in *Arabidopsis thaliana* seedlings. *Plant Physiol.* 111:243–57

142. Server AC, Mobley WC. 1991. Neuronal cell death and the role of apoptosis. See Ref. 150a, pp. 263–78

143. Steinmann F, Brändle R. 1984. Carbohydrate and protein metabolism in the rhizomes of the bulrush (*Schoenoplectus lacustris* (L.) Palla) in relation to natural development of the whole plant. *Aquat. Bot.* 19:53–63

144. Subbaiah CC, Bush DS, Sachs MM. 1994. Elevation of cytosolic calcium precedes anoxic gene expression in maize suspension-cultured cells. *Plant Cell* 6:1747–62

145. Subbaiah CC, Zhang J, Sachs MM. 1994. Involvement of intracellular calcium in anaerobic gene expression and survival of maize seedlings. *Plant Physiol.* 105: 369–76

146. Summers JE, Jackson MB. 1996. Anaerobic promotion of stem extension in *Potomogeton pectinatus:* roles for carbon dioxide, acidification and hormones. *Physiol. Plant.* 96:615–22

147. Thierfelder S, Doepner B, Gebhardt C, Hirche H, Benndorf K. 1994. ATP-sensitive K^+ channels in heart muscle cells first open and subsequently close at maintained anoxia. *FEBS Lett.* 351:356–59

148. Thomine S, Zimmermann S, Guern J, Barbier-Brygoo H. 1995. ATP-dependent regulation of an anion channel at the plasma membrane of protoplasts from epidermal cells of Arabidopsis hypocotyls. *Plant Cell* 7:2091–100

149. Thomson CJ, Atwell BJ, Greenway H. 1989. Response of wheat seedlings to low

O₂ concentrations in nutrient solution. I. Growth, O₂ uptake and synthesis of fermentative end-products by root segments. *J. Exp. Bot.* 40:985–91

150. Thomson CJ, Greenway H. 1991. Metabolic evidence for stelar anoxia in maize roots exposed to low O₂ concentrations. *Plant Physiol.* 96:1294–301

150a. Tomei LD, Lope FO, eds. 1991. *Apoptosis: The Molecular Basis of Cell Death.* New York: Cold Spring Harbor Lab. Press

151. Umeda M, Uchimiya H. 1994. Differential transcript levels of genes associated with glycolysis and alcohol fermentation in rice plants (*Oryza sativa* L.) under submergence stress. *Plant Physiol.* 106:1015–22

152. VanToai T, Bolles C. 1991. Postanoxic injury in soybean (*Glycine max*) seedlings. *Plant Physiol.* 97:588–92

153. Vartapetian BB, Andreeva IN, Kozlova GI, Agapova LP. 1977. Mitochondrial ultrastructure in roots of mesophyte and hydrophyte at anoxia and glucose feeding. *Protoplasma* 93:243–56

154. Walker JC, Howard EA, Dennis ES, Peacock WJ. 1987. DNA sequences required for anaerobic expression of the maize *alcohol dehydrogenase-1* gene. *Proc. Natl. Acad. Sci. USA* 84:6624–28

155. Wang H, Li J, Botock RM, Gilchrist DG. 1996. Apoptosis: a functional paradigm for programmed cell death induced by host-selective phytotoxin and invoked during development. *Plant Cell* 8:375–91

156. Waters I, Armstrong W, Thompson CJ, Setter TL, Adkins S, et al. 1989. Diurnal changes in radial oxygen loss and ethanol metabolism in roots of submerged and non-submerged rice seedlings. *New Phytol.* 113: 439–51

157. Waters I, Kuiper PJC, Watkin E, Greenway H. 1991. Effects of anoxia on wheat seedlings. I. Interaction between anoxia and other environmental factors. *J. Exp. Bot.* 42:1427–35

158. Waters I, Morrell S, Greenway H, Colmer TD. 1991. Effects of anoxia on wheat seedlings. II. Influence of O₂ supply prior to anoxia on tolerance to anoxia, alcoholic fermentation, and sugar levels. *J. Exp. Bot.* 42:1437–47

159. Watson NR, Peschke VM, Russell DA,

Sachs MM. 1992. Analysis of L-alanine:2-oxolutarate aminotransferase isozymes in maize. *Biochem. Genet.* 30:371–83

160. Webb T, Armstrong W. 1983. The effect of anoxia and carbohydrates on the growth and viability of rice, pea and pumpkin roots. *J. Exp. Bot.* 34:579–603

161. Webster C, Gaut RL, Browning KS, Ravel JM, Roberts JKM. 1991. Elongation and termination reactions of protein synthesis on maize root tip polyribosomes studied in a homologous cell-free system. *Plant Physiol.* 96:418–25

162. Webster C, Gaut RL, Browning KS, Ravel JM, Roberts JKM. 1991. Hypoxia enhances phosphorylation of eukaryotic initiation factor 4A in maize root tips. *J. Biol. Chem.* 266:23341–46

163. Xia JH, Roberts JKM. 1994. Improved cytoplasmic pH regulation, increased lactate efflux, and reduced cytoplasmic lactate levels are biochemical traits expressed in root tips of whole maize seedlings acclimated to a low-oxygen environment. *Plant Physiol.* 105:651–57

164. Xia JH, Roberts JKM. 1996. Regulation of H⁺ extrusion and cytoplasmic pH in maize root tips acclimated to a low-oxygen environment. *Plant Physiol.* 111:227–33

165. Xia JH, Saglio PH. 1992. Lactic acid efflux as a mechanism of hypoxic acclimation of maize root tips to anoxia. *Plant Physiol.* 100:40–46

166. Xia JH, Saglio PH, Roberts JKM. 1995. Nucleotide levels do not critically determine survival of maize root tips acclimated to a low-oxygen environment. *Plant Physiol.* 108:589–95

167. Yoshida S, Eguchi H. 1994. Environmental analysis of aerial O₂ transport through leaves for root respiration in relation to water uptake in cucumber plants (*Cucumis sativus* L.) in O₂-deficient nutrient solution. *J. Exp. Bot.* 45:187–92

168. Zhang Q, Greenway H. 1994. Anoxia tolerance and anaerobic catabolism of aged beetroot storage tissues. *J. Exp. Bot.* 45: 567–75

169. Zitomer RS, Lowry CV. 1992. Regulation of gene expression by oxygen in *Saccharomyces cerevisiae*. *Microbiol. Rev.* 56: 1–11

Annu. Rev. Plant Physiol. Plant Mol. Biol. 1997. 48:251–75

THE OXIDATIVE BURST IN PLANT DISEASE RESISTANCE

Chris Lamb[1] *and Richard A. Dixon*[2]

[1]Plant Biology Laboratory, Salk Institute for Biological Studies, 10010 North Torrey Pines Road, La Jolla, California 92037

[2]Plant Biology Division, Samuel Roberts Noble Foundation, 2510 Sam Noble Parkway, Ardmore, Oklahoma 73402

KEY WORDS: cell death, H_2O_2, superoxide

ABSTRACT

Rapid generation of superoxide and accumulation of H_2O_2 is a characteristic early feature of the hypersensitive response following perception of pathogen avirulence signals. Emerging data indicate that the oxidative burst reflects activation of a membrane-bound NADPH oxidase closely resembling that operating in activated neutrophils. The oxidants are not only direct protective agents, but H_2O_2 also functions as a substrate for oxidative cross-linking in the cell wall, as a threshold trigger for hypersensitive cell death, and as a diffusible signal for induction of cellular protectant genes in surrounding cells. Activation of the oxidative burst is a central component of a highly amplified and integrated signal system, also involving salicylic acid and perturbations of cytosolic Ca^{2+}, which underlies the expression of disease-resistance mechanisms.

CONTENTS

INTRODUCTION

Attempted infection by avirulent pathogens elicits the activation of a battery of defenses often accompanied by the collapse of challenged plant cells in the so-called hypersensitive response (HR) (130). The HR is the outcome of recognition by ligand/receptor interactions specified by paired plant resistance (*R*) and pathogen avirulence (*avr*) genes and results in a restricted lesion at the site of attack clearly delimited from surrounding healthy tissue. In addition, systemic acquired resistance (SAR) to subsequent attack by normally virulent pathogens gradually develops throughout the rest of the plant (111, 116). Many inducible defense responses including phytoalexin synthesis and deployment of pathogenesis-related (PR) proteins are regulated transcriptionally, and avirulence signals cause massive switches in host gene expression (39, 62). In 1983 Doke reported the generation of superoxide (O_2^-) and H_2O_2 as a novel response distinct from classical transcription-dependent defenses (42). Rapid generation of oxidants has now been described in many plant/ pathogen interactions and is a characteristic feature of the HR. With the cloning of *R* genes putatively encoding avirulence signal receptors there is great interest in the oxidative burst as one of the earliest responses to pathogen attack (2, 10, 40, 94). After discussing the underlying chemistry, we review recent advances in the biology, mechanism, and function of the oxidative burst.

CHEMISTRY OF ACTIVE OXYGEN IN BIOLOGICAL SYSTEMS

Although dioxygen in its ground state is relatively unreactive, partial reduction gives rise to "active oxygen species," including the superoxide anion (O_2^-), hydrogen peroxide (H_2O_2), and the hydroxyl radical ($^.OH$). O_2^- is a by-product of mitochondrial electron transport, photosynthesis, and flavin dehydrogenase reactions, and can then be converted to other active oxygen species, of which $^.OH$ is the most reactive. Much attention has been given to the mechanisms used to remove these toxic products, particularly under conditions of physical stress, e.g. high temperature and drought, which lead to increased production of active oxygen species (121).

The first reaction in the partial reduction of dioxygen is addition of a single electron to form O_2^-. This can be protonated at low pH (pKa = 4.8) to yield the perhydroxyl radical ($^\cdot HO_2$) (Equation 1), and O_2^- and $^\cdot HO_2$ undergo spontaneous dismutation to produce H_2O_2 (Equations 2 and 3).

$$H^+ + O_2^- \leftrightarrow {}^\cdot HO_2 \qquad\qquad 1.$$

$$^\cdot HO_2 + {}^\cdot HO_2 \rightarrow H_2O_2 + O_2 \qquad\qquad 2.$$

$$^\cdot HO_2 + O_2^- + H_2O \rightarrow H_2O_2 + O_2 + OH^- \qquad\qquad 3.$$

Equation 3 represents the major route for O_2^- decay at cellular pH. Because of the equilibrium in Equation 1, spontaneous radical dismutation will decrease as cellular pH increases. Metal-containing superoxide dismutase enzymes catalyze a highly efficient conversion (up to 10^{10} times faster than the spontaneous rate) of $O_2^-/{}^\cdot HO_2$ to H_2O_2.

H_2O_2 is stable and less reactive than O_2^-. However, in the presence of reduced transition metals such as Fe^{2+}, which may be free or complexed to chelating agents or proteins, H_2O_2-dependent formation of $^\cdot OH$ can occur, and O_2^- can act as the initial reducing agent for the metal (Equations 4 and 5).

$$O_2^- + Fe^{3+} \rightarrow O_2 + Fe^{2+} \qquad\qquad 4.$$

$$Fe^{2+} + H_2O_2 \rightarrow Fe^{3+} + OH^- + {}^\cdot OH \qquad\qquad 5.$$

$^\cdot OH$ is a very strong oxidant and can initiate radical chain reactions with a range of organic molecules. This can lead to lipid peroxidation, enzyme inactivation, and nucleic acid degradation.

H_2O_2 is removed by catalase or various peroxidases including ascorbate and glutathione peroxidases. The general equations for catalase and peroxidase reactions (Equations 6 and 7) are similar because the catalase reaction can be viewed as a peroxidative reaction with H_2O_2 as both substrate and acceptor. In Equation 7, the R group is often aromatic, in which case a diphenol is converted to a diquinone, a reaction important in lignin polymerization.

$$HO\text{-}OH + HO\text{-}OH \rightarrow 2H_2O + O{=}O \qquad\qquad 6.$$

$$HO\text{-}OH + HO\text{-}R\text{-}OH \rightarrow 2H_2O + O{=}R{=}O \qquad\qquad 7.$$

Peroxidase can also catalyze the formation of both O_2^- and H_2O_2 by a complex reaction in which NADH is oxidized using trace amounts of H_2O_2 first produced by nonenzymatic breakdown of NADH. The $^\cdot NAD$ radical formed then reduces O_2 to O_2^-, some of which dismutates to H_2O_2 and O_2 (62). This reaction is stimulated by monophenols and Mn^{2+} and may be a mechanism for generation of extracellular H_2O_2 for lignin polymerization (63).

BIOLOGY OF THE OXIDATIVE BURST

Doke (42) showed that inoculation of aged potato tuber tissue with an avirulent race of *Phytophthora infestans* stimulated the reduction of cytochrome *c* or the dye nitroblue tetrazolium, and addition of superoxide dismutase blocked these reactions, indicating that the potato tissue responded to the avirulent pathogen by the generation of O_2^- at the onset of the HR (42). A virulent race failed to stimulate O_2^- production. Rapid generation of O_2^- and H_2O_2 has now been reported in many HRs following inoculation with avirulent fungal, bacterial, or viral pathogens (2, 94), and a clear exposition of the underlying biology has come from studies of the response to bacterial inoculation (10) (Figure 1). Inoculation of soybean cells with either virulent or avirulent races of *Pseudomonas syringae* pv. *glycinea* causes a rapid but weak transient accumulation of oxidants (phase I) (10, 102). In cells inoculated with the avirulent but not the virulent strain there is a second, massive and prolonged oxidative burst between 3 to 6 h after inoculation (phase II). Similar two-phase kinetics are observed in tobacco cells inoculated with *Pseudomonas syringae* pv. *syringae,* which is not a tobacco pathogen (4, 79). Phase II depends on *avr* expression in the race-host cultivar interaction and expression of the *hypersensitive response and pathogenicity* (*Hrp*) gene cluster in the nonhost interaction. Transfer of the *Hrp* cluster to *Pseudomonas fluorescens* enables the saprophyte to trigger the phase II burst and hypersensitive cell death (57). Thus, the weak, transient phase I burst is a biologically nonspecific reaction, whereas the phase II burst and HR require *Hrp* action in nonhost resistance and *avr* action in race-cultivar specific resistance.

Induction of the accumulation of O_2^- and/or H_2O_2 can also be observed with specific bacterial or fungal avirulence factors (9, 11, 94). For example, a race-specific peptide elicitor derived from the *Cladosporium fulvum avr9* gene rapidly stimulates an oxidative burst when infiltrated into leaves of the tomato *Cf9* genotype, and activation of *C. fulvum avr*-mediated responses causes oxidative stress (64, 138). O_2^- formation precedes increases in glutathione and ethylene levels, lipoxygenase induction, and cell death (63, 93). Likewise, a variety of nonrace-specific fungal elicitors and endogenous oligogalacturonide elicitors derived from the plant cell wall rapidly induce the oxidative burst within 2 min of exposure to plant cells (3, 32, 34, 76, 87, 101, 115). The burst involves a massive activation of oxidative metabolism reminiscent of that following neutrophil activation in the mammalian native immune system (3,

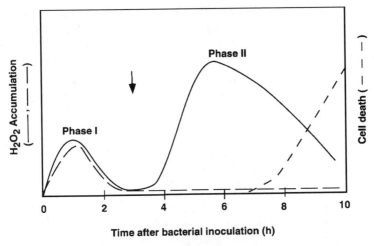

Figure 1   Kinetics for accumulation of H_2O_2 and induction of cell death in plant cells following bacterial inoculation. Cell death and H_2O_2 accumulation following inoculation with avirulent pathogen or nonhost pathogen; H_2O_2 accumulation following inoculation with virulent pathogen or nonpathogen.

24, 84). For example, in soybean cells H_2O_2 accumulates to ~1.2 mMol•liter^{-1} of packed cells within 10 min of oligogalacturonide treatment, a rate of oxidant generation similar to that in activated neutrophils (3, 6, 84, 98), and treatment of bean cells with a fungal elicitor doubles the rate of O_2 consumption (19).

MECHANISM OF THE OXIDATIVE BURST

NADPH Oxidase

The neutrophil oxidative burst involves the reaction $O_2 + NADPH \rightarrow O_2^- + NADP^+ + H^+$, followed by dismutation of O_2^- to H_2O_2 (6, 98). O_2^- production is catalyzed by a plasma membrane oxidase composed of an unusual *b*-type cytochrome with two subunits, p22 and gp91 (Figure 2), together with a putative p32 flavoprotein, which may bind NADPH (6, 98). During neutrophil activation two related cytosolic proteins, p67 and p47, the latter after multiple phosphorylation, associate with the plasma membrane to stimulate the oxidase (6, 18, 98). G-proteins including Rac2 and Rap1A are also involved in assembly of the active complex (36).

An NADPH-dependent O_2^--generating system is present in microsomes from potato tubers (44, 45). Incubation of tuber slices with an incompatible race of *P. infestans* or treatment with cell wall–derived elicitors stimulates the NADPH-dependent O_2^--generating activity, which was predominantly located in the plasma membrane fraction. No stimulation of oxidase activity was detected in plasma membrane fractions from tissues inoculated with a compatible race or from control tissues. The burst could also be observed in elicitor-

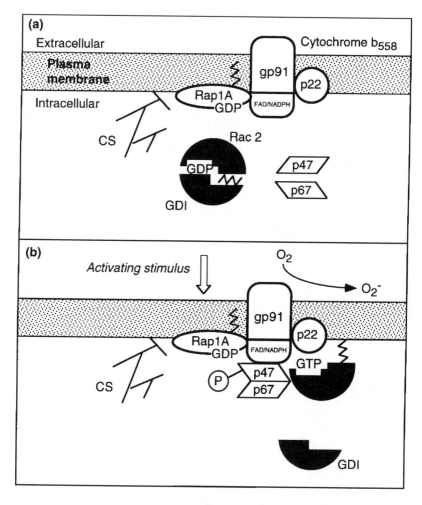

Figure 2   Schematic model of the human neutrophil NADPH oxidase in resting (*a*) and stimulated (*b*) cells leading to assembly of the oxidase on the membrane and O_2^- production. CS, cytoskeletal elements possibly associated with components of the oxidase; the zigzags represent isoprenyl groups modifying Rac2 and Rap1A. Adapted with permission from Bokoch (18).

treated protoplasts (43), indicating that oxidant generation did not require the presence of the cell wall or apoplastic enzymes. Moreover, release of O_2^- was extracellular. Because O_2^- is poorly diffusible across membranes, O_2^- is most likely released at the external surface of the plasma membrane (5, 43). In most systems H_2O_2 is the major product accumulating, but treatment of elicited rose cells with N,N-diethyldithiocarbamate, a superoxide dismutase inhibitor, blocks H_2O_2 generation while causing a dramatic accumulation of O_2^-. Thus, O_2^- generation at the external surface is followed by rapid enzyme-catalyzed dismutation to H_2O_2, and accumulation of significant levels of O_2^- in some systems, e.g. tomato and rice cells, but not others, e.g. soybean and rose cells (5, 64, 87, 133), likely reflects differences in the relative rates of O_2^- generation and dismutation. ˙OH can also be detected by EPR in elicited cells (81) and is likely generated from H_2O_2 by an Fe^{2+}-mediated reaction (Equations 4 and 5 above).

Several inhibitors of the neutrophil NADPH oxidase, including the suicide substrate inhibitor diphenylene iodonium (DPI) (6, 98), also block the fungal elicitor-stimulated oxidative burst in plant cells (5, 34, 87). In soybean cells DPI blocks H_2O_2 production in response to fungal elicitor or inoculation with avirulent *P. syringae* pv. *glycinea* at concentrations similar to those effective with the neutrophil enzyme (87). DPI also blocks H_2O_2 production from NADPH in vitro by microsomal membranes from rose cells (5). Moreover, a polyclonal antiserum to a specific peptide of the neutrophil NADPH oxidase reacts with a single polypeptide of the same size in western blots of soybean microsomal membrane preparations (133). Antibody binding to soybean p22 is blocked by coincubation with the p22 peptide immunogen. Likewise, antisera to neutrophil p47 and p67 react with plant cytosolic polypeptides of appropriate size (34, 46). A rice expressed sequence tag, RICR1091A, has been identified showing a high degree of similarity to the human gene sequence encoding the apoprotein of gp91, and the corresponding rice gene, *rbohA* isolated (61). The deduced protein sequence shows >30% amino acid sequence identity with human gp91, and the predicted topological organization closely matches that of human gp91. Identification of this rice gp91 gene sequence suggests that the functional similarities of the neutrophil and plant oxidative bursts are accompanied by mechanistic conservation at the level of the oxidase.

Alternative Mechanisms

The increase in O_2 consumption in elicitor-treated bean cells is sensitive to relatively low concentrations of KCN (19), whereas basal respiration is insensitive to KCN. While it was not demonstrated whether the elicitor-induced accumulation of H_2O_2, which accompanies the increase in O_2 consumption, is also KCN-sensitive, these observations suggest possible differences between the oxidative burst in bean cells and the KCN-insensitive neutrophil NADPH oxidase. Elicitor causes a transient alkalinization of the bean apoplast, where the pH rises to ~7.2. H_2O_2 production in alkaline conditions could be modeled in vitro with a number of peroxidases, one of which, a cell wall cationic peroxidase, can sustain H_2O_2 production at neutral pH. Based on comparative pH profiles between the cells and the purified peroxidase activity, it was proposed that elicitor stimulates a direct production of H_2O_2 by this peroxidase. However, such a model is inconsistent with the ability of isolated protoplasts to generate an oxidative burst and the involvement of O_2^- as a key mechanistic intermediate. Genes encoding peroxidase (77, 112, 128), other H_2O_2-generating enzymes, e.g. germin-like oxalate oxidases (69, 149), and enzymes of oxidative metabolism, e.g. lipoxygenase (30, 41, 103), are induced by pathogen signals. Transcription-dependent induction of such enzymes is too slow to contribute to the initial stages of the oxidative burst but may play a role in maintaining high levels of oxidants later in the response.

Signal Pathways

Activation of the oxidative burst is governed by a phosphorylation/dephosphorylation poise because the protein phosphatase 2A inhibitor cantharidin induces H_2O_2 production in soybean cells in the absence of an elicitor and potentiates the response to avirulent *P. syringae* pv. *glycinea* (87, 133). The protein-serine kinase inhibitor K252A blocks the oxidative burst, and the level of H_2O_2 accumulation can be titrated by varying the ratio of K252A to cantharidin. Such a regulatory poise may be important for rapid induction and tight control of the magnitude and duration of the oxidative burst (52, 133) (Figure 3). Protein kinase cascades have been implicated in the activation of the HR (37, 52, 87), and the *Pto* and *Xa21 R* genes encode protein-serine kinases (88, 129). Xa21 is a receptor-like protein kinase with putative extracellular leucine-rich repeat (LRR) and intracellular catalytic domains separated by a transmembrane region (129). *Pto* encodes a soluble protein-serine kinase that phosphorylates and activates a second protein-serine kinase, Pti-1, also implicated in the HR (150). However, the mechanism by which these protein kinases activate the oxidative burst and the possible involvement of putative p47 and p67 homologs as target substrates remain to be established.

In *Cf5* tomato cells, the activation of plasma membrane H^+-ATPase, NADH oxidase, and inhibition of ascorbate peroxidase by elicitor preparations containing *avr5* products depends on G-protein mediated signaling (137, 138).

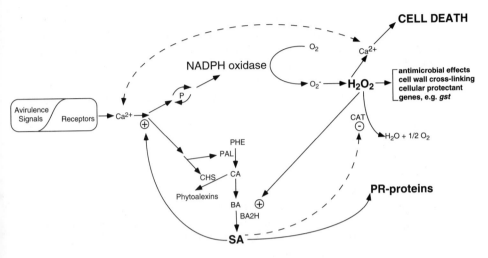

Figure 3 Signal networks in the HR. BA, benzoic acid; CA, cinnamic acid; CAT, catalase; P, a protein kinase/phosphatase phosphorylation poise; PHE, phenylalanine.

Overexpression of a Ras-related G-protein in tobacco perturbs several long-term defense responses (119), and similar effects were seen in tobacco expressing the cholera toxin A1 subunit (13), which irreversibly blocks G-protein GTPase activity. Mastoporan, a specific activator of G-proteins, induces H_2O_2 accumulation in soybean cells in the absence of elicitor, and introduction of the Fab fragment of a G-protein antibody into soybean cells perturbs elicitor stimulation of the burst (24, 83). Ca^{2+} influx and an H^+ influx/K^+ efflux exchange are important for activation of the oxidative burst (4, 11, 101, 123, 132). However, the molecular basis for G-protein action, Ca^{2+} redistribution, and Ca^{2+}-mediated signal transduction remains to be determined.

FUNCTION OF THE OXIDATIVE BURST

Direct Antimicrobial Activity

The addition of O_2^- scavengers to tobacco cells inoculated with *P. syringae* pv. *tabaci* increases recovery of bacteria (79), and H_2O_2 inhibits the germination of spores of a number of fungal pathogens (106). Moreover, a potent antimicrobial protein isolated in a screen of fungal culture fluids turned out to be a glucose oxidase and the isolated enzyme was effective against fungal and bacterial phytopathogens in vitro because of H_2O_2 generated in the reaction glucose + $O_2 \rightarrow$ gluconate + H_2O_2 (144). Potato expressing the *Aspergillus niger* glucose oxidase gene showed enhanced protection against *Erwinia caratovora,* although these in vivo effects might reflect, at least in part, indirect signal functions of H_2O_2 (see below) in addition to direct protective effects arising from several-fold elevation of H_2O_2 levels (144). Oxidizing subcellular compartments capable of myeloperoxidase-mediated generation of toxic hypochlorite from H_2O_2 equivalent to the phagosomes of activated neutrophils have not been reported in plants (6, 98).

Oxidative Cross-Linking in the Cell Wall

In a study of early events at the surface of elicitor-treated soybean and bean cells, two cell wall proteins designated p33 and p100 became insolubilized (20). Insolubilization is initiated within 2–5 min of elicitation and is completed within 20–30 min as assayed by the loss of SDS-extractable p33 and p100 against retention of strong immunofluorescence labeling when assayed in situ in the cell wall. Likewise, p33 and p100 insolubilization is observed in bean leaf tissue during a *Hrp*-induced nonhost HR to *P. syringae* pv. *tabaci* and early in an incompatible interaction between soybean and *P. syringae* pv. *glycinea,* but not at any stage during an equivalent compatible interaction (22).

p33 is a tyrosine-rich, proline-rich protein. Insolubilization of such tyrosine-rich proteins is driven by H_2O_2 from the elicitor-induced oxidative burst, because catalase or ascorbic acid blocks the loss of SDS-extractable p33 and p100, and exogenous H_2O_2 causes insolubilization within 2 min. The absence of low-molecular-weight partial degradation products and frequent detection of dimers and tetramers of p33 provide further evidence for cross-linking of these proteins. Isodityrosine formation mediates oxidative cross-linking of *Chlamydomonas* cell walls (139), and a novel di-isodityrosine residue in tomato cell walls may represent an intermolecular linkage (21). Fungal elicitor treatment increases the resistance of cell walls to digestion by microbial wall degrading enzymes as assayed by rapid protoplast release (22). Protoplast release is almost completely blocked 30 min after elicitation as an index of the increased effectiveness of the wall as a pathogen barrier. Elicitor-induced inhibition of protoplast release was not blocked by actinomycin D or cycloheximide and could be mimicked by exogenous H_2O_2. Moreover, dithiothreitol completely blocks the response to fungal elicitor, which is noteworthy because dithiothreitol inhibits elicitor-induced oxidative cross-linking, but stimulates elicitor induction of defense genes.

Rapid oxidative cross-linking in the cell wall thus resembles a self-sealing tire to slow pathogen ingress and spread before the deployment of transcriptional-dependent defenses and to trap pathogens in host cells destined to die. It is interesting to note that fungal elicitor and wounding inhibit the expression of genes encoding cell wall structural proteins with low tyrosine contents but stimulate those encoding tyrosine-rich versions (29, 77, 120, 126). This should increase the capacity of the cell wall for subsequent oxidative cross-linking, e.g. to protect against secondary infections, and increased levels of hydroxyproline and peroxidase in the cell wall can be observed systemically (112, 128). Moreover, polysaccharides in the pectic fraction often contain feruloyl residues capable of undergoing oxidative cross-linking (55). Infection of barley leaves with an incompatible race of the rust pathogen *Puccinia coronata* f.sp. *avenae* leads to changes in cell walls at the site of attempted ingress such that treatment with protoplasting enzymes strips away all cells except those undergoing the HR (70). One component of this response appears to be oxidative cross-linking of feruloyl residues because diferulic acid can be recovered from the infected leaves.

Gene Activation and Transcription-Dependent Defenses

Antioxidant enzymes or free radical scavengers can reduce phytoalexin accumulation (3, 42, 49), and exogenous H_2O_2 can stimulate phytoalexin accumulation without elicitor (3, 113). However, while catalase was reported to block elicitor induction of phytoalexin accumulation in soybean cells (3), the phytoalexin response can be dissociated from the oxidative burst (32), and antioxidant treatments fail to inhibit phytoalexin accumulation in bacterially inoculated white clover cells (35) or tobacco cells treated with an elicitin peptide from *Phytophthora cryptogea* (115). Overall the effect of antioxidant treatments on phytoalexin induction varies widely, and oxidant-independent pathways make major contributions to the regulation of phytoalexin accumulation (94, 133).

For example, in soybean cells exogenous H_2O_2 evokes only weak accumulation of transcripts encoding phenylalanine ammonia-lyase (PAL) and chalcone synthase (CHS) relative to their accumulation in response to fungal elicitor, whereas genes encoding the cellular protectant enzymes glutathione S-transferase (GST), glutathione peroxidase, and polyubiquitin are rapidly induced by exogenous H_2O_2 (87). Elicitor induction of *gst,* but not *pal* and *chs,* is perturbed by inhibition of the oxidative burst with DPI, dissipation of H_2O_2 with catalase or scavengers, or modulation of the branch signal pathway specific for the oxidative burst (87) (Figure 3). In these experiments, transcript levels were assayed 1 h after stimulation to monitor only direct signal functions and minimize potential contributions from secondary effects. H_2O_2 has been implicated in the regulation of the response to chilling stress (108) and the induction of alternative oxidase (140). Likewise oxidative stress from ozone results in the induction of various cellular protectant and defense gene systems (76). However, while UV-induced PR-1 accumulation is mediated by active oxygen species (58), this response is blocked by cycloheximide (58), and induction of PR-proteins by UV (58), ozone (76, 124, 125), or H_2O_2 (27) may involve secondary signals or a slow response to chronic oxidative stress rather than a rapid, direct oxidant-mediated signal pathway.

H_2O_2 is a small diffusible molecule, and while inoculation of soybean cells with avirulent *P. syringae* pv. *glycinea* induces cell death only in the challenged cells and not in cells separated from the challenged cells by two dialysis membranes (87), *gst* transcripts accumulate in the second, uninfected set of cells. Inclusion of catalase in the medium between the dialysis membranes completely blocks induction of *gst* and glutathione peroxidase in the uninfected cells. H_2O_2 induction of *gst* is not dependent on Ca^{2+}, whereas induction of cell death is blocked by Ca^{2+} channel inhibitors and can be mimicked by Ca^{2+} ionophores (86). In animals, H_2O_2 is a pleiotropic activator of immune and inflammatory response genes, mediated by the transcription factor NF-κB (122), but no plant homologs of components of NF-κB have been described. However, a plant homolog of Ref-1, designated Arp, has been reported (8). Ref-1 is a nuclear redox factor that modulates Fos-Jun heterodimer binding to the AP-1 *cis* element involved in oxidative stress gene activation, by the reduction/oxidation of conserved cysteines in the DNA-binding domains (145). AP-1-like binding sites (54) and the EpRE (electrophile-responsive element) are present in animal *gst* promoters, and AP-1 regulates both basal and oxidative stress–inducible expression (31). A soybean *gst* is activated by H_2O_2 (135), and the ocs *cis* element may respond to oxidative stress (48, 135, 149). Similar sequences are present in the EpER element, which consists of two adjacent AP-1 binding sites. Ocs-binding proteins (OBPs) have been isolated, and an Arabidopsis *trans* factor—OBP1—has been described that enhances OBP binding to ocs elements (147). The promoter of an H_2O_2-inducible Arabidopsis *gst* contains closely linked OBF and OBP1 binding sites (25), and an OBP1 homolog in pumpkin encodes a DNA-binding protein that binds to the ascorbate oxidase promoter (50). However, redox-sensitive transcriptional activation by these *trans* factors remains to be demonstrated.

Hypersensitive Cell Death

In tobacco cells inoculated with *P. syringae* pv. *syringae,* induction of host cell death and the oxidative burst are dependent on the *Hrp* gene cluster (10).

Mutation of any of 12 of the 13 genes in this cluster eliminates induction of both responses. However, mutation of the *hrmA* gene, which has a regulatory function, still results in H_2O_2 accumulation in tobacco cells, but gives only a weak HR based on macroscopic symptoms in tobacco leaf tissue, suggesting that the oxidative burst may not be sufficient for cell death (57). However, several reports document evidence for a causal relationship between the oxidative burst and cell death based on the manipulation of endogenous levels of oxidants. Thus, antioxidant, scavenger, or catalase treatments block the induction of cell death in several systems in response to various avirulence signals (34, 80, 84, 87, 91, 113). In soybean cells, DPI, catalase, or scavengers suppress cell death induced by an avirulent strain of *P. syringae* pv. *glycinea* (87, 133), and 3-aminotriazole potentiates the response, although inhibition of H_2O_2 turnover in the absence of an oxidative burst is not sufficient to trigger cell death. Moreover, the protein phosphatase 2A inhibitor cantharidin, which stimulates H_2O_2 production but not early defense gene transcription in the absence of avirulence signals, likewise stimulates cell death in uninfected cells and also potentiates the HR to avirulent *P. syringae* pv. *glycinea*. Conversely, the protein kinase inhibitor K252A, which blocks elicitor induction of the oxidative burst, also reduces cell death in soybean cell suspension cultures inoculated with *P. syringae* pv. *glycinea*.

Hypersensitive cell death is an active process (65, 73, 86), and the observation of mutants that spontaneously develop hypersensitive-like lesions implies genetic control (38, 59, 60, 72, 73). Certain mutations at the *Rp1* disease resistance locus of maize generate such phenotypes (68) as do the recessive *mlo* resistance alleles to powdery mildew in barley (143). However, while lesion mimic phenotypes are consistent with physiological cell death in the HR, and may in some cases reflect dysfunction of *R* gene products, mutations in other genes required for any of a range of cellular functions independent of the HR could also generate such phenotypes. This would account for the many alleles identified (72) and the observation that expression of a bacterial gene encoding a proton pump causes lesion mimic phenotypes in transgenic tobacco (97).

Selective cell deletion in animals often reflects programmed cell death manifested by extensive blebbing of the plasma membrane, DNA processing, chromatin condensation, collapse, and shrinkage of the cytosol with the retention of the integrity of cell-internal organelles, leading to the development of electron opaque bodies in a shrunken corpse (73, 110, 131, 136). This cellular deconstruction, termed apoptosis, is in sharp contradistinction to necrotic cell death, which involves cell swelling, rapid disintegration of cell-internal structures, and DNA degradation in the absence of discrete processing intermediates. Cell death triggered in leaves of two resistant cowpea cultivars by the cowpea rust fungus is accompanied by the cleavage of nuclear DNA into oligonucleosomal fragments that reflect a late stage in DNA processing during apoptosis in animals (117). Likewise, host-selective AAL toxins secreted by *Alternaria alternata* f.sp. *lycopersici* cause DNA processing and apoptotic cell death in tomato protoplasts and leaflets as revealed by both DNA ladders and development of apoptotic-like bodies (141). In soybean cells, avirulent *P. syringae* pv. *glycinea* but not the isogenic virulent race activates a physiological cell death program resulting in the generation of large (~50 kb) DNA fragments characteristic of the initial stages of DNA processing, as well as the cell corpse morphology characteristic of apoptosis (86). Apoptotic corpses were also observed at the onset of the HR in Arabidopsis leaves inoculated with avirulent *P. syringae* pv. *tomato* and tobacco cells treated with the fungal peptide cryptogein, which is involved in the induction of nonhost resistance to *Phytophthora cryptogea*. Hence apoptosis may prove to be a common, although not necessarily ubiquitous, feature of HRs. However, while a re-evaluation of cytological studies of the HR of tobacco to *Pseudomonas syringae* pv. *phaseolicola* reveals corpses similar to those in soybean, Arabidopsis, and tobacco (53), changes in the structure of lettuce leaf cells in response to injection of the same pathogen indicate necrotic rather than apoptotic cell death (15), and more than one cell death mechanism may therefore operate in the HR.

H_2O_2 stimulates a rapid influx of Ca^{2+} (109). In soybean cells this is required to activate the cell death program, and apoptosis can be induced by the Ca^{2+} ionophore A23187 in the presence of cell-external Ca^{2+} (86). Thus Ca^{2+} has a signal function downstream as well as upstream of the oxidative burst (Figure 3). However, although Ca^{2+} is a common trigger of apoptosis in animals (90) it remains to be established whether interleukin-1 cleavage enzyme-like proteases (131) also function in plants. Those systems in which apoptosis is not observed may reflect generation of toxic lipid hydroperoxides, and lipid peroxidation as a consequence of oxidant accumulation has been observed in a number of systems (1, 78, 80, 113, 115).

Functional Integration

It is likely that changes in O_2^- and H_2O_2 levels are not monitored directly by reversible binding to a receptor, but by generation of bioactive metabolic derivatives analogous to prostaglandins (17) or by perturbation of redox poises such as the glutathione/ascorbate cycle modulating redox-sensitive signal systems equivalent to OxyR (134) or NF-κB (96, 122). Hence the steady state levels of O_2^- or H_2O_2 may not be good indicators of signal strength, which for such metabolism-dependent mechanisms is more likely to reflect time-integrated flux. Jasmonic acid and other octadecanoic acids function as signals in elicitor as well as wound signal pathways (17), although the role of the oxidative burst in generating oxygenated derivatives of membrane fatty acids or their function in oxidant-mediated signaling is not clear. Glutathione treatment induces H_2O_2 production and the expression of early defense genes such as *pal* and *chs* in bean or soybean cells (142) and superoxide dismutase in tobacco cells (66). Whereas fungal elicitor causes a marked accumulation of glutathione in alfalfa cells and homoglutathione in bean cells (47), this is slow compared with *pal* and *chs* gene induction, and increases in the intracellular levels of these tripeptides caused by *L*-oxothiazolidine-4-carboxylate treatment do not elicit these defenses. However, it may be the rates of cycling between reduced and oxidized forms in redox homeostatic systems such as glutathione that are important in signaling cued by the oxidative burst, in line with emerging data on redox regulation of gene expression in animals (96).

The evidence that the oxidative burst has a causal role in both gene activation and cell death comes from the effects of manipulation of the accumulation of endogenous oxidants. However, both cellular protectant genes and cell death can be induced by exogenous oxidants. In soybean cells, *gst* and other cellular protectant genes are optimally induced by H_2O_2 applied at 1–2 mM, whereas a threshold >4 mM is required for induction of cell death (87). H_2O_2 is very rapidly metabolized and in soybean cells 10 mM H_2O_2 is cleared within ~5 min (87). Therefore, while the applied concentration is substantially higher than physiological levels during the oxidative burst, exposure to exogenous H_2O_2 following a single treatment is extremely brief, and a better mimic of the sustained pathogen-induced oxidative burst is continuous generation of H_2O_2 in situ by addition of an H_2O_2-generating system such as glucose/glucose oxidase. Nonetheless, the relative dose responses to exogenous H_2O_2 suggest a possible resolution of the paradox that the oxidative burst stimulates both cell death and cellular protectants.

Glutathione peroxidase and other peroxidative enzymes destroy active oxygen species and block oxidant-mediated programmed cell death (67). Likewise, GST detoxifies lipid hydroperoxides generated by active oxygen species (14). In addition, expression in transgenic tobacco of a mutant ubiquitin gene, which blocks ubiquitin function, leads to the generation of spontaneous HR lesions, suggesting that the ubiquitin system also functions in cell death regulation (12). Induction of *gst* and other cellular protectant genes in response to substantially lower doses of H_2O_2 than those required to trigger hypersensitive cell death thus provides a regulatory system in which H_2O_2 from the oxidative burst can act both as a local trigger of cell death in challenged cells and also as a diffusible signal for induction of cellular protectants to block oxidant-mediated programmed cell death in surrounding cells (87, 133). This arrangement would promote the strict spatial limitation of hypersensitive cell death and help maintain the capacity of surrounding cells to deploy transcription-dependent defenses. Several lesion mimic mutants exhibit spreading lesions in response to an avirulent pathogen, indicating that spatial limitation of the HR is under genetic control (37, 59, 60, 72) and that O_2^- but not H_2O_2 causes runaway cell death in the *lsd1* lesion mimic mutant (71), consistent with dysfunction of an oxidant protection mechanism.

Many flavonoids, lignans, and hydroxycinnamic acid derivatives can act as scavengers of O_2^- (82, 146), and ascorbate-dependent peroxidases utilizing phenolic substrates may remove H_2O_2 from the apoplastic space (95). Using a screen for cDNAs that, on expression in yeast, confer tolerance against the thiol-oxidizing drug diamide, an Arabidopsis NADPH reductase was identified (7) with striking sequence similarity to the isoflavone reductase (IFR) induced in legumes for phytoalexin synthesis (104). A similar IFR-like protein is induced in maize by treatments that decrease glutathione levels (107). Neither the maize nor Arabidopsis proteins has IFR activity. They appear to belong to a new class of oxidoreductases that may function in a thiol-independent response to oxidative stress under conditions of glutathione shortage. Their extensive homology to IFR suggests that the isoflavonoid phytoalexin pathway may have evolved from an oxidative defense.

H_2O_2 induction of *gst* in cells surrounding the HR lesion may also contribute to the sequestration of antimicrobial phytoalexins. *Bronze-2* in maize encodes a GST that catalyzes the formation of anthocyanin-glutathione conjugates, which are transported into the vacuole via a high-affinity, uncoupler-insensitive glutathione pump (89). This GST can also act on the alfalfa phytoalexin medicarpin, and the corresponding conjugate can be taken into mung bean vacuoles by the same mechanism (87a). Conversely, application of high concentrations of H_2O_2 to cells of *Pueraria lobata* cause rapid disappearance of vacuolar isoflavone conjugates, which then appear covalently coupled to lignocellulose, presumably by the action of peroxidases (105). Thus, high concentrations of H_2O_2 in the immediate vicinity of the HR lesion result in isoflavonoid mobilization as an early transcription-independent response, whereas lower concentrations at a distance induce mechanisms for isoflavonoid sequestration, to prevent toxic effects and help develop a preformed chemical barrier to subsequent pathogen ingress.

Systemic Responses and Interactions with Salicylic Acid

Salicylic acid (SA) accumulates both locally and to a lesser extent systemically following inoculation with avirulent pathogens, and exogenous SA, albeit often at high concentrations, induces defense genes and SAR (111, 116). Experiments with the PAL inhibitor α-aminooxy-β-phenylpropionic acid and transgenic plants expressing the bacterial *nahG* gene encoding a SA hydroxylase indicate that SA is required for both localized resistance and SAR (33, 56, 92). An SA-binding protein from tobacco (26) is a catalase (27). SA inhibits this catalase in vitro and causes an increase in the level of H_2O_2 in vivo (27). Injection of 1 mM H_2O_2 induces PR-proteins, and it was proposed that SA inhibition of catalase causes an accumulation of H_2O_2 as a signal for SAR. SA-binding proteins and SA-inhibitable catalases were also demonstrated in Arabidopsis, tomato, and cucumber (118), and structural analogs including the agrichemical dichloroisonicotinic acid, which are able to induce PR-protein expression and SAR, also inhibit catalase activity in vitro (28). However, injection of high concentrations of SA gives only a ~0.4-fold increase in H_2O_2 in vivo (27), and several observations challenge the catalase inhibitor model for SA function in SAR (16, 100). Thus, no systemic accumulation of H_2O_2 was detected in tobacco expressing SAR, and, although H_2O_2 induces PR-proteins, the effect is much weaker than with SA or dichloroisonicotinic acid. PR-protein induction by H_2O_2 is suppressed in *nahG* plants, suggesting that SA acts downstream of H_2O_2, and injection of 1 M H_2O_2 does not induce protection against tobacco mosaic virus. 3-Aminotriazole is only a weak inducer of PR-1, and in tobacco and Arabidopsis no significant changes in catalase activity are detected following immunization with *P. syringae* p.v. *syringae*. Likewise, no inhibition of catalase is observed in leaf discs preincubated with concentrations of SA that induce PR-1. These results suggest that H_2O_2 functions upstream of SA in PR-protein expression, and SA stimulation of H_2O_2 accumulation may only be relevant at the site of infection where concentrations sufficient for effective inhibition of catalase activity may be reached in the later stages of the HR (Figure 3). Transgenic tobacco expressing antisense catalase constructs shows no visible phenotype under normal conditions, and catalase suppression was not sufficient for PR-1 induction, although under high light conditions severe leaf damage was observed consistent with catalase function in protection against chronic oxidative stress (23).

SA also binds to plant, fungal, and animal catalases, and to pig liver aconitase, indicating a general affinity for iron-containing enzymes (114). Binding of SA to heme-containing enzymes may result in the generation of salicylate radicals that might have signal functions in gene activation and cell death, although no evidence for the occurrence or function of these radicals in vivo has been presented. Preincubation of parsley suspension cultures with SA enhances both spontaneous and elicitor-induced production of H_2O_2, although the optimal effect required >500 µM SA and >24 h of preincubation (75). Likewise, tobacco plants hydroponically fed with 1–2 mM SA for 1–7 days exhibit enhanced defense gene expression following wounding or pathogen infection (99). The requirement for long preincubations with high concentrations suggest that SA is mediating an inductive response, and indeed SA preconditioning of parsley cells is blocked by protein synthesis inhibitors (51).

The potency of SA is markedly enhanced when tested in the presence of pathogen (127). Thus, physiological concentrations of SA dramatically accelerate enhanced H_2O_2 accumulation in soybean cells simultaneously inoculated with avirulent *P. syringae* pv. *glycinea*, potentiating *gst* induction and cell death (127, 127a). *pal* and *chs* induction are also potentiated, indicating that SA acts before the divergence of branch signal pathways for antimicrobial defense gene induction and activation of the oxidative burst (Figure 3). SA also potently enhances the burst in response to cantharidin treatment in the absence of pathogen. This response occurs without detectable lag and is insensitive to cycloheximide, indicating that SA stimulates an agonist-dependent, highly geared intrinsic gain control mechanism, distinct from long-term inductive effects on signal pathway components or downstream effector genes. Moreover, because SA acts upstream of *pal* transcription, this gain control subsumes a feedback loop for autoamplification of SA synthesis and helps integrate transcription-dependent and oxidant-dependent responses (Figure 3). H_2O_2 stimulates the activity of benzoic acid 2-hydroxylase, which is the last step in SA biosynthesis (85), and such cross-talk between the branch pathways may be critical for signal amplification and integration (Figure 3).

Recent data indicate that inoculation of Arabidopsis with an avirulent strain of *Pseudomonas syringae* pv. *tomato* rapidly induces *gst* and low frequency systemic apoptosis in uninoculated leaves (ME Alvarez, RI Pennell, PJ Meijer, RA Dixon & C Lamb, unpublished observations). The primary oxidative burst at the site of the immunizing inoculation appears to be the signal initiating systemic responses because "micro-HRs" and SAR are not observed if the initial burst is blocked with a localized application of DPI, and the systemic responses can be induced in secondary leaves by localized generation of H_2O_2 following infiltration of glucose/glucose oxidase (but not neat H_2O_2 or a O_2^- generating system) into the primary leaf in the absence of pathogen. Although it remains to be determined whether the mobile signal is H_2O_2 or a product of H_2O_2 metabolism, these recent observations indicate that the oxidative burst plays a central role in SAR as well as localized resistance in the HR. The next few years should see elucidation of the molecular mechanisms underlying the localized activation of the oxidative burst following perception of pathogen avirulence signals and key downstream responses including gene activation, cell death, and long-distance signaling, as well as use of this information to develop novel strategies for engineering enhanced protection against pathogens by manipulation of the oxidative burst and oxidant-mediated signal pathways.

ACKNOWLEDGMENTS

We thank Cindy Doane for help in preparation of the manuscript and the Samuel Roberts Noble Foundation for encouragement and support.

> Visit the *Annual Reviews home page* at http://www.annurev.org.

Literature Cited

1. Adam A, Farkas T, Somlyai G, Hevesi M, Kiraly Z. 1989. Consequence of O_2^- generation during a bacterially induced hypersensitive response in tobacco: deterioration of membrane lipids. *Physiol. Mol. Plant Pathol.* 34:13–26

2. Alvarez ME, Lamb C. 1996. Oxidative burst-mediated defense responses in plant disease resistance. In *Oxidative Stress and the Molecular Biology of Antioxidant Defenses*, ed. J Scandalios. Cold Spring Harbor, NY: Cold Spring Harbor Lab. Press. In press

3. Apostol I, Heinstein PF, Low PS. 1989. Rapid stimulation of an oxidative burst during elicitation of cultured plants cells: role in defense and signal transduction. *Plant Physiol.* 90:109–16

4. Atkinson MM, Keppler LD, Orlandi EW, Baker CJ, Mischke CF. 1990. Involvement of plasma membrane calcium influx in bacterial induction of the K^+/H^+ exchange and hypersensitive responses in tobacco. *Plant Physiol.* 92:215–21

5. Auh C-K, Murphy TM. 1995. Plasma membrane redox enzyme is involved in the synthesis of O_2^- and H_2O_2 by *Phytophthora* elicitor-stimulated rose cells. *Plant Physiol.* 107:1241–47

6. Babior BM. 1992. The respiratory burst oxidase. *Adv. Enzymol. Relat. Areas Mol. Biol.* 65:49–95

7. Babiychuk E, Kushnir S, Bellesboix E, Van Montagu M, Inzé D. 1995. *Arabidopsis thaliana* NADPH oxidoreductase homologs confer tolerance of yeast toward the thiol-oxidizing drug diamide. *J. Biol. Chem.* 270:26224–31

8. Babiychuk E, Kushnir S, Van Montagu M, Inzé D. 1994. The *Arabidopsis thaliana* apurinic endonuclease Arp reduces human transcription factors Fos and Jun. *Proc. Natl. Acad. Sci. USA* 91:3299–303

9. Baker CJ, Mock N, Glazener J, Orlandi E. 1993. Recognition responses in pathogen/nonhost and race/cultivar interactions involving soybean (*Glycine max*) and *Pseudomonas syringae* pathovars. *Physiol. Mol. Plant Pathol.* 43:81–94

10. Baker CJ, Orlandi EW. 1995. Active oxygen in plant pathogenesis. *Annu. Rev. Phytopathol.* 33:299–321

11. Baker CJ, Orlandi EW, Mock NM. 1993. Harpin, an elicitor of the hypersensitive response in tobacco caused by *Erwinia amylovora*, elicits active oxygen production in suspension cells. *Plant Physiol.* 102:1341–44

12. Becker F, Buschfeld E, Schell J, Bachmair A. 1993. Altered response to viral infection by tobacco plants perturbed in ubiquitin system. *Plant J.* 3:875–81

13. Beffa R, Szell M, Beuwly P, Pay A, Vögeli-Lange R, et al. 1995. Cholera toxin elevates pathogen resistance and induces pathogenesis-related gene expression in tobacco. *EMBO J.* 14:5753–61

14. Berhane K, Widersten M, Engström A,

Kozarich JW, Mannervik B. 1994. Detoxification of base propenals and other β-unsaturated aldehyde products of radical reactions and lipid peroxidation by human glutathione transferases. *Proc. Natl. Acad. Sci. USA* 91:1480–84

15. Bestwick CS, Bennett MH, Mansfield JW. 1995. *Hrp* mutants of *Pseudomonas syringae* pv. *phaseolicola* induce cell wall alterations but not membrane damage leading to the hypersensitive response. *Plant Physiol.* 108:503–16

16. Bi YM, Kenton P, Mur L, Darby R, Draper J. 1995. Hydrogen peroxide does not function downstream of salicylic acid in the induction of PR protein expression. *Plant J.* 8:235–45

17. Blechert S, Brodschelm W, Hölder S, Kammerer L, Kutchan TM, et al. 1995. The octadecanoic pathway: signal molecules for the regulation of secondary pathways. *Proc. Natl. Acad. Sci. USA* 92:4099–105

18. Bokoch GM. 1994. Regulation of the human neutrophil NADPH oxidase by the Rac GTP-binding proteins. *Curr. Opin. Cell Biol.* 6:212–18

19. Bolwell GP, Butt VS, Davies DR, Zimmerlin A. 1995. The origin of the oxidative burst in plants. *Free Radic. Res.* 23:517–32

20. Bradley DJ, Kjellbom P, Lamb CJ. 1992. Elicitor- and wound-induced oxidative cross-linking of a proline-rich plant cell wall protein: a novel, rapid defense response. *Cell* 70:21–30

21. Brady JD, Sadler IH, Fry SC. 1996. Di-isodityrosine, a novel tetrameric derivative of tyrosine in plant cell wall proteins: a new potential cross-link. *Biochem. J.* 315:323–27

22. Brisson LF, Tenhaken R, Lamb CJ. 1994. Function of oxidative cross-linking of cell wall structural proteins in plant disease resistance. *Plant Cell* 6:1703–12

23. Chamnongpol S, Willekens H, Langebartels C, van Montagu M, Inzé D, van Camp W. 1996. Transgenic tobacco with a reduced catalase activity develops necrotic lesions and induces pathogenesis-related expression under high light. *Plant J.* 10:491–503

24. Chandra S, Low PS. 1995. Role of phosphorylation in elicitation of the oxidative burst in cultured soybean cells. *Proc. Natl. Acad. Sci. USA* 92:4120–23

25. Chen W, Chao G, Singh KB. 1996. The promoter of a H_2O_2 inducible Arabidopsis glutathione *S*-transferase gene contains closely linked OBF and OBP1 binding sites. *Plant J.* 10(6):955–66

26. Chen ZX, Ricigliano JW, Klessig DF. 1993.

Purification and characterization of a soluble salicylic acid-binding protein from tobacco. *Proc. Natl. Acad. Sci. USA* 90:9533–37

27. Chen ZX, Silva H, Klessig DF. 1993. Active oxygen species in the induction of plant systemic acquired resistance by salicylic acid. *Science* 262:1883–86

28. Conrath U, Chen ZX, Ricigliano JR, Klessig DF. 1995. Two inducers of plant defense responses, 2,6-dichloroisonicotinic acid and salicylic acid, inhibit catalase activity in tobacco. *Proc. Natl. Acad. Sci. USA* 92:7143–47

29. Corbin DR, Sauer N, Lamb CJ. 1987. Differential regulation of a hydroxyproline-rich glycoprotein gene family in wounded and infected plants. *Mol. Cell Biol.* 7:4337–44

30. Croft KPC, Voisey CR, Slusarenko AJ. 1990. Mechanisms of hypersensitive cell collapse: correlation of increased lipoxygenase activity with membrane damage in leaves of *Phaseolus vulgaris* (L) inoculated with an avirulent race of *Pseudomonas syringae* pv. *phaseolicola*. *Physiol. Mol. Plant Pathol.* 36:49–62

31. Daniel V. 1993. Glutathione *S*-transferases: gene structure and regulation of expression. *Crit. Rev. Biochem. Mol. Biol.* 28:173–207

32. Davis D, Merida J, Legendre L, Low PS, Heinstein P. 1993. Independent elicitation of the oxidative burst and phytoalexin formation in cultured plant cells. *Phytochemistry* 32:607–11

33. Delaney TP, Uknes S, Vernooij B, Friedrich L, Weymann K, et al. 1994. A central role of salicylic acid in plant disease resistance. *Science* 266:1247–50

34. Desikan R, Hancock JT, Coffey MJ, Neill SJ. 1996. Generation of active oxygen in elicited cells of *Arabidopsis thaliana* is mediated by a NADPH oxidase-like enzyme. *FEBS Lett.* 382:213–17

35. Devlin WS, Gustine DL. 1992. Involvement of the oxidative burst in phytoalexin accumulation and the hypersensitive reaction. *Plant Physiol.* 100:1189–95

36. Diekmann E, Abo A, Johnston C, Segal AW, Hall A. 1994. Interaction of *Rac* with p67 *Phox* and regulation of the phagocytic NADPH oxidase. *Science* 265:531–33

37. Dietrich A, Mayer JE, Hahlbrock K. 1990. Fungal elicitor triggers rapid, transient, and specific protein phosphorylation in parsley cell suspension cultures. *J. Biol. Chem.* 265:6360–68

38. Dietrich RA, Delaney TP, Uknes SJ, Ward ER, Ryals JA, Dangl JL. 1994. *Arabidopsis*

mutants simulating disease resistance response. *Cell* 77:565–77

39. Dixon RA, Harrison MJ. 1990. Activation, structure and organization of genes involved in microbial defense in plants. *Adv. Genet.* 28:165–234

40. Dixon RA, Harrison MJ, Lamb CJ. 1994. Early events in the activation of plant defense responses. *Annu. Rev. Phytopathol.* 32:479–501

41. Doehlert DC, Wicklow DT, Gardner HW. 1993. Evidence implicating the lipoxygenase pathway in providing resistance to soybean against *Aspergillus flavus*. *Phytopathology* 83:1473–78

42. Doke N. 1983. Involvement of superoxide anion generation in hypersensitive response of potato tuber tissues to infection with an incompatible race of *Phytophthora infestans*. *Physiol. Plant Pathol.* 23:345–47

43. Doke N. 1983. Generation of superoxide anion by potato tuber protoplasts during the hypersensitive response to hyphal cell wall components of *Phytophthora infestans* and specific inhibition of the reaction by suppressors of hypersensitivity. *Physiol. Plant Pathol.* 23:359–67

44. Doke N. 1995. NADPH-dependent O_2^- generation in membrane fractions isolated from wounded potato tubers inoculated with *Phytophthora infestans*. *Physiol. Plant Pathol.* 27:311–22

45. Doke N, Miura Y. 1995. In vitro activation of NADPH-dependent O_2^- generating system in a plasma membrane-rich fraction of potato tuber tissues by treatment with an elicitor from *Phytophthora infestans* or with digitonin. *Physiol. Mol. Plant Pathol.* 46:17–28

46. Dwyer SC, Legendre L, Low PS, Leto TL. 1995. Plant and human neutrophil oxidative burst complexes contain immunologically related proteins. *Biochim. Biophys. Acta* 1289:231–37

47. Edwards R, Blount J, Dixon RA. 1991. Glutathione as an elicitor of the phytoalexin response in legume cell cultures. *Planta* 184:403–9

48. Ellis JG, Tokuhisa JG, Llewellyn DJ, Bouchez D, Singh K, et al. 1993. Does the ocs element occur as a functional component of the promoters of plant genes? *Plant J.* 4:433–43

49. Epperlein MM, Noronha-Dutra AA, Strange RN. 1986. Involvement of the hydroxyl radical in the abiotic elicitation of phytoalexins in legumes. *Physiol. Plant Pathol.* 28:67–77

50. Esaka M, Fujisawa K, Goto M, Kisu Y.

51. 1992. Regulation of ascorbate oxidase expression in pumpkin by auxin and copper. *Plant Physiol.* 100:231–37

51. Fauth M, Merten A, Hahn MG, Jeblick W, Kauss H. 1996. Competence for elicitation of H_2O_2 in hypocotyls of cucumber is induced by breaching the cuticle and is enhanced by salicylic acid. *Plant Physiol.* 110:347–54

52. Felix G, Regenass M, Spanu P, Boller T. 1994. The protein phosphatase inhibitor calyculin A mimics elicitor action in plant cells and induces rapid hyperphosphorylation of specific proteins as revealed by pulse labeling with [^{33}P]phosphate. *Proc. Natl. Acad. Sci. USA* 91:952–56

53. Fett WF, Jones SB. 1995. Microscopy of the interaction of *hrp* mutants of *Pseudomonas syringae* pv. *phaseolicola* with a nonhost plant. *Plant Sci.* 107:27–39

54. Frilling RS, Bergelson S, Daniel V. 1992. Two adjacent AP-1-like binding sites form the electrophile-responsive element of the murine glutathione *S*-transferase Ya subunit gene. *Proc. Natl. Acad. Sci. USA* 89: 668–72

55. Fry SC. 1986. Cross-linking of matrix polymers in the growing cell wall of angiosperms. *Annu. Rev. Plant Physiol.* 37: 165–86

56. Gaffney T, Friedrich L, Vernooij B, Negrotto D, Nye G, et al. 1993. Requirement of salicylic acid for the induction of systemic acquired resistance. *Science* 261: 754–56

57. Glazener JA, Orlandi EW, Baker JC. 1996. The active oxygen response of cell suspensions to incompatible bacteria is not sufficient to cause hypersensitive cell death. *Plant Physiol.* 110:759–63

58. Green R, Fluhr R. 1995. UV-B-induced PR-1 accumulation is mediated by active oxygen species. *Plant Cell* 7:203–12

59. Greenberg JT, Ausubel FM. 1993. *Arabidopsis* mutants compromised for the control of cellular damage during pathogenesis and aging. *Plant J.* 4:327–41

60. Greenberg JT, Guo A, Klessig DF, Ausubel FM. 1994. Programmed cell death in plants: a pathogen-triggered response activated coordinately with multiple defense functions. *Cell* 77:551–63

61. Groom QJ, Torres MA, Fordham-Skelton AP, Hammond-Kosack KE, Robinson NJ, Jones JDG. 1996. *rbohA*, a rice homologue of the mammalian *gp91phox* respiratory burst oxidase gene. *Plant J.* 10:515–22

62. Hahlbrock K, Scheel D, Logemann E, Nürnberger T, Parniske M, et al. 1995. Oligopeptide elicitor-mediated defense gene

activation in cultured parsley cells. *Proc. Natl. Acad. Sci. USA* 92:4150–57

63. Halliwell B. 1978. Lignin synthesis: the generation of hydrogen peroxide and superoxide by horseradish peroxidase and its stimulation by manganese (II) and phenols. *Planta* 140:81–88

64. Hammond-Kosack KE, Silverman P, Raskin I, Jones JDG. 1996. Race-specific elicitors of *Cladosporium fulvum* induce changes in cell morphology and the synthesis of ethylene and salicylic acid in tomato plants carrying the corresponding *Cf* disease resistance gene. *Plant Physiol.* 110: 1381–94

65. He SY, Huang H-C, Collmer A. 1993. *Pseudomonas syringe* pv. *syringae* Harpin$_{PSS}$: a protein that is secreted by the *Hrp* pathway and elicits the hypersensitive response in plants. *Cell* 73:1255–66

66. Hérouart D, van Montagu M, Inzé D. 1993. Redox-activated expression of the cytosolic copper/zinc superoxide dismutase gene in *Nicotiana. Proc. Natl. Acad. Sci. USA* 90:3108–12

67. Hockenbery DM, Oltvai ZN, Yin X-M, Milliman CL, Korsmeyer SJ. 1993. Bcl-2 functions in an antioxidant pathway to prevent apoptosis. *Cell* 75:241–51

68. Hu G, Richter TE, Hulbert SH, Pryor T. 1996. Disease lesion mimicry caused by mutations in the rust resistance gene *Rp1. Plant Cell* 8:1367–76

69. Hurkman WJ, Tanaka CK. 1996. Germin gene expression is induced in wheat leaves by powdery mildew infection. *Plant Physiol.* 111:735–39

70. Ikegawa T, Mayama S, Nakayashiki H, Kato H. 1996. Accumulation of diferulic acid during the hypersensitive response of oat leaves to *Puccinia coronata* f. sp. *avenae* and its role in the resistance of oat leaves to cell wall degrading enzymes. *Physiol. Mol. Plant Pathol.* 48:245–55

71. Jabs T, Dietrich RA, Dangl JL. 1996. Extracellular superoxide initiates runaway cell death in an Arabidopsis mutant. *Science* 273:1853–56

72. Johal GS, Hulbert SH, Briggs SP. 1995. Disease lesion mimics of maize: a model for cell death in plants. *BioEssays* 17: 685–92

73. Jones AM, Dangl JL. 1996. Logjam at the Styx: the multiplicity of programmed cell death pathways in plants. *Trends Plant Sci.* 1:114–19

74. Kangasjärvi J, Talvinen J, Utriainen M, Karjalainen R. 1994. Plant defense systems induced by ozone. *Plant Cell Environ.* 17: 783–94

75. Kauss H, Jeblick W. 1995. Pretreatment of parsley suspension cultures with salicylic acid enhances spontaneous and elicited production of H_2O_2. *Plant Physiol.* 108: 1171–78

76. Kauss H, Jeblick W. 1996. Influence of salicylic acid on the induction of competence for H_2O_2 elicitation. Comparison of ergosterol with other elicitors. *Plant Physiol.* 111:755–63

77. Kawalleck P, Schmelzer E, Hahlbrock K, Somssich IE. 1995. Two pathogen-responsive genes in parsley encode a tyrosine-rich hydroxyproline-rich glycoprotein (hrgp) and an anionic peroxidase. *Mol. Gen. Genet.* 247:444–52

78. Keppler LD, Baker CJ. 1989. O_2^--initiated lipid peroxidation in a bacteria-induced hypersensitive reaction in tobacco cell suspensions. *Phytopathology* 79:555–62

79. Keppler LD, Baker CJ, Atkinson MM. 1989. Activated oxygen production during a bacteria induced hypersensitive reaction in tobacco suspension cells. *Phytopathology* 79:974–78

80. Keppler LD, Novacky A. 1987. The initiation of membrane lipid peroxidation during bacteria-induced hypersensitive reaction. *Physiol. Mol. Plant Pathol.* 30:233–45

81. Kuchitsu K, Kosaka H, Shiga T, Shibuya N. 1995. EPR evidence for generation of hydroxyl radical triggered by *N*-acetylchitooligosaccharide elicitor and a protein phosphatase inhibitor in suspension-cultured rice cells. *Protoplasma* 188:138–42

82. Larson RA. 1988. The antioxidants of higher plants. *Phytochemistry* 27:969–78

83. Legendre L, Heinstein PF, Low PS. 1992. Evidence for participation of GTP-binding proteins in elicitation of the rapid oxidative burst in cultured soybean cells. *J. Biol. Chem.* 267:20140–47

84. Legendre L, Reuter S, Heinstein PF, Low PS. 1993. Characterization of the oligogalacturonide-induced oxidative burst in cultured soybean (*Glycine max*) cells. *Plant Physiol.* 102:233–40

85. León J, Lawton MA, Raskin I. 1995. Hydrogen peroxide stimulates salicylic acid biosynthesis in tobacco. *Plant Physiol.* 108:1673–78

86. Levine A, Pennell RI, Alvarez ME, Palmer R, Lamb C. 1996. Calcium-mediated apoptosis in a plant hypersensitive disease resistance response. *Curr. Biol.* 6:427–37

87. Levine A, Tenhaken R, Dixon RA, Lamb C. 1994. H_2O_2 from the oxidative burst orchestrates the plant hypersensitive response. *Cell* 79:583–93

87a. Li Z-S, Alfenito M, Rea PA, Walbot V,

Dixon RA. 1997. Vacuolar uptake of the phytoalexin medicarpin by the glutathione pump. *Phytochemistry.* In press

88. Loh YT, Martin GB. 1995. The disease-resistance gene *Pto* and the fenthion-sensitivity gene *Fen* encode closely related functional protein kinases. *Proc. Natl. Acad. Sci. USA* 92:4181–84

89. Marrs KA, Alfenito MR, Lloyd AM, Walbot V. 1995. A glutathione *S*-transferase involved in vacuolar transfer encoded by the maize gene *Bronze-2. Nature* 375: 397–400

90. Martin SJ, Green DR, Cotter TG. 1994. Dicing with death: dissecting the components of the apoptosis machinery. *Trends Biochem Sci.* 19:26–30

91. Masuta C, Van Den Bulcke M, Bauw G, van Montagu M, Caplan AB. 1991. Differential effects of elicitors on the viability of rice suspension cells. *Plant Physiol.* 97: 619–29

92. Mauch-Mani B, Slusarenko AJ. 1996. Production of salicylic acid precursors is a major function of phenylalanine ammonialyase in the resistance of Arabidopsis to *Peronspora parasitica. Plant Cell* 8: 203–12

93. May MJ, Hammond-Kosack KE, Jones JDG. 1996. Involvement of reactive oxygen species, glutathione metabolism, and lipid peroxidation in the *Cf*-gene-dependent defense response of tomato cotyledons induced by race-specific elicitors of *Cladosporium fulvum. Plant Physiol.* 110: 1367–79

94. Mehdy MC. 1994. Active oxygen species in plant defense against pathogens. *Plant Physiol.* 105:467–72

95. Mehlhorn H, Lelandais M, Korth HG, Foyer CH. 1996. Ascorbate is the natural substrate for plant peroxidases. *FEBS Lett.* 378:203–6

96. Meyer M, Schreck R, Bauerle PA. 1993. H_2O_2 and antioxidants have opposite effects on activation of NF-κB and AP-1 in intact cells: AP-1 as secondary antioxidant-responsive factor. *EMBO J.* 12:2005–15

97. Mittler R, Shulaev V, Lam E. 1995. Coordinated activation of programmed cell death and defense mechanisms in transgenic tobacco plants expressing a bacterial proton pump. *Plant Cell* 7:29–42

98. Morel F, Doussiere J, Vignais PV. 1991. The superoxide-generating oxidase of phagocytic cells: physiological, molecular and pathological aspects. *Eur. J. Biochem.* 201: 523–46

99. Mur LAJ, Naylor G, Warner SAJ, Sugars JM, White RF, Draper J. 1996. Salicylic acid potentiates defence gene expression in leaf tissue exhibiting acquired resistance to pathogen attack. *Plant J.* 9:559–71

100. Neuenschwander U, Vernooij B, Friedrich L, Uknes S, Kessmann H, Ryals J. 1995. Is hydrogen peroxide a second messenger of salicylic acid in systemic acquired resistance? *Plant J.* 8:227–33

101. Nürnberger T, Nennstiel D, Jabs T, Sacks WR, Hahlbrock K, Scheel D. 1994. High affinity binding of a fungal oligopeptide elicitor to parsley plasma membranes triggers multiple defense responses. *Cell* 78: 449–60

102. Orlandi EW, Hutcheson SW, Baker CJ. 1992. Early physiological responses associate with race-specific recognition in soybean leaf tissue and cell suspensions treated with *Pseudomonas syringae* pv. *glycinea. Physiol. Mol. Plant Pathol.* 40:173–80

103. Otha H, Shida K, Pen Y-L, Furusawa I, Shishiyama J, et al. 1992. A lipoxygenase pathway is activated in rice after infection with rice blast fungus. *Magnaporthe grisea. Plant Physiol.* 97:94–98

104. Paiva NL, Edwards R, Sun Y, Hrazdina G, Dixon RA. 1991. Stress responses in alfalfa (*Medicago sativa* L.) XI. Molecular cloning and expression of alfalfa isoflavone reductase, a key enzyme of isoflavonoid phytoalexin biosynthesis. *Plant Mol. Biol.* 17:653–67

105. Park HH, Hakamatsuka T, Sankawa U, Ebizuka Y. 1995. Involvement of oxidative burst in isoflavonoid metabolism in elicited cell suspension cultures of *Pueraria lobata. Z. Naturforsch. Teil C* 50:824–32

106. Peng M, Kuc J. 1992. Peroxidase-generated hydrogen peroxide as a source of antifungal activity in vitro and on tobacco leaf disks. *Phytopathology* 82:696–99

107. Petrucco S, Bolchi A, Foroni C, Percudani R, Rossi GL, Ottonello S. 1996. A maize gene encoding an NADPH binding enzyme highly homologous to isoflavone reductases is activated in response to sulfur starvation. *Plant Cell* 8:69–80

108. Prasad TK, Anderson MD, Martin BA, Stewart CR. 1994. Evidence for a chilling-induced oxidative stress in maize seedlings and a regulatory role for hydrogen peroxide. *Plant Cell* 6:65–74

109. Price AH, Taylor S, Ripley SJ, Griffiths A, Trewavas AJ, Knight MR. 1994. Oxidative signals in tobacco increase cytosolic calcium. *Plant Cell* 6:1301–10

110. Raff MC. 1992. Social controls on cell survival and cell death. *Nature* 356:397–400

111. Raskin I. 1992. Role of salicylic acid in

plants. *Annu. Rev. Plant Physiol. Plant. Mol. Biol.* 43:439–63

112. Rasmussen JB, Smith JA, Williams S, Burkhart W, Ward E, et al. 1995. cDNA cloning and systemic expression of acidic peroxidases associated with systemic acquired resistance to disease in cucumber. *Physiol. Mol. Plant Pathol.* 46:389–400

113. Rogers KR, Albert F, Anderson A. 1988. Lipid peroxidation is a consequence of elicitor activity. *Plant Physiol.* 86:547–53

114. Rüffer M, Steipe B, Zenk MH. 1995. Evidence against specific binding of salicylic acid to plant catalase. *FEBS Lett.* 377: 175–80

115. Rustérucci C, Stallaert V, Milat M-L, Pugin A, Ricci P, Blein J-P. 1996. Relationship between active oxygen species, lipid peroxidation, necrosis, and phytoalexin production induced by elicitins in *Nicotiana*. *Plant Physiol.* 111:885–91

116. Ryals J, Uknes S, Ward E. 1994. Systemic acquired resistance. *Plant Physiol.* 104: 1109–12

117. Ryerson DE, Heath MC. 1996. Cleavage of nuclear DNA into oligonucleosomal fragments during cell death induced by fungal infection or by abiotic treatments. *Plant Cell* 8:393–402

118. Sanchez-Casas P, Klessig DF. 1994. A salicylic acid-binding activity and a salicylic acid-inhibitable catalase activity are present in a variety of plant species. *Plant Physiol.* 106:1675–79

119. Sano H, Seo S, Orudgev E, Youssefian S, Ishizuka K, Ohashi Y. 1994. Expression of the gene for a small GTP binding protein in transgenic tobacco elevates endogenous cytokinin levels, abnormally induce salicylic acid in response to wounding and increases resistance to tobacco mosaic virus infection. *Proc. Natl. Acad. Sci. USA* 91:10556–60

120. Sauer N, Corbin DR, Keller B, Lamb CJ. 1990. Cloning and characterization of a wound-specific hydroxyproline-rich glycoprotein in *Phaseolus vulgaris*. *Plant Cell. Environ.* 13:257–66

121. Scandalios JG. 1993. Oxygen stress and superoxide dismutases. *Plant Physiol.* 101: 7–12

122. Schreck R, Riber P, Bauerle PA. 1991. Reactive oxygen intermediates as apparently widely used messengers in the activation of the NF-κB transcription factor and HIV-1. *EMBO J.* 10:2247–58

123. Schwacke R, Hager A. 1992. Fungal elicitor induce a transient release of active oxygen species from cultured spruce cells that

is dependent on Ca^{2+} and protein kinase activity. *Planta* 187:136–41

124. Sharma YK, Davis KR. 1994. Ozone-induced expression of stress-related genes in *Arabidopsis thaliana*. *Plant Physiol.* 105: 1089–96

125. Sharma YK, León J, Raskin I, Davis KR. 1996. Ozone-induced responses in *Arabidopsis thaliana:* the role of salicylic acid in the accumulation of defense-related transcripts and induced resistance. *Proc. Natl. Acad. Sci. USA.* 93:5099–104

126. Sheng J, D'Ovidio R, Mehdy MC. 1991. Negative and positive regulation of a novel proline-rich protein mRNA by fungal elicitor and wounding. *Plant J.* 3:345–54

127. Shirasu K, Dixon RA, Lamb C. 1996. Signal transduction in plant immunity. *Curr. Opin. Immunol.* 8:3–7

127a. Shirasu K. Nakajima H, Rajasekhar VK, Dixon RA, Lamb C. 1996. Salicylic acid potentiation of an agonist-dependent gain control amplifying pathogen signals in the activation of defense mechanisms. *Plant Cell.* In press

128. Smith JA, Hammerschmidt R, Fulbright DW. 1991. Rapid induction of systemic resistance in cucumber by *Pseudomonas syringae* pv. *syringae*. *Physiol. Mol. Plant Pathol.* 38:232–35

129. Song W-Y, Wang G-L, Chen L-L, Kim H-S, Pi L-Y, et al. 1995. A receptor kinase-like protein encoded by the rice disease resistance gene *Xa21*. *Science* 270:1804–6

130. Staskawicz BJ, Ausubel FM, Baker BJ, Ellis JG, Jones JDG. 1995. Molecular genetics of plant disease resistance. *Science* 268:661–67

131. Steller H. 1995. Mechanisms and genes of cellular suicide. *Science* 267:1445–99

132. Tavernier E, Wendehenne D, Blein J-P, Pugin A. 1995. Involvement of free calcium in action of cryptogein, a proteinaceous elicitor of hypersensitive reaction in tobacco cells. *Plant Physiol.* 109:1025–31

133. Tenhaken R, Levine A, Brisson LF, Dixon RA, Lamb C. 1995. Function of the oxidative burst in hypersensitive disease resistance. *Proc. Natl. Acad. Sci. USA* 92: 4158–63

134. Toldeano MB, Kullik I, Trinh F, Baird PT, Schneider TD, Storz G. 1994. Redox-dependent shift of oxyR-DNA contacts along an extended DNA-binding site: a mechanism for differential promoter selection. *Cell* 78:897–909

135. Ulmasov T, Ohmiya A, Hagen G, Guilfoyle T. 1995. The soybean GH2/4 gene that encodes a glutathione *S*-transferase has a

promoter that is activated by a wide range of chemical agents. *Plant Physiol.* 108: 919–27

136. Vaux DL, Haecker G, Strasser A. 1994. An evolutionary perspective on apoptosis. *Cell* 76:777–79

137. Vera-Estrella R, Barkla BJ, Higgins VJ, Blumwald E. 1994. Plant defense response to fungal pathogens. II. G-protein-mediated changes in host plasma membrane redox reactions. *Plant Physiol.* 106:97–102

138. Vera-Estrella R, Blumwald E, Higgins VJ. 1992. Effect of specific elicitors of *Cladosporium fulvum* on tomato suspension cells. *Plant Physiol.* 99:1208–15

139. Waffenschmidt S, Woessner JP, Beer K, Goodenough UW. 1993. Isodityrosine cross-linking mediates insolubilization of cell walls in *Chlamydomonas*. *Plant Cell* 5:809–20

140. Wagner AM. 1995. A role for active oxygen species as second messengers in the induction of alternative oxidase gene expression in *Petunia hybrida* cells. *FEBS Lett.* 368: 339–42

141. Wang H, Li J, Bostock RM, Gilchrist DG. 1996. Apoptosis: a functional paradigm for programmed cell death induced by a host-selective phytotoxin and invoked during development. *Plant Cell* 8:375–91

142. Wingate VPM, Lawton MA, Lamb CJ. 1988. Glutathione causes a massive and selective induction of plant defense genes. *Plant Physiol.* 87:206–10

143. Wolter M, Hollricher K, Salamini F, Schulze-Lefert P. 1993. The *mlo* resistance alleles to powdery mildew infection in barley trigger a developmentally controlled defense mimic phenotype. *Mol. Gen. Genet.* 239:122–28

144. Wu GS, Short BJ, Lawrence EB, Levine EB, Fitzsimmons KC, Shah DM. 1995. Disease resistance conferred by expression of a gene encoding H_2O_2-generating glucose oxidase in transgenic potato plants. *Plant Cell* 7:1357–68

145. Xanthoudakis S, Curran T. 1992. Identification and characterization of Ref-1, a nuclear protein that facilitates AP-1 DNA-binding activity. *EMBO J.* 11:653–65

146. Yamasaki H, Uefusi H, Sakihama Y. 1996. Bleaching of the red anthocyanin induced by superoxide radical. *Arch. Biochem. Biophys.* 332:183–86

147. Zhang B, Chen W, Foley RC, Büttner M, Singh KB. 1995. Interactions between distinct types of DNA binding proteins enhance binding to ocs element promoter sequences. *Plant Cell* 7:2241–52

148. Zhang B, Singh KB. 1994. ocs element promoter sequences are activated by auxin and salicylic acid in *Arabidopsis*. *Proc. Natl. Acad. Sci. USA* 91:2507–11

149. Zhang ZG, Collinge DB, Thordal-Christensen H. 1995. Germin-like oxalate oxidase, a H_2O_2-producing enzyme, accumulates in barley attacked by the powdery mildew fungus. *Plant J.* 8:139–45

150. Zhou J, Loh Y-T, Bressan RA, Martin GB. 1996. The tomato gene *Pti1* encodes a serine/threonine kinase that is phosphorylated

Annu. Rev. Plant Physiol. Plant Mol. Biol. 1997. 48:277–96
Copyright © 1997 by Annual Reviews Inc. All rights reserved

THE ETHYLENE RESPONSE PATHWAY IN ARABIDOPSIS

Joseph J. Kieber

Department of Biological Sciences, Laboratory for Molecular Biology, University of Illinois at Chicago, Chicago, Illinois 60607

KEY WORDS: signal transduction, plant hormones, protein kinase, two-component system, mutants

---

ABSTRACT

The simple gas ethylene influences a diverse array of plant growth and developmental processes including germination, senescence, cell elongation, and fruit ripening. This review focuses on recent molecular genetic studies, principally in Arabidopsis, in which components of the ethylene response pathway have been identified. The isolation and characterization of two of these genes has revealed that ethylene sensing involves a protein kinase cascade. One of these genes encodes a protein with similarity to the ubiquitous Raf family of Ser/Thr protein kinases. A second gene shows similarity to the prokaryotic two-component histidine kinases and most likely encodes an ethylene receptor. Additional elements involved in ethylene signaling have only been identified genetically. The characterization of these genes and mutants will be discussed.

CONTENTS

277

INTRODUCTION

The simple gas ethylene (C_2H_4) has profound effects on a diverse array of plant growth and developmental processes, including germination, senescence, abscission, flowering, fruit ripening, and stress responses (reviewed in 1, 58). Significant progress has been made in our understanding of the ethylene biosynthetic pathway, culminating with the cloning and characterization of the genes encoding the two key enzymes, ACC synthase and ACC oxidase (reviewed in 40, 41, 87, 93, 96). Genetic and molecular analysis of ethylene signaling, mainly in Arabidopsis, has begun to provide glimpses into the molecular basis for ethylene perception and signal transduction (for other reviews see 9, 15, 17, 26, 42, 44, 96). This review is an account of the emerging picture of ethylene signaling that has been gleaned from these studies.

GENETIC ANALYSIS

When seedlings from many different dicotyledonous plant species are grown in the dark in the presence of ethylene they adopt a striking morphology, referred to as the triple response. In pea, the triple response consists of an inhibition of elongation and radial swelling of the epicotyl and an altered response to gravity (diageotropism) (46). Using this response, Neljubov first identified ethylene as a regulator of plant development by demonstrating that it was the active component of illuminating gas, which had previously been shown to have dramatic effects on plants in the late nineteenth century (66). In Arabidopsis, the triple response consists of an inhibition of root and hypocotyl elongation, radial swelling of the hypocotyl, and an exaggeration of the curvature of the apical hook. This response of Arabidopsis seedlings provides a facile screen for the isolation of ethylene response mutants and has been the key in unraveling the molecular basis of ethylene signaling. Mutants have been identified that fail to display the triple response in the presence of exogenously added ethylene (insensitive) as have mutations that constitutively display this response. These mutations affect virtually all ethylene responses in both seedlings and adult plants, suggesting that they affect central components in ethylene signaling. An additional class of mutants affect ethylene responses in only a subset of tissues, such as the root (*eir1*) or the apical hook (*hls1*) and may act late in the ethylene response pathway.

Ethylene-Insensitive Mutants

Ethylene-insensitive mutants fail to display the triple response in the presence of saturating levels of exogenously added ethylene and are readily identifiable as tall seedlings protruding above the "lawn" of short, wild-type seedlings. The first such mutant identified was *etr1* (*e*thylene-*r*esistant), which is inherited as a single gene, dominant mutation (8). This mutant is defective in a number of ethylene responses, including promotion of seed germination, enhancement of peroxidase activity, and acceleration of senescence of detached leaves. In addition, *etr1* mutants fail to display increased expression of ethylene-induced genes upon treatment with ethylene. *etr1* leaves bind only 20% as much exogenous ethylene as wild-type leaves, which is consistent with data suggesting *ETR1* encodes an ethylene receptor (see below). The *ein1* mutations, which were identified in a separate study (33), have been found to be allelic to *etr1*.

The other well-characterized ethylene-insensitive mutation is *ein2* (33), which is recessive and displays a relatively strong ethylene-insensitive phenotype as judged by the severity of the triple response in the presence of saturating levels of ethylene. Like *etr1* mutants, *ein2* mutants are defective in all ethylene responses that have been tested. Both *ein2* and *etr1* also show a 5- to 10-fold increase in the basal level of ethylene production in both etiolated seedlings and leaves from adult plants, suggesting that they are defective in the negative feedback regulation of ethylene biosynthesis (8, 33). One unique aspect of *ein2* mutant plants is that they fail to display disease symptoms, such as chlorosis and water-soaked lesions, upon infection with various strains of virulent bacteria, including *Pseudomonas syringae* and *Xanthomonas campestris* (6). However, the growth of pathogen is unaffected in *ein2* mutant plants. This reduction in disease symptoms is not observed with other ethylene-insensitive mutant plants, which suggests that the ethylene response pathway may contain a branch specific for pathogen-induced damage.

The *ein4* and *etr2* mutations, like *etr1*, are also dominant and act upstream of *ctr1* (15, 76). The absence of recessive, loss-of-function alleles at these three loci suggests that these genes are required for viability or that they are genetically redundant. The latter explanation is most likely correct in the case of *ETR1* because it has been found to be encoded by a small multigene family (see below). It is possible that *ein4* and *etr2* encode members of the ETR1 gene family because they are both dominant and act upstream of CTR1 (see below and Figure 1).

Mutations that result in a weaker ethylene-insensitive phenotype have been identified at five other genetic loci. The *ain1* (ACC-insensitive), *ein3*, *ein5*, and *ein6* mutations are recessive, whereas the *ein7* mutation is semidominant (76, 89). The weak phenotype of these mutations could be due to partial redundancy of their gene products or to mutations that are only partial loss of function. The *ein5* mutation maps very close to the recessive *ain1* mutation and thus may be allelic.

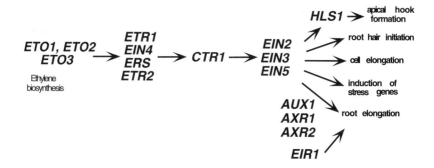

Figure 1 Proposed order of action of genes involved in ethylene production and perception in Arabidopsis. The order of the pathway was derived from the analysis of epistasis between the various mutants (15, 45, 76). Genetic analysis suggests that the pathway is primarily linear, with the exception of the *EIR1* gene product (see text), which may lie on a parallel pathway. Various responses to ethylene are indicated at the right of the Figure.

In an additional study, five ethylene-insensitive mutants were isolated that were referred to as *eti* (34). These have been only partially characterized genetically, and so it is unclear whether they represent novel loci. The ability of *eti* mutant seedlings to emerge through sand was shown to be directly proportional to the sensitivity of the seedlings to ethylene. This is consistent with the hypothesis that the triple response morphology is an adaptation that allows penetration of the seedling through soil, perhaps by protecting the delicate apical meristem from mechanical damage.

A number of Arabidopsis mutants that were isolated for resistance to auxin in the root also display resistance to other hormones, often including ethylene. The *auxl, axrl, axr2,* and *axr3* mutants display resistance to auxin, ethylene, and in some cases cytokinin or abscisic acid (37, 71, 92). These mutants all have an altered response to gravity. *axr1* and *axr2* mutants have distinct rosette and inflorescence phenotypes that are consistent with auxin insensitivity, whereas the *auxl* mutant primarily affects the root. The mechanism of multiple hormone resistance is unclear. The *AXR1* gene has been cloned and found to encode a protein similar to ubiquitin-conjugating enzyme, suggesting a role for protein degradation (51). These mutants, like the *hls1* mutant (50; also see below), highlight the interaction of multiple plant hormones in regulating various aspects of plant growth and development.

Constitutive Response Mutants

A number of mutants display the triple response in the absence of exogenous ethylene. These mutations fall into two classes. One class overproduces ethylene to such a level that the triple response is induced (33, 45). Such Eto (*e*thylene *o*verproducers) mutations are recognized because they are reverted by inhibitors of ethylene biosynthesis, such as aminoethoxyvinylglycine (AVG; 3) and α-aminoisobutyric acid (AIB; 77) and by inhibitors of ethylene binding such as *trans*-cyclooctene (81) and silver ion (7). The Eto mutations identify genes important in regulating ethylene biosynthesis. Five different Eto loci have been identified, and they produce from 10- to 100-fold more ethylene than wild-type, etiolated seedlings (33, 45; K Woeste & J Kieber, unpublished observations).

A second class of mutants display the triple response even in the presence of inhibitors of ethylene biosynthesis and binding, which suggests that these mutations affect ethylene signal transduction (45). As expected, these Ctr (*c*onstitutive *t*riple *r*esponse) mutants do not make significantly more ethylene than wild-type seedlings. The *ctr1* mutation is recessive and has dramatic effects on the morphology and development of both seedlings and adult plants. *ctr1* mutant alleles are transmitted at a reduced frequency relative to the wild-type allele, and reciprocal backcrosses have revealed that this is due to a defect in female *ctr1* gametophytes (43). Etiolated *ctr1* seedlings take longer to open the apical hook and expand their cotyledons when shifted to light as compared with wild-type seedlings. *ctr1* leaves and roots are smaller, the plants flower later, and the inflorescence is much more compact than that of wild-type plants. These phenotypes can be copied by growing wild-type plants in ethylene, suggesting that the various *ctr1* phenotypes reflect constitutive ethylene responses. This is supported by the fact that *ctr1* mutant seedlings and adult plants display a constitutively high level of expression of ethylene-regulated genes.

Leaf epidermal cells from air-grown, wild-type plants are fivefold larger than those from either air-grown *ctr1* mutant plants or ethylene-grown wild-type plants (45). This smaller cell size accounts for the majority of the decrease in leaf size observed in *ctr1* mutant plants and may underlie some of the other *ctr1* phenotypes such as the shortened hypocotyl, compacted inflorescence, and reduced root system. Ethylene has been shown to inhibit cell elongation in other plant species, perhaps because of a reorientation of the cytoskeleton and/or cell wall (4, 49, 75, 80, 84, 95). Ethylene-insensitive mutations such as *etr1* and *ein2* have 25% larger rosette leaves than wild-type plants (8, 33). This larger leaf size was correlated with an increase in cell size, consistent with the results obtained with *ctr1*. It seems likely that this increase in cell size in the ethylene-insensitive mutants is due to a failure to respond to a basal level of ethylene. These results suggest that an important role of ethylene is to regulate the size and stature of plants in response to various environmental cues.

An additional phenotype of *ctr1* is the formation of ectopic root hairs (23). Ethylene causes proliferation of root hairs in many plant species as well as the production of root hairs in certain species under conditions in which hairs are not normally observed (1, 20, 22). Arabidopsis root hairs are almost always located in epidermal cells that overlie a junction between adjacent cortical cells (24). In *ctr1* mutant roots, extra root hairs form in ectopic locations in the epidermis, suggesting that *ctr1* may act to negatively regulate the decision to adopt a hair cell fate (23). Ectopic hairs are also found in wild-type seedlings that are grown in the presence of ACC, which is readily converted to ethylene by plants (86). This indicates that ethylene may be the diffusible signal that has been proposed to regulate root hair cell fate (13). Consistent with this, seedlings grown in the presence of ethylene inhibitors (such as silver ion) display a reduction in root hair formation as do ethylene-insensitive mutant seedlings. The *axr2* mutant, which displays insensitivity to ethylene in the root, does not develop root hairs (92). A mutant (*rhd6*) has also been identified that requires exogenous ethylene to develop root hairs (57).

The recessive nature and molecular analysis of five *ctr1* mutations indicate that these are loss-of-function alleles. *ctr1* mutations result in a constitutive activation of ethylene responses in both seedlings and adult plants. These results suggest that the wild-type function of CTR1 is to negatively regulate ethylene signaling in seedlings and adult plants. The pleiotropic nature of *ctr1*, as well as of the ethylene-insensitive mutations, suggests that seedlings and adult plants share common components for responding to ethylene.

Several lines of evidence suggest that the screen for ethylene response mutations is not yet saturated. A single recessive allele of a second Ctr gene, *ctr2*, has been identified (K Woeste & J Kieber, unpublished observations), and only a single allele has been identified for several loci that result in ethylene insensitivity. Furthermore, additional genes involved in ethylene signaling may not have been detected because they are functionally redundant or their loss of function may result in lethality or infertility. Only limited attempts have been made to identify lethal or infertile ethylene response mutants. A likely possibility for functional redundancy in ethylene signaling is MAP kinase, which is encoded by a large gene family in Arabidopsis [at least ten genes (39, 59–61)] and which may be involved in ethylene signaling because in other species MAP kinases act downstream of Raf (see below). Thus, while many genes involved in ethylene signaling have been found by genetic means, it is likely that additional elements will be identified by a combination of further genetic screens and biochemical and molecular methods.

Analysis of Epistasis

The epistatic relationships among the various ethylene response mutants has been determined (15, 38, 45, 76), and the results are summarized in Figure 1. The epistasis between *ctr1* and the *etr1, ein2, ein3,* and *ein4* mutations is complete, and double mutants among the *etr1, ein2,* and *ein3* mutations do not display any additivity in their phenotype. These results are consistent with a primarily linear pathway. The order of action of the ethylene-insensitive mutations relative to the *ctr1* mutation has been determined using this approach. The *ETR1, EIN4, ETR2,* and *ERS* gene products are predicted to act upstream of *CTR1,* while the *EIN2, EIN3, EIN5, EIN6,* and *EIN7* gene products are predicted to act downstream. All the latter mutations also mask the adult phenotype of *ctr1* with the exception of the *ein7* mutation, which suggests that perhaps a developmental change in *EIN7* expression occurs in adult plants. *etr1* and *ein2* are epistatic to the ethylene-overproducing mutants (*eto1* and *eto2*) indicating that these ethylene-insensitive mutations are also resistant to endogenous ethylene.

   eir1 mutant seedlings display insensitivity to ethylene in the root, but the apical portions of the seedlings display a normal ethylene response. The epistasis between *eir1* and *ctr1* is incomplete; the double mutant has a root that is intermediate in length (76). The *eir1* mutation also displays an additive phenotype with the *ein3* and *ein5* mutations. These results suggest that *eir1* may lie on an independent signaling pathway or that *eir1* may be a "leaky" allele.

ETHYLENE PERCEPTION: ETR1

The *ETR1* gene was isolated by positional cloning (16) and found to encode a protein with similarity to bacterial two-component histidine kinases (Figure 2). The amino-terminal region of the protein contains three hydrophobic stretches, each of which is predicted to span a membrane. The carboxy terminus of ETR1 displays similarity to both the histidine kinase and response regulator domains of bacterial two-component sensing systems (reviewed in 69, 85). Two-component regulators are the major route by which bacteria sense and respond to various environmental cues such as phosphate availability, chemosensory stimuli, and osmolarity. It has been estimated that *Escherichia coli* has at least 50 different two-component systems (85). The two components

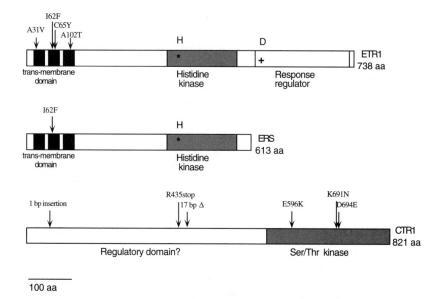

Figure 2   Schematic representations of the *ETR1, ERS,* and *CTR1* proteins. The different domains of each that have homology to known proteins are indicated. The scale in amino acids (aa) is shown in the lower left. The various mutations that have been identified are indicated above each gene, with the wild-type residue written first, followed by its position in the predicted protein and the amino acid present in the mutant version. The conserved histidine (H) and aspartic acid (D) residues that are predicted to be phosphorylated in ETR1 and ERS are indicated by a (★) and a (+) respectively.

consist of a sensor and an associated response regulator. The sensor is responsible for perceiving the signal (the input domain), which induces autophosphorylation of the histidine kinase domain on a conserved histidine residue. This phosphate is then transferred to a conserved aspartate residue on the receiver domain of the cognate response regulator, which in turn regulates the activity of the output domain. Based on its primary amino acid sequence and binding to ethylene (see below), the ETR1 protein is predicted to contain a sensor component fused to a receiver domain of a response regulator.

ETR1 is found as a membrane-associated, disulfide-linked dimer in extracts of Arabidopsis and when expressed in yeast (79). Many two-component sensors dimerize upon ligand binding, although not by disulfide linkage. The disulfide linkage in ETR1 was localized by expression of truncated forms and by in vitro mutagenesis to two cystines at the amino-terminus (Cys4 and Cys6) in a presumed extracellular domain of the protein. These cystines are present in all ETR1 homologs that have been sequenced, suggesting that dimerization may be a conserved feature in these proteins. ETR1 is also associated with membranes when extracted from Arabidopsis (79). It was unclear which membrane system ETR1 was associated with in either yeast or Arabidopsis. ETR1 does not contain a conserved signal sequence for membrane insertion. However, topological analysis predicts that the carboxy-terminal portion of the protein lies on the cytoplasmic face of the plasma membrane.

Yeast cells that express wild-type ETR1 protein were shown to bind ethylene with a high affinity, and this binding was saturable, whereas yeast expressing a mutant version of the ETR1 protein (*etr1*-1) did not display detectable saturable binding of ethylene (78). The κ_d of binding was 0.04 µl/liter, which is close to the dose required for the half-maximal response in the Arabidopsis triple response (18). This binding was reduced by *trans*-cyclooctene and 2,5-norbornadiene, two competitive inhibitors of ethylene binding in many plant species, including Arabidopsis (33, 81, 82). These results, together with the observation that *etr1* mutant seedlings display reduced ethylene binding (8) and genetic epistasis analysis placing ETR1 as the earliest acting of the ethylene mutations (45) provide compelling evidence that ETR1 encodes an ethylene receptor. An important experiment to confirm this is to demonstrate that ethylene binding alters the functionality of the ETR1 protein.

The binding of ethylene was localized to the amino-terminal, hydrophobic domain of ETR1 (78). A truncated version of ETR1 containing only the first 165 amino acids still displayed significant binding to ethylene, suggesting that this portion was sufficient for binding. The four *etr1* mutations all occur in this amino-terminal hydrophobic domain within the three transmembrane segments, and one of these mutations, *etr1*-1, has been shown to disrupt ethylene binding both in Arabidopsis and when expressed in yeast. This suggests that the transmembrane segments are required for ethylene binding. This region is the most conserved among the four ETR1 homologs that have been analyzed (see below), which further supports the notion that ethylene binds in this region. Binding in a membrane environment is not unexpected because ethylene is 14 times more soluble in lipids than in water under physiological conditions (1).

ETR1 is present as a small gene family in Arabidopsis. One such homolog, ERS was cloned by low-stringency hybridization to ETR1 (38). ERS is 67% identical to ETR1 at the amino acid level, but lacks the receiver domain of the response regulator. ERS does not map to any identified ethylene-insensitive mutation. A mutation analogous to the *etr1-4* mutation (a Ile → Phe change in the second transmembrane segment) was introduced into ERS and found to result in dominant ethylene insensitivity when expressed in Arabidopsis. Epistasis analysis indicated that the *ers* mutation acts upstream of the *ctr1* mutation, consistent with the results observed with *etr1*. Thus, ERS and ETR1 may both act as ethylene receptors and may be partially functionally redundant. This redundancy in function may explain why only dominant *etr1* mutations have been identified and why *etr1* seedlings still retain 20% of their ethylene binding when the same mutant version of ETR1 lacks detectable binding when expressed in yeast. An intriguing possibility is that heterodimers may form between the various ETR1 proteins allowing for modulation of the sensitivity of a tissue to ethylene. No ethylene-insensitive mutations have been identified at the ERS locus, which further supports the notion that the screen for ethylene-insensitive mutations has not yet been saturated.

The gene corresponding to the tomato *Never-ripe* (*Nr*) mutation has been found to encode a protein with high similarity to ETR1 (90, 94). The *Nr* mutation is a dominant ethylene-insensitive mutation that affects both seedling and adult ethylene responses (48). Like ERS, the predicted NR protein lacks the response regulator domain. The *Nr* mutation occurs in a residue identical to the residue affected in the *etr1-4* mutation. Interestingly, the *NR* gene is strongly induced by ethylene during fruit ripening, suggesting that perhaps sensitivity of ripening fruit to ethylene is modulated by the regulation of expression of NR (90). A second gene with homology to ETR1, eTAE1, has also been identified from tomato (97).

SIGNAL TRANSDUCTION

CTR1 Protein Kinase

The first gene identified in ethylene signaling was CTR1. This gene was cloned using a T-DNA tagged allele (45), and its structure is shown in Figure 2. The carboxyl terminus of CTR1 has all the hallmark features of a serine/threonine protein kinase, and expression of CTR1 in insect cells using baculoviral vectors confirms that the protein does have intrinsic Ser/Thr protein kinase activity (Y Huang & J Kieber, unpublished observations). The amino acid sequence of CTR1 is most similar to the Raf family of protein kinases. Raf was originally identified as a cellular homologue of v-raf, the transforming gene from an avian retrovirus (73). Raf-1 encodes a 73 kDa protein, the carboxy-terminal half of which encodes a Ser/Thr protein kinase (reviewed in 21, 56, 63, 64, 72, 91). The oncogenic forms of Raf are typically deleted in the 5' half of the gene, and the minimal transforming sequence was found to correspond to the kinase catalytic domain (35, 83). The various Raf proteins from animal cells share three highly conserved regions (CR1-3; 36).

The first domain, CR1, consists of a binding domain for the Ras protein followed by a cystine-rich zinc finger motif that has been shown to bind two molecules of zinc (30). CR2 contains a high proportion of serine and threonine residues that are the targets of Raf-1 autophosphorylation as well as phosphorylation by other serine/threonine kinases such as protein kinase C (47, 65). CR3, which encompasses the carboxyl-terminal half of Raf, comprises the kinase catalytic domain. The carboxyl-terminal half of CTR1 shows high homology to the CR3 domain, but the amino terminus shows only limited similarity to Raf. Notably, CTR1 lacks the zinc finger and Ras-binding motifs found in the Raf CR2 region.

Raf is part of a cascade of conserved protein kinases that are involved in the transduction of a number of external regulatory signals including mitogens, growth hormones such as PDGF and EGF, insulin and IL-2 as well as a number of developmental signaling cascades (reviewed in 14, 36, 70, 72). The receptors for these signals generally activate Raf indirectly, through the small GTP-binding protein Ras (reviewed in 5, 21). GTP-bound Ras localizes Raf from the cytoplasm to the plasma membrane where Raf then becomes activated, most likely by phosphorylation within the amino-terminal half. The current model of activation of Raf is that the amino-terminal half acts to auto-inhibit the kinase domain, and activation results from a conformation change caused by phosphorylation, which results in a relief of this autoinhibition (21). However, recent results from functional mapping of Raf indicate that deletion of the amino-terminal region does not necessarily result in increased basal kinase activity, suggesting that the regulation of Raf may be more complex (19). The lack of homology of CTR1 to the amino-terminal portion of Raf suggests that CTR1 may be regulated by different upstream factors and/or by a distinct mechanism. The apparent absence of a plant homolog to the Ras protein also supports this idea. However, it is interesting to note that the three-dimensional structure of the bacterial protein CheY, whose structure is thought to be representative of all bacterial two-component regulators, is similar to that of Ras (53), which suggests that ETR1 may directly regulate CTR1.

The substrate range of Raf-1 is extremely limited (29); the only physiological target is the dual-specificity protein kinase MEK, which is activated by phosphorylation of two conserved serine residues (2). This phosphorylation is only efficient if MEK is in its native form, suggesting that Raf recognizes a tertiary structure of MEK (29). Activated MEK, in turn, then activates MAP kinase via phosphorylation on both a tyrosine and a threonine residue. The activated MAPK then phosphorylates numerous downstream targets, including a number of transcription factors such as c-Myc and c-Jun. The region of Raf that is important for the recognition of MEK is completely contained within the protein kinase catalytic domain (88). Because CTR1 displays high similarity to the kinase domain of Raf, it seems plausible that CTR1 may phosphorylate a similar target. Consistent with this, an Arabidopsis MEK gene has recently been identified that gives a constitutive triple-response phenotype when its expression is repressed by introduction of an antisense version of the gene in transgenic Arabidopsis (P Morris, personal communication). Raz and Fluhr have found that ethylene induces a rapid and transient phosphorylation of several proteins in tobacco (74), which supports the idea that a protein kinase cascade is involved in the ethylene signal transduction pathway.

All the mutations in the CTR1 gene that have been analyzed are predicted to disrupt its kinase activity, including three single–amino acid changes in residues that are extremely conserved in protein kinases (45). Coupled with the recessive nature of *ctr1* mutations, these results indicate that the kinase activity of CTR1 is required to negatively regulate the ethylene response pathway. Furthermore, these results suggest that the protein kinase activity of CTR1 is active in air and is shut off in the presence of ethylene.

The *CTR1* gene hybridizes to genomic DNA from several other plant species, including tomato, tobacco, and maize (43), and several rice EST clones show high homology to CTR1. Two tomato homologs of ETR1 have been identified (90, 97). These results suggest that these components of the ethylene signal transduction pathway may be conserved in higher plants.

Downstream Components

One of the primary effects of ethylene is to alter the expression of various target genes. Ethylene treatment results in an increase in the level of mRNA of numerous plant genes, including cellulase, chitinase, peroxidase, chalcone synthase, a basic-type PR gene, and β1-3 glucanase, as well as ripening-related genes and ethylene biosynthetic genes (for examples see 11, 27, 28, 29a, 87; reviewed in 12). DNA sequences [(ethylene response elements (ERE)] that confer ethylene responsiveness to a minimal promoter have been identified from the ethylene-regulated pathogenesis-related (PR) genes, and a GCCGCC repeat motif was identified as both necessary and sufficient for this regulation (67). Four proteins that bind to ERE sequences were identified in tobacco, and the steady-state level of RNA for these genes increases dramatically and rapidly in response to ethylene. These ERE-binding proteins (EREBPs) may be primary targets of the ethylene response pathway and may regulate the expression of other secondary response genes such as the PR genes. The DNA-binding domain of these EREBPs has been localized to a 59–amino acid region that shows similarity to a number of uncharacterized plant proteins, as well as to the AP2 protein (26), a predicted nuclear protein that is involved in floral development in Arabidopsis.

Ethylene regulates several plant responses that require differential cell elongation, such as epinasty, tendril coiling, and apical hook formation (reviewed in 1). Growth in the presence of ethylene results in a marked exaggeration in the curvature of the apical hook in etiolated Arabidopsis seedlings. The apical hook is the result of differential cell expansion of the hypocotyl caused by the influence of ethylene and auxin, as well as other factors. The *hookless1* (*hls1*) mutation disrupts apical hook formation in both air- and ethylene-grown Arabidopsis seedlings (33). The *HLS1* gene has recently been cloned and found to show similarity to a diverse group of *N*-acetyltransferases from bacteria to mammalian cells (50). *HLS1* expression is uniform transversely across the apical hook, suggesting that it does not influence hook formation by differential expression in the adaxial and abaxial tissues of the hook. However, the steady-state level of *HLS1* mRNA is induced by ethylene, and overexpression of HLS1 resulted in seedlings that displayed a constitutively exaggerated apical hook. This suggests that increased *HLS1* expression is sufficient to induce hook formation and that ethylene may influence hook curvature by regulating the expression of *HLS1*. The pattern of expression of two primary auxin-upregulated genes, *AtAux2-11* and *SAUR-AC1*, was found to be altered in *hls1* mutants in the tissue that normally forms the apical hook, but not in other parts of the seedling, suggesting that HLS1 may play a role in auxin metabolism or transport in this tissue. The HLS1 protein may regulate auxin function in the apical hook by the acetylation of an IAA-related metabolite, thereby altering its activity (50). A second possible function of HLS1 is the acetylation of the amino terminus of a group of proteins involved in auxin transport. Amino-terminal acetylation of eukaryotic proteins is widespread and has been shown to block protein degradation (reviewed in 25). As mentioned previously, there is evidence that protein degradation plays an important role in auxin signaling, and acetylation may play a role in regulating this process.

MODEL FOR ETHYLENE SIGNALING

Ethylene signaling begins with binding to a receptor(s), which almost certainly includes members of the ETR1 gene family. This binding has been hypothesized to be mediated through a transition metal coordinated within the hydrophobic region of ETR1, which then induces a conformational change of ETR1 that may alter the rate of *trans*-phosphorylation between subunits (9). The altered conformation of ETR1 then results in the shutting off of the kinase activity of CTR1, either directly or indirectly (Figure 3).

The nature of the dominance of the *etr1* and *ers* mutations is important to the understanding of how these proteins act in ethylene signaling. One possibility is that the mutations act by a dominant negative mechanism. In this case, the protein is present in a multisubunit complex, and the mutant form of the

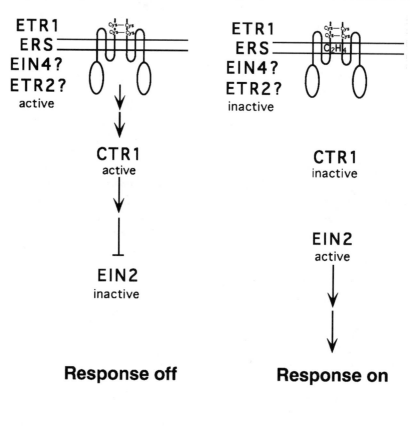

Figure 3 One model for ethylene signal transduction in Arabidopsis. Arrows represent positive regulatory steps, and the flat symbol represents a negative regulatory step. Each arrow may represent several steps in the transduction pathway as no direct interactions have yet been demonstrated. The double line represents a membrane. This model, as well as an alternative model in which ETR1 functions to inactivate CTR1 in the presence of ethylene, is discussed in the text.

enzyme acts to block the activity of the wild-type form by poisoning the entire complex. This model seems somewhat unlikely given the fact that there may be at least four members of this gene family (ETR1, ERS, and possibly EIN4 and ETR2) and the mutations result in almost complete insensitivity to ethylene. If the proteins form only dimers, then 1/4 of the receptor complexes would be poisoned (assuming approximately equal levels of expression of these four genes), which would not seem to be sufficient to eliminate ethylene perception. However, it is possible that higher-order complexes are formed and that a single mutant subunit may poison enough complexes to decrease the level of active receptors below a threshold required for a response.

A second possibility (diagrammed in Figure 3) is that the *etr1* mutations are dominant because they result in gain-of-function changes in the protein. In this case, the mutant form of the enzyme would be locked in the catalytically active state. In this context, "active" means that ETR1 has the ability to result in the inactivation of the kinase activity of CTR1. In this model, ETR1 is active in the absence of ethylene and becomes inactivated upon binding of the ligand. Thus, mutations that disrupt ethylene binding would result in an enzyme that was constitutively on, which would result in a constitutive inactivation of CTR1 and thus a lack of responsiveness to ethylene. Such a model would predict that the mutant phenotype of the *etr1* mutations would be insensitive to gene dosage, or to the presence of multiple functionally redundant genes. It is also consistent with the absence of ethylene binding seen in the mutant forms of ETR1. One two-component precedent for such a model is the FixL receptor, which regulates nitrogen fixation genes in *Rhizobium* in response to oxygen (62). The histidine kinase of FixL is active in the absence of oxygen and is inactivated upon binding of oxygen to a heme moiety (31, 52). A second intriguing example is the osmosensing pathway in yeast. This system is comprised of the *SLN1,* gene which encodes a putative two-component histidine kinase and the *SSK1* gene that encodes a response regulator (55, 68). *SSK1* then regulates the activity of a MAP kinase cascade consisting of the SSK2/SSK22 (MAPKKKs), PBS2 (MAPKK), and HOG1 (MAPK) proteins (10, 54). This combination of a two-component regulator with a MAP kinase cascade is reminiscent of ethylene signaling. The SLN1 histidine kinase is active in the noninducing conditions (low osmolarity), and it inactivates the downstream MAP kinase cascade, resulting in the osmolarity response not being turned on. In the presence of "inducer" (high osmolarity), SLN1 is inactive, which results in activation of the MAPK cascade and the induction of the osmolarity response.

The function of CTR1 is to negatively regulate ethylene responses, and genetic and molecular data suggest that the kinase activity of CTR1 is required for this repression. Thus the kinase activity of CTR1 should be active in air, and ETR1 probably regulates this activity, either directly or indirectly. Presumably, the activity of CTR1 is regulated through its amino-terminal domain, as is the case with Raf and other protein kinases with amino-terminal extensions such as STE11 and protein kinase C. CTR1 then regulates the activity of the *EIN2* and *EIN3* gene products, perhaps via a MAP kinase cascade. While this model is consistent with all the data obtained, it is almost certainly incomplete, and it is unclear if any of these interactions are direct and what other elements are involved in ethylene signaling.

CONCLUDING REMARKS

Tremendous advances have been made in the understanding of ethylene signaling in Arabidopsis, and it is now the best understood signal transduction pathway in plants. Several key genes in the pathway have been identified and cloned. The recent cloning of the *EIN2* and *EIN3* genes should soon extend our knowledge of the molecular nature of the ethylene signal transduction pathway. This work strikingly highlights the power of molecular genetic analysis in Arabidopsis. However, much remains to be learned. Certainly one area for which little information is available is the biochemical analysis of the protein products of these genes and an understanding of how these proteins interact to transduce the ethylene signal. Does ETR1 possess intrinsic histidine kinase activity, and is its activity altered upon binding of ethylene? What are the factors that regulate the kinase activity of CTR1 and what are the targets of its kinase activity? What do the other genes in the pathway encode? The near future should provide answers to some of these as well as other questions and should prove to be an exciting and enlightening period in ethylene research.

ACKNOWLEDGMENTS

I thank L Kaufman and various members of my laboratory for critically reading this manuscript. Supported in part by an NASA/NSF grant (IBN-9416017) and a USDA grant.

Visit the *Annual Reviews home page* at http://www.annurev.org.

Literature Cited

1. Abeles FB, Morgan PW, Saltveit ME Jr. 1992. *Ethylene in Plant Biology*. San Diego: Academic. 2nd ed.
2. Alessi D, Saito Y, Campbell D, Cohen P, Sithanandam G, et al. 1994. Identification of the sites in MAP kinase kinase-1 phosphorylated by p74raf-1. *EMBO J*. 13: 1610–19

3. Amrhein N, Wenker D. 1979. Novel inhibitors of ethylene production in higher plants. *Plant Cell Physiol.* 20:1635–42
4. Apelbaum A, Berg S. 1971. Altered cell microfibrillar orientation in ethylene-treated *Pisum sativum* stems. *Plant Physiol.* 48:648–52
5. Avruch J, Zhang X-f, Kyriakis J. 1994. Raf meets Ras: completing the framework of a signal transduction pathway. *Trends Biochem. Sci.* 19:279–83
6. Bent AF, Innes R, Ecker J, Staskawicz B. 1992. Disease development in ethylene-insensitive *Arabidopsis thaliana* infected with virulent and avirulent *Pseudomonas* and *Xanthomonas* pathogens. *Mol. Plant Microbe Interact.* 5:372–78
7. Beyer EM Jr. 1976. A potent inhibitor of ethylene action in plants. *Plant Physiol.* 58:268–71
8. Bleecker AB, Estelle MA, Somerville C, Kende H. 1988. Insensitivity to ethylene conferred by a dominant mutation in *Arabidopsis thaliana. Science* 241:1086–89
9. Bleecker AB, Schaller G. 1996. The mechanism of ethylene perception. *Plant Physiol.* 111:653–60
10. Brewster JL, de Valoir T, Dwyer ND, Winter E, Gustin MC. 1993. An osmosensing signal transduction pathway in yeast. *Science* 259:1760–63
11. Broglie KE, Biddle P, Cressman R, Broglie R. 1989. Functional analysis of DNA sequences responsible for ethylene regulation of a bean chitinase gene in transgenic tobacco. *Plant Cell* 1:599–607
12. Broglie R, Broglie KE. 1991. Ethylene and gene expression. See Ref. 58, pp. 101–14
13. Bünning E. 1951. Ueber die differenzierungsvorgange in der Crucifierwurzel. *Planta* 39:126–53
14. Campbell JS, Seger R, Graves JD, Graves LM, Jensen AM, Krebs EG. 1995. The MAP kinase cascade. *Recent Prog. Horm. Res.* 50:131–59
15. Chang C. 1996. The ethylene signal transduction pathway in *Arabidopsis:* an emerging paradigm? *Trends Biochem. Sci.* 21:129–33
16. Chang C, Kwok SF, Bleecker AB, Meyerowitz EM. 1993. *Arabidopsis* ethylene-response gene *ETR1:* similarity of product to two-component regulators. *Science* 262:539–44
17. Chang C, Meyerowitz EM. 1995. The ethylene hormone response in *Arabidopsis:* a eukaryotic two-component signaling system. *Proc. Natl. Acad. Sci. USA* 92:4129–33
18. Chen QHG, Bleecker AB. 1995. Analysis of ethylene signal transduction kinetics associated with seedling-growth responses and chitinase induction in wild-type and mutant *Arabidopsis. Plant Physiol.* 108:597–607
19. Chow Y-H, Pumiglia K, Jun TH, Dent P, Sturgill TW, Jove R. 1995. Functional mapping of the N-terminal regulatory domain in the human Raf-1 protein kinase. *J. Biol. Chem.* 270:14100–6
20. Cormack R. 1935. The development of root hairs by *Elodea canadensis. New Phytol.* 34:19–25
21. Daum G, Eisenmann-Tappe I, Fries H-W, Troppmair J, Rapp UR. 1994. The ins and outs of Raf kinases. *Trends Biochem. Sci.* 19:474–80
22. De Munk W, De Rooy M. 1971. The influence of ethylene on the development of 5 C-precooled Apeldoorn tulips during forcing. *HortScience* 6:40–41
23. Dolan L, Duckett CM, Grierson C, Linstead P, Schneider K, et al. 1994. Clonal relationships and cell patterning in the root epidermis of *Arabidopsis. Development* 120:2465–74
24. Dolan L, Janmaat K, Willemsen V, Linstead P, Poethig S, et al. 1993. Cellular organization of the *Arabidopsis thaliana* root. *Development* 119:71–84
25. Driessen HPC, de Jong WW, Tesser GI, Bloemendal H. 1985. The mechanism of N-terminal acetylation of proteins. *CRC Crit. Rev. Biochem.* 18:281–325
26. Ecker JR. 1995. The ethylene signal transduction pathway in plants. *Science* 268:667–75
27. Ecker JR, Davis RW. 1987. Plant defense genes are regulated by ethylene. *Proc. Natl. Acad. Sci. USA* 84:5202–6
28. Felix G, Meins F. 1987. Ethylene regulation of β–1,3-glucanase in tobacco. *Planta* 172:386–92
29. Force T, Bonventre JV, Heidecker G, Rapp U, Avruch J, Kyriakis JM. 1994. Enzymatic characteristics of the Raf-1 protein kinase. *Proc. Natl. Acad. Sci. USA* 91:1270–74
29a. Fray RG, Grierson D. 1993. Molecular genetics of tomato fruit ripening. *Trends Genet.* 9:438–43
30. Ghosh SJ, Xie WQ, Quest AFG, Mabrouk GM, Strum JC, Bell RM. 1994. The cystine-rich region of Raf-1 kinase contains zinc, translocates to the liposomes, and is adjacent to a segment that binds GTP-Ras. *J. Biol. Chem.* 269:10000–7
31. Gilles-Gonzalez MA, Gonzalez G. 1993. Regulation of the kinase activity of heme protein FixL from the two-component system FixL/FixJ of *Rhizobium meliloti. J. Biol. Chem.* 268:16293–97

32. Deleted in proof
33. Guzman P, Ecker JR. 1990. Exploiting the triple response of *Arabidopsis* to identify ethylene-related mutants. *Plant Cell* 2: 513–23
34. Harpham NVJ, Berry AW, Knee EM, Roveda-Hoyos G, Raskin I, et al. 1991. The effect of ethylene on the growth and development of wild-type and mutant *Arabidopsis thaliana* (L.) Heynh. *Ann. Bot.* 68: 55–62
35. Heidecker G, Huleihel M, Cleveland JL, Kolch W, Beck TW, et al. 1990. Mutational activation of c-Raf-1 and definition of the minimal transforming sequence. *Mol. Cell. Biol.* 10:2503–12
36. Heidecker G, Kolch W, Morrison DK, Rapp UR. 1992. The role of Raf-1 phosphorylation in signal transduction. *Adv. Cancer Res.* 58:53–73
37. Hobbie L, Estelle M. 1994. Genetic approaches to auxin action. *Plant Cell Environ.* 17:525–40
38. Hua J, Chang C, Sun Q, Meyerowitz EM. 1995. Ethylene-insensitivity conferred by *Arabidopsis ERS* gene. *Science* 269: 1712–14
39. Jonak C, Heberle-Bors E, Hirt H. 1994. MAP kinases: universal multipurpose signaling tools. *Plant Mol. Biol.* 24:407–16
40. Kende H. 1989. Enzymes of ethylene biosynthesis. *Plant Physiol.* 91:1–4
41. Kende H. 1993. Ethylene biosynthesis. *Annu. Rev. Plant Physiol. Plant Mol. Biol.* 44:283–307
42. Kieber JJ. 1996. The ethylene signal transduction pathway in *Arabidopsis*. *J. Exp. Bot.* In press
43. Kieber JJ, Ecker JR. 1994. Molecular and genetic analysis of the ethylene response mutant *ctr1*. In *NATO ASI Series: Molecular Genetic Analysis of Plant Development and Metabolism*, ed. P Puigdomenech, G Coruzzi, pp. 193–201. Heidelberg: Springer-Verlag
44. Kieber JJ, Ecker JR. 1993. Ethylene gas! It's not just for ripening anymore. *Trends Genet.* 9:356–63
45. Kieber JJ, Rothenberg M, Roman G, Feldmann KA, Ecker JR. 1993. CTR1, a negative regulator of the ethylene response pathway in *Arabidopsis*, encodes a member of the Raf family of protein kinases. *Cell* 72: 427–41
46. Knight LI, Rose RC, Crocker W. 1910. Effects of various gases and vapors upon etiolated seedlings of the sweet pea. *Science* 31:635–36
47. Kolch W, Heidecker G, Kochs G, Hummel R, Vahidi H, et al. 1993. Protein kinase Cα activates RAF-1 by direct phosphorylation. *Nature* 364:249–52
48. Lanahan MB, Yen H-C, Giovannoni JJ, Klee HJ. 1994. The *Never-ripe* mutation blocks ethylene perception in tomato. *Plant Cell* 6:521–30
49. Lang JM, Eisinger WR, Green PB. 1982. Effects of ethylene on the orientation of microtubules and cellulose microfibrils of pea epicotyl cells with polyamellate cell walls. *Protoplasma* 110:5–14
50. Lehman A, Black R, Ecker JR. 1996. *Hookless1*, an ethylene response gene, is required for differential cell elongation in the *Arabidopsis* hook. *Cell* 85:183–94
51. Leyser HMO, Lincoln CA, Timpte C, Lammer D, Turner J, Estelle M. 1993. *Arabidopsis* auxin-resistant gene *AXR1* encodes a protein related to ubiquitin-activating enzyme E1. *Nature* 364:161–64
52. Lois AF, Weinstein M, Ditta GS, Helinski DR. 1993. Autophosphorylation and phosphatase activities of the oxygen-sensing protein FixL of *Rhizobium meliloti* are coordinately regulated by oxygen. *J. Biol. Chem.* 268:4370–75
53. Lukat GS, Lee BH, Mottonen JM, Stock AM, Stock JB. 1991. Roles of the highly conserved aspartate and lysine residues in the response regulator of bacterial chemotaxis. *J. Biol. Chem.* 266:8348–54
54. Maeda T, Takekawa M, Saito H. 1995. Activation of the yeast PBS2 MAPKK by MAPKKKs or by binding of an SH3-containing osmosensor. *Science* 269:554–58
55. Maeda T, Wurgler-Murphy S, Saito H. 1994. A two-component system that regulates an osmosensing MAP kinase cascade in yeast. *Nature* 369:242–45
56. Marshall C. 1994. MAP kinase kinase kinase, MAP kinase kinase, and MAP kinase. *Curr. Opin. Genet. Dev.* 4:82–89
57. Masucci JD, Schiefelbein JW. 1994. The *rhd6* mutation of *Arabidopsis thaliana* alters root-hair initiation through an auxin- and ethylene-associated process. *Plant Physiol.* 106:1335–46
58. Mattoo A, Suttle J. 1991. *The Plant Hormone Ethylene*. Boca Raton, FL: CRC Press
59. Mizoguchi T, Gotoh Y, Nishida E, Yamaguchi-Shinozaki K, Hayashida N, et al. 1994. Characterization of two cDNAs that encode MAP kinase homologues in *Arabidopsis thaliana* and analysis of the possible role of auxin in activating such kinase activities in cultured cells. *Plant J.* 5:111–22
60. Mizoguchi T, Hayashida N, Yamaguchi-Shinozaki K, Kamada H, Shinozaki K. 1993. ATMPKs: a gene family of plant

MAP kinases in *Arabidopsis thaliana*. *FEBS Lett.* 336:440–44

61. Mizoguchi T, Ichimura K, Shinozaki K. 1996. Environmental stress response in plants: the role of mitogen-activated protein kinases (MAPKs). *Trends Biotechnol.* In press

62. Monson EK, Weinstein M, Ditta GS, Helsinki DR. 1992. The FixL protein of *Rhizobium meliloti* can be separated into a heme-binding oxygen-sensing domain and a functional C-terminal kinase domain. *Proc. Natl. Acad. Sci. USA* 89:4280–84

63. Moodie SA, Wolfman A. 1994. The 3Rs of life: Ras, Raf and growth regulation. *Trends Genet.* 10:44–48

64. Morrison DK. 1990. The Raf-1 kinase as a transducer of mitogenic signals. *Cancer Cells* 2:377–82

65. Morrison DK, Heidecker G, Rapp UR, Copeland TD. 1993. Identification of the major phosphorylation sites of the Raf-1 kinase. *J. Biol. Chem.* 268:17309–16

66. Neljubov D. 1901. Uber die horizontale Nutation der Stengel von *Pisum sativum* und einiger Anderer. *Pflanzen Beih. Bot. Zentralb.* 10:128–39

67. Ohme-Takagi M, Shinshi H. 1995. Ethylene-inducible DNA binding proteins that interact with an ethylene-responsive element. *Plant Cell* 7:173–82

68. Ota IM, Varshavsky A. 1993. A yeast protein similar to bacterial two-component regulators. *Science* 262:566–69

69. Parkinson JS. 1993. Signal-transduction schemes of bacteria. *Cell* 73:857–71

70. Pelech SL, Sanghera JS. 1992. Mitogen-activated protein kinases: versatile transducers for cell signaling. *Trends Biochem. Sci.* 17:233–38

71. Pickett FB, Wilson AK, Estelle M. 1990. The *aux1* mutation of *Arabidopsis* confers both auxin and ethylene resistance. *Plant Physiol.* 94:1462–66

72. Rapp UR. 1991. Role of Raf-1 serine/threonine protein kinase in growth factor signal transduction. *Oncogene* 6:495–500

73. Rapp UR, Goldsborough MD, Mark GE, Bonner TI, Groffen J, et al. 1983. Structure and biological activity of v-*raf*, a unique oncogene tranduced by a retrovirus. *Proc. Natl. Acad. Sci. USA* 80:4218–22

74. Raz V, Fluhr R. 1993. Ethylene signal is transduced via protein phosphorylation events in plants. *Plant Cell* 5:523–30

75. Roberts IN, Lloyd CW, Roberts K. 1985. Ethylene-induced microtubule reorientations: mediation by helical arrays. *Planta* 164:439–47

76. Roman G, Lubarsky B, Kieber JJ, Rothen-berg M, Ecker JR. 1995. Genetic analysis of ethylene signal transduction in *Arabidopsis thaliana*: five novel mutant loci integrated into a stress response pathway. *Genetics* 139:1393–409

77. Satoh S, Esashi Y. 1983. α-Aminoisobutyric-acid, propyl gallate and cobalt ion and the mode of inhibition of ethylene production by cotyledonary segments of cocklebur seeds. *Physiol. Plant.* 57:521–26

78. Schaller GE, Bleecker AB. 1995. Ethylene-binding sites generated in yeast expressing the *Arabidopsis ETR1* gene. *Science* 270:1809–11

79. Schaller GE, Ladd AN, Lanahan MB, Spanbauer JM, Bleecker AB. 1995. The ethylene response mediator ETR1 from *Arabidopsis* forms a disulfide linked dimer. *J. Biol. Chem.* 270:12526–30

80. Shibaoka H. 1994. Plant hormone-induced changes in the orientation of cortical microtubules: alterations in the crosslinking between microtubules and the plasma membrane. *Annu. Rev. Plant Physiol. Plant Mol. Biol.* 44:527–44

81. Sisler EC, Blankenship SM, Guest M. 1990. Competition of cyclooctenes and cyclooctadienes for ethylene binding and activity in plants. *Plant Growth Regul.* 9:157–64

82. Sisler EC, Yang SF. 1984. Anti-ethylene effects of *cis*-2-butene and cyclic olefins. *Phytochemistry* 23:2765–68

83. Stanton VP, Nichols DW, Laudano AP, Cooper GM. 1989. Definition of the human raf amino-terminal regulatory region by deletion mutagenesis. *Mol. Cell Biol.* 9:639–47

84. Steen DA, Chadwick AV. 1981. Ethylene effects in pea stem tissue, evidence of microtubule mediation. *Plant Physiol.* 67:460–66

85. Stock JB, Stock AM, Mottonen JM. 1990. Signal transduction in bacteria. *Nature* 344:395–400

86. Tanimoto M, Roberts K, Dolan L. 1995. Ethylene is a positive regulator of root hair development in *Arabidopsis thaliana*. *Plant J.* 8:943–48

87. Theologis A. 1992. One rotten apple spoils the whole bushel: the role of ethylene in fruit ripening. *Cell* 70:181–84

88. Van Aelst L, Barr M, Marcus S, Polverino A, Wigler M. 1993. Complex formation between RAS and RAF and other protein kinases. *Proc. Natl. Acad. Sci. USA* 90:6213–17

89. Van der Straeten D, Djudzman A, Van Caeneghem W, Smalle J, Van Montagu M. 1993. Genetic and physiological analysis of

a new locus in *Arabidopsis* that confers resistance to 1-aminocyclopropane-1-carboxylic acid and ethylene and specifically affects the ethylene signal transduction pathway. *Plant Physiol.* 102:401–8

90. Wilkinson JQ, Lanahan MB, Yen H-C, Giovannoni JJ, Klee HJ. 1995. An ethylene-inducible component of signal transduction encoded by *Never-ripe*. *Science* 270:1807–9

91. Williams NT, Roberts TM. 1994. Signal transduction pathways involving the Raf proto-oncogene. *Cancer Metastasis Rev.* 13:105–16

92. Wilson AK, Pickett FB, Turner JC, Estelle M. 1990. A dominant mutation in Arabidopsis confers resistance to auxin, ethylene, and abscisic acid. *Mol. Gen. Genet.* 222:377–83

93. Yang SF, Hoffman NE. 1984. Ethylene biosynthesis and its regulation in higher plants. *Annu. Rev. Plant Physiol.* 35:155–89

94. Yen H-C, Lee SY, Tanksley SD, Lanahan MB, Klee HJ, Giovannoni J. 1995. The tomato *Never-ripe* locus regulates ethylene-inducible gene expression and is linked to a homolog of the *Arabidopsis ETR1* gene. *Plant Physiol.* 107:1343–53

95. Yuan M, Shaw PJ, Warn RM, Lloyd CW. 1994. Dynamic reorientation of cortical microtubules, from transverse to longitudinal, in living cells. *Proc. Natl. Acad. Sci. USA* 91:6050–53

96. Zarembinski TI, Theologis A. 1994. Ethylene biosynthesis and action: a case of conservation. *Plant Mol. Biol.* 26:1579–97

97. Zhou DB, Mattoo AK, Tucker ML. 1996. The mRNA for an *ETR1* homologue in tomato is constitutively expressed in vegetative and reproductive tissue. *Plant Mol. Biol.* 30:1331–38

Annu. Rev. Plant Physiol. Plant Mol. Biol. 1997. 48:297–326

PLANT TRANSFORMATION:
Problems and Strategies for Practical Application

R. G. Birch

Department of Botany, The University of Queensland, Brisbane, 4072, Australia

KEY WORDS: plant improvement, gene transfer, transgenic plants, transgene expression, genetic engineering

ABSTRACT

Plant transformation is now a core research tool in plant biology and a practical tool for cultivar improvement. There are verified methods for stable introduction of novel genes into the nuclear genomes of over 120 diverse plant species. This review examines the criteria to verify plant transformation; the biological and practical requirements for transformation systems; the integration of tissue culture, gene transfer, selection, and transgene expression strategies to achieve transformation in recalcitrant species; and other constraints to plant transformation including regulatory environment, public perceptions, intellectual property, and economics. Because the costs of screening populations showing diverse genetic changes can far exceed the costs of transformation, it is important to distinguish absolute and useful transformation efficiencies. The major technical challenge facing plant transformation biology is the development of methods and constructs to produce a high proportion of plants showing predictable transgene expression without collateral genetic damage. This will require answers to a series of biological and technical questions, some of which are defined.

CONTENTS

INTRODUCTION

Plant transformation is at a threshold. Over 3000 field trials of transformed plants are in progress or completed in at least 30 countries. These trials involve over 40 plant species modified for various economic traits (33, and updates in *Genetic Technology News*). We are emerging from a period of plant transformation research dominated by the need to develop proven genetic transformation methods for the major experimental and economic plant species, into the era of application of transformation as a core research tool in plant biology and a practical tool for cultivar improvement. Some of the most important issues (problems and strategies) affecting these uses are very different from those foremost in our thinking in recent years while the discipline focused on the scientific understanding and technical development of reliable systems for genetic transformation of a wide range of plant species.

The capacity to introduce and express diverse foreign genes in plants, first described for tobacco in 1984 (37, 74, 120), has been extended to over 120 species in at least 35 families. Successes include most major economic crops, vegetables, ornamental, medicinal, fruit, tree, and pasture plants. The rapid and simultaneous developments in transformation technology and information technology make tabulations of transformed species quickly out of date (41, 131), and it is advisable to use computer-based searches to locate references to current transformation methods for species of interest. The process of diversification and refinement of transformation techniques for greater convenience, higher efficiency, broader genotype range, and desired molecular characteristics of transformants will continue to good effect for some time. However, gene transfer and regeneration of transgenic plants are no longer the factors limiting the development and application of practical transformation systems for many plant species. Attention is increasingly being directed to achieving the desired patterns of expression of introduced genes and to solving economic constraints on practical plant molecular improvement.

There are excellent recent reviews of the development of plant transformation systems using *Agrobacterium* (72, 146, 165), direct gene transfer into protoplasts (49, 113, 114), or particle bombardment (14, 25, 26), and the potential for their practical application (30, 73). These topics are now comprehensively addressed in recent texts (12, 54, 61, 90, 110) and methods manuals (39, 53, 57, 127, 153), to which the reader is referred for background information. In this review I aim to identify key problems remaining in the development of plant transformation systems, key issues to be resolved in the practical application of these systems, and strategies by which we may overcome these limitations.

DEFINITION AND VERIFICATION OF TRANSFORMATION

This review is concerned with the stable incorporation and expression of genes introduced into plants by means other than fusion of gametes or other cells. It focuses on transformation involving integration of introduced genes in the plant nuclear genome, although some issues are equally applicable to transformed plants in which introduced genes are expressed from an organelle genome (21), or a replicating viral vector (139).

Ingo Potrykus in 1991 (126) offered a provocative but clarifying assessment of plant transformation technologies based on a rigid definition of proof of integrative transformation, requiring a combination of genetic, phenotypic, and physical data. Unfortunately the combination specified was not useful in practice for verification of transformation in some plants. For example, analysis of sexual offspring populations is problematic in trees that are slow to reproduce sexually, and in some vegetatively propagated crops such as sugarcane with complicated (polyploid, aneuploid) genetics, and many sexually sterile cultivars. Similarly, a "tight" correlation between physical and phenotypic data is not a defining characteristic of gene transfer methods that result in integration of multiple and often rearranged copies of the transferred DNA, because many transformants have unexpressed copies of introduced sequences. Integrative transformation has nevertheless been unequivocally verified in such cases.

Most critical researchers would therefore accept a more generally applicable subset of criteria as rigorous proof of integrative transformation:

1. Southern DNA hybridization analysis of multiple independent transformants, using a probe(s) for the introduced gene(s) and restriction enzymes predicted to generate hybridizing fragments of different length at different integration sites (for a graphic representation, see 80). It is important to confirm that sizes of hybridizing fragments including flanking DNA at each integration site are reproducible within a transformed line, and that they differ between independently transformed lines. High molecular weight signals in uncut DNA, PCR-generated bands, or signals from a single putative transformed line are not acceptable substitutes, because it is more difficult to exclude the possibility of artifacts in such data.

2. Phenotypic data showing sustained expression of the introduced gene(s) exclusively in the cells of plant lines positive for the gene(s) by Southern analysis. Unambiguous phenotypic data require: (a) negative results from all untransformed controls (the tested control population size must be at least equivalent to that yielding 10 independent transformants from a parallel treated population), and (b) an assay revealing the product of transgene expression within plant cells as distinct from contaminating microbial cells, preferably from a transgene shown not to be expressed in bacterial cells. Intron-GUS (149) or anthocyanin regulatory (16, 18) reporter systems are suitable for such assays, as are in situ analyses for gene products without simple visual assays (153). Survival of lines on "escape-free" selection is not sufficient, because of the possibility of cross-protection by secreted products of contaminating microbial cells, or selection of mutants resistant to the selective agent. Enzyme assays on cell extracts are inadequate because of the possibility of contaminating transformed microbial cells.

Of course, more detailed molecular, phenotypic, and genetic characterization is likely to be undertaken on transformed lines produced for practical purposes. In some cases, target gene silencing rather than transgene expression may provide an unambiguous phenotype (15, 103). Data on co-transmission of introduced gene copies and the resulting phenotype in sexual offspring populations, where available, provide compelling confirmation of transformation, given suitable controls (126). Applicability in several independent laboratories is an important practical confirmation, because some techniques with published molecular evidence have never been repeatable.

PURPOSES OF PLANT TRANSFORMATION

An Experimental Tool for Plant Physiology

The capacity to introduce and express (or inactivate) specific genes in plants provides a powerful new experimental tool, allowing direct testing of some hypotheses in plant physiology that have been exceedingly difficult to resolve using other biochemical approaches (31). Exciting examples include the molecular genetic analysis of cellular signals controlling sexual reproduction and plant-microbe interactions (116, 143); the roles of specific enzymes in metabolic processes determining partitioning of photosynthates, and thus harvestable yield (67); and the roles of specific enzymes and hormones in plant developmental processes, including those affecting quality and storage life of marketed plant products (109, 147).

A Practical Tool for Plant Improvement

Much of the support for plant transformation research (and more broadly for plant molecular biology) has been provided because of expectations that this approach could: (*a*) generate plants with useful phenotypes unachievable by conventional plant breeding, (*b*) correct faults in cultivars more efficiently than conventional breeding, or (*c*) allow the commercial value of improved plant lines to be captured by those investing in the research more fully than is possible under intellectual property laws governing conventionally bred plants.

The first of these expectations has been met, with production of the first commercial plant lines expressing foreign genes conferring resistance to viruses, insects, herbicides, or post-harvest deterioration (54, 71, 138, 147), and accumulation of usefully modified storage products (30, 145), including several cases where there was no source of the desired trait in the gene pool for conventional breeding. The future prospects in this respect are also exciting, with preliminary indications that novel genes can be introduced to generate plant lines useful for production of materials ranging from pharmaceuticals (65) to biodegradable plastics (112).

The extent to which the other practical or commercial expectations of plant transformation can be met depends on the efficiency and predictability of production of lines with the desired phenotype, and without undesired side effects of the transformation process. As the sophistication of the physiological hypotheses to be tested by transformation increases, exactly the same factors become limiting. This occurs because of the practical difficulty of screening large numbers of plants for the desired expression pattern and the need to avoid misleading results from unrelated physiological effects of unintended genetic changes during the transformation process.

BIOLOGICAL REQUIREMENTS FOR TRANSFORMATION

The essential requirements in a gene transfer system for production of transgenic plants are: (*a*) availability of a target tissue including cells competent for plant regeneration, (*b*) a method to introduce DNA into those regenerable cells, and (*c*) a procedure to select and regenerate transformed plants at a satisfactory frequency.

One of the simplest available plant transformation systems involves infiltration of *Agrobacterium* cells into *Arabidopsis* plants before flowering, and direct selection for rare transformants in the resulting seedling populations (7, 23). Unfortunately, small plant size, rapid generation time, and high seed yield per plant are prerequisites for this method. These features are not shared by any economically important plant species. Other approaches to transformation via plant gametes have not been successful in practice (126). Therefore, the totipotency of some somatic plant cells underlies most plant transformation systems. The efficiency with which such cells can be prepared as targets for transformation is today the limiting factor in achievement of transformation in recalcitrant plant species.

Using either *Agrobacterium* or particle bombardment, it is now possible to introduce DNA into virtually any regenerable plant cell type. Only a small proportion of target cells typically receive the DNA during these treatments, and only a small proportion of these cells survive the treatment and stably integrate introduced DNA (47, 59). It is therefore generally essential to efficiently detect or select for transformed cells among a large excess of untransformed cells (12), and to establish regeneration conditions allowing recovery of intact plants derived from single transformed cells (156). Alternatively, transformed cells must contribute to the germline so that nonchimeric transformants can be obtained in the progeny from sexual reproduction (28).

PRACTICAL REQUIREMENTS OF TRANSFORMATION SYSTEMS FOR PLANT IMPROVEMENT

Beyond the biological requirements to achieve transformation and the technical requirements for verification of reproducible transformation, desired characteristics to consider in evaluating alternative techniques, or developing new ones for cultivar improvement, include:

1. Ready availability of the target tissue. The resources required to maintain a continuous supply of explants such as immature embryos at the correct developmental stage for transformation can be substantial.

2. Applicability to a range of cultivars. Genotype-specific techniques are of lower value because of the added complexity of extended breeding work to move desired genes into preferred cultivars.

3. High efficiency, economy, and reproducibility, to readily produce many independent transformants for testing.

4. Safety to operators, avoiding procedures, or substances requiring cumbersome precautions to avoid a high hazard to operators (e.g. potential carcinogenicity of silicone carbide whiskers).

5. Technical simplicity, involving a minimum of demanding or inherently variable manipulations, such as protoplast production and regeneration.

6. Frequent cotransformation with multiple genes, so that a high proportion of plant lines selected for marker gene expression will also incorporate cotransformed useful genes.

7. Unequivocal selection or efficient screening to recover transgenic plants from transformed cells.

8. Minimum time in tissue culture, to reduce associated costs and avoid undesired somaclonal variation.

9. Stable, uniform (nonchimeric) transformants for vegetatively propagated species, or fertile germline transformants for sexually propagated species.

10. Capacity to introduce defined DNA sequences without accompanying vector sequences not required for integration or expression of the introduced genes.

11. Capacity to remove reporter genes or other sequences not required following selection of transformed lines.

12. Simple integration patterns and low copy number of introduced genes, to minimize the probability of undesired gene disruption at insertion sites, or multicopy associated transgene silencing.

13. Stable expression of introduced genes in the pattern expected from the chosen gene control sequences, rather than patterns associated with the state of the cells at the time of transformation (34), or the chance site of integration (122).

14. Optionally applicable to transformation of organelle genomes.

15. Absence of valid patent claims on products.

When tested against the above criteria, most published techniques for gene transfer into plant cells must be dismissed as either disproven, unproven, or impractical for use in routine production of transgenic plants. The techniques that have been proven to produce transgenic plants from a range of species, and in many laboratories, are *Agrobacterium*-mediated transformation, bombardment with DNA-coated microprojectiles, and electroporation or PEG treatment of protoplasts. Techniques requiring protoplasts are generally avoided because of the associated inconvenience and time in culture. As a result virtually all plant transformation work aimed at direct production of improved cultivars currently uses either *Agrobacterium* or microprojectiles for gene transfer. Neither of these approaches is free of patent claims, however, and there is continuing interest in development of alternative techniques (141). The stages and time-courses for typical transformation strategies using *Agrobacterium* or DNA-coated microprojectiles shown in Figure 1.

RECALCITRANT SYSTEMS AND APPARENT CONSTRAINTS

Cereals, legumes, and woody plants are commonly categorized as recalcitrant to transformation, because these groups have included a disproportionate

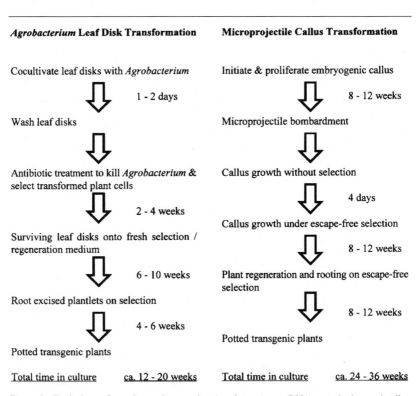

Figure 1 Typical transformation regimens using *Agrobacterium* or DNA-coated microprojectiles.

number of untransformed or difficult to transform species. However, the generalization is becoming less useful as one species after another from these groups joins the list of plants with reliable transformation systems. The hypothesis that some plants lack the biological capacity to respond to essential triggers for integrative transformation, or have cellular mechanisms preventing integrative transformation, can effectively be rejected.

It is a reasonable proposition that transgenic plants can be regenerated only from cells competent for both regeneration and integrative transformation (126). Preliminary evidence indicates that T-DNA integration may be the limiting step in maize transformation (111), but there is no evidence that actively dividing, regenerable cells are not competent to integrate introduced DNA. Where tissue culture systems have been developed to produce proliferating and regenerable cells, into which DNA can be introduced at a high frequency (as indicated by transient gene expression) without interfering with regenerability of the penetrated cells, previously recalcitrant species have become transformable. The transformation efficiency has been proportional to the efficiency of the tissue culture and gene transfer systems (18, 70, 76, 93).

STRATEGIES TO ACHIEVE TRANSFORMATION

It is instructive to consider species such as rice, which was once considered recalcitrant to transformation but can now be transformed via direct gene transfer into protoplasts (140), particle bombardment of immature embryos (27) or cell cultures (20), or *Agrobacterium* treatment of embryogenic callus (70). In each case, success seems to have followed identification (or production through tissue culture) of explants with many regenerable cells, optimization of parameters for gene transfer into those cells, and tailoring selection and regeneration procedures to recover transgenic plants. A generalized approach is illustrated in Figure 2. The nearest transformed relatives of an untransformed species of interest are an obvious reference point in initial work to develop suitable tissue culture, gene transfer, and selection regimens.

Tissue Culture Strategies

Tissue culture is not a theoretical prerequisite for plant transformation, but it is employed in almost all current practical transformation systems to achieve a workable efficiency of gene transfer, selection, and regeneration of transformants. Detailed consideration of the options for and optimization of tissue culture systems useful for plant transformation is beyond the scope of this

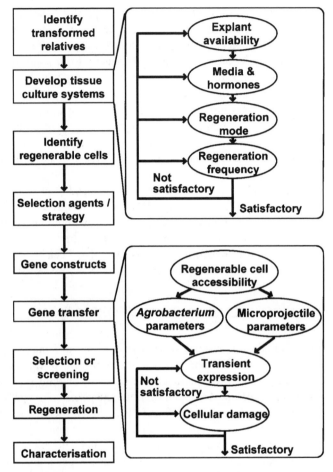

Figure 2 A generalized approach to development of transformation systems for recalcitrant plant species.

review, but the broad technology and detailed protocols are addressed in recent techniques manuals (52, 154).

In tissue culture systems for plant transformation, what is most important is a large number of regenerable cells that are accessible to the gene transfer treatment, and that will retain the capacity for regeneration for the duration of the necessary target preparation, cell proliferation, and selection treatments. A high multiplication ratio from a micropropagation system does not necessarily indicate a large number of regenerable cells accessible to gene transfer (97).

Gene transfer into potentially regenerable cells may not allow recovery of transgenic plants if the capacity for efficient regeneration is short-lived (132). There seems to be no reason to prefer embryogenic or organogenic plant regeneration. In the happy event that several tissue culture systems meeting the primary requirement above are available for a species of interest, the choice can be made based on features affecting convenience or efficiency, including ready availability of explants, and minimal time in tissue culture. Somaclonal variation, once considered a potentially useful source of genetic variation for plant improvement, is now more a bane of gene transfer programs (81). In some circumstances, particularly the direct introduction of genes for desired commercial traits into elite vegetatively propagated cultivars, the need to avoid such random genetic change may become the overriding consideration in the choice of tissue culture and gene transfer systems.

The desire to minimize somaclonal variation is one motivation for eliminating or minimizing the tissue culture phase, by gene transfer into intact tissue explants and regeneration without substantial in vitro culture. In several cereals, it has been possible to dispense with tissue culture for target preparation, by gene transfer into immature zygotic embryos, although embryogenic callus culture is still required to recover transgenic plants (27, 157, 159).

The goal of genetic transformation without tissue culture has been approached in soybean, cotton, bean, and peanut by particle bombardment into meristematic tissue of excised embryonic axes, shoot proliferation to yield some lines with transformation of germline cells, and screening for transformed sexual progeny (26). The limiting factors remain the ability to mechanically prepare the explants, transfer genes into regenerable cells, and select or screen for transformants at an efficiency sufficient for practical use in cultivar improvement. For example, the reported germline transformation rate for bean (0.06% of excised and bombarded apical meristems, 0.03% of assayed shoots) would make the process too expensive for many laboratories (133). There remains unexplained cultivar specificity for transformation via meristem bombardment within some plant species such as bean (133), impracticality of gene transfer into meristems of others such as rice (27), and uncertainty about applicability in vegetatively propagated species such as sugarcane (51).

While tissue culture remains an essential component of practical transformation systems for most plant species, research aimed at minimizing somaclonal variation deserves a high priority. Curiously little of the published research on somaclonal variation has been directed at this goal (81).

Gene Transfer Strategies

Suggested approaches to development or optimization of transformation protocols using particle bombardment (12) or *Agrobacterium* (152) have been published, based on practical experiences in laboratories working on recalcitrant crops. For particle bombardment, it is generally most efficient to first examine the available tissue culture systems, determine the modes of regeneration and the location of the cells involved, optimize tissue culture conditions to increase the number or accessibility of such cells if necessary, and then develop conditions for nonlethal transfer of DNA into large numbers of such cells per bombardment (12).

For *Agrobacterium,* it is considered more efficient to first establish the conditions for gene transfer and then work on conditions for regeneration of transformed cells (152). This contrast may be biologically well-founded, because of the greater complexity and lesser understanding of the biological interaction preceding the gene transfer event from *Agrobacterium.* Unfortunately, there is no guarantee that a transformable plant cell type will prove regenerable, even in the hands of the most successful tissue culturist.

If preliminary transformation experiments using techniques successful in similar plant systems are not successful, my advice is to establish by histological studies the precise cellular origins and timing of events leading to plant regeneration within the explants to be used as targets for gene transfer. This is likely to avoid much wasted time and frustration from optimizing gene transfer and regeneration within the same region of the explant, but potentially in different cells, so that transformed plants are unlikely to result. Work in sunflower is a fine example of the value of this relatively simple check to explain and potentially overcome difficulty in transformation of a recalcitrant species (91).

Assays for transient expression of introduced reporter genes in plant cells can provide unequivocal evidence of gene transfer. A great deal of time can be wasted unless this analysis is focused on regenerable cells, which often comprise a small and inaccessible fraction of the target tissue (12, 91). Exhaustive experiments to maximize transient expression are also futile if they involve conditions harmful to regeneration or molecular characteristics of transformed cells (8, 26). For example, particle bombardment conditions can now be arranged to give highly reproducible results in transient assays, by delivering a large number of copies of potentially transcribable DNA into the nuclei of target cells. However, a different form and concentration of DNA is likely to be optimal for efficient production of low copy number transformants (12).

There is little information on forms of DNA or sequences that may increase the frequency of stable transformation (8, 19, 142).

There is considerable batch-to-batch variation in the frequency of transient expression events following cocultivation with *Agrobacterium* (78). The correlation between transient expression and stable transformation has not been thoroughly tested. *Agrobacterium* employs a highly evolved and still incompletely understood gene transfer and integration system that appears optimized for efficient nuclear targeting and integration of a protein-complexed single-stranded DNA introduced as a small number of copies per cell (165). Therefore, stable transformation may occur at a high frequency in cells without detectable transient expression of the introduced DNA. Positive results from transient assays for *Agrobacterium*-mediated gene transfer into regenerable cells are encouraging (70, 76, 93), but until the parameters affecting such expression are better understood, it is unwise to abandon hope too quickly based on negative results from transient assays.

It may be necessary to introduce a considerable number of genes into plants for some purposes. The limits of available gene transfer techniques have not yet been defined. At least 12 separate plasmids and up to 600 kb of DNA can be introduced at once by particle bombardment (62), but the number of expressed genes has not been tested. There are some indications that large plasmids (>10 kb) may be subject to greater fragmentation during particle bombardment (12). Transposon-derived vectors have been shown to deliver an increased proportion of intact, single-copy inserts of up to 10 kb following direct gene transfer into protoplasts (92). Recent evidence indicates that use of a binary bacterial artificial chromosome vector, with helper plasmids enhancing production of VirG and VirE proteins, can allow efficient *Agrobacterium*-mediated transfer of at least 150 kb of foreign DNA into the plant nuclear genome. Furthermore, the transferred DNA appeared to be present as an intact single copy that was faithfully inherited in the progeny of several of the characterized transformants (64).

Vectors overexpressing *virG* are also a component of the thoroughly verified systems for *Agrobacterium*-mediated transformation of rice (70), maize (76), and cassava (93), and deserve wider testing in recalcitrant plants. Other key variables in *Agrobacterium*-mediated gene transfer include *Agrobacterium* and plant genotype, treatment with *vir* gene inducers such as acetosyringone, wounded cell extracts, feeder cells or sugars; pH, temperature, cell concentration, light conditions, and duration of cocultivation; explant type, quality, preculture (152), hormone treatment (50), wounding, or infiltration (11); and use of appropriate antibacterial agents (95), antioxidants (124), ethylene antagonists (35), and/or methylation inhibitors (117) to reduce damage and/or gene silencing in treated plant cells. With so many variables, and little evidence of combinations that are broadly applicable across plant species, it is evident why transient expression assays are useful, if imperfect, as indicators of suitable conditions for gene transfer (76), before more expensive studies to optimize stable transformation.

Selection Strategies

For transformation systems that generate substantial numbers of nonchimeric primary transformants, genes conferring resistance to a selective chemical agent (161), genes conferring a phenotype allowing visual or physical screening (16, 18, 129), or even PCR screening to identify plants containing transferred genes (27, 83) can all be used to recover transformants. Transformation systems that generate chimeric primary transformants including transformed germline cells, as intermediates in the production of homogeneously transformed (R_1) progeny plants, generally require screening rather than lethal selection to reveal primary transformants (28, 83, 105).

Screening approaches are expensive unless the transformation efficiency is high, and generally impractical if the proportion of transformants among regenerated lines is below 10^{-2} to 10^{-3}. In our hands, the recovery of transformed plants was 10-fold lower from visual screening compared with antibiotic selection (18). This may occur because antibiotic selection provides a continuous advantage to transformed cells, which may otherwise be overgrown by the far greater numbers of proliferating nontransformed cells (48, 78). Under these circumstances, antibiotic selection may allow a higher proportion of transformed cells to multiply and regenerate, in addition to facilitating the recognition of transformants.

An excellent review has been published on selectable marker genes, assayable reporter genes, and criteria for their use in plant transformation studies (16). Broadly applicable, simple, and robust selection regimens now exist for transgenic plants, requiring little experimentation with the timing and concentration of selective agents to match the target tissue and gene transfer system. However, it is still important to consider the physiology of antibiotic action and resistance mechanisms when choosing or modifying selection protocols (36). There are also reports of interactions between selective agent and subsequent regenerability (137), and interactions between antibiotic and gelling agents (101). Attention is increasingly being directed to introduction of multiple agronomically useful genes into plant lines, without having to pyramid selectable genes in the process (27, 32, 155, 163).

Transgene Expression Strategies

The first attempts to express foreign genes failed because of the inability of the plant transcriptional machinery to recognize some foreign gene control sequences, particularly promoter sequences of many bacterial genes. This initial hurdle was overcome by exploiting control sequences isolated from genes of *Agrobacterium* and cauliflower mosaic virus, which were known to be expressed in plant cells as part of the process of molecular subversion of host cell machinery by these pathogens. Continuing characterization of many plant genes, and analyses of transient and stable expression of foreign gene constructs in plants, have contributed to a growing understanding of features useful for the regulated expression of transgenes in plants. Features typically considered in preparation of such constructs include: (*a*) appropriate transcriptional promoters and enhancers (9); (*b*) introns (98, 100); (*c*) transcriptional terminators and 3′ enhancers (130); (*d*) polyadenylation signals (3, 162); (*e*) untranslated 5′ leader and 3′ trailer sequences (38, 56, 148); (*f*) codon usage (87, 125); (*g*) optimal sequence context around transcription and translation start sites, including absence of spurious start codons (5, 55, 66, 99); (*h*) transit sequences for appropriate subcellular compartmentation and stability of the gene product (58, 68, 82, 115); (*i*) absence of sequences such as cryptic introns (129) or polyadenylation signals resulting in inappropriate RNA processing (87, 125); and (*j*) absence of sequences resulting in undesired glycosylation or lipid anchor sites (24, 46). Recent progress with chloroplast transformation has added the possibility of expression from the plastid or nuclear genome (21, 104).

This understanding has greatly improved the ability to tailor transgenes for various strengths and patterns of expression essential for practical plant genetic engineering (10). Levels and patterns of expression generally vary to some extent, even between independent single copy transformants. This reflects the influence of different sequences flanking the integration sites upon expression of the transgene (122, 151). Although this variation may generate transformed lines which by chance have new and useful expression patterns (34), this is unlikely to occur at a frequency that is useful in practice, unless we can bias integration toward regions of the DNA that are preferentially expressed under specified conditions. Otherwise, random variation will greatly increase the expense of the varietal improvement process, because rigorous evaluation of many transformed lines will be needed to identify those with the desired phenotype. One attraction of transformation for cultivar improvement is the theoretical potential for very precise genetic change. If random variation around a desired phenotype in individual transformants necessitates extensive field characterization to select commercial lines, there may be no advantage in using transformation over conventional breeding. In these circumstances, the value of plant transformation in cultivar improvement would be restricted to traits not achievable by conventional breeding, particularly introduction of new genes to the germplasm available to breeders.

Of even greater concern is unpredictable loss of the improved phenotype in transformed cultivars because of silencing of the transgene. This phenomenon is discussed in several excellent recent reviews (45, 103, 107). In practical terms, it is fortunate that transgene silencing has to date always been detectable soon after transformation, crossing, or field testing of transformants (see 45). The problem would be much more serious if unstable lines could not be reliably detected and eliminated during the routine screening and propagation of transgenic lines before commercial release, because of the potential commercial damage from loss of an essential trait such as disease resistance. There is still much to be learned about the causes of transgene inactivation. For example, multicopy transgenes have been identified as a probable cause (see 45), indicating a disadvantage of direct gene transfer techniques which result in higher average copy numbers than *Agrobacterium*. However, our experience from several years of laboratory and field testing with hundreds of transgenic sugarcane lines is that some promoter-reporter gene constructs are silenced at high frequency, whereas others are almost invariably stably expressed, with no relationship to copy number (13). It will be important to discover what sequences within these constructs trigger or inhibit silencing, and whether matrix attachment regions (142), demethylation sequences (94), or targeted integration systems (2, 118) can be developed to protect susceptible foreign sequences from silencing in transgenic plants.

Integrating Components of Transformation Strategies

Features of representative strategies for production of transgenic plants are compared in Table 1. Other combinations exist, but the table illustrates that the choice of transformation strategy influences many secondary parameters such as time in tissue culture and number of plants processed per transformant, which often determine the practicality of the system.

It is commonly generalized that *Agrobacterium* produces simpler integration patterns than direct gene transfer, but both approaches result in a similar range of integration events, including truncations, rearrangements, and various copy numbers and insertion sites. Furthermore, the frequency distributions of copy number and rearrangements vary with transformation parameters for both gene transfer methods (25, 60, 107, 150). More careful work is required to

Table 1 Features influencing practical application of successful plant transformation strategies

Parameters	In planta	Vegetative tissue	Meristem	Embryo	Callus
			Successful Transformation Strategies		
Gene transfer method	*Agrobacterium*	*Agrobacterium*	Particle bombardment	Particle bombardment	Particle bombardment
Explant	Flowering plant	Any tissue with regenerable cells	Embryonic axes or meristems	Intact or sectioned embryos	Embryogenic or organogenic callus
Target cells for effective gene transfer	Germline cells late in floral development	Any regenerable cells	Germline cells in meristems	Epidermal cells of scutellum	Surface or subsurface cells
Regeneration	Flowering shoot/zygotic embryo	Organogenesis or embryogenesis	Shoot formation/zygotic embryo	Embryogenesis	Embryogenesis or organogenesis
1° regenerant chimeric	Few floral cells transformed	No	Germline transformed	No	No
Sexual reproduction	Essential	Unnecessary	Probably essential	Unnecessary	Unnecessary
Selection	Yes (R_1 progeny)	Yes	Not before R_1 progeny	Yes	Yes
Screenable marker	Optional	Optional	Essential	Optional	Optional
Hormonal treatment	Unnecessary	Auxin +/or cytokinin	Usually cytokinin	Auxin+/− cytokinin	Auxin+/− cytokinin
Transformants per treated explant	1–10	1–10	0.001	0.01	0.01–1
Transformants per tested regenerant	0.001	1	0.0003	1	1
Tissue culture duration	Unnecessary	10 weeks	4–6 weeks	10 weeks	20 weeks
Time from treatment of non-chimeric transformant	7–10 weeks	10 weeks	One generation time	10 weeks	10 weeks
Proven applicability (e.g.)	*Arabidopsis* (7)	Many spp. (59)	Several spp. including legumes (26, 105)	Several cereals (27, 77)	Many spp. (17, 26)

optimize methods for simple integration patterns before any reliable conclusions are drawn about the relative potential of the techniques to deliver such patterns at a satisfactory frequency.

The apparent targeting of T-DNA integration into transcribed regions is useful for gene and promoter tagging, and for transgene insertion into regions favoring subsequent expression (86). However, the observation that over 90% of T-DNA insertions may disrupt transcriptional units (96), with 15–26% of transformants showing visible mutant phenotypes resulting from T-DNA insertions (44), sounds an alarm for direct production of improved cultivars in highly selected crops, where most phenotypic changes from random mutations are likely to be adverse. For such work, integration should ideally be directed to transcribed regions without disruption of existing plant genes. To achieve this will require more research to bring gene targeting technology in plants closer to the level achieved in model animals (79, 118). Whether DNA introduced into plant cells by direct gene transfer is also preferentially integrated into transcribed regions or active genes has not been adequately tested (85).

OTHER CONSTRAINTS TO RESEARCH AND DEVELOPMENT IN PLANT TRANSFORMATION

Regulatory Environment and Public Perceptions

In most countries, planned field releases and commercial development of transgenic plants are first scrutinized and approved by regulatory authorities established by the national government, to ensure that products are safe to the environment and consumers (33, 43, 158). This process can be important to obtain the maximum social benefit from transgenic plant lines. For example, in several countries, release of transgenic insect-resistant plant varieties has been linked to mandatory programs of insect monitoring and industry responses to avoid premature loss of useful insect control genes because of a build-up of resistant insect populations (88).

Scrutiny by regulatory authorities is also an important mechanism to reassure the general public of the safety of a new technology that is not well understood by most people. However, if the process is conducted inefficiently by the regulatory authorities, it can severely slow research. For example, it is essential to characterize a substantial number of independent transformed plant lines in both physiological experiments and in selecting genetically improved cultivars. The availability of sufficient containment greenhouse space rapidly becomes limiting if the process of evaluation before approval of field releases is slow.

The conservative assumption underlying regulations in many countries is that all transgenic plants are potentially hazardous. Scientific theory and practical experience show that this is not the case. The hazards relate to the genes transferred or the phenotype produced, not to the gene transfer method used. There have been no reports of any harmful environmental effects or other hazardous unforeseen behavior of transgenic plants in the thousands of field trials conducted internationally to date (33). As public experience and understanding of plant transformation increase, it is to be hoped that regulatory processes may be streamlined, with the focus on products rather than on processes of plant genetic modification (108).

Consumer response to transgenic plant products has now been tested with the commercial release of improved varieties in a range of crops (Table 2).

Table 2 Commercial releases of transgenic plant varieties

Trait	Crop	Name	Company	Product Status
Quality (vine-ripened flavor, shelf life	Tomato	Flavr Savr	Calgene	Released 1994
Quality (vine-ripened flavor, shelf life)	Tomato	Endless Summer	DNA Plant Technology	Blocked by patent claims
Quality (paste consistency)	Tomato	—	Zeneca	Released 1995
Oil characteristics	Canola	Laurical	Calgene	Released 1994
Virus resistance	Tobacco Tomato Capsicum	—	(China)	Released 1993–1994
Virus resistance	Squash	Freedom II	Asgrow	Released 1995
Insect resistance	Cotton Potato Maize	Bollgard NewLeaf YieldGuard	Monsanto	Released 1996–1997
Insect resistance	Maize	Maximizer	Ciba Seeds	Released 1996
Herbicide resistance	Flax	Triffid	University of Saskatchewan	Released 1995
Herbicide resistance	Cotton	BXN	Calgene	Released 1995
Herbicide resistance	Canola Corn	Innovator Liberty Link	AgrEvo	Released 1995–1996
Herbicide resistance	Soybean Canola Cotton	Roundup Ready	Monsanto	Released 1995–1996
Herbicide resistance	Soybean Corn	Roundup Ready, STS Liberty Link	Pioneer	Released 1996–1997
Male sterility hybrid system	Canola	—	Plant Genetic Systems	Approved 1996 (USA, FDA)

These releases have coincided with increasing dissemination of information on transgenic plants in forms accessible to the general public (158). In each case, consumers have responded positively to quality or price advantages, and in several cases demand has outstripped supply in the first season. However, it is clear that continued work is important to provide the broad community with information to support considered responses to emerging products (63).

Intellectual Property

As technical limitations are overcome, it is possible that commercial limitations will become more serious barriers to exploitation of genetic transformation. New technologies developed in this area are effectively inventions and are therefore eligible for patent protection (84, 123). For example, patents have already been issued on most established or promising plant genetic transformation strategies (29, 69, 89, 102, 119, 134, 136) and on many isolated genes, promoters, and techniques for plant gene manipulation (see 121 and monthly patent updates in *Genetic Technology News*). The patent literature has become an important source of information in plant transformation research, albeit more difficult and expensive to search than the scientific literature (6, 40).

A patent provides the inventor or assignee with a period of exclusive ownership, or formally a right to exclude others from making, using, or selling the invention. There is no statutory exclusion for infringment when patented products or methods are used for research purposes. The widespread misconception that disclosure in the patent document allows researchers to practice the invention in order to improve on it (73) possibly arises because patent owners are generally reticent in instituting infringement proceedings until the level of damages that may accrue becomes commercially significant. There is no obligation to license and no constraint on royalty levels provided the patent holder makes active use of the intellectual property. Some patents make extremely broad claims, and patent holders are not required to develop all possible manifestations of an invention to retain broad ownership (144). Penalties for infringement of proprietary rights can be severe, so it is important to determine whether the tools or topics of proposed research are already in the public domain or subject to patent protection (160). This can be difficult to establish without periodic patent searching, or even legal challenges (75).

The position may differ between countries. For example, particle bombardment for gene transfer into plant cells has been patented in the United States (134) but not in Australia. Apparatus for particle bombardment involving a macroprojectile and stop plate has been patented in Australia (135) but not apparatus involving particle acceleration in a gas pulse. Patent coverage for many genes and promoters is similarly restricted. For example, the maize ubiquitin promoter is the subject of granted patents or applications in Europe, Japan, and the United States (128) but apparently not in many other countries, including Australia.

Under these circumstances, a transgenic plant variety produced and used commercially without infringing any patent rights in one country could infringe certain patent claims if used (even for research or other noncommercial purposes) in another country. To complicate matters further, patent applications are not available publicly in some countries (notably the United States) until the patent is granted, which can be years after the application is filed. There are moves to harmonize international practice, e.g. by providing for ownership for 20 years from the date of application and publishing 18 months after application to eliminate the practice of "submarine" patent applications that only surface after a competitor has independently made the same invention (42). The issues involved are complicated, and some important differences in patent law between countries appear unlikely to be resolved in the near future (4).

Patents are intended to encourage and reward useful invention and technical innovation, and the new technology enters the public domain after a period of 17 to 20 years (42, 84). In the interim, commercial restrictions can appear quite ruthless as patent holders adopt commercialization strategies to capture the value of protected intellectual property (75). In an era of tight public sector research funding and high research and development costs, the benefits of corporate investment to develop transformation technologies outweigh the inconvenience of patent restrictions. Debate continues on mechanisms to balance the competing interests (22).

FUTURE NEEDS AND DIRECTIONS IN PLANT TRANSFORMATION RESEARCH AND DEVELOPMENT

Transformation Efficiency

The methods for gene transfer into plant cells, particularly *Agrobacterium* and particle bombardment, are now sufficiently developed to allow transformation of essentially any plant species in which regenerable cells can be identified. Broadly applicable selection methods are well established. The key to transformation of recalcitrant species appears to be development of methods to expose many regenerable cells to nondestructive gene transfer treatments.

What currently limits the practical transformation of many plant species is the combination of a low frequency of transformation and a high frequency of undesired genetic change or unpredictable transgene expression. These problems necessitate expensive large-scale transformation and screening programs to produce useful transformants. The first constraint may be addressed by research into tissue culture systems to enrich for regenerable cells accessible to gene transfer. Contract transformation services (106) may implement economies of scale to afford robotic systems (1) for routine large scale target preparation and gene transfer treatments.

A clearer understanding of the events surrounding gene transfer by *Agrobacterium* is also required. Is transient expression a satisfactory test for *Agrobacterium*-mediated gene transfer into plant cells, or can another convenient test be developed to allow rapid detection and optimization of this key event? Does *Agrobacterium* select between cell types, and if so what features determine favored cells for gene transfer? Can these features be imparted to highly regenerable cell types? Direct gene transfer experiments indicate that if naked DNA is transferred into many actively dividing and regenerable cells, a proportion will be stably transformed. Is the same true for cells receiving typically lower doses of T-DNA, or are there additional physiological requirements for efficient T-DNA integration (111)? Is T-DNA integration targeted to potentially expressed regions of the genome, or to regions undergoing active transcription? Can the transcriptional status of target cells be manipulated to achieve a high frequency of integration into regions suitable for subsequent transgene expression, but a low frequency of insertional inactivation of genes influencing the phenotype of regenerated transformants?

There are at least as many relevant questions surrounding direct gene transfer. Is stable transformation efficiency as sensitive as transient expression to decreased DNA concentration? Does DNA concentration affect mean copy number or cotransformation frequency in resulting stable transformants? Is integration targeted to potentially transcribed regions as appears to be the case for T-DNA from *Agrobacterium*? Can artificial T-DNA complexes be manufactured, and will they influence the efficiency or integration patterns available from direct gene transfer?

Useful vs Absolute Transformation Efficiency

In the longer term, a more important goal than increased transformation efficiency is the development of transformation methods and constructs tailored for predictable transgene expression, without collateral genetic damage. We may conclude that much of the current effort in plant transformation directed toward increased transformation frequencies is naive and misdirected. We need to distinguish between absolute and useful transformation frequencies. The limiting process in the application of plant transformation for more sophisticated studies of plant physiology or for cultivar improvement is generally not the production of transformants but the screening (or subsequent breeding) required to eliminate transformants with collateral genetic damage that would interfere with meaningful physiological analysis or commercial use. Depending on the ratio of effort required for these processes a large increase in absolute transformation efficiency may be futile if accompanied by even a small decrease in the proportion of useful transformants. Conversely, a large drop in absolute transformation frequency may be more than compensated by a smaller gain in the proportion of useful transformants. These ideas are familiar to most practicing plant breeders but have understandably not been foremost in the minds of most transformation scientists while they struggled to develop reliable and efficient systems for gene transfer into target plant species.

Collateral Genetic Damage

To achieve a high proportion of useful transformants, we need to understand more clearly the factors contributing to undesired genetic change during the transformation process. To what extent is such change associated with the integration of single or multiple copies of foreign DNA, as distinct from the processes of tissue culture, selection, and plant regeneration? Is genetic change induced or selected during such processes, or is it commonly the effect of preexisting mutations in somatic cells that are simply detected when entire plants are regenerated from single (transformed) cells? If change is induced, which are the mutagenic stages in the protocols, and can they be avoided? If mutations are preexisting, can the procedures be tailored to selectively prevent regeneration of mutated cells? Compared to adventitious shoot proliferation, is somatic embryogenesis disadvantageous because of longer duration in culture, or advantageous because the complexity of the embryogenic process acts as a filter to eliminate many cells with mutations? Does the approach of germline transformation of uncultured explants followed by crossing to obtain non-chimeric transgenic progeny reduce the frequency of undesired genetic change, or just mask such change in the background of genetic variation from sexual reproduction? The relative importance of these questions varies between plant species; vegetatively vs sexually propagated crops provide an extreme example. Unfortunately, the answers to many of these questions may also be genotype specific.

Ideal and Model Transformation Systems

As the emphasis on useful transformation frequency increases, we may see a trend toward minimization or elimination of tissue culture stages, targeted integration of single copy transgenes, and direct (leaf-disc PCR) screening for transformants with useful genes to eliminate the need for reporter sequences. As our understanding of the genetic basis of agronomic traits increases, it is likely that this goal will be extended to the introduction of greater lengths of DNA encoding multiple genes. We will need to determine the capacity of available methods to introduce such lengths of DNA intact.

Although some of these questions may be answered and approaches developed with model plants, the features that make the models attractive for some genetic studies (e.g. small genome, small plant size, rapid generation time for *Arabidopsis*) generally cannot be exploited in practical transformation systems for most economically important plants. We must be prepared to select the models according to the questions, and test the answers for applicability to the practical targets.

Transformation, Breeding, and Genetic Diversity

As with conventional breeding, it is highly undesirable for plant transformation to lead to excessive genetic uniformity in current varieties of any crop. Even a single gene in all varieties can create problems. For example, the United States maize crop in 1970 was devastated because of disease susceptibility accompanying a cytoplasmic male sterility trait used to simplify hybrid seed production (164). This is another reason to aim for the capacity to transform diverse genotypes within a species, to develop diverse genes for desired phenotypes, and to eliminate unnecessary sequences from the transformation process.

When Practical Means Commercial

Plant transformation is already sufficiently developed to allow the testing and even commercialization of plants with novel phenotypes under simple genetic control. For continuing practical benefits, it will be necessary to extend our understanding of the biological basis for efficient plant transformation and develop improved technologies for predictable transgene expression without collateral genetic damage, at a pace matching the exciting scientific advances in gene cloning and characterization. This will require support from industry for the underlying research. As transformation projects are increasingly undertaken with the possibility of generating commercially useful products, transformation scientists in turn must increasingly integrate social, legal, and economic issues as well as technical issues from the earliest stages of project design.

ACKNOWLEDGMENTS

I thank my graduate students, who have contributed to many elements of this review through their critical enthusiasm for the science and technology of plant transformation. Tanya Newton contributed the concept for Figure 2.

Visit the *Annual Reviews home page* at http://www.annurev.org.

Literature Cited

1. Aitken CJ, Kozai T, Smith MAL. 1995. *Automation and Environmental Control in Plant Tissue Culture.* Dordrecht: Kluwer
2. Albert H, Dale EC, Lee E, Ow DW. 1995. Site-specific integration of DNA into wild-type and mutant *lox* sites placed in the plant genome. *Plant J.* 7:649–59
3. An GH, Mitra A, Choi HK, Costa MA, An KS, et al. 1989. Functional analysis of the 3′ control region of the potato wound-inducible proteinase inhibitor II gene. *Plant Cell* 1:115–22
4. Ardley J, Hoptroff CGM. 1996. Protecting plant 'invention': the role of plant variety rights and patents. *Trends Biotechnol.* 14: 67–69
5. Baker BF. 1993. The 5′ cap of mRNA: biosynthesis, function and structure as related to antisense drugs. In *Antisense Research and Applications,* ed. ST Crooke, B Lebleu, pp. 37–53. Boca Raton, FL: CRC Press
6. Barks AH. 1994. Patent information in biotechnology. *Trends Biotechnol.* 12:352–64
7. Bechtold N, Ellis J, Pelletier G. 1993. *In planta Agrobacterium* mediated gene transfer by infiltration of adult *Arabidopsis thaliana* plants. *C. R. Acad. Sci. III* 316: 1194–99
8. Benediktsson I, Spampinato CP, Schieder O. 1995. Studies of the mechanism of transgene integration into plant protoplasts: improvement of the transformation rate. *Euphytica* 85:53–61
9. Benfey PN, Ren L, Chua N-H. 1990. Combinatorial and synergistic properties of CaMV 35S enhancer subdomains. *EMBO J.* 9:1685–96
10. Bennett J. 1993. Genes for crop improvement. In *Genetic Engineering,* ed. JK Setlow, 15:165–89. New York: Plenum
11. Bidney D, Scelonge C, Martich J, Burrus

M, Sims L, et al. 1992. Microprojectile bombardment of plant tissues increases transformation frequency by *Agrobacterium tumefaciens. Plant Mol. Biol.* 18: 301–13
12. Birch RG, Bower R. 1994. Principles of gene transfer using particle bombardment. In *Particle Bombardment Technology for Gene Transfer,* ed. N-S Yang, P Christou, pp. 3–37. New York: Oxford Univ. Press
13. Birch RG, Bower R, Elliott AR, Potier BAM, Franks T, et al. 1996. Expression of foreign genes in sugarcane. In *Proc. Int. Soc. Sugarcane Technol. Congr., Cartegena, Sept. 1995, 22nd,* ed. JH Cock, T Brekelbaum, 2:368–73. Cali, Colombia: Tecnicana
14. Birch RG, Franks T. 1991. Development and optimization of microprojectile systems for plant genetic transformation. *Aust. J. Plant Physiol.* 18:453–69
15. Bourque JE. 1995. Antisense strategies for genetic manipulations in plants. *Plant Sci.* 105:125–49
16. Bowen B. 1993. Markers for plant gene transfer. See Ref. 90, 1:89–123
17. Bower R, Birch RG. 1992. Transgenic sugarcane plants via microprojectile bombardment. *Plant J.* 2:409–16
18. Bower R, Elliott AR, Potier BAM, Birch RG. 1996. High-efficiency, microprojectile-mediated cotransformation of sugarcane, using visible or selectable markers. *Mol. Breed.* 2:239–49
19. Buising CM, Benbow RM. 1994. Molecular analysis of transgenic plants generated by microprojectile bombardment: Effect of petunia transformation booster sequence. *Mol. Gen. Genet.* 243:71–81
20. Cao J, Duan X, McElroy D, Wu R. 1992. Regeneration of herbicide resistant transgenic rice plants following microprojectile-

mediated transformation of suspension culture cells. *Plant Cell Rep.* 11:586–91

21. Carrer H, Maliga P. 1995. Targeted insertion of foreign genes into the tobacco plastid genome without physical linkage to the selectable marker gene. *BioTechnology* 13: 791–94

22. Caskey CT. 1996. Gene patents: a time to balance access and incentives. *Trends Biotechnol.* 14:298–302

23. Chang SS, Park SK, Kim BC, Kang BJ, Kim DU, et al. 1994. Stable genetic transformation of *Arabidopsis thaliana* by *Agrobacterium* inoculation *in planta. Plant J.* 5:551–58

24. Chow M, Der CJ, Buss JE. 1992. Structure and biological effects of lipid modifications on proteins. *Curr. Opin. Cell Biol.* 4: 629–36

25. Christou P. 1992. Genetic transformation of crop plants using microprojectile bombardment. *Plant J.* 2:275–81

26. Christou P. 1995. Strategies for variety-independent genetic transformation of important cereals, legumes and woody species utilizing particle bombardment. *Euphytica* 85:13–27

27. Christou P, Ford TL, Kofron M. 1992. The development of a variety-independent gene-transfer method for rice. *Trends Biotechnol.* 10:239–46

28. Christou P, McCabe DE. 1992. Prediction of germ-line transformation events in chimeric R0 transgenic soybean plantlets using tissue-specific expression patterns. *Plant J.* 2:283–90

29. Coffee RA, Dunwel JM. 1995. Transformation of plant cells (to Zeneca Ltd.). *US Patent No. 5,464,765*

30. Collins GB, Shepherd RJ, eds. 1996. *Engineering Plants for Commercial Products and Applications.* Ann. NY Acad. Sci., Vol. 792

31. Coruzzi G, Puigdomenech P, eds. 1994. *Plant Molecular Biology: Molecular Genetic Analysis of Plant Development and Metabolism.* Berlin: Springer-Verlag

32. Dale EC, Ow DW. 1991. Gene transfer with subsequent removal of the selection gene from the host genome. *Proc. Natl. Acad. Sci. USA* 88:10558–62

33. Dale PJ. 1995. R & D regulation and field trialling of transgenic crops. *Trends Biotechnol.* 13:398–403

34. De Block M. 1993. The cell biology of plant transformation: current state, problems, prospects and the implications for plant breeding. *Euphytica* 71:1–14

35. De Block M, De Brouwer D, Tenning P. 1989. Transformation of *Brassica napus* and *Brassica oleracea* using *Agrobacterium tumefaciens* and the expression of the *bar* and *neo* genes in transgenic plants. *Plant Physiol.* 91:694–701

36. De Block M, De Sonville A, Debrouwer D. 1995. The selection mechanism of phosphinothricin is influenced by the metabolic state of the tissue. *Planta* 197: 619–26

37. De Block M, Herrera-Estrella L, van Montagu M, Schell J, Zambryski P. 1984. Expression of foreign genes in regenerated plants and their progeny. *EMBO J.* 3: 1681–89

38. De Loose M, Danthine X, Van Bockstaele E, Van Montagu M, Depicker A. 1995. Different 5′ leader sequences modulate -glucuronidase accumulation levels in transgenic *Nicotiana tabacum* plants. *Euphytica* 85:209–16

39. Draper J, Scot R, Armitage P, Walden R, eds. 1988. *Plant Genetic Transformation and Gene Expression: A Laboratory Manual.* Oxford: Blackwell

40. Electronic Data Systems Corporation. 1995. *Shadow Patent Off., http://www.spo.eds.com/spo*

41. Ellis JR. 1993. Plant tissue culture and genetic transformation. In *Plant Molecular Biology Labfax*, ed. RRD Croy, pp. 253–85. Oxford: BIOS Sci.

42. Enayati E. 1995. Intellectual property under GATT. *BioTechnology* 13:460–62

43. Engel K-H, Takeoka GR, eds. 1995. *Genetically Modified Foods: Safety Issues.* Washington, DC: Am. Chem. Soc.

44. Feldmann KA. 1991. T-DNA insertion mutagenesis in *Arabidopsis*: mutational spectrum. *Plant J.* 1:71–82

45. Finnegan J, McElroy D. 1994. Transgene inactivation: plants fight back! *BioTechnology* 12:883–88

46. Firek S, Whitelam GC, Draper J. 1994. Endoplasmic reticulum targeting of active modified beta-glucuronidase (GUS) in transgenic tobacco plants. *Transgenic Res.* 3:326–31

47. Franks T, Birch RG. 1991. Microprojectile techniques for direct gene transfer into intact plant cells. See Ref. 110, pp. 103–27

48. Fromm ME, Morrish F, Armstrong C, Williams R, Thomas J, et al. 1990. Inheritance and expression of chimaeric genes in the progeny of transgenic maize plants. *BioTechnology* 8:833–39

49. Gad AE, Rosenberg N, Altman A. 1990. Liposome-mediated gene delivery into plant cells. *Physiol. Plant.* 79:177–83

50. Gafni Y, Icht M, Rubinfeld BZ. 1995. Stimulation of *Agrobacterium tumefaciens*

virulence with indole-3-acetic acid. *Lett. Appl. Microbiol.* 20:98–101

51. Gambley RL, Bryant JD, Masel NP, Smith GR. 1994. Cytokinin-enhanced regeneration of plants from microprojectile bombarded sugarcane meristematic tissue. *Aust. J. Plant Physiol.* 21:603–12

52. Gamborg OL, Phillips GC, eds. 1995. *Plant Cell, Tissue and Organ Culture. Fundamental Methods.* Berlin: Springer-Verlag

53. Gartland KMA, Davey MR, eds. 1995. *Agrobacterium Protocols: Methods in Molecular Biology,* Vol. 44. Totowa, NJ: Humana Press

54. Gatehouse AMR, Hilder VA, Boulter D, eds. 1992. *Plant Genetic Manipulation for Crop Protection.* Wallingford, Engl: Cent. Agric. Biosci. Int.

55. Geballe AP, Morris DR. 1994. Initiation codons within 5'-leaders of mRNAs as regulators of translation. *Trends Biochem. Sci.* 19:159–64

56. Gil P, Green PJ. 1996. Multiple regions of the *Arabidopsis* SAUR-AC1 gene control transcript abundance: the 3' untranslated region functions as an mRNA instability determinant. *EMBO J.* 15:1678–86

57. Glick BR, Thompson JE, eds. 1993. *Methods in Plant Molecular Biology and Biotechnology.* Boca Raton, FL: CRC Press

58. Gomord V, Faye L. 1996. Signals and mechanisms involved in intracellular transport of secreted proteins in plants. *Plant Physiol. Biochem.* 34:165–81

59. Grant JE, Dommisse EM, Christey MC, Conner AJ. 1991. Gene transfer to plants using *Agrobacterium.* See Ref. 110, pp. 50–73

60. Grevelding C, Fantes V, Kemper E, Schell J, Masterson R. 1993. Single-copy T-DNA insertions in *Arabidopsis* are the predominant form of integration in root-derived transgenics, whereas multiple insertions are found in leaf discs. *Plant Mol. Biol.* 23:847–60

61. Grierson D, ed. 1991. *Plant Genetic Engineering.* Glasgow: Blackie

62. Hadi MZ, McMullen MD, Finer JJ. 1996. Transformation of 12 different plasmids into soybean via particle bombardment. *Plant Cell Rep.* 15:500–5

63. Hallman WK. 1996. Public perceptions of biotechnology: another look. *BioTechnology* 14:35–38

64. Hamilton CM, Frary A, Lewis C, Tanksley SD. 1996. Stable transfer of intact high molecular weight DNA into plant chromosomes. *Proc. Natl. Acad. Sci. USA* 93: 9975–79

65. Haq TA, Mason HS, Clements JD, Arntzen CJ. 1995. Oral immunization with recombinant bacterial antigen produced in transgenic plants. *Science* 268:714–16

66. Hensgens LAM, Fornerod MWJ, Rueb S, Winkler AA, van der Veen S, et al. 1992. Translation controls the expression level of a chimaeric reporter gene. *Plant Mol. Biol.* 20:921–38

67. Herbers K, Sonnewald U. 1996. Manipulating metabolic partitioning in transgenic plants. *Trends Biotechnol.* 14:198–205

68. Hicks GR, Smith HMS, Shieh M, Raikhel NV. 1995. Three classes of nuclear import signals bind to plant nuclei. *Plant Physiol.* 107:1055–58

69. Hiei Y, Komari T. 1994. Transformation of monocotyledons using *Agrobacterium. Int. Patent WO 94/00977*

70. Hiei Y, Ohta S, Komari T, Kumashiro T. 1994. Efficient transformation of rice (Oryza sativa L.) mediated by Agrobacterium and sequence analysis of the boundaries of the T-DNA. *Plant J.* 6: 271–82

71. Hinchee MAW, Padgette SR, Kishore GM, Delannay X, Fraley RT. 1993. Herbicide tolerant crops. See Ref. 90, 1:243–64

72. Hooykaas PJJ, Schilperoort RA. 1992. Agrobacterium and plant genetic engineering. *Plant Mol. Biol.* 19:15–38

73. Horsch RB. 1993. Commercialization of genetically engineered crops. *Philos. Trans. R. Soc. London Ser. B* 342:287–91

74. Horsch RB, Fraley RT, Rogers SG, Sanders PR, Lloyd A, et al. 1984. Inheritance of functional foreign genes in plants. *Science* 223:496–98

75. Hoyle R. 1996. Another salvo in the patent wars. *Nat. Biotechnol.* 14:680–82

76. Ishida Y, Saito H, Ohta S, Hiei Y, Komari T, et al. 1996. High efficiency transformation of maize (Zea mays L.) mediated by *Agrobacterium tumefaciens. Nat. Biotechnol.* 14:745–50

77. Jähne A, Becker D, Lörz H. 1995. Genetic engineering of cereal crop plants: a review. *Euphytica* 85:35–44

78. Janssen B-J, Gardner RC. 1989. Localized transient expression of GUS in leaf discs following cocultivation with *Agrobacterium. Plant Mol. Biol.* 14:61–72

79. Jasin M, Moynahan ME, Richardson C. 1996. Targeted transgenesis. *Proc. Natl. Acad. Sci. USA* 93:8804–8

80. Jenes B, Moore H, Cao J, Zhang W, Wu R. 1993. Techniques for gene transfer. See Ref. 90, 1:125–46

81. Karp A. 1995. Somaclonal variation as a tool for crop improvement. *Euphytica* 85: 295–302

82. Keegstra K, Bruce B, Hurley M, Li HM, Perry S. 1995. Targeting of proteins into chloroplasts. *Physiol. Plant.* 93:157–62

83. Kim JW, Minamikawa T. 1996. Transformation and regeneration of french bean plants by the particle bombardment process. *Plant Sci.* 117:131–38

84. Kjeldgaard RH, Marsh DR. 1994. Intellectual property rights for plants. *Plant Cell* 6:1524–28

85. Klein B, Töpfer R, Sohn A, Schell J, Steinbi H-H. 1990. Promoterless reporter genes and their use in plant gene transformation. In *Progress in Plant Cellular and Molecular Biology*, ed. HJJ Nijkamp, LHW Van Der Plas, J Van Aartrijk, pp. 79–84. Dordrecht: Kluwer

86. Koncz C, Martini N, Mayerhofer R, Koncz-Kalman Z, Körber H, et al. 1989. High-frequency T-DNA mediated gene tagging in plants. *Proc. Natl. Acad. Sci. USA* 86:8467–71

87. Koziel MG, Beland GL, Bowman C, Carozzi NB, Crenshaw R, et al. 1993. Field performance of elite transgenic maize plants expressing an insecticidal protein derived from *Bacillus thuringiensis*. *BioTechnology* 11:194–200

88. Koziel MG, Carozzi NB, Desai N, Warren GW, Dawson J, et al. 1996. Transgenic maize for the control of European corn borer and other maize insect pests. See Ref. 30, pp. 164–71

89. Kryzyzek R, Laursen CRM, Anderson PC. 1995. Stable transformation of maize cells by electroporation. *US Patent 5,472,869*

90. Kung S, Wu R, eds. 1993. *Transgenic Plants*, Vols. 1–2. San Diego: Academic

91. Laparra H, Burrus M, Hunold R, Damm B, Bravo-Angel A-M, et al. 1995. Expression of foreign genes in sunflower (*Helianthus anuus* L.): evaluation of three gene transfer methods. *Euphytica* 85:63–74

92. Lebel EG, Masson J, Bogucki A, Paszkowski J. 1995. Transposable elements as plant transformation vectors for long stretches of foreign DNA. *Theor. Appl. Genet.* 91:899–906

93. Li H-Q, Sautter C, Potrykus I, Puonti-Kaerlas J. 1996. Genetic transformation of cassava (*Manihot esculenta* Crantz). *Nat. Biotechnol.* 14:736–40

94. Lichtenstein M, Keini G, Cedar H, Bergman Y. 1994. B-cell-specific demethylation: a novel role for the intronic kappachain enhancer sequence. *Cell* 76:913–23

95. Lin JJ, Assadgarcia N, Kuo J. 1995. Plant hormone effect of antibiotics on the transformation efficiency by *Agrobacterium tumefaciens* cells. *Plant Sci.* 109:171–77

96. Lindsey K, Wei W, Clarke MC, McArdale HF, Rooke LM, et al. 1993. Tagging genomic sequences that direct transgene expression by activation of a promoter trap in plants. *Transgenic Res.* 2:33–47

97. Livingstone DM, Birch RG. 1995. Plant regeneration and microprojectile-mediated gene transfer in embryonic leaflets of peanut (*Arachis hypogaea* L.). *Aust. J. Plant Physiol.* 22:585–91

98. Luehrsen KR, Walbot V. 1991. Intron enhancement of gene expression and the splicing efficiency of introns in maize cells. *Mol. Gen. Genet.* 225:81–93

99. Luehrsen KR, Walbot V. 1994. The impact of AUG start codon context on maize gene expression in vivo. *Plant Cell Rep.* 13:454–58

100. Maas C, Laufs J, Grant S, Korfhage C, Werr W. 1991. The combination of a novel stimulatory element in the first exon of the maize Shrunken-1 gene with the following intron 1 enhances reporter gene expression up to 1000-fold. *Plant Mol. Biol.* 16:199–207

101. Maheswaran G, Welander M, Hutchinson JF, Graham MW, Richards D. 1992. Transformation of apple rootstock M26 with *Agrobacterium tumefaciens*. *J. Plant. Physiol.* 139:560–68

102. Maliga P, Maliga ZS. 1995. Method for stably transforming plastids of multicellular plants. *US Patent 5,451,513*

103. Matzke MA, Matzke AJM. 1995. How and why do plants inactivate homologous (trans)genes? *Plant Physiol.* 107:679–85

104. McBride KE, Svab Z, Schaaf DJ, Hogan PS, Stalker DM, et al. 1995. Amplification of a chimeric *Bacillus* gene in chloroplasts leads to an extraordinary level of an insecticidal protein in tobacco. *BioTechnology* 13:362–65

105. McCabe DE, Martinell BJ. 1993. Transformation of elite cotton cultivars by particle bombardment of meristems. *BioTechnology* 11:596–98

106. McElroy D. 1996. The industrialization of plant transformation. *Nat. Biotechnol.* 14:715–16

107. Meyer P. 1995. Variation of transgene expression in plants. *Euphytica* 85:359–66

108. Miller HI. 1995. Unscientific regulation of agricultural biotechnology: time to hold the policymakers accountable? *Trends Biotechnol.* 13:123–25

109. Mol JNM, Holton TA, Koes RE. 1995. Floriculture: genetic engineering of commercial traits. *Trends Biotechnol.* 13:350–55

110. Murray DR, ed. 1991. *Advanced Methods*

in *Plant Breeding and Biotechnology.* Wallingford, Engl: CAB Int.

111. Narasimhulu SB, Deng X, Sarria R, Gelvin SB. 1996. Early transcription of *Agrobacterium* T-DNA genes in tobacco and maize. *Plant Cell* 8:873–86

112. Nawrath C, Poirier Y, Somerville C. 1995. Plant polymers for biodegradable plastics: cellulose, starch and polyhydroxyalkanoates. *Mol. Breed.* 1:105–22

113. Negrutiu I, Dewulf J, Pietrzak M, Botterman J, Rietveld E, et al. 1990. Hybrid genes in the analysis of transformation conditions. II. Transient expression vs stable transformation: analysis of parameters influencing gene expression levels and transformation efficiency. *Physiol. Plant.* 79:197–205

114. Neuhaus G, Spangenberg G. 1990. Plant transformation by microinjection techniques. *Physiol. Plant.* 79:213–17

115. Neuhaus JM. 1996. Protein targeting to the plant vacuole. *Plant Physiol. Biochem.* 34:217–21

116. Newbigen E, Smyth DR, Clarke AE. 1995. Understanding and controlling plant development. *Trends Biotechnol.* 13:338–43

117. Palmgren G, Mattson O, Okkels FT. 1993. Treatment of *Agrobacterium* or leaf discs with 5-azacytidine increases transgene expression in tobacco. *Plant Mol. Biol.* 21:429–35

118. Paszkowski J, ed. 1994. *Homologous Recombination and Gene Silencing in Plants.* Amsterdam: Kluwer

119. Paszkowski J, Potrykus I, Hohn B, Shillito RD, Hohn T, et al. 1995. Transformation of hereditary material in plants. *US Patent 5,453,367*

120. Paszkowski J, Shillito RD, Saul M, Mandak V, Hohn T, et al. 1984. Direct gene transfer to plants. *EMBO J.* 3:2717–22

121. Patent bibliography: plant biotechnology. 1993. *Curr. Opin. Biotechnol.* 4:253–55

122. Peach C, Velten J. 1991. Transgene expression variability (position effect) of CAT and GUS reporter genes driven by linked divergent T-DNA promoters. *Plant Mol. Biol.* 17:49–60

123. Peet RC. 1995. *Protection of plant-related inventions in the United States.* http://biotechlaw.ari.net/plnt-ovr.html

124. Peri A, Lotan O, Abu-Abied M, Holland D. 1996. Establishment of an *Agrobacterium*-mediated transformation system for grape (*Vitis vinifera* L.): the role of antioxidants during grape-*Agrobacterium* interactions. *Nat. Biotechnol.* 14:624–28

125. Perlak FJ, Fuchs RL, Dean DA, McPherson SL, Fischhoff DA. 1991. Modification of the coding sequence enhances plant expression of insect control genes. *Proc. Natl. Acad. Sci. USA* 88:3324–28

126. Potrykus I. 1991. Gene transfer to plants: Assessment of published approaches and results. *Annu. Rev. Plant Physiol. Plant Mol. Biol.* 42:205–25

127. Potrykus I, Spangenberg G, eds. 1995. *Gene Transfer to Plants.* Berlin: Springer-Verlag

128. Quail PH, Christiansen AH, Hershey HP, Sharrock RA, Sullivan TD. 1994. Plant ubiquitin promoter system. *Eur. Patent EP 342926*

129. Reichel C, Mathur J, Eckes P, Langenkemper K, Koncz C, et al. 1996. Enhanced green fluorescence by the expression of an *Aequorea victoria* green fluorescent protein mutant in mono- and dicotyledonous plant cells. *Proc. Natl. Acad. Sci. USA* 93:5888–93

130. Richardson JP. 1993. Transcription termination. *Crit. Rev. Biochem. Mol. Biol.* 28:1–30

131. Ritchie SW, Hodges TK. 1993. Cell culture and regeneration of transgenic plants. See Ref. 90, 1:147–78

132. Ross AH, Manners JM, Birch RG. 1995. Embryogenic callus production, plant regeneration, and transient gene expression following particle bombardment, in the pasture grass *Cenchrus ciliarus* (Gramineae). *Aust. J. Bot.* 43:193–99

133. Russell DR, Wallace KM, Bathe JH, Martinell BJ, McCabe DE. 1993. Stable transformation of *Phaseolus vulgaris* via electric-discharge mediated particle acceleration. *Plant Cell Rep.* 12:165–69

134. Sanford JC, Wolf ED, Allen NK. 1990. Method for transporting substances into living cells and tissues and apparatus therefor. *US Patent 4,954,050*

135. Sanford JC, Wolf ED, Allen NK. 1992. Biolistic apparatus for delivering substances into cells and tissues in a nonlethal manner. *Aust. Patent AU 621561*

136. Schilperoort RA, Hoekema A, Hooykaas PJJ. 1990. Process for the incorporation of foreign DNA into the genome of dicotyledonous plants. *US Patent 4,940,838*

137. Schöpke C, Taylor N, Carcamo R, Konan NK, Marmey P, et al. 1996. Regeneration of transgenic cassava plants (Manihot esculenta Crantz) from microbombarded embryogenic suspension cultures. *Nat. Biotechnol.* 14:731–35

138. Shah DM, Rommens CMT, Beachy RN. 1995. Resistance to diseases and insects in transgenic plants: progress and applica-

tions to agriculture. *Trends Biotechnol.* 13: 362–68

139. Shen WH, Hohn B. 1994. Amplification and expression of the beta-glucuronidase gene in maize plants by vectors based on maize streak virus. *Plant J.* 5:227–36

140. Shimamoto K, Terada R, Izawa T, Fujimoto H. 1989. Fertile transgenic rice plants regenerated from transformed protoplasts. *Nature* 338:274–76

141. Songstad DD, Somers DA, Griesbach RJ. 1995. Advances in alternative DNA delivery techniques. *Plant Cell Tissue Organ Cult.* 40:1–15

142. Spiker S, Thompson WF. 1996. Nuclear matrix attachment regions and transgene expression in plants. *Plant Physiol.* 110: 15–21

143. Staskawicz BJ, Ausubel FM, Baker BJ, Ellis JG, Jones JDG. 1995. Molecular genetics of plant disease resistance. *Science* 268:661–67

144. Stone R. 1995. Sweeping patents put biotech companies on the warpath. *Science* 268:656–58

145. Sun SSM, Larkins BA. 1993. Transgenic plants for improving seed storage proteins. See Ref. 90, pp. 1:339–72

146. Tepfer D. 1990. Genetic transformation using *Agrobacterium* rhizogenes. *Physiol. Plant.* 79:140–46

147. Theologis A. 1994. Control of ripening. *Curr. Opin. Biotechnol.* 5:152–57

148. Turner R, Foster GD. 1995. The potential of plant viral translational enhancers in biotechnology for increased gene expression. *Mol. Biotechnol.* 3:225–36

149. Vancanneyt G, Schmidt R, O'Connor-Sanchez A, Willmitzer L, Rocha-Sosa M. 1990. Construction of an intron-containing marker gene: splicing of the intron in transgenic plants and its use in monitoring early events in *Agrobacterium*-mediated plant transformation. *Mol. Gen. Genet.* 220: 245–50

150. van der Graaff E, den Dulk-Ras A, Hooykaas PJJ. 1996. Deviating T-DNA transfer from *Agrobacterium tumefaciens* to plants. *Plant Mol. Biol.* 31:677–81

151. Van-der-Hoeven C, Dietz A, Landsmann J. 1994. Variability of organ-specific gene expression in transgenic tobacco plants. *Transgenic Res.* 3:159–66

152. van Wordragen MF, Dons JJM. 1992. Agrobacterium tumefaciens-mediated transformation of recalcitrant crops. *Plant Mol. Biol. Rep.* 10:12–36

153. Varner JE, Ye Z-H. 1995. Tissue printing to detect proteins and RNA in plant tissues. In *Methods in Plant Molecular Biology: A Laboratory Course Manual*, ed. P Maliga, DF Klessig, AR Cashmore, W Gruissem, JE Varner, pp. 79–94. Plainview, NY: Cold Spring Harbor Lab. Press

154. Vasil IK, Thorpe TA, eds. 1994. *Plant Cell and Tissue Culture.* Dordrecht: Kluwer

155. Von Bodman SB, Domier LL, Farrand SK. 1995. Expression of multiple eukaryote genes from a single promoter in *Nicotiana*. *BioTechnology* 13:587–91

156. Walden R, Wingender R. 1995. Gene-transfer and plant-regeneration techniques. *Trends Biotechnol.* 13:324–31

157. Wan Y, Lemaux PG. 1994. Generation of large numbers of independently transformed fertile barley plants. *Plant Physiol.* 104:37–48

158. Webber GD. 1995. *Biotechnology Information Series* (NCR# 483, 487, 551, 553, 557). Ames: Iowa State Univ.

159. Weeks TJ, Anderson OD, Blechl AE. 1993. Rapid production of multiple independent lines of fertile transgenic wheat (*Triticum aestivum*). *Plant Physiol.* 102:1077–84

160. Williams KM. 1994. How to avoid patent infringement. *BioTechnology* 12:297–98

161. Wilmink A, Dons JJM. 1993. Selective agents and marker genes for use in transformation of monocotyledonous plants. *Plant Mol. Biol. Rep.* 11:165–85

162. Wu L, Ueda T, Messing J. 1995. The formation of mRNA 3'-ends in plants. *Plant J.* 8:323–29

163. Yoder JI, Goldsbrough AP. 1994. Transformation systems for generating marker-free transgenic plants. *BioTechnology* 12:883–88

164. Zadoks JC, Schein RD. 1979. Southern corn leaf blight. In *Epidemiology and Plant Disease Management*, pp. 331–35. New York: Oxford Univ. Press

165. Zupan JR, Zambryski P. 1995. Transfer of T-DNA from Agrobacterium to the plant cell. *Plant Physiol.* 107:1041–47

Annu. Rev. Plant Physiol. Plant Mol. Biol. 1997. 48:327–354

CYANOBACTERIAL CIRCADIAN RHYTHMS

Susan S. Golden,[1] Masahiro Ishiura,[2] Carl Hirschie Johnson,[3] and Takao Kondo[2]

[1]Department of Biology, Texas A&M University, College Station, Texas, 77843

[2]Division of Biological Science, Graduate School of Science, Nagoya University, Chikusa, Nagoya, 464-01 Japan

[3]Department of Biology, Vanderbilt University, Nashville, Tennessee 37235

KEY WORDS: biological clock, gene expression, luciferase, luxAB, Synechococcus

ABSTRACT

Evidence from a number of laboratories over the past 12 years has established that cyanobacteria, a group of photosynthetic eubacteria, possess a circadian pacemaker that controls metabolic and genetic functions. The cyanobacterial circadian clock exhibits the three intrinsic properties that have come to define the clocks of eukaryotes: The timekeeping mechanism controls rhythms that show a period of about 24 h in the absence of external signals, the phase of the rhythms can be reset by light/dark cues, and the period is relatively insensitive to temperature. The promise of cyanobacteria as simple models for elucidating the biological clock mechanism is being fulfilled, as mutants affected in period, rhythm generation, and rhythm amplitude, isolated through the use of real time reporters of gene expression, have implicated genes involved in these aspects of the clock.

CONTENTS

327

1040-2519/97/0601-0327$08.00

INTRODUCTION

A living cell supports a harmonious pattern of interwoven metabolic processes whose cadence of ebb and flow is governed by an internal pacemaker. Unlike a metronome that merely taps out the beat, the circadian clock that times the events in a cell knows the score by heart and is sensitive to the instructions of an external conductor. As the name suggests, circadian rhythms are variations that peak and trough with a recurring period of about (circa) a day (diem). The clock that times them runs even without external stimuli, such that the period remains close to one day in constant conditions, yet it can be reset by cycles of Earth's daily environmental cues, such as the predictable alternation of light and dark. Perhaps the most remarkable property of the clock is that tempera-ture, though recognized as a cue when it varies cyclically, has little effect on the pace of the timekeeping mechanism at different constant temperatures. That is, the period is temperature compensated, and the clock ticks on reliably even as the ambient conditions change.

Plants have played a central role in the history of circadian rhythms re-search. deMairan first demonstrated in 1729 that leaves of a plant fold with a daily rhythm that is controlled by an endogenous timer (11). The study of leaf movements of plants continued when, a century after deMairan's observations, deCandolle showed that the period of the *Mimosa* leaf movement rhythm in constant conditions is not exactly 24 h (10). Other eminent nineteenth-century plant biologists who studied daily leaf movement rhythms included Darwin, Sachs, and Pfeffer [reviewed by Sweeney (90)]. Circadian rhythm research really started, however, with the pioneering studies of Erwin Bünning. In 1930, Bünning made careful measurements of the leaf movement rhythm of the common bean *Phaseolus*. He found that the movements oscillated in constant darkness with a period of 25.4 h and that exposure to a recurring 24-h light signal could entrain this "free-running" rhythm to a period of exactly 24 h [reviewed by Sweeney (90)]. Bünning also proposed that this 24-h clock might be the time-measuring process involved in photoperiodism (5). At first, "Bünning's hypothesis" for photoperiodic time measurement was ridiculed, but later experiments showed that the circadian clock is indeed the daylength (or nightlength) timer involved in photoperiodic responses (28, 90).

Since its discovery more than two centuries ago, the circadian biological clock has become widely recognized as influencing a multitude of cellular processes, ranging from medical consequences in humans when the clock goes awry [resulting in such conditions as depression, jet lag, or insomnia (9, 54, 67, 87)] to the timing of emergence of insects from their pupal cases (73, 75). The wide spectrum of organisms known to depend on the clock suggests its importance in the biological world (2, 74, 90). However, the biochemical basis of an oscillator that possesses these features—persistence in constant conditions, phase resetting by light and dark signals, and temperature compensation—is not easily modeled a priori. The nature of the clock is a fundamental question in biology today, and one that is enjoying a burst of progress as researchers using a variety of systems and approaches converge on what will be either a single instrument that evolution has customized for a diverse market, or a delightful array of solutions to a single challenge that may or may not employ similar biochemical components.

For several years the search for the biochemical oscillator has focused on using genetic approaches. Thus, organisms in which mutants could be obtained and the nature of the mutations determined have taken center stage as experimental models (13, 79). Mutants affected in what may be the central timekeeping mechanism have been reported for the eukaryotic organisms *Drosophila melanogaster* (50, 84), *Neurospora crassa* (17), *Chlamydomonas reinhardtii* (for a review, see 40), hamster (76), mouse (92, 98), and *Arabidopsis thaliana* (61). Among these organisms, some are more compliant than others. Only *Drosophila* and *Neurospora* have thus far yielded up the genes of interest (3, 60, 71, 77). The recent elucidation of the *timeless* gene of flies and the interaction of its product (TIM) with that of the *period* gene (PER) (22) have yielded a model in which a molecular feedback loop defines the properties of persistence and phase resetting that are diagnostic of a clock-controlled function. However, the sequences of the known genes in *Drosophila* and *Neurospora* do not resemble one another markedly, suggesting either that the clocks are different, that functionally similar proteins are not homologous, or that different pieces of the mechanism have been identified in each system. It is likely that the clock comprises several additional components, and that more loci must be identified to build a comprehensive view of the clock.

The potential for saturation mutagenesis to achieve the identification of all clock components could be greatly facilitated by the ability to work with an organism that has a small genome that is easily manipulated. Additional advantages would accrue from studying a microorganism in which thousands of individuals or clones could be screened for mutant circadian phenotypes, and by using automated methods to follow circadian behaviors. Until recently no organism possessed all of the desirable characteristics to provide a real advancement over the model systems that have been in place for years. The ideal organism for saturation mutagenesis would be a prokaryote, but the unambiguous demonstration of circadian behavior in prokaryotes—specifically, among the cyanobacteria—has been seen only within the past six years (8). Until 1993, no transformable cyanobacterium was known to be suitable for circadian studies. This review documents the advent and progress of a truly novel system in the quest for the circadian clock: a prokaryote with a genome size of 2.7 Mb (42), a suite of genetic tools for facile manipulation, unambiguous circadian rhythms of gene expression, and automation of analysis through a bioluminescent reporter as the circadian behavior.

THE CASE FOR A CIRCADIAN CLOCK IN CYANOBACTERIA

The Diazotrophy-Photosynthesis Paradox

The earliest reports that, in hindsight, demonstrated circadian rhythms in cyanobacteria grew out of the question of how an organism could fix nitrogen in the same cells that generate oxygen during photosynthesis. The nitrogenase enzyme is notoriously sensitive to oxygen, and the organisms that fix nitrogen, termed diazotrophs, have evolved a number of schemes to protect it (19, 20). Diazotrophic cyanobacteria have the additional challenge of shielding the sensitive enzyme from a by-product of their very livelihood, because they are the only prokaryotes that carry out oxygenic photosynthesis. Many filamentous cyanobacteria that are diazotrophic differentiate specialized cells called heterocysts for nitrogen fixation and use a division of labor strategy: Vegetative cells produce oxygen through full-chain photosynthesis but do not fix nitrogen, and heterocysts turn off photosystem II (PSII) while they express nitrogenase activity (20). The spatial segregation design cannot, however, explain the ability of unicellular or nonheterocystous filamentous species to both fix nitrogen and carry out oxygenic photosynthesis.

As early as 1969 it was known that some unicellular cyanobacteria are capable of nitrogen fixation (102). In 1974 Gallon et al reported a separation of nitrogen fixation and photosynthesis related to culture age in *Gloeocapsa* grown in continuous light, but not with a daily period; a single cycle of rising and falling activity occurred for each over the course of several days (21). In 1981 Millineaux et al (63) showed that when grown in a diurnal cycle of 12 h light–12 h darkness (LD 12:12), *Gloeocapsa* fixed nitrogen daily but only during the dark periods, whereas the heterocystous strain *Anabaena cylindrica*, in which nitrogenase and PSII activities are spatially separated, fixed nitrogen preferentially in the light (when energy for the metabolically expensive reaction would be plentiful). The first hint of circadian control in *Gloeocapsa* (in retrospect) appeared when diurnal conditions were altered to LD 16:8. Nitrogenase activity peaked 8 h after the onset of the darkness as it had in LD 12:12, but this timing placed the peak at the onset of the light period rather than during the middle of the dark period. In the same year, Weare & Benemann (99) also suggested temporal separation of oxygen production and nitrogen fixation in the nonheterocystous filamentous strain *Plectonema boryanum*, although the experiments were not designed to track the oscillations of these activities over several days.

The idea of temporal separation of nitrogenase and PSII activities as a strategy for tolerating oxygen was suggested and demonstrated repeatedly during the 1980s (41). The first report that can be interpreted in retrospect as strong evidence for a circadian rhythm of nitrogenase activity in a cyanobacterium was published by Stal & Krumbein in 1985 (85). They found that a nonheterocystous *Oscillatoria* species (filamentous) not only exhibits nitrogenase activity exclusively during the dark period of an LD 16:8 cycle, but shows two other remarkable characteristics: The nitrogenase activity begins to rise before the onset of darkness, and the oscillation of activity persists when cells are transferred from the LD cycling conditions to continuous light (LL), maintaining the pattern established by the preceding LD cycle. The authors discussed the nitrogenase pattern in terms of a metabolic trigger, with the depletion of nitrogen stores during photosynthetic growth inducing the enzyme. They later examined photosynthetic activity as well as nitrogen fixation and showed reciprocally peaking activities that persist in that pattern after cells adapted to an LD cycle are transferred to LL (86). However, they did not directly address the mechanism by which the oscillations continued in the absence of cycling environmental cues.

Two additional groups published papers in 1986 that supported the finding of daily nitrogenase activity oscillation in constant conditions. Mitsui et al (64) reported that two strains of unicellular marine cyanobacteria of the genus *Synechococcus* (spp. Miami BG 43511 and 43522) separate nitrogen fixation and photosynthesis temporally, timing each in a distinct phase of the cell division cycle. Their report was the most comprehensive at that time, measuring photosynthetic oxygen evolution, respiratory oxygen uptake, acetylene reduction (nitrogenase activity), carbohydrate synthesis, and cell division in synchronized populations. Each activity peaked once daily, with photosynthetic oxygen evolution, carbohydrate synthesis, and nitrogen fixation in different phases. Patterns established during three LD cycles continued when the cultures were exposed to 20 h of continuous light (a separate experiment showed three cycles of nitrogenase activity during 70 h of incubation in constant light). Under the conditions used, the *Synechococcus* spp. double approximately once per 24 h, and the authors concluded that the cell cycle was the timing mechanism for the other metabolic oscillations. A reassessment of these conclusions is presented in the section on "Relationship to the Cell Division Cycle."

Huang and colleagues published the first reports in 1986 describing RF-1, a strain of freshwater *Synechococcus* that figured prominently in establishing that cyanobacteria have a bona fide circadian clock (26, 31). *Synechococcus* RF-1 fixes nitrogen at varying levels with no detectable pattern if it is grown in continuous light without ever having seen darkness, but it restricts nitrogen fixation to the dark portion of an LD cycle (31), as had been reported previously for other cyanobacteria. This group recognized the persistence of the pattern in LL after an adaptive period in LD to be an endogenous rhythm; they further noted that the setting of the phase of the rhythm by the preceding LD cycle is reminiscent of the endogenous circadian rhythms reported for eukaryotes (26). Another claim for a genuine circadian rhythm in cyanobacteria was reported in 1989 by Sweeney and Borgese, who found a temperature-compensated 24-h recurrence of cell division in a marine strain of *Synechococcus,* WH7803 (91).

Convincing Evidence for a Prokaryotic Circadian Clock

The backdrop for these reports was an entrenched dogma in the circadian rhythms community that clocks are restricted to eukaryotes (6, 15, 45, 75, 83, 89). This conclusion was based primarily on an absence of evidence rather than on evidence of absence in prokaryotes. Some reports of possible prokaryotic rhythms were unconvincing (27, 88), and one concerted effort to detect circadian rhythms in cyanobacteria did not find them (93). A review in 1991 concluded that all the prokaryote evidence at that time, including the reports from cyanobacteria in the 1980s, left the door open for explanations other than circadian control (45).

Once the idea that the cyanobacterial nitrogenase rhythms might be evidence for a clock had been raised, researchers had clear hypotheses to test to determine whether the rhythmic behaviors they observed paralleled those seen in eukaryotes. Huang and colleagues working on *Synechococcus* RF-1 addressed several aspects of circadian biology (32). They found rhythms not only in nitrogen fixation, but in uptake rates of at least eight amino acids (8). The rhythms were temperature-compensated, showing approximate 24-h periods in constant conditions between 22°C and 33°C (34) or between 21°C and 37°C (8), and they could be set by temperature cycles as well as by LD cycles. The endogenous nitrogenase rhythm in RF-1 was shown to be independent of nitrogen fixation per se, because the phase of accumulation of *nifH* mRNA, which is present only under nitrogen fixing conditions, can be set by prior LD cycles imposed under non-nitrogen-fixing conditions (30). Another marine unicellular cyanobacterium was identified, *Cyanothece* sp. strain ATCC 51142, which shows persistent 24-h alternation of photosynthesis and nitrogen fixation as described previously for other strains. Cell ultrastructure data for this organism dramatize the rhythmic accumulation of stored carbohydrate (82). At this point numerous species were known that could be prokaryotic models for investigating the mechanism of the clock. However, all lacked the central requirement to exploit the virtue of the small genome: None had a means of genetic manipulation.

Using Genetics

The circadian behaviors, or detectable manifestations of the clock, that had been described in cyanobacteria before 1993 are all laborious to assay. To detect a persistent 24-h rhythm requires several days of measurement, and samples must be taken at least every few hours to observe the peaks and troughs. The prospect of a mutant hunt in which each mutagenized individual has to be assayed is daunting. In eukaryotes, schemes to automate the assay of activity have been developed, such as automatic recording of running wheel movement in the case of rodents and the interruption of an infrared beam by flying *Drosophila*. The morphological patterns of *Neurospora* as it forms conidia on agar medium provide a visible assay (80). Despite the lack of an automated system for cyanobacterial rhythms, Huang and colleagues succeeded in isolating mutants of *Synechococcus* RF-1 that show altered nitrogenase and amino acid uptake rhythms (35). They began with a screen based on the premise that growth of a mutant with a disrupted nitrogenase rhythm would be impaired under nitrogen-fixing conditions. Individual colonies from a mutagenized population were transferred to liquid and incubated in LL or LD without a combined nitrogen source for two weeks, then compared to the wild type for cell density. Rhythms of nitrogenase activity and leucine uptake were assayed in clones that had impaired growth. Among these, two mutants were identified in which both rhythms were lost, and two more in which leucine uptake remained normal although the nitrogenase rhythm was disrupted. Whether these phenotypes represent single or multiple mutations, whether they carry defects in the clock itself or in output pathways, and whether the growth phenotype is related to these has not yet been established.

Several strains of cyanobacteria have been developed for genetic research (94), but they were not among those in which the evidence for circadian rhythms was initially discovered for a simple reason: None is a unicellular diazotroph in which the question of temporal separation of metabolic functions provided the impetus for investigation. The accumulating evidence for circadian rhythms in diverse species of cyanobacteria made it likely that clocks are widespread among this major prokaryotic group. We approached the circadian problem from a different perspective—a genetic system in search of a circadian behavior to monitor. Our experimental organism is the unicellular freshwater strain known as *Synechococcus* sp. strain PCC 7942, which is not diazotrophic. It is naturally transformed by circular or linear DNA, can receive DNA by conjugation from *Escherichia coli* at high efficiency, and can express reporter genes (94). We had as a goal the ability to use the genetic flexibility of this organism to achieve saturation mutagenesis if a clock were uncovered, and thus chose to look at a behavior that would be amenable to automation: namely, light production.

The natural circadian bioluminescence rhythms of the dinoflagellate *Gonyaulax* provided the model for an ideal automation system, because the light emission can be captured and recorded in real time over several days by a variety of photodetectors (51, 69). The firefly luciferase gene, *luc,* when fused to the promoter of a *c*lock-*c*ontrolled *g*ene (*ccg*), creates a bioluminescent circadian phenotype in *Arabidopsis* (62). Artificial bioluminescence reporters have been employed in prokaryotes, including cyanobacteria, by using the *lux* genes of marine vibrios to express of the luciferase enzyme (16, 101). We fused a promoterless *luxAB* gene set encoding the luciferase enzyme from *Vibrio harveyi* to the promoter of a PSII gene (*psbAI*), which we knew provided strong expression in *Synechococcus* sp. strain PCC 7942 (55). The reporter integrated at a site in the chromosome that had been used previously for harboring heterologous genes (neutral site I[1]) (48). The resulting reporter

---

[1] Database accession U30252.

strain, AMC149, produced light when provided with the aldehyde substrate of the enzyme and showed a persistent 24-h rhythm of bioluminescence when cells were shifted from LD to LL. Moreover, the bioluminescence rhythm was reset by light and dark signals, as evidenced by a culture split into two parts that were entrained in opposite LD cycles: The two showed opposite peaks of light production after transfer to LL (Figure 1). Single pulses of 4 h of darkness in otherwise LL conditions shifted the peaks earlier or later, depending on the phase of the rhythm at which the pulse was applied; this differential sensitivity to phase-resetting cues at different times during the circadian cycle is typical of clock-controlled phenomena in eukaryotes (6, 90). As a final test, the period remained close to 24 h when measured at constant temperatures of 25°C, 30°C, and 35°C, demonstrating temperature compensation of the bioluminescence rhythm (48). We also demonstrated rhythmic expression of bioluminescence expressed from a *luxAB* reporter in two other cyanobacteria that can be genetically manipulated: the unicellular *Synechocystis* sp. strain PCC 6803 (1), a strain widely used in photosynthesis research whose genome was recently sequenced entirely (43), and the heterocystous filamentous species *Anabaena* sp. strain PCC 7120 (M Ishiura & T Kondo, unpublished data).

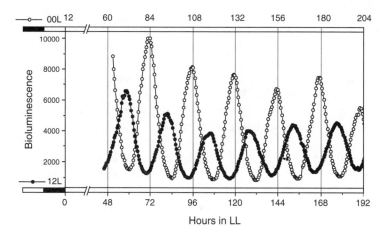

Figure 1 Circadian rhythm of bioluminescence from liquid cultures of the $P_{psbAI}::luxAB$ reporter strain AMC149 in continuous light conditions. AMC149 was cultured at 30°C under an LD 12:12 cycle, then transferred to vials held in LL for periodic measurement of bioluminescence by a photomultiplier device; the long chain aldehyde substrate decanal was provided exogenously as a vapor during measurement (48). The two traces are from cultures that were previously entrained to LD cycles that were 12 h out of phase. The last LD cycles preceding continuous light conditions (LL) are illustrated on the abscissa (*open bar* = light period, *closed bar* = dark period). [Reproduced with permission from Kondo et al (48).]

THE UTILITY OF BIOLUMINESCENCE REPORTING

Reliability of the Reporter

The utility of AMC149 as the model for genetic dissection of the cyanobacterial clock rested on the reliability of bioluminescence in reporting accurately a circadian behavior. Simultaneous measurement of cultures entrained or phase-shifted by different LD patterns confirmed that the bioluminescence rhythm was endogenous, rather than a consequence of some unrecognized cue from the measurement environment. The supposition that light emission reported activity of the *psbAI* promoter was less clear. The bacterial luciferase enzyme uses oxygen and reduced flavin mononucleotide as substrates, and these might vary in a circadian fashion as a consequence of rhythmic photosynthesis (the third substrate, a long chain aldehyde, was provided exogenously). We found that abundances of *psbAI* mRNA (48, 56), the reporter gene mRNA, and the luciferase enzyme all varied rhythmically (56). We concluded that bioluminescence is an accurate reporter of activity of the *psbAI* gene in AMC149 (56).

Automated Screening of Thousands of Colonies for Circadian Phenotypes

A prerequisite for exploiting the bioluminescence phenotype to hunt for mutants of AMC149 was to devise a technique for efficiently screening individuals for altered circadian rhythms. The first step was achieved by demonstrating that reliable rhythms could be monitored from isolated colonies on agar plates, with the decanal substrate administered continuously as a vapor from a reservoir in the petri dish (46). Colonies were monitored by both a photon-counting camera and a photomultiplier tube device, and the critical features of persistence, phase resetting, and temperature compensation were evident (46). It should be noted that neither the brief periodic incubations in darkness required for obtaining the bioluminescent signal, nor the weak light emission from the reporter itself, is sufficient to disrupt or reset the phase of the rhythms.

This led to construction of a turntable device that holds a dozen petri plates, each of which can carry approximately 1000 sufficiently separated colonies. Custom software allows a cooled charge-coupled-device (CCD) camera to monitor each plate for several minutes while it is temporarily masked from incident light, identify every light spot on the plate as a colony, and record the light emission from that colony. Approximately every 45 min, the same plate has returned through revolution of the turntable, and a new light emission can be measured from each colony. At the end of several days' monitoring, circadian profiles of bioluminescence are displayed, outliers are identified, and the relevant parameters of the rhythm of each colony are available. Development of this monitoring device marked the entry of cyanobacteria into the race with the more established eukaryotic genetic systems for the elucidation of the circadian mechanism.

Recent expansions of our automated monitoring methods for *Synechococcus* sp. strain PCC 7942 include the development of autonomously bioluminescent strains that carry the *Xenorhabdus luminescens luxCDE* genes, which direct biosynthesis of the aldehyde substrate for the *V. harveyi* luciferase (18; NV Lebedeva, J Yarrow, CR Andersson & SS Golden, unpublished results). Independence from aldehyde administration has allowed us to adopt the microplate-reading Packard TopCount as a monitoring device (Figure 2), as pioneered by the lab of Steve Kay for screening bioluminescent *Arabidopsis* and *Drosophila* (4).

RHYTHMIC GENE EXPRESSION

The robust rhythmicity of bioluminescence from AMC149, the first *lux*-containing reporter strain we engineered, called into question how many genes are controlled by the circadian clock in *Synechococcus* sp. strain PCC 7942. We made several fusions of the regulatory regions of known genes to *luxAB,* and all resulting reporter strains showed rhythmic bioluminescence, albeit at lower amplitudes and intensities than AMC149 (58). We then tried a global approach to determine what percentage of the genome is under circadian control by fusing a promoterless *luxAB* gene set randomly throughout the chromosome. This is readily performed in PCC 7942, because it can undergo homologous recombination between a segment on the chromosome and a cloned copy of the same region on a nonreplicating plasmid. The outcome is a single crossover event that incorporates the transforming plasmid into the chromosome at the site of recombination, creating a duplication of the cloned genomic segment from the plasmid (94). The result should be a fusion of *luxAB* to the segment of DNA cloned into the plasmid, which may or may not carry a promoter in the proper orientation for expression of the reporter (58).

We constructed a library of plasmids that contained random partial *Sau*3A fragments of the *Synechococcus* sp. strain PCC 7942 genome inserted upstream of *luxAB* in a pBR322-based vector. The plasmid carries a kanamycin resistance marker for selection of recombinants after integration into the genome; the vector contains the *bom* site, which allows mobilization from *E. coli* into the cyanobacterium via conjugation, a more efficient delivery system than transformation (96). As expected, only a subset of transformants showed conspicuous bioluminescence. Surprisingly, all bioluminescent strains tested

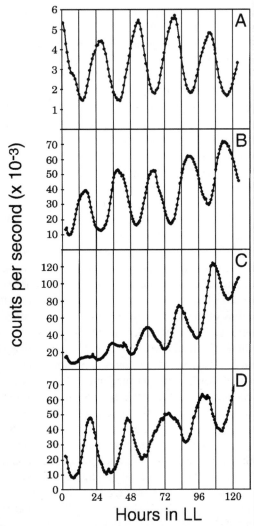

Figure 2 Rhythm traces from wild-type and mutant strains that are autonomously bioluminescent. Strains carry (*A*) *luxAB* fused to the *purF* promoter at the *purF* locus by recombination of a plasmid into the chromosome (58), and the *luxCDE* genes fused to the *psbAI* promoter in neutral site I; (*B-D*) $P_{psbAI}::luxAB$ in neutral site I, and $P_{psbAI}::luxCDE$ in neutral site II. (*B*) Wild-type circadian rhythm. (*C*) Short period (22 h) mutant. (*D*) long period (28 h) mutant. All strains were grown and measured on an agar surface; they were entrained to LD 12:12 before assay by a Packard TopCount in LL for approximately 5 days. Bioluminescence values were extracted from files with a program prepared by Martin Straume of the NSF Center for Biological Timing (University of Virginia) and plotted using Microsoft Excel.

showed a circadian rhythm of light production (58). The possibility of rhythmic substrate production (O_2 or $FMNH_2$) imposing rhythmic activity on constitutively expressed luciferase was discounted because not all clones expressed rhythms in the same phase or with the same waveform. If the rhythms reflected a rhythm of substrate(s), then we would expect them to share the same phase relationship.

Approximately 80% of the bioluminescent clones showed peaks near subjective dusk (i.e. the time during LL that corresponds to just before lights off of the entraining LD cycle) as does AMC149; we defined this phenotype as Class 1. Smaller groups of clones showed peak bioluminescence at other times, including some whose phase was almost opposite that of AMC149 (defined as Class 2). One such clone was AMC287, in which the *luxAB* gene set is integrated into the *purF* gene (Figure 2A), which encodes the key enzyme of de novo purine biosynthesis (57, 58). The encoded glutamine PRPP amidotransferase is known to be oxygen sensitive in other organisms. This result is reminiscent of nitrogenase in the diazotrophs discussed above. Thus, the teleological argument of temporal separation is supported, although its requirement has not been demonstrated. Fusion to *luxAB* of defined segments that contain the DNA upstream of *purF* demonstrated that *cis* information defining phase is present in or near the ORF. The *purF* gene immediately follows *purL* in *Synechococcus* sp. strain PCC 7942, and it is cotranscribed with *purL* as part of a large operon in other bacteria. However, the segment upstream of the *purL* gene from this organism drives bioluminescence in a Class 1 pattern. A second promoter that drives Class 2 expression of *purF* appears to lie within the C-terminal coding region of *purL* (57).

In more recent experiments we have used a Tn5-*luxAB* transposon (100) as an alternate means of creating random *luxAB* fusions throughout the *Synechococcus* sp. strain PCC 7942 genome (J Shelton, NF Tsinoremas & SS Golden, unpublished data). This approach confirmed widespread rhythmic gene expression and again revealed a minority of genes that are expressed in Class 2.

The pervasive rhythmic expression of bioluminescence among our library of random *luxAB* insertion clones suggests that there is global circadian control of gene expression in *Synechococcus* sp. strain PCC 7942. However, the fact that genes are expressed with various phase relationships indicates that there is some individuality in clock control for specific genes, as would be expected if circadian rhythmicity is to provide a means of temporal separation of biochemical processes. Our model encompasses both global and specific layers of circadian regulation (Figure 3). We propose that the clock imposes rhythmicity of transcription through a general mechanism that is not gene-specific, such as changes in supercoiling of DNA, energy charge, or RNA polymerase activity

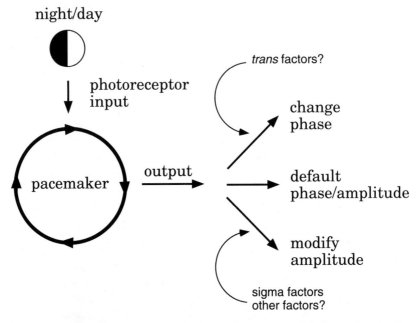

Figure 3 Model for circadian gene expression in *Synechococcus* sp. strain PCC 7942. The circadian pacemaker is set by cycles of light and darkness (and also can be entrained by temperature cycles). At least one output pathway globally affects transcription, even of genes that carry no specific *cis* information. However, additional pathways should exist to adjust the phase relationship or modify the amplitude of circadian expression from specific genes, such as Class 2 phase of *purF* (represented by proposed *trans* factors adjusting phase), and *rpoD2*-dependent high amplitude oscillation of *psbAI* (indicated as sigma factor modification of amplitude).

(58). In the absence of other layers of control, a gene would be transcribed rhythmically in this "default" phase. Support for this is provided by Class 1 rhythmic bioluminescence from strains that carry gene fusions to minimal promoters, including an *E. coli* consensus promoter and a *Synechococcus* promoter fragment that contains only the 39 bp upstream of the transcription start site (NV Lebedeva & SS Golden, unpublished data). However, some genes, including a subset of Class 1 genes and the genes in other classes, should have specific *cis* elements and *trans* acting factors that modify the default information by increasing amplitude or shifting the phase of expression.

Researchers working with other organisms have identified a small subset of genes regulated by the circadian clock (e.g. 59), whereas we found global control of gene expression. The difference may have less to do with the extent of circadian gene regulation between prokaryotes and eukaryotes than with the methods used to detect circadian cycling. Most circadian clock-controlled genes (ccgs) in eukaryotes were identified by subtraction of mRNA populations from different times of the circadian cycle. Thus, an mRNA would be identified as belonging to a ccg only if its accumulation varies considerably at different time points. We found that although steady-state mRNA levels of some genes we have identified do oscillate, the amplitude and the persistence are not as robust as the oscillations of bioluminescence, which we take to be a direct reflection of promoter activity. In particular, trough values for *psbAI* mRNA levels rise after the first full cycle, so amplitude is reduced. Likewise we see circadian oscillation of bioluminescence when the *luxAB* genes are driven by an rRNA promoter (58), but rRNA levels do not change notably throughout the circadian cycle (NV Lebedeva & SS Golden, unpublished data). We suggest that the difference between mRNA levels and bioluminescence occurs because transcription and message degradation are uncoupled in the reporter system. The heterologous *luxAB* reporter message is very unstable in *Synechococcus* sp. strain PCC 7942 (56) and is not likely to be subject to regulation in the cyanobacterium. If transcription of all genes in *Synechococcus* sp. strain PCC 7942 is controlled by the clock, but degradation of messages requires LD cycling, this would explain the weaker persistence of oscillation in LL at the RNA level. Although we have not tested the hypothesis directly, Herrin's lab reported this phenomenon for the *tufA* gene of *Chlamydomonas:* Transcription is robustly circadian, but periodicity of transcript degradation requires a dark period (37). Others may have missed widespread circadian transcription if the detection method relied on oscillations in transcript accumulation. Carré & Kay (7) recently reported weak circadian oscillation of bioluminescence from *Arabidopsis* plants in which the firefly luciferase gene, *luc,* is driven by the CaMV 35S promoter. This may be evidence of default circadian transcription in a eukaryote as well.

If transcription is globally rhythmic, but other layers of regulation (such as transcript or protein accumulation) damp this rhythm in most cases, then the ultimate outcome may be what was originally anticipated: a subset of biochemical processes actually oscillates in a circadian manner. Alternatively, the damping we see in mRNA oscillations may be considered an artefact of the requirement for maintaining the organism in constant conditions for clock studies; after all, the real environment imposes a daily cycle of light and darkness. It is likely that both are true: The clock acts more globally than previously supposed, but not every activity in the cell oscillates in a circadian fashion. Once the circadian components of cellular processes are identified in the isolation of constant conditions, it will be important to return the organism to a more natural environment to assess how they cycle in the real world.

THE CYANOBACTERIAL CIRCADIAN CLOCK

The circadian system can be described as having three necessary divisions: the clock itself (oscillator); a mechanism for setting the phase of the clock which should include one or more photoreceptors, and perhaps receptors for other environmental cues such as temperature (input pathways); and a means of relaying clock information to the various behaviors that are under circadian control (output pathways) (39). Each of these can be addressed efficiently in *Synechococcus* sp. strain PCC 7942 with the bioluminescence reporter system.

The Circadian Oscillator in Synechococcus

A prediction regarding the circadian oscillator is that it should control the rhythmicity and period of all downstream behaviors. Thus, mutants affected in circadian period are likely candidates for identification of the clock component genes. We mutagenized strain AMC149 with ethyl methanesulfonate and screened colonies for altered circadian phenotypes, identifying a wide range of mutants whose bioluminescence rhythms were altered (49; Figure 2*B–D*). Mutants exhibited periods ranging from 16 h to 60 h, and some had amplitudes that were low to the point of apparent arrhythmia. We first determined that the phenotypes did not result from mutations in the *psbAI* regulatory region that drives the reporter. This was accomplished by transforming with a neutral site I vector which carried a different selectable marker, but no *luxAB* fusion, to select for recombinants in which the reporter gene had been purged by recombination. Subsequent reintroduction of the P$_{psbAI}$:*luxAB* reporter confirmed that the altered clock phenotypes were the result of mutations in other loci.

We addressed the possibility that the mutations might affect only expression of the *psbAI* promoter (e.g. an output pathway mutation) rather than the clock itself by replacing the P$_{psbAI}$:*luxAB* fusion with a P$_{purF}$:*luxAB* fusion in two period mutant backgrounds (95). In each case the reporter strain had a bioluminescence rhythm that peaked in the expected phase for the promoter and showed the period of the mutant background. Although the mutation could be in an output pathway that affects both genes, the difference in phase suggests that at least some pathways leading from the clock to the two genes are different. Furthermore, an apparently arrhythmic mutant showed arrhythmic bioluminescence for all light-producing clones after introduction of the library that produces random insertions of *luxAB* throughout the genome (T Kondo &

M Ishiura, unpublished results). We concluded that at least some of the mutants identified in our search were likely to be affected in the central timekeeping system, and that the *psbAI* and *purF* genes receive information from the same biochemical clock (or coupled clocks).

Initial attempts to complement mutants were based on a single recombination strategy similar to that used for the random insertion of *luxAB* genes throughout the genome, and the attempts resulted in rescue of the phenotype in the 22-h period mutant SP22 (49). The segment of DNA that rescued this mutant did not restore the phenotype in other mutants, suggesting that at least one other locus is important for the clock. Subsequent experiments have indicated that this segment of DNA is a modifier of the circadian period rather than the locus that is mutant in SP22 (S Kutsuna, T Kondo & M Ishiura, unpublished data).

A new library based on double recombination to target DNA to neutral site II^2 contained plasmids that rescued the phenotypes of a number of circadian period mutants of AMC149. The maps of DNA inserts from plasmids that rescued four different long period mutants overlap (M Ishiura, S Aoki, S Kutsuna, CR Andersson, H Iwasaki, et al, unpublished data). One of these plasmids could rescue the phenotypes of the other three, and of additional mutants that showed long or short periods, or arrhythmic bioluminescence. Analysis of the segment of DNA that rescues these phenotypes is still in progress, but initial data predict three open reading frames that may form an operon. The sequences are not similar to the clock genes known from other systems—*period, timeless,* and *frequency.* Altered circadian phenotypes can arise from mutation in any of the three ORFs.

Output Pathways

One mutant that was isolated by a different strategy identified a gene that is part of a specific output pathway of the clock (95). We designed a library in which small segments of *Synechococcus* DNA will integrate the vector at the site of homology between the cloned segment and the chromosome. The size of inserts was chosen to favor integrations that frequently would cause gene truncations (12). Three mutant phenotypes were identified in the screen: short period (M2), arrhythmic (M11), and low amplitude oscillation of bioluminescence (M16). The first two were shown to have mutations unlinked to the plasmid insertion by a series of tests (95). Subsequent experiments showed that these mutants could be complemented by the same segment of DNA that

---

2

Database accession U44761.

complements the period mutants we obtained by chemical mutagenesis. There-fore, they probably also carry defects in the newly identified clock ORFs (CR Andersson, M Ishiura, T Kondo, SS Golden & CH Johnson, unpublished data).

The insertion event in M16 occurred within the *rpoD2* gene, which encodes a group 2 sigma 70-type transcription factor (95). In this case, the insertion clearly caused the mutant phenotype, because inactivation of the *rpoD2* locus in AMC149 reproduces the phenotype, and it can be complemented by a wild-type copy of *rpoD2* expressed in *trans* from neutral site II. We concluded that *rpoD2* could be a component of the oscillator, of an output pathway that affects clock control of all downstream genes, or of a specific output pathway that controls a subset of genes. In either of the first two scenarios, inactivation of *rpoD2* should have the same amplitude-reducing effect on all *Synechococ-cus* sp. strain PCC 7942 promoters. However, disrupting *rpoD2* in AMC287, a strain that carries the P_{purF}:*luxAB* reporter, had no effect on its biolumines-cence rhythm. This would not be surprising if the third scenario were correct, because AMC287 has a Class 2 phase rhythm, which is almost opposite to that of AMC149. Therefore, one might expect these strains to be connected to the clock by different output pathways. However, four Class 1 phase strains ob-tained by random insertion of *luxAB* into the chromosome were also tested, and three were unaffected by inactivation of *rpoD2*, whereas one showed the mutant phenotype. We concluded that the low amplitude defect is specific for a subset of promoters, and therefore *rpoD2* is part of an output pathway that affects circadian expression of a subset of genes in *Synechococcus* sp. strain PCC 7942.

Because overall bioluminescence from M16 is at least as high as for the parent strain AMC149, RpoD2 is not likely to be the primary sigma factor needed for expression of the P_{psbAI} promoter (95). In fact, the low amplitude phenotype appears to reflect an elevation of the trough rather than a decrease in the peak of bioluminescence. We proposed scenarios in which RpoD2 acts either directly or indirectly to affect transcription of *psbAI*. In the first, RpoD2 would be the sigma factor that is complexed with RNA polymerase during trough times of expression, and this form of holoenzyme would catalyze transcription at a lower rate than holoenzyme that uses a different form of sigma (such as RpoD1, the primary sigma 70 factor in *Synechococcus*). Clock control would occur at the level of exchange of sigma forms in the holoenzyme complex. Alternatively, RpoD2 may not interact with the *psbAI* promoter at all. It may be the sigma factor that recognizes the promoter of an undiscovered gene whose product affects (negatively) *psbAI* transcription. In this case, clock control could be exerted either at the level of interaction of RpoD2 with the regulatory gene's promoter, or through modulation of the activity of the gene's product.

Our model for control of gene expression suggests that some Class 1 genes, and all of the genes whose phases of expression are not in the default phase of the clock, should have specific regulators that modify circadian expression (Figure 3). RpoD2 appears to be a factor that increases circadian amplitude for some genes, although it is too early to conclude specificity for Class 1 genes. We anticipate elucidation of additional factors that are components of the pathways that wire other classes of genes to the clock.

Input Pathways

The analysis of input pathways that set the phase of the clock in *Synechococcus* sp. strain PCC 7942 has just begun. Clearly the alternation of light and dark is a strong stimulus for setting the rhythm, as shown by phase resetting in AMC149 (48). Weak circadian oscillations in random phases occur even in apparently unentrained cultures, although an LD cycle increases the amplitude and can set a specific phase (NV Lebedeva, SS Golden, & T Kondo, unpublished data). Schneegurt et al reported that circadian cycling of nitrogen fixation and carbohydrate granule accumulation in *Cyanothece* sp. strain ATCC 51142 is initiated by inoculation of a dilute culture from a stationary phase stock (82), and temperature cycles will entrain the clock of *Synechococcus* RF-1 (29) and AMC149 (T Kondo, J Shelton, SS Golden & CH Johnson, unpublished data). Therefore, we anticipate that photoreceptors and perhaps other environmental sensors (such as for temperature, pH, or osmotic environment) will be identified that provide phase-setting input to the cyanobacterial circadian clock.

Evidence for regulatory photoreceptors in *Synechococcus* sp. strain PCC 7942 comes from our analysis of the expression of the *psbA* gene family in a changing light environment (24). The *psbAI* gene, whose promoter we have used for circadian studies, is one of three *psbA* genes that encode the D1 protein of the PSII reaction center. The three *psbA* genes respond differently to changes in light intensity and quality. In response to an increase in intensity of white light, the *psbAII* and *psbAIII* genes are transcriptionally induced and the *psbAI* and *psbAIII* mRNAs are subject to accelerated degradation (52). We found that transfer of cells from white to low-fluence blue light causes them to behave as though they are experiencing an increase in white light: The level of the *psbAI* message decreases and levels of *psbAII* and *psbAIII* messages increase. Although red light alone has little effect on mRNA levels, a red pulse following blue exposure attenuates or reverses the blue-light response with respect to both transcriptional and posttranscriptional effects (97). Blue and red wavelengths are also the most effective at resetting the clock in AMC149, but red light does not appear to be able to reverse blue-light resetting of the clock (CH Johnson & T Kondo, unpublished data). We have not yet determined what overlap, if any, exists between these signal transduction pathways, but the potential for integration is intriguing.

Further evidence for cyanobacterial photoreceptors can be found in the genome of the cyanobacterium *Synechocystis* sp. strain PCC 6803, whose sequence revealed homologs of the plant photoreceptor phytochrome (43). A gene with sequence similarity to plant phytochrome and ethylene receptor genes, and to the phytochrome-like sequences in the *Synechocystis* sp. strain PCC 6803 genome, was recently shown to be involved in the chromatic adaptation of *Fremyella diplosiphon,* a cyanobacterium that changes the pigment composition of its light harvesting antenna to optimize absorption of the prevailing wavelengths in its environment (44). Whether homologs of known photoreceptors such as phytochrome provide input to the clock is unknown, but both reverse and forward genetic approaches are under way to identify the relevant sensors.

RELATIONSHIP TO THE CELL DIVISION CYCLE

Many cyanobacteria grow under optimal conditions with a doubling time much faster than the circadian period, which brings into question how the timing circuits for cell division and circadian rhythms are coordinated. A prevalent bias in circadian biology held that circadian rhythms could not persist under conditions in which cells double faster than once per day (15, 75). Many reports of circadian rhythms in cyanobacteria demonstrated persistence of the rhythm in cells that divided once per 24 h (64), more slowly than once a day (91), or in stationary or nongrowing conditions (33, 48, 65).

Can *Synechococcus* sp. strain PCC 7942 maintain a 24-h circadian cycle under conditions of rapid growth? Several lines of evidence indicated that the answer is yes. AMC149 was cultured on agar medium and in liquid culture under rapid growth conditions, and the bioluminescence was followed as the culture developed. In both cases, bioluminescence increased and oscillated as predicted based on the assumption of a circadian rhythm in every cell and exponential growth of the culture (47). We also observed 24-h rhythms of transcript abundance from the *psbAI* and *psbAII* genes in LL in continuous culture at light intensities that support doubling every 12 h and 5–6 h, respectively. Mori et al showed clear circadian rhythms of bioluminescence when AMC149 was diluted continuously with a doubling time of 10 h (68).

However, the experiments of Mori et al also showed that the clock and the cell division cycle are not entirely independent (68). Although the clock runs independently of cell division, the cell division cycle is "gated" by the clock. Continuously diluted cultures show circadian rhythms in the rate of cell division and in DNA content (genome equivalents), but certain phases of the circadian cycle appear to forbid cell division. Although the ability of the clock to continue ticking when cells double several times per day is at odds with what was predicted from eukaryotic organisms, circadian control over the timing of the cell division cycle is in keeping with observations from eukaryotes (14).

Some reports based on data from other species of cyanobacteria are not in complete agreement with our findings on the independence of clock-controlled functions from the cell division cycle. Mitsui et al had concluded in 1986 that temporal separation of photosynthetic O_2 production and nitrogenase activity acts by the restriction of each to a specific phase of the cell division cycle, using conditions in which the cells divided approximately once per day (64). In 1993 Mitsui and colleagues extended this study, showing that changes in growth temperature that altered the cell doubling time had the same effect on the oscillations of linked "cell cycle events" of photosynthesis and nitrogen fixation (66). This timing was not 24 h at some temperatures, and they interpreted this as confirmation that the cell division cycle times the oscillations of the other activities. However, a recent report by Mitsui & Suda (65) shows that nitrogenase-dependent H_2 production and photosynthetic O_2 evolution occur cyclically with a daily period in nongrowing cultures as well. Therefore, the cell division cycle alone cannot explain the oscillations. They concluded that both the cell cycle and a second endogenous rhythm mechanism control these processes, with a means of switching between the two control circuits. Alternatively, their observations may all reflect circadian control, with the cell cycle also governed by the clock. If so, the 1993 study shows a breakdown of the temperature compensation property that is usually associated with circadian phenomena; it is possible that the temperatures used may have been outside the compensation range for this organism.

ADAPTIVE SIGNIFICANCE OF THE CLOCK

The presence of the circadian mechanism in extant cyanobacteria implies that it fulfills a biological function and provides a selective advantage to the organism. Arguments for temporal control can be made easily, such as the separation of photosynthesis and nitrogen fixation in the organisms that led to the discovery of the cyanobacterial clock. However, the real utility of the clock may be subtle. Returning to the diazotrophy-photosynthesis paradox, two reports have shown that under some conditions nitrogen fixation and photosynthesis oscillate with a circadian period, but in the same phase, which indicates that temporal separation may not be absolutely necessary for nitrogenase function (72, 78). This underscores the fact that diazotrophs use many mechanisms to protect nitrogenase from oxygen, including clever respiratory schemes to consume oxygen, and temporal or spatial separation probably enhances these other strategies rather than substitutes for them (20). For example, *Synechococcus* RF-1 grows well in LL, under which conditions there is some photosynthetic oxygen evolution at all times. Cultures can even be bubbled with oxygen gas up to 40% (O_2/N_2 gas mixture) and still possess detectable nitrogenase activity (25).

Several arrhythmic mutants of AMC149 were identified in the screen for clock mutants, and they grow quite similarly to wild type in the lab (49). However, natural selection can magnify even small fitness advantages, and some conditions in the wild may be more affected by the presence or absence of a functioning circadian system. The fact that three cyanobacterial laboratory strains of different genera—maintained for decades with no special care to maintain a working clock—showed circadian oscillation of a bioluminescent reporter gene at least implies some level of selective advantage. Other properties, such as differentiation of heterocysts by some species and transformability, are often lost in laboratory culture when their specific functions are not frequently selected.

A potential advantage of circadian regulation of biological processes is the ability to anticipate dawn. This may be of particular utility to oxygenic photosynthetic organisms such as cyanobacteria, whose photosynthetic apparatus is particularly susceptible to damage if light catches it unprepared (23). Whether a real difference in biological selection accounts for the fact that cyanobacteria are the only prokaryotes in which a clock has yet been found, or whether yet undiscovered circadian prokaryotes are legion, remains to be determined (38).

HOW MANY CLOCKS HAVE EVOLVED?

The discovery of a circadian clock in cyanobacteria greatly increases the evolutionary distance over which such a timekeeping system is found. A key question that remains unanswered is whether a clock with similar properties has arisen several times through evolution, or a single clock has persisted and been modified, tailored to the cellular structure and metabolism of individual organisms. The sparse sequence data available for a few clock genes do not reveal homology, but we may yet be like the blind men and the elephant (81), examining trunks and legs and tails and declaring them to be snakes, trees, and ropes rather than parts of a pachyderm. New data on the *timeless* gene of *Drosophila* provide strong evidence that its product interacts with that of the *period* gene, and that migration between the nucleus and the cytoplasm is an important "delay feature" of the timekeeping mechanism (36, 53, 70, 103). One might conclude from this that the cyanobacteria, as prokaryotes, will by necessity have a different clock mechanism, because they lack a nuclear envelope. The concept of prokaryotes as zip-lock bags of enzymes is naive; compartmentalization and internal organelles exist in prokaryotes, although they are less obvious or well-studied than in eukaryotes. The coupling of transcription and translation in bacteria is, however, a fundamental difference that requires a deviation from the *period-timeless* model to provide time delay.

The evolutionary relationship between cyanobacteria and chloroplasts poses the interesting possibility that the circadian clock originated within prokaryotes and moved to plants via endosymbiosis. If so, the circadian oscillator in plants may be more readily recognizable as a homologue of the cyanobacterial clock than are the clock parts of *Drosophila* and *Neurospora*. A locus in *Arabidopsis* that may encode an oscillator component, *toc1* (61), has been mapped, and chromosome walking has narrowed the region of DNA that contains the gene (CA Strayer & SA Kay, unpublished results); many more *toc* (*timing of cab* expression) mutants have been identified and await analysis (61). Thus, direct comparison of clock components from these two groups may be possible in the near future.

CONCLUDING REMARKS

This is a very exciting time for the field of circadian biology, as the components of the biochemical oscillator are being unveiled simultaneously in several organisms through advances in genetic and molecular strategies. Although cyanobacteria have distinct technical advantages and the allure of evolutionary distance, clock structure in several eukaryotic organisms may be elucidated as soon. Bioluminescent transgenic *Arabidopsis* is nearing its entry into the clock gene fray, and *Drosophila* has unveiled a second clock part (TIM) as a partner of PER. As more components are defined, the prospect of a model that adequately explains this remarkable biological timing device becomes progressively more plausible. The evolutionary questions will yield up their answers simultaneously with the revelation of the mechanism(s), as groups compare their timepieces.

The quest for the oscillator has taken center stage because of the novelty of this biological machine, but the related mysteries of input and output mechanisms are equally as important. Input pathways are the means by which the clock is set to correspond to its environment; without this information, the clock would be out of synch with the world it measures. Output pathways are the means by which the clock does work. Without connections leading from the clock to cellular processes, the oscillator would be like an alarm clock whose bell is broken, unable to ring and thereby rouse slumbering functions at the right times. These issues, too, become more tractable with every technical and theoretical advance, and should unfold in the foreseeable future.

ACKNOWLEDGMENTS

We thank the members of our laboratories who allowed us to refer to their results before publication, and specifically CA Andersson, J Shelton, and N Tsinoremas for suggestions regarding the manuscript, and preparation of data for Figure 2. We also thank CP Wolk (Michigan State University) for advice and plasmids for constructing autonomously bioluminescent strains, and SA Kay, M Straume, J Plautz, and CA Strayer (NSF Center for Biological Timing, University of Virginia) for assistance in adopting the Packard TopCount for measurement and analysis of cyanobacterial rhythms. Our research has been supported by grants from the NIMH (MH43836 and MH01179 to CHJ), NIH (GM37040 to SSG), and NSF (MCB-9219880 and MCB 9633267 to CHJ; MCB-9311352 and MCB-9513367 to SSG); and from the Ministry of Education, Science, and Culture (07670097, 07554045, 07253228, 07558103, 08404053, and 08454244 to TK and MI). NSF and JSPS supported a joint USA-Japan Co-operative Program grant for travel (NSF INT9218744 and JSPS BSAR382). The Human Frontier Science Program funds collaborative research of the four coauthors.

> **Visit the *Annual Reviews* home page at http://www.annurev.org.**

Literature Cited

1. Aoki S, Kondo T, Ishiura M. 1995. Circadian expression of the *dnaK* gene in the cyanobacterium *Synechocystis* sp. strain PCC 6803. *J. Bacteriol.* 177:5606–11
2. Aschoff J. 1989. Temporal orientation: circadian clocks in animals and humans. *Anim. Behav.* 37:881–96
3. Bargiello TA, Jackson FR, Young MW. 1984. Restoration of circadian behavioural rhythms by gene transfer in *Drosophila*. *Nature* 312:752–54
4. Brandes C, Plautz JD, Stanewsky R, Jami-

son CF, Straume M, et al. 1996. Novel features of *Drosophila* transcription revealed by real-time luciferase reporting. *Neuron* 16:687–92
4a. Bryant DA, ed. 1994. *The Molecular Biology of Cyanobacteria*. Dordrecht: Kluwer
5. Bünning E. 1936. Die endogene Tagesrhythmik als Grundlage der photoperiodischen Reaktion. *Ber. Dtsch. Bot. Ges.* 54: 590–607
6. Bünning E. 1973. *The Physiological Clock*. New York: Springer-Verlag. 3rd ed.

7. Carré IA, Kay SA. 1995. Multiple DNA-protein complexes at a circadian-regulated promoter element. *Plant Cell* 7:2039–51
8. Chen T-H, Chen T-L, Hung L-M, Huang T-C. 1991. Circadian rhythm in amino acid uptake by *Synechococcus* RF-1. *Plant Physiol.* 97:55–59
9. Czeisler CA, Johnson MP, Duffy JF, Brown EN, Ronda JM, Kronquer RE. 1990. Exposure to bright light and darkness to treat physiologic maladaptation to night work. *N. Engl. J. Med.* 322:1253–59
10. deCandolle AP. 1832. *Physiologie vegetale, ou, Exposition des forces et des fonctions vitales des vegetaux: pour servir de suite a l'organographie vegetale, et d'introduction a la botanique geographique et agricole.* Paris: Bechet
11. deMairan JJ. 1729. *Observation Botanique.* Paris: Hist. Acad. R. Sci.
12. Dolganov N, Grossman AR. 1993. Insertional inactivation of genes to isolate mutants of *Synechococcus* sp. strain PCC 7942: isolation of filamentous strains. *J. Bacteriol.* 175:7644–51
13. Dunlap JC. 1993. Genetic analysis of circadian clocks. *Annu. Rev. Physiol.* 55:683–728
14. Edmunds LN. 1988. *Cellular and Molecular Bases of Biological Clocks.* New York: Springer-Verlag
15. Ehret CF, Wille JJ. 1970. The photobiology of circadian rhythms in protozoa and other eukaryotic microorganisms. In *Photobiology of Microorganisms,* ed. P Halldal, pp. 369–416. New York: Wiley
16. Elhai F, Wolk CP. 1990. Developmental regulation and spatial pattern of expression of the structural genes for nitrogenase in the cyanobacterium *Anabaena. EMBO J.* 9: 3379–88
17. Feldman JF, Hoyle MN. 1973. Isolation of circadian clock mutants of *Neurospora crassa. Genetics* 75:605–13
18. Fernández-Piñas F, Wolk CP. 1994. Expression of *luxCD-E* in *Anabaena* sp. can replace the use of exogenous aldehyde for in vivo localization of transcription by *luxAB. Gene* 150:169–74
19. Gallon JR. 1981. The oxygen sensitivity of nitrogenase: a problem for biochemists and micro-organisms. *Trends Biochem. Sci.* 6: 19–23
20. Gallon JR. 1992. Tansley Review No. 44: reconciling the incompatible: N_2 fixation and O_2. *New Phytol.* 122:571–609
21. Gallon JR, LaRue TA, Kurz WGW. 1974. Photosynthesis and nitrogenase activity in the blue-green alga *Gloeocapsa. Can. J. Microbiol.* 20:1633–37

22. Gekakis N, Saez L, Delahaye-Brown A-M, Myers MP, Sehgal A, et al. 1995. Isolation of *timeless* by PER protein interaction: defective interaction between *timeless* protein and long-period mutant PER^L. *Science* 270: 811–15
23. Golden SS. 1994. Light-responsive gene expression and the biochemistry of the photosystem II reaction center. See Ref. 4a, pp. 693–714
24. Golden SS. 1995. Light-responsive gene expression in cyanobacteria. *J. Bacteriol.* 177:1651–54
25. Grobbelaar N, Lin H-Y, Huang T-C. 1987. Induction of a nitrogenase activity rhythm in *Synechococcus* and the protection of its nitrogenase against photosynthetic oxygen. *Curr. Microbiol.* 15:29–33
26. Grobbelaar N, Huang T-C, Lin H-Y, Chow T-J. 1986. Dinitrogen-fixing endogenous rhythm in *Synechococcus* RF-1. *FEMS Microbiol. Lett.* 37:173–77
27. Halberg F, Conner RL. 1961. Circadian organization and microbiology: variance spectra and a periodogram on behaviour of *Escherichia coli* growing in fluid culture. *Proc. Minn. Acad. Sci.* 29:227–39
28. Hamner KC, Takimoto A. 1964. Circadian rhythms and plant photoperiodism. *Am. Nat.* 98:295–322
29. Huang T-C, Chen H-M, Pen S-Y, Chen T-H. 1994. Biological clock in the prokaryote *Synechococcus* RF-1. *Planta* 193:131–36
30. Huang T-C, Chou W-M. 1991. Setting of the circadian N_2-fixing rhythm of the prokaryote *Synechococcus* sp. RF-1 while its *nif* gene is repressed. *Plant Physiol.* 96: 324–26
31. Huang T-C, Chow T-J. 1986. New type of N2-fixing unicellular cyanobacterium (blue-green alga). *FEMS Microbiol. Lett.* 36:109–10
32. Huang T-C, Grobbelaar N. 1995. The circadian clock in the prokaryote *Synechococcus* RF-1. *Microbiology* 141:535–40
33. Huang T-C, Pen S-Y. 1994. Induction of a circadian rhythm in *Synechococcus* RF-1 while the cells are in a "suspended state". *Planta* 194:436–38
34. Huang T-C, Tu J, Chow T-J, Chen T-H. 1990. Circadian rhythm of the prokaryote *Synechococcus* sp. RF-1. *Plant Physiol.* 92: 531–33
35. Huang T-C, Wang S-T, Grobbelaar N. 1993. Circadian rhythm mutants of the prokaryotic *Synechococcus* RF-1. *Cur. Microbiol.* 27:249–54
36. Hunter-Ensor M, Ousley A, Sehgal A. 1996. Regulation of the *Drosophila* protein Timeless suggests a mechanism for reset-

ting the circadian clock by light. *Cell* 84: 677–85

37. Hwang S, Kawazoe R, Herrin DL. 1996. Transcription of *tufA* and other chloroplast-encoded genes is controlled by a circadian clock in *Chlamydomonas*. *Proc. Natl. Acad. Sci. USA* 93:996–1000

38. Johnson CH, Golden SS, Ishiura M, Kondo T. 1996. Circadian clocks in prokaryotes. *Mol. Microbiol.* 21:5–11

39. Johnson CH, Hastings JW. 1986. The elusive mechanism of the circadian clock. *Am. Sci.* 74:29–36

40. Johnson CH, Kondo T, Goto K. 1992. Circadian rhythms in *Chlamydomonas*. In *Circadian Clocks from Cell to Human: Proceedings of the 4th Sapporo Symposium on Biological Rhythms,* ed. pp. 139–55. Hokkaido: Hokkaido Univ. Press

41. Kallas T, Rippka R, Coursin T, Rebiere M-C, Tandeau de Marsac N, Cohen-Bazire G. 1983. Aerobic nitrogen fixation by nonheterocystous cyanobacteria. In *Photosynthetic Prokaryotes: Cell Differentiation and Function,* ed. GC Papageorgiou, L Packer, pp. 281–302. New York: Elsevier Biomed.

42. Kaneko T, Matsubayashi T, Sugita M, Sugiura M. 1996. Physical and gene maps of the unicellular cyanobacterium *Synechococcus* sp. strain PCC6301 genome. *Plant Mol. Biol.* 31:193–201

43. Kaneko T, Sato S, Kotani H, Tanaka A, Asamizu E, et al. 1996. Sequence analysis of the genome of the unicellular cyanobacterium *Synechocystis* sp. strain PCC6803. II. Sequence determination of the entire genome and assignment of potential protein-coding regions. *DNA Res.* 3: 109–36

44. Kehoe DM, Grossman AR. 1996. Similarity of a chromatic adaptation sensor to phytochrome and ethylene receptors. *Science.* 273:1409–12

45. Kippert F. 1991. Essential clock proteins/circadian rhythms in prokaryotes: What is the evidence? *Bot. Acta* 104:2–4

46. Kondo T, Ishiura M. 1994. Circadian rhythms of cyanobacteria: monitoring the biological clocks of individual colonies by bioluminescence. *J. Bacteriol.* 176:1881–85

47. Kondo T, Mori T, Lebedeva NV, Aoki S, Ishiura M, Golden SS. 1996. Circadian rhythms in rapidly dividing cyanobacteria. *Science.* 275:224–27

48. Kondo T, Strayer CA, Kulkarni RD, Taylor W, Ishiura M, et al. 1993. Circadian rhythms in prokaryotes: luciferase as a reporter of circadian gene expression in cy-anobacteria. *Proc. Natl. Acad. Sci. USA* 90:5672–76

49. Kondo T, Tsinoremas NF, Golden SS, Johnson CH, Kutsuna S, Ishiura M. 1994. Circadian clock mutants of cyanobacteria. *Science* 266:1233–36

50. Konopka RJ, Benzer S. 1971. Clock mutants of *Drosophila melanogaster. Proc. Natl. Acad. Sci. USA* 68:2112–16

51. Krasnow R, Dunlap JC, Taylor W, Hastings JW, Vetterling W, Gooch V. 1980. Circadian spontaneous bioluminescent glow and flashing of *Gonyaulax polyedra. J. Comp. Physiol.* 138:19–26

52. Kulkarni RD, Schaefer MR, Golden SS. 1992. Transcriptional and posttranscriptional components of *psbA* response to high light intensity in *Synechococcus* sp. strain PCC 7942. *J. Bacteriol.* 174:3775–81

53. Lee C, Parikh V, Itsukaichi T, Bae K, Edery I. 1996. Resetting the *Drosophila* clock by photic regulation of PER and a PER-TIM complex. *Science* 271:1740–44

54. Lewy AJ, Sack RL, Miller LS, Hoban TM. 1987. Antidepressant and circadian phase-shifting effects of light. *Science* 235: 352–54

55. Li R, Golden SS. 1993. Enhancer activity of light-responsive regulatory elements in the untranslated leader regions of cyano-bacterial *psbA* genes. *Proc. Natl. Acad. Sci. USA* 90:11678–82

56. Liu Y, Golden SS, Kondo T, Ishiura M, Johnson CH. 1995. Bacterial luciferase as a reporter of circadian gene expression in cyanobacteria. *J. Bacteriol.* 177:2080–86

57. Liu Y, Tsinoremas NF, Golden SS, Kondo T, Johnson CH. 1996. Circadian expression of genes involved in the purine biosynthetic pathway of the cyanobacterium *Synechococcus* sp. strain PCC 7942. *Mol. Microbiol.* 20:1071–81

58. Liu Y, Tsinoremas NF, Johnson CH, Lebedeva NV, Golden SS, et al. 1995. Circadian orchestration of gene expression in cyanobacteria. *Genes Dev.* 9:1469–78

59. Loros JJ, Denome SA, Dunlap JC. 1989. Molecular cloning of genes under control of the circadian clock in *Neurospora crassa. Science* 243:385–88

60. McClung CR, Fox BA, Dunlap JC, Jackson FR, Young MW. 1989. The *Neurospora* clock gene *frequency* shares a sequence element with the *Drosophila* clock gene *period. Nature* 339:558–62

61. Millar AJ, Carré IA, Strayer CA, Chua N-H, Kay SA. 1995. Circadian clock mutants in *Arabidopsis* identified by Luciferase imaging. *Science* 267:1161–66

62. Millar AJ, Short SR, Chua N-H, Kay SA.

1992. A novel circadian phenotype based on firefly luciferase expression in transgenic plants. *Plant Cell* 4:1075–87

63. Millineaux PM, Gallon JR, Chaplin AE. 1981. Acetylene reduction (nitrogen fixation) by cyanobacteria grown under alternating light-dark cycles. *FEMS Microbiol. Lett.* 10:245–47

64. Mitsui A, Kumazawa S, Takahashi A, Ikemoto H, Cao S, Arai T. 1986. Strategy by which nitrogen-fixing unicellular cyanobacteria grow photoautotrophically. *Nature* 323:720–22

65. Mitsui A, Suda S. 1995. Alternative and cyclic appearance of H_2 and O_2 photoproduction activities under nongrowing conditions in an aerobic nitrogen-fixing unicellular cyanobacterium *Synechococcus* sp. *Curr. Microbiol.* 30:1–6

66. Mitsui A, Suda S, Hanagata N. 1993. Cell cycle events at different temperatures in aerobic nitrogen-fixing marine unicellular cyanobacterium *Synechococcus* sp. Miami BG 043511. *J. Marine Biotechnol.* 1:89–91

67. Moore-Ede M, Czeisler C, Richardson GS. 1983. Circadian timekeeping in health and disease. *N. Engl. J. Med.* 309:469–76, 530–36

68. Mori T, Binder B, Johnson CH. 1996. Circadian gating of cell division in cyanobacteria growing with average doubling times of less than 24 hours. *Proc. Natl. Acad. Sci. USA.* 93:10183–88

69. Morse DS, Fritz L, Hastings JW. 1990. What is the clock? Translational regulation of circadian bioluminescence. *Trends Biochem. Sci.* 15:262–65

70. Myers MP, Wager-Smith K, Rothenfluh-Hilfiker A, Young MW. 1996. Light-induced degradation of TIMELESS and entrainment of the *Drosophila* circadian clock. *Science* 271:1736–40

71. Myers MP, Wager-Smith K, Wesley CS, Young MW, Sehgal A. 1995. Positional cloning and sequence analysis of the *Drosophila* clock gene, *timeless. Science* 270:805–8

72. Ortega-Calvo J-J, Stal LJ. 1991. Diazotrophic growth of the unicellular cyanobacterium *Gloeothece* sp. PCC 6909 in continuous culture. *J. Gen. Microbiol.* 137:1789–97

73. Pittendrigh CS. 1954. On temperature independence in the clock system controlling emergence time in *Drosophila. Proc. Natl. Acad. Sci. USA* 40:1018–29

74. Pittendrigh CS. 1981. Circadian Systems: General Perspective and Entrainment. In *Handbook of Behavioral Neurobiology:*

Biological Rhythms, ed. J Aschoff, pp. 57–80, 95–124. New York: Plenum

75. Pittendrigh CS. 1993. Temporal organization: reflections of a Darwinian clock-watcher. *Annu. Rev. Physiol.* 55:17–54

76. Ralph MR, Menaker M. 1988. A mutation of the circadian system in golden hamsters. *Science* 241:1225–27

77. Reddy P, Zehring WA, Wheeler DA, Pirrotta V, Hadfield C, et al. 1984. Molecular analysis of the period locus in *Drosophila melanogaster* and identification of a transcript involved in biological rhythms. *Cell* 44:21–32

78. Roenneberg T, Carpenter EJ. 1993. Daily rhythm of O_2-evolution in the cyanobacterium *Trichodesmium thiebautii* under natural and constant conditions. *Marine Biol.* 117:693–97

79. Rosbash M, Hall JC. 1989. The molecular biology of circadian rhythms. *Neuron* 3:387–98

80. Sargent ML, Kaltenborn SH. 1972. Effects of medium composition and carbon dioxide on circadian conidiation in *Neurospora. Plant Physiol.* 50:171–75

81. Saxe JG. 1952. The blind men and the elephant. In *The Family Book of Best Loved Poems,* ed. George DL, pp. 400–1. New York: Hanover House

82. Schneegurt MA, Sherman DM, Nayar S, Sherman LA. 1994. Oscillating behavior of carbohydrate granule formation and dinitrogen fixation in the cyanobacterium *Cyanothece* sp. strain ATCC 51142. *J. Bacteriol.* 176:1586–97

83. Schweiger H-G, Schweiger M. 1977. Circadian rhythms in unicellular organisms: an endeavor to explain the molecular mechanism. *Int. Rev. Cytol.* 51:315–42

84. Sehgal A, Price JL, Man B, Young MW. 1994. Loss of circadian behavioral rhythms and *per* RNA oscillations in the *Drosophila* mutant *timeless. Science* 263:1603–6

85. Stal LJ, Krumbein WE. 1985. Nitrogenase activity in the nonheterocystous cyanobacterium *Oscillatoria* sp. grown under alternating light-dark cycles. *Arch. Microbiol.* 143:67–71

86. Stal LJ, Krumbein WE. 1987. Temporal separation of nitrogen fixation and photosynthesis in the filamentous, nonheterocystous cyanobacterium *Oscillatoria* sp. *Arch. Microbiol.* 149:76–80

87. Strogatz SH, Kronauer RE, Czeisler CA. 1987. Circadian pacemaker interferes with sleep onset at specific times of each day: role in insomnia. *Am. J. Physiol.* 253:R172–78

88. Sturtevant RP. 1973. Circadian variability

in *Klebsiella* demonstrated by cosinor analysis. *Int. J. Chronobiol.* 1:141–46

89. Sweeney BM. 1974. A physiological model for circadian rhythms derived from the *Acetabularia* rhythm paradoxes. *Int. J. Chronobiol.* 2:95–110

90. Sweeney BM. 1987. *Rhythmic Phenomena in Plants.* San Diego: Academic. 2nd ed.

91. Sweeney BM, Borgese MB. 1989. A circadian rhythm in cell division in a prokaryote, the cyanobacterium *Synechococcus* WH7803. *J. Phycol.* 25:183–86

92. Takahashi JS, Pinto LH, Vitaterna MH. 1994. Forward and reverse genetic approaches to behavior in the mouse. *Science* 264:1724–33

93. Taylor WR. 1979. *Studies on the bioluminiscent glow rhythm of Gonyaulax polyedra.* PhD thesis. Univ. Mich., Ann Arbor

94. Thiel T. 1994. Genetic analysis of cyanobacteria. See Ref. 4a, pp. 581–611

95. Tsinoremas NF, Ishiura M, Kondo T, Andersson CR, Tanaka K, et al. 1996. A sigma factor that modifies the circadian expression of a subset of genes in cyanobacteria. *EMBO J.* 15:2488–95

96. Tsinoremas NF, Kutach AK, Strayer CA, Golden SS. 1994. Efficient gene transfer in *Synechococcus* sp. strain PCC 7942 and PCC 6301 by interspecies conjugation and

chromosomal recombination. *J. Bacteriol.* 176:6764–68

97. Tsinoremas NF, Schaefer MR, Golden SS. 1994. Blue and red light reversibly control *psbA* expression in the cyanobacterium *Synechococcus* sp. strain PCC 7942. *J. Biol. Chem.* 269:16143–47

98. Vitaterna MH, King DP, Chang A-M, Kornhauser JM, Lowrey PL, et al. 1994. Mutagenesis and mapping of a mouse gene, *Clock,* essential for circadian behavior. *Science* 264:719–25

99. Weare NM, Benemann JR. 1974. Nitrogenase activity and photosynthesis in *Plectonema boryanum. J. Bacteriol.* 119:258–65

100. Wolk CP, Cai Y, Panoff J-M. 1991. Use of a transposon with luciferase as a reporter to identify environmentally responsive genes in a cyanobacterium. *Proc. Natl. Acad. Sci. USA* 88:5355–59

101. Wolk CP, Elhai J, Kuritz T, Holland D. 1993. Amplified expression of a transcriptional pattern formed during development of *Anabaena. Mol. Microbiol.* 7:441–45

102. Wyatt JT, Silvey JKG. 1969. Nitrogen fixation of *Gloeocapsa. Science* 165:908–9

103. Zeng H, Qian Z, Myers MP, Rosbash M. 1996. A light-entrainment mechanism for the *Drosophila* circadian clock. *Nature* 380:129–35

Annu. Rev. Plant Physiol. Plant Mol. Biol. 1997. 48:355–81

BIOSYNTHESIS AND ACTION OF JASMONATES IN PLANTS

Robert A. Creelman and John E. Mullet

Department of Biochemistry and Biophysics, Crop Biotechnology Center, Texas A&M University, College Station, Texas 77843

KEY WORDS: jasmonic acid, chemistry, gene expression, insect and disease resistance

ABSTRACT

Jasmonic acid and its derivatives can modulate aspects of fruit ripening, production of viable pollen, root growth, tendril coiling, and plant resistance to insects and pathogens. Jasmonate activates genes involved in pathogen and insect resistance, and genes encoding vegetative storage proteins, but represses genes encoding proteins involved in photosynthesis. Jasmonic acid is derived from linolenic acid, and most of the enzymes in the biosynthetic pathway have been extensively characterized. Modulation of lipoxygenase and allene oxide synthase gene expression in transgenic plants raises new questions about the compartmentation of the biosynthetic pathway and its regulation. The activation of jasmonic acid biosynthesis by cell wall elicitors, the peptide systemin, and other compounds will be related to the function of jasmonates in plants. Jasmonate modulates gene expression at the level of translation, RNA processing, and transcription. Promoter elements that mediate responses to jasmonate have been isolated. This review covers recent advances in our understanding of how jasmonate biosynthesis is regulated and relates this information to knowledge of jasmonate modulated gene expression.

CONTENTS

1040-2519/97/0601-0355$08.00

INTRODUCTION

Jasmonic acid (JA), and its methyl ester (methyl jasmonate, MeJA) are linolenic acid (LA)-derived cyclopentanone-based compounds of wide distribution in the plant kingdom. MeJA was first identified as a component of the
essential oil of several plant species, while JA was first obtained from a fungal
culture filtrate. Early studies showed that exogenous JA or MeJA can promote
senescence and act as a growth regulator. Subsequent research revealed that JA
specifically alters gene expression and that wounding and elicitors could cause
JA/MeJA accumulation in plants. These results implied a role for jasmonate in
plant defense that has recently been confirmed. Other research described roles
for jasmonates in vegetative development, fruit development, and pollen viability. The dual role of JA in plant development and defense is examined in
this review. This review emphasizes new information in this field since the last
review in this series (103). Other excellent reviews are available that cover
JA/MeJA with respect to herbivory (7, 13), signaling (40, 106), chemistry and
biochemistry (51, 56), gene expression (89), and chromatography (117). For
the purposes of this review, (3R, 7RS)-JA/MeJA will be referred to collectively as jasmonates unless it is necessary to identify specific isomers.

CHEMISTRY AND QUANTITATION

The jasmonate molecule (Figure 1) contains two chiral centers located at C3
and C7 generating four possible stereoisomers, since either chiral center can
have an R or S absolute configuration. The mirror image isomers, (3R, 7S) and
(3S, 7R), have their side chains in a *cis* orientation. These isomers are known
as (+)– and (−)-7-iso-JA or (+)– and (−)-epi-JA. The enantiomers, (3R, 7R)-
and (3S, 7S)-JA or (−)-JA and (+)-JA, have their side chains in the *trans*
configuration. Because of increased steric hindrance, the *cis* orientation is less
stable and will epimerize to the more stable *trans* configuration. This occurs
via a keto-enol tautomerization involving the C6 ketone and the C7 proton to

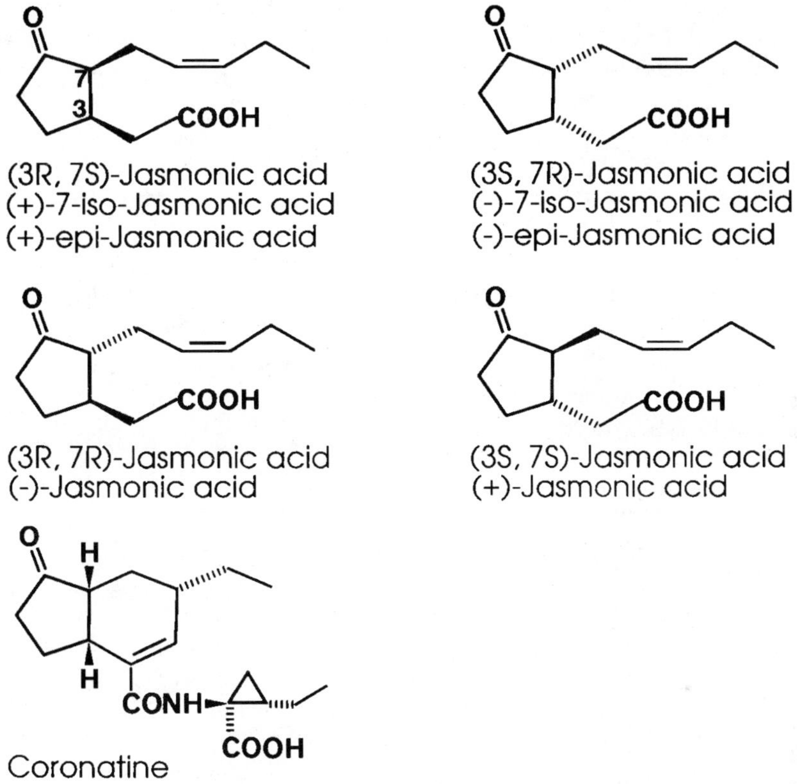

(3R, 7S)-Jasmonic acid
(+)-7-iso-Jasmonic acid
(+)-epi-Jasmonic acid

(3S, 7R)-Jasmonic acid
(-)-7-iso-Jasmonic acid
(-)-epi-Jasmonic acid

(3R, 7R)-Jasmonic acid
(-)-Jasmonic acid

(3S, 7S)-Jasmonic acid
(+)-Jasmonic acid

Coronatine

Figure 1   Chemical structures of various isomers of jasmonic acid and coronatine. The 3S, 7S isomer is the final product of the jasmonic acid biosynthetic pathway and is easily converted to its diastereomer 3R, 7S isomer during extraction. The corresponding enantiomers are found in synthetic material. Coronatine is a bacterial phytotoxin with biological activities similar to jasmonic acid.

form the corresponding diastereomers. During extraction or in the presence of acids or bases, (+)-7-iso-JA, thought to be the initial jasmonate formed in plants, is believed to epimerize to an equilibrium mixture of approximately 9:1 (−)-JA:(+)-7-iso-JA (80). The actual equilibrium concentration *in planta* is unknown. Consequently, analysis of jasmonates isolated from plants should indicate which isomers are being analyzed. Commercially available synthetic MeJA used in many experiments is composed of a 9:1 ratio (±)-MeJA:(±)-7-iso-MeJA. The methyl esters may be converted to the free acids with either basic hydrolysis or incubation with commercially available esterases.

A monoclonal antibody for the analysis of (−)-JA (analyzed as the methyl ester) has been described (1). Assuming that (−)-MeJA has a cross reaction of 100%, (+)-7-iso-MeJA had a cross reactivity of 86%. Albrecht et al (1) suggested that, when using this monoclonal antibody, it is advisable to allow endogenous jasmonates to reach their stable equilibrium before quantitation. Furthermore, as is the case with any antibody-based technique, procedures must be developed to estimate losses during extraction. In addition, antibody methods must be validated using a physical method such as GC-MS to check for possible errors due to interfering compounds present in extracts.

Synthetic JA or jasmonate analogs containing deuterium or ^{13}C have been used to quantify endogenous jasmonates by GC-MS selected ion monitoring. The jasmonate diastereomers can be resolved by gas chromatography using capillary columns enabling their separate quantitation. Furthermore, mass spectra of endogenous compounds can be to used to unequivocally identify jasmonates based on comparison with published spectra. Mueller and Brodschelm (80) derivatized jasmonates to the pentafluorobenzyl esters for quantitation by GC-MS-NICI and reported a limit of detection of ~500 fg. The presence of the fluorine atoms accounts for the increased sensitivity of this method. Creelman et al (29) used (^{13}C,^{2}H$_3$)-MeJA, whereas Gundlach et al (53) used 9,10 dihydrojasmonic acid to estimate JA levels. However, use of compounds that are structurally similar to the compound being measured may under- or overestimate endogenous levels unless the recovery efficiencies are identical. To circumvent this problem, Creelman & Mullet (28) synthesized (2-^{13}C)-JA and used it to measure JA in soybean tissue. Use of (2-^{13}C)-JA in calculating endogenous levels of JA must take into account the m/e 225 (M+1) in isotope dilution calculations because of the natural abundance of heavy isotopes. A significant improvement was the synthesis of (1,2-^{13}C)-JA with a molecular weight two mass units higher than endogenous JA (Z-P Zhang, ES McCloud & IT Baldwin, personal communication). The positive and negative ion electrospray mass spectra of several jasmonate amino acid conjugates were determined by combined HPLC-MS (102).

JA ACCUMULATION AND DISTRIBUTION

The level of JA in plants varies as a function of tissue and cell type, developmental stage, and in response to several different environmental stimuli. JA levels are low in soybean seeds but increase to 2 μg/g fresh weight in developing axes within 12 h of imbibition. In soybean seedlings, levels of JA are higher in the hypocotyl hook, a zone of cell division, and young plumules compared to the zone of cell elongation and more mature regions of the stem, older leaves, and roots. High levels of JA are also found in flowers and pericarp tissues of developing reproductive structures (28, 71).

In mature soybean leaves, high levels of jasmonate responsive gene expression are observed in paraveinal mesophyll cells and bundle sheath cells that surround veins and to a lesser extent in epidermal cells (50, 61). Little expression of jasmonate responsive genes is observed in palisade parenchyma cells unless leaves are treated with jasmonate (61). This suggests that JA levels are lower or compartmentalized differently in palisade cells or that these cells are less sensitive to JA. In illuminated plants, JA will accumulate in chloroplasts because it is a weak acid, assuming it becomes distributed in cellular compartments like ABA (58). Because palisade cells are filled with chloroplasts, this may lower the level of JA in the cytoplasm of these cells below the threshold needed to activate gene expression.

The accumulation of JA in chloroplasts may help explain why exogenous JA activates gene expression but increasing endogenous levels of JA by over-expression of allene oxide synthase does not (57). Exogenous JA will cause transient and large changes in the concentration of JA in most plant cells before reaching internal equilibrium in tissues. In contrast, over-expression of allene oxide synthase in chloroplasts provides cells with elevated rates of JA synthesis over the life of the plants. JA is synthesized in these plants in cells containing chloroplasts where it can be sequestered as it is produced. This may keep the level of JA from rising in other regions of the cell where JA receptors that modulate gene expression are presumably located.

Jasmonate levels are rapidly and transiently increased by mechanical perturbations such as those causing tendril coiling (39, 123) and turgor reduction induced by water deficit (28). Mechanical impedance during root growth might also induce JA accumulation, causing inhibition of root growth (113). Other mechanical perturbations, involving wind or touch, induce changes in growth that may be mediated in part by JA. In *Arabidopsis thaliana,* the *TCH* genes are upregulated in response to mechanical perturbation (19). The *TCH* genes encode calmodulin and two other calcium-binding proteins. Expression of these genes increases when cytoplasmic calcium levels rise (4). In animal cells, calcium stimulates eicosanoid biosynthesis by activating phospholipase and lipoxygenase. In plant cells, similar enzymes are also involved in JA biosynthesis, suggesting that a similar signal transduction pathway involving calcium may activate synthesis of JA in response to touch and turgor reduction.

Jasmonate accumulates in response to plant wounding (29). Localized wound-induced JA accumulation in injured cells could result from the mixing of compartments that contain lipases, membranes rich in LA, the precursor of

JA, lipoxygenase and the other enzymes that are involved in JA biosynthesis (described below). However, the situation is more complex because JA accumulation can be induced in cell cultures and plants by oligosaccharides derived from plant cell walls, by elicitors such as chitosans derived from fungal cell walls, and by peptide inducers such as systemin (37, 53, 82). These compounds are thought to stimulate JA biosynthesis via receptor-mediated processes. Elicitor-induced phosphorylation of a plasmalemma protein (42) and inhibition of elicitor-mediated JA accumulation by the protein kinase inhibitor staurosporin (16) are consistent with this idea. Furthermore, wound-induced accumulation of JA requires a MAP kinase (104). In some plants, ABA appears to be needed for wound, elicitor, or systemin-mediated JA accumulation (89). JA and JA responsive genes also accumulate systemically in plants in response to localized wounding (84). The systemic signal is apparently released at the wound site and migrates through the phloem to other parts of the plant. Systemin, an 18–amino acid peptide, has been shown to move in the phloem and to induce JA and JA-responsive genes throughout the apical portion of plants (84). Systemin may be released from the wound site upon hydrolysis of a precursor polypeptide (77). Electrical signals have also been proposed to mediate systemic induction of JA in response to wounding (124). In addition, systemic accumulation of JA as well as transfer of JA among plants could occur via the vapor phase in the form of MeJA (43, 49), although the significance of this latter pathway under physiological conditions is not clear.

BIOSYNTHETIC PATHWAY AND REGULATION

The biosynthesis of jasmonates begins with LA (Figure 2). This fatty acid is converted to 13-hydroperoxylinolenic acid by lipoxygenase. 13-hydroperoxylinolenic acid is a substrate for allene oxide synthase [also known as hydroperoxide dehydratase or hydroperoxide dehydrase; see Simpson & Gardner (105)] and allene oxide cyclase resulting in the formation of 12-oxo-phytodienoic acid (12-oxo-PDA). Following reduction and three steps of beta oxidation, (−)-7-iso-JA is formed. Jasmonic acid can be catabolized to form MeJA and numerous conjugates and catabolites that may have biological activity (56). The accumulation of JA in plants in response to wounding, or treatment with elicitors and systemin, can be blocked using inhibitors of lipoxygenase (8, 36, 41, 86). Therefore, increases in JA level mediated by these inducers results from de novo synthesis rather than release from JA conjugates.

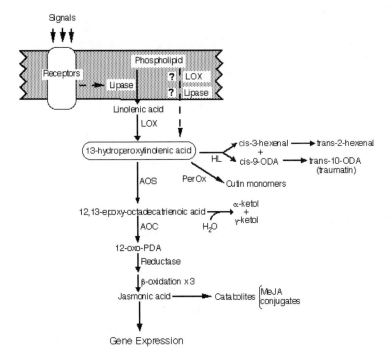

Figure 2   Biosynthetic pathway of jasmonic acid. It is postulated that signals (such as elicitors) interact with a membrane receptor, which causes the eventual production of 13-hydroperoxyli-nolenic acid. Production of 13-hydroperoxylinolenic acid is believed to occur with the release of linolenic acid via either a phospholipase or lipase followed by oxidation by lipoxygenase (LOX), but a preliminary oxidation of linolenic acid while still esterified to a phospholipid and subsequent release by a lipase cannot be ruled out. 13-hydroperoxylinolenic acid can then be catabolized by hydroperoxy lyase (HL), eventually forming volatile aldehydes and traumatic acid, or via peroxygenase pathway to cutin monomers. Jasmonic acid arises from 13-hydroperoxylinolenic acid via an allene oxide synthase (AOS) and an allene oxide cyclase (AOC)-dependent pathway with 12-oxophytodienoic acid (12-oxo-PDA) as an intermediate. Jasmonic acid then acts to modulate gene expression or can be further catabolized.

Linolenic Acid and Lipases

The *A. thaliana fad3-2 fad7-2 fad8* mutant has very low levels of linolenic acid and is unable to accumulate JA in response to wounding (76; RA Creelman, M McConn, J Browse & JE Mullet, unpublished data). Application of LA to plants results in accumulation of JA (44). This indicates that the level, distribution, or availability of LA could determine the rate of JA biosynthesis. In one study, the level of free linolenic acid measured before wounding was many times higher than the maximum JA accumulated after wounding (27). Free LA levels doubled within 1 h after wounding, while JA levels rose 10-fold (27). Therefore, the wound-induced increase in JA level could have resulted from release of linolenic acid from phospholipids, or the utilization of LA present before wounding for JA biosynthesis.

Plant membranes, especially chloroplast membranes, are a rich source of LA esterified in glycerolipids and phospholipids. This has led to the suggestion that increases in JA could result from the activation of phospholipases that release LA from membranes (44). Plant extracts also contain highly active acyl hydroxylases that can release fatty acids from lipids. In animal systems, phospholipase A_2, activated by micromolar levels of calcium, releases arachidonic acid used in the biosynthesis of eicosanoids. The eicosanoids, leukotrienes, and prostaglandins have chemical structures similar to jasmonates. These compounds mediate localized stress and inflammatory responses in animal cells. Direct evidence for the role of a specific phospholipase in JA synthesis is lacking. However, analysis of phospholipid changes and phospholipase A activity was done in tobacco cells treated with elicitors derived from *Phytophthora parasitica* var. *nicotianae* (97). Time course studies showed that the amount of phosphatidylcholine was reduced in response to elicitors and that a concomitant increase of phospholipase A activity occurred. In soybean cell culture, harpin and an extract from the pathogenic fungus *Verticillium dahliae* promoted rapid increases in phospholipase A activity (23). Upon wounding of potato tuber tissue, JA levels rose about 100-fold in 4 h; however, inhibitors of animal phospholipase A_2 (manoalide and quinacrine) did not inhibit the accumulation of JA (69). Phospholipase D, which cleaves the head group from phospholipids, could also trigger release of LA and stimulate JA biosynthesis. Plant phospholipase D has been identified in plants and is proposed to play a role in plant defense (121).

In some instances, fatty acids may be oxidized before release from lipids for JA biosynthesis. Fuessner et al (46) describe the presence of lipoxygenase (LOX) in lipid bodies which oxidizes fatty acids before further metabolism. Phospholipases preferring oxygenated fatty acids have been observed in several plant species (6, 9). Therefore, the oxidation of fatty acids during high irradiance, exposure to ozone, or as a consequence of the oxidative burst associated with plant defense (20) may stimulate the activity of phospholipases or nonspecific acyl hydrolases resulting in release of oxidized fatty acids for JA biosynthesis.

In animal cells, arachidonic acid is delivered to cells for eicosanoid biosynthesis via low density lipoproteins (54). In plants, transfer of linolenic acid among cells could be carried out by lipid transfer proteins that are localized in the extracellular space (114). Current models postulate that jasmonate biosynthesis is regulated by pathogens or herbivory through the production of elicitors or systemic signaling molecules that interact with receptors present on the plasma membrane. The observation that the enzymes of the JA biosynthetic pathway are primarily localized in plastids (10, 17) suggests that mechanisms must exist to shuttle LA released from the plasma membrane to the plastid. Alternatively, signal perception could be transmitted to plastids for subsequent release of free LA in that organelle.

Lipoxygenase (LOX)

Treatment of plants with LOX inhibitors (8, 86) and transgenic plants with reduced LOX activity (10) have reduced ability to synthesize JA. Therefore, LOX mediates an essential step in JA biosynthesis. In animals, eicosanoid biosynthesis is regulated in part through calcium and protein modulated interaction of LOX with membranes and its substrate (35, 79). In plants, the role of LOX in the regulation of JA biosynthesis has been difficult to analyze because most plants have numerous genes encoding LOX, different isoforms of LOX have different enzyme specificity, and LOX is present in more than one compartment in plant cells (i.e. 45, 64, 83, 100).

Plant lipoxygenases (EC 1.13.11.12) oxygenate linolenic acid at the 9 or 13 position to give 9- or 13-hydroperoxylinolenic acid. The role of 9-hydroperoxides and their catabolites in plants is unclear (119). The role of LOX isozymes in the production of hydroperoxide isomers also needs further investigation. LOX from tendrils of *Bryonia dioica* consists of a major isoform with pI = 6.5 and minor constituents with pI = 6.7 and 7.3 (38). These LOX isoforms primarily produce 13-hydroperoxylinolenic acid, whereas a preparation from cell cultures contains at least seven LOX isoforms in the pI range from 6.3–6.7 and 7.3–7.5 with the major reaction product 9-hydroperoxylinolenic acid (38).

LOX isozymes are found in the plasmalemma/microsomes of cucumber cotyledons (*Cucumis sativus* L.) (81). In soybean leaves, LOX accumulates in the vacuoles of paraveinal mesophyll cells (52). Elsewhere, LOX is associated with epidermal and cortical cells and is present in vacuoles and plastids. In plastids, a methyl jasmonate–induced LOX is sequestered into protein inclusion bodies (52). This may be important in restricting the interaction of LOX with fatty acids. Similarly, in barley, jasmonate–induced LOX isozymes are localized in plastids (45).

Changes in the distribution and abundance of LOX during development and in different tissues and compartments is due in part to the expression of different members of the *Lox* gene family. For example, two different LOX genes, *AtLox1* (78) and *AtLox2* (11), have been identified in *A. thaliana*. *AtLox1* is expressed in leaves, roots, inflorescences, and young seedlings, with the highest expression found in roots and young seedlings. Because *AtLox1* lacks obvious targeting sequences, this enzyme is most likely localized in the cytoplasm. In contrast, *AtLox2* is localized in chloroplasts (10). The presence of a plastid transit sequence suggests that a rice LOX that catalyzes the exclusive formation of 13-hydroperoxylinolenate is also localized in chloroplasts (90). *AtLox2* mRNA levels are high in leaves and inflorescences but low in seeds, roots, and stems. The physiological role of this chloroplast lipoxygenase was analyzed by reducing LOX2 accumulation in transgenic plants (10). The reduction of *AtLox2* expression caused no obvious changes in plant growth. However, the wound-induced accumulation of JA observed in control plants was absent in leaves of transgenic plants lacking LOX2. Therefore, plastid localized LOX2 is required for wound-induced synthesis of jasmonates in Arabidopsis leaves.

Allene Oxide Synthase and Other Steps in JA Biosynthesis

The fate of 13-hydroperoxylinolenate produced by lipoxygenase is another key branchpoint in the jasmonate biosynthetic pathway (Figure 2). Hydroperoxide lyase will cleave 13-hydroperoxylinolenate to form volatile six carbon aldehydes and 12-oxo-dodecenoic acid (119). 13-hydroperoxylinolenate can also be used by peroxygenase to produce precursors of cutin molecules (18). In contrast, production of jasmonates requires that 13-hydroperoxylinolenate be metabolized to allene oxide by allene oxide synthase (AOS; IUBMB name hydroperoxide dehydratase, EC 4.2.1.92). Other names previously used for AOS include hydroperoxide isomerase, hydroperoxide cyclase, and fatty acid hydroperoxide dehydrase. Flax and Arabidopsis AOS have been cloned and characterized (21, 107, 108; E Bell, RA Creelman & JE Mullet, unpublished data). Flax AOS is a 55-kDa hemoprotein with the spectral characteristics of a cytochrome P450 and a turnover rate of 1000 min^{-1}. The primary structure of AOS deduced from its cDNA reveals a protein of 536 amino acids containing a C-terminal domain homologous to a region in many cytochrome P450s that contain a heme-binding cysteine. The flax cDNA encodes a 58–amino acid N-terminal sequence characteristic of chloroplast transit peptides. This is consistent with localization of AOS activity in chloroplasts (120). The Arabidopsis AOS (available from the Ohio State University Arabidopsis Biological Resource Center as EST 94J16T7, GenBank Accession T20864) shares a high degree of homology with flax AOS and also contains a putative N-terminal plastid targeting sequence (RA Creelman, E Bell & JE Mullet, unpublished data). The Arabidopsis AOS gene exists as a single copy, based on analysis of total DNA Southern blots (E Bell, RA Creelman & JE Mullet, unpublished data). Allene oxide synthase activity has been localized to the plastid outer envelope in spinach (17), but given its putative plastid targeting sequence and high turnover number further localization studies are warranted.

Over-expression of flax AOS in transgenic potato plants increased JA levels (57), indicating that the amount of AOS protein limits JA biosynthesis. The high turnover rate of flax AOS (~ 1000 min^{-1}) may help this enzyme compete for substrate also used by hydroperoxy lyase and peroxygenase depending on the localization of these enzymes and their substrate accessibility. 13-hydroperoxylinolenate produced by a plastid localized LOX may be more accessible to AOS localized to plastids. Recent studies have shown that AOS activity was higher in tissues with elevated JA, suggesting that differences in JA that occur during plant development may be caused by variation in AOS abundance or activity (105). Similar analysis of AOS level and activity needs to be carried out in plants exposed to elicitors and systemin, or after wounding.

Nonenzymatic cyclization of allene oxide will yield a racemic mixture of 12-oxo-PDA. However, allene oxide cyclase (55) catalyzes the stereospecific formation of the 9S, 13S enantiomer of 12-oxo-PDA. The enzymes catalyzing reduction of 12-oxo-PDA and β-oxidation leading to JA have been demonstrated in vitro but have not been extensively characterized (119).

The tomato mutant, JL5, is inhibited in the conversion of 13-hydroperoxylinolenate to 12-oxo-PDA (60). This mutant, which could be altered in AOS or AOC activity, was identified by screening plants for reduced levels of wound-induced *Pin2* expression. Diethyldithiocarbamic acid (DIECA), an inhibitor of JA biosynthesis, was shown to efficiently reduce 13-hydroperoxylinolenate to 13-hydroxylinolenic acid (37, 41). This suggests that DIECA inhibits JA biosynthesis by reducing the precursor pool leading to allene oxide. Salicylic acid (SA), a mediator of some plant defense responses, inhibits the conversion of 13-S-hydroperoxy linolenic acid to 12-oxo-PDA (37, 41, 87, 124).

JASMONATE SIGNAL TRANSDUCTION

The JA signal transduction pathway is mainly unknown. It is presumed that jasmonate interacts with receptors in the cell that activate a signaling pathway resulting in changes in transcription, translation, and other responses mediated by JA. Lack of high specific activity JA and the lipophilic and volatile nature of JA and MeJA will make direct analysis of JA receptors difficult. Jasmonate receptors and other components of the signal transduction pathway are more likely to be discovered through analysis of mutants that are insensitive or altered in their response to JA. To date, up to four different classes of JA-in-sensitive mutants have been identified; *coi1, jar1, jin1,* and *jin4* (12, 14, 113). Genetic studies were unable to determine whether *jin4* and *jar1* were allelic (14). The mutants *jar1, jin1,* and *jin4* were recovered using a root growth screen (wild type *A. thaliana* root growth is inhibited by 1–10 mM JA). In contrast, *coi1* was identified because plants were resistant to coronatine (47). Coronatine is a chlorosis-inducing toxin that has a chemical structure (Figure 1) and biological activity similar to JA. The *coi1* mutant also shows a MeJA–insensitive root growth phenotype.

Application of jasmonate to plants causes large changes in translation, transcription, and mRNA populations (103). Treatment of barley leaves with jasmonate reduced synthesis of the large and small subunits of Rubisco and other proteins (122). Decreased translation of the large subunit in chloroplasts was correlated with a site-specific cleavage in the 5'-untranslated portion of the *rbcL* mRNA (94, 96). The altered *rbcL* mRNA 5'-end presumably reduces access to the ribosome–binding site located near the site of translation initiation. Reduced synthesis of Rubisco small subunit and other cytoplasmic proteins occurred through supression of translation initiation and reduction of mRNA levels (95).

The promoters of two jasmonate–inducible genes, *Pin2* and *VspB,* have been studied in some detail (67, 74). A 50–bp domain was identified in the promoters of both genes that conferred JA responsiveness on truncated reporter gene constructs. The *VspB* 50–bp domain did not confer JA responsiveness on a truncated −46 CaMV promoter, but induction was observed when this DNA region was added to a −90 CaMV truncated promoter. This indicates that factors binding to the −90 CaMV promoter were required to observe JA–stimulated transcription. Both JA–responsive domains contain a G-box sequence (CACGTG), which in other promoters has been shown to bind bZIP transcription factors (126). Because bZIP protein binding sites are found in numerous promoters that are not responsive to JA, it is likely that this factor, if it plays a role in JA mediated responses, interacts with other *trans* factors to modulate transcription. Mutation of the G-box in the *Pin2* promoter did not prevent JA-mediated induction of *Pin2* (72). Jasmonate induced accumulation of *VspB* and *Pin2* mRNA is blocked by cycloheximide. Bestatin, an inhibitor of aminopeptidases in plants and animals, induces *Pin2* in the absence of JA (101). This observation suggests that induction of *Pin2,* and perhaps other jasmonate modulated plant genes, is normally prevented by the action of an aminopeptidase. Induction of JA–responsive genes could therefore be mediated by inactivation of the protease or stabilization of the target protein.

JA FUNCTION AND RESPONSIVE GENES

Jasmonate modulates the expression of numerous genes and influences specific aspects of plant growth, development, and responses to abiotic and biotic stresses (see Table 1). Many of these responses were identified by application of jasmonate to plants, sometimes at nonphysiological levels. Interactions between JA and other plant growth regulators make assignment of physiological roles for JA even more complicated. In the section below, proposed actions of JA in plants are related to the level of JA in plant tissues, the activity of JA responsive genes, and insights provided by JA insensitive and JA deficient plants.

Seed Germination and Growth

JA and MeJA inhibit the germination of nondormant seeds and stimulate the germination of dormant seeds. JA, MeJA, ABA, and ethylene inhibit germination of the recalcitrant seeds of *Quercus robur* (48). When these dessication-sensitive seeds were dried, the concentration of MeJA and JA increased before

Table 1 Jasmonate functions and responsive genes

Physiological Function	JA Responsive Genes	References
Seed germination and growth		
(-) seed, pollen germination	???	47, 48, 76
(-) root growth	???	12, 14, 113
(+) tendril coiling	???	38, 39, 123
Photosynthesis/vegetative sinks		
(-) photosynthetic apparatus	(-) *rbcL, rbcS*	24, 94–96
(+) vegetative protein storage	(+) *Vsp, Lox*	11, 15, 22, 45, 50, 52, 61, 65, 73, 75, 87
(+) tuberization	(+) *Pin2*	63, 85, 87, 88, 93
Flower and fruit development		
(+) pollen viability	???	47, 76
(+) fruit ripening, pigments	(+) *EFE*	30, 71
(+) seed development	(+) cruciferin, napin, *Vsps*	110, 125
Insect and disease resistance		
(+) insect resistance	(+) *Pin2*, other genes	36, 37, 41, 43, 44, 49, 62
(+) disease resistance	(+) osmotin, thionin, RIP60	9a, 129
	(+) *Chs, Pal,* HMGR, PPO, *pdf,* ???	25, 26, 29, 49, 82, 91

the loss in seed viability. The increase in jasmonate was correlated with lipid peroxidation, which suggests that the production of jasmonate may not be regulating germination but rather is a consequence of membrane damage. In apple, jasmonate stimulated the germination of dormant embryos and increased alkaline lipase activity (92). Lipase activation may stimulate the mobilization of lipid reserves to provide sugars for seedling growth. Inhibitors of lipoxygenase-inhibited embryo germination and JA partially reversed inhibitor action. The level of jasmonate in soybean seeds 12 days post-anthesis is low (~0.1 µg/g fresh weight), whereas in older seeds JA levels are higher (0.5 ng/g fresh weight) (28). 12 h after imbibition, the level of JA increased fivefold to 2 µg/g fresh weight in axes. JA levels declined with further seedling development. The observed increase in JA levels following imbibition is correlated with seed reserve mobilization and therefore may be a consequence rather than a trigger of germination. The jasmonate-insensitive mutants, *jin4* and *jar1,* show increased sensitivity to ABA inhibition of germination (14, 113). Therefore, JA may stimulate seed germination by decreasing sensitivity to ABA. Alternatively, JA-mediated growth inhibition could block seed germination.

JA strongly inhibits root growth by a mechanism not mediated by ethylene (14). JA also inhibits IAA-stimulated coleoptile elongation possibly by blocking incorporation of glucose into cell wall polysaccharides (118). Furthermore, JA activates the differential growth involved in tendril coiling, a response that does not directly involve ethylene or IAA (39). Further work is needed to define the role of JA in growth processes.

Vegetative Sinks and Storage Proteins

Plants have the capacity to accumulate large amounts of carbon and nitrogen in specific cells and tissues and to mobilize these materials for use in other parts of the plant. This capacity is used during seed formation when nutrients are moved from the vegetative plant to developing seeds, and during seed germination when carbon and nitrogen are mobilized for seedling development. Transient storage and mobilization of nutrients also occur during vegetative growth. For example, carbon often accumulates in chloroplasts during the day and is mobilized at night to other parts of the plant. Carbon and nitrogen may also accumulate in cells located in meristematic regions for use during rapid cell growth. For osmotic reasons, cells are only able to accumulate a limited amount of sucrose and amino acids. Therefore, large amounts of carbon and nitrogen accumulate as polymers in the form of starch, fructans, and proteins.

A role for jasmonic acid in protein storage in plants was suggested, in part, because jasmonate levels are high in vegetative sinks. As noted above, jasmonate levels are higher in soybean axes, plumules, and the hypocotyl hook relative to the hypocotyl zone of cell elongation and the nonelongating portion of stems and roots (28). In six-week-old soybean seedlings, JA levels are higher in young growing leaves that are importing carbon and nitrogen than in older fully expanded leaves (28). High levels of JA are present in developing reproductive structures, especially pods, with lower levels in seeds. Jasmonate or a derivative of jasmonate, tuberonic acid, has been proposed to play a role in the formation of tubers, a special type of vegetative sink (68, 85, 93).

A second reason to suggest that jasmonate plays an important role in protein storage during plant development derives from the discovery that genes encoding vegetative storage proteins (VSPs) (111) are regulated by jasmonate (2). Vegetative storage proteins were first described in soybean (127, 128). The VSPs accumulate in the vacuoles of paraveinal mesophyll and bundle sheath cells that surround veins in soybean leaves (50). If pods are continuously removed from plants, the VSPs accumulate and can account for as much as 45% of the soluble protein in leaves (128). Other studies showed that the VSPs accumulate in pods and other parts of the developing reproductive structure but not in seeds (109). This led to the suggestion that the VSPs represent temporary deposits of amino acids derived from disassembly of Rubisco and other leaf proteins that were being mobilized for seed development. The observation that jasmonate regulates VSP accumulation was made by Anderson (2, 3) when treating soybean cell cultures with this compound. Later studies showed that the soybean VSPs accumulate in soybean axes, hypocotyl hooks, and young developing leaves and that *VspB* expression is correlated with endogenous levels of JA in plants (28, 73, 75). *VspB* expression also increases in young leaves when plants are exposed to water deficit (75). This is most likely due to inhibition of leaf growth but continued import and storage of carbon and nitrogen in the leaf. The soybean VSPs include two proteins having low acid phosphatase activity (Vspa, Vspb) (33) and lipoxygenase (65, 115). The genes encoding these proteins are regulated in a complex way by JA, sugars, phosphate, nitrogen, and auxin (34, 73, 98, 112).

Photosynthesis, Senescence, and Abiotic Stress

Application of JA to leaves decreases expression of nuclear and chloroplast genes involved in photosynthesis (22, 122). JA treatments also cause a loss of chlorophyll from leaves or cell cultures (122). Jasmonate's ability to cause chlorosis led to the suggestion that this compound plays a role in plant senescence (116). However, this suggestion is difficult to reconcile with the finding of high JA levels in zones of cell division, young leaves, and reproductive structures. Unfortunately, a complete analysis of JA levels in senescing leaves has not been carried out. A limited study of this question in soybean revealed only small changes in JA level in soybean leaves during pod fill (22). Thus although JA can induce senescence-like symptoms, the role of this hormone in mediating senescence is at present unclear.

If jasmonate-induced chlorosis and inhibition of genes encoding proteins involved in photosynthesis are not involved in senescence, then what is the physiological role of this JA activity? JA may inhibit the synthesis of chloroplast proteins during an early phase of leaf formation where cell division and import of nutrients are very active. This is consistent with higher levels of JA in young leaves compared with older leaves of soybean. In monocot leaves, meristematic cells are localized to the leaf base. This region of the leaf contains little chlorophyll, and expression of genes involved in photosynthesis is limited. Plastids in this developmental stage often contain starch grains, and these cells may accumulate vegetative storage proteins before cell enlargement and chloroplast development. Once cells stop dividing and begin to elongate, chloroplast development is initiated. If developing monocot leaves are similar to soybean hypocotyls, higher jasmonate concentrations will be present in the meristematic cells of the leaf base than in expanding and mature cells nearer the leaf apex. If this is the case, JA could act to inhibit premature accumulation of the photosynthetic apparatus in meristematic cells of the leaf base while stimulating accumulation of carbon and nitrogen reserves needed for later cell development. We speculate that treatment of excised mature leaves with JA may be recapitulating this earlier phase of development. Application of JA to mature leaves may drive the developmental program in reverse by inhibiting expression of chloroplast genes and stimulating accumulation of vegetative storage and other proteins.

The expression of many genes involved in photosynthesis is higher during plant illumination compared to darkness. JA's ability to inhibit expression of genes involved in photosynthesis could lower expression of these genes in darkness. If JA is distributed like ABA in plant cells, this compound will accumulate in chloroplasts in illuminated plants because of the increase in pH in this compartment. At night, JA accumulated in chloroplasts during the day will be released into the cytoplasm, where it could inhibit expression of genes involved in photosynthesis.

The ability of JA to inhibit expression of genes involved in photosynthesis suggests that jasmonate could help reduce the plant's capacity for carbon assimilation under conditions of excess light or carbon. The photosynthetic apparatus may absorb more light energy than can be used for photosynthesis under conditions where carbon fixation exceeds the capacity of cells to export or store carbon. Inhibition of genes encoding the photosynthetic apparatus under these conditions would eventually balance energy absorbing and using capacities. In the short term, JA-mediated induction of vegetative storage protein synthesis under conditions of high sugar accumulation creates a sink for carbon and nitrogen and releases phosphate from sugar phosphate pools for further carbon fixation.

Excess light absorption also occurs in plants exposed to water deficit, which induces stomatal closure, making plants deficient in CO_2. Under these conditions, the products of photosynthetic electron transport can no longer be used for carbon fixation, and the energy harvested by the chlorophyll antennae must be dissipated rather than used for the formation of ATP and reducing power. Some of the excess energy can be dissipated via the xanthophyll cycle or through other energy-quenching mechanisms (31). However, once the capacity of these systems is exceeded, membrane damage will occur. Lipoxygenase and other enzymes that metabolize fatty acids may protect membranes from damage by removing oxidized fatty acids. The lipoxygenase-mediated generation of JA could, in turn, induce changes in the cell that ameliorate further photochemical damage. For example, jasmonate-induced loss of chlorophyll would decrease the amount of energy absorbed by the photosynthetic apparatus. The accumulation of anthocyanins that is stimulated by JA in illuminated plants (49) could also provide some protection from excess radiation.

Flower and Fruit Development

Jasmonate might be expected to play a role in formation of flowers, fruit, and seed because of the relatively high levels of this compound in developing plant reproductive tissues. The presence of jasmonate and related volatile fatty acid derivatives may be involved in insect attraction related to pollen dispersal. Other aspects of flower, fruit, and seed development that can be modulated by jasmonate include fruit ripening, fruit carotenoid composition, and expression of genes encoding seed and vegetative storage proteins. Jasmonate-stimulated tomato and apple fruit ripening most likely occurs through activation of EFE and production of ethylene (30). It is possible that jasmonate levels gradually increase in developing fruit leading to enhanced synthesis of ethylene and subsequent fruit ripening. Application of JA to tomato fruit inhibited the accumulation of lycopene and stimulated accumulation of β-carotene (99). The biochemical basis and physiological role of this JA-mediated change needs further investigation.

Soybean *Vsps* and *A. thaliana AtVsp* show high expression in flowers and developing fruit (15, 110). The vegetative storage proteins may provide temporary storage of carbon and nitrogen arriving at the reproductive apex for use during rapid synthesis of seed storage proteins. The AtVSP proteins were missing in flowers of *coi1* mutants that are insensitive to JA but could be induced by treatment of plants with JA (12). The JA-deficient LA mutant of *A. thaliana* also does not express *AtVsp* unless plants are provided with exogenous JA (RA Creelman, M McConn, E Bell, J Browse & JE Mullet, unpublished data). Ovules of the LA-deficient mutant were viable, indicating that JA and expression of the *AtVsp* were not essential for seed formation. Therefore, although JA may modulate expression of genes encoding seed storage proteins (125), JA is not essential for production of viable ovules in *A. thaliana*. However, fatty acid mutants of *A. thaliana* that lack LA and JA and the *coi1* JA-insensitive mutant fail to produce viable pollen unless supplied with JA (76).

Insect and Disease Resistance

JA plays an important role in plant insect and disease resistance. Several lines of evidence support this conclusion. First, JA accumulates in wounded plants (29) and in plants or cell cultures treated with elicitors of pathogen defense (53). Second, JA activates genes encoding protease inhibitors that help protect plants from insect damage (62). JA also activates expression of genes encoding antifungal proteins such as thionin (9a), osmotin (129), PDF (91), and the ribosome-inactivating protein RIP60 (24). JA modulates expression of cell wall proteins such as PRP (29) that may be involved in synthesis of barriers to infection. Furthermore, JA induces genes involved in phytoalexin biosynthesis (*Chs, Pal,* HMGR) (25, 29) and phenolics (polyphenol oxidase; 37) that are involved in plant defense. The oxylipin pathway that leads to JA is also the source of other volatile aldehydes and alcohols that function in plant defense and wound healing. For example, the C6-aldehyde 2-hexenal completely inhibited growth of *Pseudomonas syringae* and *E. coli* (32), and C6-aldehydes and alcohols reduced aphid fecundity (59). These compounds are synthesized from 13-hydroperoxylinolenic acid via the action of hydroperoxy lyase (Figure 2). Jasmonate, wounding, and elicitors increase the expression of lipoxygenase and stimulate hydroperoxy lyase activity (5, 52). This response enhances the ability of plants to produce the six carbon compounds that contribute to plant protection.

A third type of evidence for JA's role in pest resistance comes from analysis of plants having modified levels of JA. For example, treatment of potato with JA increases resistance to *Phytophthora infestans* (26). The tomato mutant, JL5, which is inhibited in the conversion of 13-hydroperoxylinolenic acid to 12-oxo-PDA is more susceptible to damage by *Manduca sexta* (60). An *A. thaliana* mutant deficient in LA contains negligible amounts of JA and neither accumulates JA nor induces JA-responsive genes when wounded. These mutants are very susceptible to fungal gnats (M McConn, RA Creelman, E Bell, JE Mullet & J Browse, unpublished data). Treatment of the mutants with JA restored fungal gnat resistance, demonstrating an essential role for JA in resistance.

Although a general role for JA in plant defense is now well established, the specific way that JA is deployed relative to other defense mechanisms in response to insects and pathogens needs further investigation. The complexity of plant defense responses and JA's role was demonstrated in a recent study of two genes (*Pdf1.1, Pdf1.2*) that are involved in fungal resistance in *A. thaliana*. SA, an inducer of many genes involved in responses to pathogens, was able to induce PR-1 but not *Pdf* (91). *Pdf* expression was induced by JA, ethylene, and oxygen radical generators such as paraquat and rose bengal. The *A. thaliana* mutants, *ein2* and *coi1*, which are blocked in their responses to ethylene and MeJA, respectively, were not altered in pathogen-mediated induction of PR-1 but blocked in accumulation of PDF. These results have several important implications. First, the oxidative burst that often accompanies plant responses to pathogens may induce JA by producing oxidized fatty acids. The oxidative burst has been linked to programmed cell death (70) responses and represents one line of plant pathogen defense. JA released from injured cells could then activate further defense responses including systemic ones. Second, this study shows that pathogens such as *A. brassicola* can trigger at least two defense pathways, one involving SA and one involving JA and ethylene. This observation is consistent with reports that JA and ethylene synergistically induce genes that are involved in plant defense (129). Furthermore, SA is an inhibitor of JA biosynthesis and action (36, 86). This interaction may allow the plant to modulate the relative amount of SA- and JA-inducible defenses as a function of time after attack or in response to specific pathogens or herbivores.

JA's Dual Role in Development and Defense

In this final section, we attempt to provide a rationale for JA's dual role in plant development and defense. This discussion starts with the following questions: Why are genes encoding vegetative storage proteins and genes such as *Pin2* that are involved in insect defense regulated similarly? Why are jasmonate-inducible genes involved in plant defense also regulated by sugars, phosphate, and auxin? Why are JA levels high in young apical sinks and especially reproductive structures but inducible by wounding or elicitation in older parts of the plant?

The regulation of expression of *VspB*, which encodes a soybean vegetative storage protein, and *Pin2*, which encodes a tomato protease inhibitor involved in plant defense, is remarkably similar. Both genes are highly expressed in apical regions of vegetative plants and in flowers and reproductive tissues. Both genes are induced upon wounding and application of JA. Expression of these genes is much higher when JA/wounding treatment occurs in illuminated plants and induction of gene expression by JA is synergistically stimulated by sugars and inhibited by phosphate and auxin (34, 66). The products of these two genes are targeted to vacuoles, and large amounts of the two proteins often accumulate in plant cells. The similarity of expression of these two genes and localization of their gene products suggest that they play nearly identical roles in plants, except for the activity of the encoded proteins.

For many reasons it is not surprising that some proteins involved in plant defense also function as vegetative storage proteins. Unlike seed storage proteins that have to be stored at high density in nearly dehydrated seeds, vegetative storage proteins accumulate in fully hydrated plants containing large vacuoles. Therefore, the constraints on proteins that serve as VSPs are fewer than for seed storage proteins. Plants need to accumulate large amounts of vegetative storage proteins and proteins involved in defense without disrupting metabolism. Sequestering these proteins in vacuoles may help accomplish this. Moreover, both types of proteins are mobilized to recover amino acids when the need for storage or defense is gone. Therefore many proteins involved in plant defense, in particular those that are sequestered in vacuoles for action when ingested, are ideally suited for a role as vegetative storage proteins. Because VSPs that can also aid in plant defense have additional value for the plant, perhaps most VSPs will eventually be found to serve a role in plant defense.

Elevated expression of JA-responsive genes such as *Pin2* in vegetative apices, young leaves, and reproductive structures may simply be the consequence of their role as vegetative storage proteins. However, the accumulation of defensive proteins in these tissues also provides the plant with a preformed deterrent to herbivory and disease in regions of the plant critical to survival and reproduction. The differential accumulation of compounds involved in defense in plant tissues of high value is consistent with the "optimal defense theory" described by researchers working in the area of chemical ecology (7). This theory predicts that defense should be allocated to plant parts that contribute significantly to a plant's fitness and have a high probability of attack. A second class of chemical defense theory, the "C/N theory," emphasizes the fact that allocation of carbon and nitrogen to defense may occur in competition with the use of these resources for growth and development (7). This theory rationalizes why plants may induce carbon-rich defenses (i.e. tannins, phenols) vs nitrogen-rich defenses (i.e. proteins) depending on nutrient availability. The inducible or activated defense system mediated by JA is consistent with the C/N theory of chemical defense in several ways. In older leaves, JA levels and JA-inducible defense is activated in response to wounding or elicitors, thus limiting the allocation of nutrients to defense to situations where this is required. In addition, expression of *Pin2* is regulated not only by the presence of JA but also by sugars (63, 88). Dual regulation by sugars and JA is also observed in other genes such as *Chs* that are involved in plant defense. Furthermore, *Vsp* expression and perhaps genes such as *Pin2* is inhibited when plants are grown in limiting nitrogen conditions (112). This type of regulation would minimize the allocation of carbon and nitrogen to proteins and secondary metabolites involved in plant defense under conditions where nutrients are limiting.

CONCLUDING REMARKS

The volume of publications and reviews on jasmonate over the past decade documents the increasing interest in this compound and its role in plants. Research on this topic has solidified our understanding of the chemistry and biosynthetic pathway of jasmonates. However, additional research is needed into the mechanisms that regulate the synthesis of JA in plants during development and in response to wounding and oligosaccharides and peptides that modulate JA biosynthesis. Transgenic plants containing sense/antisense constructs of genes in the biosynthetic pathway and mutants deficient in JA will help provide definitive information. Our understanding of the JA signal transduction pathway will rapidly advance as genes identified through analysis of JA-insensitive mutants. Recent direct evidence for JA's role in plant defense confirms this role for JA originally suggested from studies of soybean cell cultures. The deployment of JA-inducible genes as part of the complex plant defense system will be a topic of intense future study. Insight gained from these studies should lead to better design of durable plant defense and improved utilization of proteins and genes from nonplant sources for plant protection.

ACKNOWLEDGMENTS

We thank our colleagues in this field for their insightful comments and for sending recent publications on this topic. Because of length restrictions and the availability of other reviews, we apologize in advance for being unable to cite all of the excellent publications on this topic. The authors' research on jasmonate has been supported by the USDA National Research Initiative, the National Science Foundation, and the Texas Agricultural Experiment Station.

> **Visit the *Annual Reviews home page* at http://www.annurev.org**

Literature Cited

1. Albrecht T, Kehlen A, Stahl K, Knöfel HD, Sembdner G, Weiler EW. 1993. Quantification of rapid transient increases in jasmonic acid in wounded plants using a monoclonal antibody. *Planta* 191:86–94
2. Anderson JM. 1988. Jasmonic acid-dependent increases in the level of specific polypeptides in soybean suspension cultures and seedlings. *J. Plant Growth Regul.* 7:203–11
3. Anderson JM. 1991. Jasmonic acid–dependent increase in vegetative storage protein in soybean tissue culture. *J. Plant Growth Regul.* 10:5–10
4. Antosiewicz DM, Polisenky DH, Braam J. 1995. Cellular localization of the Ca^{2+} binding TCH3 protein of *Arabidopsis. Plant J.* 8:623–36
5. Avdiushko S, Croft KP, Brown GC, Jackson DM, Hamilton-Kemp TR, Hildebrand D. 1995. Effect of volatile methyl jasmonate on the oxylipin pathway in tobacco, cucumber, and Arabidopsis. *Plant Physiol.* 109:1227–30
6. Bafor M, Smith MA, Jonsson L, Stobart K, Stymne S. 1993. Biosynthesis of vernoleate (cis-12-epoxyoctadeca-cis-9-enoate) in microsomal preparations from developing endosperm of *Ephorbia lagascae. Arch. Biochem. Biophys.* 303:145–51
7. Baldwin IT. 1994. Chemical changes rapidly induced by folivory. In *Insect Plant Interactions*, ed. EA Bernays, pp. 1–23. Boca Raton, FL: CRC Press
8. Baldwin IT, Schmelz EA, Zhang ZP. 1996. Effects of octadecanoid metabolites and inhibitors on induced nicotine accumulation in *Nicotiana sylvestris. J. Chem. Ecol.* 22: 61–74
9. Banas A, Johansson I, Stymne S. 1992. Plant microsomal phospholipases exhibit preference for phosphatidylcholine with oxygenated acyl groups. *Plant Sci.* 84: 137–44
9a. Becker W, Apel K. 1992. Isolation and characterization of a cDNA clone encoding

a novel jasmonate-induced protein of barley (*Hordeum vulgare* L.). *Plant Mol. Biol.* 19:1065–67
10. Bell E, Creelman RA, Mullet JE. 1995. A chloroplast lipoxygenase is required for wound-induced jasmonic acid accumulation in Arabidopsis. *Proc. Natl. Acad. Sci. USA* 92:8675–79
11. Bell E, Mullet JE. 1993. Characterization of an Arabidopsis lipoxygenase gene responsive to methyl jasmonate and wounding. *Plant Physiol.* 103:1133–37
12. Benedetti CE, Xie DX, Turner JG. 1995. *COI1*-dependent expression of an Arabidopsis vegetative storage protein in flowers and siliques and in response to coronatine or methyl jasmonate. *Plant Physiol.* 109: 567–72
13. Bennett RN, Wallsgrove RM. 1994. Secondary metabolites in plant defence mechanisms. *New Phytol.* 127:617–33
14. Berger S, Bell E, Mullet JE. 1996. Two methyl jasmonate-insensitive mutants show altered expression of *AtVsp* in response to methyl jasmonate and wounding. *Plant Physiol.* 111:525–31
15. Berger S, Bell E, Sadka A, Mullet JE. 1995. *Arabidopsis thaliana Atvsp* is homologous to soybean *VspA* and *VspB*, genes encoding vegetative storage protein acid phosphatases and is regulated similarly by methyl jasmonate, wounding, sugars, light and phosphate. *Plant Mol. Biol.* 27:933–42
16. Blechert S, Brodschelm W, Hölder S, Kammerer L, Kutchan TM, et al. 1995. The octadecanoic pathway: signal molecules for the regulation of secondary pathways. *Proc. Natl. Acad. Sci. USA* 92:4099–105
17. Blée E, Joyard J. 1996. Envelope membranes from spinach chloroplasts are a site of the metabolism of fatty acid hydroperoxides. *Plant Physiol.* 110:445–54
18. Blée E, Schuber F. 1994. Oxylipins in plants: the peroxygenase pathway. See Ref. 63a, pp. 262–64
19. Braam J, Davis RW. 1990. Rain-, wind-,

and touch-induced expression of cal-modulin and calmodulin-related genes. *Cell* 60:357–64

20. Bradley DJ, Kjellbom P, Lamb CJ. 1992. Elicitor- and wound-induced oxidative cross-linking of a proline-rich plant cell wall protein: a novel, rapid defense response. *Cell* 70:21–30

21. Brash AR, Song WC. 1995. Structure-function features of flaxseed allene oxide synthase. *J. Lipid Mediat. Cell Signal* 12:275–82

22. Bunker TW, Koetje DS, Stephenson LC, Creelman RA, Mullet JE, Grimes HD. 1995. Sink limitation induces the expression of multiple soybean vegetative lipoxygenase mRNAs while the endogenous jasmonic acid level remains low. *Plant Cell* 7:1319–31

23. Chandra S, Heinstein PF, Low PS. 1996. Activation of a phospholipase A by plant defense elicitors. *Plant Physiol.* 110:979–86

24. Chaudhry B, Mueller-Uri F, Cameron-Mills V, Gough S, Simpson D, et al. 1994. The barley 60 kDa jasmonate-induced protein (JIP60) is a novel ribosome-inactivating protein. *Plant J.* 6:815–24

25. Choi D, Bostock RM, Avdiushko S, Hildebrand DF. 1994. Lipid-derived signals that discriminate wound- and pathogen-responsive isoprenoid pathways in plants: methyl jasmonate and the fungal elicitor arachidonic acid induce different 3-hydroxy-3-methylglutaryl-coenzyme A reductase genes antimicrobial isoprenoids in *Solanum tuberosum* L. *Proc. Natl. Acad. Sci. USA* 91:2329–33

26. Cohen Y, Gisi U, Niderman T. 1993. Local and systemic protection against *Phytophthora infestans* induced in potato and tomato plants by jasmonic acid and jasmonic methyl ester. *Phytopathology* 83:1054–62

27. Conconi A, Miquel M, Browse JA, Ryan CA. 1996. Intracellular levels of free linolenic and linoleic acids increase in tomato leaves in response to wounding. *Plant Physiol.* 111:797–803

28. Creelman RA, Mullet JE. 1995. Jasmonic acid distribution and action in plants: Regulation during development and response to biotic and abiotic stress. *Proc. Natl. Acad. Sci. USA* 92:4114–19

29. Creelman RA, Tierney ML, Mullet JE. 1992. Jasmonic acid/methyl jasmonate accumulate in wounded soybean hypocotyls and modulate wound gene expression. *Proc. Natl. Acad. Sci. USA* 89:4938–41

30. Czapski J, Saniewski M. 1992. Stimulation of ethylene production and ethylene-forming enzyme in fruits of the non-ripening nor and rin tomato mutants by methyl jasmonate. *J. Plant Physiol.* 139:265–68

31. Demmig-Adams B, Adams WW III. 1992. Photoprotection and other responses of plants to high light stress. *Annu. Rev. Plant Physiol. Plant Mol. Biol.* 43:599–626

32. Deng W, Hamilton-Kemp TR, Nielsen MT, Andersen RA, Collins GB, Hildebrand DF. 1993. Effects of six-carbon aldehydes and alcohols on bacterial proliferation. *J. Agric. Food Chem.* 41:506–10

33. DeWald DB, Mason HS, Mullet JE. 1992. The soybean vegetative storage proteins VSPa and VSPb are acid phosphatases active on polyphosphates. *J. Biol. Chem.* 267:15958–64

34. DeWald DB, Sadka A, Mullet JE. 1994. Sucrose modulation of soybean *Vsp* gene expression is inhibited by auxin. *Plant Physiol.* 104:439–44

35. Dixon RAF, Diehl RE, Opas E, Rands E, Vickers PJ, et al. 1990. Requirement of a 5-lipoxygeanse-activating protein for leukotriene synthesis. *Nature* 343:282–84

36. Doares SH, Narvaez-Vasquez J, Conconi A, Ryan CA. 1995. Salicylic acid inhibits synthesis of proteinase inhibitors in tomato leaves induced by systemin and jasmonic acid. *Plant Physiol.* 108:1741–46

37. Doares SH, Syrovets T, Weiler EW, Ryan CA. 1995. Oligogalacturonides and chitosan activate plant defensive genes through the octadecanoid pathway. *Proc. Natl. Acad. Sci. USA* 92:4095–98

38. Ehret R, Schab J, Weiler EW. 1994. Lipoxygenases in *Bryonia dioica* Jacq. tendrils and cell cultures. *J. Plant Physiol.* 144:175–82

39. Falkenstein E, Groth B, Mithofer A, Weiler EW. 1991. Methyljasmonate and a-linolenic acid are potent inducers of tendril coiling. *Planta* 185:316–22

40. Farmer EE. 1994. Fatty acid signalling in plants and their associated microorganisms. *Plant Mol. Biol.* 26:1423–37

41. Farmer EE, Caldelari D, Pearce G, Walker-Simmons MK, Ryan CA. 1994. Diethyldithiocarbamic acid inhibits the octadecanoid signaling pathway for the wound induction of proteinase inhibitors in tomato leaves. *Plant Physiol.* 106:337–42

42. Farmer EE, Pearce G, Ryan CA. 1989. In vitro phosphorylation of plant plasma membrane proteins in response to the proteinase inhibitor inducing factor. *Proc. Natl. Acad. Sci. USA* 86:1539–42

43. Farmer EE, Ryan CA. 1990. Interplant communication: airborne methyl jasmon-

ate induces synthesis of proteinase inhibitors in plant leaves. *Proc. Natl. Acad. Sci. USA* 87:7713–16

44. Farmer EE, Ryan CA. 1992. Octadecanoid precursors of jasmonic acid activate the synthesis of wound-inducible proteinase inhibitors. *Plant Cell* 4:129–34

45. Feussner I, Hause B, Voros K, Parthier B, Wasternack C. 1995. Jasmonate-induced lipoxygeanse forms are localized in chloroplast of barley leaves (*Hordeum vulgare* cv. Salome). *Plant J.* 7:949–57

46. Feussner I, Wasternack C, Kindl H, Kuhn H. 1995. Lipoxygenase-catalyzed oxygenation of storage lipids is implicated in lipid mobilization during germination. *Proc. Natl. Acad. Sci. USA* 92:11849–53

47. Feys BJF, Benedetti CE, Penfold CN, Turner JG. 1994. Arabidopsis mutants selected for resistance to the phytotoxin coronatine are male sterile, insensitive to methyl jasmonate, and resistant to a bacterial pathogen. *Plant Cell* 6:751–59

48. Finch-Savage WE, Blake PS, Clay HA. 1996. Desiccation stress in recalcitrant *Quercus robur* L seeds results in lipid peroxidation and increased synthesis of jasmonates and abscisic acid. *J. Exp. Bot.* 47: 661–67

49. Franceschi VR, Grimes HD. 1991. Induction of soybean vegetative storage proteins and anthocyanins by low-level atmospheric methyl jasmonate. *Proc. Natl. Acad. Sci. USA* 83:6745–49

50. Franceschi VR, Wittenbach VA, Giaquinta RT. 1983. Paraveinal mesophyll of soybean leaves in relation to assimilate transfer and compartmentation. *Plant Physiol.* 72: 586–89

51. Gardner HW. 1995. Biological roles and biochemistry of the lipoxygenase pathway. *HortScience* 30:197–205

52. Grimes HD, Koetje DS, Franceschi VR. 1992. Expression, activity, and cellular accumulation of methyl jasmonate-responsive lipoxygenase in soybean seedlings. *Plant Physiol.* 100:433–43

53. Gundlach H, Müller MJ, Kutchan TM, Zenk MH. 1992. Jasmonic acid is a signal transducer in elicitor-induced plant cell cultures. *Proc. Natl. Acad. Sci. USA* 89: 2389–93

54. Habenicht AJR, Salbach P, Goerig M, Zeh W, Janssen-Timmen U, et al. 1990. The LDL receptor pathway delivers arachidonic acid for eicosanoid formation in cells stimulated by platelet-derived growth factor. *Nature* 345:634–36

55. Hamberg M, Fahlstadius P. 1990. Allene oxide cyclase: a new enzyme in plant lipid

metabolism. *Biochem. Biophys. Acta* 276: 518–26

56. Hamberg M, Gardner HW. 1992. Oxylipin pathway to jasmonates: biochemistry and biological significance. *Biochim. Biophys. Acta* 1165:1–18

57. Harms K, Atzorn R, Brash A, Kühn H, Wasternack C, Willmitzer L, Peña-Cortés H. 1995. Expression of a flax allene oxide synthase cDNA leads to increased endogenous jasmonic acid (JA) levels in transgenic potato plants but not to a corresponding activation of JA-Responding Genes. *Plant Cell* 7:1645–54

58. Heilmann B, Hartung W, Gimmler H. 1980. The distribution of abscisic acid between chloroplasts and cytoplasm of leaf cells and the permeability of the chloroplast envelope for abscisic acid. *Z. Pflanzenphysiol.* 97:67–78

59. Hildebrand DF, Brown GC, Jackson DM, Hamilton-Kemp TR. 1993. Effects of some leaf-emitted volatile compounds on aphid population increase. *J. Chem. Ecol.* 19: 1875–87

60. Howe GA, Lightner J, Browse J, Ryan CA. 1996. An octadecanoid pathway mutant (JL5) of tomato is compromised in signalling for defense against insect attack. *Plant Cell.* In press

61. Huang J-F, Bantroch DJ, Greenwood JS, Staswick PE. 1991. Methyl jasmonate treatment eliminates cell-specific expression of vegetative storage protein genes in soybean leaves. *Plant Physiol.* 97:1512–20

62. Johnson R, Narvaez J, An GH, Ryan C. 1989. Expression of proteinase inhibitors I and II in transgenic tobacco plants: effects on natural defense against *Manduca sexta* larvae. *Proc. Natl. Acad. Sci. USA* 86: 9871–75

63. Johnson R, Ryan CA. 1990. Wound-inducible potato inhibitor II genes: enhancement of expression by sucrose. *Plant Mol. Biol.* 14:527–36

63a. Kader J-C, ed. 1994. *Plant Lipid Metabolism.* Dordrecht: Kluwer

64. Kato T, Ohta H, Tanaka K, Shibata D. 1992. Appearance of new lipoxygenases in soybean cotyledons after germination and evidence for expression of a major new lipoxygenase gene. *Plant Physiol.* 98:324–30

65. Kato T, Shirano Y, Iwamoto H, Shibata D. 1993. Soybean lipoxygenase L-4, a component of the 94-kilodalton storage protein in vegetative tissues: expression and accumulation in leaves induced by pod removal and by methyl jasmonate. *Plant Cell Physiol.* 34:1063–72

66. Kernan A, Thornburg RW. 1989. Auxin

levels regulate the expression of a wound-inducible proteinase inhibitor II-chloramphenicol acetyl transferase gene fusion in vitro and in vivo. *Plant Physiol.* 91:73–78

67. Kim SR, Choi JL, Costa MA, An GH. 1992. Identification of a G-box sequence as an essential element for methyl jasmonate response of potato proteinase inhibitor II promoter. *Plant Physiol.* 99:627–31

68. Koda Y. 1992. The role of jasmonic acid and related compounds in the regulation of plant development. *Int. Rev. Cytol.* 135:155–99

69. Koda Y, Kikuta Y. 1994. Wound-induced accumulation of jasmonic acid in tissues of potato tubers. *Plant Cell Physiol.* 35:751–56

70. Korsmeyer SJ. 1995. Regulators of cell death. *Trends Genet.* 11:101–5

71. Lopez R, Dathe W, Bruckner C, Miersch O, Sembdner G. 1987. Jasmonic acid in different parts of the developing soybean fruit. *Biochem. Physiol. Pflanzenphysiol.* 182:195–201

72. Lorbeth R, Dammann C, Ebneth M, Amati S, Sanchez-Serrano JJ. 1992. Promoter elements involved in environmental and developmental control of potato proteinase inhibitor II expression. *Plant J.* 2:477–86

73. Mason HS, DeWald DB, Creelman RA, Mullet JE. 1992. Coregulation of soybean vegetative storage protein gene expression by methyl jasmonate and soluble sugars. *Plant Physiol.* 98:859–67

74. Mason HS, DeWald DB, Mullet JE. 1993. Identification Of a methyl jasmonate-responsive domain In the soybean *vspB* promoter. *Plant Cell* 5:241–51

75. Mason HS, Mullet JE. 1990. Expression of two soybean vegetative storage protein genes during development and in response to water deficit, wounding, and jasmonic acid. *Plant Cell* 2:569–79

76. McConn M, Browse J. 1996. The critical requirement for linolenic acid is pollen development, not photosynthesis, in an Arabidopsis mutant. *Plant Cell* 8:403–16

77. McGurl B, Ryan CA. 1992. The organization of the prosystemin gene. *Plant Mol. Biol.* 20:405–9

78. Melan MA, Dong XH, Endara ME, Davis KR, Ausubel FM, Peterman TK. 1993. An Arabidopsis thaliana lipoxygenase gene can be induced by pathogens, abscisic acid, and methyl jasmonate. *Plant Physiol.* 101:441–50

79. Miller DK, Gillard JW, Vickers PJ, Sadowski S, Léveille C, et al. 1990. Identification and isolation of a membrane protein necessary for leukotriene production. *Nature* 343:278–81

79a. Moore TS, ed. 1993. *Lipid Metabolism in Plants.* Boca Raton, FL: CRC Press

80. Mueller MJ, Brodschelm W. 1994. Quantification of jasmonic acid by capillary gas chromatography-negative chemical ionization-mass spectrometry. *Anal. Biochem.* 218:425–35

81. Nellen A, Rojahn B, Kindl H. 1995. Lipoxygenase forms located at the plant plasma membrane. *Z. Naturforsch. Teil C* 50:29–36

82. Nojiri H, Sugimori M, Yamane H, Nishimura Y, Yamada A, et al. 1996. Involvement of jasmonic acid in elicitor-induced phytoalexin production in suspension-cultured rice cells. *Plant Physiol.* 110:387–92

83. Park TK, Polacco JC. 1989. Distinct lipoxygenase species appear in the hypocotyl/radicle of germinating soybean. *Plant Physiol.* 90:285–90

84. Pearce G, Strydom D, Johnson S, Ryan CA. 1991. A polypeptide from tomato leaves induces wound-inducible proteinase inhibitor proteins. *Science* 253:895–98

85. Pelacho AM, Mingo-Castel AM. 1991. Jasmonic acid induces tuberization of potato stolons cultured in vitro. *Plant Physiol.* 97:1253–55

86. Peña-Cortés H, Albrecht T, Prat S, Weiler EW, Willmitzer L. 1993. Aspirin prevents wound-induced gene expression in tomato leaves by blocking jasmonic acid biosynthesis. *Planta* 191:123–28

87. Peña-Cortés H, Fisahn J, Willmitzer L. 1995. Signals involved in wound-induced proteinase inhibitor II gene expression in tomato and potato plants. *Proc. Natl. Acad. Sci. USA* 92:4106–13

88. Peña-Cortés H, Liu XJ, Serrano JS, Schmid R, Willmitzer L. 1992. Factors affecting gene expression of patatin and proteinase-inhibitor-II gene families in detached potato leaves: implications for their co-expression in developing tubers. *Planta* 186:495–502

89. Peña-Cortés H, Willmitzer L. 1995. The role of hormones in gene activation in response to wounding. In *Plant Hormones,* ed. PJ Davies, pp. 395–414. Dordrecht: Klewer

90. Peng Y-L, Shirano Y, Ohta H, Hibino T, Tanaka K, Shibata D. 1994. A novel lipoxygenase from rice: primary structure and specific expression upon incompatible infection with rice blast fungus. *J. Biol. Chem.* 269:3755–61

91. Penninckx IAMA, Eggermont K, Terras

FRG, Thomma BPHJ, De Samblanx GW, et al. 1996. Pathogen-induced systemic activation of a plant defense gene in *Arabidopsis* follows a salicylic acid–independent pathway involving components of the ethylene and jasmonic acid responses. *Plant Cell.* In press

92. Ranjan R, Lewak S. 1992. Jasmonic acid promotes germination and lipase activity In nonstratified apple embryos. *Physiol. Plant.* 86:335–39

93. Ravnikar M, Vilhar B, Gogala N. 1992. Stimulatory effects of jasmonic acid on potato stem node and protoplast culture. *J. Plant Growth Regul.* 11:29–33

94. Reinbothe S, Reinbothe C, Heintzen C, Seidenbecher C, Parthier B. 1993. A methyl jasmonate-induced shift in the length of the 5' untranslated region impairs translation of the plastid *rbcL* transcript in barley. *EMBO J.* 12:1505–12

95. Reinbothe S, Reinbothe C, Parthier B. 1993. Methyl jasmonate represses translation initiation of a specific set of mRNAs in barley. *Plant J.* 4:459–67

96. Reinbothe S, Reinbothe C, Parthier B. 1993. Methyl jasmonate-regulated translation of nuclear-encoded chloroplast proteins in barley (*Hordeum vulgare* L. cv. salome). *J. Biol. Chem.* 268:10606–11

97. Roy S, Pouenat ML, Caumont C, Cariven C, Prevost MC, Esquerre-Tugaye MT. 1995. Phospholipase activity and phospholipid patterns in tobacco cells treated with fungal elicitor. *Plant Sci.* 107:17–25

98. Sadka A, Dewald DB, May GD, Park WD, Mullet JE. 1994. Phosphate modulates transcription of soybean *VspB* and other sugar-inducible genes. *Plant Cell* 6: 737–49

99. Saniewski M, Czapski J. 1983. The effect of methyl jasmonate on lycopene and β-carotene accumulation in ripening red tomato. *Experientia* 39:1373–74

100. Saravitz DM, Siedow JN. 1995. The lipoxygenase isozymes in soybean [*Glycine max* (L.) Merr.] leaves. *Plant Physiol.* 107:535–43

101. Schaller A, Bergey DR, Ryan CA. 1995. Induction of wound response genes in tomato leaves by bestatin, an inhibitor of aminopeptidases. *Plant Cell* 7:1893–98

102. Schmidt J, Kramell R, Schneider G. 1995. Liquid chromatography electrospray tandem mass spectrometry of amino acid conjugates of jasmonic acid under positive and negative ionisation. *Eur. Mass Spectrom.* 1:573–81

103. Sembdner G, Parthier B. 1993. The biochemistry and the physiological and molecular actions of jasmonates. *Annu. Rev. Plant Physiol. Plant Mol. Biol.* 44:569–89

104. Seo S, Okamoto M, Seto H, Ishizuka K, Sano H, Ohashi Y. 1995. Tobacco MAP kinase: A possible mediator in wound signal transduction pathways. *Science* 270: 1988–92

105. Simpson TD, Gardner HW. 1995. Allene oxide synthase and allene oxide cyclase, enzymes of the jasmonic acid pathway, localized in *Glycine max* tissues. *Plant Physiol.* 108:199–202

106. Smith CJ. 1996. Accumulation of phytoalexins: defence mechanisms and stimulus response. *New Phytol.* 132:1–45

107. Song WC, Brash AR. 1991. Purification of an allene oxide synthase and identification of the enzyme as a cytochrome P-450. *Science* 253:781–84

108. Song WC, Funk CD, Brash AR. 1993. Molecular cloning of an allene oxide synthase: a cytochrome P450 specialized for the metabolism of fatty acid hydroperoxides. *Proc. Natl. Acad. Sci. USA* 90:8519–23

109. Staswick PE. 1989. Developmental regulation and the influence of plant sinks on vegetative storage protein gene expression in soybean leaves. *Plant Physiol.* 89: 309–15

110. Staswick PE. 1989. Preferential loss of an abundant storage protein from soybean pods during seed development. *Plant Physiol.* 90:1252–55

111. Staswick PE. 1994. Storage proteins of vegetative plant tissues. *Annu. Rev. Plant Physiol. Plant Mol. Biol.* 45:303–22

112. Staswick PE, Huang J-F, Rhee Y. 1991. Nitrogen and methyl jasmonate induction of soybean vegetative storage protein genes. *Plant Physiol.* 96:130–36

113. Staswick PE, Su W, Howell SH. 1992. Methyl jasmonate inhibition of root growth and induction of a leaf protein are decreased in an *Arabidopsis thaliana* mutant. *Proc. Natl. Acad. Sci. USA* 89:6837–40

114. Sterk P, Booij H, Schellekens GA, Van Kammen A, De Vries SC. 1991. Cell-specific expression of the carrot EP2 lipid transfer protein gene. *Plant Cell* 3:907–21

115. Tranbarger TJ, Franceschi VR, Hildebrand DF, Grimes HD. 1991. The soybean 94-kilodalton vegetative storage protein is a lipoxygenase that is localized in paravienal mesophyll cell vacuoles. *Plant Cell* 3: 973–87

116. Ueda J, Kato J, Yamane H, Takahashi N. 1981. Inhibitory effect of methyl jasmonate and its related compounds on kinetin-induced retardation of oat leaf senescence. *Physiol. Plant.* 52:305–9

117. Ueda J, Miyamoto K, Kamisaka S. 1994. Separation of a new type of plant growth regulator, jasmonates, by chromatographic procedures. *J. Chromatogr.* 658:129–42

118. Ueda J, Miyamoto K, Kamisaka S. 1995. Inhibition of the synthesis of cell wall polysaccharides in oat coleoptile segments by jasmonic acid: relevance to its growth inhibition. *J. Plant Growth Regul.* 14:69–76

119. Vick BA. 1993. Oxygenated fatty acids of the lipoxygenase pathway. See Ref. 79a, pp. 167–91

120. Vick BA. 1994. Temporal and organ-specific expression of enzymes of fatty acid hydroperoxide metabolism in developing sunflower seedlings. See Ref. 63a, pp. 280–82

121. Wang X. 1993. Phospholipases. See Ref. 79a, pp. 505–25

122. Weidhase RAE, Kramell HM, Lehmann J, Liebisch HW, Lerbs W, Parthier B. 1987. Methyljasmonate-induced changes in the polypeptide pattern of senescing barley leaf segments. *Plant Sci.* 51:177–86

123. Weiler EW, Albrecht T, Groth B, Xia ZQ, Luxem M, et al. 1993. Evidence for the involvement of jasmonates and their octadecanoid precursors in the tendril coiling response of *Bryonia dioica*. *Phytochemistry* 32:591–600

124. Wildon DC, Thain JF, Minchin PEH, Gubb IR, Reilly AJ, et al. 1992. Electrical signalling and systemic proteinase inhibitor induction in the wounded plant. *Nature* 360:62–65

125. Wilen RW, Rooijen GJH, Pearce DW, Pharis RP, Holbrook LA, Moloney MM. 1991. Effects of jasmonic acid on embryospecific processes in *Brassica* and *Linum* oilseeds. *Plant Physiol.* 95:399–405

126. Williams ME, Foster R, Chua NH. 1992. Sequences flanking the hexameric G-box core CACGTG affect the specificity of protein binding. *Plant Cell* 4:485–96

127. Wittenbach VA. 1982. Effect of pod removal on leaf senescence in soybeans. *Plant Physiol.* 70:1544–48

128. Wittenbach VA. 1983. Purification and characterization of a soybean leaf storage glycoprotein. *Plant Physiol.* 73:125–29

129. Xu Y, Chang PFL, Liu D, Narasimhan ML, Raghothama KG, et al. 1994. Plant defense genes are synergistically induced by ethylene and methyl jasmonate. *Plant Cell* 6:1077–85

Annu. Rev. Plant Physiol. Plant Mol. Biol. 1997. 48:383–98

PLANT IN VITRO TRANSCRIPTION SYSTEMS

Masahiro Sugiura

Center for Gene Research, Nagoya University, Nagoya, 464-01, Japan

KEY WORDS: plant genes, RNA polymerase I, Pol II, Pol III

ABSTRACT

In vitro transcription systems provide a powerful tool for detailed analysis of transcription reactions including initiation, elongation, and termination. Despite problems inherent to plant cells, efforts have been made to develop plant in vitro transcription systems in the past decade. These efforts have finally culminated in the development of reliable in vitro systems from suspension-cell cultures of both monocot and dicot plants. These systems can be useful in elucidating the specific mechanisms involved in the process of plant transcription and thus can potentially open a new era of transcription studies in plants.

CONTENTS

1040-2519/97/0601-0383$08.00

INTRODUCTION

The growth and development of plants and animals depend on the concerted expression of genes. Transcription is the essential regulatory step for the expression of nuclear genes encoding not only mRNAs but also stable RNAs, and this in turn involves a complex set of nucleic acid–protein and protein-protein interactions. Our current understanding of the fundamental and intricate mechanisms involved in the process and regulation of eukaryotic transcription owes a great deal to the in vitro systems prepared from animal and yeast cells.

In vitro systems provide the most significant tool for analyzing individual steps in transcription, that is, initiation, elongation, and termination. Furthermore, in vitro functional assays allow the biochemical identification or isolation of factors involved in each step. These systems also facilitate screening of functionally significant motifs through the analysis of many deletion and/or scanning-substitution mutations. This would practically be impossible using transgenic plants. Furthermore, interaction of these motifs with *trans*-acting factors, whose availability might in turn depend on the internal and/or external cues, can also be studied using specific in vitro systems. These systems can be useful for studying the role of protein-protein interaction and the possible involvement of low–molecular mass substances in transcription. Therefore, the availability of versatile in vitro transcription systems from plant cells has long been awaited (32, 37, 47).

To date, delineation of molecular aspects of transcription has mainly depended on time- and labor-consuming methods involving transgenic plants such as tobacco. In vitro transcription of several plant genes was also attempted using heterologous systems derived, for example, from HeLa cells and yeast cells. Such heterologous systems could faithfully initiate transcription from several plant genes, including those encoding maize zein (5, 6, 26), pea legumin (12), and cauliflower mosaic virus (CaMV) DNA (19). Heterologous systems could also be used to analyze basic processes of transcription and of similarity of promoter elements and protein factors between plants and animals, but they would not elucidate processes unique to plant cells. Therefore, no further advancements were made using these systems.

Recent attempts were made to develop RNA polymerase II–dependent in vitro transcription systems derived from several plant species. However, these systems were either inefficient or of limited use. Finally, a long-awaited technical advance was achieved with the development of convenient, versatile, and reliable in vitro transcription systems from dicotyledonous (tobacco) and monocotyledonous (rice) plants (13, 48).

PREPARATION OF IN VITRO TRANSCRIPTION SYSTEMS

Assay Methods

In addition to the quality of the extract, the success of any in vitro system depends largely on the sensitivity of the assay method. A good assay method can selectively distinguish de novo synthesized RNAs (from exogenously added template DNA) from the extensive background consisting of endogenous transcripts from cell extracts, transcripts resulting from endogenous DNA incompletely removed during the preparation of extracts, RNA caused by nonspecific initiation and termination on the added template DNA, and products with nonspecific incorporation of a labeled substrate by side reactions. To achieve an acceptable level of sensitivity while minimizing the effect of these nonspecific contaminants, several assay methods have been tested.

RUN-OFF ASSAY   In vitro reactions are generally carried out using linear DNA templates, and transcription is allowed to proceed up to the end of the fragment. Specific initiation sites can be calculated from the size of a transcript(s) using polyacrylamide gel electrophoresis (PAGE). This assay is simple but requires higher activities of transcription elongation; otherwise, premature termination can cause unacceptably high levels of background. One way to enhance the detection of this procedure is the use of poly (dA) tail at the end of the template and subsequent selection of de novo full-length in vitro transcripts using oligo (dT) columns (30). Lower sensitivities are possible with this method because of the limitation of using linear templates, which are less active than circular templates (33, 48). Limited success in the past might result from the use of linear DNA templates in this assay. A similar procedure with circular DNA templates encoding small stable RNAs can facilitate the analysis of the transcription process as a whole, including initiation, elongation, and termination.

TRANSCRIPTION COMPLEX ASSAY   Soon after the transcription reaction mixture is assembled—consisting of an extract, template DNA, and labeled nucleoside triphosphates, or NTPs, in a transcription buffer—initiation complexes form and synthesize short oligoribonucleotides. The reaction is stopped by addition of Sarkosyl, which removes most proteins from the DNA but leaves the stable transcription complexes intact. The amount of labeled oligoribonucleotides can be subsequently estimated by PAGE. Inefficiency of in vitro systems often results from a block in the elongation of RNA chains. This method can be used even when specific transcripts are not detected by other assays and can lead to the tentative identification of transcription initiation factors (1).

S1 MAPPING AND RNASE PROTECTION ASSAYS Both of these methods work on the principle of specific hybridization of de novo transcripts to either labeled sense DNA strand or antisense RNA followed by nuclease digestion of unhybridized nucleic acid molecules. Thereafter, the site of transcription initiation is estimated by analyzing the size of nuclease-protected DNA/RNA fragments using PAGE. These methods are extremely sensitive, but they also detect signals from endogenous RNAs, transcripts resulting from endogenous DNA, and premature termination of in vitro reaction, with equally high sensitivity. Hence, this often results in exceptionally high backgrounds. Multiple sizes of product could indicate multiple initiation sites or multiple premature termination sites. Furthermore, sites outside the region of the probe cannot be detected.

G-FREE SEQUENCE SYSTEM The template used in this assay contains a promoter region followed by a defined length of DNA (~100 bp) lacking G residues on the sense strand in a plasmid vector. When the reaction is carried out in the absence of GTP or in the presence of RNase T1, RNA synthesis terminates at the first G in the plasmid sequence. Specific initiation at the promoter is determined by analyzing the size of transcripts. This assay is fast and quantitative and provides a more enzymological approach to the analysis of specific transcription initiation, e.g. purification of transcription factors (31).

PRIMER EXTENSION ASSAY In this procedure, the in vitro products are used as templates for the synthesis of cDNAs (extended products) using reverse transcriptase (RT) and a $(5'^{32}P)$-labeled primer complementary to a portion of template DNA used (23). Specific transcripts can be detected on gel by expected size for accurate initiation (from the 5' end of primers to in vivo initiation sites). Because this assay uses only those in vitro transcripts that have elongated farther than the position of oligonucleotide primer, there can be a considerable reduction of background signals resulting from premature termination of an in vitro reaction. The use of primers complementary to a reporter gene sequence or a vector sequence (a nonplant DNA sequence) reduces background from nonspecific transcription from endogenous DNA or RNAs. This assay together with DNA sequence ladders resolves transcriptional initiation sites at the nucleotide level. Although the method is lengthier, it prevents artifactual results sometimes observed even when the G-free sequence system is used (14). One technical problem is that premature cessation of RT action results in multiple products; some plant genes are transcribed from several authentic sites but usually multiple-size fragments are an artifact.

Plant Materials

The preparation of high-quality in vitro systems critically depends on the quality of starting material. For development of plant in vitro transcription systems, two main sources have been extensively exploited: wheat germ and suspension-cell cultures.

WHEAT GERM   Wheat germ has been widely used for the preparation of in vitro translation extracts that have also been reported to support pre-tRNA processing and splicing in vitro (35). It is the only plant source whose RNA polymerases have been extensively purified and characterized (21, 22). One of the reasons it is the material of choice is relatively low levels of nuclease and protease activities. However, it is not always easy to obtain good batches of wheat germ suitable for in vitro systems (42). Wheat germ is essentially a dormant material and may be deficient or inactivated with regard to factors required for general transcription (30). It is known to contain several inhibitors of the transcription that need to be removed before active systems can be developed (16, 18, 33). Extracts of isolated wheat shoot nuclei also contain inhibitors for transcription in a HeLa cell extract (20).

SUSPENSION-CELL CULTURES   Rapidly growing established cell cultures provide a continuous source of homogenous cells with defined growth characteristics. Suspension cultures of tobacco, rice, wheat, soybean, and parsley have been used for development of in vitro transcription systems. Because these cells contain large vacuoles, it is necessary to remove vacuole constituents from in vitro systems, e.g. by isolating nuclei or by fractionating whole cell extracts. Because these undifferentiated cells are probably devoid of any cell type–specific factors they are suitable for preparing "basal" in vitro transcription systems.

Preparation of Extracts

Plant cells generally contain high levels of proteases, nucleases, and other substances that may cause inactivation of transcription machinery during the preparation of extracts. Therefore, protease and phosphatase inhibitors and other protecting reagents are routinely added in extraction buffers. Moreover, the time of extract preparation should be minimized to reduce possible damage to the active constituents involved in transcription. Furthermore, plant cells are low in protein concentration compared with animal cells; therefore, a minimal volume of extraction buffers should be used to maintain higher protein concentrations. Nuclei, the site of transcription, are obviously the best choice for extracts preparation; nevertheless, whole cells (or tissues) have also been used because they are easier to handle.

WHOLE CELL EXTRACTS   Cells are disrupted mechanically by a Waring blender or a Bead Beater so that a large amount of starting material can be processed. Crude extracts containing endogenous DNA are generally inactive and require further fractionation, e.g. by column chromatography or ammonium sulfate precipitation. These treatments remove hazardous compounds as well as endogenous nucleic acids.

NUCLEAR EXTRACTS   Using isolated nuclei as starting material avoids the potential hazard of cytoplasm (mainly vacuoles) and contamination of transcription activities from other organelles. For the isolation of intact nuclei, cells are gently disrupted, a step that is best achieved by osmotic lysis of protoplasts. Ficoll is added for adequate osmotic pressure to prevent the loss of nuclear contents by diffusion during preparation of nuclei. This is followed by extraction of nuclear proteins by high salt concentrations (30). Endogenous DNA is removed by ultracentrifugation. It is rather difficult to handle a large amount of cells using this procedure because of practical limitations.

RNA POLYMERASE II-DEPENDENT TRANSCRIPTION

RNA polymerase II (Pol II) is responsible for transcription of mRNAs and a class of U snRNAs. Purified Pol II recognizes and accurately transcribes mRNA promoters only when supplemented with additional factors present in crude cellular and nuclear extracts. In plant genes, most of the work has been devoted to the process of transcription initiation of mRNA genes, whose expression is differentially regulated by developmental and by external cues. Many cis-elements have been defined by transgenic assays, and a number of proteins interacting with DNA motifs have been isolated by affinity methods. There are a few studies elucidating detailed biochemical processes of transcriptional initiation, such as functional interaction between the cis-element and the DNA-binding protein. The analysis of these problems would be considerably facilitated by the use of in vitro transcription systems specific to plants.

Extracts Prepared from Wheat Germ

Initial attempts to develop in vitro Pol II–dependent transcription systems were made using wheat germ (1, 45). Ackerman et al (1) prepared extracts from wheat germ whole cells, nuclei, and cytosols, and found that these extracts, especially a nuclear extract, support the formation of initiation complexes on externally added plant promoters. Although these extracts were not active in the Pol II run-off assay, some factors necessary for transcription initiation could be isolated by this assay method. This extract did not contain apparent RNase or DNase but inhibited transcription when added to a HeLa cell in vitro transcription system. The nuclear extract preparations were then fractionated, and inhibitors were removed by using a HeLa system for assays. A reconstituted system thus obtained selectively transcribed plant genes but produced an extremely low level of full-sized transcripts because of the apparent block in elongation at 20–30 nucleotides (nt) (16).

A wheat protein fraction (termed KB) that substitutes for the HeLa TFIIA was isolated by substituting HeLa fractions with those of wheat using a HeLa in vitro system (7). The wheat TFIIA homologue was a single polypeptide of ~35 kDa. The CaMV 35S promoter directed accurate and efficient transcription in the wheat/HeLa as well as a pure HeLa component assay. Therefore, the wheat TFIIA seems to convert Pol II to a more processive enzyme, which suggests that TFIIA functions during elongation and is also necessary for initiation complex assembly (10). Furthermore, wheat and HeLa TATA-binding proteins were enzymatically similar (S Ackerman, personal communication). This was also the case for a TATA-binding protein fractionated from Arabidopsis (27). These observations suggest that basic mechanisms of transcription are highly conserved among eukaryotes. Using the HeLa system, the minimal promoter of the CaMV 35S gene could be defined as −35 through TATA and the initiator sequence to +5 (34).

Wheat Germ Chromatin Extract

Yamazaki & Imamoto (45) prepared a soluble chromatin fraction from wheat germ crude extract by Polymer P fractionation followed by ammonium sulfate precipitation that was found to direct accurate transcription initiation from exogenously added transcript 7 gene (TC7) promoter of the T-DNA region of Ti plasmid, as measured by run-off and primer extension assays. The run-off transcript was estimated to amount to ~80% of the total RNA synthesized, and its synthesis was completely inhibited by low concentrations of α-amanitin, a potent inhibitor for Pol II. This was the first report of accurate in vitro transcription of a plant-related gene. Using deletion experiments, they further suggested that the TATA-box is an important determinant of accurate initiation, and the region between +181 to +242 bp, in the middle of the coding region, is required for efficient initiation and elongation of TC7 transcription in vitro. A striking characteristic of the system was that accurate transcription initiation was most active in the presence of 0.5-1 mM of Mn^{2+}; far less activity was obtained using Mg^{2+} (44). The chromatin extract also supported accurate transcription initiation of a circular DNA containing the CaMV 35S promoter (44). Using this in vitro system with the 35S promoter containing the TATA-box and the upstream activation sequence, *as-1*, the tobacco DNA-binding protein TGA1a for *as-1* was reported to stimulate transcription three- to fivefold through increasing the number of active preinitiation complexes (46). TGA1a can also function as a transcription activator in a HeLa system with the 35S promoter (24), again suggesting high conservation of basic processes of transcription in eukaryotes.

Using the chromatin extracts with the G-free sequence assay, Schweizer & Mösinger (33) examined the sequence requirements for faithful and efficient transcription initiation of a series of chimeric promoter constructs from several plant genes. They found that the chromatin extract transcribes the parsley chalcone synthase promoter in an initiator sequence- (encompassing the transcriptional start site, positions −7 to +13) dependent manner, but not in a TATA-dependent manner. Their result was not in agreement with the earlier report on TATA-box–dependent in vitro transcription from a linear TC7 promoter (45) and a circular CaMV 35S DNA (44). In their analysis, requirement for circular DNA was obligatory, and presence of 2 mM Mg^{2+} was preferred over 1 mM Mn^{2+}. Moreover, the in vitro transcription initiation site of the parsley pathogenesis-related protein 2 promoter was different from the in vivo site [24 nt upstream from the in vivo site (33)]. These observations suggested that the wheat germ chromatin extract represented only a partly functional in vitro system, and in vitro systems from wheat germ would need further refinement before they could be used for transcription analysis using genuine plant promoters.

Cultured Cell Extracts

Efforts were also made to develop in vitro transcription systems from several plant cultured cells. Cooke & Penon reported that although a whole cell extract is transcriptionally inactive, an in vitro system derived from tobacco-cultured cells only after a single chromatographic separation can direct selective transcription from the CaMV 19S promoter (8, 9). Using this system, transcription from the 35S promoter led to the accumulation of short RNAs, although the 19S promoter was suggested to be weaker than the most widely used 35S promoter (19).

Roberts & Okita (30) presented a simple method for the preparation of nuclear extracts from cultured cells of rice, wheat, or tobacco. These extracts were shown to initiate transcription from the wheat gliadin and CaMV 35S promoters. They used protoplasts to prepare the nuclear extracts from cultured cells. In their procedures, high concentration of Ficoll was used during nuclear isolation, nuclear proteins were recovered using ammonium sulfate, and oligo (dT) selection of run-off transcripts was employed for their assay to reduce nonspecific transcripts.

A major problem in using cultured cell extracts is the high background levels, which hinder the identification and quantification of desired exogenous template-dependent transcription products. To avoid this problem, Arias et al (3) used immobilized DNA templates. Transcription complexes were assembled on the soybean chalcone synthase 15 promoter (CHS 15)–containing templates coupled to agarose beads using homologous whole cell and nuclear extracts from soybean-cultured cells. These beads were then washed to remove unbound materials and incubated with labeled substrates to allow transcription. While increasing the recovery of exogenous DNA-dependent transcripts, the washing of immobilized transcription complexes considerably reduced the background. This homologous in vitro system directed accurate and efficient transcription initiation from the CHS15 promoter. By using this system, it was shown that *trans* factors that bind to G-box (CACGTG, −74 to −69) and H-box (CCTACC, −61 to −56 and −121 to −126) *cis*-elements, respectively, greatly contribute to the transcription of the CHS15 promoter in vitro, and that both *cis*-element/*trans*-factor interactions in combination are required for maximal activity. Authentic transcription from the CHS15 promoter was also observed with whole cell extracts from bean-, tobacco-, and rice-cultured cells, and the soybean whole cell extract transcribed several other immobilized promoters. Although it requires additional steps (e.g. immobilization and washing) and confines transcription to a single cycle per assay, the system is useful for analysis of transcriptional initiation of plant promoters.

Yamaguchi et al (42) also reported an in vitro system from isolated nuclei of tobacco-cultured BY-2 cells after protoplast formation. Their system directs accurate transcription initiation from the TC7 promoter by using G-free sequence assay. Characteristically, the assay was again found to be most accurate and active only in the presence of 1 mM Mn^{2+}. The requirement of Mn^{2+}, therefore, seems to be unique to the TC7 or a limited class of plant promoters. However, Mn^{2+} is known to alter the specificity of templates and substrates in DNA and RNA polymerase reactions (11).

Evacuolated Protoplast Extracts

Plant cells have large vacuoles; when ruptured they dilute the extracts and contain many compounds that may damage transcription components. Frohnmeyer et al (17) reported a light-responsive in vitro transcription system from evacuolated protoplasts of parsley-cultured cells. They further showed that light-treated lysates preferentially enhance the transcription activity from a transformed parsley chalcone synthase gene compared with dark-treated lysates. Similar conditions, however, had no enhancement effect on the constitutively transcribed transformed CaMV 35S promoter. Although their system cannot be used for externally added templates in the reaction mixtures, it remarkably retains the light responsiveness for the CHS gene, indicating the maintenance of a largely intact signal transduction chain between photoreceptor(s) and the promoter. Thus if exploited, it can prove to be a valuable tool for elucidating signaling mechanisms, at least for light response.

Convenient In Vitro Systems

The in vitro transcription systems described above were either of limited use or needed special skill or laborious procedures. Recently, a long-awaited technical advance has been achieved in the development of convenient and versatile in vitro systems by two groups (37, 47). Zhu et al (48) have developed a convenient in vitro transcription system using rice and tobacco whole cell extracts and circular DNA templates. Using a procedure based on that reported by Arias et al (3), they prepared extracts from suspension cultures in the exponential to early stationary phases of growth under carefully controlled culture conditions. Breakage of cells was done using a Bead Beater homogenizer, and the homogenate was fractionated using ammonium sulfate. The transcription reactions were optimized using templates containing homologous promoters either from the rice phenylalanine ammonia-lyase (PAL) gene or from the tobacco sesquiterpene cyclase gene. Accurate initiation of transcription, using circular but not linear templates, was unambiguously detected by primer extension assays. The optimized rice in vitro system supports three to four cycles of transcription on each transcriptionally competent template per assay, enough for a wide range of applications for analyzing the Pol II–dependent transcription. Moreover, these extracts can be stored more than one year at $-80°C$ without losing significant activity.

Using this system, Zhu et al (49) pursued detailed dissection of the functional architecture of plant minimal promoters using the rice extracts with many mutagenized promoters from the rice PAL gene. This proved for the first time that the TATTTAA sequence (positions -35 to -28) is an authentic

TATA-box essential for Pol II–dependent transcription. Moreover, the −1 and +1 sites of the initiator sequence and the spacing between the TATA box and initiation site were necessary for the correct placement of the initiation site. Their key findings were confirmed by in vivo experiments using a homologous system—a rice gene, rice extracts, and rice plants. A rice basic chitinase gene and a rice tungro bacilliform virus promoter were also accurately transcribed (41, 47). Furthermore, several DNA-binding proteins that are possibly involved in the regulation of rice gene transcription have been isolated from the rice extract (47; Q Zhu, personal communication).

We have also developed a basal in vitro transcription system from rapidly grown, nongreen cultured BY-2 cells of tobacco (13, 14). Intact nuclei were isolated from protoplasts and disrupted using a high concentration of salt followed by ultracentrifugation to remove endogenous DNA. The in vitro reaction was optimized using a circular DNA template containing the tobacco β-1,3-glucanase promoter and primer extension assay. The system supported accurate transcription initiation not only from tobacco β-1,3-glucanase gene but also the CaMV 35S promoter, the adenovirus 2 major late promoter, and the simian virus 40 early major promoter. The tobacco nuclear extract supported ~1.5 cycles of transcription on each transcriptionally competent template per assay.

Nuclear genes encoding components of the photosynthesis apparatus are actively transcribed in green leaves under illumination but are poorly expressed in dark-grown plants and other nonphotosynthetic organs. The tobacco in vitro system was then used to transcribe a tomato (close to tobacco) gene encoding the small subunit of ribulose-1,5-bisphosphate carboxylase/oxygenase (rbcS3C) whose expression is tissue specific and light dependent. As expected because BY-2 cells are undifferentiated and contain nonphotosynthetic plastids, the tomato promoter was inactive in the BY-2 in vitro system, whereas accurate transcription of this promoter was observed by supplementing the BY-2 system with a nuclear extract from light-grown tomato seedlings. By using the activated in vitro system, the functional elements of the rbcS3C promoter were analyzed. A sequence 351 bp upstream from the transcription initiation site was essential for transcription, and the region between −351 to −441 bp enhances transcription. The basal/activated BY-2 system may be useful for analyzing a wide range of Pol II–dependent transcription reactions and for identifying specific signals from differentiated tissues or organelles, which modulate transcription activity. This system also supports Pol I– and Pol III–dependent transcription, making it the most versatile system so far.

POL I– AND POL III–DEPENDENT TRANSCRIPTION

Pol I Transcription

Pol I transcribes tandemly repeated rRNA gene (rDNA) clusters consisting of 17S rDNA, 5.8S rDNA, and 25S rDNA. Moreover, Pol I–dependent transcription is known to be species specific, at least in mammalian cells. Yamashita et al (43) recently reported an in vitro system for transcription of the *Vicia faba* rDNA. The soluble whole cell extract prepared from *V. faba* embryonic axes initiated transcription of a linear template containing a *V. faba* rDNA promoter in the presence of α-amanitin, which inhibits Pol II (at a low concentration) and Pol III (at a high concentration) but not Pol I. In addition, the major transcript species was determined by run-off and primer extension analyses to initiate at the same position previously determined by S1 analysis in vivo.

Using the tobacco in vitro system, we also showed transcription from a circular tobacco rDNA template as determined by primer extension (13, 15). The transcription initiation site was found at an A, which corresponds to that lying within the consensus sequence surrounding plant pre-rRNA initiation (28). Because this transcription was resistant to a high concentration of α-amanitin, the in vitro system most likely supports accurate initiation of tobacco rDNA transcription by Pol I. The in vivo transcription initiation site of tobacco rDNA has not been determined because of rapid processing of pre-rRNA (K Yakura, personal communication). The in vitro assay can help in overcoming this difficulty. The system, however, did not support transcription of *V. faba* rDNA, indicating that plant Pol I–dependent rDNA transcription is also species specific (15).

Pol III Transcription

Pol III is responsible for transcription of tRNAs, 5S rRNA, U6 snRNA, and 7SL RNA. Plant U3 snRNA genes were transcribed by Pol III and not by Pol II as in vertebrates or lower eukaryotes (25).

A gene encoding the major cytoplasmic tRNA[Tyr] from *Nicotiana rustica* contains a 13-bp intron and was transcribed efficiently by Pol III in a HeLa nuclear extract (40). The pre-tRNA was subsequently processed and spliced into its mature size tRNA[Tyr] in the HeLa extract. Wheat germ extracts were then shown to accurately process, splice, and modify the pre-tRNA[Tyr] transcribed in HeLa extracts (35). Human and yeast cell extracts were also shown to transcribe several other plant tRNA genes (2, 4, 29, 36, 38, 39).

We estimated Pol III activity to be ~10% of the total RNA polymerase activities in the tobacco in vitro system (13). Therefore, in vitro transcription from a circular DNA template containing an *Arabidopsis* U6 snRNA promoter was conducted, and the in vitro products were assayed by primer extension. Transcription initiated at the same site as in vivo, and it was resistant to a low concentration of α-amanitin, which inhibits Pol II, indicating that the tobacco in vitro system also supports accurate initiation of U6 transcription. We can improve the tobacco system with respect to its Pol III activity, which now allows us to carry out run-off assay with full-size genes. The improved system supported accurate transcription initiation and termination from *Arabidopsis* U3, U6, and 7SL RNA genes, and also from Arabidopsis tRNA[Ser] genes (Y Yukawa, unpublished data). In vitro transcripts from these Pol III–dependent genes are small (~100 nt) and stable. Therefore we are now in a position to investigate the transcription process in its totality, i.e. initiation, elongation, and termination.

PROBLEMS AND PROSPECTS

Because of the lack of reliable plant in vitro transcription systems, our understanding of transcription processes and their regulation in plants has not been at par with that in animals and yeast. Hence, the gap has often been filled using the knowledge obtained from animal and yeast systems. This might not always be right; for example, plant U3 snRNA genes were transcribed by Pol II and not by Pol III, as in animals and yeast (25).

   Recently, in vitro transcription systems, which are potentially useful for studying a wide range of transcription processes, from suspension-cultured cells of rice and tobacco have been reported. Results obtained by in vitro experiments are often necessary to be confirmed by in vivo analysis, or vice versa. Rice and tobacco have been the best choices because these have also been favorites for in vivo studies regarding gene expression, including transgenic assays. Moreover, these plant cells can easily and economically be cultured in a large amount necessary for isolating transcription factors.

   The plant in vitro transcription system is an essential tool to understand plant-specific processes, such as light-response and chloroplast signals. For wide use, the in vitro system should be reproducible in other laboratories, and should be as simple and versatile as possible. Therefore, further improvement and optimization of existing plant in vitro systems are still necessary. Because of the complex nature of transcription (also splicing and translation), protocols for extract preparation are not as complete and elaborate as those of DNA sequencing or gene cloning. Therefore, personal experience is often necessary to obtain active extracts. Fractionation of extracts and reconstitution of subfractions or purified components are the next steps in understanding transcription in plants. Nevertheless, plant in vitro transcription systems have presented encouraging possibilities for the analysis of biochemical processes of transcription from plant genes and may provide new insights into processes that are unique to plant cells.

ACKNOWLEDGMENTS

I thank Sanjay Kapoor for critical reading of the text. I am also grateful to M Sugita, Y Yukawa, K Yakura, S Ackerman, Q Zhu, K Yamazaki, H Beier, R Cooke, and P Schweizer for providing information.

> **Visit the *Annual Reviews* home page at http://www.annurev.org.**

Literature Cited

1. Ackerman S, Flynn PA, Davis EA. 1987. Partial purification of plant transcription factors. I. Initiation. *Plant Mol. Biol.* 9: 147–58
2. Arends S, Kraus J, Beier H. 1996. The tRNA[Tyr] multigene family of *Triticum aestivum*: genome organization, sequence analyses and maturation of intron-containing pre-tRNAs in wheat germ extract. *FEBS Lett.* 384:222–26
3. Arias JA, Dixon RA, Lamb CJ. 1993. Dissection of the functional architecture of a plant defense gene promoter using a homologous in vitro transcription initiation system. *Plant Cell* 5:485–96
4. Beier D, Beier H. 1992. Expression of variant nuclear *Arabidopsis* tRNA[Ser] genes and pre-tRNA maturation differ in HeLa, yeast and wheat germ extracts. *Mol. Gen. Genet.* 233:201–8
5. Boston RS, Goldsbrough PB, Larkins BA. 1985. Transcription of a zein gene in heterologous plant and animal systems. In *Plant Genetics*, ed. M Freeling, pp. 629–39. New York: Liss
6. Boston RS, Larkins BA. 1986. Specific transcription of a 15-kilodalton zein gene in HeLa cell extracts. *Plant Mol. Biol.* 7: 71–79
7. Burke C, Yu XB, Marchitelli L, Davis EA, Ackerman S. 1990. Transcription factor IIA of wheat and human function similarly with plant and animal viral promoters. *Nucleic Acids Res.* 18:3611–20
8. Cooke R, Penon P. 1990. In vitro transcription from cauliflower mosaic virus promot-

ers by a cell-free extract from tobacco cells. *Plant Mol. Biol.* 14:391–405
9. Cooke R, Penon P. 1990. In vitro transcription of class II promoters in higher plants. *Methods Mol. Biol.* 49:271–89
10. de Mercoyrol L, Job C, Ackerman S, Job D. 1989. A wheat-germ nuclear fraction required for selective initiation in vitro confers processivity to wheat-germ RNA polymerase II. *Plant Sci.* 64:31–38
11. Dixon M, Webb EC. 1980. Enzyme biosynthesis. In *Enzymes,* pp. 570–621. London: Longman. 3rd ed.
12. Evans IM, Bown D, Lycett GW, Croy RRD, Boulter D, Gatehouse JA. 1985. Transcription of a legumin gene from pea (*Pisum sativum* L.) in vitro. *Planta* 165:554–60
13. Fan H, Sugiura M. 1995. A plant basal in vitro system supporting accurate transcription of both RNA polymerase II- and III-dependent genes: supplement of green leaf component(s) drives accurate transcription of a light-responsive *rbcS* gene. *EMBO J.* 14:1024–31
14. Fan H, Sugiura M. 1996. Basal and activated in vitro transcription in plants by RNA polymerase II and III. *Methods Enzymol.* 273:268–77
15. Fan H, Yakura K, Miyanishi M, Sugita M, Sugiura M. 1995. In vitro transcription of plant RNA polymerase I-dependent rRNA genes is species-specific. *Plant J.* 8:295–98
16. Flynn PA, Davis EA, Ackerman S. 1987. Partial purification of plant transcription factors. II. An in vitro transcription system is inefficient. *Plant Mol. Biol.* 9:159–69

17. Frohnmeyer H, Hahlbrock K, Schäfer E. 1994. A light-responsive in vitro transcription system from evacuolated parsley protoplasts. *Plant J.* 5:437–49

18. Furter R, Hall BD. 1991. Substances in nuclear wheat germ extracts which interfere with polymerase III transcriptional activity in vitro. *Plant Mol. Biol.* 17:773–85

19. Guilley H, Dudley RK, Jonard G, Balàzs E, Richards KE. 1982. Transcription of cauliflower mosaic virus DNA: detection of promoter sequences, and characterization of transcripts. *Cell* 30:763–73

20. Henfrey RD, Proudfoot LMF, Covey SN, Slater RJ. 1989. Identification of an inhibitor of transcription in extracts prepared from wheat shoot nuclei. *Plant Sci.* 64: 91–98

21. Jendrisak J. 1981. Purification and subunit structure of DNA- dependent RNA polymerase III from wheat germ. *Plant Physiol.* 67:438–44

22. Jendrisak JJ, Burgess RR. 1975. A new method for the large-scale purification of wheat germ DNA-dependent RNA polymerase II. *Biochemistry* 14:4639–44

23. Kadonaga JT. 1990. Assembly and disassembly of the *Drosophila* RNA polymerase II complex during transcription. *J. Biol. Chem.* 265:2624–31

24. Katagiri F, Yamazaki K, Horikoshi M, Roeder RG, Chua NH. 1990. A plant DNA-binding protein increases the number of active preinitiation complexes in a human in vitro transcription system. *Genes Dev.* 4:1899–909

25. Kiss T, Marshallsay C, Filipowicz W. 1991. Alteration of the RNA polymerase specificity of U3 snRNA genes during evolution and in vitro. *Cell* 65:517–26

26. Langridge P, Feix G. 1983. A zein gene of maize is transcribed from two widely separated promoter regions. *Cell* 34:1015–22

27. Mukumoto F, Hirose S, Imaseki H, Yamazaki K. 1993. DNA sequence requirement of a TATA element-binding protein from *Arabidopsis* for transcription in vitro. *Plant Mol. Biol.* 23:995–1003

28. Perry KL, Palukaitis P. 1990. Transcription of tomato ribosomal DNA and the organization of the intergenic spacer. *Mol. Gen. Genet.* 221:102–12

29. Reddy PS, Padayatty JD. 1988. Effects of 5' flanking sequences and changes in the 5' internal control region on the transcription of rice tRNA^{Gly}GCC gene. *Plant Mol. Biol.* 11:575–83

30. Roberts MW, Okita TW. 1991. Accurate in vitro transcription of plant promoters with nuclear extracts prepared from cultured plant cells. *Plant Mol. Biol.* 16:771–86

31. Sawadogo M, Roeder RG. 1985. Factors involved in specific transcription by human RNA polymerase II: analysis by a rapid and quantitative in vitro assay. *Proc. Natl. Acad. Sci. USA* 82:4394- 98

32. Schweizer P. 1994. In vitro transcription of plant nuclear genes. In *Results and Problems in Cell Differentiation,* ed. L Nover, 20:105–21. Berlin: Springer-Verlag

33. Schweizer P, Mösinger E. 1994. Initiator-dependent transcription in vitro by a wheat germ chromatin extract. *Plant Mol. Biol.* 25:115–30

34. Sif S, Cummings A, Davis EA, Ackerman S. 1993. Interaction of human transcription factors IIA and IID with the cauliflower mosaic virus 35S promoter. *Mol. Biol.* 12: 53–61

35. Stange N, Beier H. 1987. A cell-free plant extract for accurate pre tRNA processing, splicing and modification. *EMBO J.* 6: 2811-18

36. Stange N, Beier D, Beier H. 1991. Expression of nuclear tRNA^{Tyr} genes from *Arabidopsis thaliana* in HeLa cell and wheat germ extracts. *Plant Mol. Biol.* 16:865–75

37. Sugiura M. 1996. Plant in vitro transcription: the opening of a new era. *Trends Plant Sci.* 1:41

38. Teichmann T, Urban C, Beier H. 1994. The tRNA^{Ser}-isoacceptors and their genes in *Nicotiana rustica:* genome organization, expression in vitro and sequence analyses. *Plant Mol. Biol.* 24:889–901

39. Ulmasov B, Folk W. 1995. Analysis of the role of 5' and 3' flanking sequence elements upon in vivo expression of the plant tRNA^{Trp} genes. *Plant Cell* 7:1723–34

40. van Tol H, Stange N, Gross HJ, Beier H. 1987. A human and a plant intron-containing tRNA^{Tyr} gene are both transcribed in a HeLa cell extract but spliced along different pathways. *EMBO J.* 6:35–41

41. Xu Y, Zhu Q, Panbangred W, Shirasu K, Lamb C. 1996. Regulation, expression and function of a new basic chitinase gene in rice (*Oryza sativa* L.). *Plant Mol. Biol.* 30:387–401

42. Yamaguchi Y, Mukumoto F, Imaseki H, Yamazaki K. 1994. Preparation of an in vitro transcription system of plant origin, with methods and templates for assessing its fidelity. In *Plant Molecular Biology Manual,* ed. SB Gelvin, RA Schilperoot, pp. 1–15. Dordrecht: Kluwer. 2nd ed.

43. Yamashita J, Nakajima T, Tanifuji S, Kato A. 1993. Accurate transcription initiation of

Vicia faba rDNA in a whole cell extract from embryonic axes. *Plant J.* 3:187–90

44. Yamazaki K, Chua NH, Imaseki H. 1990. Accurate transcription of plant genes in vitro using a wheat germ-chromatin extract. *Plant Mol. Biol. Rep.* 8:114–23

45. Yamazaki K, Imamoto F. 1987. Selective and accurate initiation of transcription at the T-DNA promoter in a soluble chromatin extract from wheat germ. *Mol. Gen. Genet.* 209:445–52

46. Yamazaki K, Katagiri F, Imaseki H, Chua NH. 1990. TGA1a, a tobacco DNA-binding protein, increases the rate of initiation in a plant in vitro transcription system. *Proc. Natl. Acad. Sci. USA* 87:7035–39

47. Zhu Q. 1996. RNA polymerase II-dependent plant in vitro transcription systems. *Plant J.* 10:185–88

48. Zhu Q, Chappell J, Hedrick SA, Lamb C. 1995. Accurate in vitro transcription from circularized plasmid templates by plant whole cell extracts. *Plant J.* 7:1021–30

49. Zhu Q, Dabi T, Lamb C. 1995. TATA box and initiator functions in the accurate transcription of a plant minimal promoter in vitro. *Plant Cell* 7:1681–89

Annu. Rev. Plant Physiol. Plant Mol. Biol. 1997. 48:399–429
Copyright © 1997 by Annual Reviews Inc. All rights reserved

AQUAPORINS AND WATER PERMEABILITY OF PLANT MEMBRANES

Christophe Maurel

Institut des Sciences Végétales, CNRS, Avenue de la Terrasse, F-91198 GIF-SUR-YVETTE Cedex, France

KEY WORDS: hydraulic conductivity, lipid bilayer, osmosis, turgor, water channel

ABSTRACT

The mechanisms of plant membrane water permeability have remained elusive until the recent discovery in both vacuolar and plasma membranes of a class of water channel proteins named aquaporins. Similar to their animal counterparts, plant aquaporins have six membrane-spanning domains and belong to the MIP superfamily of transmembrane channel proteins. Their very high efficiency and selectivity in transporting water molecules have been mostly characterized using heterologous expression in *Xenopus* oocytes. However, techniques set up to measure the osmotic water permeability of plant membranes such as transcellular osmosis, pressure probe measurements, or stopped-flow spectrophotometry are now being used to analyze the function of plant aquaporins in their native membranes. Multiple mechanisms, at the transcriptional and posttranslational levels, control the expression and activity of the numerous aquaporin isoforms found in plants. These studies suggest a general role for aquaporins in regulating transmembrane water transport during the growth, development, and stress responses of plants. Future research will investigate the integrated function of aquaporins in long-distance water transport and cellular osmoregulation.

This review is dedicated to Professor Jean Guern on the occasion of his retirement.

399

CONTENTS

INTRODUCTION

Because of their lack of mobility in an often challenging environment, plants depend on a supply of water for their growth and development and have to tightly control water balance. Numerous studies have addressed the overall physiological and biophysical mechanisms of plant water relations (7, 14, 104) and pointed to a variety of physiological processes, from long-distance water transport to single cell expansion and osmoregulation, that require the transport of water across cellular membranes (7, 104). However, the mechanisms of plant membrane water permeability have remained elusive until the recent discovery in animals and plants of a class of water-transport proteins named aquaporins (3, 11). These water channel proteins facilitate the passive exchange of water across membranes, and their discovery in plants emphasized the limitation that transmembrane water transport may exert on a number of physiological processes (12). Their discovery also suggested the existence of yet to be discovered regulatory processes that might be critical to plant water relations.

The first aim of this review is to bridge the gap between the new molecular insights made possible by the discovery of aquaporins and a wealth of information reported over the years by plant physiologists and biophysicists. The significance of aquaporins for our understanding of plant water relations and the experimental and theoretical perspectives that they open are also discussed.

BACKGROUND

In plants, transmembrane water movements are primarily determined by hydrostatic and osmotic pressure gradients, respectively noted ΔP and $\Delta \pi$ (14, 105).

The buildup of a hydrostatic pressure or turgor is made possible by the mechanical resistance of the cell wall whose relation to cell volume is characterized by the elastic modulus (ε) (105).

The osmotic driving force $\Delta \psi_{osm}$ can be expressed as

$$\Delta \psi_{osm} = \sigma \Delta \pi = \sigma RT \Delta C. \qquad 1.$$

It is determined by the solute concentration gradient (ΔC) but also by the solute reflection coefficient (σ) for the membrane (R is the universal gas constant, T is absolute temperature) (22). This σ coefficient quantifies the membrane selectivity for the solute and determines its osmotic efficiency. In most cases, the solute cannot significantly permeate the membrane or at most at a rate several orders of magnitude less than water, and σ is maximal and equal to unity. However, σ may take lower values and is null (no osmotic effect) when the solute permeates the membrane as efficiently as water (22).

The basic flow equation (Equation 2) that governs transmembrane water transport has been derived from the theory of irreversible thermodynamics (14, 53, 105). It dictates that, if only one solute is considered, the net transmembrane volume flow J_v is proportional to the motive force (ΔP-$\sigma \Delta \pi$), and the surface area (A) and the hydraulic conductivity (L_p) of the membrane.

$$J_v = L_p A (\Delta P - \sigma \Delta \pi). \qquad 2.$$

This equation emphasizes the major role played, besides σ and ε, by L_p, which describes the intrinsic permeability of the membrane to water. L_p can be converted into P_f, the osmotic water permeability or filtration coefficient according to

$$P_f = L_p RT / V_w, \qquad 3.$$

where R and T have their usual meaning and V_w is the partial molar volume of water (22). Examining the biophysical, molecular, and physiological significance of these parameters, P_f and L_p, is the object of this review. Reviews that cover all aspects of plant water relations can be found elsewhere (7, 104).

MEMBRANE WATER PERMEABILITY MEASUREMENTS

While P_f describes overall water movement in response to hydrostatic or osmotic pressure gradients, the diffusional permeability coefficient (P_d) describes the unidirectional flux (diffusion) of water molecules that still occurs across the membrane without any driving force (22). Although of lesser physiological significance, this parameter gives an estimate of the membrane water transport capacity and, when compared with P_f, provides valuable information about the biophysical mechanisms of water permeation (see below).

Diffusional Water Permeability

All early measurements of plant cell and tissue diffusional water permeability were performed using heavy (HDO) or tritiated (HTO) water (28, 40, 101). However, the rapid diffusional water exchange rates and, above all, unstirred layer effects render diffusional water permeability (P_d) measurements using these tracer flow techniques very difficult (14, 22, 105). Unstirred layer effects were especially prominent when P_d was determined in multicellular plant systems or in cells with high water permeability; most of the reported values (101) may well be spurious. Recently, Henzler & Steudle (34) measured the P_d of *Chara* internodal cells taking advantage of the osmotic (solute-like) effects of HDO and following its diffusion by means of a pressure probe (see below). Proton nuclear magnetic resonance (^{1}H-NMR) provides approaches to circumvent the extracellular unstirred layer artifacts and permits probing of water compartments in intact plant tissues (77, 100, 110, 135a). Membrane water permeability can be derived from saturation transfer (77) or more commonly from relaxation times of intracellular water protons. For this, the ^{1}H-NMR signal of the water molecules residing in the extracellular space or diffusing from the intracellular space is doped by an extracellular paramagnetic ion such as Mn^{2+}. However, the P_d values deduced from these measurements greatly depend on assumptions about the penetration and compartmentation characteristics of the paramagnetic agent into the plant cell or tissue (100, 110, 135a) and can still be dominated by intracellular unstirred layers (110).

Osmotic Water Permeability

The overall exchange of water induced osmotically in intact tissues or organs can provide indications on cell water permeability (25, 32). More direct estimates of osmotic water permeability (P_f) in single plant cells were first obtained by the plasmometric method (reviewed in 102). A preliminary plasmolysis separates the plant protoplast from any interaction with the cell wall, and P_f values are derived from the rate of protoplast swelling observed upon a subsequent hypotonic treatment. This technique has been criticized (51, 137) mostly because of its low resolution due to the restricted diffusion of the osmoticum and/or unstirred layers in the plant cell walls. These drawbacks have recently been avoided through the direct micromanipulation of isolated protoplasts (91).

The giant internodal cells of the Characeae provide convenient materials for studying membrane transport and allowed in particular P_f measurements by means of transcellular osmosis (51). For this, the two ends of a cell are brought into contact with external media of different osmolarities. This creates a transcellular water flow from the less to the most concentrated compartment, whose rate indicates the cell P_f. Despite pitfalls due to intracellular osmotic polarization (15, 51), this methods provides P_f measurements with little change in cell volume and turgor. It also allows an easy distinction between endo- (inward) and exo- (outward) osmotic water movements.

In contrast with transcellular osmosis, the pressure probe technique can be applied to giant algal cells but also to normal-sized higher plant cells (reviewed in 103, 136). An oil-filled microcapillary is inserted into the cell and its coupling to a mechanical device allows to primarily measure or clamp the cell turgor pressure. Water movement is generated upon imposition of a sudden osmotic or hydrostatic pressure gradient and can be followed through the associated turgor pressure relaxation. Alternatively, volume relaxations can be followed under pressure-clamp conditions (135). These types of measurements allow the successive determination of the cell volumetric elastic modulus, the cell hydraulic conductivity, and the membrane solute reflection coefficients, which together define the plant cell water relation (103, 136). Another great potential of the pressure probe is that it allows in situ measurements on individual cells under osmotic or hydrostatic conditions (108, 112). However, owing to technical difficulties, this technique has only been applied to a restricted number of higher plant cell types. Two other drawbacks hamper its application to the direct analysis of membrane water transport. First, the respective contributions of the membranes and cell-to-cell connections (plasmodesmata) can hardly be distinguished in the overall cell L_p. Second, because the probe is inserted into the vacuole, the cell L_p integrates the resistance in series of the tonoplast and the plasma membrane. This restriction applies to nearly all techniques described above and has long been recognized as a main difficulty in interpreting water transport measurements in plant cells (14, 28, 124). In *Chara*, these difficulties could be circumvented by the selective disruption of the tonoplast (58) or by a three-compartment analysis of turgor relaxation kinetics (124, 125).

More recently, optical methods (reviewed in 117) have been applied to measuring the P_f of purified plant membrane vesicles (81, 96). The mixing of a vesicle suspension with an anisoosmotic solution is accomplished in <5 ms in a stopped-flow apparatus. The subsequent osmotic adjustment of the sealed vesicle volume can be followed, for instance, from the simultaneous change in scattered light intensity. A P_f value can be derived from both the time course of vesicle volume adjustment and the vesicle size, determined by an independent technique such as electron microscopy.

A Wide Range of Measured Permeabilities

Early estimates (14, 101) of plant membrane water permeability ranged over more than three orders of magnitude. Because of methodological uncertainties, this variability could hardly be interpreted at that time (14). Since then, novel methods have been developed and their reliability assessed. For instance, consistent P_f values of *Chara* cells can be measured using either transcellular osmosis or a pressure probe (Table 1). The latter technique indicated that L_p values of different plant cell types range from 2×10^{-8} to 10^{-5} m • s^{-1} • MPa^{-1} (i.e. P_f ranging from 3×10^{-4} to 10^{-1} cm • s^{-1}) (Table 1; 105). Stopped flow measurements also indicated that membrane subfractions isolated from the same tobacco cells can display strikingly different water permeabilities, from 6×10^{-4} to 6×10^{-2} cm • s^{-1} (Table 1; C Maurel, F Tacnet, J Güclü, J Guern & P Ripoche, unpublished manuscript). Thus, the wide range of water permeabilities measured in plant cells appears to reflect a genuine property of their membranes. The reported values also correspond to the range of permeabilities determined for animal cell membranes (Table 1; 22, 116).

EMERGENCE OF WATER CHANNELS IN PLANT PHYSIOLOGY

The Red Blood Cell Paradigm

The lipid phase of membranes was long considered the major path for water exchange in living cells. Transport along this path is based on the solubility/diffusion of individual water molecules into the phospholipid bilayer and is characterized by equal P_f and P_d and a high Arrhenius activation energy (E_a = 14–16 kcal • mol^{-1}) (22, 30). The determination of a P_f/P_d greater than unity in the erythrocyte and other animal cell membranes has been the first experimental evidence for transmembrane aqueous pores that would mediate, as theoretically predicted, an apparent positive interaction during osmotic water flow (59, 83). The low E_a (<5 kcal • mol^{-1}) of this process in erythrocytes further suggested the existence of a bulk flow of water across membrane pores (66, 116). That the red blood cell water channels might contain a proteinaceous component was inferred from their reversible blockade by mercury sulfhydryl reagents (66).

Early Evidence for Plant Water Channels

The possible existence of aqueous pores in plant membranes was discussed in the early 1960s by Ray (92) and Dainty (14). To investigate this possibility, Gutknecht (28) determined the P_f/P_d ratio of *Valonia* cells. Its value close to unity and the low measured value of P_f of these cells (2.4 × 10^{-4} cm • s^{-1}; Table 1) suggested the absence of aqueous membrane pores. Probably because of the well-recognized artifacts caused by unstirred layers in such an experiment (14), this was, until recently (34), the only report of P_f/P_d in a plant membrane (Table 1).

The dependence of water transport on temperature was also characterized in several plant systems (15, 26, 40). The derived E_a values, higher than the E_a for free water diffusion or water diffusion in dead tissues, were interpreted to mean that plant membranes, and more specifically aqueous pores (15), may create an impediment to water flow. However, E_a has not been used until recently to distinguish between lipid- and channel-mediated water transports. The search for inhibitors of water transport in plants led to the eventual use of mercury derivatives (26, 40). Although these compounds were suggested to directly alter the permeability of plant membranes (40), they were used in most cases as general metabolic inhibitors (26).

Actually, it was generally assumed that, if present, aqueous pores in plant membranes should allow the simultaneous flow of water and solutes. This led to the search of possible frictional interactions between solute and water in the membrane that yield σ values inferior to unity. Now that the very high transport selectivity of aquaporins has been demonstrated (see below), it appears that the absence of such friction (σ = 1) (29) is not evidence against the presence of aqueous pores. In contrast, Dainty & Ginzburg (16) and others (107) noticed—and this argument still remains valid (106)—that the σ values

Table 1 Water permeabilities of cells and membranes from plants and animals

Species	Tissue	System[a]	Permeability $(10^{-4}$ cm • $s^{-1})$[b]	Method	Reference
		Materials and measurement techniques			
Chara corallina	Internode	Cell	P_f: 260 ± 72 P_f: 243 ± 15	Pressure probe Transcellular osmosis	34 121
Elodea nuttallii	Leaf	Cell	P_d >300	^{1}H-NMR	110
Elodea densa	Leaf lower epidermis	Cell	P_f: 19.0 ± 3.1 (P<4 bar) P_f: 7.5 ± 2.7 (P>4 bar)	Pressure probe	109
Pisum sativum	Epicotyl epidermis	Cell	P_f: 3–30	Pressure probe	13
	Epicotyl cortex	Cell	P_f: 50–1200	Pressure probe	13
		Cell, plasma membrane and endomembranes			
Chara australis	Internode	Cell	P_f: 97 ± 17	Transcellular osmosis	58
		PM	P_f: 94 ± 13	Transcellular osmosis	
Allium cepa	Bulb inner epidermis	Cell Vacuole	P_f: 6–8 P_f: 40–540	Deplasmolysis Deplasmolysis	113
Nicotiana tabacum	Cell suspension	PM ves. TP ves.	P_f: 6.2 ± 0.4 P_f: 659 ± 83	Stopped-flow Stopped-flow	c
Apple	Fruit parenchyma	Vacuole	P_d: 24.4	^{1}H-NMR	100
Liriodendron tulipifera	Leaf	Chloroplast envelope	P_d: 9 ± 2	^{1}H-NMR	77
		Animal cell membranes			
Hog	Stomach	PM ves.	P_f: 2.8 ± 0.3	Stopped-flow	116
Toad	Bladder	Apical membrane ves.	P_f: 3.9 ± 0.4 (– vasopressin) P_f: 450 (+ vasopressin)	Stopped-flow Stopped-flow	116
Human	Erythrocyte	PM ves.	P_f: 230 ± 30	Stopped-flow	71
		P_f/P_d of plant membranes			
Valonia ventricosa	—	Cell	P_f: 2.4 ± 0.3 P_d: 2.4 ± 0.2	Transcellular osmosis Tracer flow	28
Chara corallina	Internode	Cell	P_f: 260 ± 72 P_d: 7.7 ± 3	Pressure probe Pressure probe	34

for organic molecules in Characeae cells are so low that this must reflect the presence of water-filled pores.

Nevertheless, the characterization of membrane permeability in Characeae cells progressively led to the idea that solute transport, but also most of the osmotic flow of water, occurs across membrane pores in these cells (55, 121, 122). For instance, Wayne & Tazawa (121, 122), prior to the molecular identification of animal and plant aquaporins, presented most of the now classical arguments favoring membrane water channels and clearly demonstrated their major contribution to osmotic water transport.

Surprisingly, this view remained marginal in plant physiology. High P_f values ($>10^{-2}$ cm • s^{-1}) had been measured in several higher plant cell types and might have suggested water-channel-mediated transport. Yet the prevailing idea was that the reported range of P_f could be accounted for by water diffusion across plant lipid membranes of distinct composition (102). This view can indeed be supported by studies with artificial membranes that reported P_f values ranging from 2×10^{-5} to 1×10^{-2} cm • s^{-1} (21, 22, 61) but was scarcely investigated (27, 96) using plant lipid membranes.

Molecular Identification of Plant Aquaporins

The abundance of aquaporins in plants is undoubtedly the main reason that led to their redundant identification by biochemical and molecular biological approaches, well before their function was clearly identified. The high hydrophobicity of these intrinsic membrane proteins also facilitated their biochemical purification. The α-Tonoplast Intrinsic Protein (α-TIP), for instance, represents 2% of the total extractable proteins of bean cotyledons and could be easily purified by Triton X-114 extraction and partitioning into the detergent-rich phase (46). Surprisingly, the epitopes carried by aquaporin-like proteins in plant membranes are so prominent that several polyclonal antibodies raised to total purified plasma membranes or tonoplasts were found to mostly react with these proteins (52, 70). A large class of Arabidopsis plasma membrane aquaporins was thus isolated using such antibodies, by immuno-selection from a mammalian COS cell expression system (52). Several aquaporins such as γ-TIP are also some of the most abundantly expressed sequence tags (ESTs) in Arabidopsis and were repeatedly identified in systematic cDNA sequencing

---

[a]PM, Plasma membrane; TP, tonoplast; ves., vesicles.

[b]P_d: Diffusional water permeability coefficient; P_f: osmotic water permeability coefficient. When not provided by the authors, P_f values were derived, for comparison with other P_f values, from the indicated L_p values according to Equation 3, with T = 293 K. P_f and P_d values are indicated ± SD.

[c]C Maurel, F Tacnet, J Güçlü, J Guern & P Ripoche, unpublished manuscript

programs (35). Finally, the modulation of aquaporin gene expression in response to various physiological stimuli has also led to their identification by procedures such as the differential screening of cDNA libraries (49, 86, 126, 127). γ-TIP and RD28 were thus identified as genes up-regulated by, respectively, gibberellic acid (GA_3) and drying treatments (86, 127). More recently, the duplicated Asn-Pro-Ala (NPA) motifs conserved in the aquaporin amino acid sequences (see below) have provided a unique region for the PCR amplification of putative aquaporin genes.

The archetypal animal water channel CHIP28, subsequently renamed aquaporin-1 (AQP-1), was also first identified by biochemical means in the erythrocyte membrane (reviewed in 116). The work of Preston et al (87), who hypothesized from the expression pattern of CHIP28/AQP-1 that it might correspond to one of the long sought after water channels of the red blood cell and kidney epithelia, has proved to be a major breakthrough for animal and plant biologists (11). Evidence that plant aquaporins facilitate the transport of water across membranes comes from the observation that these proteins, similar to AQP-1, induce a specific and mercury-sensitive increase in the P_f of *Xenopus* oocyte membranes (75). The two other criteria that functionally define water channels, i.e. a low E_a and a $P_f/P_d > 1$, have been demonstrated for AQP-1 (71, 87). This and the reconstitution of functional water channels in proteoliposomes containing the purified protein (115, 130, 131) have provided unequivocal evidence that AQP-1, and by extension its functional animal and plant homologs, have an intrinsic water transport activity.

MOLECULAR FEATURES OF WATER CHANNELS

The aquaporins define a functional class of water-transport proteins that belong to the larger Major Intrinsic Protein (MIP) family of transmembrane channels (3). This ancient family was named after its archetype, the MIP of mammalian lens fiber, and includes members in a wide variety of living organisms (93). At present, aquaporins have been identified in vertebrates, higher plants, bacteria, and insects (8, 63, 75, 87).

MIP Homologs, Aquaporins and Other Water-Transport Proteins

Sequences available in the gene data banks indicate that the large MIP family of membrane channels is even more extended in plants than in animals (93). For instance, 23 MIP-like unique ESTs have been found in Arabidopsis (A Zweig & MJ Chrispeels, personal communication) while only 7 MIP-like sequences have been identified in humans (118). However, most of the plant MIP proteins and cDNAs remain to be functionally characterized. While they potentially represent novel aquaporins, they may also encode transmembrane channels for other substrates. The bacterial MIP homolog, GlpF, for instance, facilitates the diffusion of small polyhydric alcohols but is not permeable to water (33, 75, 76). The function of mammalian MIP, as an ion or a water channel, remains controversial (54, 118). The water transport activity of soybean nodulin-26 remains to be determined, but the purified protein was shown to form large conductance ion channels in artificial membranes (123).

Protein-mediated transmembrane transport of water is not necessarily due to aquaporins. Because they contain a hydrated path, all membrane transport proteins have a finite permeability to water. It was found, for instance, that in most ion channels the transport of one ion is coupled to that of 5–10 water molecules, which correspond to the number of water molecules filling the constriction of the pore (4, 22). Higher water transport activities have been assigned to animal membrane proteins such as the cystic fibrosis transmembrane conductance regulator (CFTR), glucose transporters, or inorganic ion cotransporters (reviewed in 132). In plants, the tonoplast of *Chara* internodal cells contains a channel that allows the coupling of 25 molecules of water with a single K^+ ion (39). The inhibition of water transport in *Chara* cells by K^+ channel blockers (122) provides evidence that either nonclassical aquaporins or other proteins can significantly contribute to the P_f of plant membranes. Nevertheless, with a unit permeability of 3–12 $cm^3 \cdot s^{-1}$ (115, 130, 131), the water transport efficiency of aquaporins is at least 20 times greater than that of any other known protein.

Aquaporins Are Expressed in the Plant Vacuolar and Plasma Membranes

The sequence relationship between all plant MIP-like cDNAs (126) indicates that the encoded proteins fall into three sequence subclasses. The first two classes contain, respectively, the Tonoplast Intrinsic Proteins (TIP) and the Plasma membrane Intrinsic Proteins (PIP) that have been localized in the tonoplast and in the plasma membrane. This was shown by several groups using immunocytochemistry (36, 46, 48) and protein immunodetection in isolated organelles (36) or purified membrane fractions (18, 48, 52, 68). The third class comprises nodulin-26, which is expressed in the peribacteroid membrane of root symbiotic nodules (23). However, the localization of nodulin-26 homologues in nonlegumes remains to be determined.

Molecular Structure of the Aquaporin Water Channel

All of the known aquaporin cDNAs typically encode 25–30 kDa polypeptides (116, 118). Although the overall identity among MIP homologs might fall under 25%, their amino acid sequences contain very well-conserved residues including a SGxHxNPAVT motif (NPA box) (Figure 1; 93). Sequence analyses also suggest that the MIP gene family arose after an early intragenic duplication. Thus the amino- and carboxy-terminal halves of the proteins share sequence homology, each possess a NPA box, and each is oriented in opposite directions in the membrane (Figure 1; 93).

The conserved hydropathy profiles deduced from MIP-like cDNAs suggest a typical structure, with six putative membrane-spanning segments and cytoplasmic amino- and carboxy-terminal domains (Figure 1; 93). Experimental evidence for this topological model mostly comes from the analysis of vectorial proteolysis at defined epitopes introduced in recombinant AQP-1 (89). Phosphorylation and glycosylation studies with other animal and plant MIP homologues are consistent with such a model (6, 74, 79).

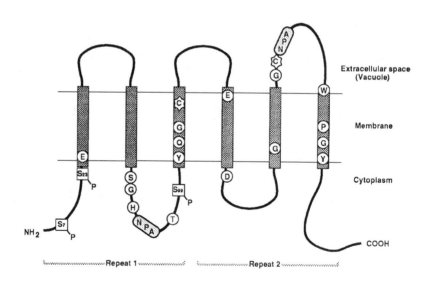

Figure 1 Proposed structure of plant aquaporins based on topological studies on plant and animal MIP homologs (see text). The most conserved residues in the MIP family are encircled. The two NPA boxes are shaded. Phosphorylation sites identified in α-TIP are squared off. Stars indicate positions for cysteine residues that confer mercury sensitivity on plant aquaporins.

The low electrophoretic mobility of native plant aquaporins (68) and their propensity to aggregate in vitro suggested that they may occur as oligomers in plant membranes (17, 18, 46, 52, 68). This view is consistent with size exclusion chromatography and hydrodynamic analyses of native AQP-1 that determined a stoichiometry of 4 subunits per oligomer (99). Recently, X-ray and electron diffraction of two-dimensional crystals of AQP-1 yielded high resolution projection maps. These maps depict a structural motif of 6–7 nm in diameter with a central depression and four trapezoid subunits, each displaying 6–7 putative transmembrane segments (43, 80, 119).

Although aquaporins assemble in homotetrameric or even more organized structures, radiation inactivation analyses indicated a functional size of 30 kDa for animal water channels (114). In addition, the coexpression in *Xenopus* oocytes of wild type AQP-1 monomers with inactive permeability mutants indicated no dominant negative effects of the latter (88, 97, 133). These data suggested that in animals, and probably in plants as well, each aquaporin monomer forms a functionally independent pore (118).

Mercury derivatives have been extensively used to probe the molecular structure of water channels. They are thought to block these channels via oxidation of cysteine residue(s) proximal to the aqueous pore and subsequent occlusion of the latter by the large mercury ion. The mercury-sensitive sites of AQP-1 and vasopressin-regulated AQP-2 have been identified at, respectively, Cys189 and Cys181, each in the third extracytoplasmic loop next to the second NPA box (6, 88, 133). The functional importance of this and of the symmetrical loop carrying the other NPA box (see Figure 1) was also suggested by the converse observation that animal and plant mercury-resistant proteins can be sensitized to the inhibitor by introduction of target cysteine residues in either of these domains (18, 47, 98). However, the cysteine residues that confer mercury sensitivity to plant γ-TIP and δ-TIP sit in the third transmembrane segment, on the hydrophilic side of a putative amphiphilic α-helix (17). This points to another domain that possibly lines the aqueous transmembrane pore.

FUNCTIONAL STUDIES OF PLANT WATER CHANNELS

Heterologous Expression of Plant Aquaporins in Xenopus Oocytes

The capacity of *Xenopus* oocytes to translate intracellularly injected mRNAs, their large size, and the relatively low P_f ($\leq 1 \times 10^{-3}$ cm • s^{-1}) of their native membrane make these cells convenient for the functional expression of water-transport proteins (134). Oocytes are classically incubated in isoosmotic conditions for few days after mRNA injection to allow exogenous protein expression. The initial rate of cell swelling observed upon subsequent transfer of the oocytes into a hypoosmotic solution indicates their membrane P_f (134). Similar to animal AQP-1 (87), several plant MIP homologs of either the tonoplast or the plasma membrane (18, 52, 75) were found to induce a 2-20-fold increase in the P_f of the oocyte membrane and were thus identified as novel aquaporins (Table 2). Oocyte expression also provides powerful approaches to study aquaporin topology (89), oligomer assembly (97), posttranslational regulation (60, 74), transport selectivity (75), and structure-function relationships (47).

Contribution of Aquaporins to Membrane Water Transport

Because most studies on plant aquaporins have been performed after heterologous expression of proteins in oocytes, several groups are now concentrating on the study of water channels in plant membranes (34, 81, 122) (C Maurel, F Tacnet, J Güclü, J Guern & P Ripoche, unpublished manuscript). In all these experiments, aquaporin function was probed through mercury inhibition. Because some plant and animal aquaporins are resistant to mercury action (18, 31), this approach only indicates the minimal contribution of water channels to the total water transport capacity of the membrane. Nevertheless, 1 mM mercuric chloride ($HgCl_2$) can reduce by more than 80% the P_f of tonoplast vesicles purified from tobacco suspension cells (C Maurel, F Tacnet, J Güclü, J Guern & P Ripoche, unpublished manuscript). The high intrinsic water permeability of these membranes ($P_f = 6 \times 10^{-2}$ cm • s^{-1}) and the low E_a of transport (~2.5 kcal • mole^{-1}) provide additional evidence that water transport in these vesicles occurs predominantly through membrane channels. In contrast, plasma membrane vesicles isolated from the same tobacco cells exhibited a low P_f (6×10^{-4} cm • s^{-1}) and a high E_a (~14 kcal • mole^{-1}), both indicative of a low density of functional aquaporins, if any. The membranes of *Chara* and *Nitellopsis* cells are also highly permeable to water, and mercurials reduced their P_f by up to 70 and 30%, respectively (34, 122). This inhibition was associated with either a decrease in P_f/P_d (34) or a twofold increase in E_a (122), both indicating the blockade of pore-mediated transport. Although these studies indicate that the high water permeability of certain plant membranes is mainly accounted for by water channel activities, they provide no molecular identification of the corresponding channel proteins. To address this question, Kaldenhoff et al (48) have used transgenic Arabidopsis that constitutively expressed antisense PIP1b aquaporin transcripts and displayed an associated 40–80% reduction in PIP1a and PIP1b mRNA levels (R Kaldenhoff, personal communication). The osmotic swelling of isolated protoplasts was three- to

Table 2 Cell localization and expression of plant aquaporins

Name	Species	Aquaporin function[a]	Localization[b]	Organ-specific	Stimulus-induced
				Expression	
	Tonoplast Intrinsic Protein (TIP) subfamily				
α-TIP	*Phaseolus vulgaris*	74[c]	TP (36, 46)[c]	Seeds and seedlings (46, 78)c	
γ-TIP	*Arabidopsis thaliana*	75	TP (37)	Roots & shoots (elong. zone)— flowers (37, 65)	GA$_3$ (86)
δ-TIP	*Arabidopsis thaliana*	17	TP (17)	Shoots > roots (vasc. tissues)— flowers (17)	
pRB7	*Nicotiana tabacum*	82	n.d.	Roots (meristem, immature central cylinder) (128)	Root-knot nematode infection (82)
	Plasma membrane Intrinsic Protein (PIP) subfamily				
PIP1a, PIP1c	*Arabidopsis thaliana*	52	PM (52)	Roots (central cylinder>cortex)— leaves (vasc. tissues> mesophyll) (52; AR Schäffner, personal communication)	
PIP1b (AthH2)	*Arabidopsis thaliana*	48, 52	PM (48, 52)	Roots & shoots (differentiating and elong. tissues, guard cells) (48, 52)	Blue light, ABA, GA$_3$ (48–50)
PIP2a, PIP2b	*Arabidopsis thaliana*	52	PM (52)	Roots & shoots (52)	
RD28 (PIP2)	*Arabidopsis thaliana*	18	PM (18)	All organs except seeds (18)	Desiccation (up-regulation) (127)
MIPA	*Mesembryanthemum crystallinum*	126	n.d.	Leaves (vasc. tissues, meristem)— roots (epiderm., develop. xylem) (126)	Salt stress (down-regulation) (126)
MIPB	*Mesembryanthemum crystallinum*	126	n.d.	Root tip (126)	

[a]Aquaporin function was determined by oocyte expression in all cases, except in Reference 48 where transgenic Arabidopsis expressing an antisense gene construct was used.

[b]Cell localization: TP, tonoplast; PM, plasma membrane; n.d., not determined.

[c]Number(s) indicate(s) the corresponding reference(s).

fourfold slower in transgenic protoplasts than in controls, suggesting that PIP1 aquaporins significantly contribute to water transport in these cell membranes (48).

Transport Selectivity

Well before the molecular identification of aquaporins, indirect evidence suggested a reasonably high selectivity for animal and plant water channels (55, 66). For instance, the water permeability of *Chara* cell membranes can be modulated without change in electrical conductance, suggesting the existence of two distinct paths, for water and ions, respectively (55). More recently, the selectivity of plant aquaporins has been directly supported by the observation that neither γ-TIP nor α-TIP has detectable effects on glycerol and ion transport in oocytes (74, 75). In particular, current measurements during osmosis did not indicate any detectable passive ion permeability or any solvent-drag of ions in conditions where the water channels were known to operate, indicating the eventual passage of at most 1 ion for every 2×10^5 water molecules (75). This view has recently been challenged by the finding that when stimulated by forskolin in the oocyte membrane, animal AQP-1 exhibits a high permeability to cations (129).

Using the pressure probe technique, Henzler & Steudle (34) investigated the selectivity of plant water channels directly in the membrane of *Chara* cells. Mercury was found to reduce the membrane permeability to water but not to solutes such as dimethyl formamide, ethanol, or acetone. This indicates that the major transport path for these solutes is not through water channels but rather through the lipid bilayer. Using a theoretical treatment of membrane organization as a composite structure with arrays of water channels and others with lipid phase, Henzler & Steudle (34) further calculated reflection coefficients for these solutes along the water channel path. These coefficients were less than unity and were interpreted to mean that these molecules can permeate the channel pore, though at a very slow rate, and develop strong frictional interactions with water molecules. Recent investigations using oocyte expression have also revealed basal permeabilities of animal aquaporins to small solutes such as urea, ethylene glycol, and glycerol but not to larger molecules (2, 20, 42). α-TIP and γ-TIP displayed a selectivity in these assays higher than that of their animal counterparts (1). It remains to be determined whether plant cells possess other aquaporins that, similar to animal AQP-3, discriminate poorly between water and small neutral solutes (20, 42). Nevertheless, our current knowledge of plant aquaporin selectivity, and their assumed capacity to exclude protons (75, 131), points to the remarkable capacity of these proteins in improving the semipermeable characteristics of plant membranes.

Mechanisms of Water Permeation

The molecular bases of water channel selectivity are currently unknown (118). These channels have been assumed to form narrow pores of 0.3–0.4 nm in diameter that would exclude any molecule larger than water (55, 66). Water molecules would flow one by one in the constriction of the pore, and the number of molecules in a single file would determine the P_f/P_d ratio, typically greater than unity in water channels (22, 59, 83). The exclusion of small ions, and protons in particular (75, 131), further suggests that size exclusion, but also other mechanisms such as electrical filtering, determine aquaporin selectivity. The comparative structure/function analysis of aquaporins with distinct selectivity profiles, such as plant TIPs and animal AQP-3, will be of great interest to address these issues.

Water is thought to flow across the water channel pore in either direction, down its potential gradient. However, early reports (quoted in 57) mentioned a difference in the rates of plasmolysis and deplasmolysis of plant protoplasts. This might well be explained by a solute sweep-away or other unstirred-layer effects during osmosis (14). However, more precise analyses of hydrostatic or osmotic water movements in *Nitella* suggest that these cell membranes at least possess an intrinsic capacity to rectify water flow, by a factor of 1.1 to 2.7 (15, 51, 57, 108, 121). It remains to be determined whether this polarity results, as first proposed, from a water flow–induced dehydration of the lipid membrane (15) or, in view of the most recent studies, from intrinsic features of water permeability in membrane channels or from their differential expression and/or regulation in subcellular membrane domains (see below).

THE REGULATION OF AQUAPORIN ACTIVITY

Although not fully understood, the large variability of water permeability values reported in different plant cell types (Table 1) suggests the importance of species specific, developmental, and physiological factors in determining water transport in plant membranes. In *Chara,* the L_p of internodal cells depends on their distance from the apex and on the vegetative and reproductive status of the plant (120). L_p can also be modulated by the external solute (15, 56) or CO_2 (120) concentration. In addition, in higher plants numerous physiological processes might involve the modulation of transmembrane water transport. For instance, the L_p of roots varies in response to a variety of environmental factors including diurnal cycling, drought and salinity, low temperature, and nutrient and O_2 deficiency (5, 10, 85, 104, 135). Phytochrome (Pfr) (9) and auxin (64) were shown to increase the deplasmolysis rate of epidermal cells of *Taraxacum* and *Allium,* respectively, while abscisic acid (ABA) down-regulates water transport in carrot disks (26).

Changes in membrane lipid composition and fluidity are known to occur during plant development (111) or in response to specific physiological conditions (10, 72). These changes might provide plant cells with distinct baseline membrane water permeabilities (61, 96). However, water channel proteins appear as the most powerful means to mediate rapid and large amplitude changes in the water transport capacity of membranes.

Gene Expression

A detailed description of plant cell types and tissues known to express aquaporins is presented in Table 2. Although aquaporin expression can be found in most tissues, the distribution of each individual aquaporin isoform is under tight developmental and physiological control and suggests specific roles for each of these proteins. For instance, several aquaporins (Arabidopsis δ-TIP, tobacco pRB7) are preferentially expressed in the parenchyma cells of vascular tissues and are possibly involved in long-distance water transport (17, 128), whereas other isoforms (Arabidopsis PIP1b/AthH2, γ-TIP) have patterns associated with elongating and differentiating cells (48, 65). Detailed analyses of seed-specific expression of α-TIP indicated that developmental control of plant aquaporins can be exerted at the levels of gene transcription, translation, or subcellular routing of protein (reviewed in 73). Table 2 also shows that aquaporin expression can be controlled by various hormonal (ABA, GA3) or environmental factors (blue light, water stress, pathogen infection) (48–50, 82, 86, 126, 127). Of particular interest is the up- or down-regulation of aquaporin genes from Arabidopsis, ice plant, and sunflower in response to drought or salinity that supports a role for these proteins in regulating overall plant water balance under stress conditions (126, 127) (X Sarda, D Tousch & T Lamaze, personal communication). All these factors can interact in providing each aquaporin with a physiologically relevant pattern of expression. The γ-TIP gene, for instance, is preferentially expressed in elongating tissues or after stimulation by GA3 (65, 86). The TobRB7 promoter contains two distinct regulatory *cis*-acting elements that direct root-specific expression, under normal developmental conditions or after root-knot nematode infection (82, 128).

Although more work is needed to integrate the spatial and temporal expression of all plant aquaporin isoforms, the variety of these isoforms and expression patterns might well reflect the need for distinct cell types and tissues to accommodate various hydric requirements in specific physiological conditions.

Cell Localization

It is well known that water reabsorption in the mammalian kidney is regulated by vasopressin via the rapid (<5 min) insertion and removal, by a vesicle shuttle mechanism, of AQP-2 proteins in the apical membrane of collecting duct epithelial cells (reviewed in 54). Although similar dynamic regulation has not been reported for plant aquaporins, their very strict subcellular localization points to a critical role in the functional specialization of membrane compartments or even domains. For instance, γ-TIP and α-TIP can coreside in the same seed parenchyma or root tip cells, but are confined to, respectively, the protein storage vacuole or the nascent vegetative vacuole (38, 84), two functionally distinct vacuolar compartments (84). In Arabidopsis mesophyll cells, PIP1 aquaporins accumulate in highly convoluted invaginations of the plasma membrane called plasmalemmasomes (94). This might indicate local regulation in water permeability for membrane domains in close contact with the tonoplast.

Posttranslational Modifications

α-TIP can be phosphorylated in bean seeds by a tonoplast-bound calcium-dependent protein kinase (45). Putative aquaporins in the spinach leaf plasma membrane are also phosphorylated, in response to changes in apoplastic water potential (44). The protein kinase(s) and protein phosphatase(s) that target plant aquaporins may thus provide an efficient link between water transport and the signaling cascades involved in plant cell osmoregulation (44, 73).

The functional significance of α-TIP phosphorylation was investigated in *Xenopus* oocytes (74). For this, a set of site-specific mutants of α-TIP where three putative phosphorylation sites were disrupted, individually or in combination, was used for in vitro and in vivo phosphorylation by animal cAMP-dependent protein kinase (PKA). Parallel functional assay of these mutant proteins showed that direct phosphorylation of α-TIP at these three sites stimulates by 100–150% its water channel activity. From these and similar results obtained on animal AQP-2 (60), it has been proposed that aquaporin phosphorylation can provide a mechanism to control in situ opening and closing of water channels (60, 73, 74). Alternatively, this might represent a signal for subcellular protein transport, as suspected in the specific case of AQP-2 regulation by vasopressin (62).

Other posttranslational modifications occur in aquaporins, such as the proteolytic maturation from a larger precursor of a 23-kDa TIP in pumpkin seeds (41). They possibly represent novel mechanisms for regulating aquaporin expression and activity.

Other Regulatory Mechanisms

The endo- and exo-osmotic L_p measured in *Chara* cells by transcellular osmosis can be differentially inhibited by, respectively, pCMBS or the K^+ channel blocker nonyltriethylammonium (122). This suggests the presence of distinct types and/or regulations of water channels at the two poles of the cell, which may explain the observed cell polarity for water transport. In particular, the specific sensitivity of endoosmotic water flow to cytochalasins B and E points to a role for the actin cytoskeleton in regulating the activity or localization of some of these water-transport proteins (121).

The L_p of giant-celled algae can also be modulated in response to hydrostatic or osmotic pressure gradients (15, 56, 107, 108, 137). For instance, the intracellular and extracellular osmotic pressures can induce a linear increase by up to 100% and 30%, respectively, in the hydraulic resistance of *Nitella flexilis* membranes (56). A pressure-dependent membrane behavior has also been observed in the leaf cells of the higher plant *Elodea* (Table 1; 109), but was found in neither *Tradescantia* nor *Mesembryanthemum* (see references in 105). Turgor- and concentration-dependent water permeability might be of prime importance to plant cell water relations, but its molecular mechanisms remain unknown. This phenomenon seems to be related to the polarity of water transport observed in certain cells, because polarity was stronger at high tonicity (57), and similar mechanisms involving lipid membrane extension or folding, compression, or dehydration have been proposed to explain the two phenomena (reviewed in 103). None of the cloned plant or animal aquaporins has been reported to be directly gated by hydrostatic or osmotic pressure gradients.

THE INTEGRATED FUNCTION OF AQUAPORINS IN PLANTS

Studies in animals clearly indicate a role for aquaporins in renal functions and more generally in epithelial fluid absorption/secretion. In contrast, the function of aquaporins in red blood cells or as putative osmoreceptors in the brain remains unclear (54). The physiological and developmental processes that involve aquaporin function in plants can be inferred from the expression pattern of these proteins, but this provides no direct indication on their basic cellular function(s). For instance, water-stress regulated aquaporins might contribute to the tolerance of plants to drought or salinity, but it is not clear whether they are involved in adjusting the overall L_p of the plant to its growth capacity (126), locally facilitate water mobilization toward critical cells and organs during drought (12), or help in preadapting dessicated tissues to a sudden rehydration stress after drought (75).

Transcellular and Long-Distance Water Transport

In plants, most of the long-distance longitudinal transport occurs through phloem and xylem vessels that pose no or poorly resistant membrane barriers to water flow. In contrast, the intense water flow that radially traverses roots and leaves is mediated through nonvascular living tissues (7). The respective roles of the apoplasmic (across cell walls) and cell-to-cell paths followed by water in these tissues are still a matter of debate and may well depend on the physiological conditions, the organ or the plant species considered (Figure 2; 7, 104). Although it has been interpreted that cell-to-cell water transport mostly occurs by the symplasmic path (across plasmodesmata) (Figure 2; 85, 135), it has not been experimentally possible to distinguish its contribution

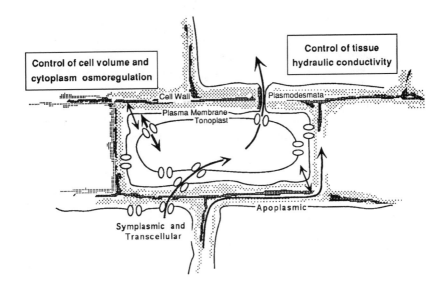

Figure 2   Putative functions of aquaporins in plant cells.

from that of the transcellular path (across membranes). Aquaporins, in particular those at the plasma membrane (PIPs), might play a major role in controlling the latter (12). However, cell types where transmembrane water flow might be limiting remain to be clearly identified. In roots, the Casparian band of endodermal cell walls is thought to create an impediment to radial water flow toward the stele and the vascular tissues (85, 104). This means that either in the root cortex or in the endodermis, apoplastic water flow has to enter the symplasm. The expression of PIP1 aquaporins higher in the root endodermis than in the root cortex of Arabidopsis (A Schäffner, personal communication) rather fits with the latter possibility. The xylem parenchyma cells are also a site of high aquaporin expression (Table 2) and surely play a key role in controlling osmotic and hydrostatic water movements.

Although integrated models of water transport in roots have been proposed (85, 104), a direct answer to these questions will require the precise quantification of membrane P_f in the different root cell types. The combination of the cell and root pressure probes, together with the use of mechanically modified roots bring a first hint at these questions (5, 24, 104). The use of plants altered in the expression or activity of root aquaporins will be helpful to complement these approaches.

The blockade of aquaporins by mercury has recently been taken as a basis to investigate the overall role of cellular membranes in long-distance water transport (10, 32, 69). Maggio & Joly (69) measured pressure-induced sap flows in excised tomato roots and found that treating these roots with 0.5 mM $HgCl_2$ induced a 57% decrease in their hydrostatic L_p, without any change in global K^+ transport. Carvajal et al (10) demonstrated an effect of a lower $HgCl_2$ concentration (50 µM) on sap exudation in excised wheat roots. Although their measurements of osmotic L_p in roots are more questionable than the hydrostatic L_p measurements performed by others (69, 103), their results suggest that the down-regulation of mercury-sensitive water channels might account for the decrease in root L_p observed upon nitrogen or phosphorus deprivation. A simultaneous reduction in plasma membrane fluidity points to a complementary role for lipids in regulating membrane permeability (10). The osmotic extension and shrinkage of peeled sunflower hypocotyl segments was also slowed down after mercury treatment (32). Although mercury-induced change in L_p has been identified in these three approaches, it remains unclear whether aquaporins were the target of the inhibitor. Mercury may target other membrane transport proteins or act as a metabolic poison, thus greatly altering local water potential gradients. However, mercury inhibition could be reversed in all cases by reducing agents, suggesting that the blocker did not alter intratissue organization.

Cell Volume and Osmoregulation

The abundance of aquaporins in the tonoplast of different cell types (46, 67, 70) and the few reports on water transport in vacuoles (58, 100, 113, 125; C Maurel, F Tacnet, J Güclü, J Guern & P Ripoche, unpublished manuscript), both suggest that most plant cells exhibit a high tonoplast water permeability. This might provide the cells with a reduced vacuolar resistance to transcellular water flow (12). The observation that in tobacco suspension cells and in onion epidermal cells the tonoplast can be up to 100-fold more permeable to water than the plasma membrane (Table 1) suggests a second role for water transport at the tonoplast (75). In many plant cells, the vacuole(s) occupies most of the interior, and the cytoplasm is restricted to a thin layer between the tonoplast and the plasma membrane (Figure 2). Nonlimiting water transport at the tonoplast might allow plant cells to efficiently use their vacuolar space to buffer osmotic fluctuations occurring in the cytoplasm (75). This would, in particular, avoid the collapse or the swelling of the latter, in case of a sudden osmotic challenge originating from the extracellular space. The clustering of Arabidopsis plasma membrane aquaporins, in plasmalemmasomes that protrude deep into the vacuole (94), might also contribute to optimizing water exchange between the vacuole and the apoplast, with minor osmotic perturbation in the cytoplasm. All together, these studies strongly suggest an original osmoregulatory function of aquaporins in plant cells directly related to the typical compartmentation of these cells.

The role of cell L_p (water supply) in controlling extension growth is still unclear (7, 104). A baseline water permeability at the plasma membrane of individual elongating cells could be a priori sufficient, even under a reduced water potential gradient, to account for the overall flow of water that mediates their relatively slow expansion. Such hydraulic resistance might also explain the disequilibria in water potential observed in growing tissues (7). However, a high L_p has been reported for some elongating cells and might be necessary for a cell-to-cell supply of water, from xylem to growing tissues (13, 104). Accordingly, the plasma membrane aquaporin PIP1b is preferentially expressed in the elongating cells of Arabidopsis root and hypocotyl (48). The parallel expression of γ-TIP in these cells (65) might reflect a critical role for tonoplast aquaporins during vegetative vacuole differentiation and expansion. This would permit, in particular, water uptake into the vacuole or transvacuolar water flow in quasi-isoosmotic conditions.

Aquaporins are surely involved in some other aspects of plant cell turgor and volume regulation: for instance, the re-imbibition of dessicated plant cells that can occur during seed germination (73, 74) or after exposure to drought.

The observation that pollen grain hydration in *Brassica* is regulated through protein synthesis in the stigma and that this control possibly relates to the self-incompatibility mechanisms (95) might also suggest a key role for aquaporins in plant reproduction. Finally, plant infection by root knot nematodes leads to the swelling of two adjacent root cells that serve as nutrient reservoirs and feeding site for the parasites. The *trans*-induction of aquaporin pRB7 by the pathogen in these cells provides a striking example of the role that aquaporins may play in adjusting membrane P_f to cell size and water demand (82).

In conclusion, aquaporins seem to fulfill two main functions in plants, in individual cell osmoregulation, on the one hand, and in the control of trans-cellular and -tissue water transport, on the other hand. These two functions appear to be intrinsically related to the dual presence of these proteins, respectively, on the tonoplast and the plasma membrane (Figure 2).

PERSPECTIVES

The discovery of aquaporins in plants and other organisms provides a molecular basis for the passive permeability of membranes to water. Beyond its immediate implications in molecular biophysics, this discovery has raised a novel emphasis on the notion of hydraulic conductivity of plant membranes, cells, and tissues and opens new perspectives to understand the role and the complexity of this parameter. Two major points have received strong molecular support from the most recent work on plant aquaporins. First, the hydraulic conductivity of membranes appears to be crucially determined at the spatial level in the plant. This control is exerted down to the level of subcellular compartments and maybe even membrane domains and determines fundamental features of cell organization and polarity. In particular, water transport in and through the plant cell is intrinsically related to its compartmentation into vacuole and cytoplasm. The spatial control of membrane hydraulic conductivity also contributes to determining water exchange between adjacent cells. This is especially important because, due to plant tissue organization, any single cell type has to cope with specific and local water potential gradients. Finally, aquaporins likely participate in the complex integration of water movement in the whole plant. For instance, the radial flow of water in roots that involves distinct parallel paths converging toward the vascular tissue of the stele illustrates the expected complexity of this integration and can now be addressed in molecular terms. The second idea that emerges is that a large variety of regulatory mechanisms can now be envisaged that can control the intensity of transmembrane water flow and its relative contribution to total water transport. These mechanisms allow a dynamic adjustment of membrane properties and appear to be integrated in most developmental programs, from seed to flower, and in a variety of physiological responses to biotic and abiotic stimuli. The discovery of aquaporins has directed our attention to the importance of water transport regulation in most of these processes. Conversely, the need for such an extended spatial and temporal control of water transport through plant membranes may well provide a basis to explain the wealth of aquaporin isoforms and associated regulatory mechanisms that are being discovered.

The description of as yet unidentified plant aquaporins and their expression patterns will surely reveal new physiological situations that involve aquaporin function. Nevertheless, numerous molecular probes are already available that allow accurate aquaporin mRNA and protein quantification and localization in plant cells. These tools, coupled to accurate water permeability measurements by means of the cell pressure probe or stopped-flow spectrophotometry, should enable the development of novel cell biological approaches for investigating the variety of physiological and developmental regulations that have already been identified in algal and higher plant cells. The most exciting developments opened by aquaporins discovery might yet be expected in the field of genetics. In human beings, individuals mutated in the AQP-1 and AQP-2 genes have provided valuable insights into the physiological significance of these aquaporins (19, 71, 90). In plants, reverse genetics using antisense aquaporin transgenes seems to be equally promising (48) (R Kaldenhoff, personal communication). The development of other genetical strategies, such as transposon or T-DNA insertion mutagenesis, should allow a more accurate targeting of aquaporin genes and functions.

In conclusion, aquaporins provide a molecular basis to the passive transport of water across plant membranes. However, it remains crucial to dissect the complementary mechanisms that determine the driving forces responsible for water transport and balance at the different organization levels of the plant. We are just at the beginning of understanding the entire significance of plant aquaporins, but the characterization of these proteins has already inseminated original physiological and theoretical approaches to the study of plant water relations (10, 69, 106).

ACKNOWLEDGMENTS

I thank J Dainty, MJ Chrispeels, J Guern, and P Ripoche for discussions; MJ Chrispeels, R Kaldenhoff, X Sarda, and AR Schäffner for communicating unpublished results; and H Barbier-Brygoo, MJ Chrispeels, D Geelen, J Guern, C Lurin, and P Ripoche for critical reading of the manuscript.

Literature Cited

1. Abrami L, Berthonaud V, Deen PMT, Rousselet G, Tacnet F, et al. 1996. Glycerol permeability of mutant aquaporin 1 and other AQP-MIP proteins: inhibition studies. *Pflügers Arch.* 431:408–14
2. Abrami L, Tacnet F, Ripoche P. 1995. Evidence for a glycerol pathway through aquaporin 1 (CHIP28) channels. *Pflügers Arch.* 430:447–58
3. Agre P, Sasaki S, Chrispeels MJ. 1993. Aquaporins: a family of water channel proteins. *Am. J. Physiol.* 261:F461
4. Alcayaga C, Cecchi X, Alvarez O, Latorre R. 1989. Streaming potential measurements in Ca^{2+}-activated K^+ channels from skeletal and smooth muscle. *Biophys. J.* 55:367–71
5. Azaizeh H, Gunse B, Steudle E. 1992. Effects of NaCl and $CaCl_2$ on water transport across root cells of maize (*Zea mays* L.) seedlings. *Plant Physiol.* 99:886–94
6. Bai L, Fushimi K, Sasaki S, Marumo F. 1996. Structure of aquaporin-2 vasopressin water channel. *J. Biol. Chem.* 271:5171–76
7. Boyer JS. 1985. Water transport. *Annu. Rev. Plant Physiol.* 36:473–516
8. Calamita G, Bishai WR, Preston GM, Guggino WB, Agre P. 1995. Molecular cloning and characterization of AqpZ, a water channel from *Escherichia coli*. *J. Biol. Chem.* 270:29063–66
9. Carceller MS, Sánchez RA. 1972. The influence of phytochrome in the water exchange of epidermal cells of *Taraxacum officinale*. *Experientia* 28:364
10. Carvajal M, Cooke DT, Clarkson DT. 1996. Responses of wheat plants to nutrient deprivation may involve the regulation of water-channel function. *Planta* 199:372–81
11. Chrispeels MJ, Agre P. 1994. Aquaporins: water channel proteins of plant and animal cells. *Trends Biochem. Sci.* 19:421–25
12. Chrispeels MJ, Maurel C. 1994. Aquaporins: the molecular basis of facilitated water movement through living plant cells? *Plant Physiol.* 105:9–13
13. Cosgrove D, Steudle E. 1981. Water relations of growing pea epicotyl segments. *Planta* 153:343–50

14. Dainty J. 1963. Water relations of plant cells. *Adv. Bot. Res.* 1:279–326
15. Dainty J, Ginzburg BZ. 1964. The measurement of hydraulic conductivity (osmotic permeability to water) of internodal characean cells by means of transcellular osmosis. *Biochim. Biophys. Acta* 79:102–11
16. Dainty J, Ginzburg BZ. 1964. The reflection coefficient of plant cell membranes for certain solutes. *Biochim. Biophys. Acta* 79:129–37
17. Daniels MJ, Chaumont F, Mirkov TE, Chrispeels MJ. 1996. Characterization of a new vacuolar membrane aquaporin sensitive to mercury at a unique site. *Plant Cell* 8:587–99
18. Daniels MJ, Mirkov TE, Chrispeels MJ. 1994. The plasma membrane of *Arabidopsis thaliana* contains a mercury-insensitive aquaporin that is a homolog of the tonoplast water channel protein TIP. *Plant Physiol.* 106:1325–33
19. Deen PMT, Verdijk MAJ, Knoers NVAM, Wieringa B, Monnens LAH, et al. 1994. Requirement of human renal water channel aquaporin-2 for vasopressin-dependent concentration of urine. *Science* 264:92–95
20. Echevarria M, Windhager EE, Tate SS, Frindt G. 1994. Cloning and expression of AQP3, a water channel from the medullary collecting duct of rat kidney. *Proc. Natl. Acad. Sci. USA* 91:10997–11001
21. Fettiplace R, Haydon DA. 1980. Water permeability of lipid membranes. *Physiol. Rev.* 60:510–50
22. Finkelstein A. 1987. Water movement through lipid bilayers, pores, and plasma membranes. Theory and reality. In *Distinguished Lecture Series of the Society of General Physiologists,* Vol. 4. New York: Wiley. 228 pp.
23. Fortin MG, Morrison NA, Verma DPS. 1987. Nodulin-26, a peribacteroid membrane nodulin is expressed independently of the development of the peribacteroid compartment. *Nucleic Acids Res.* 15:813–24
24. Frensch J, Hsiao TC, Steudle E. 1996. Water and solute transport along developing maize roots. *Planta* 198:348–55

25. Glinka Z, Reinhold L. 1962. Rapid changes in permeability of cell membranes to water brought about by carbon dioxide and oxygen. *Plant Physiol.* 37:481–86

26. Glinka Z, Reinhold L. 1972. Induced changes in permeability of plant cell membranes to water. *Plant Physiol.* 49:602–6

27. Graziani Y, Livne A. 1972. Water permeability of bilayer lipid membranes: sterol-lipid interaction. *J. Membr. Biol.* 7:275–84

28. Gutknecht J. 1967. Membranes of *Valonia ventricosa:* apparent absence of water-filled pores. *Science* 158:787–88

29. Gutknecht J. 1968. Permeability of *Valonia* to water and solutes: apparent absence of aqueous membrane pores. *Biochim. Biophys. Acta* 163:20–29

30. Haines TH. 1994. Water transport across biological membranes. *FEBS Lett.* 346:115–22

31. Hasegawa H, Ma T, Skach W, Matthay MA, Verkman AS. 1994. Molecular cloning of a mercurial-insensitive water channel expressed in selected water-transporting tissues. *J. Biol. Chem.* 269:5497–500

32. Hejnowicz Z, Sievers A. 1996. Reversible closure of water channels in parenchymatic cells of sunflower hypocotyl depends on turgor status of the cells. *J. Plant Physiol.* 147:516–20

33. Heller KB, Lin ECC, Wilson TH. 1980. Substrate specificity and transport properties of the glycerol facilitator of *Escherichia coli. J. Bacteriol.* 144:274–78

34. Henzler T, Steudle E. 1995. Reversible closing of water channels in *Chara* internodes provides evidence for a composite transport model of the plasma membrane. *J. Exp. Bot.* 46:199–209

35. Höfte H, Desprez T, Amselem J, Chiapello H, Caboche M, et al. 1993. An inventory of 1152 expressed sequence tags obtained by partial sequencing of cDNAs from *Arabidopsis thaliana. Plant J.* 4:1051–61

36. Höfte H, Faye L, Dickinson C, Herman EM, Chrispeels MJ. 1991. The protein-body proteins phytohemagglutinin and tonoplast intrinsic protein are targeted to vacuoles in leaves of transgenic tobacco. *Planta* 184:431–37

37. Höfte H, Hubbard L, Reizer J, Ludevid D, Herman EM, et al. 1992. Vegetative and seed-specific forms of tonoplast intrinsic protein in the vacuolar membrane of *Arabidopsis thaliana. Plant Physiol.* 99:561–70

38. Hoh B, Hinz G, Jeong B-K, Robinson DG. 1995. Protein storage vacuoles form de novo during pea cotyledon development. *J. Cell Sci.* 108:299–310

39. Homblé F, Véry AA. 1992. Coupling of water and potassium ions in K^+ channels of the tonoplast of *Chara. Biophys. J.* 63:996–99

40. House CR, Jarvis P. 1968. Effect of temperature on the radial exchange of labelled water in maize roots. *J. Exp. Bot.* 19:31–40

41. Inoue K, Takeuchi Y, Nishimura M, Hara-Nishimura I. 1995. Characterization of two integral membrane proteins located in the protein bodies of pumpkin seeds. *Plant Mol. Biol.* 28:1089–1101

42. Ishibashi K, Sasaki S, Fushimi K, Uchida S, Kuwahara M, et al. 1994. Molecular cloning and expression of a member of the aquaporin family with permeability to glycerol and urea in addition to water expressed at the basolateral membrane of kidney collecting duct cells. *Proc. Natl. Acad. Sci. USA* 91:6269–73

43. Jap BK, Li H. 1995. Structure of the osmoregulated H_2O-channel, AQP-CHIP, in projection at 3.5 Å resolution. *J. Mol. Biol.* 251:413–20

44. Johansson I, Larsson C, Ek B, Kjellbom P. 1996. The major integral proteins of spinach leaf plasma membranes are putative aquaporins and are phosphorylated in response to Ca^{2+} and apoplastic water potential. *Plant Cell* 8:1181–91

45. Johnson KD, Chrispeels MJ. 1992. Tonoplast-bound protein kinase phosphorylates tonoplast intrinsic protein. *Plant Physiol.* 100:1787–95

46. Johnson KD, Herman EM, Chrispeels MJ. 1989. An abundant, highly conserved tonoplast protein in seeds. *Plant Physiol.* 91:1006–13

47. Jung JS, Preston GM, Smith BL, Guggino WB, Agre P. 1994. Molecular structure of the water channel through aquaporin CHIP: the hourglass model. *J. Biol. Chem.* 269:14648–54

48. Kaldenhoff R, Kölling A, Meyers J, Karmann U, Ruppel G, et al. 1995. The blue light–responsive *AthH2* gene of *Arabidopsis thaliana* is primarily expressed in expanding as well as in differentiating cells and encodes a putative channel protein of the plasmalemma. *Plant J.* 7:87–95

49. Kaldenhoff R, Kölling A, Richter G. 1993. A novel blue light– and abscisic acid-inducible gene of *Arabidopsis thaliana* encoding an intrinsic membrane protein. *Plant Mol. Biol.* 23:1187–98

50. Kaldenhoff R, Kölling A, Richter G. 1997. Regulation of the *Arabidopsis thaliana* aquaporin-gene AthH2 (PIP1b). *J. Photobiol. Photobiochem.* In press

51. Kamiya N, Tazawa M. 1956. Studies on water permeability of a single plant cell by means of transcellular osmosis. *Protoplasma* 46:394–422

52. Kammerloher W, Fischer U, Piechottka GP, Schäffner AR. 1994. Water channels in the plant plasma membrane cloned by immunoselection from a mammalian expression system. *Plant J.* 6:187–99

53. Kedem O, Kachalsky A. 1958. Thermodynamic analysis of the permeability of biological membranes to non-electrolytes. *Biochim. Biophys. Acta* 27:229–46

54. King LS, Agre P. 1996. Pathophysiology of the aquaporin water channels. *Annu. Rev. Physiol.* 58:619–48

55. Kiyosawa K, Ogata K. 1987. Influence of external osmotic pressure on water permeability and electrical conductance of *Chara* cell membrane. *Plant Cell Physiol.* 28: 1013–22

56. Kiyosawa K, Tazawa M. 1972. Influence of intracellular and extracellular tonicities on water permeability in characean cells. *Protoplasma* 74:257–70

57. Kiyosawa K, Tazawa M. 1973. Rectification characteristics of *Nitella* membranes in respect to water permeability. *Protoplasma* 78:203–14

58. Kiyosawa K, Tazawa M. 1977. Hydraulic conductivity of tonoplast-free *Chara* cells. *J. Membr. Biol.* 37:157–66

59. Koefoed-Johnsen V, Ussing HH. 1953. The contributions of diffusion and flow to the passage of D_2O through living membranes. Effect of neurohypophyseal hormone on isolated anuran skin. *Acta Physiol. Scand.* 28:60–76

60. Kuwahara M, Fushimi K, Terada Y, Bai L, Marumo F, et al. 1995. cAMP-dependent phosphorylation stimulates water permeability of aquaporin-collecting duct water channel protein expressed in *Xenopus* oocytes. *J. Biol. Chem.* 270:10384–87

61. Lande MB, Donovan JM, Zeidel ML. 1995. The relationship between membrane fluidity and permeabilities to water, solutes, ammonia, and protons. *J. Gen. Physiol.* 106: 67–84

62. Lande MB, Jo I, Zeidel ML, Somers M, Harris HW. 1996. Phosphorylation of aquaporin-2 does not alter the membrane water permeability of rat papillary water channel–containing vesicles. *J. Biol. Chem.* 271:5552–57

63. Le Cahérec F, Bron P, Verbavatz J-M, Garret A, Morel G, et al. 1996. Incorporation of proteins into (*Xenopus*) oocytes by proteoliposome microinjection: functional characterization of a novel aquaporin. *J. Cell Sci.* 109:1285–95

64. Loros J, Taiz L. 1982. Auxin increases the water permeability of *Rhoeo* and *Allium* epidermal cells. *Plant Sci. Lett.* 26:93–102

65. Ludevid D, Höfte H, Himelblau E, Chrispeels MJ. 1992. The expression pattern of the tonoplast intrinsic protein γ-TIP in *Arabidopsis thaliana* is correlated with cell enlargement. *Plant Physiol.* 100:1633–39

66. Macey RI. 1984. Transport of water and urea in red blood cells. *Am. J. Physiol.* 246:C195–203

67. Maeshima M. 1992. Characterization of the major integral protein of vacuolar membrane. *Plant Physiol.* 98:1248–54

68. Maeshima M, Hara-Nishimura I, Takeuchi Y, Nishimura M. 1994. Accumulation of vacuolar H^+-pyrophosphatase and H^+-ATPase during reformation of the central vacuole in germinating pumpkin seeds. *Plant Physiol.* 106:61–69

69. Maggio A, Joly RJ. 1995. Effects of mercuric chloride on the hydraulic conductivity of tomato root systems: evidence for a channel-mediated water pathway. *Plant Physiol.* 109:331–35

70. Marty-Mazars D, Clémencet M-C, Dozolme P, Marty F. 1995. Antibodies to the tonoplast from the storage parenchyma cells of beetroot recognize a major intrinsic protein related to TIPs. *Eur. J. Cell Biol.* 66:106–18

71. Mathai JC, Mori S, Smith BL, Preston GM, Mohandas N, et al. 1996. Functional analysis of aquaporin-1 deficient red cells: the Colton-null phenotype. *J. Biol. Chem.* 271: 1309–13

72. Mathieu C, Motta C, Hartmann M-A, Thonat C, Boyer N. 1995. Changes in plasma membrane fluidity of *Bryonia dioica* internodes during thigmomorphogenesis. *Biochim. Biophys. Acta* 1235: 249–55

73. Maurel C, Chrispeels MJ, Lurin C, Tacnet F, Geelen D, et al. 1997. Function and regulation of seed aquaporins. *J. Exp. Bot.* 48:In press

74. Maurel C, Kado RT, Guern J, Chrispeels MJ. 1995. Phosphorylation regulates the water channel activity of the seed-specific aquaporin α-TIP. *EMBO J.* 14:3028–35

75. Maurel C, Reizer J, Schroeder JI, Chrispeels MJ. 1993. The vacuolar membrane protein γ-TIP creates water specific channels in *Xenopus* oocytes. *EMBO J.* 12: 2241–47

76. Maurel C, Reizer J, Schroeder JI, Chrispeels MJ, Saier MHJ. 1994. Functional

characterization of the *Escherichia coli* glycerol facilitator, *GlpF*, in *Xenopus* oocytes. *J. Biol. Chem.* 269:11869–72

77. McCain DC, Markley JL. 1985. Water permeability of chloroplast envelope membranes. In vivo measurement by saturation-transfer NMR. *FEBS Lett.* 183:353–58

78. Melroy DL, Herman EM. 1991. TIP, an integral membrane protein of the protein-storage vacuoles of the soybean cotyledon undergoes developmentally regulated membrane accumulation and removal. *Planta* 184:113–22

79. Miao G-H, Hong Z, Verma DPS. 1992. Topology and phosphorylation of soybean nodulin-26, an intrinsic protein of the peribacteroid membrane. *J. Cell Biol.* 118:481–90

80. Mitra AK, van Hoek AN, Wiener MC, Verkman AS, Yeager M. 1995. The CHIP28 water channel visualized in ice by electron crystallography. *Nat. Struct. Biol.* 2:726–29

81. Niemietz C, Tyerman SD. 1995. Physiology of aquaporins in native plant membranes. *Int. Workshop Plant Membr. Biol.*, 10th, Regensburg, Ger. C76 (Abstr.)

82. Opperman CH, Taylor CG, Conkling MA. 1994. Root-knot nematode-directed expression of a plant root-specific gene. *Science* 263:221–23

83. Paganelli CV, Solomon AK. 1957. The rate of exchange of tritiated water across the human red cell membrane. *J. Gen. Physiol.* 41:259–77

84. Paris N, Stanley CM, Jones RL, Rogers JC. 1996. Plant cells contain two functionally distinct vacuolar compartments. *Cell* 85:563–72

85. Passioura JB. 1988. Water transport in and to roots. *Annu. Rev. Plant Physiol. Plant Mol. Biol.* 39:245–65

86. Phillips AL, Huttly AK. 1994. Cloning of two gibberellin-regulated cDNAs from *Arabidopsis thaliana* by subtractive hybridization: expression of the tonoplast water channel, γ-TIP, is increased by GA3. *Plant Mol. Biol.* 24:603–15

87. Preston GM, Carroll TP, Guggino WB, Agre P. 1992. Appearance of water channels in *Xenopus* oocytes expressing red cell CHIP28 protein. *Science* 256:385–87

88. Preston GM, Jung JS, Guggino WB, Agre P. 1993. The mercury-sensitive residue at cysteine 189 in the CHIP28 water channel. *J. Biol. Chem.* 268:17–20

89. Preston GM, Jung JS, Guggino WB, Agre P. 1994. Membrane topology of aquaporin CHIP. Analysis of functional epitope-scanning mutants by vectorial proteolysis. *J. Biol. Chem.* 269:1668–73

90. Preston GM, Smith BL, Zeidel ML, Moulds JJ, Agre P. 1994. Mutations in *aquaporin-1* in phenotypically normal humans without functional CHIP water channels. *Science* 265:1585–87

91. Ramahaleo T. 1996. *Conductances ioniques, élasticité et perméabilité osmotique de cellules racinaires de colza* (Brassica napus). PhD thesis. Univ. Rouen, France. 162 pp.

92. Ray PM. 1960. On the theory of osmotic water movement. *Plant Physiol.* 35:783–95

93. Reizer J, Reizer A, Saier MHJ. 1993. The MIP family of integral membrane channel proteins: sequence comparisons, evolutionary relationships, reconstructed pathway of evolution, and proposed functional differentiation of the two repeated halves of the proteins. *Crit. Rev. Biochem. Mol. Biol.* 28:235–57

94. Robinson DG, Sieber H, Kammerloher W, Schäffner AR. 1996. PIP1 aquaporins are concentrated in plasmalemmasomes of *Arabidopsis thaliana* mesophyll. *Plant Physiol.* 11:645–49

95. Sarker RH, Elleman CJ, Dickinson HG. 1988. Control of pollen hydration in *Brassica* requires continued protein synthesis, and glycosylation is necessary for intraspecific incompatibility. *Proc. Natl. Acad. Sci. USA* 85:4340–44

96. Schuler I, Milon A, Nakatani Y, Ourisson G, Albrecht A-M, et al. 1991. Differential effects of plant sterols on water permeability and on acyl chain ordering of soybean phosphatidylcholine bilayers. *Proc. Natl. Acad. Sci. USA* 88:6926–30

97. Shi L-b, Skach WR, Verkman AS. 1994. Functional independence of monomeric CHIP28 water channels revealed by expression of wild-type mutant heterodimers. *J. Biol. Chem.* 269:10417–22

98. Shi L-b, Verkman AS. 1996. Selected cysteine point mutations confer mercurial sensitivity to the mercurial-insensitive water channel MIWC/AQP-4. *Biochemistry* 35:538–44

99. Smith BL, Agre P. 1991. Erythrocyte M_r 28,000 transmembrane protein exists as a multisubunit oligomer similar to channel proteins. *J. Biol. Chem.* 266:6407–15

100. Snaar JEM, van As H. 1992. Probing water compartments and membrane permeability in plant cells by ^{1}H NMR relaxation measurements. *Biophys. J.* 63:1654–58

101. Stadelmann EJ. 1969. Permeability of the

428   MAUREL

plant cell. *Annu. Rev. Plant Physiol.* 20: 585–606

102. Stadelmann EJ, Lee-Stadelmann OY. 1989. Passive permeability. *Methods Enzymol.* 174:246–66

103. Steudle E. 1993. Pressure probe techniques: basic principles and application to studies of water and solute relations at the cell, tissue and organ level. In *Water Deficits: Plant Responses From Cell to Community*, ed. JAC Smith, H Griffiths, pp. 5–36. Oxford: Bios Sci.

104. Steudle E. 1994. The regulation of plant water at the cell, tissue, and organ level: role of active processes and of compartmentation. In *Flux Control in Biological Systems: From Enzymes to Populations and Ecosystems*, ed. ED Schultze, pp. 237–99. San Diego: Academic

105. Steudle E. 1989. Water flow in plants and its coupling to other processes: an overview. *Methods Enzymol.* 174:183–225

106. Steudle E, Henzler T. 1995. Water channels in plants: do basic concepts of water transport change? *J. Exp. Bot.* 46:1067–76

107. Steudle E, Tyerman SD. 1983. Determination of permeability coefficients, reflection coefficients, and hydraulic conductivity of *Chara corallina* using the pressure probe: effects of solute concentrations. *J. Membr. Biol.* 75:85–96

108. Steudle E, Zimmermann U. 1974. Determination of the hydraulic conductivity and of reflection coefficients in *Nitella flexilis* by means of direct cell-turgor pressure measurements. *Biochim. Biophys. Acta* 332: 399–412

109. Steudle E, Zimmermann U, Zillikens J. 1982. Effect of cell turgor on hydraulic conductivity and elastic modulus of *Elodea* leaf cells. *Planta* 154:371–80

110. Stout DG, Cotts RM, Steponkus PL. 1977. The diffusional water permeability of *Elodea* leaf cells as measured by nuclear magnetic resonance. *Can. J. Bot.* 55: 1623–31

111. Strzalka K, Hara-Nishimura I, Nishimura M. 1995. Changes in physical properties of vacuolar membrane during transformation of protein bodies into vacuoles in germinating pumpkin seeds. *Biochim. Biophys. Acta* 1239:103–10

112. Tyerman SD, Steudle E. 1982. Comparison between osmotic and hydrostatic water flows in a higher plant cell: determination of hydraulic conductivities and reflection coefficients in isolated epidermis of *Tradescantia virginiana. Aust. J. Plant Physiol.* 9:461–79

113. Url WG. 1971. The site of penetration resistance to water in plant protoplasts. *Protoplasma* 72:427–47

114. van Hoek AN, Hom ML, Luthjens LH, de Jong MD, Dempster JA, et al. 1991. Functional unit of 30 kDa for proximal tubule water channels as revealed by radiation inactivation. *J. Biol. Chem.* 266:16633–35

115. van Hoek AN, Verkman AS. 1992. Functional reconstitution of the isolated erythrocyte water channel CHIP28. *J. Biol. Chem.* 267:18267–69

116. van Os CH, Deen PMT, Dempster JA. 1994. Aquaporins: water selective channels in biological membranes. Molecular structure and tissue distribution. *Biochim. Biophys. Acta* 1197:291–309

117. Verkman AS. 1995. Optical methods to measure membrane transport processes. *J. Membr. Biol.* 148:99–110

118. Verkman AS, van Hoek AN, Ma T, Frigeri A, Skach WR, et al. 1996. Water transport across mammalian cell membranes. *Am. J. Physiol.* 270:C12–30

119. Walz T, Typke D, Smith BL, Agre P, Engel A. 1995. Projection map of aquaporin-1 determined by electron crystallography. *Nat. Struct. Biol.* 2:730–32

120. Wayne R, Mimura T, Shimmen T. 1994. The relationship between carbon and water transport in single cells of *Chara corallina. Protoplasma* 180:118–35

121. Wayne R, Tazawa M. 1988. The actin cytoskeleton and polar water permeability in characean cells. *Protoplasma* (Suppl. 2): 116–30

122. Wayne R, Tazawa M. 1990. Nature of the water channels in the internodal cells of *Nitellopsis. J. Membr. Biol.* 116:31–39

123. Weaver CD, Shomer NH, Louis CF, Roberts DM. 1994. Nodulin 26, a nodule-specific symbiosome membrane protein from soybean, is an ion channel. *J. Biol. Chem.* 269:17858–62

124. Wendler S, Zimmermann U. 1985. Compartment analysis of plant cells by means of turgor pressure relaxation. I. Theoretical considerations. *J. Membr. Biol.* 85:121–32

125. Wendler S, Zimmermann U. 1985. Compartment analysis of plant cells by means of turgor pressure relaxation. II. Experimental results on *Chara corallina. J. Membr. Biol.* 85:133–42

126. Yamada S, Katsuhara M, Kelly WB, Michalowski CB, Bohnert HJ. 1995. A family of transcripts encoding water channel proteins: tissue-specific expression in the common ice plant. *Plant Cell* 7: 1129–42

127. Yamaguchi-Shinozaki K, Koizumi M, Urao S, Shinozaki K. 1992. Molecular

cloning and characterization of 9 cDNAs for genes that are responsive to desiccation in *Arabidopsis thaliana:* sequence analysis of one cDNA clone that encodes a putative transmembrane channel protein. *Plant Cell Physiol.* 33:217–24

128. Yamamoto YT, Taylor CG, Acedo GH, Cheng C-L, Conkling MA. 1991. Characterization of *cis*-acting sequences regulating root-specific gene expression in tobacco. *Plant Cell* 3:371–82

129. Yool AJ, Stamer WD, Regan JW. 1996. Forskolin stimulation of water and cation permeability in aquaporin1 water channels. *Science* 273:1216–18

130. Zeidel ML, Ambudkar SV, Smith BL, Agre P. 1992. Reconstitution of functional water channels in liposomes containing purified red cell CHIP28 protein. *Biochemistry* 31: 7436–40

131. Zeidel ML, Nielsen S, Smith BL, Ambudkar SV, Maunsbach AB, et al. 1994. Ultrastructure, pharmacologic inhibition, and transport selectivity of aquaporin channel–forming integral protein in proteoliposomes. *Biochemistry* 33:1606–15

132. Zeuthen T. 1995. Molecular mechanisms for passive and active transport of water. *Int. Rev. Cytol.* 160:99–161

133. Zhang R, van Hoek AN, Biwersi J, Verkman AS. 1993. A point mutation at cysteine-189 blocks the water permeability of rat kidney water channel CHIP28k. *Biochemistry* 32:2938–41

134. Zhang R, Verkman AS. 1991. Water and urea permeability properties of *Xenopus* oocytes: expression of mRNA from toad urinary bladder. *Am. J. Physiol.* 260: C26–34

135. Zhang WH, Tyerman SD. 1991. Effect of low O_2 concentration and azide on hydraulic conductivity and osmotic volume of the cortical cells of wheat roots. *Aust. J. Plant Physiol.* 18:603–13

135a. Zhang WH, Jones GP. 1996. Water permeability in wheat root protoplasts determined from nuclear magnetic resonance relaxation times. *Plant Sci.* 118:97–106

136. Zimmermann U. 1989. Water relations of plant cells: pressure probe technique. *Methods Enzymol.* 174:338–66

137. Zimmermann U, Steudle E. 1974. The pressure-dependence of the hydraulic conductivity, the membrane resistance and membrane potential during turgor pressure regulation in *Valonia utricularis. J. Membr. Biol.* 16:331–52

Annu. Rev. Plant Physiol. Plant Mol. Biol. 1997. 48:431–60

GIBBERELLIN BIOSYNTHESIS:
Enzymes, Genes and Their Regulation

Peter Hedden

IACR-Long Ashton Research Station, Department of Agricultural Science, University of Bristol, Bristol, BS18 9AF, United Kingdom

Yuji Kamiya

Frontier Research Program, The Institute of Physical and Chemical Research (RIKEN), Hirosawa 2-1, Wako-shi, Saitama 351-01, Japan

KEY WORDS: terpene cyclases, monooxygenases, 2-oxoglutarate-dependent dioxygenases, feedback, light and temperature regulation

ABSTRACT

The recent impressive progress in research on gibberellin (GA) biosynthesis has resulted primarily from cloning of genes encoding biosynthetic enzymes and studies with GA-deficient and GA-insensitive mutants. Highlights include the cloning of *ent*-copalyl diphosphate synthase and *ent*-kaurene synthase (formally *ent*-kaurene synthases A and B) and the demonstration that the former is targeted to the plastid; the finding that the *Dwarf-3* gene of maize encodes a cytochrome P450, although of unknown function; and the cloning of GA 20-oxidase and 3β-hydroxylase genes. The availability of cDNA and genomic clones for these enzymes is enabling the mechanisms by which GA concentrations are regulated by environmental and endogenous factors to be studied at the molecular level. For example, it has been shown that transcript levels for GA 20-oxidase and 3β-hydroxylase are subject to feedback regulation by GA action and, in the case of the GA 20-oxidase, are regulated by light. Also discussed is other new information, particularly from mutants, that has added to our understanding of the biosynthetic pathway, the enzymes, and their regulation and tissue localization.

431

CONTENTS

INTRODUCTION

The gibberellin (GA) hormones act throughout the life cycle of plants, influencing seed germination, stem elongation, flower induction, anther development, and seed and pericarp growth. Furthermore, they mediate environmental stimuli, which modify the flux through the GA-biosynthetic pathway. Regulation of GA biosynthesis is therefore of fundamental importance to plant development and its adaptation to the environment.

The last review of GA biosynthesis in this series, by Graebe (38), was followed by several reviews covering this topic (68, 75, 121). Graebe's article focused on GA-biosynthetic pathways in cell-free systems, characteristics of biosynthetic enzymes, and factors affecting GA production. He predicted accurately that the main topic of the next review in this series would be the cloning and characterization of genes for the GA-biosynthetic enzymes. Several of the enzymes have now been cloned, and the availability of their cDNAs is providing new and often unexpected information on the nature of these proteins. For example, some of the enzymes catalyze multiple steps in the pathway. It is also now possible to investigate the mechanisms by which GA biosynthesis is regulated in response to environmental and endogenous signals.

In light of these exciting advances, a new review on GA biosynthesis is appropriate. As is usual practice, we discuss the biosynthetic pathway, shown in Figures 1 and 2, in three sections according to the nature of the enzymes: terpene cyclases involved in *ent*-kaurene synthesis, monooxygenases, and di-

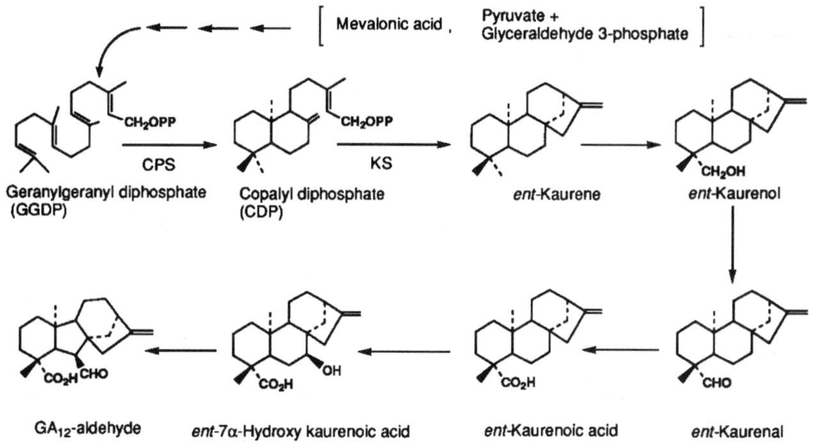

Figure 1 Early GA-biosynthetic pathway to GA_{12}-aldehyde. GGDP is produced in plastids by the isoprenoid pathway, originating from mevalonic acid or, possibly, pyruvate/glyceraldehyde 3-phosphate.

oxygenases. We also include discussions on regulation and sites of synthesis. We restrict ourselves to higher plants, because GA biosynthesis in other organisms, principally fungi and ferns, is not as well understood. However, future phylogenetic comparisons between GA biosynthesis genes in all organisms should be instructive in determining the origin of GAs. Much of the current progress on GA biosynthesis has come from work with mutants, which has been covered comprehensively in several reviews (98, 99, 101, 105) and is not dealt with specifically here. Some of the better characterized GA-deficient mutants, with the position of lesions, are given in Table 1, which also lists cDNA clones for biosynthetic enzymes.

ent-KAURENE SYNTHESIS

ent-Kaurene is synthesized by the two-step cyclization of geranylgeranyl diphosphate (GGDP) via the intermediate, *ent*-copalyl diphosphate (CDP). The enzymes that catalyze these reactions are referred to as the A and B activities, respectively, of *ent*-kaurene synthase (formerly *ent*-kaurene synthetase). However, we adopt the more logical nomenclature proposed by MacMillan (75). Thus, the conversion of GGDP to CDP is catalyzed by *ent*-copalyl diphosphate synthase (CPS) and of CDP to *ent*-kaurene by *ent*-kaurene synthase (KS). The biosynthesis of GGDP from mevalonic acid is common to many terpenoid pathways and is covered in the review by Chappell

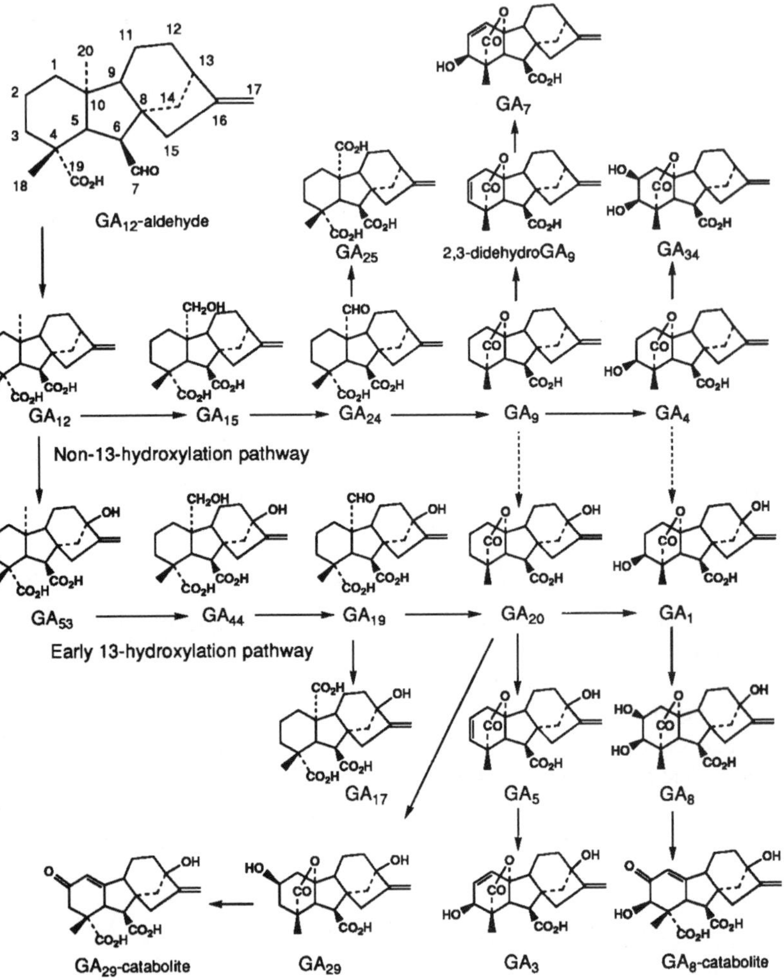

Figure 2 Gibberellin-biosynthetic pathway from GA_{12}-aldehyde.

(13). Recently, a nonmevalonate pathway to isoprenoids, involving pyruvate and glyceraldehyde-3-phosphate, has been proposed in green algae (112). Such a pathway may operate in plastids of higher plants, given the difficulty in demonstrating the incorporation of mevalonate into isoprenoids in these organelles.

Table 1 Mutants and cDNA clones for GA-biosynthetic enzymes

Enzyme	Plant	Mutant	References	cDNA cloning	Data base
CPS	Arabidopsis thaliana	gal	63	124	U11034
	Zea mays	Anl	53	9	L37750
	Pisum sativum	ls-1	128	2	U63652
	Lycopersicon esculentum	gib-1	10	—	
KS	Cucurbita maxima	—	—	147	U43904
	A. thaliana	ga2	150	—	
	Z. mays	d5	48	—	
	L. esculentum	gib-3	10	—	
ent-Kaurene	P. sativum	lh^i	127	—	
oxidase	A. thaliana	ga3	150	—	
	Oryza sativa	dx	89	—	
Monoxy-	Z. mays	d3	28	144	U32579
genase	P. sativum	na	49	—	
GA 20-	C. maxima	—	—	71	X73314
oxidase	A. thaliana	ga5	129	146	U20872
					U20873
					U20901
	A. thaliana	—	—	92	X83379
					X83380
					X83381
	P. sativum	—	—	77	X91658
				31	U70471
	P. sativum	—	—	73	U58830
	Phaseolus vulgaris	—	—	31	U70530
					U70531
					U70532
	O. sativa	—	—	137	U50333
	Spinacia oleracea	—	—	145	U33330
GA 3β-hy-	A. thaliana	ga4	129	14	L37126
droxylase	Z. mays	dl	28	—	
	O. sativa	dy	59	—	
	P. sativum	le	50	—	
	Lathyrus odoratus	l	106	—	
GA 2-oxidases	P. sativum	sln	108	—	

CPS and KS were first separated by anion-exchange chromatography on extracts of *Marah macrocarpus* endosperm (23). There were also indications for the involvement of two enzymes from studies on GA-deficient mutants; work with cell-free extracts of young fruits suggested that the dwarf tomato mutants, *gib-1* and *gib-3,* have lesions at CPS and KS, respectively (10). Using an *ent*-kaurene oxidase inhibitor to estimate rates of *ent*-kaurene biosynthesis, a method first used with germinating barley grain (40), Zeevaart & Talon (150) demonstrated that the nonallelic *ga1* and *ga2* mutants of Arabidopsis were both defective in *ent*-kaurene production, whereas *ga4* and *ga5* are blocked later in the pathway (129).

ent-Copalyl Diphosphate Synthase

Koornneef et al (63) constructed a fine structure genetic map of the Arabidopsis *GA1* locus using nine independent *ga1* alleles, three of which were made by fast neutron bombardment and the rest by treatment with ethylmethanesulfonate (EMS). Among the fast-neutron-generated mutants, *ga1-3* contains a large deletion (5 kb) and failed to recombine with the EMS-treated mutants (63). Sun et al (123) used *ga1-3* to clone the *GA1* locus by genomic subtraction. Cosmid clones containing wild-type DNA inserts spanning the deletion in *ga1-3* complemented the dwarf phenotype when integrated into the *ga1-3* genome by T-DNA transformation. *GA1* cDNA contains a 2.4-kb open reading frame, which was shown by functional analysis to encode CPS (124). *Escherichia coli* co-transformed with a bacterial GGDP synthase gene (12), and the *GA1* cDNA produced CDP, from which copalol was identified by combined gas chromatography–mass spectrometry (GC-MS) after alkaline hydrolysis. Although *ga1-3* contains a large deletion and genomic Southern analysis indicated that *GA1* is a single-copy gene, the mutant produces low amounts of GAs (150), suggesting that there are *GA1* homologues in Arabidopsis or there is an alternative pathway for *ent*-kaurene synthesis. A similar situation exists in maize, from which the *An1* (*Anther ear-1*) locus was cloned by transposon tagging (9). The predicted amino acid sequence of *An1* cDNA shares high sequence identity (51%, without transit peptide sequence) with that of the GA1 protein. A homozygous deletion mutant of *An1, an1-bz2-6923*, accumulated *ent*-kaurene to 20% of the wild-type content, indicating the presence of isoenzymes. Furthermore, a putative homologous cDNA, *An2*, was cloned by RT-PCR (8). At least two different *GA1* homologues have been obtained from tomato seedlings by RT-PCR using oligonucleotide primers based on Arabidopsis and maize CPS sequences (R Imai, personal communication). It appears, therefore, that leakiness of the *ga1-3* and *an1* deletion mutants is due to the presence of other CPSs.

The *Ls* locus of pea was shown to encode CPS. The *ls-1* dwarf mutant had reduced CPS activity in a cell-free system from immature seeds (127). Confirmation was obtained after cloning a CPS from pea by RT-PCR, its identity being confirmed by expression in *E. coli* of a glutathione S-transferase fusion protein with CPS activity (2). The *ls-1* mutation, produced by EMS treatment, is due to a G-to-A substitution at an intron-exon border that causes impaired splicing and a frameshift in the transcript (2).

ent-Kaurene Synthase

ent-Kaurene synthase (KS) was purified from endosperm of pumpkin (*Cucurbita maxima*) (110), which is a rich source of GA-biosynthetic enzymes (38). The enzyme, which had a predicted M_r of 81,000, required divalent cations, such as Mg^{2+}, Mn^{2+}, and Co^{2+}, for activity and had an optimal pH range of 6.8–7.5 (110). The K_m for CDP was 0.35 µM. Purification of KS was quickly followed by its molecular cloning (147). PCR was used with degenerate oligonucleotides, designed from amino acid sequences of the purified protein, to produce a cDNA fragment for library screening. The isolated full-length cDNA was expressed in *E. coli* as a fusion protein, with maltose-binding protein, which converted [³H]CDP to ent-[³H]kaurene. The KS transcript is abundant in growing tissues, such as apices and developing cotyledons, and is present in every organ in pumpkin seedlings. Although it is difficult to compare mRNA abundance across species, it appears that KS is expressed at much higher levels than is CPS. Whereas CPS transcripts are undetectable in leaves of Arabidopsis (124) and pea (2) by northern blot analysis, requiring RNase protection assays (A Silverstone & TP Sun, personal communication) or RT-PCR, respectively, KS mRNA, though of low abundance, can be assayed in pumpkin leaves by northern hybridization (147). This is consistent with strict regulation of the first step of ent-kaurene synthesis from the abundant GGDP.

The deduced amino acid sequence of KS shares significant homology with other terpene cyclases (Figure 3A), with highest homology (51% amino acid similarity) with CPS from Arabidopsis and maize. It contains the DDXXD motif, which is conserved in casbene synthase (81), 5-*epi*-aristolochene synthase (24), and limonene synthase (17), and which is proposed to function as a binding site for the divalent metal ion-diphosphate complex (13). CPS lacks the DDXXD motif, consistent with its catalytic activity not involving cleavage of the diphosphate group.

Subcellular Localization of ent-Kaurene Synthesis

Although there has been evidence for ent-kaurene synthesis in plastids for over 20 years (reviewed in 38), unequivocal confirmation of this has been provided only recently. By precise use of marker enzymes to assess plastid purity and GC-MS to identify enzyme products, Aach et al (1) clearly demonstrated that CPS/KS activity (GGDP to ent-kaurene) is localized in developing chloroplasts from wheat seedlings and leucoplasts from pumpkin endosperm. Mature chloroplasts contained little activity. These results were supported by the recent cDNA cloning of CPS (124) and KS (147). The first 50 N-terminal amino acids of the GA1 protein (Arabidopsis CPS) are rich in serine and

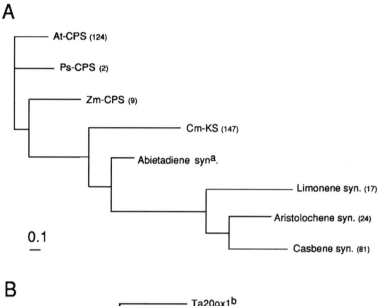

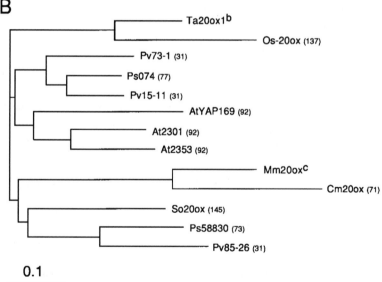

Figure 3 Phylogenetic trees, produced using the PHYLIP package (J Felsenstein, University of Washington, Seattle), for (*A*) terpene cyclases, including *ent*-copalyl diphosphate synthase (CPS) and *ent*-kaurene synthase (KS), and (*B*) GA 20-oxidases. Reference numbers are shown in parentheses. (Sources: [a]R Croteau et al, unpublished data; [b]NEJ Appleford, JR Lenton, AL Phillips & P Hedden, unpublished data; [c]DA Ward, J MacMillan, AL Phillips & P Hedden, unpublished data.)

threonine with an estimated pI of 10.2 (124). Such properties are common features of precursors of many chloroplast-localized proteins, such as the small subunit of Rubisco (54). The transit peptide is cleaved on entry into the plastid to produce a functional mature protein. Incubation of a ^{35}S-labeled Arabidopsis pre-CPS of 86 kDa with isolated pea chloroplasts resulted in transport into the chloroplasts and processing to a 76-kDa protein (124). The deduced amino acid sequence of KS also contains a putative transit peptide, although import into chloroplasts could not be demonstrated (147).

Plastids are the major site of production of GGDP, and most of the GGDP synthases cloned in plants have transit peptides for plastid transport (6). Localization of GGDP synthase from *Capsicum annum* in plastids has been demonstrated immunocytochemically (64). GGDP is a common precursor for many plastid-localized terpenoids, including carotenoids and the phytol side-chain of chlorophyll. Overexpression of phytoene synthase, which converts GGDP to phytoene, in transgenic tomato resulted in a lower chlorophyll content than in wild-type plants and a dwarf phenotype that was partially reversed by applying GA$_3$ (27). The endogenous GA concentrations in apical shoots of the transgenic plants were reduced to about 3% of that in wild-type shoots. If, as suggested (27), overproduction of phytoene has depleted GGDP content, resulting in reduced synthesis of GA and chlorophyll, the three pathways must share the same pool of GGDP and thus be interdependent.

MONOOXYGENASES

The highly hydrophobic *ent*-kaurene is oxidized by membrane-bound monooxygenases to GA$_{12}$. The enzymes require NADPH and oxygen and, on the basis of the early demonstration that *ent*-kaurene and *ent*-kaurenal oxidation are inhibited by carbon monoxide with reversibility by light at 450 nm (86), are all assumed to involve cytochrome P450. The involvement of cytochrome P450 in *ent*-kaurenoic acid 7β-hydroxylase in the fungus *Gibberella fujikuroi* has now also been shown (51). At least one of the enzymes (GA$_{12}$-aldehyde synthase) is associated with the endoplasmic reticulum in pea embryos and pumpkin endosperm (37), requiring transport of *ent*-kaurene, or perhaps a later intermediate, from the plastid.

Several GA-deficient dwarf mutants are defective in *ent*-kaurene oxidase activity. In pea, the lh^i (*lh-2*) mutation affects stem elongation and seed development (125–127). Cell-free extracts from immature *lh-2* seeds were deficient in *ent*-kaurene oxidase activity relative to wild-type seeds; the three steps from *ent*-kaurene to *ent*-kaurenoic acid were affected, suggesting that a single enzyme might catalyze these reactions (127). However, unequivocal verification of this must await the availability of pure enzyme because a regulatory function for *Lh* cannot be excluded. The Tan-ginbozu mutant of rice (*dx*) is probably also deficient in *ent*-kaurene oxidase activity (89). Application of uniconazole, an *ent*-kaurene oxidase inhibitor, mimics the phenotype of Tan-ginbozu and produces endogenous GA concentrations similar to those in the mutant (89). Because of growth responses to applied *ent*-kaurene, Tan-ginbozu (85) and also *lh* (49) were thought previously to have lesions before *ent*-kaurene synthesis. The recent findings (89, 127) indicate that such application experiments may give misleading results.

Although less progress has been made with the monooxygenases than with the other enzymes, the recent cloning of the *Dwarf-3* (*D3*) gene of maize (144) should enable rapid progress in characterizing this group of enzymes. The *D3* gene, obtained by transposon tagging, was found, on the basis of its deduced amino acid sequence, to encode a member of a new class of cytochrome P450 monooxygenases with closest homology to sterol hydroxylases. Unfortunately, there is uncertainty about the step catalyzed by the D3 protein (BO Phinney, personal communication), although this should now be revealed by functional expression of the cDNA in a suitable heterologous system.

DIOXYGENASES

The enzymes involved in the third stage of the pathway are soluble oxidases that use 2-oxoglutarate as a co-substrate. These 2-oxoglutarate–dependent dioxygenases belong to a family of nonheme Fe-containing enzymes that have been the subject of several recent reviews (20, 94, 95). The enzymes show considerable diversity of function, but they are clearly related on the basis of conserved amino acid sequences.

The reactions known to be catalyzed by 2-oxoglutarate–dependent dioxygenases are shown as a network of pathways in Figure 2. Although they were originally delineated in developing seeds (38), both 13-hydroxylation and non-13-hydroxylation pathways have now been demonstrated in vegetative tissues (41, 60). The individual steps between GA_{12}-aldehyde and GA_3 and GA_8 were demonstrated in intact maize shoots by applying each isotopically labeled intermediate and identifying its immediate metabolite by GC-MS (30, 60). Both pathways were observed in a cell-free system from embryos/scutella of two-day-old germinating barley grain (41), although GA_4 was not identified. The dioxygenases in GA biosynthesis will now be discussed in detail, with particular emphasis on newer aspects not covered in the review by Lange & Graebe (68).

7-Oxidase

Oxidation at C-7 from an aldehyde to a carboxylic acid may be catalyzed by either dioxygenases or monooxygenases. Pumpkin endosperm contains both 7-oxidase activities (46), as does barley embryos/scutella (41). The dioxygenase activity from pumpkin has been partially purified and shown to have a very low pH optimum (72), whereas the monooxygenase is most active above pH 7 (46). The two types of activity also differ in their substrate specificities; the monooxygenase is specific for GA_{12}-aldehyde, whereas the soluble activity oxidizes several hydroxylated GA_{12}-aldehyde derivatives (46). The presence of both types of enzyme in a single tissue may indicate subcellular compartmentation of GA-biosynthetic pathways.

13- and 12α-Hydroxylases

As for GA 7-oxidase, both dioxygenase and monooxygenase forms of these hydroxylases have been described. The soluble, 2-oxoglutarate–dependent 13-hydroxylase detected in cell-free extracts from spinach leaves (36) is still the only example of a dioxygenase with this activity. 13-Hydroxylases in pumpkin endosperm (46, 69), developing pea embryos (52), and barley embryos/scutella (41) are of the monooxygenase type. The preferred substrate for the 13-hydroxylases is probably GA_{12}, although other GAs are hydroxylated to some extent. GA_{12}-aldehyde is 13-hydroxylated in embryo cell-free systems from *Phaseolus coccineus* (140) and *P. vulgaris* (128a), indicating that GA_{53}-aldehyde is an intermediate in the 13-hydroxylation pathway in these tissues. In the barley embryos/scutella system, GA_{15} and GA_{24} were 13-hydroxylated at very low rates compared with GA_{12}, while GA_9 was not metabolized (41). This result is similar to that found previously with microsomes from immature pea embryos (52), but in this case GA_{15} and GA_9 were hydroxylated only when their lactones were opened by hydrolysis. Although "late" 13-hydroxylation (on GA_9 or GA_4) can often be demonstrated, it may be relatively inefficient and accompanied by hydroxylation at other positions on the C and D rings (56). However, in some species, such as *Picea abies* (84), it would appear to be the major pathway.

Both forms of the 12α-hydroxylases are present in pumpkin seed, the monooxygenase hydroxylating GA_{12}-aldehyde (GA_{12} is not a substrate) (46), whereas the dioxygenase uses a variety of GA tricarboxylic acid substrates (69, 70). The monooxygenase has a low pH optimum and may thus catalyze part of the same pathway as the soluble 7-oxidase, for which 12α-hydroxyGA_{12}-aldehyde is a substrate. The soluble 12α-hydroxylase is sensitive to the presence of phosphate (69) and was, therefore, undetected in previous studies in which phosphate, rather than Tris, buffer was used to extract the enzymes (45). Because phosphate removes Fe by precipitation, it would appear that the 12α-hydroxylase has an unusually high requirement for Fe^{2+}.

20-Oxidases

Formation of the C_{19}-GA skeleton requires successive oxidation of C-20 from a methyl group, as in GA_{12} or GA_{53}, through the alcohol and aldehyde, from which this C atom is lost as CO_2 (Figure 2). As discussed below, a single enzyme (GA 20-oxidase) can catalyze this reaction sequence, although the number of enzymes that are actually involved in vivo is unknown. A 2-oxoglutarate-dependent dioxygenase that converted GA_{53} to GA_{44} and GA_{19} was partially purified from 20-day-old developing embryos of *P. sativum* (67). Although the proportion of the two products remained constant throughout purification, it was uncertain whether a single enzyme catalyzed both steps. Clear evidence that GA 20-oxidases are multifunctional was obtained after purification of the enzyme to homogeneity from pumpkin endosperm (66). The enzyme converted GA_{12} to GA_{15}, GA_{24} and GA_{25}, and GA_{53} to GA_{44}, GA_{19} and GA_{17}, with a small amount of putative GA_{20} produced at high protein concentrations. GA_{12} was converted more efficiently than was GA_{53}. The production of the tricarboxylic acids, GA_{25} and GA_{17}, is characteristic of the pumpkin endosperm cell-free system (69) and indicates that the 20-oxidase in this tissue is functionally different from that encountered in other systems, which produce predominantly C_{19}-GAs (52, 128a).

The purification of the GA 20-oxidase from pumpkin was quickly followed by the cloning of a cDNA that encoded this enzyme (71). The cDNA was selected from an expression library, derived from developing embryos, using antiserum against a peptide sequence from the purified protein. As well as binding the antibodies, the expressed fusion protein was functionally active and catalyzed the same reactions as the native enzyme. The aldehyde intermediate, GA_{24}, was converted to both GA_{25} and GA_9, although the latter was obtained in less than 1% yield. The presence of hydroxyl groups reduced the efficiency of conversion, by about 50% in the case of GA_{19} (13-hydroxylated) and 95% for GA_{23} (3β, 13-dihydroxylated), although small amounts of the C_{19}-GA product were detected in each case. The derived amino acid sequence corresponds to a protein of 43.3 kDa, which is close to that estimated for the native enzyme from gel filtration (44 kDa), and contains the conserved regions found in other plant dioxygenases.

The cloning of the GA 20-oxidase cDNA from pumpkin seeds enabled the isolation of homologous clones from other species. The first examples, from

Arabidopsis, were obtained independently in two laboratories (92, 146). Two GA 20-oxidase cDNAs were cloned from the *gal-3* mutant utilizing PCR with degenerate primers designed from the conserved amino acid sequences; a third cDNA clone was found after scrutiny of the Data Base of Expressed Sequence Tags (92). Confirmation that all three cDNAs encoded GA 20-oxidases was obtained by demonstrating that the products of heterologous expression in *E. coli* converted GA_{12} to GA_9 and GA_{53} to GA_{20}, with GA_{12} the preferred substrate. Small amounts of the tricarboxylic acids, GA_{25} and GA_{17}, respectively, were formed but, in contrast to the pumpkin enzyme, the C_{19}-GAs were the major products. Thus, these enzymes appeared to be involved in the biosynthesis of active GAs. It was confirmed that at least one of the isozymes is active in vivo when a genomic clone encoding one of the GA 20-oxidases was isolated from Arabidopsis by probing a genomic library with the pumpkin 20-oxidase cDNA (146). The clone mapped tightly to the *GA5* locus, mutation of which results in semidwarfism (62) and a reduction in the concentrations of C_{19}-GAs (129). Expression of the GA 20-oxidase genes is tissue-specific, with transcripts detected, respectively, in stems/floral apices, floral apices/siliques, and siliques (92). The silique-specific 20-oxidase transcript is much more abundant than the others. The stem-specific gene corresponds to the *GA5* locus, mutation of which, in *ga5,* is due to a G to A substitution that introduces a premature stop codon (146). Although the mutant protein would be highly truncated and unlikely to be catalytically active, the *ga5* plant is semidwarfed and contains low amounts of C_{19}-GAs (129). It must be assumed, therefore, that other GA 20-oxidases, such as that expressed in the floral apex, supply GAs to the stem.

Gibberellin 20-oxidase cDNAs have been cloned from at least seven species, with multiple genes found in several of them. Their encoded amino acid sequences share a relatively low degree of sequence conservation, with amino acid identities ranging from 50–75%. The relationship between the sequences is shown in Figure 3*B*. With the exception of the enzyme from pumpkin seed, the proteins have very similar functions, converting 20-methyl GAs to the corresponding C_{19} lactones. It is notable that the enzyme cloned from developing cotyledons of *Marah macrocarpus* (J MacMillan & DA Ward, unpublished information) produces C_{19}-GAs despite its being most closely related to the pumpkin enzyme on the basis of sequence (Figure 3*B*). The structural differences that determine whether the 20-oxo intermediates are oxidized to C_{19}-GAs or to tricarboxylic acids are likely to be subtle. The 20-oxidases from pumpkin, *Marah,* and Arabidopsis prefer nonhydroxylated substrates to the 13-hydroxylated analogues. This is consistent with the types of GAs found in the tissues in which these enzymes are present. For example, GA_4, which is not 13-hydroxylated, is the major GA in Arabidopsis shoots (129). In contrast, a GA 20-oxidase cloned from shoots of rice, in which 13-hydroxy C_{20}-GAs are the predominant forms (61), oxidizes GA_{53} more efficiently than it does GA_{12} (137).

There is evidence for the presence in shoot tissues of GA 20-oxidases with properties different from those that have been cloned so far. GA_{44} oxidase activity from spinach leaves was separated by anion-exchange chromatography from GA_{53} oxidase and GA_{19} oxidase activities, which co-eluted (35). These last two activities are induced by transfer of plants to long days, whereas GA_{44} oxidase activity is not photoperiod-sensitive (36). The spinach GA_{44} oxidase converts the lactone form of this GA (36), as do cell-free systems from pea shoots (38) and germinating barley embryos (41), whereas GA 20-oxidases from immature seeds require a free alcohol at C-20 for oxidation to occur (45, 52, 128a). Detailed studies with recombinant GA 20-oxidase, produced by expression of one of the Arabidopsis cDNAs in *E. coli*, revealed that this enzyme also required a free alcohol function and that oxidation of the alcohol was much slower than that of the methyl and aldehyde substrates (47). It seems likely, therefore, that a separate enzyme(s) with a high affinity for the 20-alcohols, perhaps as the lactones, exists in shoot tissues. An enzyme with similar properties to the Arabidopsis GA 20-oxidase has been cloned from spinach leaves (145). On the basis of the activity of the protein after expression in *E. coli* and its higher expression in long days than in short days, it would appear to correspond to the GA_{53} and GA_{19} oxidases that were observed in the leaf homogenates.

An unexpected difference between the spinach GA_{44} oxidase and the recombinant Arabidopsis GA 20-oxidase is in the stereospecific removal of a hydrogen atom during oxidation of the C-20 alcohol intermediates. It was shown, using GA_{15} or GA_{44} labeled stereospecifically with deuterium, that the Arabidopsis enzyme removes the *pro-R* H atom on conversion of the free alcohol to the aldehyde (142). In contrast, GA_{44}, as the lactone, is oxidized with loss of the *pro-S* H, by cell-free extracts of spinach leaves. This observation provides further evidence for the existence of a distinct lactone oxidase; the different stereochemistry of the reactions is presumably due to the fixed orientation of C-20 in the lactone as opposed to it assuming an energetically more favored conformation as the free alcohol.

3β-Hydroxylases and Related Enzymes

3β-Hydroxylation results in the conversion of the C_{19}-GAs GA_{20} and GA_9 to GA_1 and GA_4, respectively, in the final step in the formation of physiologically active GAs. There is now increasing evidence that, in common with GA 20-oxidases, certain GA 3β-hydroxylases may be multifunctional. An enzyme purified from developing embryos of *P. vulgaris* catalyzed 2,3-desaturation and 2β-hydroxylation reactions, in addition to 3β-hydroxylation (115, 116). GA_{20} and GA_9 were about equally reactive as substrates. A 3β-hydroxylase from the same source also epoxidized GA_5 to GA_6 (65). The enzyme could use non-2β-hydroxylated C_{19} (γ-lactone) GAs or 19–20 δ- lactone C_{20}-GAs as substrates (65), the latter presumably acting as structural analogues of the former, which are the natural substrates. An enzyme that 3β-hydroxylated GA_{15} to give GA_{37} was partially purified from pumpkin endosperm (72). It did not possess desaturase, although it was not tested with C_{19}-GAs.

The pumpkin GA 3β-hydroxylase has the typical properties of a 2-oxoglutarate-dependent dioxygenase (72). In particular, it was possible to demonstrate a 1:1 stoichiometry between the formation of hydroxy GA and succinate, once uncoupled oxidation of 2-oxoglutarate was subtracted. In contrast, although the *P. vulgaris* enzyme requires 2-oxoglutarate for activity, Smith et al (116) could find no evidence that this compound functioned as a substrate. They suggested that ascorbate may serve as the cosubstrate, as it does in the related enzyme, 1-aminocyclopropane-1-carboxylic acid (ACC) oxidase (22). However, 2-oxoglutarate is essential for full 3β-hydroxylase activity, whereas it serves no function for ACC oxidase activity. The nature of the *P. vulgaris* 3β-hydroxylase is unresolved.

The desaturase activity of 3β-hydroxylases provides the first step in the production of GA_3. After applying $[^3H]GA_5$ to immature seeds of apricot, de Bottini et al (19) obtained chromatographic evidence for the presence of GA_1, GA_3, and GA_6 in the products. Unequivocal evidence for conversion of GA_{20} to GA_3 via GA_5 was obtained in shoots of *Zea mays* (30), thus establishing a new biosynthetic pathway. There was no indication that GA_5 was converted to GA_1, a reduction that is without precedent in GA biosynthesis. The equivalent pathway for non-13-hydroxylated GAs was demonstrated in cell-free systems from immature seeds of *Marah* and apple (3). Enzyme activity present in both endosperm and developing embryos of *Marah* results in the conversion of GA_9 to GA_4 and 2,3-dehydroGA_9, which is further oxidized to GA_7 (75a). The *Marah* and apple systems have a marked preference for non-13-hydroxylated substrates; although both systems converted GA_5 to GA_3, GA_{20} was metabolized to GA_1 (3β-hydroxylation), GA_{29} (2β-hydroxylation), and GA_{60} (1β-hydroxylation), but not to GA_5, by the *Marah* system and was unmetabolized by the apple preparation. The branch pathway from GA_{20} to GA_3 occurs also in barley embryos (41) and may be common, although not ubiquitous, in higher plants. The conversion of GA_5 to GA_3 has also been demonstrated in pea shoots (93) and in a cell-free system from rice anthers (58), although because GA_5 is not formed in these systems (50, 57), the function of this activity is unclear.

The conversions of GA_5 to GA_3, and of 2,3-didehydroGA_9 to GA_7, are unusual reactions that are initiated by loss of the 1β-H (3). Hydrogen abstraction is accompanied by rearrangement of the 2,3 double bond to the 1,2 position and hydroxylation on C-3β. This enzymatic activity may also result in the 1β-hydroxylation of GA_{20} and GA_5, also observed in *Marah* (3). The enzyme that converts GA_5 to GA_3 requires 2-oxoglutarate, but not added Fe^{2+}, for activity and, in contrast with most other related dioxygenases, it is not inhibited by iron chelators (116). Its activity, however, is reduced by Mn^{2+} and other metal ions, the inhibition by Mn^{2+} being reversed by Fe^{2+}. It would appear that Fe is bound very tightly at the active site.

GA_3 and GA_1 formation in maize is apparently catalyzed by one enzyme (122). As well as affecting the conversion of GA_{20} to GA_1, the *dwarf-1* mutation reduces formation of GA_5 and the conversion of GA_5 to GA_3. If *Dwarf-1* is a structural gene, a single enzyme must catalyze all three reactions. In contrast, the *le* (3β-hydroxylation) mutation of pea was purported not to affect the conversion of GA_5 to GA_3 (93). This conclusion was based on equal growth responses of *Le* and *le* plants to applied GA_5, which was assumed to have no intrinsic biological activity. However, GA_5 is as active as GA_1 and GA_3 on *dwarf-1* maize shoots, despite no metabolism to GA_3 by this genotype (122).

The *ga4* mutation of Arabidopsis results in low amounts of GA_1 and GA_4 and an accumulation of GA_{20} and GA_9 in flowering shoots, indicating reduced 3β-hydroxylase activity (129). This conclusion was supported by an 85% reduction in the amount of GA_1 produced from labeled GA_{20} in *ga4* seedlings compared with those of Landsberg *erecta* or the *ga5* (20-oxidase) mutant (55). The *GA4* locus has been cloned by T-DNA insertion (14). The gene encodes a dioxygenase that has relatively low amino acid sequence identity with the GA 20-oxidases; it has 30% identity (50% similarity) with the Arabidopsis stem-specific 20-oxidase (GA5). Expression of the *ga4* cDNA in *E. coli* has confirmed that it encodes a 3β-hydroxylase (J Williams, AL Phillips & P Hedden, unpublished information). The preferred substrate for the recombinant enzyme is GA_9, for which the K_m is tenfold lower than that for GA_{20}. Thus, as with the Arabidopsis GA 20-oxidases, the presence of a 13-hydroxyl group reduces substrate affinity for the enzyme. The enzyme also epoxidizes the 2,3-double bond in GA_5 and 2,3-didehydroGA_9, and hydroxylates certain C_{20}-GAs, albeit with low efficiency. This activity could account for the presence of 3β-hydroxy C_{20}-GAs in Arabidopsis (129). However, there are undoubtedly other

GA 3β-hydroxylases active in this species. The original EMS-induced mutation (*ga4-1*) results in semidwarfism (62), and a reduction to about 30% of the normal content of 3β-hydroxy GAs (129). The mutant enzyme has an amino acid substitution, cysteine to tyrosine (14), that might allow a low level of activity. The mutant with the T-DNA insertion (*ga4-2*) is also a semidwarf, phenotypically similar to *ga4-1*, with very little likelihood of the mutant gene encoding an active 3β-hydroxylase. Residual growth in this mutant must, therefore, result from the action of other enzymes.

2β-*Hydroxylases and Related Enzymes*

Hydroxylation on C-2β results in the formation of inactive products and is, therefore, important for turnover of the physiologically active GAs. The natural substrates for these enzymes are normally C_{19}-GAs, although 2β-hydroxy C_{20}-GAs are also found in plant tissues, particularly where the concentration of C_{20}-GAs is high (76). 2β-Hydroxylases have been partially purified from cotyledons of *P. sativum* (118) and *Phaseolus vulgaris* (39, 117). There is evidence that, for both sources, at least two enzymes with different substrate specificities are present. Two activities from cotyledons of imbibed *P. vulgaris* seeds were separable by cation-exchange chromatography and gel-filtration (39). The major activity, corresponding to an enzyme of M_r 26,000 by size-exclusion HPLC, hydroxylated GA_1 and GA_4 in preference to GA_9 and GA_{20}, while GA_9 was the preferred substrate for the second enzyme (M_r 42,000).

Formation of 2-keto derivatives (GA catabolites) by further oxidation of 2β-hydroxy GAs (Figure 2) occurs in several species, but it is particularly prevalent in developing seeds (119) and roots (108) of pea. The conversion of GA_{29} to GA_{29}-catabolite in pea seeds was inhibited by prohexadione-calcium, an inhibitor of 2-oxoglutarate–dependent dioxygenase (87), indicating that the reaction is catalyzed by an enzyme of this type (108). Although the slender (*sln*) mutation of pea blocks both the conversion of GA_{20} to GA_{29} and of GA_{29} to GA_{29}-catabolite in seeds, the inability of unlabeled GA_{20} to inhibit oxidation of radiolabeled GA_{29}, and vice versa, indicated that the steps are catalyzed by separate enzymes (108). Furthermore, in shoot tissues, the slender mutation inhibits 2β-hydroxylation of GA_{20}, but not the formation of GA_{29}-catabolite. It was, therefore, proposed that *Sln* encodes a regulatory protein (108). Formation of the GA catabolites could be initiated by oxidation either at C-1 or C-2α. Although the reaction sequence is unknown, it is of interest that, after application of labeled GA_{20} to leaves of pea, labeled GA_{81} (2α-hydroxy GA_{20}) accumulates in the roots, together with the catabolites of GA_8 and GA_{29} (108).

GA_{81} is not formed from GA_{29} and must, therefore, be formed directly as a result of 2α-hydroxylase activity.

REGULATION OF GA BIOSYNTHESIS

The role of GAs as mediators of environmental stimuli is well established. Factors, such as photoperiod and temperature, can modify GA metabolism by changing the flux through specific steps in the pathway. More recent work has shown that GA biosynthesis is modified by the action of GA itself in a type of feedback regulation. The mechanisms underlying these regulatory processes can now be investigated as a result of the current advances in the molecular biology of GA biosynthesis.

Feedback Regulation

The presence of abnormally high concentrations of C_{19}-GAs in certain GA-insensitive dwarf mutants, such as *Rht3* wheat (5), *Dwarf-8* maize (29), and *gai* Arabidopsis (130), indicate a link between GA action and biosynthesis. In maize, there is a gene-dosage effect, with a 60-fold increase in the GA_1 content of homozygous *Dwarf-8* shoots, compared with wild-type and a 33-fold increase in the heterozygote. It was suggested that GA action results in the production of a transcriptional repressor that limits the expression of GA-biosynthetic enzymes (111). Mutants with impaired response to GA would lack this repressor and have elevated rates of GA production. Such plants normally contain semidominant mutations, which may result in a gain of function (91). It has been proposed that GA 20-oxidase is a primary target for feedback regulation (5, 18, 44). In addition to an elevated C_{19}-GA content, the GA-insensitive dwarfs often contain lower amounts of C_{20}-GAs than their corresponding wild-types, suggesting increased GA 20-oxidase activity. Conversely, overgrowth mutants, such as slender (*sln*) barley and *la cry^s* pea, that grow as if saturated with GA, even in its absence, contain reduced amounts of C_{19}-GAs and elevated content of C_{20}-GA GAs (18, 77). It appears that the slender mutation activates the GA signal transduction pathway, even in the absence of GA, and may thereby cause constitutive downregulation of GA 20-oxidase activity.

Further support for feedback regulation of GA 20-oxidase activity was provided by work with GA-biosynthesis mutants. Reduced concentrations of C_{20}-GAs, as well as a highly elevated GA_{20} content, are characteristics of most 3β-hydroxylase-deficient mutants, including *dwarf-1* maize (28), *le* pea (96), and *l* sweet pea (109). Treatment of the maize (44) or pea (77) mutants with 2,2-dimethylGA$_4$, a synthetic and highly bio-active GA, restored the concentrations of GA_{20} and GA_{19} to those of wild types. Although the change in GA contents in the foregoing examples is associated with altered growth rates, the effect of GA action on GA metabolism is not a consequence of the change in growth rate. For example, there are numerous GA-insensitive dwarf mutants with normal GA contents, in which the mutation, normally recessive, is likely to affect processes that are not part of the primary response to GAs (99, 105).

With the availability of GA 20-oxidase cDNA clones, it has been possible to begin a molecular analysis of the feedback mechanism. Transcript levels for each of the three Arabidopsis GA 20-oxidase genes are much higher in the *ga1-3* (CPS-deficient) mutant than in Landsberg *erecta* and are very substantially reduced by treating the mutant with GA_3 (92). The reduction occurs within 1–3 h, long before a growth response is discernible (AL Phillips, D Valero & P Hedden, unpublished information), confirming that it is not related to growth-rate and indicating that the message is turned over rapidly. The level of mRNA for the stem-specific GA 20-oxidase is also higher in the *ga5* (GA-deficient) and *gai* (GA-insensitive) mutants than in the wild type (146). Treatment of *ga5* and, to a lesser extent, wild-type with GA_4, caused a reduction in GA 20-oxidase transcript levels, whereas treatment of *gai* resulted in a slight increase in 20-oxidase mRNA. Strong downregulation of GA 20-oxidase transcript levels by GA has also been observed in pea (77) and rice (137). Low endogenous GA concentration, as in mutants or after treatment with a biosynthesis inhibitor, consistently resulted in increased mRNA levels. Conversely, these levels were substantially reduced by application of GA. Furthermore, leaves of the slender (*la cry^s*) pea mutant contained only small amounts of 20-oxidase transcript, consistent with strong downregulation of gene expression (77).

Other enzymes in the pathway, particularly the GA 3β-hydroxylase, may also be subject to feedback regulation. Treatment of seedlings of the GA-deficient *na* mutant of pea with 2,2-dimethylGA$_4$ caused a slight reduction in the conversion of GA_{19} to GA_{20}, but a much greater reduction in GA_{20} metabolism (77). Furthermore, Chiang et al (14) found much more *GA4* (3β-hydroxylase) transcript in rosette leaves of *ga4* mutants than in Landsberg *erecta,* and that treatment with GA_3 reduced the amount of transcript within 8 h. It is possible that several enzymes are subject to regulation by GA; the identities of others may emerge when more enzymes of the pathway have been cloned.

Regulation by Light

The involvement of GAs in the photoperiod-induced bolting of long-day rosette plants is well documented (38). Transfer of *Silene armeria* plants from short days (SD) to long days (LD) causes the GA_1 content to increase several-fold, particularly in the subapical region, with a decrease in GA_{53} content consistent with increased GA_{53} metabolism (131, 133, 134). In spinach (*Spinacia oleracea*), changes in GA concentrations (135) and enzyme activity in cell-free systems (36) on transfer from SD to LD are consistent with enhanced oxidation of GA_{53} and GA_{19} in LD. The activities of GA_{53} and GA_{19} 20-oxidases, now known to be the same enzyme (145), increase in the light and decrease in the dark (36). Furthermore, there are higher amounts of GA 20-oxidase mRNA in plants grown in LD than those in SD or in total darkness (145). It has been suggested that, in LD, there is sufficient GA 20-oxidase activity to raise the GA_1 concentration above the threshold required for stem extension (135). In fact, light appears to increase the total flux through the pathway, because *ent*-kaurene synthesis is also enhanced in LD in spinach and in *Agrostemma githago* (149). Although GA_{53} 20-oxidase activity is regulated by light, oxidation of GA_{44}, in the lactone form, remains at high, constant levels irrespective of light or dark treatment (36). As discussed earlier, this latter activity is probably because of another enzyme, which is not under light regulation.

Despite many attempts to implicate GA metabolism in phytochrome-mediated changes in growth rate, supporting evidence is sparse. Enhancement of GA_{20} 3β-hydroxylation by far-red light has been observed in lettuce (138, 139) and cowpea epicotyls (25, 31, 78, 79). In the latter case, higher GA_1 concentrations in plants grown in far-red light were due also to reduced 2β-hydroxylation and were accompanied by heightened tissue responsiveness to GA (78, 80). However, in peas, enhanced shoot elongation by treatment with far-red-rich light was not associated with increased GA_1 content (100). There is also no evidence to suggest that dark-grown peas (34, 120, 143) or sweet peas (109) contain more GA_1 than light-grown plants. In fact, work with phytochrome- and GA-deficient mutants of pea indicates that growth inhibition by red light, which is mediated by phytochrome B, is due to altered responsiveness to GA, rather than to changes in the concentration of GA_1 (143). Several phytochrome-deficient mutants, such as the *ein* mutant of *Brassica rapa* (21) and the ma_3^R mutant of *Sorghum* (15, 16), have an overgrowth phenotype and were originally thought to contain abnormally high GA levels. Although the GA_1 content of these plants may be elevated (7, 104), this is apparently not the cause of their altered phenotype. Phytochrome B-deficient mutants of cucumber (74) and pea (143) contain comparable amounts of active GAs to the wild types, but show an enhanced response to GAs. In *Sorghum,* altered GA content is due to a shift in the phase of a diurnal fluctuation in GA concentrations (26). It was proposed that phytochrome deficiency disrupted diurnal regulation of the conversion of GA_{19} to GA_{20}, resulting in a 12-h shift in the peaks of GA_{20} and GA_1 concentrations, whereas the pattern of fluctuation in the levels of GA_{12} and GA_{53} was unaffected.

Gibberellin metabolism is sensitive to light quantity. When pea seedlings were grown in low irradiance (40 μmol • m^{-2} • s^{-1}), GA_{20} concentration increased sevenfold compared with plants grown in high irradiance (386 μmol • m^{-2} • s^{-1}) (34), whereas in plants grown in the dark, the GA_{20} content was reduced to 25% of that in high irradiance. Moreover, the response of the seedling to exogenous GA_1 was heightened in the dark. These results indicate that the rate of GA 20-oxidation is sensitive to light fluence; with the cloning of GA 20-oxidases now reported for many species, the mechanisms underlying this process as well as the diurnal fluctuation in GA biosynthesis can be probed.

Regulation by Temperature

Induction of seed germination (stratification) or of flowering (vernalization) by exposure to low temperatures are processes in which GAs have been implicated. There are, however, few examples in which GAs have been shown unequivocally to mediate the temperature stimulus. The most extensively studied system is *Thlaspi arvense,* in which stem extension and flowering are induced by exposure to low temperatures followed by a return to higher temperatures (82). The same effect can be obtained without cold induction by application of GAs, the most active of those tested being GA_9 (83). In noninduced plants, *ent*-kaurenoic acid accumulates to high concentrations in the shoot tip, the site of perception of the cold stimulus, whereas after vernalization and return to high temperatures, the level of this intermediate falls within days to relatively low values (43). Metabolism of labeled *ent*-kaurenoic acid to GA_9 could be demonstrated in thermo-induced shoot tips, but not in noninduced material (42). Furthermore, microsomes from induced shoots metabolized *ent*-kaurenoic acid and *ent*-kaurene, but microsomes from noninduced shoots were much less active for both activities (43). Leaves, or microsomes extracted from leaves, from thermo-induced and noninduced plants metabolized *ent*-kaurenoic acid to the same extent. These results are consistent with regulation of *ent*-kaurenoic acid 7β-hydroxylase and, to a lesser degree, *ent*-kaurene oxidase by cold treatment in shoot tips of *Thlaspi.*

A change in GA content following vernalization was found in shoot tips of *Brassica napus,* in which GAs, including GA_1 and GA_3, accumulated during the cold period (148). The higher rates of GA production in vernalized, relative to nonvernalized, plants appeared to persist for 1–2 weeks after the return to high temperatures. In contrast to these findings, there was no evidence to suggest that GAs were the signal for cold-induced flowering in *Raphanus sativus* (88) or *Tulipa gesneriana* (97).

Although the mechanism for thermo-induction of GA biosynthesis is not yet known, it has been suggested that cold treatment may allow increased rates of gene expression, possibly via demethylation of the promoters (11). Some circumstantial support for this theory was obtained by reducing DNA methylation in *Thlaspi* and late-flowering ecotypes of Arabidopsis by treatment with 5-azacytidine. Flowering times in noninduced plants were reduced in both cases. Confirmation of this theory must await the isolation and characterization of the 7β-hydroxylase gene.

SITES OF GA BIOSYNTHESIS

Developmental regulation of GA biosynthesis is determined mainly by changes in plant ontogeny during development and the tissue distribution of individual enzymes of the pathway. Work with legumes indicates that GA-biosynthesis occurs mainly in actively growing tissues, with leaves and internodes important sites (113, 114). Gibberellin 20-oxidase transcript levels are much higher in pea leaves than in internodes (32). The same 20-oxidase gene is expressed in shoots, young seeds, and expanding pods (32), but a different gene is expressed in developing cotyledons (73). Orthologues of the pea GA 20-oxidases, with similar patterns of gene expression, were cloned from French bean (32), and a third gene, which is expressed in developing cotyledons, leaves, and roots, was also detected in this species.

The tissue-specific expression of the GA 20-oxidase genes, also noted in Arabidopsis (92), has not been found for CPS (2, 124), KS (147), or GA 3β-hydroxylase (14). Whereas different GA 20-oxidase genes are expressed during pea seed development, the same CPS gene is expressed in a biphasic manner, corresponding with the two stages of GA production (2). The first phase, which results in the production of GA_1 and GA_3 (33, 103), is associated with seed development (126–128) and pod growth (33, 103). Both endosperm and testa are potential sites of synthesis (103). The second phase occurs in the developing embryo and has no known physiological function. As seeds approach maturity there is often an increase in 2-oxidation activities in embryos and testa (4, 119). The physiological significance of this deactivation mechanism is vividly demonstrated by the slender (*sln*) mutant of pea, which accumulates high concentrations of GA_{20} in the mature seed due to a lack of 2-oxidase activities (102, 107, 108). On germination, the GA_{20} is 3β-hydroxylated to GA_1, resulting in overgrowth of the first 10–12 internodes (102). An intriguing phenotype was obtained by crossing slender with *na*, a GA-deficient dwarf, blocked at an intermediate step in the biosynthetic pathway, but not expressed in the developing seed (49). The double mutant is phenotypically slender to the six-leaf stage, but thereafter is severely dwarfed (108). Crossing slender with *le* (3β-hydroxylase-deficient) produced a dwarf from emergence, as did crossing with lh^i, which blocks at *ent*-kaurene oxidase in seeds (127) and prevents the accumulation of GA_{20} (108).

Gibberellin production in the pericarp of developing fruit may be important for fruit growth, particularly in the absence of seeds (132). Work with pea indicates that the presence of seeds stimulates GA biosynthesis in the pericarp (90, 141). Both seeds and 4-chloroindole-3-acetic acid, the proposed seed-derived signal, stimulated the conversion of GA_{19} to GA_{20} in pea pods. However, the finding that GA 20-oxidase transcript levels in pericarp from seeded pea fruit are much lower than in seedless fruit (32) does not support a regulatory role for GA 20-oxidation in fruit development in this species.

CONCLUDING REMARKS

Research on GA biosynthesis has reached an exciting stage. cDNA and genomic clones have been obtained for five types of enzyme, including members of each of the three enzymatic classes: cyclases, monooxygenases, and dioxygenases. The cloning of each enzyme in the pathway should be achieved within the next three to five years. The availability of clones is enabling significant advances in several directions. Enzymes can be prepared by heterologous expression in sufficient quantities for detailed studies on their structure and function, and for the production of antibodies. Promoter-reporter gene fusions, in situ hybridization, and immunolocalization can be used to determine cellular and subcellular sites of synthesis. It is possible to examine the regulation of individual enzymes at the transcript and protein levels. Such studies are already yielding important information on the regulation of GA 20-oxidase transcript abundance by GA action and by light, and on the developmental control of CPS and KS gene expression.

From a practical standpoint, it will be possible to manipulate GA production in transgenic plants by altering the expression of individual genes. This technology offers an alternative to the use of chemical growth regulators for the control of plant development. It also provides a means to alter the abundance of specific enzymes and thereby determine their contributions to the flux through the biosynthetic pathway. For example, overexpression of GA 20-oxidase cDNAs in Arabidopsis results in accelerated bolting, confirming that the activity of this enzyme is rate-limiting for this developmental process (53).

Experiments of this type may provide new insights into the role of GAs in plant development.

ACKNOWLEDGMENTS

We thank colleagues for their critical reading of the manuscript. IACR receives grant-aided support from the Biotechnology and Biological Sciences Research Council of the United Kingdom.

> **Visit the *Annual Reviews* home page at http://www.annurev.org**

Literature Cited

1. Aach H, Böse G, Graebe JE. 1995. *ent*-Kaurene biosynthesis in a cell-free system from wheat (*Triticum aestivum* L.) seedlings and the localization of *ent*-kaurene synthetase in plastids of three species. *Planta* 197:333–42
2. Ait-Ali T, Swain SM, Reid JB, Sun TP, Kamiya Y. 1997. The *Ls* locus of pea encodes the gibberellin biosynthesis enzyme *ent*-kaurene synthase A. *Plant J.* 77:443–54
3. Albone KS, Gaskin P, MacMillan J, Phinney BO, Willis CL. 1990. Biosynthetic origin of gibberellin A_3 and gibberellin A_7 in cell-free preparations from seeds of *Marah macrocarpus* and *Malus domestica*. *Plant Physiol.* 94:132–42
4. Albone K, Gaskin P, MacMillan J, Smith VA, Weir J. 1989. Enzymes from seeds of *Phaseolus vulgaris* L.: hydroxylation of gibberellins A_{20} and A_1 and 2,3-dehydrogenation of gibberellin A_{20}. *Planta* 117: 108–15
5. Appleford NEJ, Lenton JR. 1991. Gibberellins and leaf expansion in near-isogenic wheat lines containing *Rht1* and *Rht3* dwarfing alleles. *Planta* 183:229–36
6. Bartley GE, Scolnik PA, Giuliano G. 1994. Molecular biology of carotenoid biosynthesis in plants. *Annu. Rev. Plant Physiol. Plant Mol. Biol.* 45:287–301
7. Beal FD, Morgan PW, Mander LN, Miller FR, Babb KH. 1991. Genetic regulation of development in *Sorghum bicolor* V. The ma_3^R allele results in gibberellin enrichment. *Plant Physiol.* 95:116–25
8. Bensen RJ, Crane VC, Wang X, Duvick J, Briggs SP. 1995. Maize GA-mutants: isolation, characterization, and gene cloning and expression. In *Abstract of International Conference on Plant Growth Substances.*

(Abstr. 096). Minneapolis: Int. Plant Growth Substances Assoc.
9. Bensen RJ, Johal GS, Crane VC, Tossberg JT, Schnable PS, et al. 1995. Cloning and characterization of the Maize *An1* gene. *Plant Cell* 7:75–84
10. Bensen RJ, Zeevaart JAD. 1990. Comparison of *ent*-kaurene synthetase A and B activities in cell-free extracts from young tomato fruits of wild-type and *gib-1, gib-2,* and *gib-3* tomato plants. *J. Plant Growth Regul.* 9:237–42
11. Burn JE, Bagnell DJ, Metzger JD, Dennis ES, Peacock WJ. 1993. DNA methylation and the initiation of flowering. *Proc. Natl. Acad. Sci. USA* 90:287–91
12. Chamovitz D, Misawa N, Sandmann G, Hirschberg J. 1992. Molecular cloning and expression in *Escherichia coli* of a cyanobacterial gene coding for phytoene synthase, a carotenoid biosynthesis enzyme. *FEBS Lett.* 296:305–10
13. Chappell J. 1995. Biochemistry and molecular biology of the isoprenoid biosynthetic pathway in plants. *Annu. Rev. Plant Physiol. Plant Mol. Biol.* 46:521–47
14. Chiang HH, Hwang I, Goodman HM. 1995. Isolation of the Arabidopsis *GA4* locus. *Plant Cell* 7:195–201
15. Childs KL, Cordonnier-Pratt MM, Pratt LH, Morgan PW. 1992. Genetic regulation of development in *Sorghum bicolor* VII. The ma_3^R flowering mutant lacks a phytochrome that predominates in green tissue. *Plant Physiol.* 99:765–70
16. Childs KL, Pratt LH, Morgan PW. 1991. Genetic regulation of development in *Sorghum bicolor* VI: the ma_3^R allele results in abnormal phytochrome physiology. *Plant Physiol.* 97:714–19

17. Colby SM, Alonso WR, Katahira EV, McGarvey DJ, Croteau R. 1993. 4S-Limonene synthase from the oil glands of spearmint (*Mentha spicata*): cDNA isolation, characterization, and bacterial expression of the catalytically active monoterpene cyclase. *J. Biol. Chem.* 268:23016–24

18. Croker SJ, Hedden P, Lenton JR, Stoddart JL. 1990. Comparison of gibberellins in normal and slender barley seedlings. *Plant Physiol.* 94:194–200

19. de Bottini GA, Bottini R, Koshioka M, Pharis RP, Coombe BG. 1987. Metabolism of [³H]gibberellin A₅ by immature seeds of apricot (*Prunus armeniaca* L.). *Plant Physiol.* 83:137–42

20. De Carolis E, De Luca V. 1994. 2-Oxoglutarate-dependent dioxygenases and related enzymes: biochemical characterization. *Phytochemistry* 36:1093–107

21. Devlin PF, Rood SB, Somers DE, Quail P, Whitelam GC. 1992. Photophysiology of the elongated internode (ein) mutant of *Brassica rapa. Plant Physiol.* 100:1442–47

22. Dong JG, Fernández-Maculet JC, Yang SF. 1992. Purification and characterization of 1-aminocyclopropane-1-carboxylate oxidase from apple fruit. *Proc. Natl. Acad. Sci. USA* 89:9789–93

23. Duncan JD, West CA. 1981. Properties of kaurene synthetase from *Marah macrocarpus* endosperm: evidence for the participation of separate but interacting enzymes. *Plant Physiol.* 68:1128–34

24. Facchini PJ, Chappell J. 1992. Gene family for an elicitor-induced sesquiterpene cyclase in tobacco. *Proc. Natl. Acad. Sci. USA* 89:11088–92

25. Fang N, Bonner BA, Rappaport L. 1991. Phytochrome mediation of gibberellin metabolism and epicotyl elongation in cow pea, *Vigna sinensis* L. See Ref. 128b, pp. 280–88

26. Foster KR, Morgan PW. 1995. Genetic regulation of development in *Sorghum bicolor* IX: the ma3^R allele disrupts diurnal control of gibberellin biosynthesis. *Plant Physiol.* 108:337–43

27. Fray RG, Wallace A, Fraser PD, Valero D, Hedden P, et al. 1995. Constitutive expression of a fruit phytoene synthase gene in transgenic tomatoes causes dwarfism by redirecting metabolites from the gibberellin pathway. *Plant J.* 8:693–701

28. Fujioka S, Yamane H, Spray CR, Gaskin P, MacMillan J, et al. 1988. Qualitative and quantitative analyses of gibberellins in vegetative shoots of normal, *dwarf-1, dwarf-2, dwarf-3* and *dwarf-5* seedlings of *Zea mays* L. *Plant Physiol.* 88:1367–72

29. Fujioka S, Yamane H, Spray CR, Katsumi M, Phinney BO, et al. 1988. The dominant nongibberellin-responding dwarf mutant (*D8*) of maize accumulates native gibberellins. *Proc. Natl. Acad. Sci. USA* 85:9031–35

30. Fujioka S, Yamane H, Spray CR, Phinney BO, Gaskin P, et al. 1990. Gibberellin A₃ is biosynthesized from gibberellin A₂₀ via gibberellin A₅ in shoots of *Zea mays* L. *Plant Physiol.* 85:9031–35

31. García-Martínez JL, Keith B, Bonner BA, Stafford A, Rappaport L. 1987. Phytochrome regulation of the response to exogenous gibberellins by epicotyls of *Vigna sinensis* L. *Plant Physiol.* 85:212–16

32. García-Martínez JL, López-Díaz I, Sánchez-Beltrán MJ, Phillips AL, Ward DA, et al. 1997. Expression of gibberellin 20-oxidase genes in pea and bean in relation to fruit development. *Plant Mol. Biol.* In press

33. García-Martínez JL, Santes C, Croker SJ, Hedden P. 1991. Identification, quantitation and distribution of gibberellins in fruits of *Pisum sativum* cv. Alaska during pod development. *Planta* 184:53–60

34. Gawronska H, Yang YY, Furukawa K, Kendrick RE, Takahashi N, et al. 1995. Effects of low irradiance stress on gibberellin levels in pea seedlings. *Plant Cell Physiol.* 36:1361–67

35. Gilmour SJ, Bleecker AB, Zeevaart JAD. 1987. Partial purification of gibberellin oxidases from spinach leaves. *Plant Physiol.* 85:87–90

36. Gilmour SJ, Zeevaart JAD, Schwenen L, Graebe JE. 1986. Gibberellin metabolism in cell-free extracts from spinach leaves in relation to photoperiod. *Plant Physiol.* 82:190–95

37. Graebe JE. 1982. Gibberellin biosynthesis in cell-free systems from higher plants. In *Plant Growth Substances 1982*, ed. PF Wareing, pp. 71–80. London: Academic

38. Graebe JE. 1987. Gibberellin biosynthesis and control. *Annu. Rev. Plant Physiol.* 38:419–65

39. Griggs DL, Hedden P, Lazarus CM. 1991. Partial purification of two gibberellin 2β-hydroxylases from cotyledons of *Phaseolus vulgaris. Phytochemistry* 30:2507–12

40. Grosselindemann E, Graebe JE, Stöckl D, Hedden P. 1991. *ent*-Kaurene biosynthesis in germinating barley (*Hordeum vulgare* L., cv Himalaya) caryopses and its relation to α-amylase production. *Plant Physiol.* 96:1099–104

41. Grosselindemann E, Lewis MJ, Hedden P,

Graebe JE. 1992. Gibberellin biosynthesis from gibberellin A_{12}-aldehyde in a cell-free system from germinating barley (*Hordeum vulgare* L., cv. Himalaya) embryos. *Planta* 188:252–57

42. Hazebroek JP, Metzger JD. 1990. Thermoinductive regulation of gibberellin metabolism in *Thlaspi arvensi* L. I. Metabolism of [²H]kaurenoic acid and [¹⁴C]gibberellin A_{12}-aldehyde. *Plant Physiol.* 94: 154–65

43. Hazebroek JP, Metzger JD, Mansager ER. 1993. Thermoinductive regulation of gibberellin metabolism in *Thlaspi arvensi* L. II. Cold induction of enzymes in gibberellin biosynthesis. *Plant Physiol.* 102:547–52

44. Hedden P, Croker SJ. 1992. Regulation of gibberellin biosynthesis in maize seedlings. See Ref. 52a, pp. 534–44

45. Hedden P, Graebe JE. 1982. Cofactor requirements for the soluble oxidases in the metabolism of the C_{20}-gibberellins. *J. Plant Growth Regul.* 1:105–16

46. Hedden P, Graebe JE, Beale MH, Gaskin P, MacMillan J. 1984. The biosynthesis of 12α-hydroxylated gibberellins in a cell-free system from *Cucurbita maxima* endosperm. *Phytochemistry* 23:569–74

47. Hedden P, Phillips AL, Jackson GS, Coles JP. 1995. Molecular cloning of gibberellin biosynthetic enzymes, a route to the genetic manipulation of plant growth. In *Proceedings of the Ninth Forum for Applied Biotechnology,* pp. 1559–66. Ghent: Med. Fac. Landbouww. Univ. Gent

48. Hedden P, Phinney BO. 1979. Comparison of *ent*-kaurene and *ent*-isokaurene synthesis in cell-free systems from etiolated shoots of normal and *dwarf-5* maize seedlings. *Phytochemistry* 18:1475–79

49. Ingram TJ, Reid JB. 1987. Internode length in *Pisum* gene *na* may block gibberellin synthesis between *ent*-7α- hydroxykaurenoic acid and gibberellin A_{12}-aldehyde. *Plant Physiol.* 83:1048–53

50. Ingram TJ, Reid JB, Murfet IC, Gaskin P, Willis CL, et al. 1984. Internode length in *Pisum*. The *Le* gene controls the 3β-hydroxylation of gibberellin A_{20} to GA_1. *Planta* 160:455–63

51. Jennings JC, Coolbaugh RC, Nakata DA, West CA. 1993. Characterization and solubilization of kaurenoic acid hydroxylase from *Gibberella fujikuroi*. *Plant Physiol.* 101:925–30

52. Kamiya Y, Graebe JE. 1983. The biosynthesis of all major pea gibberellins in a cell-free system from *Pisum sativum*. *Phytochemistry* 22:681–89

52a. Karssen CM, Van Loon LC, Vreugenhil D,

eds. 1992. *Progress in Plant Growth Regulation*. Dordrecht: Kluwer

53. Katsumi M. 1964. Gibberellin-like activities of certain auxins and diterpenes. PhD thesis. Univ. Calif., Los Angeles

54. Keegstra K, Olsen LD, Theg SM. 1989. Chloroplastic precursors and their transport across the envelope membranes. *Annu. Rev. Plant Physiol. Plant Mol. Biol.* 40:471–501

55. Kobayashi M, Gaskin P, Spray CR, Phinney BO, MacMillan J. 1994. The metabolism of gibberellin A_{20} to gibberellin A_1 by tall and dwarf mutants of *Oryza sativa* and *Arabidopsis thaliana*. *Plant Physiol.* 106: 1367–72

56. Kobayashi M, Gaskin P, Spray CR, Suzuki Y, Phinney BO, et al. 1993. Metabolism and biological activity of gibberellin A_4 in vegetative shoots of *Zea mays, Oryza sativa,* and *Arabidopsis thaliana*. *Plant Physiol.* 102:379–86

57. Kobayashi M, Kamiya Y, Sakurai A, Saka H, Takahashi N. 1990. Metabolism of gibberellins in cell-free extracts of anthers from normal and dwarf rice. *Plant Cell Physiol.* 31:289–93

58. Kobayashi M, Kwak SS, Kamiya Y, Yamane H, Takahashi N, et al. 1991. Conversion of GA_5 to GA_6 and GA_3 in cell-free systems from *Phaseolus vulgaris* and *Oryza sativa*. *Agric. Biol. Chem.* 55: 249–51

59. Kobayashi M, Sakurai A, Saka H, Takahashi N. 1989. Quantitative analysis of endogenous gibberellins in normal and dwarf cultivars of rice. *Plant Cell Physiol.* 30: 963–69

60. Kobayashi M, Spray CR, Phinney BO, Gaskin P, MacMillan J. 1996. Gibberellin metabolism in maize—the stepwise conversion of gibberellin A_{12}-aldehyde to gibberellin A_{20}. *Plant Physiol.* 110:413–18

61. Kobayashi M, Yamaguchi I, Murofushi N, Ota Y, Takahashi N. 1988. Fluctuation and localization of endogenous gibberellins in rice. *Agric. Biol. Chem.* 52:1189–94

62. Koornneef M, van der Veen JH. 1980. Induction and analysis of gibberellin sensitive mutants in *Arabidopsis thaliana* (L.) Heynh. *Theor. Appl. Genet.* 58:257–63

63. Koornneef M, van Eden J, Hanhart CJ, de Jongh AMM. 1983. Genetic fine-structure of the *GA-1* locus in the higher plant *Arabidopsis thaliana* (L.) Heynh. *Genet. Res.* 41:57–68

64. Kuntz M, Römer S, Suire C, Hugueney P, Weil JH, et al. 1992. Identification of a cDNA for the plastid-located geranylgeranyl pyrophosphate synthase from *Cap-*

sicum annuum: correlative increase in enzyme activity and transcript level during fruit ripening. *Plant J.* 2:25–34

65. Kwak SS, Kamiya Y, Sakurai A, Takahashi N, Graebe JE. 1988. Partial purification nad characterization of gibberellin 3β-hydroxylase from immature seeds of *Phaseolus vulgaris* L. *Plant Cell Physiol.* 29:935–43

66. Lange T. 1994. Purification and partial amino-acid sequence of gibberellin 20-oxidase from *Cucurbita maxima* L. endosperm. *Planta* 195:108–15

67. Lange T, Graebe JE. 1989. The partial purification and characterization of a gibberellin C-20 hydroxylase from immature *Pisum sativum* L. seeds. *Planta* 179:211–21

68. Lange T, Graebe JE. 1993. Enzymes of gibberellin synthesis. In *Methods in Plant Biochemistry,* ed. PJ Lea, 9:403–30. London: Academic

69. Lange T, Hedden P, Graebe JE. 1993. Biosynthesis of 12α- and 13-hydroxylated gibberellins in a cell-free system from *Cucurbita maxima* endosperm and the identification of new endogenous gibberellins. *Planta* 189:340–49

70. Lange T, Hedden P, Graebe JE. 1993. Gibberellin biosynthesis in cell-free extracts from developing *Cucurbita maxima* embryos and the identification of new endogenous gibberellins. *Planta* 189:350–59

71. Lange T, Hedden P, Graebe JE. 1994. Expression cloning of a gibberellin 20-oxidase, a multifunctional enzyme involved in gibberellin biosynthesis. *Proc. Natl. Acad. Sci. USA* 91:8522–66

72. Lange T, Schweimer A, Ward DA, Hedden P, Graebe JE. 1994. Separation and characterization of three 2-oxoglutarate-dependent dioxygenases from *Cucurbita maxima* L. endosperm involved in gibberellin biosynthesis. *Planta* 195:98–107

73. Lester DR, Ross JJ, Ait-Ali T, Martin DN, Reid JB. 1996. A gibberellin 20-oxidase cDNA (Accession No. U58830) from pea (*Pisum sativum* L.) seed (PGR 96-050). *Plant Physiol.* 111:1353

74. López-Juez E, Kobayashi M, Sakurai A, Kamiya Y, Kendrick RE. 1995. Phytochrome, gibberellin, and hypocotyl growth. A study using cucumber (*Cucumis sativus* L.) long hypocotyl mutant. *Plant Physiol.* 107:131–40

75. MacMillan J. 1997. Biosynthesis of the gibberellin plant hormones. *Nat. Prod. Rep.* In press

75a. MacMillan J, Ward DA, Phillips AL, Sánchez-Beltrán MJ, Gaskin P, et al. 1997.

Gibberellin biosynthesis from gibberellin A₁₂-aldehyde in endosperm and embryos of *Marah macrocarpus.* *Plant Physiol.* In press

76. Mander LN, Owen DJ, Croker SJ, Gaskin P, Hedden P, et al. 1996. Identification of three C₂₀ gibberellins: GA₉₇ (2β-hydroxy-GA₅₃), GA₉₈ (2β-hydroxy-GA₄₄ and GA₉₉ (2β-hydroxy-GA₁₉). *Phytochemistry* 43: 23–28

77. Martin DN, Proebsting WM, Parks TD, Dougherty WG, Lange T, et al. 1996. Feedback regulation of gibberellin biosynthesis and gene expression in *Pisum sativum* L. *Planta.* 200:159–66

78. Martínez-García JF, García-Martínez JL. 1992. Phytochrome modulation of gibberellin metabolism in cow pea epicotyls. See Ref. 52a, pp. 585–90

79. Martínez-García JF, García-Martínez JL. 1992. Interaction of gibberellins and phytochrome in the control of cowpea epicotyl elongation. *Physiol. Plant.* 86:236–44

80. Martínez-García JF, García-Martínez JL. 1995. An acylcyclohexanedione retardant inhibits gibberellin A₁ metabolism, thereby nullifying phytochrome-modulation of cowpea epicotyl explants. *Physiol. Plant.* 94:708–14

81. Mau CJD, West CA. 1994. Cloning of casbene synthase cDNA: evidence for conserved structural features among terpenoid cyclases in plants. *Proc. Natl. Acad. Sci. USA* 91:8497–501

82. Metzger JD. 1985. Role of gibberellins in the environmental control of stem growth in *Thlaspi arvensi* L. *Plant Physiol.* 78: 8–13

83. Metzger JD. 1990. Comparison of biological activities of gibberellins and gibberellin-precursors native to *Thlaspi arvensi* L. *Plant Physiol.* 94:151–56

84. Moritz T. 1995. Biological activity, identification and quantification of gibberellins in seedlings of Norway spruce (*Picea abies*) grown under different photoperiods. *Physiol. Plant.* 95:67–72

85. Murakami Y. 1972. Dwarfing genes in rice and their relation to gibberellin biosynthesis. In *Plant Growth Substances, 1970,* ed. DJ Carr, pp. 166–74. Berlin: Springer-Verlag

86. Murphy PJ, West CA. 1969. The role of mixed function oxidases in kaurene metabolism in *Echinocystis macrocarpa* Greene endosperm. *Arch. Biochem. Biophys.* 133: 395–407

87. Nakayama I, Kamiya Y, Kobayashi M, Abe H, Sakurai A. 1990. Effects of a plant growth regulator, prohexadione, on the bio-

synthesis of gibberellins in cell-free systems derived from immature seeds. *Plant Cell Physiol.* 31:1183–90

88. Nakayama M, Yamane H, Nojiri H, Yokota T, Yamaguchi Y, et al. 1995. Qualitative and quantitative analysis of endogenous gibberellins in *Raphanus sativus* L. during cold treatment and the subsequent growth. *Biosci. Biotech. Biochem.* 59:1121–25

89. Ogawa S, Toyomasu T, Yamane H, Murofushi N, Ikeda R, et al. 1996. A step in the biosynthesis of gibberellins that is controlled by the mutation in semi-dwarf rice cultivar Tan-ginbozu. *Plant Cell Physiol.* 37: 363–68

90. Ozga JA, Brenner ML, Reinecke DM. 1992. Seed effects on gibberellin metabolism in pea pericarp. *Plant Physiol.* 100: 88–94

91. Peng J, Harberd NP. 1993. Derivative alleles of the Arabidopsis gibberellin-insensitive (*gai*) mutation confer a wild-type phenotype. *Plant Cell* 5:351–60

92. Phillips AL, Ward DA, Uknes S, Appleford NEJ, Lange T, et al. 1995. Isolation and expression of three gibberellin 20-oxidase cDNA clones from Arabidopsis. *Plant Physiol.* 108:1049–57

93. Poole AT, Ross JJ, Lawrence NL, Reid JB. 1995. Identification of gibberellin A4 in *Pisum sativum* L. and the effects of applied gibberellins A9, A4, A5 and A3 on the *le* mutant. *Plant Growth Regul.* 16:257–62

94. Prescott AG. 1993. A dilemma of dioxygenases (or where biochemistry and molecular biology fail to meet). *J. Exp. Bot.* 44:849–61

95. Prescott AG, John P. 1996. Dioxygenases: molecular structure and role in plant metabolism. *Annu. Rev. Plant Physiol. Plant Mol. Biol.* 47:245–71

96. Proebsting WM, Hedden P, Lewis MJ, Croker SJ, Proebsting N. 1992. Gibberellin concentration and transport in genetic lines of pea: effects of grafting. *Plant Physiol.* 100:1354–60

97. Rebers M, Vermeer E, Knegt E, Shelton CJ, van der Plas LHW. 1995. Gibberellin levels and cold-induced floral stalk elongation in tulip. *Physiol. Plant.* 94:687–91

98. Reid JB. 1990. Phytohormone mutants in plant research. *J. Plant Growth Regul.* 9: 97–111

99. Reid JB. 1993. Plant hormone mutants. *J. Plant Growth Regul.* 12:207–26

100. Reid JB, Hasan O, Ross JJ. 1990. Internode length in *Pisum*. Gibberellins and the response to far-red-rich light. *J. Plant Physiol.* 137:46–52

101. Reid JB, Ross JJ. 1993. A mutant-based approach, using *Pisum sativum*, to understanding plant growth. *Int. J. Plant Sci.* 145:22–34

102. Reid JB, Ross JJ, Swain SM. 1992. Internode length in *Pisum*. A new, slender mutant with elevated levels of C19 gibberellins. *Planta* 188:462–67

103. Rodrigo MJ, García-Martínez JL, Santes CM, Gaskin P, Hedden P. 1997. The role of gibberellins A1 and A3 in fruit growth of *Pisum sativum* L. and the identification of gibberellins A4 and A7 in young seeds. *Planta.* In press

104. Rood SB, Williams PH, Pearce D, Murofushi N, Mander LN, et al. 1990. A mutant gene that increases gibberellin production in *Brassica. Plant Physiol.* 93:1168–74

105. Ross JJ. 1994. Recent advances in the study of gibberellin mutants. *Plant Growth Regul.* 15:193–206

106. Ross JJ, Davies NW, Reid JB, Murfet IC. 1990. Internode length in *Lathyrus odoratus*. Effects of mutants *l* and *lb* on gibberellin metabolism and levels. *Physiol. Plant.* 79:453–58

107. Ross JJ, Reid JB, Swain SM. 1993. Control of stem elongation by gibberellin A1: evidence from genetic studies including the slender mutant *sln. Aust. J. Plant Physiol.* 20:585–99

108. Ross JJ, Reid JB, Swain SM, Hasan O, Poole AT, et al. 1995. Genetic regulation of gibberellin deactivation in *Pisum. Plant J.* 7:513–23

109. Ross JJ, Willis CL, Gaskin P, Reid JB. 1992. Shoot elongation in *Lathyrus odoratus* L.: gibberellin levels in light and dark-grown tall and dwarf seedlings. *Planta* 187: 10–13

110. Saito T, Abe H, Yamane H, Sakurai A, Murofushi N, et al. 1995. Purification and properties of *ent*-kaurene synthase B from immature seeds of pumpkin. *Plant Physiol.* 109:1239–45

111. Scott IM. 1990. Plant hormone response mutants. *Physiol. Plant.* 78:147–52

112. Schwender J, Seemann M, Lichtenthaler HK, Rohmers M. 1996. Biosynthesis of isoprenoids (carotenoids, sterols, prenyl side-chains of chlorophylls and plastoquinone) via a novel pyruvate/glyceraldehyde 3-phosphate nonmevalonate pathway in the green alga *Scenedesmus obliquus. Biochem. J.* 316:73–80

113. Sherriff LJ, McKay MJ, Ross JJ, Reid JB, Willis CL. 1994. Decapitation reduces the metabolism of gibberellin A20 to A1 in *Pisum sativum* L., decreasing the *Le/le* difference. *Plant Physiol.* 104:277–80

114. Smith VA. 1992. Gibberellin A1 biosynthe-

sis in *Pisum sativum* L. II. Biological and biochemical consequences of the *le* mutation. *Plant Physiol.* 99:372–77

115. Smith VA, Albone KS, MacMillan J. 1990. Enzymatic 3β-hydroxylation of gibberellins A_{20} and A_5. See Ref. 128b, pp. 62–71

116. Smith VA, Gaskin P, MacMillan J. 1990. Partial purification and characterization of the gibberellin A_{20} 3β-hydroxylase from seeds of *Phaseolus vulgaris*. *Plant Physiol.* 94:1390–401

117. Smith VA, MacMillan J. 1984. Purification and partial characterization of a gibberellin 2β-hydroxylase from *Phaseolus vulgaris*. *J. Plant Growth Regul.* 2:251–64

118. Smith VA, MacMillan J. 1986. The partial purification and characterization of gibberellin 2β-hydroxylases from seeds of *Pisum sativum*. *Planta* 167:9–18

119. Sponsel VM. 1983. The localization, metabolism and biological activity of gibberellins in maturing and germinating seeds of *Pisum sativum* cv. Progress No. 9. *Planta* 159:454–68

120. Sponsel VM. 1986. Gibberellins in dark- and red-light-grown shoots of dwarf and tall cultivars of *Pisum sativum:* the quantification, metabolism and biological activity of gibberellins in Progress No. 9 and Alaska. *Planta* 168:119–29

121. Sponsel VM. 1995. Gibberellin biosynthesis and metabolism. In *Plant Hormones: Physiology, Biochemistry and Molecular Biology*, ed. PJ Davies, pp. 66–97. Dordrecht: Martinus Nijhoff

122. Spray CR, Kobayashi M, Suzuki Y, Phinney BO, Gaskin P, et al. 1996. The *dwarf-1 (d1)* mutant of *Zea mays* blocks three steps in the gibberellin biosynthetic pathway. *Proc. Natl. Acad. Sci. USA* 93: 10515–18

123. Sun TP, Goodman HM, Ausubel FM. 1992. Cloning the Arabidopsis *GA1* locus by genomic subtraction. *Plant Cell* 4:119–28

124. Sun TP, Kamiya Y. 1994. The Arabidopsis *GA1* locus encodes the cyclase *ent*-kaurene synthetase A of gibberellin biosynthesis. *Plant Cell* 6:1509–18

125. Swain SM, Reid JB. 1992. Internode length in *Pisum:* a new allele at the *Lh* locus. *Physiol. Plant.* 86:124–30

126. Swain SM, Reid JB, Ross JJ. 1993. Seed development in Pisum. The lh^i allele reduces gibberellin levels in developing seeds, and increases seed abortion. *Planta* 191:482–88

127. Swain SM, Ross JJ, Kamiya Y, Reid JB. 1995. Gibberellin and seed development. *Plant Cell Physiol.* 36:S110

128. Swain SM, Ross JJ, Reid JB, Kamiya Y. 1995. Gibberellin and pea seed development. *Planta* 195:426–33

128a. Takahashi M, Kamiya Y, Takahashi N, Graebe JE. 1986. Metabolism of gibberellins in a cell-free system from immature seeds of *Phaseolus vulgaris* L. *Planta* 168: 190–99

128b. Takahashi N, Phinney BO, MacMillan J, eds. 1991. *Gibberellins*. New York: Springer-Verlag

129. Talon M, Koornneef M, Zeevaart JAD. 1990. Endogenous gibberellins in *Arabidopsis thaliana* and possible steps blocked in the biosynthetic pathways of the semidwarf *ga4* and *ga5* mutants. *Proc. Natl. Acad. Sci. USA* 87:7983–87

130. Talon M, Koornneef M, Zeevaart JAD. 1990. Accumulation of C_{19}-gibberellins in the gibberellin-insensitive dwarf mutant *gai* of *Arabidopsis thaliana* (L.) Heynh. *Planta* 182:501–5

131. Talon M, Tadeo FR, Zeevaart JAD. 1991. Cellular changes induced by exogenous and endogenous gibberellins in shoot tips of the long-day plant *Silene armeria*. *Planta* 185:487–93

132. Talon M, Zacarias L, Primo-Millo E. 1992. Gibberellins and parthenocarpic ability in developing ovaries of seedless mandarins. *Plant Physiol.* 99:1575–81

133. Talon M, Zeevaart JAD. 1990. Gibberellin and stem growth as related to photoperiod in *Silene armeria* L. *Plant Physiol.* 92: 1094–100

134. Talon M, Zeevaart JAD. 1992. Stem elongation and changes in the levels of gibberellins in shoot tips induced by differential photoperiodic treatments in the long-day plant *Silene armeria*. *Planta* 188:457–61

135. Talon M, Zeevaart JAD, Gage DA. 1991. Identification of gibberellins in spinach and effects of light and darkness on their levels. *Plant Physiol.* 97:1521–26

136. Deleted in proof

137. Toyomasu T, Kawaide H, Sekimoto H, von Numers C, Phillips AL, et al. 1997. Cloning and characterization of a cDNA encoding gibberellin 20-oxidase from rice (*Oryza sativa* L.) seedlings. *Physiol. Plant.* 99: 111–18

138. Toyomasu T, Tsuji H, Yamane H, Nakayama M, Yamaguchi I, et al. 1993. Light effects on endogenous levels of gibberellins in photoblastic lettuce seeds. *J. Plant Growth Regul.* 12:85–90

139. Toyomasu T, Yamane H, Yamaguchi I, Murofushi N, Takahashi N, et al. 1992. Control by light of hypocotyl elongation and levels of endogenous gibberellins in

seedlings of *Lactuca sativa* L. *Plant Cell Physiol.* 33:695–701

140. Turnbull CGN, Crozier A, Schwenen L, Graebe JE. 1985. Conversion of [^{14}C]gibberellin A$_{12}$-aldehyde to C$_{19}$- and C$_{20}$-gibberellins in a cell-free system from immature seeds of *Phaseolus coccineus* L. *Planta* 165:108–13

141. van Huizen R, Ozga JA, Reinecke DM, Twitchin B, Mander LN. 1995. Seed and 4-chloroindole-3-acetic acid regulation of gibberellin metabolism in pea pericarp. *Plant Physiol.* 109:1213–17

142. Ward JL, Jackson GS, Beale MH, Gaskin P, Hedden P, et al. 1997. Stereochemistry of oxidation of the gibberellin 20-alcohols, GA$_{15}$ and GA$_{44}$, to 20-aldehydes by gibberellin 20-oxidases. *Chem. Commun,* pp. 13–14

143. Weller JL, Ross JJ, Reid JB. 1994. Gibberellins and phytochrome regulation of stem elongation in pea. *Planta* 192:489–96

144. Winkler RG, Helentjaris T. 1995. The maize *dwarf-3* gene encodes a cytochrome P450-mediated early step in gibberellin biosynthesis. *Plant Cell* 7:1307–17

145. Wu K, Gage DA, Zeevaart JAD. 1996. Molecular cloning and photoperiod-regulated expression of gibberellin 20-oxidase from the long-day plant spinach. *Plant Physiol.* 110:547–54

146. Xu YL, Wu K, Peeters AJM, Gage D, Zeevaart JAD. 1995. The *GA5* locus of *Arabidopsis thaliana* encodes a multifunctional gibberellin 20-oxidase: molecular cloning and functional expression. *Proc. Natl. Acad. Sci. USA* 92:6640–44

147. Yamaguchi S, Saito T, Abe H, Yamane H, Murofushi N, et al. 1996. Molecular cloning and characterization of a cDNA encoding the gibberellin biosynthetic enzyme *ent*-kaurene synthase B from pumpkin (*Cucurbita maxima* L.). *Plant J.* 10:203–13

148. Zanewich KP, Rood SB. 1995. Vernalization and gibberellin physiology of winter canola. Endogenous gibberellin (GA$_1$) content and metabolism of [^{3}H]GA$_1$ and [^{3}H]GA$_{20}$. *Plant Physiol.* 108:615–21

149. Zeevaart JAD, Gage DA. 1993. *ent*-Kaurene biosynthesis is enhanced by long photoperiods in the long-day plants *Spinacia oleracea* L. and *Agrostemma githago* L. *Plant Physiol.* 101:25–29

150. Zeevaart JAD, Talon M. 1992. Gibberellin mutants in *Arabidopsis thaliana*. see Ref. 52a, pp. 34–42

Annu. Rev. Plant Physiol. Plant Mol. Biol. 1997. 48:461–91

POLLEN GERMINATION AND TUBE GROWTH

Loverine P. Taylor

Department of Genetics and Cell Biology, Washington State University, Pullman, Washington 99164-4234

Peter K. Hepler

Biology Department, University of Massachusetts, Amherst, Massachusetts 01003

KEY WORDS: pollen germination, pollen tube, tip growth

ABSTRACT

Many aspects of Angiosperm pollen germination and tube growth are discussed including mechanisms of dehydration and rehydration, in vitro germination, pollen coat compounds, the dynamic involvement of cytoskeletal elements (actin, microtubules), calcium ion fluxes, extracellular matrix elements (stylar arabinogalactan proteins), and control mechanisms of gene expression in dehydrating and germinating pollen. We focus on the recent developments in pollen biology that help us understand how the male gamete survives and accomplishes its successful delivery to the ovule of the sperm to effect sexual reproduction.

CONTENTS

461

INTRODUCTION

Pollen develops within the anther and at maturity contains the products of sporophytic gene expression, arising from the tapetal layer of the anther wall, and gametophytic gene expression from the vegetative and generative nuclei. The progamic phase of development begins with pollen dehydration, a survival aid during dispersal. When a pollen grain falls on a receptive stigma, the stored RNA, protein, and bioactive small molecules allow rapid germination and outgrowth of a tube that penetrates and grows within the style. Experiencing the most rapid growth of any plant cell known, which is restricted exclusively to the tip of the tube, the pollen tubes eventually deposit the two sperm cells in the embryo sac where they fuse with the egg and central cell to form the zygote and endosperm. Understanding this critical developmental process is not only important in our efforts to decipher the basic mechanism of sexual reproduction in flowering plants but also has value for the potential manipulation of crop plant production.

This review focuses on many aspects of pollen germination and tube growth beginning with dehydration and ending just before tube entry into the ovary. There have been many recent reviews on pollen development, but most have focused on special subjects. Heslop-Harrison (54) and Mascarenhas (100) reviewed this topic broadly; however, the subject is large and progress continues in several areas at a rapid pace, justifying the present effort. Nevertheless, space constraints have forced us to curtail expanded discussions on many points, and to greatly restrict the number of literature citations.

IN VITRO GERMINATION

Pollen of most species will germinate and grow a tube when placed in a solution of calcium, boron, and an osmoticant. Although it provides a controlled experimental system, germination in vitro does not completely mimic growth in vivo. Even with highly optimized germination media (GM), in vitro tubes reach only 30–40% of in vivo lengths, and structural anomalies are frequently observed (125). In addition, the pollen of some species, e.g. *Arabidopsis thaliana,* does not germinate well in vitro no matter what conditions are used (123). If only a fraction of the viable grains produce a tube, the resulting studies do not provide an accurate reflection of "germinating" pollen. Even though it is not possible to duplicate the dynamic interaction between the pollen and the pistil, nevertheless in many species germination and tube growth are robust under experimentally defined conditions, rendering in vitro–based studies of relevance to the in vivo situation.

The composition of the GM can dramatically affect pollen metabolism. Under osmotically equivalent conditions, PEG-based GM increased germination frequency and prevented tube bursting when compared to sucrose (e.g. 125). PEG is relatively inert metabolically and cannot enter cells (138), whereas sucrose and/or monosaccharides enter the pollen and augment the already high internal concentrations (61, 71). High sucrose levels in the GM may alter the permeability of the growing pollen tube, resulting in leaching of metabolites and ions into the media (27, 49, 138, 178). Sucrose also fuels ethanolic fermentation during in vitro growth accumulating to levels in the GM (100 mM) that may inhibit growth (12). Whether ethanol is produced during in vivo growth is unknown.

DEHYDRATION AND HYDRATION

Water Loss and Phase Transition

Dehydration, which usually occurs just before anthesis, induces a metabolically quiescent state that confers a tolerance to the environmental stresses encountered during dispersal by water, wind, insects, or animals and may be a necessary prerequisite for pollen viability and subsequent germination (89). When released from the anther, pollen is partially dehydrated, with water contents ranging from 6 to 60%; the degree of hydration is species-specific (72). Despite desiccation, pollen is viable if the structural changes that occur during dehydration are reversible upon rehydration. Therefore, the conditions under which water loss occurs can significantly affect the subsequent adhesion and germination of pollen (72, 89, 157).

Given the importance of pollen as a germplasm source (24), considerable effort has been vested in identifying the cellular targets of dehydration (e.g. 148). Water is a major determinant of the structural integrity and stability of cellular membranes, and the loss of this barrier is a major cause of decreased pollen viability. During dehydration water becomes increasingly bound and reaches a transition point where the phospholipid structure changes from a lamellar or liquid crystalline form to a gel state (27). The temperature at which this transition occurs depends on the water content (27). If conditions are suboptimal when the gel state is formed, e.g. extreme fluctuation in temperature and humidity during the dispersal phase, irreversible membrane injury can occur (72, 133). Free radicals, which are formed more readily in the dehydrated state, also contribute to membrane damage (72, 124, 157).

As water is removed it is replaced by molecules that protect cellular structures. In other dehydrating tissues, various molecules accumulate, e.g. trehalose in nematodes and yeast and the hydrophilic LEA proteins in dry seeds. Neither of these molecules has been identified in pollen, but conditions in mature pollen foster expression of *Dc3*, a *lea*-class gene involved in water stress responses. In transformed pollen the *Dc3* promoter was activated at the latest state of pollen development (149) and supported a twofold higher level of expression than *LAT52*, a late pollen-specific gene that functions during tomato pollen hydration (113). Sucrose may function as a desiccation protectant in pollen, because dehydration survival was correlated to its concentration (61).

Imbibition

Pollen hydration proceeds in a controlled manner characterized by distinct plateaus of increasing water content (13, 48, 180) and can begin in the anther before pollen release (180). Hydration plateaus can be detected in vitro: Prehydrated petunia pollen increased threefold in volume compared to a twofold increase when desiccated pollen was exposed directly to bulk water (48). This study also found that the more hydrated the grain initially, the larger the final volume. The osmoticum also influences water uptake; tube volume was greater on sucrose-based medium than on a PEG medium (178).

The damaging effects of immediate exposure of the desiccated grain to bulk water are demonstrated by the increased osmotic potential, ion concentration (K^+, NAD^+, and H^+) and enzyme activity in the GM upon addition of pollen (e.g. 27, 49). Presumably this "imbibitional damage" results from altered membrane integrity; this could be caused by incomplete or abnormal reconstitution of the lipid bilayer upon rapid hydration, possibly involving a transition from gel to liquid crystalline phase (27) or the persistence of patches of gel phase lipids in the hydrated state (72). One model of imbibitional damage predicts that the leakage can be completely eliminated by prehydrating pollen before exposing it to bulk water (27). This effectively lowers the transition point temperature so that the partially hydrated phospholipids are already in the liquid crystalline phase when rehydrated (24, 49, 61).

Ultrastructural and physiological studies of pollen hydration in *Brassica* detected two distinct phases of hydration (33, 40, 59). During the initial phase, putative signals are reciprocally exchanged between pollen and stigma (33). The second phase proceeds with an invagination of the intine in the colpial zone (aperture where pollen tube will emerge) and formation of a "foot" of pollen coating that contacts the stigma papilla. This "foot" may differ chemically from the rest of the pollen coat because a cyclohexane wash removes the coating except in this region (40). Freeze-etch preparations show microchannels at the papilla-pollen boundary through which water may flow. Water moves from stigma to pollen grain but not between grains and requires protein synthesis (131). The area around the site of pollen tube emergence is rich in pectins, and one of the earliest visible alterations of macromolecules upon hydration is a loss of protein and pectic material from the length of the colpial slit (57). Upon emergence from the germination pore, the tube grows through the foot layer and penetrates the papilla cell (40), possibly utilizing an active cutinase that is located in the intine layer and translocated to the pollen tube tip upon germination (58).

POLLEN COMPONENTS INVOLVED IN GERMINATION

Pollen Coat

Surface molecules provide contact and initiate the signaling necessary for successful adhesion (dry stigmas) and germination (wet and dry stigmas). The pollen wall consists of two layers, the inner pectic-cellulosic and the outer, highly sculpted exine, composed of sporopollenin (168). In the later stages of pollen development the tapetal layer of the anther wall disintegrates, and the cellular contents are deposited onto the surface of the pollen grain forming the tryphine (or pollenkitt) layer. The composition of this material is highly heterogeneous and includes waxes, lipid droplets, small aromatic molecules, and proteins.

Waxes

Several alleles of the epicuticular wax mutants (*cer*) of Arabidopsis (76) have an altered tryphine coating and impaired male fertility (1, 64, 123). Microscopic and biochemical analysis of *cer*6-2 pollen showed that an aberrant tryphine layer deposited during development visually disappeared by dehiscence (123). The mutant pollen failed to hydrate but stimulated callose deposition in the underlying stigma cell, a sign of an incompatible pollination. *cer*6-2 pollen was viable and could germinate in vitro, an observation that was exploited to reverse the male sterility (MS) by pollinating at high humidity. Chemical analysis of the *cer*6-2 pollen coat showed that almost none of the C29 and C30 waxes characteristic of wild-type (WT) pollen was present, suggesting that long-chain alkanes (C29) may be the bioactive molecule (123). Not all *cer* mutants with altered male fertility lack a tryphine (1, 64, 123); *cer*3-2186 has a visually normal pollen coating but exhibits a strong MS phenotype (64). Thus subtle changes in tryphine composition may have dramatic consequences for fertility.

The failure of the *cer* mutants to stimulate release of water from the stigma to the pollen is interpreted as evidence that lipids in the pollen coat are involved in the cell-cell recognition required for hydration. By microscopic observation, elements in the coat show limited mobility and thus may move from grain to grain (40). This movement could explain the mentor pollen rescue of the MS *cer* phenotype by WT pollen (64, 123). Some *cer* mutants are temperature sensitive and can be successfully pollinated at lower temperatures (18°C versus 25°C) (64). Temperature has a profound effect on lipid viscosity, which could in turn affect the mobility and/or accessibility of substances suspended or dissolved in the tryphine layer. An important issue is whether lipid-mediated hydration is specific to species with dry stigmas (or even more narrowly to the *Brassicaceae*) or whether some elements also operate in pollinations on wet stigmas.

Oleosins

Oleosin-like proteins have been detected in the tryphine layer (127). In seeds oleosins are associated with oil bodies; a model based on the presumed secondary structure of the oleosin protein proposes that the highly conserved central hydrophobic domain penetrates the oil body and that the N- and C-termini interact on the surface to stabilize the structure (62). The predicted protein product of the tapetum-specific oleosin-like cDNAs have an identical central hydrophobic domain but differ significantly from the majority of seed oleosins in the C-terminal region (e.g. 62, 126, 127). *cer* mutants lacking a tryphine (*cer*6-2) do not accumulate oleosin-like proteins (122, 127). Other *cer* mutants, which have an abnormal tryphine (e.g. *cer*6-1), have reduced amounts (D Preuss, personal communication). Even though the structure of the oleosin-like proteins recovered from cyclohexane washes of *Brassica* pollen do not contain the hydrophobic domain (127), there appears to be a connection between tryphine lipids and tapetal oleosins. Further analysis of mutant tryphine coatings for the presence of oleosin-like proteins will be required to establish a relationship.

Flavonoids

In addition to lipids and proteins, small aromatic molecules such as carotenoids, flavonoids, jasmonates, phenolic acids, brassinosteroids, and most of the classic phytohormones have been detected in pollen (73, 96, 134, 168). Of these compounds only flavonols, a specific class of flavonoid, have a demonstrated role in pollen germination (110, 174). Virtually all pollens accumulate flavonols, often to very high levels (168). By genetic analysis, pollen flavonols are sporophytic in origin; they are synthesized in the tapetum, released into the locule, and taken up and modified by the developing gametophyte (161, 171).

Early suggestions that flavonols stimulated pollen tube growth were confirmed with the isolation of flavonol-deficient plants, of which several now exist (23, 110, 174; C Napoli, LP Taylor, personal communication). Flavonol-deficient plants are self-sterile because the pollen fails to germinate (110, 142) or produce a functional tube (120, 174). The reproductive defect is conditional and can be biochemically complemented by adding flavonols at pollination or by WT stigma exudate (142, 160, 174). The bioactive compound from the stigma exudate was identified as kaempferol, a flavonol aglycone (110, 160). An in vitro pollen rescue assay established that the response to exogenous flavonols was sensitive (0.4 µM), specific for flavonol aglycones (110, 162, 174) and rapid: Tube outgrowth occurs within 2 min (110). Although flavonols accumulate before pollen maturity (28, 120, 121), all the biochemical and histological evidence indicates that they function only at germination.

Although the mechanism of flavonol action remains to be determined, a structural role within the cell wall (53) has been excluded by feeding studies using a radiolabeled flavonol (171); rather, it seems that flavonols may be localized to cytoplasmic compartments (173). Feeding experiments also showed that flavonols were not catabolized during germination but were rapidly (<1 min) conjugated to specific sugars (161), forming a pollen-unique class of flavonols (161, 171, 177). The sensitivity and specificity of the germination requirement as well as the rapid response to added flavonols suggests they may function as signal molecules, a role they perform in legume-*Rhizobium* interactions (93). If flavonols are germination signals, it is not clear when they act or how the active aglycone is generated in WT pollen, which accumulates only flavonol diglycosides (121, 161, 171). Formation of the diglycoside, which cannot stimulate germination in vitro, occurs by step-wise addition of sugars, and the final step is irreversible (161, 171). However, the aglycone can be generated from the intermediate monoside by an activity associated with intact pollen (LP Taylor, unpublished data). There are several plausible explanations for the inability to detect the aglycone or the monoside in WT pollen, including the formation of a second signal that is stored until germination. The synthesis of tagged flavonols and the isolation of mutations in key activities will be important tools to probe the mechanism of flavonol action.

The universality of the flavonol requirement is debatable. The flavonol pathway evolved in parallel with plant evolution; the more complex compounds are present in more advanced species (135). Flavonols are present in the spores of mosses and ferns (173) and gymnosperm pollen (66) (LP Taylor, unpublished data). One of the earliest descriptions of white pollen was an ancient bristle cone pine that showed decreased pollen germination frequency compared with yellow pollen from surrounding trees (66). The taxonomic distance, together with structural and physiological differences such as wet versus dry stigmas and bicellular versus tricellular pollen in maize and petunia, argues that the germination requirement may be widespread in the angiosperms. Flavonols enhanced reproductive success in tomato, forsythia, *Brassica oleracea,* wheat, and several species of *Nicotiana* (132, 173, and references therein; B McClure, personal communication). There is no flavonol requirement in an Arabidopsis CHS mutant that lacks flavonols but produces functional pollen (14). Did the requirement for flavonols evolve independently in maize and petunia (and perhaps other species), or has Arabidopsis pollen simply lost the need for flavonols? The appearance of multiple independent events that require exactly the same flavonols is more difficult to envision than the loss of the function in a single species. Until more species are examined and more flavonol-deficient mutants isolated, the extent of the requirement remains an unanswered question.

POLLEN TUBE GROWTH

Overview of the Pollen Tube Structure

Microscopic examination of growing pollen tubes reveals the characteristic zonation in which the apical region of the tube possesses a clear cap with more granular elements behind (26, 30, 54, 100, 115, 117, 137). These internal components exhibit vigorous "reverse fountain" cytoplasmic streaming (55, 56, 117). However, within the tip itself the motion is chaotic and turbulent, with vesicles appearing to move in a random, diffusive manner from the base of the clear zone to the extreme apex (117). More detail is provided at the electron microscope (EM) level (26, 115, 137), but this often necessitates

chemical fixation, a process that has been shown to produce severe disorgani-
zation of cytoplasmic components in both pollen tubes (32, 52) and fungal
hyphae (69). For this reason, the comments in this section describe studies of
pollen tubes fixed by rapid freeze fixation and freeze substitution (29, 30, 79,
83, 147) in which vital processes are halted by vitrification within a few
milliseconds, preventing cytoplasmic rearrangements.

A striking feature of the pollen tube is the marked accumulation of secretory vesicles, often aggregated in the shape of an inverted cone, at the tube
apex. The vesicles contain components for cell wall expansion. Because more
vesicles are secreted than are required to support the increased area of the
plasma membrane (PM), there appears to be a very active process of endocytosis (137). This conclusion is supported by the apical localization of clathrin
(10), and coated pits and vesicles (29, 80) as well as the uptake of FITC-dex-
tran from the medium (114). Threaded among the secretory vesicles are elements of endoplasmic reticulum (29, 30, 83), as well as a few unorganized
actin microfilaments (MFs) (79, 147). Other cytoplasmic inclusions such as
mitochondria and dictyosomes are distal to the zone of secretory vesicles;
some species, e.g. tobacco, show a uniform distribution throughout the tube
(30), while others, such as lily, show longitudinal zonation (83).

Actin Microfilaments

Actin MFs are involved in the transport of secretory vesicles essential for cell
elongation (30). The actin accumulates in substantial amounts in mature pollen
grains and possesses properties that are closely similar to muscle actin (172).
Thus, pollen-derived actin can activate myosin ATPase activity and bind muscle heavy meromyosin, yielding the characteristic arrow head decoration
(172). Support for a role of actin MFs in intracellular motility comes from
studies using the fungal toxins, cytochalasin B and D (56, 81, 100). The drugs
bind to the barbed end and prevent further assembly (25), and this presumably
leads to disassembly of the MFs. However, cytochalasin-treated tobacco (81)
and narcissus pollen tubes (56) form enlarged actin bundles, suggesting that
the drug might induce aggregation or even polymerization of F-actin, rather
than depolymerization.

MFs may play a role in the organization of the apical clear zone (30).
Several reports show that actin MFs accumulate in the apical region of the
growing pollen tube and may form a dense matrix there (for a review, see 30).
From these observations it was assumed that the MFs direct the secretory
vesicles to apical docking sites, and further that the so-called "dense matrix" of
actin becomes a structural element of such proportions that it, rather than the

cell wall, is the primary component that resists the turgor pressure of the growing tube (30, 137). These conclusions are based on observations of chemically fixed or stabilized pollen tubes; the dense matrix is not observed in pollen tubes prepared by rapid freeze fixation (79, 83, 147). More convincing are recent studies on living cells injected with fluorescent phalloidin, which reveal actin cables along the shank of the tube flaring outward to the PM (106). During tube growth, the MFs advance but do not enter the apical clear zone. When growth is stopped with caffeine and the tip-focused Ca^{2+} gradient is dissipated, the MFs move closer to the apex (106), a phenomenon that may occur during fixation (32). Thus while actin MFs drive the longitudinal flow, they do not necessarily participate in vesicle motion within the clear zone. Rather, the flow of vesicles through the clear zone may be governed by the tip-focused Ca^{2+} gradient (119) that appears to create a sink for vesicle fusion at the place of ion entry on the apical PM. Other factors may be involved including annexin, a Ca^{2+}-binding membrane-associated protein that localizes to the clear zone of lily pollen tubes (7).

Actin-Binding Proteins

MYOSIN   An early demonstration of myosin activity in pollen tubes showed that isolated lily pollen tube organelles could move on bundles of Characean actin, and at rates substantially faster than beads coated with muscle myosin (75). Subsequently, a 175-kDa myosin heavy chain was detected in *Nicotiana* pollen, and epitope mapping using two monoclonal antibodies (MAb) against myosin II indicated that antifast reacted with the head or S-1 portion of the molecule, while antipan reacted with the tail (140). Both MAbs produced a punctate immunofluorescence pattern at the tip and along the tube, with anti-S-1 reacting in some instances with the surface of the vegetative nucleus and the generative cell (140). Another myosin II antibody, this one against smooth muscle myosin, has been reported to label the generative cell wall, as well as vesicles, mitochondria, and ER in tobacco pollen tubes (146). Further work with heterologous probes revealed that Abs to myosin I (A,B), II, and V, differentially stain components within the lily pollen tube (107). Anti-myosin I localized to large organelles, the PM, and the surface of the generative cell and vegetative nucleus; myosin II appeared to stain larger organelles; and infrequently the male germ unit and myosin V were restricted to smaller organelles. The subcellular distribution of the different myosins suggests they may be functionally distinct.

Recent progress on the endogenous pollen tube myosins includes the identification of a 170-kDa peptide that binds to muscle F-actin and releases with exogenously added ATP (176). This putative myosin protein supports the movement of phalloidin-stained muscle F-actin MFs at rates (7.7 mm/s) as rapid as cytoplasmic streaming in vivo. The 170-kDa protein appears to be an unconventional myosin because a specific antibody against this protein does not cross-react with conventional myosin II from muscle (176). The immuno fluorescent image indicates that the polyclonal Ab produces both a punctate and amorphous pattern of stain, often with greater intensity at the tip of the tube (175).

PROFILIN   Profilin, a monomeric G-actin-binding protein, was originally identified in birch pollen as a potent human allergen (154) but is found in virtually all cell types and in all species examined. It is a small (14-kDa) globular protein that may contribute to F-actin assembly and certain aspects of the phosphoinositide signaling pathway. Even though they show considerable sequence diversity from their animal counterparts, plant profilins bind actin (155), poly L-proline (136, 154), and phosphatidyl 4,5-bisphosphate (34), suggesting that they might participate in actin MF regulation and membrane-associated signal transduction. Plant profilins function in nonplant cells; expression of a maize pollen profilin in a *Dictyostelium* mutant lacking the protein fully restored the WT phenotype including normal cytokinesis (70). Although there is increasing evidence for pollen-specific profilin isoforms (63), attempts to localize the protein in pollen tubes have yielded different results (50, 108). A recent approach using both immunofluorescence and immunogold probes on fixed tissues, and fluorescent analog cytochemistry of living cells injected with tagged profilin, shows uniform distribution in the cytoplasm without any evidence of accumulation or association with organelles or inclusions including actin MFs (159).

Other actin-binding proteins identified in pollen include spectrin, localized to the tip of pollen tubes (30), and a 135 kDa actin-bundling protein isolated from lily pollen tubes (139). Pollen-specific cDNAs encoding a putative actin depolymerizing factor (ADF) (22) have been isolated from lily and *Brassica napus*. The protein was immunolocalized to dense storage bodies in the mature pollen and dissipates into the cytoplasm during germination (22). Finally, a member of the Rho family of GTPases has been identified in pollen and pollen tubes (90). Because of the common association of members of this family with actin-associated processes in a variety of eukaryotic cells, Lin et al (90) argued that it may be associated with and regulate actin MFs in the pollen tube.

Microtubules

Of the three principal arrays of microtubules (MTs) in the pollen tube, only the cortical MTs in the vegetative cytoplasm are involved in pollen tube growth

(79, 147). There is controversy concerning their presence in the apical clear zone (15). Again MTs, like MFs, might be displaced during conventional fixation, and therefore future work should apply cryo-methods and/or fluorescent analog cytochemistry with derivatized tubulin to authoritatively establish whether MTs are present in the apical zone of older, elongated pollen tubes. In cryo-fixed preparations, the MTs are oriented parallel to the long axis of the pollen tube (79, 147) and are closely associated with and cross-bridged to the PM. Often each MT has one or a few MFs, presumably actin, alongside, oriented parallel to the MT and positioned at a regular distance therefrom (82). Occasionally tubular elements of ER in the immediate subjacent cytoplasm run parallel to the MTs (79).

Recent studies on elongating pollen tubes treated with anti-MT agents demonstrated a role for MTs in creating and/or maintaining cytoplasmic zonation. The position of the male germ unit and vacuoles was markedly altered in tubes lacking MTs (3, 67, 109). Distally located vacuoles moved toward the tip, virtually obliterating the normal cytoplasmic zonation of the tube (67, 68). By contrast, the male germ unit failed to move forward and could actually be cut off from the tip by a callose plug (67). In parallel with the cytological studies, a series of biochemical investigations showed evidence of the MT motor proteins, kinesin and dynein in pollen tubes (for a review, see 15). The plant homologue from *Nicotiana tabacum* pollen tubes possessed certain kinesin properties such as an MT-activated ATPase activity and ATP-dependent MT binding. The protein has been detected along the shank of the pollen tube as well as in the apical region (145). Immunogold staining of membrane-bound organelles and possibly vesicles has led to the suggestion that MT motile systems also contribute to transport within the pollen tube.

Calcium Ions

The Ca^{2+} requirement for germination and tube growth has been amply demonstrated through direct modulation of ion concentration or the use of agents that interfere with transport (30, 42, 137). Recent studies with an ion-selective vibrating probe reveal an extracellular flux of Ca^{2+} (5.4 $pmol/cm^2/s$) that is inwardly directed at the tip of the growing tube (77, 118). Parallel studies on intracellular Ca^{2+} using fluorescent indicator dyes reveal that there is a steep, tip-focused gradient, with values above 3 μM at the extreme apex dropping to less than 0.2 μM within 20 μm from the tip (94, 95, 118, 119). Both direct ion imaging and manganese-dependent quenching of dye fluorescence (95) support the contention that ion entry is restricted to a small region at the extreme apex where growth occurs and membrane deformation is maximal. Ca^{2+} influx appears to be regulated by channels in this region, perhaps stretch activated (95, 119). Ca^{2+} levels at the high point of the gradient fluctuate, with concentration changes as much as fourfold; growth rates also oscillate, being positively correlated with changes in $[Ca^{2+}]$ (119).

Conditions known to inhibit growth also eliminate the extracellular inward flux and dissipate the intracellular tip-focused Ca^{2+} gradient (118, 119). Upon return to normal conditions, the intracellular gradient and the extracellular influx reappear; first the $[Ca^{2+}]$ elevates throughout the apical zone and the tip swells. Thereafter the sector in the swollen apex with the highest level of Ca^{2+} becomes the point where growth restarts (119). Growth can also be inhibited by conditions that elevate Ca^{2+} such as direct iontophoretic injection of Ca^{2+}, photolysis of caged Ca^{2+}, or application of weak electric fields (94). During recovery there is often a reorientation in the direction of tube elongation, emphasizing the importance of elevated Ca^{2+} for tube guidance (94). Imaging confirms elevated Ca^{2+} throughout the tube; however, the use of Calcium Green, a single wavelength indicator, has prevented a detailed analysis of ion dynamics at the very tip of the tube where elongation and reorientation occur. In recent work R Malhó & AJ Trewavas (95a) showed that localized increases in intracellular $[Ca^{2+}]$ attracted pollen tubes and release of a caged Ca^{2+}-buffer, which generates a transient local reduction in the ion, repelled pollen tubes.

Elevated Ca^{2+} may play a role in the regulation of growth that occurs during the self-incompatibility response in poppy pollen tubes. Application of the S-factor, a glycoprotein, causes the incompatible, but not the compatible, pollen tubes to elevate their internal Ca^{2+}, with the increase appearing to emanate from the shank of the tube, possibly in the vicinity of the male germ unit (for a review, see 44). To elucidate the underlying cause, Franklin-Tong et al (44) found that photoactivated release of caged IP3 can generate a similar increase in Ca^{2+}, which travels as a wave toward the tip of the tube. Unfortunately, the use of Calcium Green did not permit an analysis of the ionic dynamics in the growing tip; additional studies with ratiometric dyes are required for this.

Evidence that the tip-focused gradient dictates the internal structural zonation as well as the site and degree of vesicle fusion comes from studies showing that elevated Ca^{2+} blocks cytoplasmic streaming, leads to the fragmentation of F-actin (74), and depolymerizes MTs. These observations account for the lack of an organized system of MFs and MTs in the pollen tube apex and for the lack of directed particle motion. However, when the Ca^{2+} gradient is dissipated, the actin MFs (106) and streaming lanes move toward the tip (118).

Extracellular Matrix Components

Following germination the emerging pollen tube is in intimate contact with the extracellular matrix (ECM) of the pistil. This phase is characterized by the rapid synthesis of PM and cell wall as the growing tube traverses the transmitting tract (TT) to deliver the male gametes to the ovary. The metabolic demands are enormous; a maize pollen tube can traverse a 50-cm silk at the rate of 1 cm/h (100, 129). Stylar components may serve several functions during pollen tube growth including adhesion, guidance, nutrition, and structural integrity (e.g. 20, 65, 85, 169). The structural features of the pollen and tube walls are well documented at the light and EM levels (54, 137). More recently immunolocalization with monoclonal (MAb) and polyclonal antibodies to callose, pectins, and AGP have confirmed the chemical composition of various structures (e.g. 105, 156).

Pollen Wall and Pectin

The pollen tube wall is bipartite: an inner sheath of callose (β 1→3 glucan) is covered by an outer fibrillar layer composed largely of pectin with hemicellulose and cellulose. The callose lining is absent from the tube tip. In some species the pectin and cellulose components of the fibrillar layer are separate, forming a three-layered wall (54, 86). Microscopically the outer layer of the tube wall appears to be an extension of the pectinaceous inner wall (intine) of the pollen grain (54). In the ungerminated grain, pectin is most enriched at the apertures, the sites of pollen tube emergence (86). Antibodies that can discriminate between acidic pectin and methyl esterified pectin localized the esterified form to the growing tip and to the secretory vesicles that fuse at the apex (45, 65, 86). Unesterified pectins were also tip-localized but were more prominent in the outer tube wall in regions distal to the tip. Both in ungerminated pollen and in the pollen tube there appears to be a radial gradient of methyl esterified pectin, with the highest levels at the inner surface of the outer wall layer (45).

In the current view of wall formation, pectins are polymerized and methyl esterified within the Golgi. From a structural perspective methyl conjugation prevents rigidification (Ca^{2+}-mediated cross-linking) and thus permits a more flexible structure (19). The pectin esters are transported in Golgi-derived secretory vesicles to the growing wall. Following deposition, they are progressively de-esterified, and cross-linked by Ca^{2+}. The rigidity of the resulting framework provides support for the tube as it penetrates the cell files of the TT.

Arabinogalactan Proteins

AGPs are a class of hydroxyproline-rich glycoproteins (HRGPs) characterized by a high carbohydrate content (up to 98% of their total mass) that includes arabinose and galactose residues linked in a particular pattern (4). AGPs are also distinguished from other HRGPs by binding to Yariv's reagent (4). AGPs have been identified in the pollen and the stylar ECM of many species, and there is intense interest in determining their role in plant reproduction (e.g. 20, 35, 47, 91).

There is debate about the location of AGPs in the growing pollen tubes, likely resulting from technical and physiological differences (65, 85). For example, MAC 207 detected AGPs in the tips of in vivo grown lily pollen tubes, but the same MAb produced no signal in the tips of in vitro–grown tobacco tubes (65, 85). The lack of a signal in the tip of tobacco tubes and the absence of a callose layer in this area were taken to indicate that the AGPs were located in the inner layer of the pollen tube wall (85). However, AGPs were localized to the PM of the tubes and cytoplasmic vesicles in lily (65). When lily pollen tubes were treated with pectinase, a method required to visualize the immuno-complex in tobacco (85), the fluorescent pattern changed from an overall diffuse signal to a periodic ring-like form reminiscent of tobacco (65). As more genes encoding AGPs are cloned and antibodies are generated to the protein backbones, some of these discrepancies will disappear. In the meantime, binding to Yariv's reagent provides a complementary detection method (65). One proposed function of pollen and stylar AGPs is to provide adhesion to facilitate the movement of the tube cell through the TT. AGPs and pectins may interact, thus providing a mechanism to mediate or facilitate tube growth (5). In this context, it is significant that lily pollen tubes grown in vivo, but not in vitro, aggregated during growth and that this association could be disrupted by pectinase (65). This finding underscores the importance of interactions between AGPs and pectins in the pollen with their counterparts in the style.

STYLAR AGPs   Some AGPs and/or the corresponding cDNAs have been isolated and characterized from stylar tissues (e.g. 35, 36, 92). One of the most exciting findings is that stylar AGPs are incorporated into the growing pollen tube (92, 169). Immunocytochemistry techniques detected an association of a style-specific 120-kDa glycoprotein with pollen tubes grown either in vivo or in vitro. The AGP was localized to a subapical region of the cytoplasm rich in pectin-containing secretory vesicles and to inclusions in the callose layer (inner) of older (distil) portions of the pollen tube (92). The association with elements of cell wall synthesis suggests that stylar proteins destined for incorporation in the growing pollen tube use the existing tip-associated transport machinery.

A detailed study of the transmitting tissue-specific (TTS) protein has identified a reproductive role for this particular AGP and has highlighted the central role of the glycomoiety in protein function. TTS proteins from the stylar ECM of *N. tabacum* are heterogeneous (50–100 kDa); sugars can account for up to 35% of molecular mass (20). The chemically deglycosylated protein backbone is 30 kDa (165) and has a basic PI that becomes acidic with glycosylation (169). In vitro, the TTS protein promotes pollen tube growth and acts as a chemoattractant for pollen tubes (169), but only the glycosylated form is active. Analysis of transgenic plants with reduced TTS protein reveals that pollen tube growth rate is reduced, and this in turn leads to reduced seed set (20). Immunolocalization of pollinated pistils detected TTS protein along the pollen tube surface and in the wall. Biochemical analysis suggests that association of TTS at the tip of the pollen tube is unique and may involve differences in the chemical form of the protein or type of interaction (169).

There is a developmental gradient of TTS protein in the style. In immature flowers it is higher in the stigma end but becomes more abundant in the ovary end with maturity and pollination (165). The underglycosylated form accumulates following pollination and presumably is generated by a deglycosylase activity associated with pollen tubes (169). Thus there are three gradients in the style that are all TTS related: (*a*) increasing levels of TTS protein, (*b*) TTS-bound sugars, and (*c*) decreasing pH caused by glycosylated TTS protein. All of these may contribute to the guidance of the pollen tubes to the ovary.

The expression and glycosylation state of the TTS protein is controlled by a homeotic regulatory gene that acts in a reproductive–tissue-specific manner. This is of interest because very few of the downstream targets in the floral identity cascade pathway have been identified. The regulatory effects of the tobacco *Agamous* homologue (*NAG*) were demonstrated by comparing the accumulation and chemical form of the TTS protein in two types of transgenic plants (21). TTS protein expressed from a CaMV 35S-TTS construct was present in both vegetative and reproductive tissues, but fully glycosylated TTS only accumulated in the style. However, expression of the CaMV 35S-NAG construct led to the accumulation of fully glycosylated TTS protein in all tissues, including sepals that supported ectopic pollen germination and tube growth. Obviously other factors are involved, but this result is striking because it shows that pollen germination is sensitive to the level and/or type of glycosylation of the TTS protein. It also identifies a reproductive-tissue-specific glycosylating activity that controls formation of the germinating-enhancing species of TTS protein. This is reminiscent of the dependence of pollen germination on the glycosylation state of flavonols, which is controlled by a pollen-specific glycosylating activity (110, 162; LP Taylor, unpublished data).

POLLEN GLYCOPROTEINS   Pollen glycoproteins involved in pollen tube growth and/or guidance include Pex1, a HRGP expressed exclusively in starch-filled (mature) and germinating pollen of maize (128). The proposed structure of the Pex protein (128) resembles the sexual agglutinins of *Chlamydomonas reinhardtii* which are involved in recognition of opposite mating types. Sugar analysis of Pex is unavailable, so it is not known if it is also an AGP. However, Pex1 differs from most AGPs because of its unusually large size in the unglycosylated state (~300 kDa) and by the predominance of the unglycosylated form in vivo (128). Immunolocalization studies using an antibody to the globular domain detected Pex protein in the intine of ungerminated pollen and in the callose layer of the tube wall in regions distil from the tip (129).

Pulsatory and Oscillatory Growth

Pollen tubes do not grow uniformly, but rather in bursts or pulses (54), and direct measurement at short time intervals has detected two distinct patterns (119, 141). In petunia and tobacco, the tube cell elongates with alternating bursts of fast and slower growth (46, 116), while lily tubes, especially those longer than 700 µM, grow with a periodic oscillatory pattern in which the rate changes in a smooth sine wave (119). From studies of cell wall deposition, it seems clear that the periods of slow growth coincide with the thickened deposits of pectins (86) and AGPs (85), whereas the periods of rapid growth correspond to regions where the cell wall is thin (116). However, lily pollen tubes show a uniform deposition of these wall components along their length even though they display a marked oscillation in growth rate.

To probe the underlying cause, Li et al (88) modulated turgor and were able to generate thickened bands in lily pollen tubes. However, application of caffeine under conditions that do not affect turgor generated similar bands (88). Although turgor changes are not ruled out, the ability to generate pulsatory behavior in their absence encourages consideration of other explanations. The role of the cell wall in pollen tube growth is speculative. Changes in the yielding properties at the tip, brought about by enzymes that break polymer linkages, or through the insertion of new wall material, might locally loosen and weaken the wall and permit rapid expansion (19). Both Ca^{2+} and H^+ in the cell wall space could have a crucial regulatory role in controlling the yield properties of the wall. Ca^{2+} would be expected to stiffen the unesterified pectin, whereas high levels of H^+ would have a loosening effect. Deciphering the contribution of the different components and factors constitutes important research problems for the future.

Saunders & Lord (130) have proposed that pollen tube growth is an example of "cell movement" in higher plants. Inert beads, introduced into cut styles of three different species, moved at the same rate, to the same extent and in exactly the same path as pollen tubes, providing evidence that the transmitting tract plays an active role in the extension or movement of the pollen tube to the ovary. Proteins immunologically related to vitronectin, an adhesion molecule involved in animal cell movement, were also detected in the TT. It is also postulated that focal adhesions between the ECM and actin MFs within the pollen tube participate in pulling the tube through the TT (130).

GENE EXPRESSION DURING THE PROGAMIC PHASE

Translational Control

There are several excellent reviews devoted to gene expression during the formation and maturation of the male gametophyte (99, 104, 152). This section focuses on the genes expressed during the final stages of pollen development, germination, and tube growth in the style. Despite rapid progress in molecular cloning of anther and pollen sequences, only a fraction of the estimated 20,000 sequences involved in pollen development has been characterized (99, 152). Pollen mitosis I (PMI) represents a key switch in gametophyte formation, defining the transition from microspore development to the pollen maturation phase (8, 163). Two groups of genes have been characterized according to whether maximal expression occurs before (early genes), or after (late genes) PMI (99). The late genes include many pollen-specific sequences thought to function during pollen germination and tube growth. Maturing pollen synthesizes and stores large amounts of proteins, rRNA, and mRNA that are available for rapid germination during the progamic phase (98, 152). Early studies used RNA and protein synthesis inhibitors to show that in some species, but not all, germination and initial tube growth were dependent on translation but not transcription (reviewed in 98, 100). Presumably the stored mRNA in mature pollen is translated on the polyribosomes that are formed upon hydration.

Posttranscriptional regulation is compatible with the speed of pollen hydration and germination, but basic information about translational mechanisms is sparse. For example, it is not known when stored transcripts are translated whether germination is triggered by protein synthesis or whether the required proteins are present in the dehydrated pollen and activated by hydration. In vitro translation of stored transcripts in maize generated a protein profile similar to that produced during germination (102), and a few proteins have been detected in both pre- and postgerminated pollen (37, 41, 111, 129). However, there is convincing evidence that novel proteins are produced in germinating pollen and that many of these proteins are glycosylated and/or phosphorylated (17, 59). The 1D- and 2D-gel pattern of newly synthesized proteins in growing *N. tabacum* pollen tubes showed that the majority were in low abundance because their presence was detected only by ^{14}C-fluorography, not by coomassie staining (17). Some of these proteins may be encoded by the numerous transcripts that are present in low abundance in the pollen (60).

Discrepancies have been noted in the pattern of in vitro- and in vivo-synthesized cell wall proteins (87). For obvious reasons, the protein pattern from an in vivo labeling system provides a more relevant physiological picture of protein synthesis (59). About 40 pollen proteins, including 11 that were newly phosphorylated, were synthesized within 1 h of a compatible pollination in *Brassica oleracea*. Only five of the proteins were present in ungerminated pollen. By correlating the appearance of particular proteins with cytological changes, the in vivo labeling technique may be able to identify proteins involved in the earliest events of pollen-pistil interaction.

Translational Mechanisms in Dehydrated Pollen

During dehydration the levels of intracellular K^+ increase, reaching levels of over 280 mM in *Tradescantia* pollen grains (6). Concentrations above 220 mM completely inhibit in vitro translation probably by inhibiting the formation of initiation complexes (6). Another possible effect of the increased salt concentration in dehydrating pollen is the stabilization of higher order structures within the RNA molecule. An attractive hypothesis for translational control during dehydration is based on the predicted structure of the 5' untranslated region (UTR) of pollen transcripts. Twell (152) showed that the LAT52 5' UTR acts as a translational enhancer in in vitro translation systems. In subsequent transient expression assays the LAT 5' UTR enhanced luciferase (LUC) activity three- to fivefold over a synthetic polylinker UTR, to levels comparable with known viral enhancer sequences. Most significantly the enhancement was tissue-specific: LUC activity was 20-fold higher in pollen than in leaves. A survey of the 5' UTR sequences of cloned late genes revealed a low potential for secondary structure, suggesting that the transcripts may be translated under conditions of increasing salt concentration. One can envision that the late pollen-specific transcripts comprise a graded series of 5' UTR structures that would function at different levels of hydration. Thus translation may be regulated by a rheostat mechanism rather than an off/on switch. Those involved in germination would be the last, or first, to be translated.

Translational control of pollen transcripts may be mediated by specific proteins that bind to the regulatory regions, including 5′ UTRs and poly(A) tails, of the stored mRNA. *NeIF*-4A8, a gene encoding a putative translation factor, was recently isolated from tobacco pollen (11) and shown to have homology to eukaryotic translation factors that have RNA helicase activity. Although a functional role is still to be determined, the pollen-specificity (Table 1) and temporal expression suggest that *NeIF*-4A8 may be involved in the selective translation of germination sequences. PAB proteins (PABP) that bind to the poly(A) tails of most, if not all, eukaryotic transcripts have been demonstrated to function in translational control and deadenylation-mediated mRNA decay in yeast. A recently described plant homologue, exclusively expressed in Arabidopsis flowers, was shown to functionally complement a yeast PABP mutant (9). PAB5 functioned in several aspects of translation including initiation and poly(A)-tail shortening, and it interacted with a yeast protein required for translation initiation. Transgenic Arabidopsis expressing the GUS reporter gene from the PAB5 promoter showed activity in the tapetum, in developing pollen, and in pollen tubes from germination until entry into the ovule (Table 1).

Pollination induced the deadenylation of three classes of TT mRNAs (164), and although the level of two mRNA classes subsequently declined in the style, transcripts from the third class, TTS, remained high despite substantial poly(A)-tail shortening. Thus it appears that this developmentally expressed translational mechanism can distinguish transcripts such as TTS that are required for pollen tube growth. Characterization of posttranscriptional control of pollen germination and tube growth may identify plant-unique mechanisms as well as those shared with animal reproductive development.

Gene Expression During Pollen Germination and Tube Growth

The repertoire of gene expression, and the regulatory mechanisms that act during pollen germination, must be flexible to accommodate the species-specific differences in pollen physiology and in the environmental conditions at pollination. For example, rRNA is not transcribed in germinating *Tradescantia* and lily pollen (99), but high levels are produced for several hours of germination in *Pinus taeda* (43), reflecting a requirement for RNA synthesis during the prolonged (several days) hydration and tube growth in gymnosperm pollen. To date, genes specifically associated with germination have not been identified. Even sequences isolated from cDNA libraries to germinating pollen are transcribed in nongerminating pollen (111, 112; V Guyon & LP Taylor, personal communication). However, the large number of unidentified proteins (>230) whose appearance is coincident with germination (101) suggests that it may be

Table 1   Genes expressed in germinating pollen

Gene/species	Function/homology	Location	How determined	Reference
PAB5 A. thaliana	poly(A) binding protein	germinating pollen	Transgenic plant, GUS activity	9
tua1 A. thaliana	tubulin	germinating pollen	Transgenic plant, GUS activity	18
Bcp1 B. campestris	AGP-like	germinating pollen	Northern	144
Gs15 Glycine max	Glutamine synthase	germinating pollen	Transgenic lotus, GUS activity	97
CDPK maize	Ca^{+2}-dependent protein kinase	germinating pollen	Northern, western and antisense oligonucleotide suppression	41
pex1 maize	chimeric extensin-like	pollen tube wall, callose layer	Immunolocalization (FTIC and TEM)	129
PGc9 maize	polygalacturonase	pollen tube wall	Immunolocalization (FTIC); Transgenic plant (tobacco), GUS activity	37
W2247 maize	polygalacturonase	pollen tube tip	Transgenic plant (tobacco) and pollen bombardment, GUS activity	2, 143
Zm13 maize	protease inhibitor/LAT52	vegetative cell cytoplasm	In situ hybridization	51
neIF-4A8 N. tabacum	RNA helicase	germinating pollen	Transgenic plants and pollen bombardment, GUS activity	11
NTP303 N. tabacum	ascorbate oxidase/LAT51, Bcp10	vegetative cell cytoplasm, tube tip	Northern; Pulse labelling	166
chiA (P$_{A2}$) Petunia hybrida	chalcone isomerase	germinating pollen	Transgenic plants, GUS activity	158
PGP76 P. hybrida	leucine-rich repeat Cf-2, Cf-9	germinating pollen	Northern	V Guyon & LP Taylor, unpubl data
PGP177 P. hybrida	serine-rich	germinating pollen	Northern	V Guyon & LP Taylor, unpubl data
PGP 220 P. hybrida	AGP-like	germinating pollen	Northern	V Guyon & LP Taylor, unpubl data
ppe1 Petunia inflata	pectin esterase/Bp19	pollen tubes	Northern	112
PRK1 P. inflata	receptor-like protein kinase	pollen tubes	Northern and western	111
Ps1 Rice	protease inhibitor/LAT52/pZm13	potten tube	Northern; Transgenic plants, GUS activity	179

Table 1 *(continued)*

Gene/species	Function/ homology	Location	How determined	Reference
LAT51 tomato	ascorbate oxidase/ Bp10/NTP303	tube tip and cytoplasm	In situ hybridization	153
LAT52 tomato	Kunitz trypsin inhibitor/pZm13	tube tip and cytoplasm vegetative cell nucleus	In sity hybridization transgenic plant, nuclear targeted GUS activity	151, 153
LAT56 tomato	pectate lyase/ LAT59/TP10/ Zm58	tube tip and cytoplasm	In situ hybridization tissue print	31, 153
LAT58 tomato	None	tube tip and cytoplasm	In situ hybridization	153
LAT59 tomato	pectate lyase/ LAT56/TP10/ Zm58	tube tip and cytoplasm	In situ hybridization	153

premature to conclude that none of them arises from transcripts activated specifically at germination.

Some early gene products such as alcohol dehydrogenase, actin (99), and a heat-shock protein cognate from tomato (38) persist in germinating pollen. The issue of protein turnover must be examined in parallel with the posttranscriptional control of pollen gene expression. Ubiquitin-mediated protein degradation has not been demonstrated in pollen, although elements of the system are present in virtually all gymnosperm and angiosperm pollens examined (78). Some maize inbreds show a progressive loss of ubiquitin and ubiquitinated conjugates during pollen maturation (16), but the significance of this for pollen development is unclear because it is not shown by the vast majority of species (78).

LATE GENES   Although the late genes are transcriptionally activated before dehydration, their persistence during germination and growth argues for a functional role at this stage (see Table 1). Because of the method of isolation, most of the genes in Table 1 represent abundant transcripts, and currently their function in pollen tube growth is predicated on the shared homology with genes of known function (103, 152, 173). Some sequences defy classification because they have no homology to known genes, and others may serve multiple functions as suggested by differential localization (e.g. *Pex*1, PG) and/or expression (e.g. *Bcp*1) before and after germination.

KINASE SEQUENCES IN POLLEN TUBES   A pollen-specific calcium-dependent calmodulin-independent protein kinase (CDPK) isolated from maize suggests the presence of posttranslational control mechanisms involving Ca^{2+} and phosphorylation (41). The gene is transcribed in mature and germinating pollen and is required for germination. When antisense oligonucleotides to the kinase sequences were added to the GM, germination was abolished or severely diminished. The same effect was seen if an antagonist of calmodulin or a CDPK inhibitor was present at germination. Transcripts to a receptor-like serine-threonine protein kinase, *PRK*1, were detected for at least 16 h after pollen germination in petunia (111). Although *PRK*1 was isolated from germinating pollen, transgenic plants expressing the cDNA in an antisense orientation showed arrested pollen development at the uninucleate microspore stage (84). Because the premature pollen abortion precluded testing a role in germination, another approach, perhaps the use of antisense *PRK*1 oligonucleotides at germination, will be necessary. If *PRK*1 functions at two different stages it will be interesting to know whether the kinase substrates are identical in developing and in germinating pollen.

Genes Involved in Pollen Metabolism

Some genes are expressed in both the sporophyte and the gametophyte, usually at different times in development (e.g. 144). *Bcp*1 was expressed in the tapetum of *Brassica campestris* anthers and in germinating pollen up to 5 h postgermination (144). Promoter analysis of *Bcp*1 showed that tapetal and pollen expression were controlled by different regulatory sequences (170) and that perturbation of expression in the tapetum resulted in plants that aborted all pollen at the late uninuclear stage. To test the role of *Bcp*1 expression in pollen, an antisense copy of the cDNA was expressed from the pollen-specific LAT52 promoter from tomato. *Bcp*1 expression was not perturbed in the tapetum of these plants, but 50% of the pollen aborted at the bicellular stage (170). Thus expression of *Bcp*1 is required in both sporophyte and gametophyte with the tapetal requirement preceding the pollen.

    The most complete analysis of transcription during pollen germination is that of the *NTP*303 gene from *N. tabacum*. Northern analysis and pulse-labeling of in vitro–germinated pollen detected transcripts at 10 h but not 20 h after germination (166). However, subsequent analysis by in situ hybridization detected *NTP*303 transcripts in pollen tubes growing for 72 h in the stylar transmitting tract. By this time, the tube had reached the ovary. The RNA was localized to the vegetative cell and to the tip of pollen tube (167). In situ hybridization and immunocytochemical techniques showed that pectin-degrading polygalacturonase (PG) is differentially localized, being present in the cytoplasm of ungerminated grains and in the wall of the elongating tube (e.g.

2, 37). Expression in germinating pollen was confirmed in both stably and transiently transformed pollen expressing a PG promoter-GUS marker gene fusion (37, 143).

The resolution of in situ hybridization does not allow the detection of any contribution to this signal by the relatively small generative cell. To demonstrate cell-specific transcriptional activation of a GUS reporter gene, Twell et al (151, 152) added a nuclear targeting sequence to the LAT52 GUS construct. Transgenic plants expressing GUS activity showed staining only in the vegetative nucleus, confirming that the LAT52 promoter was active only in the vegetative cell. The construct was subsequently used to show vegetative-cell-specific expression of LAT52 in tricellular pollen (Arabidopsis), although in this instance it was activated before PMI rather than after, as in bicellular pollen (39, 152).

LAT52 is a well-characterized example of a late pollen gene. It is expressed from PMI onward, and transcripts have been detected after 18 h of in vitro germination (113, 153). LAT52 has sequence homology to genes encoding proteinase inhibitor proteins, but the activity of the expressed protein has not been determined (113). To define a role for LAT52 in pollen function, antisense LAT52 tomato plants were analyzed. Segregation analysis of two classes of the transgenic plants confirmed that LAT expression was required for viable pollen (113). Primary transformants with reduced LAT52 mRNA and protein levels developed normally and showed no pollen abnormalities until the hydration phase. The transgenic grains did not appear to imbibe water from a PEG-based GM but hydrated normally when placed in an aqueous solution without PEG. The water flux was reversible because the abnormal shape was reconstituted by placing the hydrated pollen in PEG media. The mediating effect of PEG on the mutant phenotype may be related to its known effect on slowing water uptake (150). More importantly, the mutant phenotype was expressed when the pollen was hydrated on the stigma: In vivo pollination showed that many grains did not germinate and those that did grew knotted and twisted pollen tubes.

ACKNOWLEDGMENTS

We thank members of the plant reproductive community as well as members of the Taylor and Hepler research groups who were generous with their time for helpful discussions. The authors research presented in this review was supported by grants USDA NRICGP 9337304-9435 and NSF IBN-9405361 to LPT and NSF MCB-9304953 and MCB-9601087 to PKH.

Visit the *Annual Reviews home page* at http://www.annurev.org.

Literature Cited

1. Aarts MGM, Keijzer CJ, Stiekema WJ, Pereira A. 1995. Molecular characterization of the *CER1* gene of *Arabidopsis* involved in epicuticular wax biosynthesis and pollen fertility. *Plant Cell* 7:2115–27

2. Allen RL, Lonsdale DM. 1993. Molecular characterization of one of the maize polygalacturonase gene family members which are expressed during late pollen development. *Plant J.* 3:261–71

3. Astrom H, Sorri O, Raudaskoski M. 1995. Role of microtubules in the movement of the vegetative nucleus and generative cell in tobacco pollen tubes. *Sex. Plant Reprod.* 8:61–69

4. Bacic A, Gell AC, Clarke AE. 1988. Arabinogalactan proteins from stigmas of *Nicotiana alata*. *Phytochemistry* 27:679–84

5. Baldwin TC, McCann MC, Roberts K. 1993. A novel hydroxyproline-deficient arabinogalactan protein secreted by suspension-cultured cells of *Dacus carota*. *Plant Physiol.* 103:115–23

6. Bashe D, Mascarenhas JP. 1984. Changes in potassium ion concentrations during pollen dehydration and germination in relation to protein synthesis. *Plant Sci. Lett.* 35:55–60

7. Battey NH, Blackbourn HD. 1993. The control of exocytosis in plant cells. *New Phytol.* 125:307–38

8. Bedinger PA, Edgerton MD. 1990. Developmental staging of maize microspores reveals a transition in developing microspore proteins. *Plant Physiol.* 92:474–79

9. Belostotsky DA, Meagher RB. 1996. A pollen-, ovule-, and early embryo-specific poly(A) binding protein from arabidopsis complements essential functions in yeast. *Plant Cell* 8:1261–75

10. Blackbourn HD, Jackson AP. 1996. Plant clathrin heavy chain: sequence analysis and restricted localization in growing pollen tubes. *J. Cell Sci.* 109:777–87

11. Brander KA, Kuhlemeier C. 1995. A pollen-specific DEAD-box protein related to translation initiation factor eIF-4A from tobacco. *Plant Mol. Biol.* 27:637–49

12. Bucher M, Brander KA, Sbicego S, Mandel T, Kuhlemeier C. 1995. Aerobic fermentation in tobacco pollen. *Plant Mol. Biol.* 28:739–50

13. Buitink J, Walters-Vertucci C, Hoekstra FA, Leprince O. 1996. Calorimetric properties of dehydrating pollen. *Plant Physiol.* 111:235–42

14. Burbulis IE, Iacobucci M, Shorley BW. 1996. A null mutation in the first enzyme of flavonoid biosynthesis does not affect male fertility in arabidopsis. *Plant Cell* 8:1013–25

15. Cai G, Moscatelli A, Del Casino C, Cresti M. 1996. Cytoplasmic motors and pollen tube growth. *Sex. Plant Reprod.* 9:59–64

16. Callis J, Bedinger P. 1994. Developmentally regulated loss of ubiquitin and ubiquinitated proteins during pollen maturation in maize. *Proc. Natl. Acad. Sci. USA* 91:6074–77

17. Căpková V, Zbrozek J, Tupy J. 1994. Protein synthesis in tobacco pollen tubes: preferential synthesis of cell-wall 69-kDa and 66-kDa glycoproteins. *Sex. Plant Reprod.* 7:57–66

18. Carpenter JL, Ploense SE, Snustad DP, Silflow CD. 1992. Preferential expression of an α-tubulin gene of arabidopsis in pollen. *Plant Cell* 4:557–71

19. Carpita NC, Gibeaut DM. 1993. Structural models of primary cell walls in flowering plants: consistency of molecular structure with the physical properties of the walls during growth. *Plant J.* 31:1–30

20. Cheung AY, Wang H, Wu H-M. 1995. A floral transmitting tissue-specific glycoprotein attracts pollen tubes and stimulates their growth. *Cell* 82:383–93

21. Cheung AY, Zhan XY, Wang H, Wu H-M. 1996. Organ-specific and Agamous-related expression and glycosylation of a pollen tube growth-promoting protein. *Proc. Natl. Acad. Sci. USA* 93:3853–58

22. Chung YY, Magnuson NS, An GH. 1995. Subcellular localization of actin depolymerizing factor in mature and germinating pollen. *Mol. Cells* 5:224–29

23. Coe EH, McCormick SM, Modena SA. 1981. White pollen in maize. *J. Hered.* 72:318–20

24. Connor KF, Towill LE. 1993. Pollen-handling protocol and hydration/dehydration characteristics of pollen for application to long-term storage. *Euphytica* 68:77–84

25. Cooper JA. 1987. Effects of cytochalasin

and phalloidin on actin. *J. Cell Biol.* 105: 1473–78

26. Cresti M, Tiezzi A. 1990. Germination and pollen-tube Formation. In *Microspores: Evolution and Ontogeny,* ed. S Blackmore, RB Knox, pp. 239–63. New York: Academic

27. Crowe JH, Hoekstra FA, Crowe LM. 1989. Membrane phase transitions are responsible for imbibitional damage in dry pollen. *Proc. Natl. Acad. Sci. USA* 86:520–23

28. Deboo GB, Albertsen MC, Taylor LP. 1995. Flavanone 3-hydroxylase transcripts and flavonol accumulation are temporally coordinate in maize anthers. *Plant J.* 7:703–13

29. Derksen J, Rutten T, Lichtscheidl IK, de Win AHN, Pierson ES, Rongen G. 1995. Quantitative analysis of the distribution of organelles in tobacco pollen tubes: implications for exocytosis and endocytosis. *Protoplasma* 188:267–76

30. Derksen J, Rutten T, Van Amstel T, de Win A, Doris F, Steer M. 1995. Regulation of pollen tube growth. *Acta Bot. Neerl.* 44: 93–119

31. Dircks LK, Vancanneyt G, McCormick S. 1996. Biochemical characterization and baculovirus expression of the pectate lyase-like LAT56 and LAT59 pollen proteins of tomato. *Plant Physiol. Biochem.* 34:1–12

32. Doris FP, Steer MW. 1996. Effects of fixatives and permeabilisation buffers on pollen tubes: implications for localisation of actin microfilaments using phalloidin staining. *Protoplasma.* 195:25–36

33. Doughty J, Hedderson F, McCubbin A, Dickinson H. 1993. Interaction between a coat-borne peptide of the *Brassica* pollen grain and stigmatic S (self-incompatibility)-locus-specific glycoproteins. *Proc. Natl. Acad. Sci. USA* 90:467–71

34. Drøbak BK, Watkins PAC, Valenta R, Dove S, Lloyd CW, Staiger CJ. 1994. Inhibition of plant plasma membrane phosphoinositide phospholipase C by the actin-binding protein, profilin. *Plant J.* 6:389–400

35. Du H, Simpson RJ, Clarke AE, Bacic A. 1996. Molecular characterization of a stigma-specific gene encoding an arabinogalactan-protein (AGP) from *Nicotiana alata. Plant J.* 9:313–23

36. Du H, Simpson RJ, Moritz RL, Clarke AE, Bacic A. 1994. Isolation of the protein backbone of an arabinogalactan-protein from the styles of *Nicotiana alata* and characterization of a corresponding cDNA. *Plant Cell* 6:1643–53

37. Dubald M, Barakate A, Mandaron P, Mache R. 1993. The ubiquitous presence of

exopolygalacturonase in maize suggests a fundamental cellular function for this enzyme. *Plant J.* 4:781–91

38. Duck NB, Folk WR. 1994. Hsp 70 heat shock protein cognate is expressed and stored in developing tomato pollen. *Plant Mol. Biol.* 26:1031–39

39. Eady C, Lindsey K, Twell D. 1994. Differential activation and conserved vegetative cell-specific activity of a late pollen promoter in species with bicellular and tricellular pollen. *Plant J.* 5:543–50

40. Elleman CJ, Franklin-Tong V, Dickinson HG. 1992. Pollination in species with dry stigmas: the nature of the early stigmatic response and the pathway taken by pollen tubes. *New Phytol.* 121:413–24

41. Estruch JJ, Kadwell S, Merlin E, Crossland L. 1994. Cloning and characterization of a maize pollen-specific calcium-dependent calmodulin-independent protein kinase. *Proc. Natl. Acad. Sci. USA* 91:8837–41

42. Feijó JA, Malhó R, Obermeyer G. 1995. Ion dynamics and its possible role during *in vitro* pollen germination and tube growth. *Protoplasma* 187:155–67

43. Frankis RCJ. 1990. RNA and protein synthesis in germinating pine pollen. *J. Exp. Bot.* 41:1469–73

44. Franklin-Tong VE, Drøbak BK, Allan AC, Watkins PAC, Trewavas AJ. 1996. Growth of pollen tubes of *Papaver rhoeas* is regulated by a slow-moving calcium wave propagated by inositol 1,4,5-triphosphate. *Plant Cell* 8:1305–21

45. Geitmann A, Hudak J, Vennigerholz F, Walles B. 1995. Immunogold localization of pectin and callose in pollen grains and pollen tubes of *Brugmansia suaveolens:* implications for the self-imcompatibility reaction. *J. Plant Physiol.* 147:225–35

46. Geitmann A, Li Y-Q, Cresti M. 1996. The role of the cytoskeleton and dictyosome activity in the pulsatory growth of *Nicotiana tabacum* and *Petunia hybrida* pollen tubes. *Bot. Acta* 109:102–9

47. Gerster J, Allard S, Robert LS. 1996. Molecular characterization of two *Brassica napus* pollen-expressed genes encoding putative arabinogalactan proteins. *Plant Physiol.* 110:1231–37

48. Gilissen LJW. 1977. The influence of relative humidity on the swelling of pollen grains *in vitro. Planta* 137:299–301

49. Golan-Goldhirsh A, Schmidhalter U, Muller M, Oertli JJ. 1991. Germination of *Pistacia vera* L. pollen in liquid medium. *Sex. Plant Reprod.* 4:182–87

50. Grote M, Swoboda I, Meagher RB, Valenta R. 1995. Localization of profilin- and actin-

like immunoreactivity in *in vitro*-germinated tobacco pollen tubes by electron microscopy after special water-free fixation techniques. *Sex. Plant Reprod.* 8:180–86

51. Hanson DD, Hamilton DA, Travis JL, Bashe DM, Mascarenhas JP. 1989. Characterization of a pollen-specific cDNA clone from *Zea mays* and its expression. *Plant Cell* 1:173–79

52. He Y, Wetzstein HY. 1995. Fixation induces differential tip morphology and immunolocalization of the cytoskeleton of pollen tubes. *Plant Physiol.* 93:757–73

53. Heilmann J, Strack D. 1990. Incorporation of kaempferol 3-*O*-glucoside into the cell walls of Norway spruce needles. *Planta* 181:599–603

54. Heslop-Harrison J. 1987. Pollen germination and pollen-tube growth. In *Pollen: Cytology and Development*, ed. GH Bourne, KW Jeon, M Friedlander, pp. 1–78. New York: Academic

55. Heslop-Harrison J, Heslop-Harrison Y. 1990. Dynamic aspects of apical zonation in the angiosperm pollen tube. *Sex. Plant Reprod.* 3:187–94

56. Heslop-Harrison J, Heslop-Harrison Y. 1991. Restoration of movement and apical growth in the angiosperm pollen tube following cytochalasin-induced paralysis. *Philos. Trans. R. Soc. London Ser. B.* 331: 225–35

57. Heslop-Harrison Y, Heslop-Harrison J. 1992. Germination of monocolpate angiosperm pollen: evolution of the actin cytoskeleton and wall during hydration, activation and tube emergence. *Ann. Bot.* 69: 385–94

58. Hiscock SJ, Dewey FM, Coleman JOD, Dickinson HG. 1994. Identification and localization of an active cutinase in the pollen of *Brassica napus* L. *Planta* 193:377–84

59. Hiscock SJ, Doughty J, Dickinson HG. 1995. Synthesis and phosphorylation of pollen proteins during the pollen-stigma interaction in self-compatible *Brassica napus* L. and self-incompatible *Brassica oleracea* L. *Sex. Plant Reprod.* 8:345–53

60. Hodge R, Paul W, Scott R. 1992. Cold-plaque screening: a simple technique for the isolation of low abundance, differentially expressed transcripts from conventional cDNA libraries. *Plant J.* 2:257–60

61. Hoekstra FA, Crowe LM, Crowe JH. 1989. Differential dessication sensitivity of corn and *Pennisetum* pollen linked to their sucrose contents. *Plant Cell Environ.* 12: 83–91

62. Huang AH. 1996. Oleosins and oil bodies in seeds and other organs. *Plant Physiol.* 110:1055–61

63. Huang S, McDowell JM, Weise MJ, Meagher RB. 1996. The *Arabidopsis* profilin gene family. Evidence for an ancient split between constitutive and pollen-specific profilin genes. *Plant Physiol.* 111: 115–26

64. Hulskamp M, Kopczak SD, Horejsi TF, Kihl BK, Pruitt RE. 1995. Identification of genes required for pollen-stigma recognition in *Arabidopsis thaliana. Plant J.* 8: 703–14

65. Jauh GY, Lord EM. 1996. Localization of pectins and arabinogalactan-proteins in lily (*Lilium longiflorum* L.) pollen tube and style, and their possible roles in pollination. *Planta* 199:251–61

66. Johnson LC, Critchfield WB. 1974. A white-pollen variant of bristlecone pine. *J. Hered.* 65:123

67. Joos U, van Aken J, Kristen U. 1994. Microtubules are involved in maintaining cellular polarity in pollen tubes of *Nicotiana sylvestris. Protoplasma* 179:5–15

68. Joos U, van Aken J, Kristen U. 1995. The anti-microtubule drug carbetamide stops *Nicotiana sylvestris* pollen tube growth in the style. *Protoplasma* 187:182–91

69. Kaminskyj SG, Jackson SL, Heath IB. 1992. Fixation induces differential polarized translocation of organelles in hyphae of *Saprolegnia ferax. J. Microsc.* 167: 153–68

70. Karakesisoglou I, Schleicher M, Gibbon BC, Staiger CJ. 1996. Plant profilins rescue the aberrant phenotype of profilin-deficient *Dictyostelium* cells. *Cell Motil. Cytoskel.* 34:36–47

71. Kawaguchi K, Shibuya N, Ishii T. 1996. A novel tetrasaccharide, with a structure similar to the terminal sequence of an arabinogalactan-protein, accumulates in rice anthers in a stage-specific manner. *Plant J.* 9:777–85

72. Kerhoas C, Gay G, Dumas C. 1987. A multidisciplinary approach to the study of the plasma membrane of *Zea mays* pollen during controlled dehydration. *Planta* 171: 1–10

73. Knöfel H-D, Sembdner G. 1995. Jasmonates from pine pollen. *Phytochemistry* 38: 569–71

74. Kohno T, Shimmen T. 1988. Mechanism of Ca^{2+} inhibition of cytoplasmic streaming in lily pollen tubes. *J. Cell Sci.* 95:501–9

75. Kohno T, Shimmen T. 1988. Accelerated sliding of pollen tube organelles along Characeae actin bundles regulated by Ca^{2+}. *J. Cell Biol.* 106:1539–43

76. Koornneef M, Hanhart CJ, Thiel F. 1989. A genetic and phenotypic description of *Eceriferum* (*cer*) mutants in *Arabidopsis thaliana*. *J. Hered.* 80:118–22

77. Kühtreiber W, Jaffe LF. 1990. Detection of extracellular calcium gradients with a calcium-specific vibrating electrode. *J. Cell Biol.* 110:1565–73

78. Kulikauskas R, Hou A, Muschietti J, McCormick S. 1995. Comparisons of diverse plant species reveal that only grasses show drastically reduced levels of ubiquitin monomer in mature pollen. *Sex. Plant Reprod.* 8:326–32

79. Lancelle SA, Cresti M, Hepler PK. 1987. Ultrastructure of the cytoskeleton in freeze-substituted pollen tubes of *Nicotiana alata*. *Protoplasma* 140:141–50

80. Lancelle SA, Cresti M, Hepler PK. 1996. Growth inhibition and recovery in freeze-substituted *Lilium longiflorum* pollen tubes: structural effects of caffeine. *Protoplasma*. In press

81. Lancelle SA, Hepler PK. 1988. Cytochalasin-induced ultrastructure alterations in *Nicotiana* pollen tubes. *Protoplasma* (Suppl. 2):65–75

82. Lancelle SA, Hepler PK. 1991. Association of actin with cortical microtubules revealed by immunogold localization in *Nicotiana* pollen tube. *Protoplasma* 165:167–72

83. Lancelle SA, Hepler PK. 1992. Ultrastructure of freeze-substituted pollen tubes of *Lilium longiflorum*. *Protoplasma* 167: 215–30

84. Lee H-S, Karunanandaa B, McCubbin A, Gilroy S, Kao T-H. 1996. PRK1, a receptor-like kinase of *Petunia inflata*, is essential for postmeiotic development of pollen. *Plant J.* 9:613–24

85. Li Y-Q, Bruun L, Pierson ES, Cresti M. 1992. Periodic deposition of arabinogalactan epitopes in the cell wall of pollen tubes of *Nicotiana tabacum* L. *Planta* 188: 532–38

86. Li Y-Q, Chen F, Linskens HF, Cresti M. 1994. Distribution of unesterfied and esterfied pectins in cell walls of pollen tubes of flowering plants. *Sex. Plant Reprod.* 7: 145–52

87. Li Y-Q, Linskens HF. 1983. Wall-bound proteins of pollen tubes after self-and cross-pollination in *Lilium longiflorum*. *Theor. Appl. Genet.* 67:11–16

88. Li Y-Q, Zhang H-Q, Pierson ES, Huang F-Y, Linskens HF, Hepler PK, Cresti M. 1996. Enforced growth-rate fluctuation causes pectin ring formation in the cell wall of *Lilium longiflorum* pollen tubes. *Planta* 200:41–49

89. Lin JJ, Dickinson DB. 1984. Ability of pollen to germinate prior to anthesis and effect of desiccation on germination. *Plant Physiol.* 74:746–48

90. Lin Y-K, Wang Y-L, Zhu J-K, Yang Z-B. 1996. Localization of a Rho GTPase implies a role in tip growth and movement of the generative cell in pollen tubes. *Plant Cell* 8:293–303

91. Lind JL, Bacic A, Clarke AE, Anderson MA. 1994. A style-specific hydroxyproline-rich glycoprotein with properties of both extensins and arabinogalactan proteins. *Plant J.* 6:491–502

92. Lind JL, Bonig I, Clarke AE, Anderson MA. 1996. A style-specific 120-kDa glycoprotein enters pollen tubes of *Nicotiana alata* in vivo. *Sex. Plant Reprod.* 9: 75–86

93. Long SR. 1989. Rhizobium-legume nodulation: life together in the underground. *Cell* 56:203–14

94. Malhó R, Read ND, Pais MS, Trewavas AJ. 1994. Role of cytosolic free calcium in the reorientation of pollen tube growth. *Plant J.* 5:331–41

95. Malhó R, Read ND, Trewavas AJ, Pais MS. 1995. Calcium channel activity during pollen tube growth and reorientation. *Plant Cell* 7:1173–84

95a. Malhó R, Trewavas AJ. 1996. Localized apical increases of cytosolic free calcium control pollen tube orientation. *Plant Cell* 8:1935–49

96. Mandava NB. 1988. Plant growth-promoting brassinosteroids. *Annu. Rev. Plant Physiol. Plant Mol. Biol.* 39:23–52

97. Marsolier MC, Carrayol E, Hirel B. 1993. Multiple functions of promoter sequences involved in organ-specific expression and ammonia regulation of a cytosolic soybean glutamine synthetase gene in transgenic *Lotus corniculatus*. *Plant J.* 3:405–14

98. Mascarenhas JP. 1975. The biochemistry of angiosperm pollen development. *Bot. Rev.* 41:259–314

99. Mascarenhas JP. 1990. Gene activity during pollen development. *Annu. Rev. Plant Physiol. Plant Mol. Biol.* 41:317–38

100. Mascarenhas JP. 1993. Molecular mechanisms of pollen tube growth and differentiation. *Plant Cell* 5:1303–14

101. Mascarenhas JP, Mermelstein J. 1981. Messenger RNAs: their utilization and degradation during pollen germination and tube growth. *Acta Soc. Bot. Pol.* 50:13–20

102. Mascarenhas NT, Basche D, Eisenberg A, Willing RP, Xiao C-M, Mascarenhas JP. 1984. Messenger RNAs in corn pollen and protein synthesis during germination and

pollen tube growth. *Theor. Appl. Genet.* 68:323–26

103. McCormick S. 1991. Molecular analysis of male gametogenesis in plants. *Trends Genet.* 7(9):298–303

104. McCormick S. 1993. Male gametophyte development. *Plant Cell* 5:1265–75

105. Meikle PJ, Bonig I, Hoogenraad NJ, Clarke AE, Stone BA. 1991. The location of (1→3)-β-glucans in the walls of pollen tubes of *Nicotiana alata* using a (1→3)-β-glucan-specific monoclonal antibody. *Planta* 185:1–8

106. Miller DD, Lancelle SA, Hepler PK. 1996. Actin microfilaments do not form a dense meshwork in Lilium longiflorum pollen tube tips. *Protoplasma.* 195:123–32

107. Miller DD, Scordilis SP, Hepler PK. 1995. Identification and localization of three classes of myosins in pollen tubes of *Lilium longiflorum* and *Nicotiana alata. J. Cell Sci.* 198:2549–53

108. Mitterman I, Swoboda I, Pierson E, Eller N, Kraft D, et al. 1995. Molecular cloning and characterization of profilin from tobacco (*Nicotiana tabacum*): increased profilin expression during pollen maturation. *Plant Mol. Biol.* 27:137–46

109. Miyake T, Kuroiwa H, Kuroiwa T. 1995. Differential mechanisms of movement between generative cell and vegetative nucleus in pollen tubes of Nicotiana tabacum as revealed by additions of colchicine and nonanoic acid. *Sex. Plant Reprod.* 8: 228–30

110. Mo Y, Nagel C, Taylor LP. 1992. Biochemical complementation of chalcone synthase mutants defines a role for flavonols in functional pollen. *Proc. Natl. Acad. Sci. USA* 89:7213–17

111. Mu JH, Lee HS, Kao TH. 1994. Characterization of a pollen-expressed receptor-like kinase gene of *Petunia inflata* and the activity of its encoded kinase. *Plant Cell* 6:709–21

112. Mu JH, Stains JP, Kao TH. 1994. Characterization of a pollen-expressed gene encoding a putative pectin esterase of *Petunia inflata. Plant Mol. Biol.* 25:539–44

113. Muschietti J, Dircks L, Vancanneyt G, McCormick S. 1994. LAT52 protein is essential for tomato pollen development: pollen expressing antisense LAT52 RNA hydrates and germinates abnormally and cannot achieve fertilization. *Plant J.* 6:321–38

114. O'Driscoll D, Hann C, Read SM, Steer MW. 1993. Endocytotic uptake of florescent dextrans by pollen tubes grown *in vitro. Protoplasma* 175:126–30

115. Pierson ES, Cresti M. 1992. Cytoskeleton and cytoplasmic organization of pollen and pollen tubes. *Int. Rev. Cytol.* 140:73–125

116. Pierson ES, Li Y-Q, Zhang H-Q, Willemse MTM, Linskens HF, Cresti M. 1995. Pulsatory growth of pollen tubes: investigation of a possible relationship with the periodic distribution of cell wall components. *Acta Bot. Neerl.* 44:121–28

117. Pierson ES, Lichtscheidl IK, Derksen J. 1990. Structure and behavior of organelles in living pollen tubes of *Lilium longiflorum. J. Exp. Bot.* 41:1461–68

118. Pierson ES, Miller DD, Callaham DA, Shipley AM, Rivers BA, et al. 1994. Pollen tube growth is coupled to the exracellular calcium ion influx and the intracellular calcium gradient: effect of BAPTA-type buffers and hypertonic media. *Plant Cell* 6: 1815–28

119. Pierson ES, Miller DD, Callaham DA, van Aken J, Hackett G, Hepler PK. 1996. Tip-localized calcium entry fluctuates during pollen tube growth. *Dev. Biol.* 174:160–73

120. Pollak PE, Hansen K, Astwood JD, Taylor LP. 1995. Conditional male fertility in maize. *Sex. Plant Reprod.* 8:231–41

121. Pollak PE, Vogt T, Mo Y, Taylor LP. 1993. Chalcone synthase and flavonol accumulation in stigmas and anthers of *Petunia hybrida. Plant Physiol.* 102:925–32

122. Preuss D, Davis RW. 1995. Cell signaling during fertilization: Interactions between the pollen coat and the stigma. *J. Cell Biochem.* 19:140

123. Preuss D, Lemieux B, Yen G, Davis RW. 1993. A conditional sterile mutation eliminates surface components from *Arabidopsis* pollen and disrupts cell signaling during fertilization. *Genes Dev.* 7:974–85

124. Priestly DA, Werner BG, Leopold AC, McBride MB. 1985. Organic free radical levels in seeds and pollen: The effects of hydration and aging. *Physiol. Plant.* 64: 88–94

125. Read SM, Clarke AE, Bacic A. 1993. Stimulation of growth of cultured *Nicotiana tabacum* W38 pollen tubes by poly (ethylene glycol) and $Cu_{(II)}$ salts. *Protoplasma* 177:1–14

126. Roberts MR, Hodge R, Scott R. 1995. *Brassica napus* pollen oleosins posses a characteristic C-terminal domain. *Planta* 195:469–70

127. Ross JHE, Murphy DJ. 1996. Characterization of anther-expressed genes encoding a major class of extracellular oleosin-like proteins in the pollen coat of Brassicaceae. *Plant J.* 9:625–37

128. Rubinstein AL, Broadwater AH, Lowrey

KB, Bedinger PA. 1995. *Pex1*, a pollen-specific gene with an extensin-like domain. *Proc. Natl. Acad. Sci. USA* 92:3086–90

129. Rubinstein AL, Marquez J, Suarez-Cervera M, Bedinger PA. 1995. Extensin-like glycoproteins in the maize pollen tube wall. *Plant Cell* 7:2211–25

130. Sanders LC, Lord EM. 1992. A dynamic role for the stylar matrix in pollen tube extension. *Int. Rev. Cytol.* 140:297–318

131. Sarker RH, Elleman CJ, Dickinson HG. 1988. Control of pollen hydration in *Brassica* requires continued protein synthesis, and glycosylation is necessary for intraspecific incompatiblity. *Proc. Natl. Acad. Sci. USA* 85:4340–44

132. Sedgley M. 1975. Flavonoids in pollen and stigma of *Brassica oleraceae* and their effects on pollen germination *in vitro*. *Ann. Bot.* 39:1091–95

133. Shivanna KR, Linskens HF, Cresti M. 1991. Responses of tobacco pollen to high humidity and heat stress: viability and germinability in vitro and in vivo. *Sex. Plant Reprod.* 4:104–9

134. Singh S, Sawhney VK. 1992. Plant hormones in *Brassica napus* and *Lycopersicon esculetum* pollen. *Phytochemistry* 31: 4051–53

135. Stafford HA. 1991. Flavonoid evolution: an enzymatic approach. *Plant Physiol.* 96: 680–85

136. Staiger CJ, Goodbody KC, Hussey PJ, Valenta R, Drøbak BK, Lloyd CW. 1993. The profilin multigene family of maize: differential expression of three isoforms. *Plant J.* 4:631–41

137. Steer MW, Steer JM. 1989. Pollen tube tip growth. *New Phytol.* 111:323–58

138. Subbaiah CC. 1984. A polyethylene glycol based medium for *in vitro* germination of cashew pollen. *Can. J. Bot.* 62:2473–75

139. Takahara K, Yokota E, Shimmen T. 1995. Isolation of 135-kDa actin bundling protein from pollen tubes of lily. *Plant Cell Physiol.* 36:S132

140. Tang XJ, Hepler PK, Scordilis SP. 1989. Immunochemical and immunocytochemical identification of myosin heavy chain polypeptide in *Nicotiana* pollen tubes. *J. Cell Sci.* 92:569–74

141. Tang XW, Liu GQ, Yang Y, Zheng WL, Wu WL, Nie DT. 1992. Quantitative measurement of pollen tube growth and particle movement. *Acta Bot. Sin.* 34:893–98

142. Taylor LP, Jorgensen R. 1992. Conditional male fertility in chalcone synthase-deficient petunia. *J. Hered.* 83:11–17

143. Tebbutt SJ, Lonsdale DM. 1995. Deletion analysis of a tobacco pollen-specific polygalacturonase promoter. *Sex. Plant Reprod.* 8:224–46

144. Theerakulpisut P, Xu H, Singh MB, Pettitt JM, Knox RB. 1991. Isolation and development expression in Bcp 1, an anther-Specific cDNA clone in *Brassica campestris*. *Plant Cell* 3:1073–84

145. Tiezzi A, Moscatelli A, Cai G, Bartalesi A, Cresti M. 1992. An immunoreactive homolog of mammalian kinesin in *Nicotiana* pollen tube. *Cell Motil. Cytoskel.* 21: 132–37

146. Tirlapur UK, Cai G, Faleri C, Moscatelli A, Scali M, et al. 1995. Confocal imaging and immunogold electron microscopy of changes in distribution of myosin during pollen hydration, germination and pollen tube growth in *Nicotiana tabacum* L. *Eur. J. Cell Biol.* 67:209–17

147. Tiwari SC, Polito VS. 1988. Organization of the cytoskeleton in pollen tubes of *Pyrus communis*: a study employing conventional and freeze-substitution electron microscopy, immunofluorescence, and rhodamine-phalloidin. *Protoplasma* 147:100–12

148. Tiwari SC, Polito VS, Webster BD. 1990. In dry pear (*Pyrus comminus* L.) pollen, membranes assume a tightly packed multilamellate aspect that disappears rapidly upon hydration. *Protoplasma* 153:157–68

149. Touraev A, Fink CS, Stoger E, Heberle-Bors E. 1995. Pollen selection: a transgenic reconstruction approach. *Proc. Natl. Acad. Sci. USA* 92:12165–69

150. Tully RE, Musgrave ME, Leopold AC. 1981. The seed coat as a control of imbibitional chilling injury. *Crop Sci.* 21:312–17

151. Twell D. 1992. Use of a nuclear-targeted β-glucuronidase fusion protein to demonstrate vegetative cell-specific gene expression in developing pollen. *Plant J.* 6: 887–92

152. Twell D. 1994. The diversity and regulation of a gene expression in the pathway of male gametophyte development. In *Molecular and Cellular Aspects of Plant Reproduction*, ed. RJ Scott, AD Stead, pp. 83–135. Cambridge: Cambridge Univ. Press

153. Ursin VM, Yamaguchi J, McCormick S. 1989. Gametophytic and sporophytic expression of anther-specific genes in developing tomato anthers. *Plant Cell* 1:727–36

154. Valenta R, Duchene M, Pettenburger K, Sillaber C, Valent P, et al. 1991. Identification of profilin as a novel pollen allergen; IgE autoreactivity in sensitized individuals. *Science* 253:557–60

155. Valenta R, Ferriera F, Grote M, Swoboda I,

Vrtala S, et al. 1993. Identification of profilin as an actin-binding protein in higher
plants. *J. Biol. Chem.* 268:22777–81
156. Van Aelst AC, Van Went JL. 1992. Ultrastructural immuno-localization of pectins
and glycoproteins in *Arabidopsis thaliana*
pollen grains. *Protoplasma* 168:14–19
157. van Bilsen DG, van Roekel T, Hoekstra FA.
1994. Declining viability and lipid degradation during pollen storage. *Sex. Plant
Reprod.* 7:303–10
158. van Tunen AJ, Mur LA, Brouns GS, Rien-
stra J-D, Koes RE, Mol JNM. 1990. Pollen-
and anther-specific *chi* promoters from petunia: tandem promoter regulation of the
chiA gene. *Plant Cell* 2:393–401
159. Vidali L, Hepler PK. 1996. Characterization and localization of profilin in pollen grains and tubes of *Lilium longiflorum*.
Cell Motil. Cytoskel. In press
160. Vogt T, Pollak P, Tarlyn N, Taylor LP. 1994.
Pollination- or wound-induced kaempferol
accumulation in petunia stigmas enhances
seed production. *Plant Cell* 6:11–23
161. Vogt T, Taylor LP. 1995. Flavonol 3-*O*-gly-
cosyltransferases associated with petunia
pollen produce gametophyte-specific flavonol diglycosides. *Plant Physiol.* 108:
903–11
162. Vogt T, Wollenweber E, Taylor LP. 1995.
The structural requirements of flavonols
that induce pollen germination of condi-
tionally male fertile *Petunia*. *Phytochemistry* 38:589–92
163. Wang C-S, Walling LL, Eckard KJ, Lord
EM. 1992. Patterns of proteins accumulation in developing anthers of *Lilium longiflorum* correlate with histological events.
Am. J. Bot. 79:118–27
164. Wang H, Wu H-M, Cheung AY. 1996. Pollination induces mRNA poly(A) tail-short-
ening and cell deterioration in flower transmitting tissue. *Plant J.* 9:715–27
165. Wang H, Wu H-M, Cheung AY. 1993. Development and pollination regulated accumulation and glycosylation of a stylar
transmitting tissue-specific proline-rich
protein. *Plant Cell* 5:1639–50
166. Weterings K, Reijnen W, van Aarssen R,
Kortstee A, Spijkers J, et al. 1992. Characterization of a pollen-specific cDNA clone
from *Nicotiana tabacum* expressed during
microgametogenesis and germination.
Plant Mol. Biol. 18:1101–11
167. Weterings K, Reijnen W, Wijn G, van de
Heuvel K, Appeldoorn N, et al. 1995. Molecular characterization of the pollen-specific genomic clone NTPg303 and in situ
localization of expression. *Sex. Plant Reprod.* 8:11–17
168. Wiermann R, Gubatz S. 1992. Pollen wall
and sporopollenin. *Int. Rev. Cytol.* 140:35–
72
169. Wu HM, Wang H, Cheung AY. 1995. A
pollen tube growth stimulatory glycoprotein is deglycosylated by pollen tubes and
displays a glycosylation gradient in the
flower. *Cell* 82:395–403
170. Xu H-L, Davies SP, Kwan BYH, O'Brien
AP, Singh M, Knox RB. 1993. Haploid and
diploid expression of a *Brassica campestris*
anther-specific gene promoter in *Arabidopsis* and tobacco. *Mol. Gen. Genet.* 239:
58–65
171. Xu P, Vogt T, Taylor LP. 1996. Uptake and
metabolism of flavonols during in vitro germination and tube growth of *Petunia hybrida* pollen. *Planta.* In press
172. Yen L-F, Liu X-O, Cai ST. 1995. Polymerization of actin from maize pollen. *Plant
Physiol.* 107:73–76
173. Ylstra B. 1995. *Molecular Control of Fertilization in Plants.* Amsterdam: Acad.
Proefskr. Vrije Univ.
174. Ylstra B, Busscher J, Franken J, Hollman
PCH, Mol JNM, van Tunen AJ. 1994. Flavonols and fertilization in *Petunia hybrida*:
localization and mode of action during pollen tube growth. *Plant J.* 6:201–12
175. Yokota E, McDonald AR, Liu B, Shimmen
T, Palevitz BA. 1995. Localization of a
170-kDa myosin heavy chain in plant cells.
Protoplasma 185:178–87
176. Yokota E, Shimmen T. 1994. Isolation and
characterization of plant myosin from pollen tubes of lily. *Protoplasma* 177:153–62
177. Zerback R, Bokel M, Geiger H, Hess D.
1989. A kaempferol 3-glucosylgalactoside
and further flavonoids from pollen of *Petunia hybrida*. *Phytochemistry* 28:897–99
178. Zhang H-Q, Croes AF. 1982. A new medium for pollen germination *in vitro. Acta
Bot. Neerl.* 31:113–19
179. Zou J-T, Zhan X-Y, Wu H-M, Wang H,
Cheung AY. 1994. Characterization of a
rice pollen-specific gene and its expression.
Am. J. Bot. 81:552–61
180. Zuberi MI, Dickinson HG. 1985. Pollen-
stigma interaction in *Brassica*. III. hydration of the pollen grains. *Cell Sci.* 76:
321–36

Annu. Rev. Plant Physiol. Plant Mol. Biol. 1997. 48:493–523

METABOLITE TRANSPORT ACROSS SYMBIOTIC MEMBRANES OF LEGUME NODULES

Michael K. Udvardi and David A. Day

Division of Biochemistry and Molecular Biology, Faculty of Science, Australian National University, Canberra ACT, 0200, Australia

KEY WORDS: rhizobia, legumes, nitrogen fixation, bacteroids, peribacteroid membrane, symbiosome

ABSTRACT

Infection of legume roots or stems with soil bacteria of the Rhizobiaceae results in the formation of nodules that become symbiotic nitrogen-fixing organs. Within the infected cells of these nodules, bacteria are enveloped in a membrane of plant origin, called the peribacteroid membrane (PBM), and divide and differentiate to form nitrogen-fixing bacteroids. The organelle-like structure comprised of PBM and bacteroids is termed the symbiosome, and is the basic nitrogen-fixing unit of the nodule. The major exchange of nutrients between the symbiotic partners is reduced carbon from the plant, to fuel nitrogenase activity in the bacteroid, and fixed nitrogen from the bacteroid, which is assimilated in the plant cytoplasm. However, many other metabolites are also exchanged. The metabolic interaction between the plant and the bacteroids is regulated by a series of transporters and channels on the PBM and the bacteroid membrane, and these form the focus of this review.

CONTENTS

1040-2519/97/0601-0493$08.00

INTRODUCTION

Endocytobiotic associations (associations between two organisms in which one organism lives in the cell cytoplasm of the other) are widespread in nature. They include associations between algae and invertebrates, bacteria and insects, bacteria and invertebrates, and bacteria and plants. One of the most studied, and perhaps the best understood, of all endocytobiotic associations is that which involves plants and diazotrophic (nitrogen-fixing) soil bacteria of the family Rhizobiaceae. In these symbioses, the bacteria reduce N_2 to ammonia and exchange this for reduced carbon compounds from the plant. The reduced carbon is used as fuel for bacterial metabolism, including nitrogen fixation, and the ammonia enables legumes and many nonlegumes to grow in soils that have little or no available nitrogen. Symbiotic nitrogen fixation has, therefore, both ecological and agricultural significance.

A common feature of many endocytobiotic associations is a host-derived membrane surrounding the microsymbiont, which forms an organelle-like structure termed the symbiosome (134). This membrane, called the symbiosome membrane, separates the microsymbiont from the host cell cytoplasm and is not only a structural barrier between the symbionts but is also a functional barrier. The symbiosome membrane in nitrogen-fixing legume root nodules, more commonly called the peribacteroid membrane (PBM) because it surrounds bacteroids (the symbiotic, nitrogen-fixing form of rhizobia; Figure 1), regulates the transport of metabolites between the symbiotic partners and is essential for a stable symbiosis. The peribacteroid and bacteroid membranes govern the nature of the exchanges between plant and bacteroid and facilitate the metabolic coupling between the two. This review focuses on current

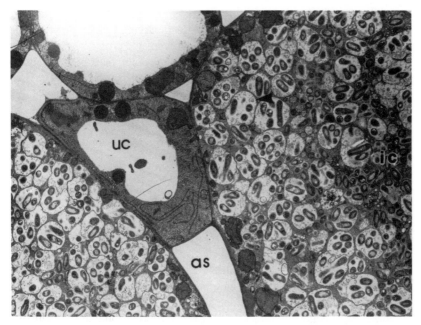

Figure 1   Electron micrograph showing cells in the infected zone of a soybean root nodule. The infected cells containing large numbers of symbiosomes are clearly seen on either side of an uninfected cell (*uc*) and an air space (*as*). The peribacteroid membrane of a symbiosome is indicated (*arrow head*) in the right-hand infected cell.

knowledge about the biochemistry and molecular biology of these important membranes.

Development of Nitrogen-Fixing Symbioses Between Legumes and Rhizobia

The interaction between rhizobia[1] in the soil and the legume root begins with an exchange of chemical signals resulting in colonization of root hairs, initiation of root cell division, and nodule formation. The initial signaling events confer specificity on the interaction, such that a given rhizobia species interacts with a limited range of plant species. The molecular basis of these interactions has been the focus of much exciting and fruitful research in recent years

1
    The term rhizobia refers to the nitrogen-fixing genera of the family Rhizobiaceae, which include *Rhizobium*, *Bradyrhizobium*, *Sinorhizobium*, and *Azorhizobium*.

(28, 46). Physical contact between rhizobia and the root hair cell is followed by root hair curling and formation of an infection thread, an outgrowth of the plant cell wall that ramifies through the root cortex and provides access for the infecting rhizobia (for a review, see 128). At some point, in response to signals that are poorly understood, the infection thread degrades and the bacteria are taken into the cells of the developing nodule via endocytosis of the plasma membrane; the symbiosome so formed becomes the fundamental nitrogen-fixing unit of the nodule (12, 134). Once inside the root cell, the bacteria divide and differentiate to form the nitrogen-fixing bacteroids. Simultaneously, the PBM proliferates and differentiates until the infected cell is filled with symbiosomes (Figure 1).

In the mature legume nodule, the enlarged infected cells house many thousands of symbiosomes (10). In some legumes, PBM division is coordinated with bacteroid division, and symbiosomes contain a single bacteroid (124). In other legumes, PBM division is not so tightly coupled to that of the bacteria, and symbiosomes may consist of many bacteroids within a single envelope (10). However, in both cases the PBM represents the physical interface between the symbionts, and all exchanges and communication must take place across this membrane. As mentioned above, the PBM is essential for a stable symbiosis, and degradation of the PBM leads to senescence of the symbiosis (177).

Proliferation of the PBM within the infected cells requires massive synthesis of lipids and proteins, with at least some of the latter being unique to the PBM (166). This synthesis seems to involve the golgi and endoplasmic reticulum in the usual manner, but with the involvement of nodule-specific isoforms of some of the enzymes (167). In the mature nodule, the PBM represents a novel membrane with properties resembling a mixture of the plasma membrane and the tonoplast (18, 98). The entire infected cell of legume nodules differentiates during symbiosome proliferation: Infected cells are generally much larger than normal, and their carbon and nitrogen metabolism becomes highly specialized (24, 149, 164).

The primary metabolic exchange between the plant and bacteroids is reduced carbon for reduced nitrogen. The nature of this exchange has been the focus of much research in the past decade, and a simple model of carbon and nitrogen transport across the PBM has emerged. However, the complete picture of metabolite exchange between the symbionts is likely to be quite complicated, and we have much yet to learn about this aspect of the symbiosis. In the following sections we focus first on recent advances in our understanding of carbon/nitrogen exchange and its regulation, and then consider the transport of other metabolites between legumes and bacteroids.

CARBON SUPPLY TO THE BACTEROIDS

Sucrose from the shoot is the principal source of reduced carbon for the nodule (84). It is metabolized via the action of sucrose synthase and glycolytic enzymes, mainly in the plant cytoplasm of the infected zone of the nodule (150). Many carbon compounds are plentiful in legume nodules, and it is possible that more than one source of carbon is provided to support nitrogen fixation. It is also likely that the nature of these compounds changes with the development of the nodule and may vary between species. However, it is generally accepted that C_4-dicarboxylates are the main products of sucrose degradation supplied to bacteroids to support nitrogen fixation in most legumes. Evidence comes from studies with both plants and bacteroids.

Inferences from Studies of Bacteroid Carbon Metabolism

Biochemical and genetic experiments have led to insights into the physicochemical environment that bacteroids inhabit inside nodules. For example, investigations of the activities of substrate-inducible enzymes in bacteroids have provided a qualitative measure of substrate concentrations in the nodule, and studies with specific bacterial mutants have highlighted the involvement of an enzyme/pathway in symbiotic nitrogen fixation.

The Entner-Doudoroff pathway is the major route for glucose catabolism in rhizobia in culture, although the Embden-Meyerhof and pentose phosphate pathways also operate in *Rhizobium* (30, 39, 93, 120, 145). However, bacteroids cannot utilize di- or monosaccharides effectively. *Bradyrhizobium japonicum* bacteroids isolated from soybean root nodules cannot transport glucose (121) and appear to lack a complete glycolytic pathway (22). *Rhizobium leguminosarum* bacteroids can neither transport nor metabolize hexoses and disaccharides (29, 51, 66). In addition, many of the inducible enzymes necessary for sugar metabolism by the fast-growing cowpea *Rhizobium* strain NGR234 are absent from isolated bacteroids of snake bean nodules, suggesting that the plant does not supply bacteroids with sugars (30, 139). The corollary to this, that sugars from the plant are not essential to bacteroid metabolism and symbiotic nitrogen fixation, is demonstrated by the fact that mutants of *Rhizobium* affected in sugar uptake or metabolism retain the ability to reduce nitrogen symbiotically (31, 36, 51, 53, 63, 96, 131).

Organic acids, including C_4-dicarboxylic acids, however, are effective substrates for supporting both bacteroid respiration (8, 20, 154) and nitrogen fixation (9, 13, 102, 115). Enzyme activities and radiorespirometry have revealed the operation of the TCA cycle in bacteroids (13, 56, 85, 95, 146), though it has been argued that the reductive part of the cycle may be partially inhibited (74). Although C_4-dicarboxylates support bacteroid respiration and nitrogen fixation, their support of the latter is not always simple or direct. For example, when ^{14}C-malate or ^{14}C-succinate are supplied to suspensions of bacteroids prepared from soybean nodules, in excess of immediate requirements, ^{14}C is stored as poly-β-hydroxybutyrate (PHB), and nitrogen fixation may be depressed to lower rates (14). If exogenous C-substrates are then withdrawn, stored PHB is utilized, $^{14}CO_2$ is evolved, and enhanced nitrogen fixation ensues (14). Thus, delivery of substrates to the bacteroids must be strictly regulated. In vivo, transport across the PBM and bacteroid inner membrane are likely to do this.

Perhaps the most convincing evidence for the involvement of C_4-dicarboxylic acids in nitrogen fixation comes from genetic studies. Bacterial mutants defective in dicarboxylate transport elicit nodules that are ineffective, or Fix⁻ (2, 17, 37, 38, 40, 41, 44, 129, 165, 180). Mutants defective in TCA cycle enzymes are also Fix⁻. A succinate dehydrogenase mutant of R. meliloti is affected in symbiotic development and nitrogen fixation (49, 50). R. meliloti isocitrate dehydrogenase and citrate synthase mutants are also Fix⁻ (74, 94). The metabolism of C_4-dicarboxylates as the sole carbon source via the TCA cycle is dependent upon generation of acetyl-CoA. Bacteroid malic enzyme converts malate to pyruvate which is subsequently converted to acetyl-CoA by pyruvate dehydrogenase. Recently, it was shown that a mutant of R. meliloti lacking NAD-malic enzyme forms ineffective nodules (34).

Gluconeogenesis can produce sugars for biosynthetic purposes if they are not supplied exogenously. The observations that gluconeogenic enzymes are active in R. leguminosarum bacteroids (95, 96), and mutants defective in gluconeogenesis are Fix⁻ (34, 42), support the conclusion that sugars are not supplied in significant amounts to the bacteroids.

There has been speculation that amino acids may also play a role in carbon and/or energy supply to the bacteroids (73, 81). Glutamate and proline can enhance respiration and support nitrogen fixation in isolated bacteroids (14, 181), although this support is apparently a consequence of prior deamination of the amino acid (14). Various amino acid auxotrophs have been tested for their ability to form nitrogen-fixing symbioses with plants. Glutamate, glutamine, glycine, serine, and most tryptophan auxotrophs produce effective nodules (5, 27, 77). In contrast, asparagine, arginine, leucine, methionine, tyrosine (and anthranilate) auxotrophs produce ineffective nodules, which suggests that these compounds are not supplied in sufficient amounts by the plant to support growth or development of mature bacteroids (5, 77, 101).

Although production of effective nodules by specific amino acid auxotrophs is consistent with the interpretation that the amino acid is supplied in planta, such studies do not prove that the amino acid is provided by the plant (they may not be needed for bacteroid development and nitrogen fixation), nor do they clarify the role of the amino acid in bacteroid metabolism. The potential roles of specific amino acids in bacteroid metabolism can be partly clarified by measuring the flux of amino acids across the peribacteroid and bacteroid membranes when the amino acids are supplied at concentrations similar to those present in the cytoplasm of infected cells. In order to play a significant role in energy supply for nitrogen fixation, rates of amino acid transport across both membranes must at least match known rates of nitrogen fixation in bacteroids.

Transport of Carbon Compounds into Isolated Bacteroids

Rhizobia are gram-negative bacteria and possess both an inner and outer cell membrane. As in other gram-negative bacteria, the inner membrane (IM) is energized by the activities of the respiratory electron transport chain and cell membrane ATPases which generate an electrochemical gradient (outside positive) across the membrane (107). It is this membrane that largely governs the transport of compounds between the bacteria and their environment.

DICARBOXYLATE TRANSPORT   The C_4-dicarboxylate carrier is the best characterized transporter in the bacteroid inner membrane and its genetics, biochemistry, and physiology have been reviewed recently (72). Isolated bacteroids take up dicarboxylic acids rapidly from the surrounding medium via an energy-dependent dicarboxylate transport (Dct) system encoded by the *dctA* gene (17, 38, 43, 44, 54, 62, 67, 95, 97, 121, 129, 136, 162, 169, 171). As mentioned above, the C_4-dicarboxylate transport system is essential for symbiotic nitrogen fixation (2, 17, 37, 38, 40, 41, 44, 49, 129, 165, 180). The Dct system employs three genes: *dctA, dctB,* and *dctD.* The *dctA* gene encodes the dicarboxylate transporter, DctA, which is a cytoplasmic membrane protein with 12 putative membrane-spanning domains. The products of the other two genes, DctB and DctD, form a two-component regulatory system that monitors the availability of dicarboxylates and interacts with the alternative sigma factor NtrA (also called RpoN) to regulate transcription of *dctA* (72). DctB is a sensor protein that resides in the bacteroid IM and activates the DctD protein, possibly by phosphorylation, in the presence of C_4 dicarboxylates and related compounds. Activated DctD interacts with NtrA to stimulate transcription of *dctA.*

The Dct system is dependent on an energized membrane, and uncouplers and metabolic poisons inhibit the activity of the system (40, 43, 62, 111). The carrier appears to catalyze H^+/malate$^-$ symport with a stoichiometry of $2H^+$

per malate (J Hendriks, L Whitehead & DA Day, unpublished results). The Dct system has a high affinity for the substrates succinate, fumarate, and malate, and K_m values range from 2–15 µM, and V_{max} from 10–80 nmol/min/mg protein in different genera (40, 43, 44, 62, 136, 162). The Dct system in *R. meliloti* can also transport aspartate, albeit with a much lower affinity (97, 172). Dct⁻ mutants cannot grow on aspartate as the sole carbon source, although a second high-affinity aspartate transport system exists in *R. meliloti* that enables cells to utilize aspartate as the sole nitrogen source (170, 172). There is evidence that the rate of nitrogen fixation limits the rate of C_4-dicarboxylate transport in bacteroids, rather than vice versa (72), which may explain why efforts to enhance symbiotic nitrogen fixation by genetically manipulating the Dct system have been largely unsuccessful (119).

SUGAR TRANSPORT   Rhizobia can grow on a number of different sugars, including mono and disaccharides, as the sole source of carbon in free-living culture. Under these conditions they express specific transport systems that enable them to accumulate the sugar(s) from the external medium (29, 51, 66). However, bacteroids are generally unable to transport most sugars. Bacteroids isolated from soybean root nodules were unable to transport glucose or sucrose (26, 121, 161). Interestingly, *B. japonicum* bacteroids express a fructose transporter (161) even though they apparently lack a complete glycolytic pathway (22). *Rhizobium leguminosarum* bacteroids are unable to transport hexoses and disaccharides (29, 51, 66). The absence of transporters for most sugars in bacteroids, together with other data summarized above, supports the conclusion that the plant does not supply significant amounts of sugars to the bacteroids. This conclusion is supported further by studies of the transport properties of the peribacteroid membrane, described below.

AMINO ACIDS   Isolated bacteroids are capable of accumulating a range of amino acids from the external medium, and these acids could provide carbon for the bacteroid. Schemes linking amino acid transport to transport of other organic acids have also been suggested. These are discussed below in the section on "Transport of Fixed Nitrogen Across Symbiotic Membranes."

Transport of Carbon Compounds Across the PBM

The development of techniques for the large-scale isolation of intact and pure symbiosomes from legume nodules has enabled the direct measurement of metabolite transport across the PBM. Bergersen & Appleby (11) first reported isolation of symbiosomes from soybean nodules, but these were unstable.

Later attempts using rate zonal centrifugation of nodule homogenates of both soybean (117) and french bean (60) were more successful. It is worth noting that in soybean the choice of strain of *B. japonicum* used to inoculate plants can be an important determinant of the stability of isolated symbiosomes. We have found (LF Whitehead, FJ Bergersen & DA Day, unpublished observation) that while strain USDA110 allows easy isolation of symbiosomes, those containing strain CB1809 are much more fragile and rupture during centrifugation. This presumably explains the difference between the results of Bergersen & Appleby (11) and Price et al (117). Isolation of symbiosomes from pea (126) had preceded the soybean work, but these symbiosomes were not used in transport assays until much later (133, 151). Transport studies with alfalfa (97), lupin (118), broad bean (153), and siratro (110) symbiosomes have also been published.

Symbiosomes isolated from this diverse range of legumes can take up a variety of reduced carbon substrates, such as glucose in french bean (60), oxalate in beans (153), and glutamate in lupin (118). However, without exception, symbiosomes isolated from all legumes can accumulate dicarboxylates very quickly and with high affinity. With a wealth of data from plant metabolism and bacteroid transport studies pointing to dicarboxylates as the major carbon source for bacteroids, it was no surprise that a dicarboxylate transporter was found on the PBM (60, 162). Kinetic and inhibitor studies with soybean symbiosomes have shown that the PBM possesses a carrier with a preference for malate and succinate, which appears to act as a uniport for the monovalent dicarboxylate anion (162). This is essentially a passive transport mechanism with malate uptake into the symbiosomes being driven by the electrical potential ($\Delta\Psi$) across the PBM (see below), and subsequent uptake and metabolism by the enclosed bacteroids (111, 159). The properties of the soybean PBM transporter show that it is quite distinct from its counterparts on other cell membranes (25). Because of substrate specificity, inhibitor sensitivity, and pH optimum, the dicarboxylate transporter on the PBM is also distinct from the DctA protein of the bacteroid (25, 60, 162). The maximum velocity of the PBM dicarboxylate transport system is less than that of the bacteroid Dct system (162), perhaps indicating that the PBM transporter will be rate-limiting when substrate is plentiful. Nonetheless, measured rates of malate transport across the PBM appear more than adequate to support rates of nitrogen fixation estimated from measurements with isolated bacteroids (26). The rate of sugar and amino acid transport in soybean symbiosomes, however, is too slow to support estimated nitrogen fixation rates (26, 161).

The PBM protein nodulin 26 has recently been shown to have channel activity when isolated and inserted into a lipid bilayer, transporting malate among other anions (175). Nodulin 26 is phosphorylated by a PBM protein kinase (173), and in isolation this renders the channel sensitive to the voltage across the membrane (88). Phosphorylation or dephosphorylation of PBM proteins is correlated with stimulation or inhibition of malate uptake by isolated soybean symbiosomes (112). Because nodulin 26 is the major PBM protein phosphorylated in soybeans (166, 173) it is possible that it is the dicarboxylate transporter. However, the channel activity of isolated nodulin 26 is rather nonspecific, and more work is required to establish the identity of the PBM dicarboxylate transporter. It is also worth noting that french bean PBM, which has a malate transporter similar to that of soybean PBM (60), does not react with nodulin 26 antibodies. This is also true of pea symbiosomes (21a). Further, incubation of french bean PBM or symbiosomes with ATP does not result in a labeled 26 kDa protein and does not provide long-term stimulation of malate transport, as it does in soybean (S Moreau, DA Day & A Puppo, unpublished results).

TRANSPORT OF FIXED NITROGEN ACROSS SYMBIOTIC MEMBRANES

Bacteroids use the enzyme nitrogenase to catalyze the reduction of N_2 to ammonia, as shown in Equation 1:

$$N_2 + 8H^+ + 8e^- + 16ATP \rightarrow 2NH_3 + H_2 + 16ADP + 16Pi \qquad 1.$$

As discussed above, the reductant and ATP required for this reaction are derived from reduced carbon provided by the plant, probably in the form of C_4-dicarboxylates. The ammonia produced is released by the bacteroid and is assimilated in the cytoplasm of the infected cells (13, 76). The first step in this assimilation is catalyzed, in all legumes, by glutamine synthetase (GS), which converts ammonia to glutamine. Subsequently, glutamine is converted to glutamate by the glutamate synthase (GOGAT) reaction (3, 23, 164). These enzymes are very active in nodules and maintain ammonia at very low concentrations in the plant cytoplasm (147). Consequently, there is a very strong "sink" driving ammonia efflux from the bacteroid. The assimilated nitrogen is translocated out of the nodules to other plant parts, but the form in which it is transported depends on the legume species. Temperate legumes export amides whereas tropical legumes export largely ureides (3, 141). The mechanism by which ammonia reaches the plant cytosol, however, remains controversial and is discussed in detail below.

Nitrogen Fixation and Nitrogen Assimilation Are Uncoupled in Bacteroids

The nitrogenase reaction consumes a considerable amount of chemical energy in the form of ATP and reductant. In addition, synthesis of the nitrogenase enzyme complex requires the coordinated expression of many genes and a concomitant investment of energy. Not surprisingly then, the synthesis and activity of nitrogenase is tightly regulated in all nitrogen-fixing organisms, including rhizobia. In most diazotrophs, deployment of nitrogenase is the ultimate response to nitrogen limitation. Rhizobia are an important and perhaps unique exception to this rule. In rhizobia, nitrogen fixation is regulated by oxygen concentration rather than nitrogen availability (7, 45), and this has important implications, considered below. First, however, it is worthwhile to consider how nitrogen assimilation is regulated in rhizobia and other bacteria.

Regulation of nitrogen assimilation in rhizobia and other bacteria is achieved at the genetic level by a central nitrogen regulatory (Ntr) system. The Ntr system is orchestrated by three genes: *ntrA, ntrB,* and *ntrC* (90, 122). The proteins NtrB and NtrC constitute a two-component regulatory system that responds to the nitrogen status of the cell (122). Under conditions of low nitrogen availability, the sensor protein NtrB phosphorylates and thereby activates NtrC, which in turn activates the transcription of genes involved in nitrogen metabolism, including itself (61, 68, 90). The NtrA protein is a unique sigma factor that combines with the RNA polymerase core enzyme and the regulatory protein NtrC to activate gene transcription (61, 68, 99). Genetic analysis has shown that the Ntr system is necessary for expression of several genes involved in nitrogen assimilation in rhizobia, including nitrate reductase, a high-affinity ammonium transporter, and GS II (4, 21, 92, 105, 114, 130, 142, 144, 152, 160). However, the Ntr system is generally not required for transcription of nitrogen fixation (*nif*) genes in rhizobia. Mutations that impair the Ntr system have little or no effect on the symbiotic nitrogen fixation by rhizobia (58, 114, 152, 160). This is an important difference between rhizobia and free-living nitrogen-fixing bacteria such as *Klebsiella* and *Azotobacter,* which require the Ntr system for *nif* gene expression (90).

Nitrogen fixation in rhizobia is regulated by oxygen concentration (7, 45). The fact that nitrogen fixation is primarily under oxygen control, whereas nitrogen assimilation is under nitrogen control in rhizobia, implies that these processes can be uncoupled. This is what occurs in bacteroids where nitrogen fixation occurs without significant ammonia assimilation. Biochemical and genetic evidence indicates that bacteroids are not nitrogen starved in nodules. Thus, the ammonia assimilation pathway is not induced in bacteroids, and activities of the high-affinity ammonium transporter, GS, and GOGAT are all low relative to nitrogenase activity (19, 52, 65, 86, 87, 108, 123). These results are consistent with the fact that mutations in the *ntr* genes that are required for induction of these enzymes have little or no effect on the symbiotic perform-ance of rhizobia (58, 114, 152, 160).

It is intriguing, and of profound importance to the symbiosis, that bac-teroids fix nitrogen at high rates even though they are apparently not nitrogen starved and cannot assimilate much of the ammonia they produce. The enor-mous energy cost of nitrogen fixation has already been pointed out. So why do bacteroids fix nitrogen if it is not to satisfy a metabolic requirement for nitrogen? A number of hypotheses have been proposed to account for this apparently unique behavior (158). Whatever the reason, nitrogen fixation by bacteroids obviously benefits the host plant and, therefore, both partners of the symbiosis. Evolution of the novel regulation of *nif* genes in rhizobia may have been an important aspect of the emergence of rhizobia as nitrogen-fixing microsymbionts. The end result of the regulatory mechanisms outlined above is that the N_2 fixed by the bacteroids is exported to the plant. The possible mechanisms of this transport are discussed below.

Transport of Ammonia Across Bacteroid Membranes

Free-living rhizobia, like other bacteria, transport ammonia in two ways. First, when concentrations of ammonia in the surrounding medium are low, NH_4^+ transporters allow rapid accumulation of ammonium in the cell (79). Ammo-nium transporters (Amt) have been characterized in a number of rhizobia grown in culture (52, 55, 65, 70, 108, 113) and they appear to be under the control of the Ntr system (160). The second general mechanism by which ammonia enters cells is by diffusion of NH_3. At high concentrations of ammo-nia, the Amt system is repressed, and diffusion of NH_3 apparently provides the cell with sufficient ammonia for growth (65).

Bacteroids isolated from nitrogen-fixing nodules do not express the Amt system; ammonia transport occurs by simple diffusion of NH_3 alone (65, 70, 91, 108). Ammonia assimilation also is repressed in bacteroids (15, 19, 64). Consequently, most of the ammonia that is produced by nitrogen fixation is not assimilated by the bacteroids but rather lost from the cell by diffusion of NH_3 down a concentration gradient (13). In the absence of an NH_4^+ trans-porter, the lost ammonia cannot be effectively recovered. Repression of the Amt system in nitrogen-fixing bacteroids, therefore, avoids a futile cycle that would otherwise arise (79). This regulation of Amt in bacteroids is compatible with their role as ammonia-producing and -exporting organisms.

Amino Acid Transport Across Bacteroid Membranes

The ability of some rhizobial amino acid auxotrophs to establish nitrogen-fixing symbioses suggests that the necessary amino acid may be supplied to bacteroids by their host plants (73). Transport studies with isolated bacteroids confirm that they can accumulate certain amino acids quite rapidly (32, 51, 135, 163). B. japonicum bacteroids isolated from soybean nodules transport glutamate, aspartate, and alanine in an energy-dependent manner (163, 178). Alanine transport exhibited a K_m of 15 μM and a V_{max} of 3 nmol mg protein^{-1} min^{-1}. Competitive inhibition between glutamate and aspartate indicates that these amino acids share a common carrier. The transporter has a high affinity for these substrates (K_m of 0.8 μM and 27.5 μM for glutamate and aspartate, respectively), and the V_{max} ranges from 3–23 nmol mg protein^{-1} min^{-1} (163, 178). Inhibition studies suggest that the same transporter also transports asparagine and glutamine but not leucine, alanine, glycine, or serine (163). An amino acid transporter that transports glutamate and other amino acids has also been described in free-living R. leguminosarum (116). In R. meliloti, glutamate and aspartate share a common transporter that has a K_m for aspartate of 1.5 μM (172). Aspartate can be transported by a second transport system in R. meliloti, namely the Dct system described above. The Dct system is necessary for growth of R. meliloti on aspartate as a sole carbon source (172). Phenylalanine, methionine, leucine, proline, and glycine are also taken up by isolated bacteroids, albeit at a much slower rate than glutamate, but transport mechanisms have not been investigated (161).

Transport of Amino Acids Across the PBM

Several models that integrate carbon and nitrogen exchange across the PBM via the movement of amino acids (73, 78, 80) have been proposed. These hypotheses include: (a) a malate/aspartate shuttle, in which malate and glutamate enter and 2-oxoglutarate and aspartate leave the bacteroid (73); (b) a proline shuttle linking supply of reductant to the bacteroid with purine synthesis and the pentose phosphate pathway in the plant (80); and (c) a malonamate (the aminated form of malonate) shuttle proposed to take advantage of abundant malonate in some legume nodules (78, 140).

However, in the majority of cases, transport studies with isolated symbiosomes from a number of different legumes have failed to identify amino acid transporters on the PBM (60, 110, 149, 153, 163, 178). These studies of amino acid transport into isolated symbiosomes demonstrate more than any others the importance of the PBM in determining the flow of metabolites between plant and bacteroid. The malate/aspartate shuttle is unlikely to operate in view of the impermeability of the PBM to glutamate, at least in soybean, french bean, and siratro symbiosomes (60, 110, 163). There is a report that lupin symbiosomes can take up glutamate (118), but the rate of uptake in this study was not calculated, and percentage integrity of the PBM was not assessed. It is difficult, therefore, to put these results in context. Likewise, no transport mechanism for proline has been identified on the PBM of soybean or siratro, although this amino acid diffuses into symbiosomes at a significant rate (110, 161). Malonamate transport across the PBM has not been measured directly to our knowledge. Malonate uptake across the PBM, which is also required in this scheme (78), can occur via the dicarboxylate transporter (111), but malonate uptake across the PBM is severely inhibited by malate, and though the malonate concentration of nodule homogenates is very high, so is that of malate (148). Malonate uptake by bacteroids appears to be restricted to passive diffusion across the bacteroid membrane (176). We consider, therefore, all of the above schemes to be unlikely mechanisms for the major flux of fixed nitrogen to the plant (or reductant to the bacteroid).

Efflux of amino acids from the bacteroid does not necessarily require operation of a coordinated exchange with other metabolites as in the schemes mentioned above. For example, it has been reported (82) that both alanine and aspartate, which can be readily formed in the bacteroid from malate (93), efflux rapidly from isolated soybean bacteroids in the absence of glutamate. Efflux of these amino acids has been seen also in long-term experiments with isolated, nitrogen-fixing symbiosomes (133). Bacteroids isolated from various legumes have a number of active transport mechanisms for the transport of amino acids, especially aspartate and glutamate, as mentioned above. When free bacteroids are loaded with ^{14}C-aspartate or glutamate, the amino acid can exchange for external cold aspartate, glutamate, and malate. If these amino acids are synthesized in the bacteroid when malate is metabolized by isolated symbiosomes, then they could exchange for malate in the peribacteroid space. The exchanged amino acid may then diffuse from the symbiosome down a concentration gradient (which will be large when isolated symbiosomes are suspended in a relatively large volume of reaction buffer), perhaps explaining the results of Rosendahl et al (133). Such efflux is, however, very slow (178).

The question arises about why bacteroids maintain active transport mechanisms for amino acids when the PBM appears to restrict availability of their substrates in planta. It has been suggested (178) that this reflects the regulatory pathways controlling synthesis of amino acid transporters in rhizobia. For example, free-living *B. japonicum* express a glutamate transporter when grown without glutamate, but when glutamate is available at high concentrations in the growth medium, the transporter is repressed (178). Similar results have been obtained with *Rhizobium trifolii* (71). The kinetics of glutamate uptake by free-living *B. japonicum* cells grown with NH_4Cl as the nitrogen source are very similar to those observed with isolated bacteroids (178). These results suggest that a glutamate transporter is maintained in bacteroids in response to the low amino acid concentration in the peribacteroid space. That is, the bacteroids are in a scavenging mode with respect to amino acids. Some amino acids diffuse across the PBM at significant rates, and this, coupled with the active transporters on the bacteroid membrane, may allow uptake of amino acids as supplementary nitrogen or carbon sources for the bacteroid (178). This may explain why some amino acid auxotrophs can form effective nodules (see above). Certainly labeling of bacteroid proteins occurs when isolated intact symbiosomes are incubated over an hour with ^{14}C-methionine, which apparently crosses the PBM by simple diffusion (75, 161).

Note that most of the above mentioned studies have been performed with symbiosomes from mature nodules. In developing nodules, before the bacteroids begin to fix atmospheric nitrogen, there must be some mechanism for delivering nitrogen to the differentiating rhizobia. It is possible, therefore, that the permeability of the PBM to amino acids may change during nodule development. It is also possible that rectification of the PBM ammonium channel (see below) changes so that ammonia is delivered from the plant to the bacteroid.

Ammonium Transport Across the PBM

Because the ammonium assimilatory enzymes in the bacteroid are repressed in the symbiotic state, NH_3 will diffuse rapidly into the PBS, driven by its concentration gradient and the expected ΔpH across the bacteroid IM (see below). The acidic pH in the PBS, the result of proton pumping by both the PBM ATPase and the bacteroid respiratory chain (see below), strongly favors protonation of NH_3 to NH_4^+. Repression of the bacteroid ammonium carrier prevents reuptake of NH_4^+, which is therefore available to the plant. However, a transporter is required on the PBM for the rapid movement of this ion.

Initial studies, using the ammonium analogue methylammonium, failed to find any evidence for such a carrier (157), but this technique has limitations when measuring efflux rather than uptake. More recently, the movement of NH_4^+ in isolated symbiosomes has been reinvestigated using the patch clamp technique, and a novel voltage-gated monovalent cation channel capable of transporting NH_4^+ across the PBM has been identified (155). This channel opens to allow NH_4^+ efflux from the symbiosome when a membrane potential is generated across the PBM. This is likely to occur in vivo by the action of the

H^+ pumping ATPase on the PBM (see below). In PBM patches, NH_4^+ currents were rectified so that the movement of NH_4^+ was unidirectional—effectively out of the peribacteroid space (155). Subsequent experiments have indicated that this rectification depends on the divalent cation concentration on either side of the PBM such that calcium or magnesium on either face of the PBM inhibit ion flow from that side of the membrane (LF Whitehead, SD Tyerman & DA Day, unpublished results). Because magnesium concentrations in the plant cytosol are expected to be in the mM range, it is probable that the channel will be outwardly rectified (from the PBS) in vivo, preventing back-flow of NH_4^+ into the symbiosome. The role of calcium in regulating the channel is less clear: Calcium concentrations in the plant cytosol are usually very low, but we (LF Whitehead, AH Millar & DA Day, unpublished results) have found substantial quantities of calcium in the PBS of soybean nodules. How much of this is free or bound is not known.

The PBM channel is selective for small monovalent cations, with K^+ and NH_4^+ the most active, but does not transport divalent cations, larger monovalent cations such as choline, or anions (155). However, at physiological concentrations (10–20 mM), the channel showed a preference for ammonium over potassium. In some patches, the conductance of the channel showed saturation kinetics with an apparent K_m of about 30 mM for NH_4^+. This is similar to the estimated concentration of NH_4^+ inside symbiosomes during nitrogen fixation (147). Analysis of ion currents in patches suggested that the channel is present in the PBM at very high density—800–1000 channels per μm^2—and that the conductance is extremely small (155). A high concentration of low conductance channel on the PBM may have some advantages for the regulation of ammonium flux to the plant. For a small particle such as the symbiosome, a large number of low-conductance channels would allow finer control of fluxes both in the long term, via protein synthesis, and in the short term via the percentage of channels in the open configuration.

Although the channel described above has properties that one would expect of a transporter delivering ammonium to the plant, its operation during nitrogen fixation remains to be demonstrated. Experiments with symbiosomes isolated under micro-aerobic conditions should confirm this. Nonetheless, the magnitude of the ammonium currents across the patch, extrapolated to the whole symbiosome, indicates that the channels could easily cope with the rate of ammonium produced by symbiosomes in vivo (155). Identification of the channel does not rule out the possibility that some diffusion of NH_3 occurs across the PBM [in vivo this will depend on the pH of the peribacteroid space (59, 151, 157)].

REGULATION OF CARBON AND NITROGEN EXCHANGE ACROSS SYMBIOTIC MEMBRANES

Energization of Symbiotic Membranes

ATPase activity is associated with the PBM from a number of legumes (16, 33, 35, 125, 151, 156). This ATPase pumps protons into the PBS and generates a $\Delta\Psi$ across the PBM. In the presence of permeant anions, this potential is readily converted to a ΔpH (35, 151, 159). In addition, the bacteroid respiratory electron transport chain pumps protons out of the bacteroid and into the PBS (107). The end result is acidification of the PBS. The precise pH of the PBS will be determined by the relative activities of the proton pumps and counter-ion movements, but could be up to two pH units more acidic than the plant cytosol (159). Thus in vivo the PBM and bacteroid IM delineate three compartments of contrasting pH: a slightly alkaline plant cytosol, an acidic PBS, and an alkaline bacteroid interior. Further, the PBM and bacteroid IM have opposite electrical polarities. These relationships are illustrated in Figure 2 and will partly determine the nature of transport systems involved in the exchange of metabolites between plant and bacteroid.

The dicarboxylate and ammonium transport mechanisms described above seem ideally suited to take advantage of the symbiosome membrane electrochemical gradients (Figure 2). The PBM dicarboxylate transporter is probably a uniport mechanism that allows movement of the malate anion down its

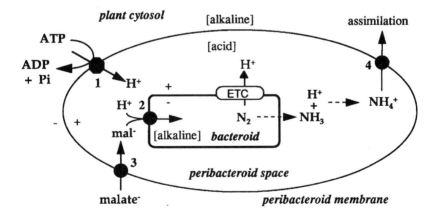

Figure 2 Scheme summarizing the exchange of carbon and nitrogen thought to occur in soybean symbiosomes. 1, PBM ATPase; 2, bacteroid IM dicarboxylate carrier; 3, PBM dicarboxylate carrier; 4, ammonium channel.

electrical and concentration gradients, into the PBS. The Dct system of the bacteroid, however, catalyzes H^+/malate$^-$ cotransport and is driven by the ΔpH across the bacteroid IM. An acidic PBS ensures that NH_3 arriving from the bacteroid is protonated to form NH_4^+, and this serves as a trap that drives further diffusion of NH_3 out of the bacteroid. The NH_4^+ channel on the PBM is voltage gated and opens to allow NH_4^+ movement into the plant cytosol upon PBM energization. Again, this is a passive process per se, but the high GS activity of the plant cytosol ensures rapid assimilation and maintenance of a strong concentration gradient out of the symbiosome (147).

Thus, the proton pumps of the PBM and bacteroid serve to coordinate and facilitate the exchange of malate for ammonium between plant and bacteroid. Conversely, the exchange of malate and ammonium across the PBM and bacteroid membranes helps to maintain pH balance in the PBS. This may be important for the stability of the symbiosis because excessive acidification of the PBS may inhibit nitrogen fixation and/or lead to early senescence of infected cells (18). In this context, it is interesting to note that Szafran & Haaker (151) found that acidification of the PBS upon adding ATP to isolated pea symbiosomes inhibited nitrogenase activity—but in this case nitrogenase activity was assayed by acetylene reduction, and ammonia was not formed. We would expect that if these experiments were repeated using N_2 instead of acetylene, and malate as a source of reductant for the symbiosomes, then ATP would stimulate nitrogen fixation.

Protein Kinases

Adding ATP to isolated symbiosomes can stimulate malate uptake across the PBM (60, 111, 112), but this may be only partly due to energization of the PBM. In soybean symbiosomes, the ATP effect remains after reisolation of the symbiosomes, and malate transport is inhibited by phosphatase treatment of the symbiosomes (112). The stimulation/inhibition of malate uptake coincided with phosphorylation/dephosphorylation of some PBM proteins, notably nodulin 26 (112). The PBM contains a protein kinase that is calcium dependent (6, 166, 173, 174) and that may help to regulate metabolite transport across that membrane. The major protein phosphorylated is nodulin 26 (166, 173), but whether this protein is responsible for malate transport across the PBM has not been established (see above).

ATP and Oxygen Concentrations

Cytosolic ATP is a substrate for both the protein kinase and the ATPase on the PBM and is therefore likely to play a key role in the regulation of metabolite exchange across that membrane. Ammonium efflux from the symbiosome is principally a passive process, but N-demand in the plant will create a major driving force by establishing a concentration gradient. Here, ATP concentration will also be important, because large quantities are required in the plant cytoplasm for ammonium assimilation. Because most of the plant cell's ATP synthesis occurs in the mitochondria, oxygen concentrations in the infected zone will be important, with sudden decreases likely to lead to decreased cytosolic ATP concentration (100) and, possibly, inhibition of malate/ammonium exchange. Consistent with this idea, decreasing oxygen concentration around soybean nodules leads to a decrease in adenylate energy charge and an accumulation of malate in the plant fraction (109). Oxygen is important in another context as well. Malate transport into isolated symbiosomes is strongly inhibited by KCN and other poisons of bacteroid respiration (60, 111), indicating that sustained uptake of malate across the PBM depends greatly on its uptake and subsequent metabolism in the bacteroid. In vivo, the rate of nitrogen fixation will largely determine this demand by the bacteroid, and this is probably, in turn, largely determined by oxygen concentration (164).

Calcium

Calcium has been implicated in the regulation of both malate and ammonium transport across the PBM. It is required for activity of the protein kinase which, in soybean at least, stimulates malate uptake by symbiosomes (112, 173). It also inhibits ion movement through the ammonium channel (155). In preliminary experiments, we (AH Millar & DA Day, unpublished results) have grown nodulated soybeans hydroponically in varying calcium concentrations and subsequently measured malate uptake by isolated symbiosomes. Low calcium during growth resulted in low rates of malate transport into symbiosomes and lowered sensitivity to phosphatase treatment. Adding ATP to these symbiosomes stimulated malate uptake substantially, suggesting that the phosphorylation status of the PBM had been altered by growth on low calcium. The symbiosome may be a store of calcium in infected cells, and transport of this ion across symbiotic membranes warrants detailed investigation.

TRANSPORT OF INORGANIC NUTRIENTS TO THE BACTEROID

A range of inorganic nutrients are required by the bacteroids both during nodule development and during steady state nitrogen fixation (127). Our knowledge of how these are acquired in planta is scant. Inorganic cations required include vanadium, iron, molybdenum, nickel, and cobalt (132). In addition, Na and K transport is likely to be important for ion and osmotic balance across the PBM and bacteroid IM. There are indications that the symbiosome contains significant concentrations of Ca, which is required, among other things, for cell wall structure in rhizobia (168). This suggests that a transporter for this ion exists on the PBM. Pi is also required by the bacteroid, but nothing is known about movement of this ion across the PBM. However, the rapidity with which some inorganic anions, such as nitrate and chloride, collapse the $\Delta\Psi$ across the PBM (156) implies that at least one anion channel exists on the PBM, and it may be that Pi shares this. One important inorganic nutrient, whose transport in nodules and rhizobia has been studied in a little more detail, is iron.

Transport of Iron Across Symbiotic Membranes

There is a large demand for iron in the bacteroid for synthesis of iron-containing proteins necessary for nitrogen fixation, such as nitrogenase itself and the cytochromes of the respiratory chain. Iron is also required in large quantities for the synthesis of heme for leghemoglobin (Lb), the oxygen-carrying protein so abundant in nodules. It was thought for some time that this heme was synthesized in the bacteroids, but recent reconsideration of early data indicates that this is unlikely and that heme for Lb is synthesized in the plant fraction (106). When one considers that each infected cell of a nodule may contain many thousands of bacteroids and that Lb is present in concentrations up to several mM (1), then it becomes obvious that relatively large amounts of iron are needed in nodules. Nutrition studies have shown that lack of iron can limit the symbiosis (127, 132). The question then arises about how this iron is accessed and how it is stored and transported within infected cells, especially because excess iron can be very toxic to cells (57).

There are two general strategies by which bacteroids could acquire iron within host plants. One is the direct utilization of host iron compounds such as ferric [Fe^{3+}]-citrate. Alternatively, the micro-organisms could dissociate iron from host iron complexes such as Fe^{3+}-organic acid complexes and ferritin, by reduction or by secretion of siderophores (57). Siderophores are small molecular weight, high-affinity Fe^{3+} chelators synthesized and secreted by microorganisms and some plants under conditions of iron limitation. In all gram-negative bacteria studied to date, internalization of iron from Fe^{3+}-siderophore and chelate complexes involves several outer- and inner-membrane receptor proteins and is absolutely dependent on the presence of a functional TonB protein

(57). Several bradyrhizobial strains, including *B. japonicum* USDA 110, have been shown to release a variety of compounds that have the ability to chelate Fe^{3+} (57).

Recently, Fe acquisition by isolated soybean symbiosomes has been investigated (89, 103). Together these two studies have shown that: (*a*) the PBM possesses a mechanism for the uptake of ferric citrate; (*b*) bacteroids can also take up ferric citrate but do so less readily than do intact symbiosomes (though this varies with the age of the nodule); (*c*) a large part of the Fe taken up as ferric citrate by isolated symbiosomes remains in the PBS from which it cannot be readily leached; and (*d*) the PBM also possesses a ferric chelate reductase that uses NADH to reduce Fe^{3+} to Fe^{2+}; added NADH stimulates Fe uptake into symbiosomes, suggesting that Fe^{2+} ions are transported across the PBM and maybe also the bacteroid membranes.

It thus appears that Fe from the plant cytosol may enter symbiosomes by two different methods (see Figure 3*A*), but the relative importance of and interaction between the two mechanisms has not been assessed.

In a separate study, Wittenberg et al (179) found that the PBS contained large quantities of Fe bound to low molecular weight compounds whose spectra resembled those of known bacterial siderophores. Taken together the results suggest that ferric citrate can be transported across the PBM into the PBS where high-affinity binding compounds, released by the bacteroids, sequester the Fe. If and how the Fe is released from these compounds is not known, but it is possible that the ferric chelate reductase identified on the PBM (89) could be involved. In this model (depicted in Figure 3*B*), the PBM reductase would function as it does on the plasma membrane, oxidizing NADH on the plant cytosolic side and Fe^{3+} on the PBS side, either for release back into the plant cytosol (e.g. for Lb synthesis) or uptake into the bacteroid. The precise identity of the compounds that bind Fe in the PBS is also unknown, although the compounds clearly differ from the siderophores produced and excreted by free-living rhizobia (57). Notwithstanding this, the Wittenberg et al (179) study suggests that the PBS may be a major storage compartment for Fe in the nodule.

MOLECULAR STRUCTURE OF PBM PROTEINS

The primary structure of several confirmed or putative PBM proteins have been deduced from cDNA sequences. These include soybean nodulins 16 (104), 23 (69), 24 (48), and 26 (47). The cDNAs encoding these proteins were isolated using nodule-specific antibodies (47, 48, 69), or by differential screening of a cDNA library using ^{32}P-labeled cDNA probes derived from root or

A.

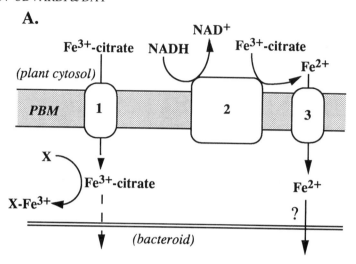

B.

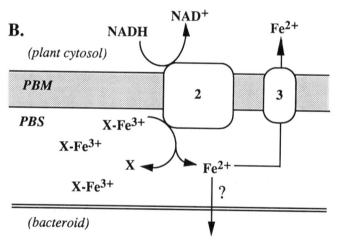

Figure 3 Putative schemes of iron transport across membranes of soybean symbiosomes. 1, PBM ferric-citrate transporter; 2, PBM ferric chelate reductase; 3, putative ferrous ion transporter; X, iron-binding compound released by bacteroids (179); PBS, peribacteroid space. In A, Fe enters the PBS either directly as ferric citrate, in which case the ferric ions can be sequestered by bacteroid chelators (X), or after reduction in the plant cytosol via a ferric chelate reductase. In B, the ferric chelate reductase is shown as oxidizing NADH on the plant cytosolic side of the PBM and reducing ferric chelates, formed following ferric citrate uptake across the PBM, in the PBS. The ferrous ions thus formed may persist under the low oxygen and acidic environment of the PBS (see text) and could be either transported back into the plant cytosol (e.g. for leghemoglobin synthesis) or into the bacteroid, as required.

nodule poly(A)+ RNA (104). This accounts for the fact that they all encode nodule-specific or nodule-enhanced proteins. Little is known about the function of these proteins. The one exception is nodulin 26, which belongs to the MIP family of integral membrane channel proteins (137). Different members of this family have different substrate specificities. As mentioned above, nodulin 26 was shown to function as a high conductance ion channel when expressed in a lipid bilayer system but showed little ion specificity (175). Phosphorylation of nodulin 26 on serine 262 affects its voltage-sensitive channel activity in planar lipid membranes (88). More recently, transformation of Xenopus oocytes with nodulin 26 resulted in the expression of two transport activities, water and anion uptake, which could be separated by their inhibitor sensitivities (DM Roberts, personal communication). It is possible, therefore, that nodulin 26 is a bifunctional protein. However, its precise physiological role remains to be determined. Recently, Kouchi & Hata (83) isolated a soybean nodulin cDNA that encodes a putative sulphate transporter (138), and it has been suggested that the transporter may be located on the PBM. However, this remains to be proven.

Many of the proteins in the PBM are unlikely to be nodule specific. These may include the H^+-ATPase, NADH-ferric reductase, and the protein kinase(s), but none of these proteins has been characterized at the molecular level. There is an obvious gap between our knowledge of the biochemical activity of PBM proteins on the one hand, and their molecular structure on the other. Closing this gap and understanding the structure/function relations of PBM proteins remains one of the greatest challenges in this area of research.

FUTURE DIRECTIONS

The peribacteroid membrane of legume root nodules is probably the best characterized of all symbiosome membranes. A number of features of the PBM are likely to be shared by other symbiosome membranes and, to a limited extent at least, what we have learned about the PBM now serves as a bench mark by which to compare and understand other symbiosome membranes. Central to the bioenergetics of the PBM is a host-encoded H^+-ATPase that is a primary-active transporter that generates an electrochemical gradient across the membrane. This electrochemical gradient provides a driving force for the transport of important metabolites such as malate and ammonium between the plant and the nitrogen-fixing bacteroids. Although the types of compounds that are traded between symbionts differ in different endocytobioses, ATPases on other symbiosome membranes are likely to play similar roles in providing the driving force for secondary transport processes. Another feature of the PBM that may be conserved among symbiosome membranes is the membrane-bound protein kinase(s) that phosphorylates and thereby regulates the activity of other membrane proteins. Of course, such features are not unique to symbiotic membranes, and it is to be expected that the activity of transporters in symbiosome membranes will also be regulated by ions such as Ca^{2+}, as they are in other biological membranes. Although we are beginning to understand how some of these factors affect the activity of specific PBM proteins, elucidating other regulatory mechanisms and understanding their significance in vivo remains one of the challenges in the study of the PBM. In addition, most of the proteins of the PBM remain to be characterized at the biochemical level, and investment of research effort in this area will almost certainly reward us with a better understanding of the role of the PBM in coupling the metabolic activities of the plant and the resident bacteroids.

Finally, it should be apparent from this review that our knowledge of the molecular structure of PBM proteins is, at best, scant, and there is a clear need for work on the isolation and molecular characterization of PBM proteins. The comparative ease with which relatively large quantities of PBM can be obtained (26) gives us hope that approaches such as protein sequencing coupled to PCR-based strategies will enable the molecular cloning and characterization of many of the proteins in the PBM. It is our expectation that while biochemistry will continue to uncover novel and important functions of the PBM, molecular biology will play a major role in developing our understanding of the biology of the perbacteroid membrane in the future.

ACKNOWLEDGMENTS

We thank Dean Price for providing the electron micrograph, and Fraser Bergersen, Lynne Whitehead, Hans Lambers, Lis Rosendahl, and Megan McKenzie for critical reading of the manuscript.

Visit the *Annual Reviews home page* at http://www.annurev.org.

Literature Cited

1. Appleby CA. 1984. Leghemoglobin and *Rhizobium* respiration. *Annu. Rev. Plant Physiol.* 35:443–78
2. Arwas R, McKay IA, Rowney FRP, Dilworth MJ, Glenn AR. 1985. Properties of organic acid utilization mutants of *Rhizobium leguminosarum* strain 300. *J. Gen. Microbiol.* 131:2059–66
3. Atkins CA. 1991. Ammonia assimilation

and export of nitrogen from the legume nodule. See Ref. 32a, pp. 293–319
4. Ausubel FM, Buikema WJ, Earl CD, Klingensmith JA, Nixon BT, Szeto WW. 1985. Organization and regulation of *Rhizobium meliloti* and *Parasponia Bradyrhizobium* nitrogen fixation genes. See Ref. 41a, pp. 165–79
5. Barsomian GD, Urzainqui A, Lohman K,

Walker GC. 1992. *Rhizobium meliloti* mutants unable to synthesize anthranilate display a novel symbiotic phenotype. *J. Bacteriol.* 174:4416–26

6. Bassarab S, Werner D. 1987. A Ca^{2+}-dependent protein kinase activity in the peribacteroid membrane from soybean root nodules. *J. Plant Physiol.* 130:233–41

7. Batut J, Boistard P. 1994. Oxygen control in *Rhizobium. Antonie Van Leeuwenhoek* 66:129–50

8. Bergersen FJ. 1958. The bacterial component of soybean root nodules; changes in respiratory activity, cell dry weight and nucleic acid content with increasing nodule age. *J. Gen. Microbiol.* 19:312–23

9. Bergersen FJ. 1977. Physiological chemistry of dinitrogen fixation by legumes. In *A Treatise on Dinitrogen Fixation, Section III: Biology*, ed. RWF Hardy, WS Silver, pp. 519–55. New York: Wiley

10. Bergersen FJ. 1982. *Root Nodules of Legumes: Structure and Functions.* Chichester: Wiley. 164 pp.

11. Bergersen FJ, Appleby CA. 1981. Leghaemoglobin within bacteroid-enclosing membrane envelopes from soybean root nodules. *Planta* 152:534–43

12. Bergersen FJ, Briggs MJ. 1958. Studies on the bacterial component of soybean root nodules: cytology and organization in the host tissue. *J. Gen. Microbiol.* 19:482–90

13. Bergersen FJ, Turner GL. 1967. Nitrogen fixation by the bacteroid fraction of breis of soybean root nodules. *Biochim. Biophys. Acta* 141:507–15

14. Bergersen FJ, Turner GL. 1990. Bacteroids from soybean root nodules: accumulation of poly-β-hydroxybutyrate during the supply of malate and succinate in relation to N_2 fixation in flow chamber reactions. *Proc. R. Soc. London Ser. B* 240:39–59

15. Bishop PE, Guevara JG, Engelke JA, Evans HJ. 1976. Relation between glutamine synthetase and nitrogenase activities in the symbiotic association between *Rhizobium japonicum* and *Glycine max. Plant Physiol.* 57:542–46

16. Blumwald E, Fortin MG, Rea PA, Verma DPS, Poole RJ. 1985. Presence of host plasma membrane type H^{+}-ATPase in the membrane envelope enclosing the bacteroids in soybean root nodules. *Plant Physiol.* 78:665–72

17. Bolton E, Higgisson B, Harrington A, O'Gara F. 1986. Dicarboxylic acid transport in *Rhizobium meliloti*: isolation of mutants and cloning of dicarboxylic acid transport genes. *Arch. Microbiol.* 144:142–46

18. Brewin NJ. 1991. Development of the legume root nodule. *Annu. Rev. Cell Biol.* 7: 191–226

19. Brown CM, Dilworth MJ. 1975. Ammonia assimilation by *Rhizobium* cultures and bacteroids. *J. Gen. Microbiol.* 86:39–48

20. Buris RH, Wilson PW. 1939. Respiratory enzyme systems in symbiotic nitrogen fixation. *Cold Spring Harbor Symp. Quant. Biol.* 7:349–61

21. Carlson TA, Martin GB, Chelm BK. 1987. Differential transcription of the two glutamine synthetase genes of *Bradyrhizobium japonicum. J. Bacteriol.* 169:5861–66

21a. Christiansen JH, Rosendahl L, Widell S. 1995. Preparation and characterization of sealed inside-out vesicles from *Pisum sativum* L. and *Glycine max* L. root nodules by aqueous polymer two-phase partitioning. *J. Plant Physiol.* 147:175–81

22. Copeland L, Vella J, Hong Z. 1989. Enzymes of carbohydrate metabolism in soybean nodules. *Phytochemistry* 28:57–61

23. Cullimore JV, Bennett MJ. 1988. The molecular biology and biochemistry of plant glutamine synthetase from root nodules of *Phaseolus vulgaris* L. and other legumes. *J. Plant Physiol.* 132:387–93

24. Day DA, Copeland L. 1991. Carbon metabolism and compartmentation in nitrogen fixing legume nodules. *Plant Physiol. Biochem.* 29:185–201

25. Day DA, Ou Yang L-J, Udvardi MK. 1990. Nutrient exchange across the peribacteroid membrane of isolated symbiosomes. In *Nitrogen Fixation: Achievements and Objectives*, ed. PM Gresshoff, LE Roth, G Stacey, WE Newton, pp. 219–26. New York: Chapman & Hall

26. Day DA, Price GD, Udvardi MK. 1989. Membrane interface of the *Bradyrhizobium japonicum–Glycine max* symbiosis: peribacteroid units from soybean nodules. *Aust. J. Plant Physiol.* 16:69–84

27. de Bruijn FJ, Rossbach S, Schneider M, Ratet P, Messmer S, et al. 1989. *Rhizobium meliloti* 1021 has three differentially regulated loci involved in glutamine biosynthesis, none of which is essential for symbiotic nitrogen fixation. *J. Bacteriol.* 171:1673–82

28. Denarie J, Cullimore J. 1993. Lipo-oligosaccharide nodulation factors: a new class of signaling molecules mediating recognition and morphogenesis. *Cell* 74:951–54

29. de Vries GE, van Brussel AAN, Quispel A. 1982. Mechanism of regulation of glucose transport in *Rhizobium leguminosarum. J. Bacteriol.* 149:872–79

30. Dilworth M, Glenn A. 1984. How does a legume nodule work? *Trends Biol. Sci.* 9: 519–23
31. Dilworth MJ, Arwas R, McKay IA, Saroso S, Glenn AR. 1986. Pentose metabolism in *Rhizobium leguminosarum* MNF300 and in cowpea *Rhizobium* NGR234. *J. Gen. Microbiol.* 132:2733–42
32. Dilworth MJ, Glenn A. 1981. Control of carbon substrate utilization by rhizobia. In *Current Perspectives in Nitrogen Fixation,* ed. AH Gibson, WE Newton, pp. 244–51. Aust. Acad. Sci.
32a. Dilworth MJ, Glenn AR, eds. 1991. *Biology and Biochemistry of Nitrogen Fixation.* Amsterdam: Elsevier
33. Domigan NM, Farnden KJF, Robertson JG, Monk BC. 1988. Characterization of the peribacteroid membrane ATPase of lupin root nodules. *Arch. Biochem. Biophys.* 264: 564–73
34. Driscoll BT, Finan TM. 1993. NAD$^+$-dependent malic enzyme of *Rhizobium meliloti* is required for symbiotic nitrogen fixation. *Mol. Microbiol.* 7:865–73
35. Dubrovo PN, Krylova VV, Livanova GI, Zhiznevskaya GY, Izmailov SF. 1992. Properties of ATPases of the peribacteroid membrane in root nodules of yellow lupine. *Sov. Plant Physiol.* 39:318–24
36. Duncan MJ. 1979. L-Arabinose metabolism in rhizobia. *J. Gen. Microbiol* 113: 177–79
37. Duncan MJ, Fraenkel DG. 1979. α-Ketoglutarate dehydrogenase mutant of *Rhizobium meliloti. J. Bacteriol.* 137:415–19
38. el Din AKYG. 1992. A succinate transport mutant of *Bradyrhizobium japonicum* forms ineffective nodules on soybeans. *Can. J. Microbiol.* 38:230–34
39. Emerich DW, Anthon GE, Hayes RR, Karr DB, Liang R, et al. 1988. Metabolism of *Rhizobium*-leguminous plant nodules with an emphasis on bacteroid carbon metabolism. In *Nitrogen Fixation: Hundred Years After,* ed. H Bothe, FJ de Bruijn, WE Newton, pp. 539–46. Stuttgart: Fischer
40. Engelke TH, Jagadish MN, Puhler A. 1987. Biochemical and genetical analysis of *Rhizobium meliloti* mutants defective in C₄-dicarboxylate transport. *J. Gen. Microbiol.* 133:3019–29
41. Engelke T, Jording D, Kapp D, Pühler A. 1989. Identification and sequence analysis of the *Rhizobium meliloti dctA* gene encoding the C₄-dicarboxylate carrier. *J. Bacteriol.* 171:5551–60
41a. Evans HJ, Bottomley PJ, Newton WE, eds. 1985. *Nitrogen Fixation Research Progress.* Dordrecht: Nijhoff

42. Finan TM, McWhinnie E, Driscoll B, Watson RJ. 1991. Complex symbiotic phenotypes result from gluconeogenic mutations in *Rhizobium meliloti. Mol. Plant-Microbe Interact.* 4:386–92
43. Finan TM, Wood JM, Jordan DC. 1981. Succinate transport in *Rhizobium leguminosarum. J. Bacteriol.* 148:193–202
44. Finan TM, Wood JM, Jordan DC. 1983. Symbiotic properties of C₄-dicarboxylic acid transport mutants of *Rhizobium leguminosarum. J. Bacteriol.* 154:1403–13
45. Fischer HM. 1994. Genetic regulation of nitrogen fixation in rhizobia. *Microbiol. Rev.* 58:352–86
46. Fisher RF, Long SR. 1992. *Rhizobium*–plant signal exchange. *Nature* 357: 655–60
47. Fortin MG, Morrison NA, Verma DPS. 1987. Nodulin-26, a peribacteroid membrane nodulin is expressed independently of the development of the peribacteroid compartment. *Nucleic Acids Res.* 15: 813–24
48. Fortin MG, Zelechowska M, Verma DPS. 1985. Specific targeting of membrane nodulins to the bacteroid-enclosing compartment in soybean nodules. *EMBO J.* 4: 3041–46
49. Gardiol A, Arias A, Cervenansky C, Martinez-Drets G. 1982. Succinate dehydrogenase mutant of *Rhizobium meliloti. J. Bacteriol.* 151:1621–23
50. Gardiol AE, Truchet GL, Dazzo FB. 1987. Requirement of succinate dehydrogenase activity for symbiotic bacteroid differentiation of *Rhizobium meliloti* in alfalfa nodules. *Appl. Environ. Microbiol.* 53:1947–50
51. Glenn AR, Dilworth MJ. 1981. The uptake and hydrolysis of disaccharides by fast- and slow-growing species of *Rhizobium. Arch. Microbiol.* 129:233–39
52. Glenn AR, Dilworth MJ. 1984. Methylamine and ammonium transport systems in *Rhizobium leguminosarum* MNF3841. *J. Gen. Microbiol.* 130:1961–68
53. Glenn AR, McKay IA, Arwas R, Dilworth MJ. 1984. Sugar metabolism and the symbiotic properties of carbohydrate mutants of *Rhizobium leguminosarum. J. Gen. Microbiol.* 130:239–45
54. Glenn AR, Poole PS, Hudman JF. 1980. Succinate uptake by free living and bacteroid forms of *Rhizobium leguminosarum. J. Gen. Microbiol.* 119:267–71
55. Gober JW, Kashket ER. 1983. Methylammonium uptake by *Rhizobium* sp. strain 32H1. *J. Bacteriol.* 153:1196–201
56. Grimes PM, Fottrell PF. 1966. Enzymes

involved in glutamate metabolism in legume root nodules. *Nature* 212:295–96

57. Guerinot ML, Yi Y. 1994. Iron: nutritious, noxious, and not readily available. *Plant Physiol.* 104:815–20

58. Gussin GN, Ronson CW, Ausubel FM. 1986. Regulation of nitrogen fixation genes. *Annu. Rev. Genet.* 20:567–91

59. Haaker H, Szafran MM, Wassink HJ, Appels MA. 1995. Malate, aspartate and proton exchange between *Rhizobium leguminosarum* symbiosomes and its symbiotic partner *Pisum sativum*. See Ref. 152a, pp. 565–72

60. Herrada G, Puppo A, Rigaud J. 1989. Uptake of metabolites by bacteroid-containing vesicles and by free bacteroids from French bean nodules. *J. Gen. Microbiol.* 135: 3165–71

61. Hirschman J, Wong P-K, Sei K, Keener J, Kustu S. 1985. Products of nitrogen regulatory genes *ntrA* and *ntrC* of enteric bacteria activate *glnA* transcription *in vitro:* evidence that the *ntrA* product is a sigma factor. *Proc. Natl. Acad. Sci. USA* 82: 7525–29

62. Hornez JP, El Guezzar M, Derieux JC. 1989. Succinate transport in *Rhizobium meliloti:* characteristics and impact on symbiosis. *Curr. Microbiol.* 19:207–12

63. Hornez JP, Timinouni M, Defives C, Derieux JC. 1994. Unaffected nodulation and nitrogen-fixation in carbohydrate pleiotropic mutants of *Rhizobium meliloti*. *Curr. Microbiol.* 28:225–29

64. Howitt SM, Gresshoff PM. 1985. Ammonia regulation of glutamine synthetase in *Rhizobium* sp. ANU289. *J. Gen. Microbiol.* 131:1433–40

65. Howitt SM, Udvardi MK, Day DA, Gresshoff PM. 1986. Ammonia transport in free-living and symbiotic *Rhizobium* sp. ANU289. *J. Gen. Microbiol.* 132:257–61

66. Hudman JF, Glenn AR. 1980. Glucose uptake by free living and bacteroid forms of *Rhizobium leguminosarum*. *Arch. Microbiol.* 128:72–77

67. Humbeck C, Werner D. 1987. Two succinate uptake systems in *Bradyrhizobium japonicum*. *Curr. Microbiol.* 14:259–62

68. Hunt TP, Magasanik B. 1985. Transcription of *glnA* by purified *Escherichia coli* components: core RNA polymerase and the products of *glnF, glnG* and *glnL. Proc. Natl. Acad. Sci. USA* 82:8453–57

69. Jacobs FA, Zhang M, Fortin MG, Verma DPS. 1987. Several nodulins of soybean share structural domains but differ in their subcellular locations. *Nucleic Acids Res.* 15:1271–80

70. Jin HN, Glenn AR, Dilworth MJ. 1988. Ammonium uptake by cowpea *Rhizobium* strain MNF 2030 and *Rhizobium trifolii* MNF 1001. *Arch. Microbiol.* 149:308–11

71. Jin HN, Glenn AR, Dilworth MJ. 1990. How does L-glutamate transport relate to selection of mixed nitrogen sources in *Rhizobium leguminosarum* biovar *trifolii* MNF1000 and cowpea *Rhizobium* MNF2030? *Arch. Microbiol.* 153:448–54

72. Jording D, Uhde C, Schmidt R, Pühler A. 1994. The C4-dicarboxylate transport-system of *Rhizobium meliloti* and its role in nitrogen fixation during symbiosis with alfalfa *(Medicago sativa)*. *Experientia* 50: 874–83

73. Kahn ML, Kraus J, Somerville JE. 1985. A model of nutrient exchange in the *Rhizobium*-legume symbiosis. See Ref. 41a, pp. 193–99

74. Kahn ML, Mortimer M, Park KS, Zhang W. 1995. Carbon metabolism in the *Rhizobium*-legume symbiosis. See Ref. 152a, pp. 525–32

75. Katinakis P, Lankhorst RMK, Louwerse J, van Kammen A, van den Bos RC. 1988. Bacteroid-encoded proteins are secreted into the peribacteroid space by *Rhizobium leguminosarum*. *Plant Mol. Biol.* 11: 183–90

76. Kennedy IR. 1966. Primary products of symbiotic nitrogen fixation. II. Pulse-labeling of serradella nodules with $^{15}N_2$. *Biochim. Biophys. Acta* 130:295–303

77. Kerppola TK, Kahn ML. 1988. Symbiotic phenotypes of auxotrophic mutants of *Rhizobium meliloti* 104A14. *J. Gen. Microbiol.* 134:913–19

78. Kim YS, Chae HZ. 1990. A model of nitrogen flow by malonamate in *Rhizobium japonicum*-soybean symbiosis. *Biochem. Biophys. Res. Commun.* 169:692–99

79. Kleiner D. 1985. Bacterial ammonium transport. *FEMS Microbiol. Rev.* 32: 87–100

80. Kohl DH, Schubert KR, Carter MB, Hagedorn CH, Shearer G. 1988. Proline metabolism in N_2-fixing root nodules: energy transfer and regulation of purine synthesis. *Proc. Natl. Acad. Sci. USA* 85:2036–40

81. Kohl DH, Straub PF, Shearer G. 1994. Does proline play a special role in bacteroid metabolism? *Plant Cell Environ.* 17:1257–62

82. Kouchi H, Fukai K, Kihara A. 1991. Metabolism of glutamate and aspartate in bacteroids isolated from soybean root nodules. *J. Gen. Microbiol.* 137:2901–10

83. Kouchi H, Hata S. 1993. Isolation and characterization of novel cDNAs representing genes expressed at early stages of soybean

nodule development. *Mol. Gen. Genet.* 238:106–19

84. Kouchi H, Yoneyama T. 1984. Dynamics of carbon photosynthetically assimilated in nodulated soya plants grown under steady state conditions. *Ann. Bot.* 53:883–96

85. Kurz WGW, LaRue TA. 1977. Citric acid cycle enzymes and nitrogenase in nodules of *Pisum sativum. Can. J. Microbiol.* 23: 1197–200

86. Kurz WGW, Rokosh DA, LaRue TA. 1975. Enzymes of ammonia assimilation in *Rhizobium leguminosarum* bacteroids. *Can. J. Microbiol.* 21:1009–12

87. Laane C, Krone W, Konings W, Haaker H, Veeger C. 1980. Short-term effect of ammonium chloride on nitrogen fixation by *Azotobacter vinelandii* and by bacteroids of *Rhizobium leguminosarum. Eur. J. Biochem.* 103:39–46

88. Lee JW, Zhang Y, Weaver CD, Shomer NH, Louis CF, Roberts DM. 1995. Phosphorylation of nodulin 26 on serine 262 affects its voltage-sensitive channel activity in planar lipid bilayers. *J. Biol. Chem.* 270: 27051–57

89. LeVier K, Day DA, Guerinot ML. 1996. Iron uptake by symbiosomes from soybean root nodules. *Plant Physiol.* 111:893–900

90. Magasanik B. 1982. Genetic control of nitrogen assimilation in bacteria. *Annu. Rev. Genet.* 16:135–68

91. Marsh SD, Wyza RE, Evans WR. 1984. Uptake of ammonia and methylamine by free-living and symbiotic *Rhizobium. Plant Physiol.* 75 (Suppl. 155):28 (Abstr.)

92. Martin GB, Chapman KA, Chelm BK. 1988. Role of the *Bradyrhizobium japonicum ntrC* gene product in differential regulation of the glutamine synthetase II gene (*glnII*). *J. Bacteriol.* 170:5452–59

93. McDermott TR, Griffith SM, Vance CP, Graham PH. 1989. Carbon metabolism in *Bradyrhizobium japonicum* bacteroids. *FEMS Microbiol. Rev.* 63:327–40

94. McDermott TR, Kahn ML. 1992. Cloning and mutagenesis of the *Rhizobium meliloti* isocitrate dehydrogenase gene. *J. Bacteriol.* 174:4790–97

95. McKay IA, Dilworth MJ, Glenn AR. 1988. C_4-dicarboxylate metabolism in free-living and bacteroid forms of *Rhizobium leguminosarum* MNF3841. *J. Gen. Microbiol.* 134:1433–40

96. McKay IA, Glenn AR, Dilworth MJ. 1985. Gluconeogenesis in *Rhizobium leguminosarum* MNF3841. *J. Gen. Microbiol.* 131:2067–73

97. McRae DG, Miller RW, Berndt WB, Joy K. 1989. Transport of C_4-dicarboxylates and amino acids by *Rhizobium meliloti* bacteroids. *Mol. Plant Microbe Interact.* 2: 273–78

98. Mellor RB. 1989. Bacteroids in the *Rhizobium*-legume symbiosis inhabit a plant internal lytic compartment: implications for other microbial endosymbioses. *J. Exp. Bot.* 40:831–39

99. Merrick MJ, Gibbins JR. 1985. The nucleotide sequence of the nitrogen-regulation gene *ntrA* of *Klebsiella pneumoniae* and comparison with conserved features in bacterial RNA polymerase sigma factors. *Nucleic Acids Res.* 13:7607–19

100. Millar AH, Day DA, Bergersen FJ. 1995. Microaerobic respiration and oxidative phosphorylation by soybean nodule mitochondria: implications for nitrogen fixation. *Plant Cell Environ.* 18:715–26

101. Miller JH. 1972. *Experiments in Molecular Genetics.* New York: Cold Spring Harbor Lab. Press

102. Miller RW, McRae DG, Al-Jobore A, Berndt WB. 1988. Respiration supported nitrogenase activity of isolated *Rhizobium meliloti* bacteroids. *J. Cell. Biochem.* 38: 35–49

103. Moreau S, Meyer J-M, Puppo A. 1995. Uptake of iron by symbiosomes and bacteroids from soybean nodules. *FEBS Lett.* 361:225–28

104. Nirunsuksiri W, Sengupta-Gopalan C. 1990. Characterization of a novel nodulin gene in soybean that shares sequence similarity to the gene for nodulin-24. *Plant. Mol. Biol.* 15:835–49

105. Nixon BT, Ronson CW, Ausubel FM. 1986. Two-component regulatory systems responsive to environmental stimuli share strongly conserved domains with the nitrogen assimilation regulatory genes *ntrB* and *ntrC. Proc. Natl. Acad. Sci. USA* 83: 7850–54

106. O'Brian MR. 1996. Heme synthesis in the *Rhizobium*-legume symbiosis: a palette for bacterial and eukaryotic pigments. *J. Bacteriol.* 178:2471–78

107. O'Brian MR, Maier RJ. 1989. Molecular aspects of the energetics of nitrogen fixation in *Rhizobium*-legume symbioses. *Biochim. Biophys. Acta* 974:229–46

108. O'Hara GW, Riley IT, Glenn AR, Dilworth MJ. 1985. The ammonium permease of *Rhizobium leguminosarum* MNF3841. *J. Gen. Microbiol.* 131:757–64

109. Oresnik IJ, Atkins CA, Layzell DB. 1995. The legume symbiosis: C-limited bacteria living within O_2 limited plant cells? See Ref. 152a, pp. 601

110. Ou Yang L-J, Day DA. 1992. Transport

properties of symbiosomes isolated from siratro nodules. *Plant Physiol. Biochem.* 30:613–23

111. Ou Yang L-J, Udvardi MK, Day DA. 1990. Specificity and regulation of the dicarboxylate carrier on the peribacteroid membrane of soybean nodules. *Planta* 182:437–44

112. Ou Yang L-J, Whelan J, Weaver CD, Roberts DM, Day DA. 1991. Protein phosphorylation stimulates the rate of malate uptake across the peribacteroid membrane of soybean nodules. *FEBS Lett.* 293: 188–90

113. Pargent W, Kleiner D. 1985. Characteristics and regulation of ammonium (methylammonium) transport in *Rhizobium. meliloti. FEMS Microbiol. Lett.* 30:257–59

114. Pawlowski K, Ratet P, Schell J, de Bruijn FJ. 1987. Cloning and characterization of *nifA* and *ntrC* genes of the stem nodulating bacterium ORS571, the nitrogen fixing symbiont of *Sesbania rostrata:* regulation of nitrogen fixation (*nif*) genes in the free living versus symbiotic state. *Mol. Gen. Genet.* 206:207–19

115. Peterson JB, La Rue TA. 1981. Utilization of aldehydes and alcohols by soybean bacteroids. *Plant Physiol.* 68:489–93

116. Poole PS, Franklin M, Glenn AR, Dilworth MJ. 1985. The transport of L-glutamate by *Rhizobium leguminosarum* involves a common amino acid carrier. *J. Gen. Microbiol.* 131:1441–48

117. Price GD, Day DA, Gresshoff PM. 1987. Rapid isolation of intact peribacteroid envelopes from soybean nodules and demonstration of selective permeability to metabolites. *J. Plant Physiol.* 130:157–64

118. Radyukina NL, Bruskova RK, Izmailov SF. 1992. Transport of ^{14}C substrate through peribacteroidal membrane of yellow-lupine nodules. *Dokl. Akad. Nauk* 323:603–6

119. Rastogi V, Labes M, Finan T, Watson R. 1992. Overexpression of the *dctA* gene in *Rhizobium meliloti:* effect on transport of C_4 dicarboxylates and symbiotic nitrogen fixation. *Can. J. Microbiol.* 38:555–62

120. Rawsthorne S, Minchin FR, Summerfield RJ, Cookson C, Coombs J. 1980. Carbon and nitrogen metabolism in legume root nodules. *Phytochemistry* 19:341–55

121. Reibach PH, Streeter JG. 1984. Evaluation of active versus passive uptake of metabolites by *Rhizobium japonicum* bacteroids. *J. Bacteriol.* 159:47–52

122. Reitzer LJ, Magasanik B. 1986. Transcription of *glnA* in *E. coli* is stimulated by activator bound to sites far from the promoter. *Cell* 45:785–92

123. Robertson JG, Farden KJF. 1980. Ultra structure and metabolism of the developing root nodule. In *The Biochemistry of Plants,* ed. BJ Miflin, 5:65–113. New York: Academic

124. Robertson JG, Lyttleton P. 1984. Division of peribacteroid membranes in root nodules of white clover. *J. Cell Sci.* 69:147–57

125. Robertson JG, Lyttleton P, Bullivant S, Grayston GF. 1978. Membranes in lupin root nodules I. The role of Golgi bodies in the biogenesis of infection threads and peribacteroid membranes. *J. Cell Sci.* 30: 129–49

126. Robertson JG, Warburton MP, Lyttleton P, Fordyce AM, Bullivant S. 1978. Membranes in lupin root nodules II. Preparation and properties of peribacteroid membranes and bacteroid envelope inner membranes from developing lupin nodules. *J. Cell Sci.* 30:151–74

127. Robson AD. 1978. Mineral nutrients limiting nitrogen fixation in legumes. In *The Mineral Nutrition of Legumes in Tropical and Subtropical Soils,* ed. CS Andrew, EJ Kamprath, pp. 277–93. Melbourne: CSIRO

128. Rolfe BG, Gresshoff PM. 1988. Genetic analysis of legume nodule initiation. *Annu. Rev. Plant Physiol. Plant Mol. Biol.* 39: 297–319

129. Ronson CW, Lyttleton P, Robertson JG. 1981. C_4-dicarboxylate transport mutants of *Rhizobium trifolii* form ineffective nodules on *Trifolium repens. Proc. Natl. Acad. Sci. USA* 78:4284–88

130. Ronson CW, Nixon BT, Albright LM, Ausubel FM. 1987. *Rhizobium meliloti ntrA* (*rpoN*) gene is required for diverse metabolic functions. *J. Bacteriol.* 169: 2424–31

131. Ronson CW, Primrose SB. 1979. Carbohydrate-metabolism in *Rhizobium trifolii:* identification and symbiotic properties of mutants. *J. Gen. Microbiol.* 112:77–88

132. Rosendahl L, Glenn AR, Dilworth MJ. 1991. Organic and inorganic inputs into legume root nodule nitrogen fixation. See Ref. 32a, pp. 259–92

133. Rosendahl L, Dilworth MJ, Glenn AR. 1992. Exchange of metabolites across the peribacteroid membrane in pea root nodules. *J. Plant. Physiol.* 139:635–38

134. Roth E, Jeon K, Stacey G. 1988. Homology in endosymbiotic systems: The term "symbiosome." In *Molecular Genetics of Plant-Microbe Interactions,* ed. R Palacios, DPS Verma, pp. 220–25. St. Paul: Am. Phytopathol. Soc.

135. Salminen SO, Streeter JG. 1987. Involvement of glutamate in the respiratory meta-

bolism of *Bradyrhizobium japonicum* bacteroids. *J. Bacteriol.* 169:495–99

136. San Francisco MJD, Jacobson GR. 1985. Uptake of succinate and malate in cultured cells and bacteroids of two slow-growing species of *Rhizobium. J. Gen. Microbiol.* 131:765–73

137. Sandal NN, Marcker KA. 1988. Soybean nodulin-26 is homologous to the major intrinsic protein of the bovine lens fiber membrane. *Nucleic Acids Res.* 16:9347

138. Sandal NN, Marcker KA. 1994. Similarities between a soybean nodulin, *Neurospora crassa* sulphate permease II and a putative human tumour suppressor. *Trends Biol. Sci.* 19:19

139. Saroso S, Dilworth MJ, Glenn AR. 1986. The use of activities of carbon catabolic enzymes as a probe for the carbon nutrition of snake bean nodule bacteroids. *J. Gen. Microbiol.* 132:243–49

140. Schramm RW. 1992. Proposed role of malonate in legume nodules. *Symbiosis* 14:103–13

141. Schubert KR. 1986. Products of biological nitrogen fixation in higher plants: synthesis, transport and metabolism. *Annu. Rev. Plant Physiol.* 37:539–74

142. Shatters RG, Somerville JE, Kahn ML. 1989. Regulation of glutamine synthetase II activity in *Rhizobium meliloti* 104A14. *J. Bacteriol.* 171:5087–94

143. Deleted in proof

144. Stanley J, van Slooten J, Dowling DN, Finan T, Broughton WJ. 1989. Molecular cloning of the *ntrA* gene of the broad host-range *Rhizobium* sp. NGR234, and phenotypes of a site-directed mutant. *Mol. Gen. Genet.* 217:528–32

145. Stowers MD. 1985. Carbon metabolism in *Rhizobium* species. *Annu. Rev. Microbiol.* 39:89–108

146. Stowers MD, Elkan GH. 1983. The transport and metabolism of glucose in cowpea rhizobia. *Can. J. Microbiol.* 29:398–406

147. Streeter JG. 1989. Estimation of ammonia concentration in the cytosol of soybean nodules. *Plant Physiol.* 90:779–82

148. Streeter JG. 1991. Transport and metabolism of carbon and nitrogen in legume nodules. *Adv. Bot. Res.* 18:129–87

149. Streeter JG. 1995. Integration of plant and bacterial metabolism in nitrogen fixing systems. See Ref. 152a, pp. 67–76

150. Streeter JG. 1995. Recent developments in carbon transport and metabolism in symbiotic systems. *Symbiosis* 19:175–96

151. Szafran MM, Haaker H. 1995. Properties of the peribacteroid membrane ATPase of pea root nodules and its effect on the nitro-

genase activity. *Plant Physiol.* 108:1227–32

152. Szeto WW, Nixon BT, Ronson CW, Ausubel FM. 1987. Identification and characterization of the *Rhizobium meliloti ntrC* gene: *R. meliloti* has separate regulatory pathways for activation of nitrogen fixation genes in free-living and symbiotic cells. *J. Bacteriol.* 169:1423–32

152a. Tikhonovich IA, Provorov NA, Romanov VI, Newton WE, eds. 1995. *Nitrogen Fixation: Fundamentals and Applications.* Dordrecht: Kluwer

153. Trinchant J-C, Guerin V, Rigaud J. 1994. Acetylene reduction by symbiosomes and free bacteroids from broad bean (*Vicia faba* L.) nodules: role of oxalate. *Plant Physiol.* 105:555–61

154. Tuzimura K, Meguro H. 1960. Respiration substrate of *Rhizobium* in the nodules. *J. Biochem.* 47:391–97

155. Tyerman SD, Whitehead LF, Day DA. 1995. A channel-like transporter for NH_4^+ on the symbiotic interface of N_2-fixing plants. *Nature* 378:629–32

156. Udvardi MK, Day DA. 1989. Electrogenic ATPase activity on the peribacteroid membrane of soybean (*Glycine max* L.) root nodules. *Plant Physiol.* 90:982–87

157. Udvardi MK, Day DA. 1990. Ammonia (^{14}C-methylamine) transport across the bacteroid and peribacteroid membranes of soybean root nodules. *Plant Physiol.* 94:71–76

158. Udvardi MK, Kahn ML. 1993. Evolution of the (*Brady*) *Rhizobium*-legume symbiosis: why do bacteroids fix nitrogen? *Symbiosis* 14:87–101

159. Udvardi MK, Lister DL, Day DA. 1991. ATPase activity and anion transport across the peribacteroid membrane of isolated soybean symbiosomes. *Arch. Microbiol.* 156:362–66

160. Udvardi MK, Lister DL, Day DA. 1992. Isolation and characterization of a *ntrC* mutant of *Bradyrhizobium* (*Parasponia*) sp. ANU289. *J. Gen. Microbiol.* 138:1019–25

161. Udvardi MK, Ou Yang L-J, Young S, Day DA. 1990. Sugar and amino acid transport across symbiotic membranes from soybean nodules. *Mol. Plant-Microbe Interact.* 3:334–40

162. Udvardi MK, Price GD, Gresshoff PM, Day DA. 1988. A dicarboxylate transporter on the peribacteroid membrane of soybean nodules. *FEBS Lett.* 231:36–40

163. Udvardi MK, Salom CL, Day DA. 1988. Transport of L-glutamate across the bacteroid membrane but not the peribacteroid

membrane from soybean root nodules. *Mol. Plant-Microbe Interact.* 1:250–54

164. Vance CP, Heichel GH. 1991. Carbon in N_2 fixation: limitation or exquisite adaptation. *Annu. Rev. Plant Physiol. Plant Mol. Biol.* 42:373–92

165. van Slooten JC, Bhuvanasvari TV, Bardin S, Stanley J. 1992. Two C_4-dicarboxylate transport systems in *Rhizobium* sp. NGR234: rhizobial dicarboxylate transport is essential for nitrogen fixation in tropical legume symbioses. *Mol. Plant-Microbe Interact.* 5:179–86

166. Verma DPS. 1992. Signals in root nodule organogenesis and endocytosis of *Rhizobium. Plant Cell* 4:373–82

167. Verma DPS, Gu X, Hong Z. 1995. Biogenesis of the peribacteroid membrane in root nodules: roles of dynamin and phosphatidyl-inositol-3-kinase. See Ref. 152a, pp. 467–70

168. Vincent JM, Humphrey BA. 1963. Partition of divalent cations between bacterial wall and cell contents. *Nature* 199:149–51

169. Wang Y-P, Birkenhead K, Boesten B, Manian S, O'Gara F. 1989. Genetic analysis and regulation of the *Rhizobium meliloti* genes controlling C_4-dicarboxylic acid transport. *Gene* 85:135–44

170. Watson RJ. 1990. Analysis of the C_4-dicarboxylate transport genes of *Rhizobium meliloti*: nucleotide sequence and deduced products of *dctA, dctB,* and *dctD. Mol. Plant-Microbe Interact.* 3:174–81

171. Watson RJ, Chan Y-K, Wheatcroft R, Yang A-F, Han SH. 1988. *Rhizobium meliloti* genes required for C_4-dicarboxylate transport and symbiotic nitrogen fixation are located on a megaplasmid. *J. Bacteriol.* 170:927–34

172. Watson RJ, Rastogi VK, Chan Y-K. 1993. Aspartate transport in *Rhizobium meliloti. J. Gen. Microbiol.* 139:1315–23

173. Weaver CD, Crombie B, Stacey G, Roberts DM. 1991. Calcium-dependent phosphory-

lation of symbiosome membrane proteins from nitrogen-fixing soybean nodules. *Plant Physiol.* 95:222–27

174. Weaver CD, Roberts DM. 1992. Determination of the site of phosphorylation of nodulin 26 by the calcium-dependent protein kinase from soybean nodules. *Biochemistry* 31:8954–59

175. Weaver CD, Shomer NH, Louis CF, Roberts DM. 1994. Nodulin 26, a nodule-specific symbiosome membrane protein from soybean, is an ion channel. *J. Biol. Chem.* 269:17858–62

176. Werner D, Dittrich W, Thierfelder H. 1982. Malonate and Krebs cycle intermediates utilization in the presence of other carbon sources by *Rhizobium japonicum* and soybean bacteroids. *Z. Naturforsch. Teil C* 37: 921–26

177. Werner D, Mellor RB, Hahn MG, Greisebach H. 1985. Soybean root response to symbiotic infection. Glyceollin I accumulation in an ineffective type of soybean nodule with an early loss of peribacteroid membrane. *Z. Naturforsch. Teil C* 40: 179–81

178. Whitehead LF, Tyerman SD, Salom CL, Day DA. 1995. Transport of fixed nitrogen across symbiotic membranes of legume nodules. *Symbiosis* 19:141–54

179. Wittenberg JB, Wittenberg BA, Day DA, Udvardi MK, Appleby CA. 1996. Siderophore-bound iron in the peribacteroid space of soybean root nodules. *Plant and Soil* 178:161–69

180. Yarosh OK, Charles TC, Finan TM. 1989. Analysis of C_4-dicarboxylate transport genes in *Rhizobium meliloti. Mol. Microbiol.* 3:813–23

181. Zhu Y, Shearer G, Kohl DH. 1992. Proline fed to intact soybean plants influences acetylene reducing activity and content and metabolism of bacteroids. *Plant Physiol.* 98:1020–28

Annu. Rev. Plant Physiol. Plant Mol. Biol. 1997. 48:525–45

PROGRAMMED CELL DEATH IN PLANT-PATHOGEN INTERACTIONS

Jean T. Greenberg

Department of Molecular, Cellular and Developmental Biology, University of Colorado at Boulder, Campus Box 347, Boulder, Colorado 80309

KEY WORDS: cell death, hypersensitive response, apoptosis

ABSTRACT

Plants cope with pathogen attacks by using mechanisms of resistance that rely both on preformed protective defenses and on inducible defenses. The latter are the most well studied, and progress is being made in determining which induced responses are responsible for limiting pathogen growth. Many plant-pathogen interactions are accompanied by plant cell death. Recent evidence suggests that this cell death is often programmed and results from an active process on the part of the host. The review considers the roles and possible mechanisms of plant cell death in response to pathogens.

CONTENTS

525

INTRODUCTION

Plants' defense mechanisms against pathogens divide into two classes: those that are present constitutively [so-called preformed defenses (58 and references therein)] and those that are induced upon exposure to a pathogen. The latter class of defense mechanisms has been the subject of intense interest because of the possibility of exploiting these natural inducible defenses to engineer broad-spectrum pathogen resistance. Although some pathogens do not appear to elicit a robust defense response on their hosts (e.g. 13), some pathogens do elicit plant defenses but can still parasitize the host, possibly because some pathogens can grow or develop faster than the host can elaborate its defenses or because the pathogen can tolerate the induced defenses or detoxify them (e.g. 87). In this context, it is useful to consider that plant-pathogen interactions are dynamic: There may be much communication between the host and the parasite, and the outcome of the interaction may reflect the attempts of the host and the parasite to adapt to defenses and virulence factors, respectively. The most robust resistance to pathogens often occurs when a plant can specifically recognize the pathogen and rapidly induce a variety of potential defenses that limit its growth and/or development.

Often the interaction between a plant and a pathogen results in plant cell death. Plant cell death can occur when the pathogen unsuccessfully parasitizes the host as well as when the pathogen successfully causes disease. The effects of host cell death on a pathogen may vary greatly depending on the lifestyle of the parasite. For example, some pathogens are obligate parasites that depend on living cells to grow and develop. Other pathogens may benefit from the release of nutrients from dead cells. The purpose of this review is to reevaluate the role and regulation of plant cell death during plant-pathogen interactions and to examine the relationship between cell death and the activation of other defenses.

THE ROLE OF CELL DEATH DURING PATHOGENESIS

When a plant is infected by a pathogen that can grow extensively, the interaction is called "susceptible." Cell death during such a susceptible interaction may benefit pathogens that do not rely on living tissue to grow and develop. Cell death can be useful for obtaining nutrients and providing a reservoir for pathogen dissemination if the pathogen is released onto the surface of the plant. However, is cell death really beneficial to the pathogen? If, for example, the cell death is accompanied by dehydration in a dry environment, it may be deleterious to the pathogen. To answer this question, it is important to look at the whole life cycle of a pathogen as well as the environmental conditions in which it grows. For example, many foliar bacterial pathogens that cause the death of the host leaf cells are thought to gain access to plant tissue through wound sites. These pathogens, such as *Xanthomonads* and *Pseudomonads*, then multiply between the plant cells in the apoplasm. However, it is unlikely that cell death per se is required for the multiplication of these bacteria. This is because there are natural plant isolates (84) and mutants (8) that are "tolerant" to bacteria; that is, they show little or no symptoms after pathogen attack and yet they support growth of the pathogen that is equal to that seen on plants where symptoms occur. These natural and induced mutations benefit the plant because less cell death occurs during the infection. However, note that bacteria may persist longer in such tolerant tissue, and when the leaves senesce and fall into the soil they may provide a higher reservoir of bacteria than leaves that show cell death. Release of pathogens onto the surface of these natural variants and mutants and long-term survival of pathogens within these tissues need to be tested to clarify the role of cell death in susceptible interactions.

CELL DEATH MECHANISMS DURING SUSCEPTIBLE INTERACTIONS

How cell death occurs during susceptible interactions is not well understood for most plant-pathogen combinations. Tolerant mutants and natural variants that show decreased cell death during susceptible interactions but normal pathogen growth may provide clues about the mechanisms of cell death. One of three possibilities seems likely to explain the basis of pathogen-induced cell death: A toxin from the microbe might directly kill the plant cells, a secreted virulence factor might cause the plant to kill itself by causing a metabolic catastrophe (such as the disruption of membrane integrity), or an endogenous cell death program might be triggered. In one case, a mutant of Arabidopsis that was tolerant of both *Pseudomonas syringae* (the causal agent of leaf spot disease) and *Xanthomonas campestris* (the causal agent of black rot), *ein2*, was identified among mutants that were isolated because of their insensitivity to the hormone ethylene (8). Mutations in other loci that also caused ethylene insensitivity showed normal symptom development. These observations suggest that either the *ein2* mutation is less leaky than the other ethylene-insensitive mutations or that the *ein2* mutations are pleiotropic and affect both the perception of ethylene and the perception of an additional pathogen-derived signal. The possibility that ethylene has a role in symptom development is intriguing because this hormone is required for the timing of the onset of senescence in Arabidopsis (34) and fruit ripening in tomato (57, 70), two processes that result in the death of cells in processes that require de novo protein synthesis. The possible involvement of hormones and other host factors in modulating cell death responses to pathogens suggests that symptom development is not necessarily due to toxic microbial products or toxic plant products. Additional evidence that cell death during susceptible interactions is genetically programmed comes from the observation that many mutants of maize mimic pathogenic diseases at the level of the visible symptoms (45, 88, 89). However, these mutants have not been characterized extensively to determine whether they exhibit the expected biochemical markers that appear during a pathogen infection. The recent isolation of mutants of Arabidopsis that mimic leaf spot disease and do express typical biochemical defense markers seen during this disease suggests that the cell death associated with this disease is genetically determined (JT Greenberg, unpublished information).

Susceptibility to pathogens is often genetically conditioned. In the case of the tomato fungal pathogen *Alternaria alternata,* susceptibility of tomato to stem canker disease is conditioned by the *Asc* gene (17). This same gene is required for susceptibility to the *Alternaria* AAL-toxin, which causes cell death (17). Experiments with purified toxin on susceptible plants have implicated ethylene in symptom development. First, toxin-induced cell death and leaf epinasty. The latter symptom is a typical ethylene response of tomato plants. Measurement of ethylene levels of toxin-treated plants were increased, and application of ethylene inhibitors partially blocked symptom production (67). Because ethylene on its own does not induce cell death, it should be considered a modulator of cell death processes. The metabolic state of the plant also conditions susceptibility to the toxin. Thus, elevated pyrimidine biosynthetic intermediates blocked symptom development, whereas inhibiting pyrimidine biosynthesis alone gave symptoms similar to those induced by the toxin (67). Although the molecular basis of the cell death induced by toxin has not been determined, recently it was shown that the induced cell death is an example of programmed cell death (pcd) in plants (91). This is one of the first examples of pcd in plants that has many of the same characteristics as those seen in animal cells undergoing a common form of pcd called apoptosis. Thus, tomato protoplasts treated with toxin exhibit internucleosomal cleavage, DNA breaks with 3′OH ends indicative of endonucleolytic cleavage, membrane blebbing, and apoptotic bodies. As would be expected for cell death that is an active process, the induction of internucleosomal DNA fragments by toxin was dependent on protein synthesis (91). Intact tomato leaves treated with toxin also exhibited DNA cleavage (91). The same toxin also induced pcd in African green monkey kidney cells, suggesting that the mechanism of activation of pcd in plants and animal cells may be conserved (90).

Cell death that appears in many different susceptible interactions may be caused by pathogen-derived factors taking advantage of endogenous plant cell death programs to effect their functions, but this remains to be determined. In the case of AAL-toxin, the pathogen may have evolved to take advantage of pcd caused by low pyrimidine levels, which on its own caused cell death (see above). However, cells with inhibited pyrimidine synthesis have not been examined for apoptotic features. Pyrimidine levels may be limiting in senescing tissue and may serve as an endogenous trigger for cell death. Senescence itself is likely a form of pcd where nutrients are reapportioned from older tissues to reproductive ones and the older cells die under the control of newly synthesized protein (35). The strategy of a pathogen causing the host to commit suicide is common in animal-pathogen interactions (94), particularly with viruses (79). It will be interesting to determine if many plant pathogens harbor virulence factors that act by regulating host pcd or if cell death arises by many different mechanisms.

CELL DEATH DURING A RESISTANT RESPONSE

Overview of the Resistant Response

Sometimes when a plant is infected with a pathogen, it can mount a rapid defense response and effectively prevent the pathogen from growing and/or developing. Such an outcome of an interaction between a pathogen that normally causes a susceptible interaction and its host plant can occur if two conditions are met: First, the pathogen must harbor a gene called an avirulence (*avr*) gene, and second, the plant must harbor a single gene called a resistance (*R*) gene that allows the plant to specifically recognize a pathogen carrying the avirulence gene. Both functions must be present to limit pathogen growth and/or development, a phenomenon termed "gene-for-gene" resistance [reviewed by Keen (47)]. The mode of recognition of *avr* functions is not known in many cases. However, in some cases avr gene products are proteins that are either known to be directly secreted into the apoplasm (e.g. 38) or thought to be secreted directly into the plant cell using a specialized bacterial secretion apparatus (type III) common to both animal and plant pathogens (32). Not all avr products are proteins; for example, the *avrD* operon encodes proteins that synthesize a c-glycosyl lipids called a syringolides (81). Several *R* genes have been cloned that are each specific for a variety of different pathogens including *Pseudomonas syringae, Tobacco Mosaic Virus* (*TMV*), *Cladosporium fulvum,* and *Melamspora lini.* The remarkable finding about these *R* genes is that they are all similar in structure even though they are specific for pathogens with radically different modes of pathogenicity, from an intracellular virus to an extracellular bacteria to fungi that are both intra- and extracellular. The notable features of several cloned R genes are that they have leucine-rich repeats that are suggestive of a domain that interacts with other proteins and a sequence that may encode a nucleotide-binding domain (82). Other plant genes involved in resistance include typical signal transduction molecules such as the Pto kinase and the Pti kinase of tomato (61, 97).

As alluded to above, resistance to pathogens that is conditioned by a gene-for-gene interaction is an active process. Attempts to understand the basis of resistance in many different plant-pathogen systems have revealed that many potential defense functions are induced during a resistance response. These include the de novo induction of many mRNA transcripts and proteins termed defense-related proteins. Often these transcripts and proteins are induced during a susceptible interaction as well as a resistant interaction, but the timing and abundance of the transcript or protein is either faster or higher, respectively, when resistance occurs. The same phenomenon has been observed for the production of plant antibiotics (phytoalexins). For example, in Arabidopsis plants undergoing a resistance response, phytoalexins accumulate to a higher level than in a susceptible interaction (30). Defense-related proteins and phytoalexins are proposed to be important for directly limiting pathogen growth. In some cases, there is direct evidence for antimicrobial activity of inducible defense-related genes and phytoalexins. For example, overproduction of the defense-related gene $PR1a$ confers resistance to oomycete fungi (1), and loss of phytoalexin synthesis in Arabidopsis causes plants to become more susceptible to $P.$ $syringae$ pathogens (30). Some defense-related proteins require the production of the signal molecule salicylic acid (SA) to be induced during a resistance response. For example, this is the case for the TMV-induced resistance response in tobacco (19). Plants that cannot accumulate SA because of the presence of a transgene called $nahG$ from $Pseudomonas$ $putida,$ whose product converts SA into the inactive catechol, have a reduced capacity to limit pathogen growth during a resistance response (29).

Other potential defenses that are induced during a resistance response are alterations of plant cell walls: At least one protein is crosslinked in the cell walls possibly by dityrosine bridges in a process that is likely to depend on H_2O_2 (11, 12). Phenolic compounds also become crosslinked in the cell walls, which become lignified during the resistant response. Modifications of the cell wall induced during the resistant response protect the cell wall from digestion by pathogens (12). The H_2O_2 required for cell wall modifications may be supplied by the activation of a membrane-localized NADPH-dependent oxidase in a process termed the oxidative burst [reviewed by Mehdy (63)]. In some cases, it appears that the H_2O_2 comes from the dismutation of superoxide in the apoplasm (3) generated from the oxidase. It is unclear whether this superoxide is dismutated spontaneously or enzymatically because no extracellular superoxide dismutase has yet been identified.

Additional changes that occur during the resistant response include the transient accumulation of Ca^{2+} into cells (55) and the activation of H^+/K^+ exchange (2, 5), processes that may be involved in signaling (see below). Finally, most characterized resistance responses are accompanied by rapid cell death of the infected cells and to a limited degree the neighboring cells. This rapid cell death is termed the hypersensitive response (HR). Sometimes the HR is referred to interchangeably with a resistance response, but for the purpose of this review the HR refers only to cell death that is associated with the resistant response. Plants that have undergone a resistance response that includes the HR on one or a few leaves develop immunity to many other pathogens in the leaves that have not previously been exposed to pathogens. This immunity is called systemic acquired resistance, and it requires the signal SA for its induction (29).

Many of the diverse biochemical and cell biological changes that occur during the resistant response do not require the continuous exchange of information between the pathogen and the host, but rather are the result of an initial stimulus that sets in motion a cascade of plant responses. Evidence for this comes from two types of observations: A number of pathogen-derived molecules when applied to plants as purified products (called elicitors) appear to induce all aspects of the resistant response (4, 83). In addition, there are plant mutants that also bypass pathogen exposure to activate an apparent resistant response spontaneously (21, 37). The cascade of responses likely involves the activation of multiple signal transduction pathways because in the case of gene expression, many defense-related genes are activated with different kinetics in response to the same stimulus (e.g. see references 26, 37, 72). One of these transduction pathways is regulated by SA because this molecule is required for full resistance during avirulent *Pseudomonas* infection of Arabidopsis (19). However, SA is not responsible for regulating phytoalexin production (JT Greenberg, unpublished information) or *LOX1* gene expression (64), two defense-related responses that are differentially regulated during the resistant response (30, 64).

Hypersensitive Cell Death Function

Because the resistance response is correlated with diverse biochemical and cell biological changes, it is often unclear whether all these changes are necessary for the limitation of pathogen growth and/or development. It is possible that different pathogens are affected by distinct aspects of the resistant response or that some potential defenses act in a redundant fashion. In the case of the HR, it has long been speculated that the cell death is directly responsible for limiting pathogen growth and development. Evaluation of this hypothesis in the absence of a way to block only the HR during a resistant response is difficult. In the case of fungal organisms that must develop within live plant cells to grow, it is possible to make limited inferences about the role of the HR in the resistance response. This is because it is possible to detect the stage at which a fungus is arrested and determine if it occurs before, concomitant with, or after the onset of cell death. In many cases this level of detail has not been characterized. However, one elegant study by Gorg et al (33) examined the timing of cell death in the interaction between powdery mildew and barley. Different *R* genes cause barley to arrest powdery mildew at distinct steps in development. In the case of the *Mlg* resistance locus, the fungus is arrested before the establishment of the haustorium, and the percentage of germinated spores that are arrested is dependent on the gene dosage of *Mlg*. Thus, one copy of *Mlg* causes 80% arrest, and two copies causes 100% arrest. However, upon close examination, infections of heterozygous plants were not accompanied by cell death at the time of the arrest of fungal development. Homozygous plants showed cell death concomitant with the developmental arrest. The authors suggested that because arrest could occur before cell death in heterozygous plants that cell death was not an obligatory step in achieving resistance. However, they suggested that cell death might be a back-up defense that ensures quantitative resistance to the fungus by the creation of a highly antimicrobial environment. It is worth noting in this context that powdery mildew is an obligate biotroph that causes disease by establishing a haustorium in a living plant cell and then undergoes further development and reproduction. This pathogen cannot establish an infection if cell death occurs. In contrast to the *Mlg* mechanism, resistance to powdery mildew conditioned by the *Mla* locus causes arrest later in the development of the fungus at the haustorium stage and is always accompanied by cell death. However, a gene dosage experiment similar to that done with *Mlg* plants has not been performed.

Because fungal arrest can occur at different stages of development, it is difficult to generalize about the role of cell death in directly affecting fungal growth. Cell death patterns can also vary greatly during a resistance response triggered by obligate biotrophic fungi. For example, *Peronospora parasitica* can cause at least six distinct cell death patterns depending on which *R* gene the host carries (42). In some cases cell death occurs only in the infected cell, and in some cases the cell death occurs distal to the site of infection. These different patterns of cell death may reflect different roles that cell death plays in pathogen limitation. Implicit in this discussion is the idea that R genes may not all activate exactly the same distribution of potential defenses judging from the cell death pattern as well as the developmental stage at which the pathogen is arrested. This is further supported by the observation that some defense-related genes are induced during a resistant response only when a specific R gene is present (72, 73), and different physiological changes can be mediated by different R genes (39). It is likely that there are a core set of events that occur during a resistant response and that these are added upon by the particular R gene that has evolved to recognize the appropriate avr product. For example, even though the Arabidopsis R genes $Rpm1$ and $Rpt2$ recognize $avrRpm1$ and $avrRpt2$ genes, respectively, and activate some defense-related genes in an $Rpm1$- or $Rpt2$-specific manner, both R genes share at least one signal transduction step. This is evidenced by the phenotype of Arabidopsis ndr mutants, which are unable to mount effective resistance against bacteria carrying either $avrRpt2$ or $avrRpm1$ (15).

It is straightforward to imagine that cell death may play a role in quantitative resistance to biotrophic fungi because they need living host cells to achieve a successful level of parasitism. How might cell death influence growth of pathogens other than obligate biotrophic fungi? In the case of intracellular pathogens such as viruses, cell death might affect the plasmadesmata through which the virus travels from one cell to another (20). In the case of a bacterial pathogen, dehydration, which often accompanies the HR, might cause a decrease in growth or long-term survival within the plant tissue. Cell death may also cause the release of defense-related proteins and metabolites into the apoplasm where bacterial and some fungal pathogens reside.

The HR: An Example of Programmed Cell Death

Pcd is the formal term used to describe cell death that is genetically programmed and thus requires active participation by the host. The term pcd implies nothing about the mechanism by which the cell death occurs. That the HR is likely a form of pcd and not due to the action of pathogen-secreted toxins that directly kill the host is bolstered by several observations. First, plants must have active protein synthesis to show an HR induced by bacteria (18, 48). Second, purified bacterial elicitors of the HR require plants to have active metabolism to show cell death (40). These same elicitors induce the expected array of defense-related biochemical markers (83). Third, overproduction of a component of the R gene Pto signal transduction pathway of tomato expressed in tobacco caused an amplification of the HR. Finally,

Arabidopsis mutants that show an apparent HR spontaneously or in response to aberrant stimuli argue that the HR is genetically controlled and coordinately regulated with other defense-related biochemical events typically seen during the resistant response (21, 37). Some maize mutations at the complex resistance locus *rp1* also result in spontaneous cell death lesions, indicating that cell death is genetically controlled and can be caused by mutations in a signal transduction component of the resistance response (43). These observations prompted a number of groups to examine whether the pcd observed in plants might be similar to pcd observed in animals when they undergo a common type of pcd called apoptosis. Typical apoptotic events include the fragmentation of DNA into ~50-kDa pieces and/or internucleosomal-sized fragments with 3'OH ends indicative of endonucleolytic cleavage (so-called DNA ladders), membrane blebbing, nuclear condensation, and fragmentation with nucleic acids found in membrane-bound vessels (apoptotic bodies) and cytoplasmic condensation. One or more of these parameters has been examined in a few plant-pathogen systems when the HR is occurring. Tomato protoplasts treated with the HR elicitor arachidonic acid showed DNA ladders and 3'OH DNA ends (91). Soybean culture cells, intact soybean, and Arabidopsis tissue all showed DNA fragmented into large pieces and a morphology suggestive of apoptotic bodies (55). In this case, the authors found that the apparent apoptosis was only elicited by bacteria harboring an *avr* gene that could be recognized by plant cells with the corresponding *R* gene. In soybean suspension cultures, induction of the HR by bacteria was an active process that required Ca^{2+} transport and kinase activity (55).

Ryerson & Heath (74) found that cowpea infected with the obligate fungal parasite *Uromyces vignae* showed in situ evidence of 3'OH DNA ends as well as DNA laddering. Killing the cells by abiotic treatment generally did not elicit DNA laddering (74), except for KCN (74, 91). Tomato protoplasts treated with the kinase inhibitor staurosporine showed DNA fragmentation, whereas soybean suspension cells treated with this drug blocked elicitor-induced oxidative burst (56). Finally, tobacco infected with TMV showed in situ evidence of 3'OH ends in the DNA. Although diverse pathogens exhibited one or more markers of apoptosis on a variety of different plants, the HR of lettuce after bacterial infection did not show apoptotic morphological features (9). This may indicate that there are two or more mechanisms of effecting pcd during the HR. Alternatively, pcd in lettuce may share some features with the other systems listed above but then diverges at a step that determines cell morphology. For example, DNA fragmentation mediated by endonucleases may occur in lettuce, although it has not yet been examined. In animals, though apoptosis is a common way for pcd to occur, there is at least one other alternative mechanism of pcd that has been observed during development (77).

Regulation and Execution of the HR

The HR is correlated with the oxidative burst, membrane damage, ion fluxes, endonuclease activation, DNA cleavage, and gene expression. It is not clear a priori which of these events may be involved specifically in the regulation or execution of the HR and which may be involved in the other aspects of the resistance response. To sort out these questions, a number of laboratories have tried to determine the role of individual changes that occur during the resistant response that might be specifically related to the regulation or execution of the HR. In particular, much emphasis has been placed on understanding the role and regulation of the oxidative burst and its relationship to the HR. In some systems, it is possible to detect superoxide in the apoplasm of cells undergoing an HR (3, 22), whereas in other systems only H_2O_2 accumulates to detectable degrees (e.g. 56). The oxidative burst (superoxide and/or hydrogen peroxide) proceeds cell death, which also makes it a candidate for a source of signaling molecules. In animal cells, oxidative stress activates apoptosis by two independent signaling mechanisms (76), making the oxidative burst in plant cells a prime candidate for the HR trigger in plants.

To determine whether the oxidative burst is responsible for triggering the HR, Glazener et al (31) used bacterial mutants, which are defective in eliciting the HR because of a mutation in the *hrp* locus (a locus important for the secretion of virulence and avirulence factors), to infect tobacco cells in culture and monitored plant responses. They found the production of active oxygen, detected as H_2O_2, using this hrp^- strain was indistinguishable from that elicited by a wild-type strain. However, plant cell death did not occur with the hrp^- strain. Thus, using the hrp^- strain, the oxidative burst was uncoupled from the cell death response (the HR). However, Levine et al (56) have found that the oxidative burst, manifest as H_2O_2 production in their soybean suspension culture system, might play a role in triggering the HR. The oxidative burst induced by the elicitor Pmg, a glucan from the mycelial walls of the fungal pathogen *Phytophthora megasperma f.* sp. *glycinea,* is completely blocked by the kinase inhibitors K252A and staurosporine. When H_2O_2 production was enhanced by inhibiting catalase during an HR elicited by avirulent bacteria, the amount of cell death was greatly increased (56). In addition, if the oxidative burst was blocked by an inhibitor of NADPH oxidase (which generates the oxidative burst) or the kinase activity was blocked by K252A, the amount of cell death of the plant cells induced by avirulent *Pseudomonas* was decreased by a factor of two. The fact that K252A completely blocked the oxidative burst by Pmg elicitor but did not completely inhibit cell death induced by avirulent bacteria suggests that H_2O_2 may not act alone to trigger cell death. However, since the oxidative burst was not monitored after avirulent bacteria inoculation in the presence of K252A, it is also possible that K252A completely inhibits the oxidative burst by Pmg but not by avirulent bacteria. Thus, the residual active oxygen may remain when soybean cells are treated with K252A and avirulent bacteria. The results of Levine et al (56) and Glazener et al (31) indicate that H_2O_2 may not be sufficient to account for all the cell death that is observed during the resistant response. However, it is not clear whether activation of all resistance responses results in the same amount of H_2O_2 production. One possibility is that if the concentration of H_2O_2 is high enough, additional signals are not necessary. This might account for the observation that 8 mM H_2O_2 was required to induce apoptosis in suspension cultures of soybean when it was estimated that the amount of H_2O_2 that is generated during the oxidative burst of tobacco cells in response to bacteria is much lower, about 12 μmol • min^{-1} (31 and references therein). It has been suggested that because catalase may be inhibited by SA (16), a rise in H_2O_2 at the site of lesion formation might contribute to the coordinated activation of cell death (54).

It is also possible that superoxide is involved in triggering cell death. Infiltration of superoxide dismutase into tobacco leaves infected with TMV compromised the development of the HR (25). Recently it was shown that the lsd1 mutant of Arabidopsis, which shows an apparent HR after shifting uninfected plants from short-day to long-day growth conditions, accumulates superoxide in the apoplasm of leaf tissue (44). These observations point to a possible function for superoxide in regulating the initiation and/or extent of cell death during the HR. Because both plants and animals show apoptosis in response to oxidative signals, it will be interesting to determine whether there is any similarity to the mechanism of apoptotic activation in these highly diverged systems.

Other potential signals for HR induction are the flux and exchange of ions. Early during the resistance response, there is an efflux of K^+. A hint that ion fluxes do play an important role in regulating the HR comes from tobacco plants that constitutively express the bacterio-opsin (bO) protein, a bacterial proton pump from Halobacterium halobium that requires rhodopsin for active proton pumping in bacteria. Such transgenic plants showed an apparent spontaneous HR accompanied by 3′OH DNA ends indicative of endonucleolytic processing and expressed many defenses normally seen during a resistance response (66). In addition, bO-expressing tobacco showed elevated levels of DNA endonucleases that were also activated during a TMV-induced resistance response on tobacco (66). Tobacco plants expressing a mutant form of the bO that should not form a functional ion channel did not show any features of the defense response. A key question in interpreting these experiments is whether the bO protein is acting by promoting passive or active translocation of protons. To do the latter, it would need to bind rhodopsin, which is not known to be present in higher plants. To translocate protons even passively, the bO protein would need to be localized to the plasma membrane to function as an ion channel and activate the resistance response downstream of pathogen recognition.

Ca^{2+} fluxes may also play a role in the execution of the HR. Blocking Ca^{2+} ion channels reduces cell death of soybean suspension cells in response to avirulent bacteria or H_2O_2 (55). Treatment of such plant cells with a Ca^{2+} ionophore also induced pcd. In the case of soybean suspension cells, Ca^{2+} fluxes were not associated with the expression of defense related genes that were induced by H_2O_2 (55). Because Ca^{2+} was also required for activation of DNA endonucleases that are associated with the resistance response (65), it is possible that DNA cleavage is a necessary step in the suicide process. Conversely, DNA breakdown may occur after the cell has committed to a cell death program and may facilitate recycling of cellular constituents.

Because the membrane changes its properties during the resistant response, it is possible that lipid-based signals are responsible for regulating the HR. Because of the differential activation and/or localization of lipoxygenase and phospholipase D, respectively, during the resistance response, it has been suggested that the products of these particular enzymatic reactions might be active as signal molecules in the activation of the HR (18, 96). Interestingly, lipoxygenase enzymes generate hydroperoxidated lipids that have been seen to induce apoptosis in animals (75).

Using pharmacological, genetic, and transgenic approaches, it seems clear that ion channels are involved in signaling the onset of cell death. Other steps in the regulation of the HR include phosphorylation and dephosphorylation (27, 55, 97) and protease activity (56). In most cases, the particular molecules involved in these activities have not been identified, but their requirement for HR regulation has been inferred using pharmacological agents that may not be specific for only one target. Many questions remain open about the regulation and execution of the HR. Are H_2O_2 and O_2^- obligatory signals in the induction of the HR? If other signals exist, what are they? Could they be lipid peroxides? What is the role of DNA cleavage? Are there HR-specific proteases as there are apoptosis-specific proteases in animals? What are the targets that become phosphorylated, dephosphorylated, or proteolyzed?

Answering these questions should contribute to a full understanding of the regulation and function of pcd resulting from plant-pathogen interactions.

CELL DEATH DURING SYSTEMIC ACQUIRED RESISTANCE

Plants that have undergone a resistance response on one part of a plant become more resistant at distal sites to subsequent attack by pathogens that would normally cause a susceptible interaction. As mentioned above, this type of resistance is called systemic acquired resistance (SAR) and requires the signal molecule SA (29). In addition, some pathogens that normally cause a resistance response on a particular host will apparently be limited to an even greater extent if these plants have SAR. Direct application of SA, a synthetic analog of SA, 2,6-dichloroisonicotinic acid (INA), or benzothiadiazole (BTH) to plants triggers the synthesis of the same biochemical markers that appear during induction of SAR by a biological agent (28, 52, 85, 92). It is possible that SA, INA, and BTH activate SAR at the same target (52, see 93 and references therein). These chemicals have been used to study basic aspects of the regulation of SAR. The observation that pathogens that normally can cause disease on a host can be prevented from doing so because of SAR has prompted several laboratories to investigate the possible mechanism of pathogen growth and/or developmental arrest. Remarkably, susceptible barley plants that are treated with INA and then are infected with powdery mildew show developmental arrest of the fungi accompanied by rapid plant cell death. These arrested fungi apparently look identical to those that are arrested because of infection of plants carrying the *Mlg* resistance gene (49). This type of observation is not unique to barley: Arabidopsis plants treated with INA or a pathogen that induces SAR show rapid cell death when infected by a normally pathogenic downy mildew isolate that does not cause cell death on control plants (62, 86). Furthermore, the induction of rapid cell death is not limited to infections by biotrophic pathogens because Arabidopsis with SAR induced by avirulent *P. syringae* also shows rapid cell death when infected with virulent *P. syringae* (14). Tobacco plants treated with SA also show a rapid cell death response when infected with the soft-rot pathogen *Erwinia carotovara* subsp. *carotovora* (71). Other cases of rapid cell death induced by normally virulent pathogens on immunized plants were noted by Kuc (50).

What might be the basis of this rapid cell death? One possibility is that induction of SAR causes either the induction or the modification of latent plant receptors that can recognize products secreted by normally virulent pathogens. If this is the case, then there may be a range of such receptors that can recognize products from all kinds of pathogens such as one finds for R genes. An alternative model is that some plant product that is generated during virulent pathogen attack is perceived by the plant differently if the plant has elevated levels of SA. In this view, SA induces a "receptor" to a plant product instead of a pathogen-derived product. A role for SA in modifying the regulation of a defense response other than cell death has been reported. For example, SA treatment or bacterial infection of tobacco induces the defense-related transgene fusion *AoPR1-GUS* only modestly, but pretreatment of the plants with SA before bacterial infection caused a 10-fold increase in the induction of the fusion (68). SA itself is also subject to the potentiating stimulus ethylene. Thus, treatment of plants with low levels of SA that did not cause defense-related gene activation caused the defense-related gene *PR1* to become inducible by ethylene. Ethylene alone did not induce *PR1* (53).

The cell death that occurs on plants with activated SAR may be triggered in a similar way as the HR. In this regard, it is intriguing that treating parsley suspension cells with SA causes them to undergo an amplified oxidative burst when exposed to fungal elicitor (46). This amplified oxidative burst is blocked by the same kinase inhibitor, K252A, that blocks the oxidative burst that occurs because of a resistant response (56). An alternative hypothesis is that cell death induced by pathogens is a default response of plants that pathogens have evolved to suppress except in special cases, such as when *avr* genes are present. Suppressors of the HR derived from virulent fungi have been documented (23, 24). Cell death in immune plants could be caused by an ability of plants to inhibit cell death suppressors.

Although SAR causes plants to undergo a rapid cell death response with normally virulent bacterial and fungal pathogens, it causes a suppression of cell death when plants are infected with avirulent viruses. These viruses normally cause localized cell death during a resistance response. When plants with a functional R gene that conditions resistance to an avirulent virus have also undergone SAR, the cell death response of the plants is decreased, as shown by a decrease in lesion size. This observation has been interpreted to mean that plants become even more resistant to the virus. Plants that cannot accumulate SA show an increase in lesion size upon viral infection (29). How should a change in lesion size be interpreted? First it is important to establish the underlying assumptions about what regulates lesion size. Lesion size may reflect at least two things: the rate of viral movement relative to the onset of cell death and/or the amount of cell death after viral recognition. Cell death during a resistant response is not entirely cell autonomous: Maize plants mosaic for the *RP1* resistance gene showed some cell death in the neighboring *rp1* tissue after infection with *Puccinia sorgum,* if the recognition of pathogen occurred in the *RP1* tissue sectors (7). Thus, lesion size may not be a reflection of pathogen replication as much as it is one of the activation of plant cell-cell communication that results in cell death. Arabidopsis plants that cannot accumulate SA exhibit more cell death when exposed to ozone (78). Sharma et al (78) suggested that this may be because of the absence of antioxidants that are regulated by SA. At least one catalase gene was activated by SA in potato plants (69). Arabidopsis also showed increased catalase in SA-treated tissue (A Castillo, P Hradecky & JT Greenberg, unpublished information). SA also induces Mn-superoxide dismutase in tobacco (10). It appears that SA may serve to both condition the activation of a cell death response and regulate the extent of cell death at the same time. These roles for SA are also consistent with the observation that some mutants of Arabidopsis (*lsd6* and *lsd7, l*esions *s*imulating *d*isease) that spontaneously accumulate cell death patches show suppressed cell death if SA levels are decreased (93).

CELL DEATH AND THE INDUCTION OF DEFENSES

Because SAR occurs at or after the time of HR development and occasionally after susceptible interactions that involve cell death, it has been widely assumed that there is a causal relationship between cell death and SAR establishment. It has been found, for example, that more numerous *TMV* lesions are correlated with more SA production (95). However, in some cases, the amount of cell death did not correlate with the amount of resistance observed in plants (14). Excision of a cucumber leaf undergoing a resistant response 6 h after treatment with *P. syringae* still allowed development of SAR even though cell death would not occur on the primary infected leaf until 24 h after inoculation. Thus, SAR can be triggered before any cell death (80). It is also unlikely that cell death per se causes SAR, because many situations can arise where plant cells die yet no SA or SAR ensues. Senescent leaves did not trigger SAR in young tissues, even though senescence is a form of pcd (see reference 35; JT Greenberg, unpublished information). Cell death occurs throughout plant development without triggering SAR (35). Plant cells killed by wounding (51, 60), freezing (59), or leaf spot disease (14) do not elicit SAR. Several Arabidopsis mutants that showed spontaneous cell death (21; JT Greenberg, unpublished information) did exhibit SAR. The connection between cell death and SAR may simply be that cell death occurs fortuitously at the same time as the development of SAR. One possibility is that when a resistance response is triggered, a number of pathways are coordinately regulated, including one for cell death and one for SAR, and one or more for other local defenses. Thus, one event in the infected leaf might lead to SAR development at distal sites and also result in a signal to generate localized cell death. In this scenario, cell death and SAR are coupled because of an early event, but cell death does not cause SAR.

CONCLUDING REMARKS

Cell death occurs in response to diverse pathogens and appears to occur by at least two pathways: one that resembles apoptosis that is seen in animal systems and one that resembles necrotic cell death morphologically, but that may also be an example of programmed cell death. Recognition of pathogens in a gene-for-gene interaction is only one of several ways that programmed cell death is triggered. It is likely that many factors influence the control of cell death during pathogenesis as exemplified by the diversity of cell death patterns seen after pathogen attack and by the numerous cell death defense mutants in maize and Arabidopsis (21, 36, 37, 45, 93). An emerging area of interest is how the metabolic state of the plant influences the cell death rate and patterns during plant-pathogen interactions. Indeed, alterations in sugar levels or modification of the ubiquitin pathway appear to trigger the HR spontaneously and alter cell death responses to pathogens (6, 41). Herbers et al (41) suggested that this may reflect a role for cell wall invertases in controlling pathogen-triggered cell death. What is clear is that cell death during plant-pathogen interactions is a common event that may be important for robust resistance as well as susceptibility to pathogens depending on the lifestyle of the pathogen in question.

ACKNOWLEDGMENTS

I thank Adam Driks for valuable discussions. JTG is a Pew Scholar and an American Cancer Society Junior Faculty Fellow. Research in JTG's laboratory is funded by grant 1R29GM54292-01 from the National Institutes of Health.

> Visit the *Annual Reviews home page* at http/:www.annurev.org.

Literature Cited

1. Alexander D, Goodman RM, Gut-Rella M, Glascock C, Weymann K, et al. 1993. Increased tolerance to two oomycete pathogens in transgenic tobacco expressing pathogenesis-related protein 1a. *Proc. Natl. Acad. Sci. USA* 90:7327–31

2. Atkinson MM, Huang JS, Knopp JA. 1985. The hypersensitive reaction of tobacco to *Pseudomonas syringae* pv. *pisi:* activation of a plasmalemma K^+/H^+ exhange in tobacco. *Phytopathology* 77:843–47

3. Auh CK, Murphy TM. 1995. Plasma membrane redox enzyme is involved in the sythesis of O_2^- and H_2O_2 by *Phytophthora*

elicitor-stimulated rose cells. *Plant Physiol.* 107:1241–47

4. Baillieul F, Genetet I, Kopp M, Saindrenan P, Fritig B, Kauffmann S. 1995. A new elicitor of the hypersensitive response in tobacco: A fungal glycoprotein elicits cell death, expression of defense genes, production of salicylic acid, and induction of systemic acquired resistance. *Plant J.* 8:551–60

5. Baker CJ, Atkinson MM, Collmer A. 1987. Concurrent loss in Tn5 mutants of *Pseudomonas syringae* pv. *syringae* of the ability to induce the HR and host plasma membrane K⁺/H⁺ exchange mechanism. *Plant Physiol.* 79:843–47

6. Becker F, Buschfeld E, Schell J, Bachmair A. 1993. Altered response to viral infection in tobacco plants perturbed in ubiquitin system. *Plant J.* 3:875–81

7. Bennetzen JL, Blevins WE, Ellingboe AH. 1988. Cell-autonomous recognition of the rust pathogen determines *Rp-1*-specified resistance in maize. *Science* 241:208–10

8. Bent AF, Innes RW, Ecker JR, Staskawicz BJ. 1992. Disease development in ethylene-insensitive *Arabidopsis thaliana* infected with virulent and avirulent *Pseudomonas* and *Xanthomonas* pathogens. *Mol. Plant-Microbe Interact.* 5:372–78

9. Bestwick CS, Bennett MH, Mansfield JW. 1995. *Hrp* mutant of *Pseudomonas syringae* pv *phaseolicola* induces cell wall alterations but not membrane damage leading to the hypersensitive reaction in lettuce. *Plant Physiol.* 108:503–16

10. Bowler C, Alliote T, DeLoose M, Van Montagu M, Inze D. 1989. The induction of manganese superoxide dismutase in response to stress in *Nicotiana plumbaginifolia. EMBO J.* 8:31–38

11. Bradley DJ, Kjellbom P, Lamb CJ. 1992. Elicitor- and wound-induced oxidative cross-linking of a proline-rich plant cell wall protein: a novel, rapid defense response. *Cell* 70:21–30

12. Brisson LF, Tenhaken R, Lamb C. 1994. The function of oxidative cross-linking of cell wall structural proteins in plant disease resistance. *Plant Cell* 6:1703–12

13. Buell CR, Somerville SC. 1995. Expression of defense-related and putative signaling genes during tolerant and susceptible interactions of Arabidopsis with *Xanthomonas campestris* pv. *campestris. Mol. Plant-Microbe Interact.* 8:435–43

14. Cameron RK, Dixon RA, Lamb CJ. 1994. Biologically induced systemic acquired resistance in *Arabidopsis thaliana. Plant J.* 5:715–25

15. Century KS, Holub EB, Staskawicz BJ. 1995. *NDR1*, a locus of *Arabidopsis thaliana* that is required for disease resistance to both a bacterial and a fungal pathogen. *Proc. Natl. Acad. Sci. USA* 92:6597–601

16. Chen Z-X, Silva H, Klessig DF. 1993. Active oxygen species in the induction of plant systemic acquired resistance by salicylic acid. *Science* 262:1883–86

17. Clouse SD, Gilchrist DG. 1987. Interaction of the Asc locus in F8 paired lines of tomato with *Alternaria alternata f. sp. lycopersici* and AAL-toxin. *Phytopathology* 77:80–82

18. Croft KPC, Voisey CR, Slusarenko AJ. 1990. Mechanism of hypersensitive cell collapse: correlation of increased lipoxygenase activity with membrane damage in leaves of *Phaseolus syringae* pv. *phaseolicola. Physiol. Mol. Plant Pathol.* 36:49–62

19. Delaney TP, Uknes S, Vernooij B, Friedrich L, Weymann K. 1994. A central role of salicylic acid in plant disease resistance. *Science* 266:1247–50

20. Deom CM, Lapidot M, Beachy RN. 1992. Plant virus movement proteins. *Cell* 69:221–24

21. Dietrich RA, Delaney TP, Uknes SJ, Ward ER, Ryals JA, Dangl JL. 1994. *Arabidopsis* mutants simulating disease resistance responses. *Cell* 77:565–77

22. Doke N. 1983. Involvement of superoxide anion generation in the hypersensitive response of potato tuber tissues to infection with an incompatible race of *Phytophthora infestans* and to the hyphal wall components. *Physiol. Plant Pathol.* 23:345–57

23. Doke N, Cahi HB, Kawaguchi A. 1987. Biochemical basis of triggering and suppression of hypersensitive cell response. In *Molecular Determinants of Plant Diseases*, ed. S Nishimura, CP Vance, N Doke, pp. 235–51. Tokyo: Jpn. Sci. Soc. Press

24. Doke N, Garas NA, Kuc J. 1980. Effect on host hypersensitivity of suppressors released during the germination of *Phytophthora infestans* cystopores. *Phytopathology* 70:35–39

25. Doke N, Ohashi Y. 1988. Involvement of O₂⁻ generating systems in the induction of necrotic lesions on tobacco leaves infected with TMV. *Physiol. Mol. Plant Pathol.* 32:163–75

26. Dong X-N, Mindrinos M, Davis KR, Ausubel FM. 1991. Induction of *Arabidopsis* defense genes by virulent and avirulent *Pseudomonas syringae* strains and by a cloned avirulence gene. *Plant Cell* 3:61–72

27. Dunigan DD, Madlener JC. 1995. Ser-

ine/threonine protein phosphatase is required for tobacco mosaic virus-mediated programmed cell death. *Virology* 207: 460–66

28. Friedrich L, Lawton K, Ruess W, Masner P, Specker N, et al. 1996. A benzothiadiazole derivative induces systemic acquired resistance in tobacco. *Plant J.* 10:61–70

29. Gaffney T, Friedrich L, Vernooij B, Negrotto D, Nye G, et al. 1993. Requirement of salicylic acid for the induction of systemic acquired resistance. *Science* 261: 754–56

30. Glazebrook J, Ausubel FM. 1994. Isolation of phytoalexin-deficient mutants of *Arabidopsis thaliana* and characterization of their interactions with bacterial pathogens. *Proc. Natl. Acad. Sci. USA* 91:8955–59

31. Glazener JA, Orlandi EW, Baker CJ. 1996. The active oxygen response of cell suspensions to incompatible bacteria is not sufficient to cause hypersensitive cell death. *Plant Physiol.* 110:759–63

32. Gopalan S, Bauer DW, Alfano JR, Loniello AO, He SY, Collmer A. 1996. Expression of the *Pseudomonas syringae* avirulence protein *avrB* in plant cells alleviates its dependence on the hypersensitive response and pathogenicity (Hrp) secretion system in eliciting genotype-specific hypersensitive cell death. *Plant Cell* 8:1095–105

33. Gorg R, Hollricher K, Schulze-Lefert P. 1993. Functional analysis and RFLP-mediated mapping of the *Mlg* resistance locus in barley. *Plant J.* 3:857–66

34. Grbic V, Bleecker AB. 1995. Ethylene regulates the timing of leaf senescence in *Arabidopsis. Plant J.* 8:595–602

35. Greenberg JT. 1996. Programmed cell death: a way of life for plants. *Proc. Natl. Acad. Sci. USA* 93:12094–97

36. Greenberg JT, Ausubel FM. 1993. *A. thaliana* mutants compromised for the control of cellular damage during pathogenesis and aging. *Plant J.* 4:327–41

37. Greenberg JT, Guo A, Klessig DF, Ausubel FM. 1994. Programmed cell death in plants: a pathogen-triggered response activated with multiple defense functions. *Cell* 77:551–63

38. Hammond-Kosack KE, Jones JDG. 1994. Incomplete dominance of tomato *Cf* genes for resistance to *Cladosporium fulvum. Mol. Plant-Microbe Interact.* 7:58–70

39. Hammond-Kosak KE, Jones JDG. 1995. Plant disease resistance genes: unravelling how they work. *Can J. Bot.* 73S:S495–505

40. He SY, Huang HC, Collmer A. 1993. *Pseudomonas syringae* pv. *syringae* Har-

pinP$_{ss}$: a protein that is secreted via the Hrp pathway and elicits the hypersensitive response in plants. *Cell* 73:1255–66

41. Herbers K, Meuwly P, Frommer WB, Metraux JP, Sonnewald U. 1996. Systemic acquired resistance mediated by the ectopic expression of invertase: possible hexose sensing in the secretory pathway. *Plant Cell* 8:793–803

42. Holub E, Benyon JL, Crute IR. 1994. Phenotypic and genotypic characterization of interactions between isolates of *Peronospora parasitica* and accessions of *Arabidopsis thaliana. Mol. Plant-Microbe Interact.* 7:223–49

43. Hu G, Richter TE, Hulbert SH, Pryor T. 1996. Disease lesion mimicry caused by mutations in the rust resistance gene *rp1. Plant Cell* 8:1367–76

44. Jabs T, Dietrich RA, Dangl JL. 1996. Initation of runaway cell death in an *Arabidopsis* mutant by extracellular superoxide. *Science* 273:1853–56

45. Johal GS, Hulbert SH, Briggs SP. 1995. Disease lesion mimics of maize: a model for cell death in plants. *BioEssays* 17:685–692

46. Kauss H, Jeblick W. 1995. Pretreatment of parsley suspension cultures with salicylic acid enhances spontaneous and elicited production of H$_2$O$_2$. *Plant Physiol.* 108: 1171–78

47. Keen NT. 1990. Gene-for-gene complementarity in plant-pathogen interactions. *Annu. Rev. Genet.* 24:447–63

48. Keen NT, Ersek T, Long M, Bruegger B, Holliday M. 1981. Inhibition of the hypersensitive reaction of soybean leaves to incompatible *Pseudomonas* spp. by blastocidin S, strepomycin or elevated temperature. *Physiol. Plant Pathol.* 18:325–37

49. Kogel KH, Beckhove U, Dreschers J, Munch S, Romme Y. 1994. Acquired resistance in barley. The resistance mechanism induced by 2,6-dichloroisonicotinic acid is a phenocopy of a genetically based mechanism governing race-specific powdery mildew resistance. *Plant Physiol.* 106:1269–77

50. Kuc J. 1982. Induced immunity to plant diseases. *BioScience* 32:854–60

51. Kuc J. 1983. Induced systemic resistance in plants to diseases caused by fungi and bacterial. In *The Dynamics of Host Defense,* ed. J Bailey, B Deverall, pp. 191–221. Sydney: Academic

52. Lawton KA, Friedrich L, Hunt M, Weymann K, Delaney T, et al. 1996. Benzothiadiazole induces disease resistance in

544 GREENBERG

Arabidopsis by activation of the systemic acquired resistance signal transduction pathway. *Plant J.* 10:71–82

53. Lawton KA, Potter SL, Uknes S, Ryals J. 1994. Acquired resistance signal transduction in *Arabidopsis* is ethylene independent. *Plant Cell* 6:581–88
54. Leon J, Lawton MA, Raskin I. 1995. Hydrogen peroxide stimulates salicylic acid biosynthesis in tobacco. *Plant Physiol.* 108:1671–78
55. Levine A, Pennell RI, Alvarez ME, Palmer R, Lamb C. 1996. Calcium-mediated apoptosis in a plant hypersensitive disease resistant response. *Curr. Biol.* 6:427–37
56. Levine A, Tenhaken R, Dixon R, Lamb C. 1994. H2O2 from the oxidative burst orchestrates the plant hypersensitive disease resistance response. *Cell* 79:583–93
57. Lincoln JE, Cordes S, Read E, Fischer RL. 1987. Regulation of gene expression by ethylene during *Lycopersicon esculentum* (tomato) fruit development. *Proc. Natl. Acad. Sci. USA* 84:2793–97
58. Maher EA, Bate NJ, Ni W, Elkind Y, Dixon RA, Lamb CJ. 1994. Increased disease susceptibility of transgenic tobacco plants with suppressed levels of preformed phenylpropanoid products. *Proc. Natl. Acad. USA* 91:7803–6
59. Malamy J, Klessig DF. 1992. Salicylic acid and plant disease resistance. *Plant J.* 2: 643–54
60. Malamy J, Sanchez-Casas P, Hennig J, Guo A, Klessig DF. 1996. Dissection of the salicylic acid signaling pathway in tobacco. *Mol. Plant-Microbe Interact.* 9: 474–82
61. Martin GB, Brommonschenkel S, Chunwongse J, Frary A, Ganal MW, et al. 1993. Map-based cloning of a protein kinase gene conferring disease resistance in tomato. *Science* 262:1432–36
62. Mauch-Mani B, Slusarenko AJ. 1994. Systemic acquired resistance in *Arabidopsis thaliana* induced by a predisposing infection with a pathogenic isolate of *Fusarium oxysporum. Mol. Plant-Microbe Interact.* 7:378–83
63. Mehdy MC. 1994. Active oxygen species in plant defense against pathogens. *Plant Physiol.* 105:467–72
64. Melan MA, Dong XN, Endara ME, Davis KR, Ausubel FM, Peterman TK. 1993. An *Arabidopsis thaliana* lipoxygenase gene can be induced by pathogens, abscisic acid, and methyl jasmonate. *Plant Physiol.* 101: 441–50
65. Mittler R, Lam E. 1995. Identification, characterization, and purification of a to-

bacco endonuclease activity induced upon hypersensitive response cell death. *Plant Cell* 7:1951–62
66. Mittler R, Shulaev V, Lam E. 1995. Coordinated activation of programmed cell death and defense mechanisms in transgenic tobacco plants expressing a bacterial proton pump. *Plant Cell* 7:29–42
67. Moussatos VV, Yang SF, Ward B, Gilchrist DG. 1994. AAL-toxin induced physiological changes in *Lycopersicon esculentum* Mill: roles for ethylene and pyrimidine intermediates in necrosis. *Physiol. Mol. Plant Pathol.* 44:455–68
68. Mur LA, Naylor G, Warner SAJ, Sugars JM, White RF, Draper J. 1996. Salicylic acid potentiates defence gene expression in tissue exhibiting acquired resistance to pathogen attack. *Plant J.* 9:559–71
69. Niebel A, Heungens K, Barthels N, Inze D, Montagu MV, Gheysen G. 1995. Characterization of a pathogen-induced potato catalase and its systemic expression upon nematode and bacterial infection. *Mol. Plant-Microbe Interact.* 8:371–78
70. Oeller PW, Wong LM, Taylor LP, Pike DA, Theologis A. 1991. Reversible inhibition of tomato fruit senescence by antisense RNA. *Science* 254:437–39
71. Palva TK, Hurtig M, Saindrenan P, Palva ET. 1994. Salicylic acid induced resistance to *Erwinia carotovora* subsp. *carotovora* in tobacco. 1994. *Mol. Plant-Microbe Interact.* 7:356–63
72. Reuber TL, Ausubel FM. 1996. Isolation of *Arabidopsis* genes that differentiate between resistance responses mediated by the *RPS2* and *RPM1* disease resistance genes. *Plant Cell* 8:241–49
73. Ritter C, Dangl JL. 1996. Interference between two specific pathogen recognition events mediated by distinct plant disease resistance genes. *Plant Cell* 8:251–57
74. Ryerson DE, Heath MC. 1996. Cleavage of nuclear DNA into oligonucleosomal fragments during cell death induced by fungal infection or by abiotic treatments. *Plant Cell* 8:393–402
75. Sandstrom PA, Pardi D, Tebbey PW, Dudek RW, Terrain DM, et al. 1995. Lipid hydroperoxide-induced apoptosis: lack of inhibition by bcl-2 over-expression. *FEBS Lett.* 365:66–70
76. Santana P, Pena LA, Haimovitz-Friedman A, Martin S, Green D, et al. 1996. Acid sphingomyelinase-deficient human lymphoblasts and mice are defective in radiation-induced apoptosis. *Cell* 86:189–99
77. Schwartz LM, Smith SW, Jones MEE, Osborne BA. 1993. Do all programmed cell

deaths occur via apoptosis? *Proc. Natl. Acad. Sci. USA* 90:980–84

78. Sharma YK, Leon J, Raskin I, Davis KR. 1996. Ozone-induced responses in *Arabidopsis thaliana:* the role of salicylic acid in the accumulation of defense-related transcripts and induced resistance. *Proc. Natl. Acad. Sci. USA* 93:5099–104

79. Shen Y, Shenk TE. 1995. Viruses and apoptosis. *Curr. Biol.* 5:105–11

80. Smith JA, Hammerschmidt R, Fulbright DW. 1991. Rapid induction of systemic resistance in cucumber by *Pseudomonas syringae pv. syringae. Physiol. Mol. Plant Pathol.* 38:223–35

81. Smith MJ, Mazzola EP, Sims JJ, Midland SL, Keen NT, et al. 1993. The syringolides: bacterial c-glycosyl lipids that trigger plant disease resistance. *Tetrahedron Lett.* 34: 223–26

82. Staskawicz BJ, Ausubel FM, Baker BJ, Ellis JG, Jones JDG. 1995. Molecular genetics of plant disease resistance. *Science* 268:661–67

83. Strobel NE, Ji C, Gopalan S, Kuc JA, He SY. 1996. Induction of systemic acquired resistance in cucumber by *Pseudomonas syringae* pv. *syringae* 61 HrpZ protein. *Plant J.* 9:431–39

84. Tsuji J, Somerville S, Hammerschmidt R. 1991. Identification of a gene in *Arabidopsis thaliana* that controls resistance to *Xanthomonas campestris pv. campestris. Physiol. Mol. Plant Pathol.* 38:57–65

85. Uknes S, Mauch-Mani B, Moyer M, Potter S, Williams S, et al. 1992. Acquired resistance in *Arabidopsis. Plant Cell* 4:645–56

86. Uknes S, Winter A, Delaney T, Vernooij B, Morse A, et al. 1993. Biological induction of systemic acquired resistance in *Arabidopsis. Mol. Plant-Microbe Interact.* 6: 680–85

87. VanEtten HD, Matthews DE, Matthews PS. 1989. Phytoalexin detoxification: importance for pathogenicity and practical implications. *Annu. Rev. Phytopathol.* 27: 143–64

88. Walbot V. 1991. Maize mutants for the 21st century. *Plant Cell* 3:851–56

89. Walbot V, Hoisington DA, Neuffer MG. 1983. Disease lesion mimic mutations. In *Genetic Engineering of Plants,* ed. T Kosuge, CP Meredith, A Hollaender, pp. 431–42. New York: Plenum

90. Wang H, Jones C, Ciacci-Zanella J, Holt T, Gilchrist DG, Dickman MB. 1996. Fumonisins and AAL toxins: sphinganine analog mycotoxins induce apoptosis in monkey kidney cells. *Proc. Natl. Acad. Sci. USA* 93:3461–65

91. Wang H, Li J, Bostock RM, Gilchrist DG. 1996. Apoptosis: a functional paradigm for programmed plant cell death induced by host-selective phytotoxin and invoked during development. *Plant Cell* 8:375–91

92. Ward ER, Uknes SJ, Williams SC, Dincher SS, Wiederhold, et al. 1991. Coordinate gene activity in response to agents that induce systemic acquired resistance. *Plant Cell* 3:1085–94

93. Weymann K, Hunt M, Uknes S, Neuenschwander U, Lawton K, et al. 1995. Suppression and restoration of lesion formation in *Arabidopsis lsd* mutants. *Plant Cell* 7:2013–22

94. Williams GT. 1994. Programmed cell death: a fundamental protective response to pathogens. *Trends Microbiol.* 2:462–64

95. Yalpani N, Silverman P, Wilson TMA, Kleier DA, Raskin I. 1991. Salicylic acid is a systemic signal and an inducer of pathogenesis-related protein in virus-infected tobacco. *Plant Cell* 3:809–18

96. Young SA, Wang XM, Leach JE. 1996. Changes in the plasma membrane distribution of rice phospholipase D during resistant interactions with *Xanthomonas oryzae* pv *oryzae. Plant Cell* 8:1079–90

97. Zhou J, Loh YT, Bressan RA, Martin GB. 1995. The tomato gene *Pti1* encodes a serine/threonine kinase that is phosphorylated by *Pto* and is involved in the hypersensitive response. *Cell* 83:925–35

Annu. Rev. Plant Physiol. Plant Mol. Biol. 1997. 48:547–74

POLLINATION REGULATION OF FLOWER DEVELOPMENT

Sharman D. O'Neill

Section of Plant Biology, Division of Biological Sciences, University of California, Davis, California 95616

KEY WORDS: auxin, ethylene, gametophyte development, ovule development, senescence

ABSTRACT

Pollination regulates a syndrome of developmental responses that contributes to successful sexual reproduction in higher plants. Pollination-regulated developmental events collectively prepare the flower for fertilization and embryogenesis while bringing about the loss of floral organs that have completed their function in pollen dispersal and reception. Components of this process include changes in flower pigmentation, senescence and abscission of floral organs, growth and development of the ovary, and, in certain cases, pollination also triggers ovule and female gametophyte development in anticipation of fertilization. Pollination-regulated development is initiated by the primary pollination event at the stigma surface, but because developmental processes occur in distal floral organs, the activity of interorgan signals that amplify and transmit the primary pollination signal to floral organs is implicated. Interorgan signaling and signal amplification involves the regulation of ethylene biosynthetic gene expression and interorgan transport of hormones and their precursors. The coordination of pollination-regulated flower development including gametophyte, embryo, and ovary development; pollination signaling; the molecular regulation of ethylene biosynthesis; and interorgan communication are presented.

CONTENTS

547

INTRODUCTION

Pollination regulates a complex syndrome of developmental events in many flowers. Because perianth senescence is the most visible manifestation of this pollination-regulated flower development and because of its horticultural importance, a preponderance of the research focused on this phase of flower development has been on wilting and abscission of the corolla and calyx. However, this review focuses on the broader context of pollination-regulated developmental responses that collectively lead to the shedding of some floral organs that have served their function in pollen dispersal and reception while simultaneously preparing other organs for fertilization, embryogenesis, and fruit development. Thus, pollination delineates prepollination flower development—during which the flower is specialized for pollen dispersal and reception—from postpollination development—whereby the flower becomes specialized to ensure fertilization and nourishment of the developing embryo and seed. Developmental processes associated with this functional transition include senescence of the perianth, pigmentation changes, ovary maturation, ovule differentiation, and female gametophyte development. It has been proposed that perianth senescence and color changes of floral organs serve as signals for insect pollinators to discriminate receptive flowers from those that have advanced to later stages of reproductive development, whereas pollination regulation of ovule maturation serves to coordinate development of the male and female gametophyte after a point where fertilization is all but assured. In addition to the proposed ecological significance of pollination-regulated flower development, processes associated with pollination-regulated flower development are important horticulturally and often result in a reduction of the commercial quality of flowers. Several previous reviews have addressed senescence and postharvest physiology of flowers (13, 50, 51), and pollination-regulated flower development has recently been reviewed from a horticultural perspective (116).

The transition from prepollination to postpollination development can occur in the absence of pollination as part of a temporal program of flower development. However, in many flowers this developmental transition is either strictly regulated by pollination or is accelerated by pollination. Flowers that demonstrate strict pollination regulation have served as excellent model systems to dissect the signals that regulate the developmental transition. A major research goal has been to understand the signals that coordinate diverse developmental responses set in motion as the flower undergoes the transition from a primary role in pollen dispersal and reception to that of fertilization and nurturing the developing embryo and seed. Because these diverse developmental programs occur in distinct floral parts, interorgan signals that coordinate the overall process are implicated. This review focuses on the current state of knowledge of the processes and signals that coordinate pollination-regulated flower development.

POLLINATION REGULATES A SYNDROME OF DEVELOPMENTAL EVENTS

In many flowers, pollination regulates a syndrome of development events that collectively prepare the flower for fertilization and embryogenesis while shedding organs that have completed their function in pollen dispersal and reception. Components of this pollination-regulated syndrome include changes in flower pigmentation, senescence and abscission of floral organs, and growth and development of the ovary. Pollination-regulated developmental events are initiated by a single event—pollination—but because they occur in distinct floral organs the activity of interorgan signals that amplify and transmit the primary signal to distal floral organs is also implicated.

Pollination Regulation of Perianth Senescence

A predominant feature of pollination-regulated development is perianth senescence, although this process is strictly regulated by pollination in only a minority of species. More typically, senescence occurs gradually as part of a temporal program of flower development that may be accelerated by pollination, although there are extreme examples, such as daylily flowers, in which the perianth senesces within 12–18 h after flower opening regardless of their pollination status (79, 85). In flowers that exhibit either pollination-dependent or pollination-accelerated senescence, pollination leads to a rapid increase in ethylene production resembling the climacteric response observed in ripening fruits (16, 53, 117). In these flowers, which include *Petunia,* carnation, cyclamen, and orchids, perianth senescence occurs in concert with the increase in endogenous ethylene production and can be prevented by treatments that inhibit ethylene production or perception (18, 53, 54, 106, 107, 111, 124, 159).

Very early studies first noted the sensitivity of the perianth of certain flowers to ethylene (26). In these studies, treatment of *Cattleya* flowers with as little as 2 ppb of ethylene for 24 h induced senescence symptoms that were identical to "dried-sepal injury," a physiological disorder that developed in the perianth of orchid flowers. This disorder was inferred to be caused by exogenous ethylene in greenhouses. The association of endogenous ethylene production with perianth senescence was made somewhat later during study of the floral color changes that accompany pollination-regulated development in another orchid species, *Vanda* cv. Miss Agnes Joaquim (2). It was noted that these flowers produced ethylene endogenously, and it was correlated with the floral color changes. Furthermore, the floral color change was accelerated by exogenous ethylene, suggesting that endogenous ethylene production triggered color fading and, by inference, also triggered perianth senescence (2). The effects of exogenous ethylene in this *Vanda* cultivar could be mimicked by emasculation or by pollination, and this further indicated that this initial description of ethylene-regulated flower senescence reflected a pollination-regulated process (SD O'Neill, unpublished observations). It is now well known that ethylene plays an important role in coordinating pollination-regulated perianth senescence in many flower species, and a number of orchid species have been particularly well characterized in this regard. For example, emasculation, pollination, auxin treatment, or wounding of orchid flowers stimulates ethylene production and induces perianth senescence in several orchid genera, including *Arachnis, Aranda, Cattleya, Cymbidium, Dendrobium, Paphiopedilum, Phalaenopsis,* and *Vanda* (2, 7, 20, 26, 27, 43, 58, 102–104, 117, 118, 165, 168, 175). Comprehensive studies have demonstrated large variability in sensitivity to ethylene among different orchid species with *Vanda* cv. Miss Joaquim reported to be the most sensitive to ethylene (43). *Cymbidium, Cattleya,* and *Paphiopedilum* were moderately sensitive to ethylene, whereas *Dendrobium* and *Oncidium* were relatively insensitive. Many other flower species, such as carnation, are also similarly sensitive to ethylene, with exogenous propylene (an ethylene analog) promoting a pattern of sustained ethylene production and symptoms of perianth senescence similar to endogenous ethylene production and senescence symptoms induced by pollination (110). These results support the conclusion that ethylene is sufficient to promote the pattern of pollination-regulated perianth senescence that is observed in a number of flowers, including orchid, carnation, and *Petunia*.

Perianth senescence is an active process that is accompanied by changes in gene expression (13, 77, 78, 170). Three cDNAs (SR5, SR8, and SR12) were initially isolated from senescing carnation petals. Two were regulated by ethylene and the third by both ethylene and by temporal cues (77, 78). Two of the senescence-related cDNAs are likely to encode β-galactosidase (SR12) and glutathione S-transferase (SR5) (90). The potential role of β-galactosidase in senescing flower petals is likely to be in cell wall disassembly that accompanies most senescence processes. The role of glutathione S-transferase was suggested to be in the detoxification of lipid and DNA hydroperoxides associated with senescence-induced oxidative processes (146). As discussed below, genes encoding key ethylene biosynthetic enzymes are also regulated by pollination and associated pollination signals in several flower species, thus providing the signaling linkage between pollination, ethylene, and the regulation of genes that contribute to perianth senescence (101, 117, 119, 147, 149). Overall, these results indicate that specific biochemical events are regulated at the level of gene expression during perianth senescence and that these genes, like the overall process, are regulated by ethylene.

Pollination Regulation of Floral Pigmentation Changes

It has been reported that over 74 angiosperm families exhibit floral pigment changes in response to pollination or flower aging and proposed that pollinators recognize color changes and preferentially visit previously unpollinated flowers (44, 157). Pollination can induce diverse patterns of pigmentation changes including color fading, enhanced pigmentation, or intensification of pigmentation in discreet spots. In *Cymbidium* orchid flowers, pollination induces anthocyanin formation, and this process has been well studied as the first visible manifestation of pollination-regulated flower development that also leads to perianth senescence (4–9, 166, 167). The change in *Cymbidium* lip coloration can be accelerated by pollination, emasculation, or treatment of the stigma with auxin (7). Subsequent studies in *Cymbidium* also demonstrated that a small incision at the base of the lip prevented emasculation-induced coloration and that treatment of completely excised lips with 1-aminocyclopropane-1-carboxylic acid (ACC) or ethylene promoted anthocyanin accumulation, suggesting that lip coloration resulted from the translocation of ACC to the lip where it was converted to ethylene in situ (167). Pigmentation changes in *Vanda* orchid flowers are also associated with pollination and ethylene production but, in contrast with *Cymbidium, Vanda* flowers undergo rapid color fading (2). In lupine flowers, the color of the banner spot changes from yellow to magenta as the flower ages (140). This change in banner spot color also appears to be regulated by ethylene and is most likely accelerated by pollination. The change in banner spot color precedes flower wilting by several days and may serve to maintain the attractiveness of a large floral display from a distance but still provide a signal for pollinators at close range.

Relatively little is known about the specific biochemical events that underlie the process of floral pigmentation changes, although certain floral organ pigmentation changes have been associated with carotenoid or anthocyanin biosynthesis (95), anthocyanin degradation (126), or changes in tissue pH (10). Because of the diversity in patterns of pigmentation changes that are pollination regulated, it is likely that different biochemical mechanisms contribute to the changes observed in different flowers. It has been suggested that pollination-regulated color changes evolved independently in angiosperms many different times (157), suggesting that the underlying biochemical mechanisms are likely to differ among taxa.

Pollination Regulation of Female Gametophyte Development

The end result of pollination is fertilization, which leads to zygote formation and subsequent embryogenesis, all of which have been reviewed extensively (14, 62, 68, 86, 88, 94, 114, 128, 130). In some species, female gametophyte development is incomplete before pollination, and the pollination event itself regulates ovule and gametophyte initiation, development, and maturation in preparation for subsequent fertilization. In certain orchids such as *Cattleya, Sophronitis, Epidendron, Laelia, Phalaenopsis, Dendrobium,* and *Doritis,* ovules are completely absent in unpollinated ovaries, and their development is triggered by pollination (29, 30, 65, 133, 173, 174, 176). In many other orchid genuses, such as *Cypripedium, Paphiopedilum, Phragmipedilum, Herminium, Epipactis,* and *Platanthera,* ovule primordia are present before pollination but remain suspended at a premeiotic stage until pollination triggers further ovule development (29, 36, 37, 138). In still other orchid species, immature ovule primordia are present at anthesis that have not yet progressed to the stage of archesporial cell differentiation (38, 76).

Pollination-regulated ovule initiation and development has been characterized in detail in *Phalaenopsis* (176). Within two days after pollination and before pollen tube germination, cell proliferation is initiated along the placental ridges of the ovary. The placental ridges continue to elongate and branch to form thousands of finger-like ovule primordia by approximately 40 days after pollination, a time still well in advance of fertilization. At this stage, the inner integument appears as a collarlike growth near the tip of the primordia, and the outer integument is initiated shortly thereafter. The archesporial cell enlarges further to form the megasporocyte that, following meiosis, gives rise to the expanding megaspore. Subsequent mitotic divisions result in the development of a *Polygonum*-type embryo sac in this orchid species (14, 163, 176). The elapsed time between pollination and maturation of the ovule and final fertilization is approximately 80 days in *Phalaenopsis*. The progression of pollination-regulated ovule development in *Phalaenopsis* is illustrated in Figure 1. The initial stages of ovule differentiation, namely cell proliferation in the placental ridge, occurs before pollen tube germination, suggesting that the initial signals for ovule initiation arise from the pollination event and are not dependent on pollen germination or tube growth. The strict regulation of ovule development by pollination results in the synchronous development of thousands of ovules in the *Phalaenopsis* ovary (176). The synchronous ovules have provided the basis to dissect the process of ovule development at the molecular level and to isolate a number of genes whose expression is correlated with discrete stages of ovule differentiation (100).

The unique regulation of megagametophyte development by pollination in a number of orchid species has been related to their reproductive ecology. A generally held view is that orchid ovule development is not initiated before pollination because of the low probability of pollination by highly specific

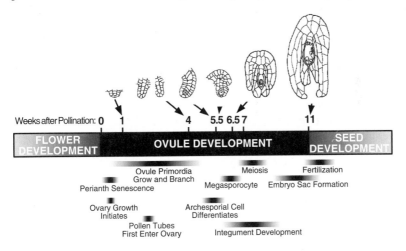

Figure 1   Pollination regulation of ovule development in *Phalaenopsis*. The timeline of ovule development beginning with the event of pollination and ending with fertilization is presented. The span of time during which various developmental changes occur is indicated below the timeline. Anatomical diagrams across the top depict the developmental stage of the ovule primordia, immature ovule, and finally, the mature ovule containing the female gametophyte (modified from Reference 100).

pollinators, compounded by the large investment required for megagametophyte and ovary maturation. The orchid reproductive strategy only invests in female reproductive development after pollination, when fertilization is all but assured. The variation observed in the stage of ovule development at the time of anthesis in different orchid taxa has also been proposed to relate to the environment in which they grow (144). For example, epiphytic orchids that grow in areas where the growing season is long can afford the extended time necessary for complete postpollination ovule development, whereas terrestrial species growing in temperate regions with a short growing season must achieve partial female development before anthesis to complete the reproductive cycle within the season.

Pollination regulation of megagametophyte development is not confined to orchids but is a feature of many higher plant reproductive systems. In many species megagametophyte development is almost complete before pollination, and the final differentiation of cells of the megagametophyte is triggered by pollination. For example, in *Prunus dulcis* (almond), ovule development is arrested at the megasporocyte stage before anthesis, with further development and enlargement of the megagametophyte occurring only after compatible pollination (123). Studies of ovule development in other *Prunus* species in which the megagametophyte is immature at anthesis indicates that the polar nuclei do not fuse before pollination and that this event is pollination regulated (31, 32, 123). Experiments with cotton ovule culture have shown that the addition of 5.0 μM IAA to the culture media can induce polar nuclei fusion, suggesting that auxin supplied by the pollen may be a natural signal inducing polar nuclei fusion and perhaps other biochemical events in preparation for fertilization (66).

Recently, and unexpectedly, it was shown that the egg cell of most ovules in *Zea mays* is not morphologically mature at the time of pollination (96). The final events of egg cell maturation are completed in the majority of ovules after pollination but before fertilization, suggesting that pollen signals are responsible for inducing the final stages of megagametophyte development in this species as well.

In barley, pollination triggers synergid degeneration before the pollen tubes reached the ovule, which allows the pollen tube to penetrate the megagametophyte during fertilization (93). It has also been shown that calcium accumulates in the synergids in response to pollination. It has been suggested that this plays a role in establishing a chemotropic gradient to attract the pollen tube and/or in causing the pollen tube to burst after it enters the synergid to release the sperm cells (22, 23, 56). It has also been suggested that a signal from the pollen tube is communicated over a short range to the ovule to promote synergid degeneration in *Nicotiana* (63). A possible candidate for the pollen tube–derived signal in this case was suggested by experiments using cotton ovule culture in which 0.5 µM gibberellin induced synergid degeneration in a manner similar to that after pollination (66), although other hormones associated with the pollen tube may also contribute to this signaling function (89).

Pollination Regulation of Embryo Development

Several studies have demonstrated that apomictic embryo development in some species is dependent on pollination. For example, in the orchid *Zygopetalum,* pollen from an incompatible species, *Oncidium,* that was ineffective in fertilization, nevertheless triggered apomictic seed production (143). This requirement for pollination to induce apomictic embryo development was confirmed in another orchid species, *Orchis* (49). A series of these and related observations led to an early interpretation that pollen-borne chemical substances induced ovary growth that then indirectly promoted embryo development even in the absence of fertilization (48). Other examples of apomictic embryo development that are dependent on pollination include the apomictic grass *Pennisetum setaceum,* which increases the set of apomictic embryos when pollinated with *P. ciliare* pollen, even though this pollen does not result in fertilization (12, 136). Gamma-irradiated pollen that is inefficient at fertilization induces a low percentage of haploid embryos to develop in apple that are presumed to be apomictic (177). Overall, these observations indicate that while pollination is not sufficient to induce embryogenesis in all species, it is required for apomictic embryo development in several plant species that are predisposed to produce apomictic embryos.

Pollination Regulation of Ovary Development

Ovary development after pollination depends upon a supply of auxin that is typically derived from developing ovules and seeds (47). In the absence of pollination and embryogenesis, certain solanaceous species such as tomato and *Petunia* could be induced to form mature seedless fruit by treatment of the ovaries with auxin (45). Extracts of pollen could mimic the effect of auxin, leading to the proposal that pollen contained auxin and that contributed to the initiation of ovary growth (46, 72–74). Unlike the role of developing seeds in promoting ovary growth, the role of pollination as a primary event regulating ovary and fruit development is less firmly established. Early studies of pollination-regulated orchid development identified changes in the curvature of the ovary to be one of the earliest pollination-regulated developmental events (29). More recently, pollination regulation of cell division in the placental ridge has been implicated in the formation of hair cells that expand the central ovary cavity (176). These morphological changes occur approximately three days before pollen germination and clearly demonstrate that pollination itself, rather than fertilization, triggers the initial stages of ovary development. Exogenously applied auxin and inhibitors of ethylene biosynthesis demonstrated that the elaboration of ovary wall hair cells, the earliest morphological change, was dependent on both auxin and ethylene (176), suggesting that pollen-borne auxin can be translocated to the ovary (141). Pollination effects on ovary development are not restricted to orchids and have been reported in many other plant species, including muskmelon ovaries, which doubled in size within 48 h after pollination (80), as well as in *Brodiaea* and carnation, where ovary expansion was promoted by ethylene treatment, which also promoted perianth senescence (55, 108, 109, 112, 113). It is possible that ovary growth responses were directly related to mobilization of carbohydrate from the senescing petals to the ovary and were thus only indirectly related to pollination.

POLLINATION SIGNALS

Pollination is first perceived at the stigmatic surface, and pollination-regulated developmental events are initiated before pollen germination or penetration of the style by the growing pollen tube. This suggests that a physical event closely associated with the pollen-stigma interaction or a pollen-borne substance is responsible for initiating pollination-regulated flower development. In most species, the primary pollination event is accompanied by an increase in ethylene evolution in the stigma and style within hours after pollination and well before pollen germination (39, 40, 42, 82–84, 97, 110). In self-incompatible species, pollination responses differ between compatible and incompatible pollinations, and this has provided some insight to the primary and secondary signals that operate in pollination-regulated flower development. Compatible pollination in *Petunia* results in two distinct phases of ethylene evolution occurring after approximately 3 and 20 h, respectively, whereas self-pollination triggers only the first phase of ethylene production (81, 137). The early phase of ethylene evolution was attributed to direct conversion of pollen-borne 1-aminocyclopropane-1-carboxylic acid (ACC) to ethylene and the second phase of ethylene evolution to endogenous synthesis of ACC in floral organs distal to the stigma. These results implicate the involvement of a primary and a secondary signal in the pollination response, the first of which is perceived in the stigma followed by a subsequent secondary signal that transmits and amplifies the primary pollination signal to distal floral organs (115). Although it is likely that the primary and secondary pollination signals are distinct, they are both linked to ethylene evolution.

Primary Pollination Signals

The primary pollination signal has been proposed to result from physical contact between the pollen and stigma, from pollen tube penetration of the stigma (40, 41), or from pollen-borne chemical messengers (60, 137, 159). In *Petunia,* carnation, and orchid flowers, the initial response to pollination is rapid ethylene evolution by the stigma. In each case, pollination-induced ethylene evolution precedes germination of the pollen tube, suggesting that penetration of the stigmatic surface by the pollen tube is not required for the induction of ethylene biosynthesis (20, 42, 54, 59, 110, 117). Furthermore, mock pollination of orchid flowers with latex beads, a pollen surrogate that has been used to study the role of the stylar matrix in pollen tube extension (135), failed to trigger ethylene production or any other pollination responses, indicating that physical contact alone is insufficient to induce any components of the pollination-regulated developmental syndrome (176). The preponderance of results suggests that physical contact between the pollen and stigma, or wounding reactions associated with pollen tube growth, are not the primary pollination signals, and considerable experimental attention has thus focused on the identification of a pollen-borne chemical that may serve as a primary pollination signal. Chemicals that have been identified in pollen include ACC, auxin, pectic oligosaccharides, brassinosteroids, and methyl jasmonate, all of which are known inducers of ethylene biosynthesis, making them attractive candidates for the primary pollen signal molecule (3, 28, 33, 67, 71, 92, 131, 134, 151, 152, 154, 156, 162, 172).

Shortly after the identification of ACC as the immediate biochemical precursor of ethylene (1), several reports identified its presence in pollen and suggested that ACC may be the primary pollen signal (127, 158). The role of pollen-borne ACC in triggering the initial burst of ethylene production and its potential translocation to other floral organs has now been extensively studied, especially in model systems such as *Petunia* and carnation (158–161). In spite of its presence in relatively large quantities in some pollen (127, 137, 158), the role of ACC in supporting the initial ethylene production in the stigma has been controversial. For example, Hoekstra & van Roekel (60) reported that ACC content of pollen from various sources is not well correlated with the level of pollination-induced ethylene production, and several reports indicated that treatment of the stigma with an inhibitor of ACC synthase, aminoethoxyvinylglycine (AVG), before pollination prevented pollination-induced ethylene production (61, 117, 169, 176). Thus, these reports suggested that pollination-induced ethylene production is derived from endogenous production of ACC rather than from exogenous pollen-borne ACC. In contrast, Singh et al

(137) reported that wide variations in pollen-borne ACC content in different *Petunia* genotypes were well correlated with the initial peak of pollination-induced ethylene production and that this early ethylene production was not inhibited by a different inhibitor of ACC synthase, aminooxyacetic acid (AOA). Thus, these data suggested that pollen-borne ACC was the substrate for initial pollination-induced ethylene production in *Petunia* (137). Other reports have suggested that the quantity of pollen-borne ACC would be vastly insufficient to support the amount of ethylene produced in the stigma following pollination (139), and two reports indicate that diffusion of ACC from the pollen is likely to be restricted under conditions that prevail in vivo, which would further restrict its availability to support ethylene production in the stigma (60, 121). Although the conflicting data indicate that pollen-borne ACC cannot account for the initial pollination-induced ethylene production in the stigma of all flowers, it is possible that exogenous pollen-borne ACC may be responsible for initiating ethylene production in the stigma of at least some flowers, and this early ethylene production may be subsequently enhanced by autocatalytic production of ACC in the stigma.

The contribution of pollen-borne ACC to pollination regulation of flower development has also been tested by exogenous application of ACC to the stigma (61, 117, 127). In each case, ACC promoted an initial burst of ethylene production but did not accelerate wilting or perianth senescence unless extremely high concentrations were used. ACC is not detectable in orchid pollen, yet pollination induces rapid and high levels of ethylene production in the stigma (117). In *Petunia,* compatible pollination elicited two phases of ethylene production and caused rapid perianth senescence, whereas self-pollination only triggered the first phase of ethylene production, and perianth senescence was delayed even though both compatible and incompatible pollen contained ACC (137). Collectively, the data suggest that pollen-borne ACC is not a universal primary pollen signal that is sufficient to elicit the full pollination-regulated developmental response. Nevertheless, even small amounts of ACC may trigger autocatalytic ethylene production in the stigma and thus may play a role in triggering the initial burst of ethylene production in some flowers.

A number of potential primary pollen signals have been tested for their capacity to induce ethylene production and stigma closure, two early pollination-regulated responses in *Phalaenopsis* orchid flowers (JA Nadeau & SD O'Neill, unpublished observations). Pollen-derived proteins, systemin (120), pollen-derived lipids, flavonoids, methyl jasmonate, and jasmonic acid were all eliminated as likely candidates because they failed to elicit either ethylene production and/or stigma closure over a time frame consistent with a role in pollination signaling. Only auxin (both IAA and NAA) was active in triggering ethylene production and stigma closure in this orchid bioassay (176). This result is consistent with very early reports of a substance, termed pollenhormon, that was present in orchid pollen that caused wilting of the flower and was later shown to be auxin (34, 35, 99). Even before auxin was identified as a component of orchid pollen, it was demonstrated that exogenous application of auxin could mimic pollination and initiate most pollination-regulated developmental events (25, 64). Since then, the capacity of auxins to initiate pollination-regulated developmental events in orchids has been repeatedly demonstrated (4, 7–9, 20, 21, 25, 142). Auxin was proposed to serve as the primary pollination signal by the direct transfer of pollen-borne auxin to the stigma, from where it diffused to distal floral organs and promoted autocatalytic ethylene production throughout the flower, leading to perianth senescence (20). Auxin has been identified as a component of *Nicotiana, Antirrhinum, Cyclamen, Petunia,* and *Datura* pollen, as well as of germinated pollen of *Pinus radiata,* suggesting that this putative primary pollination signal is present in a wide range of higher plant pollen (11, 89, 98, 145). While the role of auxin as the primary pollen signal has been consistently supported, other reports argue against the idea that auxin diffuses to the perianth where it directly stimulates ethylene production, suggesting that a distinct signal transmits the primary pollination event to distal organs (141).

Although auxin is present in orchid pollen and may act as the primary pollen signal, the levels of auxin in orchid pollen may be insufficient to be the sole primary pollen signal. It is possible that pollen contains other forms of auxin, such as auxin conjugates, or other pollen-borne factors, which may participate synergistically with auxin to elicit pollination-regulated responses in orchid. Arditti (4) suggested that the "pollenhormon" described by Fitting may be a mixture of biologically active molecules. A number of candidate signal molecules should be examined more intensively in this regard, including pectic cell wall fragments that have been shown to elicit ethylene production (150) and that are potentially produced by pollen-derived polygalacturonase and pectate lyase (17, 164).

Secondary Pollination Signals

Developmental changes such as ovule initiation and floral color changes are initiated in floral organs distal to the stigma shortly after pollination, implicating the role of secondary signals that transduce and amplify the primary pollination signal. The interorgan regulation of the pollination-regulated response suggests that the secondary signal is transmissible and moves from the stigma, through the style, to other floral organs. Early evidence for such a secondary pollination signal came from surgical experiments demonstrating that a mobile wilting factor is transmitted through the style to the corolla of *Petunia* flowers within 6 h after pollination (42). Stylar exudates also promote perianth senescence in *Petunia* and carnation, implicating the role of a chemical messenger either produced in the style or translocated through it (42, 132). Because auxin, ethylene, and ACC have been implicated in primary pollen signaling, these same molecules have been extensively evaluated as potential transmissible signals in pollinated flowers.

The model of Burg & Dijkman (20) proposed that the primary pollination signal, auxin, diffused to the labellum where it triggered ethylene biosynthesis. Although early experiments demonstrated that ^{14}C-IAA applied to the stigma was mobilized to the column and labellum, subsequent research indicated that ^{14}C-IAA applied to *Angraecum* and *Cattleya* stigmas was largely immobilized at the point of application with some translocation to the ovary (141). Because the pollination signal spreads to distal floral organs (petals and sepals) much faster than exogenously applied ^{14}C-IAA, it was concluded that auxin is unlikely to be the secondary pollination signal that regulates perianth senescence. However, the data that auxin is translocated primarily to the ovary is consistent with other evidence that auxin translocated from the stigma may specifically contribute to the regulation of the initiation of ovule differentiation in that organ (176).

Ethylene itself is a potential transmissible signal to floral parts distal from the stigma, and it has recently been suggested that diffusion of ethylene within the intercellular spaces (interstitial ethylene) may function in interorgan communication (75, 166, 168). Internal ethylene in the *Cymbidium* flower central column was measured at levels up to 15 ppm, and treatment of the column with exogenous ethylene increased ethylene concentration in the perianth, consistent with the proposal that interstitial ethylene from the column is translocated directly to the perianth (75). In contrast, Reid et al (127) demonstrated that aspiration of ethylene produced in the gynoecium did not delay petal senescence in carnation, indicating that volatile ethylene produced because of the primary pollination signal was not the transmissible signal responsible for regulating perianth senescence. Although these results suggest the possibility that interstitial ethylene acts as the secondary transmissible signal between floral organs, this does not appear to be the case in all flowers.

In spite of the suggestions that ethylene, rather than ACC, is translocated between floral organs, there is a large body of research implicating the ethylene precursor, ACC, as an important secondary pollination signal that coordinates interorgan pollination-regulated responses. Translocation of ACC was first demonstrated in waterlogged tomato plants (15). This result illustrated the potential for transport of this water-soluble hormone precursor to effect interorgan regulation of growth responses rather than transport of ethylene itself, which is less amenable to targeted translocation processes because of its gaseous state. Similarly, ACC translocation in pollinated flowers provides a potential mechanism for the targeted translocation of a secondary pollination signal. Following pollination of carnation flowers, ACC levels increased in all flower parts, which was proposed to result from ACC translocation (111). It was subsequently shown that ^{14}C-ACC applied to the stigma of carnation flowers resulted in the production of ^{14}C-labeled ethylene in the gynoecium and perianth, which provided compelling evidence that ACC was translocated from the stigma to the gynoecium and perianth, where it served as a substrate for ethylene biosynthesis (127). Emasculation-induced senescence of *Cymbidium* flowers demonstrated that changes in lip coloration, an early morphological marker of ethylene-regulated senescence, were triggered by ACC translocated from the central column (166). In *Phalaenopsis* orchid flowers, ethylene production by the intact flower significantly exceeded the sum of ethylene production by excised floral organs, which was interpreted to indicate that ethylene production by some floral organs is dependent on ACC import from other floral parts (117). Collectively, there is strong experimental support from several flower systems that ACC is translocated between floral organs, where it serves as a substrate for ethylene biosynthesis. This translocation of ACC also implicates it as a secondary pollination signal in the interorgan coordination of pollination-regulated developmental responses.

A very early response of floral tissues to pollination is increased sensitivity to ethylene (125). Substantial research has focused on the identification of transmissible "sensitivity factors" that render the floral tissue more sensitive to senescence-inducing effects of ethylene. Following pollination of *Phalaenopsis* flowers, there was a significant increase in the endogenous content of short-chain saturated free fatty acids in the column and perianth. Furthermore, exogenous application of these free fatty acids to the stigma increased sensitivity of the flower to ethylene, leading to the suggestion that these short-chain fatty acids may be ethylene sensitivity factors (52). It is likely that elucidation of the molecular components involved in ethylene perception and signal transduction will contribute to better understanding this phenomenon.

REGULATION OF ETHYLENE BIOSYNTHESIS IN POLLINATED FLOWERS

Ethylene is central to the control of pollination-regulated flower development, and the regulation of its endogenous synthesis appears to be an important component of the interorgan coordination of discrete developmental processes.

Examination of the regulation of expression of genes encoding the major biosynthetic steps in ethylene biosynthesis has provided the basis for analyzing the temporal and spatial regulation of ethylene biosynthesis in flowers and in response to both primary and secondary pollination signals. In addition, the cloning of ethylene biosynthetic genes has provided the means to genetically engineer flowers for enhanced longevity (91).

The penultimate enzyme in the ethylene biosynthetic pathway, ACC synthase, is widely regarded as the rate-controlling step. This step is highly regulated at the level of enzyme activity and at the level of gene expression (69, 70). In addition, while considered to be constitutive in many tissues, the final enzyme in ethylene biosynthesis, ACC oxidase, is also regulated at the level of enzyme activity and at the level of gene expression (69, 70). The spatial and temporal expression of both ACC synthase and ACC oxidase in response to pollination have provided significant insight into the interorgan coordination of pollination-regulated flower development.

Pollination Regulation of ACC Synthase

In all species examined to date, ACC synthase is encoded by multiple genes that exhibit differential tissue specificity and/or differential regulation by environmental or hormonal stimuli (70). Many cDNA clones encoding ACC synthase have been isolated from a number of flowers including tomato, carnation, *Petunia,* geranium, and orchid (24, 57, 91, 117, 119, 129). Because of its potential role in the regulation of ethylene production in response to pollination, ACC synthase has been the subject of intense interest regarding pollination-regulated flower development. ACC synthase cDNA clones have been isolated from senescing carnation flowers and from pollinated orchid flowers and their expression at the level of mRNA abundance characterized in detail in relation to pollination and perianth senescence (57, 117, 119, 171). In carnation flowers, ACC synthase mRNA was undetectable in all floral organs immediately after harvest but increased dramatically after 5–6 days in petals and styles coincident with the increase in ethylene production. ACC synthase activity was approximately sixfold higher in senescing styles than in petals, whereas ACC synthase mRNA accumulated to similar levels in both tissues (171). Subsequently, a second divergent ACC synthase mRNA was identified that was predominantly expressed in styles and may account for the apparent discrepancy between stylar ACC synthase mRNA and activity levels (57). Expression of both carnation ACC synthase cDNAs (termed CARACC3 and CARAS1) were themselves ethylene regulated, suggesting that they contribute to autocatalytic ethylene production. Relatively low levels of both ACC synthase mRNAs were detected in ovaries, in spite of high levels of ethylene production in this organ, suggesting the existence of a third ACC synthase gene predominantly expressed in the ovary. The discrepancies noted in attributing ACC synthase enzyme levels to the expression of particular ACC synthase illustrate the complexity in understanding the detailed regulation of ethylene production by multiple ethylene biosynthetic genes in floral tissues (57).

Expression of ACC synthase genes has also been characterized in pollinated orchid flowers (117). In this flower system, three ACC synthase cDNAs have been characterized that collectively account for the observed ACC synthase enzyme activity that has been characterized in the various organs of the flower. mRNAs corresponding to two highly homologous ACC synthase cDNAs (OAS1 and OAS2) accumulated to high levels in the gynoecium approximately 18 h after pollination, accumulated also in labellum tissue over the same period, but were undetectable in the perianth (excluding the labellum) even at 72 h after pollination (117). These results were consistent with the absence of ACC synthase enzyme activity in the perianth tissues, even though the perianth is the site of substantial ethylene evolution (19, 117).

Because auxin has been implicated as a primary signal in pollination-regulated responses in orchid flowers and because auxin has been reported to induce expression of certain ACC synthase gene family members in several plant species (105, 122, 153), the regulation of the orchid ACC synthase genes by auxin was carefully evaluated (117). Although exogenous application of NAA to the stigma strongly promoted accumulation of ACC synthase mRNA, this induction was completely reversed by pretreatment with AVG, an inhibitor of ethylene biosynthesis. This result indicated that OAS1 and OAS2 mRNAs were regulated by ethylene and, as with carnation ACC synthases, participated in autocatalytic ethylene production. A third ACC synthase cDNA was cloned from pollinated orchid flowers that was divergent from OAS1 and OAS2 and shown to be pollination regulated in the stigma and to be directly regulated by auxin (19). Taken together, in pollinated orchid flowers there appear to be at least two distinct types of ACC synthase genes that differ markedly in their spatial and hormonal regulation. The first type is responsive to a primary pollination signal, auxin; the second type is ethylene regulated and may serve to amplify or sustain the primary pollination signal by regulating autocatalytic ethylene production. None of the ACC synthase genes in pollinated orchid flowers are expressed in the perianth despite high levels of ethylene biosynthesis in those tissues (117). This implies that a source of ACC, other than endogenous production, exists for conversion to ethylene in perianth tissue.

Pollination Regulation of ACC Oxidase

ACC oxidase is the final enzyme in ethylene biosynthesis that converts ACC to ethylene and in most instances is considered to be constitutive (70). However, increases in ACC oxidase have been reported in senescing carnation flower petals (87), and ACC oxidase mRNA levels are pollination induced in *Phalaenopsis* and carnation flowers (101, 155). In carnation, ACC oxidase mRNA was undetectable in presenescent petals, ovaries, and receptacles but was present at significant levels in presenescent styles and increased dramatically in abundance following pollination (171). These results suggest that ACC oxidase may be present in carnation stigmas before pollination. Similarly, it was reported that ACC oxidase levels in *Petunia* flowers were maximal shortly after flower opening and that ACC oxidase enzyme activity was most abundant in the *Petunia* flower stigma before pollination (121). Recently, a family of *Petunia* ACC oxidase genes have been cloned and their expression in *Petunia* flowers studied in detail (147, 148). In agreement with Pech et al (121), each of the three expressed ACC oxidase mRNAs accumulated in the gynoecium during early flower development, reaching their maximal levels at the time of flower opening (147). In addition, ethylene treatment enhanced the accumulation of all three ACC oxidase mRNAs in the pistil, and ACC oxidase mRNA accumulation was also induced in the floral transmitting tissue by pollination (149). Collectively, these results suggest that ACC oxidase is likely to be present in the stigma of both carnation and *Petunia,* and these stigmas can likely convert pollen-borne ACC to ethylene as a part of the primary perception of pollen by the stigma. In addition, it appears that ACC oxidase genes expressed in flowers are ethylene regulated and so are likely to also participate, with ACC synthase, in autocatalytic ethylene production in pollinated flowers.

ACC oxidase activity and ACC oxidase mRNA accumulation have also been characterized in orchid flowers (101, 117). Unlike *Petunia,* ACC oxidase activity was not detected in mature flowers before pollination but was induced rapidly following pollination (101). Following pollination, orchid ACC oxidase (OAO1) mRNA accumulated to high levels within 48 h in all floral organs, including the perianth (101). This result demonstrated a striking disparity between the accumulation of ACC synthase mRNA, which failed to accumulate in the perianth, and suggested that the sepals and petals developed the capacity to convert ACC to ethylene but could not synthesize ACC endogenously (101, 117). In light of the high level of ethylene evolution from perianth tissues and the absence of ACC synthase enzyme activity or a corresponding mRNA, it has been suggested that ACC must be translocated from other floral organs to the perianth where it is converted to ethylene by the activity of resident ACC oxidase (117).

MODEL FOR INTERORGAN COORDINATION OF POLLINATION-REGULATED FLOWER DEVELOPMENT

Ethylene with auxin plays important roles in pollination-regulated developmental responses and, in conjunction with ACC, participates in the interorgan coordination of diverse components of pollination-regulated flower development. The model outlined in Figure 2 attempts to summarize data that address the initiation of ethylene biosynthesis by pollination and the interorgan regulation of ethylene biosynthesis in pollinated flowers because these two processes appear to be central to the coordination of pollination-regulated flower development. Much of the model relies on the recent elucidation of the patterns of expression of genes encoding ethylene biosynthetic enzymes in pollinated flowers. In this model, a pollen-associated factor is prominent as the primary stimulus in initiating ethylene biosynthesis as well as the overall process of pollination-regulated flower development. In orchid flowers, an important component of the primary pollen signal is auxin, whereas in other flowers, ACC or other as-yet-unidentified pollen-borne substances may also participate in primary pollen signaling. The earliest response to pollination is the induction of ethylene production in the stigma and style, which may use pollen-borne ACC directly as substrate or, more likely, relies on the rapid induction of ACC synthase in the stigma. In orchid flowers, expression of an auxin-regulated ACC synthase is rapidly induced in the stigma by pollination, and this is likely to account for the rapid transduction of the putative primary pollen signal, auxin, to the initial burst of ethylene production. In other flowers such as *Petunia,* a trace amount of pollen-borne ACC may be converted directly to ethylene, which leads to localized autocatalytic ethylene production in the stigma. The presence of ACC oxidase mRNA in presenescent *Petunia* stigmas and its rapid ethylene-dependent induction following pollination is consistent with this mechanism of initial pollination-induced ethylene production.

The interorgan transmission and amplification of the primary pollen signal requires secondary signals that emanate from the stigma, and an important component of this signal appears to be ACC and its subsequent conversion to ethylene in distal floral organs. In orchid flowers, the role of translocated ACC was implicated by the failure of petals and sepals to accumulate ACC synthase mRNA or enzyme activity, in spite of high levels of ethylene production by these floral organs, suggesting that ethylene production in these organs is entirely dependent on translocated ACC. This situation is less pronounced in

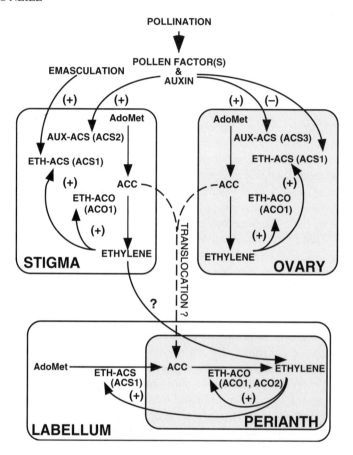

Figure 2 Model for the interorgan regulation of ethylene biosynthetic genes in pollinated flowers, based primarily on data from the *Phalaenopsis* orchid flower, with other flower systems viewed as variations on this basic theme (19, 101, 117). The primary signaling event for pollination-regulated development occurs with the transfer of pollen-borne factors, including auxin, to the stigma surface, the site of initial perception. An auxin-regulated ACC synthase gene (AUX-ACS) is induced by the primary pollination signal(s). ACC synthesized in the stigma is converted to ethylene by ACC oxidase, initially present at a low level but induced to high levels soon after pollination to initiate autocatalytic ethylene production in the stigma. Expression of other ethylene-regulated ACC synthase gene(s) (ETH-ACS) are also induced by ethylene to further amplify autocatalytic ethylene production. ACC translocated from the stigma to the perianth supports the production of ethylene in that organ, initiating its senescence. This mechanism of interorgan regulation relies on translocation of ACC to distal organs to support ethylene production. In addition to its role as a gaseous signal coordinating ethylene production at the molecular level, ethylene itself may also be translocated within the flower as a translocated signal. Abbreviations: ACC, 1-aminocyclopropane-1-carboxylic acid; ACO, ACC oxidase; ACS, ACC synthase; AdoMet, S-adenosylmethionine; AUX, auxin-regulated; ETH, ethylene-regulated.

flowers such as *Petunia* where ACC synthase is induced in petal tissues. However, the *Petunia* petal ACC synthase is ethylene induced, and the initial production of ethylene in petals required to stimulate autocatalytic ethylene production may be derived from translocated ACC. The model shown in Figure 2 is idealized in that it reflects the situation in *Phalaenopsis* orchid flowers, one of the most extreme examples of pollination-regulated flower development. Data from other species suggest that there are variations on the themes presented in Figure 2.

CONCLUSIONS AND FUTURE DIRECTIONS

Pollination regulates a suite of developmental responses that transform the flower from a structure dedicated to pollen dispersal and reception to one that is dedicated to fertilization, embryogenesis, seed development, and ultimately seed dispersal. One important aspect of this process is the shedding of floral organs, such as the petals and sepals that have completed their service in pollinator attraction. In addition, pollination demands that certain reproductive structures rapidly mature in anticipation of fertilization. Thus, the signals that regulate pollination responses are involved in simultaneously coordinating processes of programmed cell death while initiating the differentiation and/or maturation of important reproductive structures.

Research directed at understanding the nature of the pollination signals has a rich history dating to the beginnings of this century, and it has revealed important elements of the interorgan signaling needed to coordinate the full syndrome of pollination-regulated developmental processes. Although it is clear that auxin, ethylene, and ACC are important actors in this regulatory process, it is likely that this is not the complete cast. More research is needed to more fully elucidate primary pollination signals and how they interact with each other and with the stigma and style to propagate the signal to distal floral organs.

This review focuses on the early events in pollen signaling. However, the growing pollen tube continues to exchange information with the style and ovary, and the nature of these signals also represents a rich area of further research. Much research has characterized a number of model flower systems that are physiologically suited to studying pollination-regulated flower development, but these model systems do not generally overlap with model genetic systems, such as *Arabidopsis thaliana*. Despite the recognized differences between plants in terms of their pollination responses, future progress requires that information from powerful physiological systems be transferred to species that are well suited to genetic dissection and transgenic manipulation and analysis.

ACKNOWLEDGMENTS

I thank Drs. Anhthu Bui, Jeanette Nadeau, Ron Porat, and Xiansheng Zhang for discussions and research contributions described in this review and Professor Abraham H Halevy (The Hebrew University of Jerusalem) for his friendship and inspiration. I also thank the USDA-NRICGP (Plant Growth and Development Program), the National Science Foundation (Developmental Mechanisms Program), and the Binational Agricultural Research and Development Fund for providing support for my laboratory to study many of the issues discussed here.

Visit the *Annual Reviews home page* at http//:www.annurev.org.

Literature Cited

1. Adams DO, Yang SF. 1979. Ethylene biosynthesis: Identification of 1-aminocyclopropane-1-carboxylic acid as an intermediate in the conversion of methionine to ethylene. *Proc. Natl. Acad. Sci. USA* 83: 7755–59

2. Akamine EK. 1963. Ethylene production in fading Vanda orchid blossoms. *Science* 140:1217–18

3. Anderson JD, Mattoo AK, Lieberman M. 1982. Induction of ethylene biosynthesis in tomato leaf disks by cell wall-digesting enzymes. *Biochem. Res. Commun.* 107: 588–96

4. Arditti J. 1971. Orchids and the discovery of auxin. *Am. Orchid Soc. Bull.* 40:211–14

5. Arditti J. 1979. Aspects of the physiology of orchids. *Adv. Bot. Res.* 7:421–655

6. Arditti J, Flick BH. 1976. Post-pollination phenomena in orchid flowers. VI. Excised floral segments of *Cymbidium. Am. J. Bot.* 63:201–11

7. Arditti J, Hogan NM, Chadwick AV. 1973. Post-pollination phenomena in orchid flowers. IV. Effects of ethylene. *Am. J. Bot.* 60:883–88

8. Arditti J, Jeffrey DC, Flick BH. 1971. Post-pollination phenomena in orchid flowers. III. Effects and interactions of auxin, kinetin or gibberellin. *New Phytol.* 70: 1125–41

9. Arditti J, Knauft RL. 1969. The effects of auxin, actinomycin D, ethionine and puromycin on post-pollination behavior in *Cymbidium* (Orchidaceae) flowers. *Am. J. Bot.* 56:620–28

10. Asen S, Stewart RN, Norris KH. 1977. Anthocyanin and pH involved in the color of 'Heavenly Blue' morning glory. *Phytochemistry* 16:1118–19

11. Barendse GWM, Rodrigues-Pereira AS, Berkers PA, Driessen FM, Eyden-Emons van A, Linskens HF. 1970. Growth hormones in pollen, styles and ovaries of *Petunia hybrida* and of *Lilium* species. *Acta Bot. Neerl.* 19:175–86

12. Bashaw EC, Hanna WW. 1990. Apomictic reproduction. In *Reproductive Versatility in the Grasses,* ed. GP Chapman, pp. 100–30. Cambridge: Cambridge Univ. Press

13. Borochov A, Woodson WR. 1989. Physiology and biochemistry of flower petal senescence. *Hortic. Rev.* 11:15–43

14. Bouman F. 1984. The ovule. See Ref. 66a, pp. 123–57

15. Bradford KJ, Yang SF. 1980. Xylem transport of 1-aminocyclopropane-1-carboxylic acid, an ethylene precursor, in waterlogged tomato plants. *Plant Physiol.* 65:322–26

16. Brady CJ. 1987. Fruit ripening. *Annu. Rev. Plant Physiol.* 38:155–78

17. Brown SM, Crouch ML 1990. Characterization of a gene family abundantly expressed in *Oenothera organensis* pollen that shows sequence similarity to polygalacturonase. *Plant Cell* 2:263–74

18. Bufler G, Mor Y, Reid MS, Yang SF. 1980. Changes in 1-aminocyclopropane-1-carboxylic acid content of cut carnation flowers in relation to their senescence. *Planta* 150:439–42

19. Bui A. 1996. *Temporal and spatial regulation of ACC synthase genes in pollinated orchid flowers.* PhD thesis. Univ. Calif., Davis

20. Burg SP, Dijkman MJ. 1967. Ethylene and auxin participation in pollen induced fading of *Vanda* orchid blossoms. *Plant Physiol.* 42:1648–50

21. Chadwick AV, Hogan NM, Arditti J. 1980. Postpollination phenomena in orchid flowers. IX. Induction and inhibition of ethylene evolution, anthocyanin synthesis, and perianth senescence. *Bot. Gaz.* 141:422–27

22. Chaubal R, Reger BJ. 1990. Relatively high calcium is localized in synergid cells of wheat ovaries. *Sex. Plant Reprod.* 3: 98–102

23. Chaubal R, Reger BJ. 1992. The dynamics of calcium distribution in the synergid cells of wheat after pollination. *Sex. Plant Reprod.* 206–13

24. Clark D, Lind-Iversen S, Richards C, Evensen K. 1994. Ethylene-regulated gene expression in geranium flowers. See Ref. 140a, pp. 291–95

25. Curtis JT. 1943. An unusual pollen reaction in *Phalaenopsis. Am. Orchid Soc. Bull.* 21: 98–100

26. Davidson OW. 1949. Effects of ethylene on orchid flowers. *Am. Soc. Hortic. Sci.* 53: 440–46

27. Dijkman MJ, Burg SP. 1970. Auxin-induced spoiling of *Vanda* blossoms. *Am. Orchid Soc. Bull.* 39:799–804

28. Dobson HEM. 1989. Pollenkitt in plant reproduction. In *The Evolutionary Ecology of Plants*, ed. JH Bock, YB Linhart, pp. 227–46. Boulder, CO: Westview

29. Duncan RE, Curtis JT. 1942. Intermittent growth of fruits of *Cypripedium* and *Paphiopedilum*. A correlation of the growth of orchid fruits with their internal development. *Bull. Torrey Club* 69:353–59

30. Duncan RE, Curtis JT. 1943. Growth of fruits in Cattleya and allied genera in the Orchidaceae. *Bull. Torrey Club* 70:104–19

31. Eaton GW. 1959. A study of the megagametophyte in *Prunus avium* and its relation to fruit setting. *Can. J. Plant Sci.* 39:466–76

32. Eaton GW, Jamont AM. 1964. Embryo sac development in the Apricot, *Prunus armeniaca* L. CV constant. *Am. Soc. Hortic. Sci.* 86:95–101

33. Felix G, Grosskopf DG, Regenass M, Basse CW, Boller T. 1991. Elicitor-induced ethylene biosynthesis in tomato cells. *Plant Physiol.* 97:19–25

34. Fitting H. 1909. Die Beeinflussung der Orchideenblüten durch die Bestäubung und durch andere Umstände. *Z. Bot.* 1:1–86

35. Fitting H. 1910. Weitere entwicklungsphysiologishe Untersuchungen an Orchideenblüten. *Z. Bot.* 2:225–67

36. Fredrikson M. 1991. An embryological study of *Platanthera bifolia* (Orchidaceae). *Plant Syst. Evol.* 714:213–20

37. Fredrikson M. 1992. The development of the female gametophyte of *Epipactis* (Orchidaceae) and its inference for reproductive ecology. *Am. J. Bot.* 79:61–8

38. Fredrikson M, Carlsson K, Franksson O. 1988. Confocal scanning laser microscopy, a new technique used in an embryological study of *Dactylorhiza maculata* (Orchidaceae). *Nord. J. Bot.* 8:369–74

39. Gard DL. 1994. γ-tubulin is asymmetrically distributed in the cortex of *Xenopus* oocytes. *Dev. Biol.* 161:131–40

40. Gilissen LJW. 1976. The role of the style as a sense-organ in relation to wilting of the flower. *Planta* 131:201–2

41. Gilissen LJW. 1977. Style-controlled wilting of the flower. *Planta* 133:275–80

42. Gilissen LJW, Hoekstra FA. 1984. Pollination-induced corolla wilting in *Petunia hybrida*. Rapid transfer through the style of a wilting-inducing substance. *Plant Physiol.* 5:496–98

43. Goh CJ, Halevy AH, Engel R, Kofranek AM. 1985. Ethylene evolution and sensitivity in cut orchid flowers. *Sci. Hortic.* 26: 57–67

44. Gori DF. 1983. Post-pollination phenomena and adaptive floral color change. In *Handbook of Experimental Pollination Biology*, ed. CE Jones, RJ Little, pp. 31–49. New York: Van Nostrand Reinhold

45. Gustafson FG. 1936. Inducement of fruit development by growth-promoting chemicals. *Proc. Natl. Acad. Sci. USA* 22:628–36

46. Gustafson FG. 1937. Parthenocarpy induced by pollen extracts. *Am. J. Bot.* 24: 102–7

47. Gustafson FG. 1939. Auxin distribution in fruits and its significance in fruit development. *Am. J. Bot.* 26:189–94

48. Gustafsson A. 1946. Apomixis in higher plants. Part 1. The mechanism of apomixis. Lunds Univ. Årsskrift, 2:1–66.

49. Hagerup O. 1944. On fertilisation, polyploidy and haploidy in *Orchis maculatus* L. sens. lat. *Dansk Bot. Arkiv.* 11:1–25

50. Halevy AH, Mayak S. 1979. Senescence and postharvest physiology of cut flowers. *Hortic. Rev.* 1:204–36

51. Halevy AH, Mayak S. 1981. Senescence and postharvest physiology of cut flowers. *Hortic. Rev.* 3:59–143

52. Halevy AH, Porat R, Spiegelstein H, Borochov A, Botha L, Whitehead CS. 1996. Short-chain saturated fatty acids in the regulation of pollination-induced ethylene sensitivity of *Phalaenopsis* flowers. *Physiol. Plant.* 97:469–74

53. Halevy AH, Whitehead CS, Kofranek AM. 1984. Does pollination induce corolla abscission of cyclamen flowers by promoting ethylene production? *Plant Physiol.* 75: 1090–93

54. Hall IV, Forsyth FR. 1967. Production of ethylene by flowers following pollination and treatments with water and auxin. *Can. J. Bot.* 45:1163–66

55. Han SS, Halevy AH, Reid MS. 1991. The role of ethylene and pollination in petal senescence and ovary growth of Brodiaea. *J. Am. Soc. Hortic. Sci.* 116:68–72

56. He C-P, Yang H-Y. 1992. Ultrastructural localization of calcium in the embryo sac of sunflower. *Chin. J. Bot.* 4:99–106

57. Henskens JAM, Rouwendal GJA, Have AT, Woltering EJ. 1994. Molecular cloning of two different ACC synthase PCR fragments in carnation flowers and organ-specific expression of the corresponding genes. *Plant Mol. Biol.* 26:453–58

58. Hew CS, Tan SC, Chin TY, Ong TK. 1989. Influence of ethylene on enzyme activities and mobilization of materials in pollinated *Arachnis* orchid flowers. *J. Plant Growth Regul.* 8:121–30

59. Hill SE, Stead AD, Nichols R. 1987. Pollination-induced ethylene and production of 1-aminocyclopropane-1-carboxylic acid by pollen of *Nicotiana tabacum* cv White Burley. *J. Plant Growth Regul.* 6:1–13

60. Hoekstra FA, van Roekel T. 1988. Effects of previous pollination and stylar ethylene on pollen tube growth in *Petunia hybrida* styles. *Plant Physiol.* 86:4–6

61. Hoekstra FA, Weges R. 1986. Lack of control by early pistillate ethylene of the accelerated wilting of *Petunia hybrida* flowers. *Plant Physiol.* 80:403–8

62. Huang B-Q, Russell SD. 1992. Female germ unit: organization, isolation, and function. *Int. Rev. Cytol.* 140:233–93

63. Huang BQ, Russell SD. 1992. Synergid degeneration in *Nicotiana*: a quantitative, fluorochromatic and chlorotetracycline study. *Sex. Plant Reprod.* 5:151–55

64. Hubert B, Maton J. 1939. The influence of synthetic growth-controlling substances and other chemicals on post-floral in tropical orchids. *Biol. Jarrb.* 6:244–85

65. Israel HW, Sagawa Y. 1965. Post-pollination ovule development in *Dendrobium* orchids III. Fine structure of meiotic prophase I. *Caryologia* 18:15–33

66. Jensen WA, Ashton ME, Beasley CA. 1983. Pollen tube-embryo sac interaction in cotton. In *Pollen: Biology And Implications for Plant Breeding*, ed. DL Mulcahy, E Ottaviano, pp. 67–72. New York: Elsevier

66a. Johri BM, ed. 1984. *Embryology of Angiosperms*. Berlin: Springer-Verlag

67. Kandasamy MK, Dwyer KG, Paolillo DJ. 1990. *Brassica* s-proteins accumulate in the intercellular matrix along the path of pollen tubes in transgenic tobacco pistils. *Plant Cell* 2:39–49

68. Kapil RN, Bhatnagar AK. 1981. Ultrastructure and biology of female gametophyte in flowering plants. *Int. Rev. Cytol.* 70: 291–341

69. Kende H. 1989. Enzymes of ethylene biosynthesis. *Plant Physiol.* 91:1–4

70. Kende H. 1993. Ethylene biosynthesis. *Annu. Rev. Plant Physiol. Plant Mol. Biol.* 44:283–307

71. Knox RB, Heslop-Harrison J. 1970. Pollen wall proteins: localization and enzymatic activity. *J. Cell Sci.* 6:1–27

72. Laibach F. 1932. Pollenhormon und wuchsstoff. *Ber. Deutsch. Bot. Ges.* 50: 383–92

73. Laibach F. 1933. Versuche mit wuchsstoffpaste. *Ber. Deutsch. Bot. Ges.* 51:386–92

74. Laibach F. 1933. Wuchsstoffversuche mit lebenden orchideenpollinien. *Ber. Deutsch. Bot. Ges.* 51:336–40

75. Larsen PB, Woltering EJ, Woodson WR. 1993. Ethylene and interorgan signaling in flowers following pollination. In *Plant Signals in Interactions with Other Organisms*, ed. J Schultz, I Raskin, pp. 171–81. Rockville, MD: Am. Soc. Plant Physiol.

76. Law SK, Yeung EC. 1989. Embryology of Calypso bulbosa. I. Ovule development. *Am. J. Bot.* 76:1668–74

77. Lawton KA, Huang B, Goldsbough PB, Woodson WR. 1989. Molecular cloning and characterization of senescence-related genes from carnation flower petals. *Plant Physiol.* 90:690–96

78. Lawton KA, Raghothama KG, Goldsbrough PB, Woodson WR. 1990. Regulation of senescence-related gene expression in carnation flower petals by ethylene. *Plant Physiol.* 93:1370–75

79. Lay-Yee M, Stead AD, Reid MS. 1992. Flower senescence in daylily (*Hemerocallis*). *Physiol. Plant.* 86:308–14

80. Lingle SE, Dunlap JR. 1991. Sucrose metabolism and IAA and ethylene production

in muskmelon ovaries. *J. Plant Growth Regul.* 10:167–71

81. Linskens HF. 1974. Translocation phenomena in the petunia flower after cross- and self-pollination. In *Fertilization in Higher Plants,* ed. HF Linskens, pp. 285–92. Amsterdam: North-Holland

82. Lipe JA, Morgan PW. 1973. Location of ethylene production in cotton flowers and dehiscing fruits. *Planta* 115:93–96

83. Lovell PJ, Lovell PH, Nichols R. 1987. The control of flower senescence in petunia (*Petunia hybrida*). *Ann. Bot.* 60:49–59

84. Lovell PJ, Lovell PH, Nichols R. 1987. The importance of the stigma in flower senescence in petunia (*Petunia hybrida*). *Ann. Bot.* 60:41–47

85. Lukaszewski TA, Reid MS. 1989. Bulb-type flower senescence. *Acta Hortic.* 261: 59–62

86. Maheshwari P. 1950. An Introduction to the Embryology of Angiosperms. New York: McGraw-Hill

87. Manning K. 1985. The ethylene forming enzyme system in carnation flowers. In *Ethylene and Plant Development,* JA Roberts, GA Tucker, pp. 83–92. London: Butterworths

88. Mascarenhas JP. 1993. Molecular mechanisms of pollen tube growth and differentiation. *Plant Cell* 5:1303–14

89. Mascarenhas JP, Canary D. 1985. Pollen. Symbionts and symbiont-induced structures. In *Encyclopedia of Plant Physiology,* ed. RP Pharis, DM Reid, 2:579–98. Berlin/Heidelberg/New York/Tokyo: Springer-Verlag

90. Meyer RC, Goldsbrough PB, Woodson WR. 1991. An ethylene-responsive flower senescence-related gene from carnation encodes a protein homologous to glutathione *s*-transferases. *Plant Mol. Biol.* 17:277–81

91. Michael MZ, Savin KW, Baudinette SC, Graham MW, Chandler SF, et al. 1993. Cloning of ethylene biosynthetic genes involved in petal senescence of carnation and petunia, and their antisense expression in transgenic plants. In *Cellular and Molecular Aspects of the Plant Hormone Ethylene,* ed. JC Pech, A Latché, C Balogué, pp. 298–303. The Netherlands: Kluwer

92. Mo Y, Nagel C, Taylor LP. 1992. Biochemical complementation of chalcone synthase mutants defines a role for flavonols in functional pollen. *Proc. Natl. Acad. Sci. USA* 89:7213–17

93. Mogensen HL. 1984. Quantitative observations on the pattern of synergid degeneration in barley. *Am. J. Bot.* 71:1448–51

94. Mogensen HL. 1992. The male germ unit:

concept, composition and significance. *Int. Rev. Cytol.* 140:129–47

95. Mohan Ram HY, Mathur G. 1984. Flower colour changes in *Lantana camara.J. Exp. Bot.* 35:1656–62

96. Mòl R, Matthys-Rochon E, Dumas C. 1994. The kinetics of cytological events during double fertilization in *Zea mays* L. *Plant J.* 5:197–206

97. Mor Y, Halevy AH, Spiegelstein H, Mayak S. 1985. The site of 1-aminocyclopropane-1-carboxylic acid synthesis in senescing carnation petals. *Physiol. Plant.* 65: 196–202

98. Muir RM. 1947. The relationship of growth hormones and fruit development. *Proc. Natl. Acad. Sci. USA* 33:303–12

99. Müller R. 1953. Zur quantitativen bestimmung von indolylessigsäure mittels papierchromatographie und papierelektrophorese. *Beitr.zur Biol. Pflanz.* 30:1–32

100. Nadeau JA, Zhang XS, Li J, O'Neill SD. 1996. Ovule development: identification of stage- and tissue-specific cDNAs. *Plant Cell* 8:213–39

101. Nadeau JA, Zhang XS, Nair H, O'Neill SD. 1993. Temporal and spatial regulation of 1-aminocyclopropane carboxylate oxidase in the pollination induced senescence of orchid flowers. *Plant Physiol.* 103:31–39

102. Nair H. 1990. Postharvest physiology and handling of orchids. *Malayan Orchid Rev.* 18:62–68

103. Nair H, Fong TH. 1987. Ethylene production and 1-aminocyclopropane-1-carboxylic acid levels in detached orchid flowers of *Dendrobium* 'Pampadour'. *Sci. Hortic.* 32:145–51

104. Nair H, Idris ZM, Arditti J. 1991. Effects of 1-aminocyclopropane-1-carboxylic acid on ethylene evolution and senescence of *Dendrobium* (Orchidaceae) flowers. *Lindleyana* 6:49–58

105. Nakajima N, Mori H, Yamazaki K, Imaseki H. 1990. Molecular cloning and sequence of a complementary DNA encoding 1-aminocyclopropane-1-carboxylate synthase induced by tissue wounding. *Plant Cell Physiol.* 31:1021–29

106. Nichols R. 1966. Ethylene production during senescence of flowers. *J. Hortic. Sci.* 41:279–90

107. Nichols R. 1968. The response of carnations (*Dianthus caryophyllus*) to ethylene. *J. Hortic. Sci.* 43:335–49

108. Nichols R. 1971. Induction of flower senescence and gynoecium development in the carnation (*Dianthus caryophyllus*) by ethylene and 2-chloroethylphosphonic acid. *J. Hortic. Sci.* 46:323–32

109. Nichols R. 1976. Cell enlargement and sugar accumulation in the gynoecium of the glasshouse carnation (*Dianthus caryophyllus* L.) induced by ethylene. *Planta* 130: 47–52

110. Nichols R. 1977. Sites of ethylene production in the pollinated and unpollinated senescing carnation (*Dianthus caryophyllus*) inflorescence. *Planta* 135:155–59

111. Nichols R, Bufler G, Mor Y, Fujino DW, Reid MS. 1983. Changes in ethylene production and 1-aminocyclopropane-1-carboxylic acid content of pollinated carnation flowers. *J. Plant Growth Regul.* 2:1–8

112. Nichols R, Ho LC. 1975. An effect of ethylene on the distribution of ^{14}C-sucrose from the petals to other flower parts in the senescent cut inflorescence of *Dianthus caryophyllus. Ann. Bot.* 39:433–38

113. Nichols R, Ho LC. 1975. Effects of ethylene and sucrose on translocation of dry matter and ^{14}C-sucrose in the cut flower of the glasshouse carnation (*Dianthus caryophyllus*) during senescence. *Ann. Bot.* 39: 287–96

114. Noher de Halac I, Harte C. 1985. Cell differentiation during megasporogenesis and megagametogenesis. *Phytomorphology* 35: 189–200

115. O'Neill SD. 1994. Pollen signaling and interorgan regulation of the postpollination syndrome of flowers. See Ref. 140a, pp. 161–77

116. O'Neill SD, Nadeau JA. 1997. Postpollination flower development. *Hortic. Rev.* 19: 1–58

117. O'Neill SD, Nadeau JA, Zhang XS, Bui AQ, Halevy AH. 1993. Interorgan regulation of ethylene biosynthetic genes by pollination. *Plant Cell* 5:419–32

118. Oertli JJ, Kohl JHC. 1960. *Der Einfluss der Bestaubung auf die Stoffbewegung in Cymbidiumbluten.* Munchen/Bonn/Wien: BLV

119. Park KY, Drory A, Woodson WR. 1992. Molecular cloning of an 1-aminocyclopropane-1-carboxylate synthase from senescing carnation flower petals. *Plant Mol. Biol.* 18:377–86

120. Pearce G, Strydom D, Johnson S, Ryan CA. 1991. A polypeptide from tomato leaves induces wound-inducible proteinase inhibitor protein. *Science* 253:895–98

121. Pech J-C, Latché A, Larrigaudière C, Reid MS. 1987. Control of early ethylene synthesis in pollinated petunia flowers. *Plant Physiol. Biochem.* 25:431–37

122. Peck SC, Kende H. 1995. Sequential induction of the ethylene biosynthetic enzymes by indole-3-acetic acid in etiolated peas. *Plant Mol. Biol.* 28:293–301

123. Pimienta E, Polito VS. 1983. Embryo sac development in almond [*Prunus dulcis* (Mill.) DA Webb] as affected by cross-, self- and nonpollination. *Ann. Bot.* 51: 469–79

124. Porat R, Borochov A, Halevy AH. 1994. Pollination-induced changes in ethylene production and sensitivity to ethylene in cut Dendrobium orchid flowers. *Sci. Hortic.* 58: 215–21

125. Porat RA, Halevy AH, Serek M, Borochov A. 1995. An increase in ethylene sensitivity following pollination is the initial event triggering an increase in ethylene production and enhanced senescence of Phalaenopsis orchid flowers. *Physiol. Plant.* 93: 778–84

126. Proctor JTA, Creasy LL. 1969. An anthocyanin-decolorizing system in florets of *Cichorium intybus. Phytochemistry* 8: 1401–3

127. Reid MS, Fujino DW, Hoffman NE, Whitehead CS. 1984. 1-aminocyclopropane-1-carboxylic acid: the transmitted stimulus in pollinated flowers? *J. Plant Growth Regul.* 3:189–96

128. Reisler L, Fischer RL. 1993. The ovule and embryo sac. *Plant Cell* 5:1291–301

129. Rottmann WH, Peter GF, Oeller PW, Keller JA, Shen NF et al. 1991. 1-aminocyclopropane-1-carboxylate synthase in tomato is encoded by a multigene family whose transcription is induced during fruit and floral senscence. *J. Mol. Biol.* 222: 937–61

130. Russell SD. 1993. The egg cell: development and role in fertilization and early embryogenesis. *Plant Cell* 5:1349–59

131. Ryan CA, Farmer EE. 1991. Oligosaccharide signals in plants: a current assessment. *Annu. Rev. Plant Physiol.* 42:651–74

132. Sacalis J, Wulster G, Janes H. 1983. Senescence in isolated carnation petals: differential response of various petal portions to ACC, and effects of uptake of exudate from excised gynoecia. *Z. Pflanzenphysiol.* 112: 7–14

133. Sagawa Y, Israel HW. 1964. Post-pollination ovule development in *Dendrobium* orchids. I. Introduction. *Caryologia* 17: 53–64

134. Sakurai A, Fujioka S. 1993. The current status of physiology and biochemistry of brassinosteroids. *Plant Growth Regul.* 13: 147–59

135. Sanders LC, Lord EM. 1989. Directed movement of latex particles in the gynoecia of three species of flowering plants. *Science* 243:1606–8

136. Simpson CE, Bashaw EC. 1969. Cytology

and reproductive characteristics of *Pennisetum ciliare*. *Am. J. Bot.* 56:31–36

137. Singh A, Evensen KB, Kao T. 1992. Ethylene synthesis and floral senescence following compatible and incompatible pollinations in *Petunia inflata*. *Plant Physiol.* 99: 38–45

138. Sood SK, Mohana Rao PR. 1986. Development of male and female gametophytes in *Herminium angustifolium* (Orchidaceae). *Phytomorphology* 36:11–5

139. Stead AD. 1992. Pollination-induced flower senescence: a review. *Plant Growth Regul.* 11:13–20

140. Stead AD, Reid MS. 1990. The effect of pollination and ethylene on the colour change of the banner spot of *Lupinus albifrons* (Bentham) flowers. *Ann. Bot.* 66: 655–63

140a. Stephenson AG, Kao TH, eds. 1994. *Pollen-Pistil Interactions and Pollen Tube Growth.* Rockville, MD: Am. Soc. Plant Physiol.

141. Strauss M, Arditti J. 1982. Postpollination phenomena in orchid flowers. X. Transport and fate of auxin. *Bot. Gaz.* 143:286–93

142. Strauss MS, Arditti J. 1984. Postpollination phenomena in orchid flowers. XII. Effects of pollination, emasculation, and auxin treatment on flowers of *Cattleya* Porcia 'Cannizaro' and the rostellum of *Phalaenopsis*. *Bot. Gaz.* 145:43–49

143. Suessenguth K. 1923. Über die pseudogamie bei Zygopetalum Mackayi Hook. Ber. Duetsche. *Bot. Ges.* 41:16–23

144. Swamy BGL. 1943. Embryology of the Orchidaceae. *Curr. Sci.* 12:13–17

145. Sweet GB, Lewis PN. 1969. A diffusible auxin from *Pinus radiata* and its possible role in stimulating ovule development. *Planta* 89:380–84

146. Sylvestre I, Droillard M-J, Bureau J-M, Paulin A. 1989. Effects of the ethylene rise on the peroxidation of membrane lipids during the senescence of cut carnations. *Plant Physiol. Biochem.* 27:407–13

147. Tang X, Gomes AMTR, Bhatia A, Woodson WR. 1994. Pistil-specific and ethylene-regulated expression of 1-aminocyclopropane-1-carboxylate oxidase genes in petunia flowers. *Plant Cell* 6:1227–39

148. Tang X, Wang H, Brandt AS, Woodson WR. 1993. Organization and structure of the 1-aminocylopropane-1-carboxylate oxidase gene family from *Petunia hybrida*. *Plant Mol. Biol.* 23:1151–64

149. Tang X, Woodson WR. 1996. Temporal and spatial expression of 1-aminocyclopropane-1-carboxylate oxidase mRNA following pollination of immature and mature

Petunia hybrida flowers. *Plant Physiol.* 112: 503–11

150. Tong CB, Labavitch JM, Yang SF. 1986. The induction of ethylene production from pear cell culture by cell wall fragments. *Plant Physiol.* 81:929–30

151. Ueda J, Kato J. 1981. Promotive effect of methyl jasmonate on oat leaf senescence in the light. *Z. Pflanz.physiol. Bd.* 103:357–59

152. van der Donk JAWM. 1975. Recognition and gene expression during the incompatibility reaction in *Petunia hybrida* L. *Mol. Gen. Genet.* 141:305–16

153. Van der Straeten D, Van Wiemeersch L, Goodman HM, Van Montagu M. 1990. Cloning and sequencing of two different cDNAs encoding 1-aminocyclopropane-1-carboxylate synthase in tomato. *Proc. Natl. Acad. Sci. USA* 87:4859–63

154. Vogt T, Pollak P, Tarlyn N, Taylor LP. 1994. Pollination- or wound-induced kaempferol accumulation in petunia stigmas enhances seed production. *Plant Cell* 6:11–23

155. Wang H, Woodson WR. 1991. A flower senescence-related mRNA from carnation shows sequence similarity with fruit ripening-related mRNAs involved in ethylene biosynthesis. *Plant Physiol.* 96:1000–1

156. Weidhase RA, Kramell H-M, Lehmann J, Liebisch H-W, Lerbs W, Parthier B. 1987. Methyljasmonate-induced changes in the polypeptide pattern of senescing barley leaf segments. *Plant Sci.* 51:177–86

157. Weiss MR. 1991. Floral colour changes as cues for pollinators. *Nature* 354:227–29

158. Whitehead CS, Fujino DW, Reid MS. 1983. Identification of the ethylene precursor 1-aminocyclopropane-1-carboxylic acid (ACC) in pollen. *Sci. Hortic.* 21:291–97

159. Whitehead CS, Fujino DW, Reid MS. 1983. The roles of pollen ACC and pollen tube growth in ethylene production by carnations. *Acta Hortic.* 141:221–27

160. Whitehead CS, Halevy AH, Reid MS. 1984. Roles of ethylene and 1-aminocyclopropane-1-carboxylic acid in pollination and wound-induced senescence of *Petunia hybrida* flowers. *Physiol. Plant.* 61:643–48

161. Whitehead CS, Halevy AH, Reid MS. 1984. Control of ethylene synthesis during development and senescence of carnation petals. *J. Am. Soc. Hortic. Sci.* 109:473–75

162. Wiermann R, Vieth K. 1983. Outer pollen wall, an important accumulation site for flavonoids. *Protoplasma* 118:230–33

163. Willemse MTM, van Went JL. 1984. The female gametophyte. See Ref. 66a, pp. 159–96

164. Wing RA, Yamaguchi J, Larabell SK, Ursin VM, McCormick S. 1989. Molecular and

genetic characterization of two pollen-expressed genes that have sequence similarity to pectate lyases of the plant pathogen *Erwinia*. *Plant Mol. Biol.* 14:17–28

165. Woltering EJ. 1989. Lip coloration in *Cymbidium* flowers by emasculation and by lip-produced ethylene. *Acta Hortic.* 261: 145–50

166. Woltering EJ. 1990. Interorgan translocation of 1-aminocyclopropane-1-carboxylic acid and ethylene coordinates senescence in emasculated *Cymbidium* flowers. *Plant Physiol.* 92:837–45

167. Woltering EJ. 1990. Interrelationship between the different flower parts during emasculation-induced senescence in *Cymbidium* flowers. *J. Exp. Bot.* 41:1021–29

168. Woltering EJ. 1994. Ethylene and orchid flower senescence: roles of ethylene and ACC in inter-organ communication in *Cymbidium* flowers. *Proc. Nagoya Int. Orchid Congr.*, pp. 64–70. Nagoya, Jpn: Naganae Print. Co.

169. Woltering EJ, Ten Have A, Larsen PB, Woodson WR. 1994. Ethylene biosynthetic genes and inter-organ signalling during flower senescence. In *Molecular and Cellular Aspects of Plant Reproduction*, ed. RJ Scott, AD Stead, pp. 285–307. Cambridge: Cambridge Univ. Press

170. Woodson WR, Lawton KA. 1988. Ethyl- ene-induced gene expression in carnation petals. Relationship to autocatalytic ethylene production and senescence. *Plant Physiol.* 87:498–503

171. Woodson WR, Park KY, Drory A, Larsen PB, Wang H. 1992. Expression of ethylene biosynthetic pathway transcripts in senescing carnation flowers. *Plant Physiol.* 99: 526–32

172. Yamane H, Abe H, Takahashi N. 1982. Jasmonic acid and methyl jasmonate in pollens and anthers of three camellia species. *Plant Cell Physiol.* 23:1125–27

173. Yasugi S. 1983. Ovule and embryo development in *Doritis pulcherrima* (Orchidaceae). *Am. J. Bot.* 70:555–60

174. Yeung EC, Law SK. 1989. Embryology of *Epidendron ibaguense*. I. Ovule development. *Can. J. Bot.* 67:2219–26

175. Yip KC, Hew CS. 1988. Ethylene production by young *Aranda* orchid flowers and buds. *Plant Growth Regul.* 7:217–22

176. Zhang XS, O'Neill SD. 1993. Ovary and gametophyte development are coordinately regulated following pollination by auxin and ethylene. *Plant Cell* 5:403–18

177. Zhang YX, Lespinasse Y. 1991. Pollination with gamma-irradiated pollen and development of fruits, seeds and parthenogenetic plants in apple. *Euphytica* 54:101–9

Annu. Rev. Plant Physiol. Plant Mol. Biol. 1997. 48:575–607

PLANT DISEASE RESISTANCE GENES

Kim E. Hammond-Kosack and Jonathan D. G. Jones

The Sainsbury Laboratory, John Innes Center, Colney Lane, Norwich, Norfolk NR4 7UH, United Kingdom

KEY WORDS: plant disease resistance gene, plant pathogen, avirulence gene, leucine rich-repeat protein, receptors

ABSTRACT

In "gene-for-gene" interactions between plants and their pathogens, incompatibility (no disease) requires a dominant or semidominant resistance (*R*) gene in the plant, and a corresponding avirulence (*Avr*) gene in the pathogen. Many plant/pathogen interactions are of this type. *R* genes are presumed to (*a*) enable plants to detect *Avr*-gene-specified pathogen molecules, (*b*) initiate signal transduction to activate defenses, and (*c*) have the capacity to evolve new *R* gene specificities rapidly. Isolation of *R* genes has revealed four main classes of *R* gene sequences whose products appear to activate a similar range of defense mechanisms. Discovery of the structure of *R* genes and *R* gene loci provides insight into *R* gene function and evolution, and should lead to novel strategies for disease control.

CONTENTS

575

1040-2519/97/0601-0575$08.00

INTRODUCTION

Plants need to defend themselves against attack from viruses, microbes, invertebrates, and even other plants. Because plants lack a circulatory system, each plant cell must possess a preformed and/or inducible defense capability, so distinguishing plant defense from the vertebrate immune system (100). Following the rediscovery of Mendel's work, plant breeders recognized that resistance to disease was often inherited as a single dominant or semi-dominant gene (44). Considerable knowledge has since accumulated on the biochemical and genetic basis of disease resistance (27, 64, 73), while the use of resistant cultivars has become a valuable strategy to control crop disease (10). Only within the past four years have disease resistance (R) genes against distinct pathogen types been isolated. Intriguingly, the proteins encoded by R genes from different species against different pathogens have many features in common. Here we review this work and consider how R gene products may function and how the recognition of novel pathogen specificities could evolve. Several other recent reviews on isolated R genes are available (4, 37, 64, 87).

AN OVERVIEW OF PLANT-PATHOGEN ASSOCIATIONS AND THE GENETIC BASIS OF PLANT DEFENSE

Pathogens deploy one of three main strategies to attack plants: necrotrophy, biotrophy, or hemibiotrophy. Necrotrophs first kill host cells and then metabolize their contents. Some have a broad host range, and cell death is often induced by toxins and/or enzymes targeted to specific substrates (101). *Pythium* and *Botrytis* species are examples of fungal necrotrophs. Other necrotrophs produce host-selective toxins that are effective over a very narrow range of plant species. For this class of pathogens, plant resistance can be achieved via the loss or alteration of the toxin's target or through detoxification. Pathogen virulence is dominant because of the need to produce a functional toxin and/or enzyme, whereas avirulence, the inability to cause disease, is inherited as a recessive trait (Figure 1A). The first R gene to be isolated was

Hml from maize, which confers resistance to the leaf spot fungus *Cochliobolus carbonum. Hml* codes for a reductase enzyme that detoxifies the *C. carbonum* HC-toxin. This toxin inhibits histone deacetylase activity (35, 101), and the *Hml* gene product is thought to inactivate the toxin.

Biotrophic and hemibiotrophic pathogens invade living cells and subvert metabolism to favor their growth and reproduction (1). The frequent formation of "green-islands" on senescing leaves surrounding the biotrophic infection sites of fungal rusts and mildews attests to the importance of keeping host cells alive throughout this intimate association (1). Biotrophs tend to cause disease on only one or a few related plant species. In contrast, hemibiotrophic fungi such as *Phytophthora* and *Colletotrichum* kill surrounding host cells during the later stages of the infection. Due to the specialized nature of these plant-biotrophic/hemibiotrophic pathogen associations, it is not surprising that minor differences in either organism can upset the balance. Incompatibility frequently results in the activation of plant defense responses, including localized host cell death, the hypersensitive response (HR) (27).

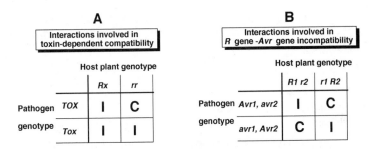

Figure 1 Various types of genetic interactions between plants and pathogenic microbes. In each panel, I denotes an incompatible interaction, where the plant is resistant to the pathogen, and C denotes a compatible interaction where the plant is susceptible to pathogen attack and disease occurs. (*A*) Interactions involved in toxin-dependent compatibility. The wild-type pathogen *TOX* gene is required for the synthesis of a toxin that is crucial for pathogenesis. *Tox* is the corresponding recessive, nonfunctional allele. The host *R* gene is required for detoxification, although resistance can also occur through expression of a toxin-insensitive form of the toxin target. Disease only occurs when the plant cannot detoxify the toxin produced by the pathogen. (*B*) Interactions involved in *R-Avr*-dependent incompatibility. *R1* and *R2* are two dominant plant resistance genes, where *r1* and *r2* are their respective recessive (nonfunctional) alleles. *R1* and *R2* confer recognition of pathogens carrying the corresponding pathogen avirulence genes, *Avr1* and *Avr2*, respectively, but not the respective recessive (nonfunctional) alleles, *avr1* and *avr2*. Disease (compatibility) occurs only in situations where either the resistance gene is absent or nonfunctional (*r1, r2*) or the pathogen lacks or has altered the corresponding avirulence gene (*avr1, avr2*). The interactions depicted in this panel are frequently called "the quadratic check" to indicate the presence of two independently acting *R-Avr* gene combinations (13, 19).

In the 1940s, using flax (*Linum usitatissimum*) and its fungal rust pathogen *Melampsora lini*, HH Flor studied the inheritance not only of plant resistance, but also of pathogen virulence (19). His work revealed the classic "gene-for-gene" model that proposes that for resistance to occur, complementary pairs of dominant genes, one in the host and the other in the pathogen, are required. A loss or alteration to either the plant resistance (*R*) gene or the pathogen avirulence (*Avr*) gene leads to disease (compatibility) (Figure 1*B*). This simple model holds true for most biotrophic pathogens, including fungi, viruses, bacteria, and nematodes (10, 44). The discovery that plants have centers of origin, where the greatest genetic diversity resides, and have co-evolved with pathogens, spurred a series of breeding programs to identify resistant germplasm in wild relatives of crop species and then introgress this for agricultural benefit (55). The spin-off for plant pathology was the development of several model "gene-for-gene" systems, ideal for intensive scrutiny because resistant and susceptible near-isogenic lines were available to minimize experimental differences due to background genetic variation. It is from these interactions that some of the first *R* genes and *Avr* genes have been isolated. The other *R* genes have been isolated from *Arabidopsis thaliana,* which has emerged in the past eight years as an excellent model system for plant-pathogen interaction studies (51).

Pathogen Avirulence (Avr) Genes

Although identified as the genetic determinants of incompatibility toward specific plant genotypes, the function of avirulence genes for the pathogen remains obscure. Plant viruses provide the only exception, where genes encoding either the coat protein, replicase, or movement protein have been demonstrated as the Avr determinant (62, 71, 93). Viral Avr specificity is altered by amino acid substitutions that do not significantly compromise the protein's function in pathogenesis. For the other microbial types, there often appears to be a fitness penalty associated with mutations from avirulence to virulence, and this suggests that the gene products have important roles for pathogenicity (11, 58). This view is reinforced by the fact that some *Avr* genes are always maintained within a pathogen population.

The molecular identities of a few fungal and bacterial Avr-generated signals are known. For fungi whose colonization is restricted to the plant's intercellular spaces (apoplast), small secreted peptides can elicit *R*-dependent defense responses in the pathogen's absence, e.g. Avr9 and Avr4 of *Cladosporium fulvum* and NIP1 of *Rhynchosporium secalis* (46). However, for biotrophic fungi that form intracellular haustoria, the nature of the Avr-derived signal is unknown.

For pathogenic bacteria, two distinct types of Avr-generated signals now appear to exist. Exported syringolides (C-glucosides with a novel tricyclic ring) are produced by enzymes encoded by the *avrD* locus of *Pseudomonas syringae* pv. *glycinea,* and these induce an HR on soybean cultivars that carry the *Rpg4* resistance gene (45). For other bacterial species, the Avr protein itself is now thought to be the signal. These *avr* gene products have no signal peptide, and yet they are recognized by *R* gene products that are likely to be cytoplasmic (see below). How do Avr products get into the plant cell? Bacterial *hrp* genes are required for both *h*ypersensitive *r*esponse induction and *p*athogenesis. *Hrp* genes code for a protein complex with strong homologies to the type III secretory system that is known to be used by some bacterial pathogens of mammalian cells (11). Recent work has conclusively shown that for the HR conditioned by genes for *Pseudomonas* resistance, i.e. *Pto, RPS2,* and *RPM1,* the corresponding Avr proteins must be delivered directly into the plant cell cytoplasm (22, 54a, 92). Although the *Xanthomonas avrBs3* family of *Avr* genes is very different in sequence from *Pseudomonas Avr* genes, delivery of the *avr* gene product into plant cells also appears necessary for their function. Members of the *avrBs3* family encode proteins with a highly reiterated internal motif of 34 amino acids in length, for example, *avrBs3* of *X. campestris* pv. *vesicatoria* and *avrXa10* of *X. oryzae* pv. *oryzae* (30, 32). By altering the number of repeats in *avrBs3* or the sequence within these repeats, both bacterial host range and *R*-mediated specificity was altered (30). Because *hrp* genes are essential for HR induction, and nuclear localization signals have been identified in the *avr* gene sequences, this indicates that gene product targeting to the plant cell nucleus may also be required for function (99a, 108).

Three Predicted Properties of R Genes and Their Products

The dominant nature of *R* and *Avr* genes has led to the inference that *R* genes encode proteins that can recognize *Avr*-gene-dependent ligands. Following pathogen recognition, the R protein is presumed to activate signaling cascade(s) that coordinate the initial plant defense responses to impair pathogen ingress. Implicit in this view is the notion that R proteins would be expressed in healthy, unchallenged plants in readiness for the detection of attack. A third requirement of R proteins is the capacity for rapid evolution of specificity. Frequently new virulent races of pathogens regularly evolve that evade specific *R* gene–mediated resistance (10, 13, 44, 64). Thus a mechanism is required by which plants can rapidly evolve new *R* genes to resist virulent isolates.

ISOLATED DISEASE RESISTANCE GENES

In the absence of a known biochemical role for R gene products, the *R* gene isolation strategies relied predominantly upon defining the gene's chromosomal location using segregating populations, and then identifying the correct sequence by either transposon insertion to destroy biological activity or cosmid complementation to restore the resistance phenotype. This technical challenge was solved simultaneously in several laboratories. A summary of the reported *R* genes is given in Table 1. Figure 2 provides an overview of the predicted structure of each R protein and their percent identity in specific regions. Figure 3 shows the alignment of amino acid sequences of particular common motifs and domains.

R Genes Predicted to Encode Cytoplasmic Proteins

ARABIDOPSIS *RPS2* AND *RPM1* GENES   *RPS2* confers resistance to strains of *Pseudomonas syringae* bacteria that carry the plasmid-borne *avrRpt2* gene. *RPM1* provides resistance against *P. syringae* strains that express either of two nonhomologous *avr* genes, *avrB* or *avrRpm1* (5, 23, 65). The predicted gene products, 909 amino acids for RPS2 and 926 amino acids for RPM1, carry in their amino termini a possible leucine zipper region (LZ), a potential nucleotide binding site (NBS), and an internal hydrophobic domain. The carboxy-terminal halves are comprised of at least 14 imperfect leucine-rich repeats (LRRs). Overall, the two predicted sequences share 23% identity and 51% similarity (Figure 2). The LZs of RPS2 and RPM1 have 4 and 6 contiguous heptad sequences, respectively, that match the consensus sequence (I/R) XDLXXX (52). It is proposed that this domain facilitates the formation of a coiled-coil structure to promote either dimerization or specific interactions with other proteins. The NBS is found in numerous ATP- and GTP-binding proteins (98). The sequence GPGGVGKT of RPS2 matches the generalized consensus

---

→

Figure 2 Comparison of the predicted primary structure of R gene products and Prf (which is required for Pto function). Each protein has been drawn to scale, and the bar at the figure's foot indicates length in amino acids. Identified protein domains and motifs are shown either within boxes or as distinct shapes. Regions encoded by directly repeated DNA sequences in L6 and Prf are indicated by arrows above each protein. The percentage values placed between some R proteins reveal the amino acid sequence identity between either corresponding regions or exons, as determined by the GAP sequence alignment program (Genetics Computer Group, University of Wisconsin). For the comparison between RPS2 and RPM1, the regions aligned were 1-135 with 1-155, 135-418 with 155-442, and 418-909 with 442-926, respectively. For the comparisons between L6, N, and RPP5 the individual exons were aligned. The extracellular LRR proteins are divided into domains A to G for Cf proteins, and domains A to I for Xa-21 (38, 86). For Cf-9, domain A is the putative signal peptide; the domains B and D flank the LRRs that comprise domain C, and the

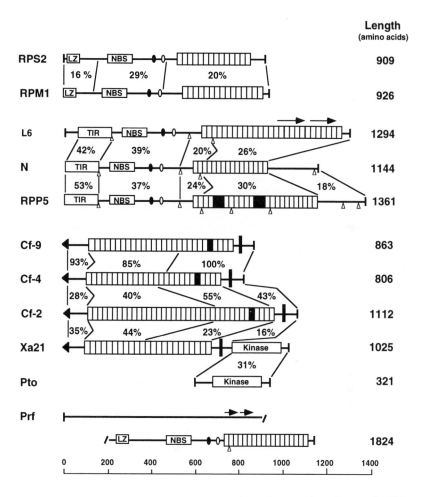

putative membrane anchor comprises domains E, F, and G. For the comparisons between the Cf-9, Cf-4, Cf-2, and Xa21, the regions aligned were grouped accordingly: domains A and B, the amino terminal 18 LRRs, 16 LRRs, 26 LRRs, and 11 LRRs, respectively; the carboxy terminal 12 LRRs, and domains D, E, F, and G, respectively. Full details of each gene product are given in the text. Abbreviations: LZ, Leucine zipper motif; NBS, Nucleotide binding site; TIR, Drosophila Toll/Human Interleukin-1 resistance gene cytoplasmic domain; ▯▯▯▯, Leucine-rich repeat domain, where the number of LRR motifs is indicated by the number of segments, and filled segments represent regions where the LRR motif is not conserved; Kinase, Serine/threonine kinase domain; filled circles, the GLPL(A/T)ax(V/S)aaG(S/G)aa motif, where a is an aliphatic amino acid; open circles, the L(R/K)xCFLY(C/I)(A/S)xF motif; +, transmembrane spanning region; ←, signal peptide; Δ, intron position.

GXGXXG(R/k)V for the kinase 1a, phosphate-binding loop (P-loop). This is followed by a kinase 2 domain, where an invariant aspartate is believed to coordinate the metal ion binding required for phospho-transfer reactions, and then a kinase 3a domain containing an arginine that in other proteins interacts with the purine base of ATP (98). These three domains have collectively been termed the NBS region in R proteins and are distinct from those found in protein

TABLE 1   THE FIVE CLASSES OF CLONED PLANT DISEASE RESISTANCE GENES

Class	Gene	Plant	Pathogen	Infection type/ organ attacked	Predicted Features of R protein	Reference
1.	Hm1	Maize	Helminthosporium maydis (race 1)	Fungal necrotroph / leaf	Detoxifying enzyme HC-toxin reductase	[35]
2.	Pto	Tomato	Pseudomonas syringae p.v. tomato (avrPto)	Extracellular bacteria / leaf	Intracellular serine/ threonine protein kinase	[59]
3a	RPS2	Arabidopsis	Pseudomonas syringae p.v. tomato (avrRpt2)	Extracellular bacteria / leaf	L. Zip / NBS / LRR	[5] [65]
	RPM1	Arabidopsis	Pseudomonas syringae p.v. maculicola (avrRpm1/ avrB)	Extracellular bacteria / leaf	Intracellular protein with amino terminal leucine zipper domain, and nucleotide binding site (NBS) and leucine rich repeat (LRR) domains	[23]
	I2 [B]	Tomato	Fusarium oxysporum f.sp. lycopersicon	Necrotrophic fungus/root and vascular tissue		[A]
3b	N	Tobacco	Mosaic virus	Intracellular virus / leaf and phloem	Toll / NBS / LRR	[105]
	L6 M	Flax	Melampsora lini (AL6, AM)	Biotrophic fungal rust with haustoria / leaf	Intracellular protein with amino terminal domain homology with Drosophila Toll protein, and NBS and LRR domains	[53] [C]
	RPP5	Arabidopsis	Peronospora parasitica	Biotrophic downy mildew fungus with haustoria / leaf		[D]
4	Cf-9, Cf-2, Cf-4 Cf-5	Tomato	Cladosporium fulvum (Avr9, Avr2, Avr4 Avr5)	Biotrophic extracellular fungus without haustoria / leaf	Extracellular LRR protein with single membrane spanning region and short cytoplasmic carboxyl terminus	[38] [14] [41] [E]
5.	Xa-21	Rice	Xanthomonas oryzae pv. oryzae (all races)	Extracellular bacteria / leaf	Extracellular LRR protein with single membrane spanning region and cytoplasmic kinase domain	[86]

[A] G. Simons and R. Fluhr, pers. comm.; [B] a very tightly linked marker to the wheat Cre3 gene that confers resistance to the root invading cereal cyst nematode Heterodera avenae is highly homologous to this R gene class, E. Lagudah and S. Anderson, pers. comm.; [C] P. Anderson, G. Lawrence and J. Ellis, pers. comm.; [D] J. Parker, M. Coleman, V. Szabo, M. Daniels and J. Jones, unpublished; [E] M. Dixon, K. Hatzixanthis and J. Jones, unpublished.

kinases (94). The presence of the NBS suggests possible activation of a kinase or a role as a G-protein, though no biochemical evidence shows that the NBS actually binds ATP or GTP. The LRRs with an average repeat unit length of 23 amino acids show a good match to the cytoplasmic LRR consensus (LxxLxxLxxLxLxx(N/C/T)x(x)LxxIPxx) (37). LRRs have been implicated in protein-protein interactions and ligand binding in a diverse array of proteins (48). Collectively, the above features suggest both the *RPS2* and *RPM1* genes code for cytoplasmically localized proteins. This is intriguing because bacterial colonization is exclusively extracellular, but as stated above the Avr gene product may be delivered into plant cells via the bacterial Hrp secretory system. Comparison of the *avrRpt2*, *avrRpm1*, and *avrB* gene sequences reveals only minimal homology between them (11).

TOBACCO *N*, FLAX *L6*, AND ARABIDOPSIS *RPP5* GENES The tobacco (*Nicotiana tabacum*) *N* gene was originally introgressed from *N. glutinosa* and confers resistance to most strains of tobacco mosaic virus. Alternative splicing of the *N* gene transcript gives rise to two sizes of mRNA (105). The larger transcript codes for a 1144–amino acid protein (N), with an NBS, an internal hydrophobic domain, and 14 LRRs (23 amino acid type) present in the carboxyl terminal half. The less abundant truncated transcript codes for a 652–amino acid protein (N^{tr}) that possesses the amino terminal 616 amino acids of N including the NBS, the hydrophobic domain, and the first 1.5 LRRs followed by 36 amino acids. Although N shows similar structural organization to RPS2 and RPM1, the amino terminal domain of N is distinct, exhibiting homology with the cytoplasmic domains of the Drosophila Toll protein and the mammalian interleukin-1 receptor (IL-R) protein (Figure 3A) (20% and 16% amino acid identity and 42% and 41% amino acid similarity, respectively) (28, 85), and by inference another seven members of the growing Toll/IL-1R superfamily (66). This region in plant *R* genes has been designated the TIR (*T*oll/*I*nterleukin-1 *R*esistance) domain (B Baker, personal communication). Because the amino terminal domain of the N protein has homology to the cytoplasmic signaling domains of these receptors, it is probably involved in signaling and not ligand binding. Direct interaction between the tobacco N protein and the probable viral avr determinant, the replicase protein (71), is plausible because TMV replication is exclusively intracellular.

The flax (*Linum usitatissimum*) *L6* gene confers resistance to strains of the rust fungus *Melampsora lini* that carry the *AL6* avirulence gene (53). Like *N*, the *L6* gene gives rise to two mRNAs via alternative splicing. The larger and predominant transcript codes for a 1294–amino acid protein that like N contains within its amino terminus a TIR domain with homology to the cytoplas-

A. Comparison of TIR homologous domains

```
            d    t                 r*r  i   ***         d  di  *r **r *
N      12  YDVFISFR..  ....GeDtRK  TFTSHIyEVI .  .NDKGIKTF  qDDKrLEYCA
RPP5   12  YDVFLSFs..  ....GvDvRK  TFlSHILkAI .  .DGKSInTF  IDhG.IERSR
L6     62  MLVFLSFR..  ....GpDtRe  QFTDFLyQSI .  .RRYKIHTF  rDDDELLKCK
TOLL  860  KDAFISY...  ....SHKDQS  FIEDYIVPQI  EHGPQKFQLC  VHERDWLVCG
IL-1R 372  YDAYILYPKT  VGEGSTSDCD  ILVFKVLPEV  LEKQCGYKLF  IYGRDDYVCE
            o                                              o
           region 1                 region 2      region 3
```

```
            tdt    t       t *    * t  tr     d        ir  *r  **
N      62  TFPGELCKAI  EESqfAIVVf  SeNYATSHVC .II....NELV  kIMeC.ktrf
RPP5   62  TIAPELIsAI  rEARISIVIf  SkNYASStWC .II....NFLV  EIhKC.fNDl
L6    112  EIGPNLLRAI  DGSKIyVpII  SsgYAGSIWC .II....mFLa  EIVRrQeEDp
TOLL  910  HIPENIMRSV  ADSRRTITVL  SQNFIKSIWA RI...EFRA  AHRSALNEGR
IL-1R 422  DIVEVINENV  KRSRRLIIIL  VRE..TSGFS  WIGGSSEIQI  AMYNALVQDG
                                          o
                         R
           region 4
```

```
             dttr    rrr  td r    d r*d*             i r   r d  **d
N     112  KqtVIPIFYD  VDPShVRnQk  EsFAKAFEe. .heTKYKDDV  EGICRWRIA  LNEAAANLKGSC
RPP5  112  gqMVIPVFYD  VDPSEVRKQT  geFGKvFEKt  cEvSKdKQPG  DQKCRWvQA  LTDIANIAGED
L6    162  RRIILPIFYm  VDPSDVRHQT  gcYKKAFrKh  aN..KFDG   QTICNWKDA  LKKVGDLKwHI
TOLL  960  SRIIVIIYSD  IG..DVEKLD  EELKAYLKMN  TYL KW GD          PW  FWDKLRFALPH
IL-1R 472  IKVVLLELEK  IQ..DYEKMP  ESIK FIKQk  HGAIRWSGD  FTQGHCSAKTR  FWKNVRYHMPV
                                                                     FWK  RY   P
           region 5                                        region 6
```

B. Comparison of serine/threonine kinase domains

```
            **            **         *  *  **
Fen   129  DLP..SMSWE  QRLEICICAA  RGLHYLEk.n ..AVIHRDVK  ctNILLIENF
Pto   129  DLPtmSMSWE  QRLEICICAA  RGLHYLEt.r ..AiIHRDVK  sINILLIENF
Irak  303  TQACPPLSWP  QRLLILLCTA  RAIQKLEQ.D  SPSLIHGDIK  SSNVLLIERL
Pelle 309  QNPLPALTWQ  QRFSISLCTA  RGIYFLHTAR  GTPLIEGDIK  PANILLIQCL
c                        G   YL         H DIK peNI
                         domain VI
```

```
            ***                       *  * *     * * * *
Fen   174  VPKIIDFGIS  K..tmpELDQ  THLSTV.... .VrGniGYia  HEYaLwGdITEKSDV
Pto   176  VPKIIDFGIS  K..kgtELDQ  THLSTV.... .VkctlGYid  HEYfIkGriTEKSDV
Irak  352  TPKICDFGIS  RFSRFAGSSP  SQSSMVARTQ  TVRGTLAYLP  EDYIKTGRIAVDTDT
Pelle 359  QPKICDFGLV  R......EGP  KSLDAVVEVN  KVFGTKIYLP  HEFRNFRQISTGVDV
c             I DFG                        gt  Y a PE          D
           domain VII                domain VIII       d IX
```

C. Comparison of other conserved motifs in NBS/LRR proteins

```
N     344  EVTALPDHES  IQLFKQHAFG  K...EVPNEN  FEKLSLEVVN  YAKGLPIALK VWGSLLHNLR
RPP5  340  EVKLPSQGLA  LKMISQYAFG  K...DSPPDD  FKEIFEVAE  LVGSLPIGLS VLGSSLKGRD
L6    395  EVGSMSKPRS  LELESKHAFK  K...NTPPSY  YETLANDVVD  TTAGLPITLK VIGSLLFKQE
PRF  1247  HLRLFRDDES  WTLLQKEVFQ  G...ESCPPE  LEDVGFEISK  SCRGLPIAVV LVAGVLKQKK
RPM1  332  EIELLKEDEA  WVLLSNKEAP  ASLEQCRTQN  LEPIFRKLVE  RCQGLPIAIA SLGSMMSTKK
RPS2  306  RVEFLEKKHA  WEDECSKVWR  KDLLESSS..  IRRLFEIIVS  KCGGLPIALI TLGGAMAHRE
                                                       Motif 3          HD
```

```
N     401  L...TEWKSA  IEHMK....N  N.SYSGIIDK  IKISYDGIEF  KQ.QEMFIDI ACFLR G
RPP5  397  K...DEWVKM  MPFIR....N  D.SDDKIEET  IRVGYIRINK  KN.REIFKCI ACFFN G
L6    452  I...AWVEDT  LEQIR....R  TLNLDEVYDR  IKISYIAIEF  EA.KEIFDI ACFFI G
PRF  1303  KTLDS.IKVV  EQSISSQRIG  SLEESI..SI  IGFSYKNLEF  HYLKPCFLYF GGFLQ G
RPM1  392  ..FESEIKKV  YSTINWELNN  NHELKIVRSI  MFLSFNDLP  YPLKRCFIYC SLF... .
RPS2  364  T..EEIWIHA  SEVITRFPAE  MKGMNYVFAL  IKFSYDNLES  DLLRSCFIYC ALFPE E
                                            Motif 2          Motif 1
```

mic domains of Toll and IL-1R (21% and 16% amino acid identity and 50% and 41% amino acid similarity, respectively). This TIR domain is followed by an NBS, a hydrophobic domain, and 27 LRRs that fit the 23–amino acid consensus but are highly imperfect in length (37). The carboxyl terminal 40% of the leucine-rich region is encoded by two directly repeated DNA sequences of 438 and 447 base pairs. L6 also possesses at its extreme amino terminus an additional 60 amino acids, which includes a potential signal anchor sequence, suggesting the protein might enter, but not pass through, the secretory pathway. The smaller transcript that arises via alternative splicing, codes for a 705–amino acid protein (L6^{tr}) that is identical to L6 for the first 676 amino acids but has a novel sequence of 29 amino acids and loses most of the LRR domain. The first 18 amino acids of the 29–amino acid C-terminal extension of L6^{tr} is predicted to be a possible membrane-spanning region. Although the N/N^{tr} and L6/L6^{tr} proteins appear structurally similar they may located in distinct subcellular compartments. Different cellular locations for these R proteins would fit well with the distinct biology of the two pathogens and the

◄ ──

Figure 3   Amino acid sequence alignments between specific regions of R proteins, and where appropriate, related proteins of known biological function. In each Prettybox alignment (UWGCG program), the amino acids shown in white on a black background indicate identical residues, whereas those shown in black on a grey background indicate similar residues. Amino acid similarities are defined as follows: I=V=L=M, D=E=Q=N, F=Y=W, H=K=R, and G=A=S=T=P (67). (*A*) Comparison of the TIR (Toll/IL-1 Receptor cytoplasmic) domain in the tobacco N, Arabidopsis RPP5, and flax L6 R proteins with the corresponding cytoplasmic domains in the Drosophila Toll receptor and the human interleukin-1 receptor (IL-1R) proteins (28, 85). The asterisks above the alignment indicate amino acids in the three R proteins that do not conform to the Toll/IL-1R consensus. r indicates the 14 conserved amino acids specific to the three R proteins where there is no conserved consensus between Toll and IL-1R. d indicates the 11 amino acids where the three R proteins have the same consensus and this differs from the Toll/IL-1R consensus. t indicates the eight amino acids specifically conserved between the R proteins and Toll. i indicates the four amino acids specifically conserved between IL-1R and the three R proteins. Below the sequence alignment the locations of six conserved regions identified in the Toll/IL-1R superfamily (66) are shown, and the single amino acids that when mutated compromise protein function are highlighted with o for Toll function and with individual letters for IL-1R function (29, 82). (*B*) Comparison of the portion of the serine/threonine protein kinase domains of tomato R protein Pto (59), which confers AvrPto specificity (92), with the same region in the tomato Fen protein that confers Fenthion insecticide sensitivity (60), the human interleukin-receptor associated kinase (IRAK) (84), and the Drosophila Pelle kinase (8). The asterisks above the alignment indicate the 17 amino acids that differ between Pto and Fen (conserved substitutions are also underlined). Below the alignment is the eukaryotic consensus sequence for serine-threonine kinases (98). (*C*) Comparison of a portion of the region between the NBS and LRR regions in the tobacco N, flax L6, tomato Prf, and Arabidopsis RPP5, RPM1, and RPS2 proteins. The four conserved motifs of unknown function indicated are the HD motif; GLPL(A/T)ax(V/S)aaG(S/G)aa, where a is an aliphatic amino acid; motif 1 with the consensus L(R/K)xCFLY(C/I)(A/S)xF; motif 2, consensus Lx(I/L/F)SYxxL(N/E)P; and motif 3 (L/F)ExaAxxaV.

potential perception of the Avr ligand. The penetration peg of the rust fungus spore passes through the host cell wall, but the haustorium does not breach the plasma membrane and eventually becomes entirely surrounded by it (1). Membrane localization of the L6/L6^{tr} proteins would therefore facilitate interception of the fungal avr-derived signal.

The Arabidopsis *RPP5* gene was isolated from the resistant accession Landsberg-*erecta* and confers resistance to the biotrophic downy mildew fungus *Peronospora parasitica,* which is a natural pathogen of Arabidopsis (39). The predicted gene product of 1361 amino acids possesses a TIR domain at the amino terminus, followed by an NBS and 21 LRRs in the carboxyl terminus. Each LRR motif varies in length from 21 to 24 amino acids, but this domain also contains two regions with less homology to the LRR consensus. Because the RPP5 sequence predicts neither the presence of a signal peptide or membrane spanning region, the protein is probably cytoplasmically localized. The *RPP5* gene is more closely related to *N* and *L6* than to *RPS2* and *RPM1* (described above), because of the TIR domain and the similarity in the positions of the intron/exon splice junctions that give rise to exons 1, 2, and 3 (Figure 2).

TOMATO *Pto, Fen,* AND *Prf* GENES   The tomato *Pto* gene confers resistance to races of *Pseudomonas syringae* pv. *tomato* that carry the *avrPto* gene. This was the first race-specific *R* gene to be isolated (59). *Pto* codes for a 321–amino acid protein and has been shown to be a serine/threonine-specific protein kinase, capable of autophosphorylation (57). Pto possesses 27 serine and 13 threonine residues, and is in the same protein kinase class as the cytoplasmic domain of the *Brassica* self-incompatibility gene *SRK,* the mammalian signaling factor Raf, the *Drosophila* pelle kinase, and the human IRAK kinase (Figure 3*A;* 7, 8, 84, 88). The protein does not possess an LRR domain or an NBS. Thus Pto appears to possess a signal transduction but no obvious recognition capacity. However, recent experiments using both *Pto* and *avrPto* sequences in a yeast 2-hybrid system indicate that the AvrPto and Pto proteins do directly interact (83, 92). Pto autophosphorylation is also required for the Pto-avrPto interaction to occur.

Pto is a member of a clustered family of five genes (60). One of the other family members is *Fen,* which specifies sensitivity to the insecticide fenthion and codes for a 318–amino acid serine/threonine protein kinase (60). The Fen protein shares 80% identity (87% similarity) with Pto but does not confer avrPto-dependent bacterial resistance. Both protein kinase activity and a putative N-terminal myristoylation site, proposed to be involved in membrane tareting, are required to confer fenthion sensitivity (77). The Pto myristoylation site is not required for resistance to *P. syringae*. The Fen kinase does not interact with avrPto in the yeast 2-hybrid system (92). The analysis of a series of Pto/Fen chimeric genes in both yeast and transgenic plants has identified a 95–amino acid stretch of Pto, between residues 129 and 224, that is required for interaction with avrPto and for disease resistance (92). Within these 95 amino acids, Fen and Pto differ by only 13 nonconservative changes (Figure 3*B*). The Fen specificity is localized to the carboxy terminal 186 amino acids (77).

Mutagenesis of *Pto*-containing tomato has revealed an additional gene, *Prf*, that is required for both *Pto* and *Fen* to function (80). *Prf* is located within the *Pto* gene family, 24 kb from the *Pto* gene but just 500 bp from the *Fen* gene (81). *Prf* encodes a 1824–amino acid protein with leucine zipper, NBS, and leucine-rich repeat motifs of the 23–amino acid type, which identifies it as a member of the resistance gene class that includes *RPS2* and *RPM1*, and is more distantly related to the *N* and *L6* genes. Prf also possesses a large amino-terminal region, 720 residues in length, with no homology to any known protein. At the end of this region are two direct repeats of 70 and 71 amino acids with 49% sequence identity. Because both *Pto* and *Prf* are essential for resistance, this demonstrates that both LRR-containing proteins and protein kinases can be components of the same signaling pathway. However, the functional relationship between Pto and Prf proteins is not yet known (see below).

R Genes Predicted to Encode Proteins with Extracytoplasmic Domains

TOMATO *Cf-9, Cf-2, Cf-4,* AND *Cf-5* GENES   Resistance to the leaf mould pathogen *Cladosporium fulvum* is conferred by distinct *Cf* genes, which have been introgressed from various wild *Lycopersicon* species or land races into cultivated tomato *Lycopersicon esculentum* (27). Two *C. fulvum Avr* genes, *Avr9* and *Avr4,* that confer avirulence on *Cf-9* and *Cf-4* expressing tomato, respectively, have been cloned (43, 99). Their secreted cysteine-rich peptide products of 28 (Avr9) and 88 (Avr4) amino acids are potentially ligands for the Cf-9 and Cf-4 proteins. Four tomato *Cf* genes have been isolated.

Cf-9 encodes an 863–amino acid membrane-anchored, predominantly extracytoplasmic glycoprotein containing 27 imperfect LRRs with an average length of 24 amino acids. The LRRs show a good match to the extracytoplasmic LRR consensus of LxxLxxLxxLxLxxNxLxGxIPxx (37). The LRR domain is interrupted by a short region, originally designated as LRR 24, which has only minimal LRR homology. This domain, now designated C2, divides the LRR domain into 23 amino terminal LRRs (domain C1) and 4 carboxy terminal LRRs (domain C3). This C2 "loop out" domain appears to be absent from most other extracytoplasmic LRR proteins, except the other Cf proteins described below. It could act as a molecular hinge that connects the C1 and C3 regions or as an extended loop that interacts with other proteins that participate in signal transduction. Flanking both ends of the LRR domain are two regions (domains B and D) that contain several cysteine residues, conserved in other LRR proteins, that may be important in maintaining the overall protein structure (37). The 21–amino acid cytoplasmic terminus of Cf-9 (domain G) concludes with the motif KKxx, which in mammals or yeast would be expected to localize the protein to the endoplasmic reticulum (97).

Cf-4, which is tightly linked to *Cf-9,* encodes an 806–amino acid protein very similar to that of Cf-9 (41). Cf-4 differs from Cf-9 by possessing two fewer LRRs and by having one other small deletion and a number of amino acid substitutions in the amino-terminal half of the protein (Figure 2). The carboxy-terminal halves of both proteins, from LRR 18 of Cf-9 onward, are identical, suggesting that resistance specificity resides in their amino-terminal portions whereas the carboxyl-terminal portion probably interacts with common signaling/regulatory component(s).

The *Cf-2* locus, unlinked to the *Cf-4/Cf-9* locus, contains two functional genes that each independently confer resistance. Each *Cf-2* gene encodes a 1112–amino acid protein, which has a similar overall structure to Cf-9 but possesses 37 LRRs (14). Both Cf-2s lack the KKxx motif of Cf-9, suggesting either a different cellular location for Cf-2, which might account for some of the differences between Cf-9- and Cf-2-mediated defense responses activated by their respective Avr gene product (27), or that this motif has no relevance to Cf protein function. The LRRs of Cf-2 are nearly all exactly 24 amino acids in length, and 20 of these have a highly conserved alternating repeat motif. A similar arrangement is not evident in the Cf-9 and Cf-4 proteins. However, like the other two Cf proteins, a short C2 domain divides the LRRs into an amino terminal block of 33 LRRs and a carboxyl terminal block of 4 LRRs. Like the other Cf proteins, both Cf-2s have many predicted NxS/T glycosylation sites. As the highest homology between the Cf-2 and Cf-9 and Cf-4 proteins resides in the carboxyl terminal 360 amino acids of Cf-2 (Figure 2), this again suggests this region plays a similar role in all three proteins.

The *Cf-5* gene, tightly linked to *Cf-2,* encodes a 968–amino acid protein very similar to that of *Cf-2* (Table 1). The two proteins differ by the exact deletion of six LRRs within the alternating repeat region of Cf-5 and by several amino acid changes in the amino terminal two thirds of each protein. The carboxyl terminal halves of Cf-2 and Cf-5 are also highly conserved.

RICE *Xa21* GENE   *Xa21* confers resistance to over 30 distinct strains of the bacterium *Xanthomonas oryzae* pv. *oryzae,* which causes leaf blight in rice.

Xa21 encodes a 1025–amino acid protein that possesses a putative signal peptide, 23 extracytoplasmic LRRs with numerous potential glycosylation sites, a single transmembrane domain, and an intracellular serine/threonine kinase domain (86). The LRR domain cannot be classified into C1, C2, and C3 domains, unlike in the Cf proteins. The Xa21 protein shows pronounced overall homology with the Arabidopsis receptor-like serine/threonine kinase RLK5, whose function is currently unknown (7). However, because Xa21 possesses both the LRR feature of the Cf-9 protein and a Pto-like serine/threonine kinase domain, this protein provides the first potential clue to the link between R proteins predicted to encode solely a receptor function and potential downstream signaling capacity.

It is somewhat surprising that resistance to strains of *Xanthomonas oryzae* pv. *oryzae* expressing *avrXa21* is conferred by an R protein structurally distinct from the other bacterial resistance proteins, i.e. RPS2, RPM1, or Pto. Although the Hrp secretory system is required by some *X. oryzae avr* genes to induce the resistance response, e.g. *avrXa10* on rice plants carrying *Xa10* (32), this has not yet been established for *avrXa21*, and neither has this gene been isoolated. The avrXa21-derived ligand might have a novel molecular identity, because *Xanthomonas oryzae* pv. *oryzae* is predominantly a xylem vessel colonizing bacterium. Conceivably, it is delivered extracellularly, unlike other bacterial Avr products, in which case the Xa21 LRRs might be involved in the recognition.

Are There Other R Gene Classes?

Various research groups have either isolated or are at the final stages of isolating *R* genes to a diverse array of additional microbes. These include the fungal resistance genes *I2* from tomato, the rust *M* gene from flax, and the cyst nematode resistance gene *Cre3* from wheat (Table 1). The majority of this second wave of isolated *R* genes can be recognized as highly related to members of existing *R* gene subclasses. In addition, based on conserved features found in the NBS/LRR class of R proteins (see below), candidate *R* genes of potato linked to the *Gro1* gene that confers resistance to the cyst nematode *Globodera rostochiensis* and the *R7* gene that gives resistance to the hemibiotrophic fungus *Phytophthora infestans,* have been identified (54). This suggests that perhaps plants use only a limited number of recognition/signal transduction systems to combat microbial attack. These new findings also clearly highlight the fact that R protein structure cannot be predicted from the nature of the pathogen or vice versa. It is possible (but not certain) that we have seen all the kinds of gene-for-gene *R* genes there are.

R PROTEIN MOTIFS AND THEIR POTENTIAL FUNCTION

Pathogen Recognition

Mechanistically, the simplest interpretation of Flor's gene-for-gene hypothesis is that the *Avr*-gene dependent ligand binds directly to the R gene product which then activates downstream signaling events to induce various defense responses (20). As the majority of the isolated *R* genes encode proteins that possess domains characteristic of authentic receptor proteins found in mammals, Drosophila, and yeast, a receptor-like function for plant R proteins appears likely. The most obvious candidate for providing the recognition specificity is the LRR domain. LRRs have been demonstrated to bind the corresponding ligand, for example in the porcine RNase inhibitor protein (PRI) and the receptors for gonadotropin and follicle-stimulating hormone (48, 69). Although the contact points between PRI and its RNase ligand have been accurately determined by co-crystallization studies (49), it is difficult to extrapolate these data to identify potential residues involved in protein binding in plant LRR proteins because of the unique nature of the PRI LRR motifs. PRI is comprised of alternating LRR motifs of 28 and 29 amino acids in length that form a horseshoe structure (49). The plant LRRs with motifs of 23 and 24 amino acids in length may form a β-helix, which is a more linear structure, as found in pectate lyase, P22 tailspike protein, and pertactin (18, 50, 89). For those R proteins predicted to be extracytoplasmic, where the LRR motifs and the integrity of the entire LRR domain are best conserved, a hydrophobic face could form to facilitate multiple interactions with other proteins or ligands. A key future goal is to elucidate the crystal structure of several plant LRR proteins.

If the tomato Cf proteins are localized to the plant plasma membrane, these could each directly bind a different extracellular peptide ligand derived from *Cladosporium fulvum*. The amino-terminal 18 LRR of Cf-9, 16 LRRs of Cf-4, and 28 LRRs of Cf-2 are likely to contain the binding region because these regions possess the greatest sequence divergence. However, it seems unlikely that all the LRRs would be involved in binding such a small ligand as Avr9, unless Avr9 multimers bind. Instead, some of the LRRs may be required to provide the correct structure surrounding the binding site. Alternatively, the Cf proteins may bind to a larger plant protein, and each Avr peptide may modulate this interaction by binding either to the Cf protein or the other protein. This model could also apply to the rice Xa21 protein if the avrXa21 gene product is secreted from bacteria but does not enter plant cells. For several extracellular LRR proteins, the glycosylation pattern within the LRR domain is crucial for ligand binding (109). Glycosylation sites are absent from the β-strand of the LRR motif in the C1 domain but are present within the β-strand of the C3 LRRs (37). It remains to be established whether glycosylation patterns influence R protein function.

For the LRR proteins predicted to be cytoplasmically localized, it is unclear whether the function of the LRR domain is to confer the specificity of recognition. By default this domain is apparently accepted as the recognition domain, primarily because all the other motifs appear to have an obvious signaling capacity. However, the role of the LRR in these R proteins could be dimerization or interaction with either upstream or downstream signaling components. The LRR domain of yeast adenylate cyclase is required to interact with ras protein (91). Clearly the LRR domain is of functional importance, because some alleles of RPS2 and RPM1 with just single amino acid changes in the LRR domain do not confer resistance (5, 23, 65). Domain swap experiments between RPS2 and RPM1, or between different alleles of the *L* locus, should provide insight into the specificity domains of *R* gene products.

The tomato Pto protein binds directly to the *Pseudomonas syringae* avrPto gene product (83, 92). Although the actual amino acids specifying binding are undetermined, this interaction entirely conforms to the biochemical interpretation of Flor's gene-for-gene hypothesis.

Signal Transduction

For the NBS/LRR class of *R* genes, the nucleotide binding site and either the leucine zipper or TIR homologous domains are the most likely to be involved in signaling. The presence of the NBS, which is found in numerous ATP and GTP binding proteins (98), suggests that although these R proteins do not possess intrinsic kinase activity, they could activate kinases or G proteins. Mutagenesis of amino acids known to be required for NBS function destroys R protein and Prf biological activity (81). The NBS domains found in R proteins are most similar to those found in ras proteins, adenylate kinases, ATP synthase β-subunits, and ribosomal elongation factors (25, 98). An important future goal is to characterize the nature of the nucleotide triphosphate binding and its significance to R protein function.

The leucine zipper regions found in RPM1, RPS2, and Prf, which in each protein precede the NBS, potentially could facilitate homodimerization of the proteins themselves or heterodimerization with other proteins (52). R proteins could exist as monomers before pathogen challenge and then undergo dimerization or oligomerization upon activation. Alternatively, they could exist initially as dimers or multimers that disassociate upon activation. Computer data-base searches with these leucine zipper regions reveal the greatest similarity to the coiled-coil regions in myosin and paramyosin proteins (5, 52).

The TIR domain in N, L6, and RPP5, though exhibiting only moderate homology to the Toll/IL-R cytoplasmic domain (Figure 3A), is tantalizing. The Drosophila Toll receptor protein, which also has an extracytoplasmic LRR domain, controls dorsal-ventral polarity in embryos (21, 28, 68). Toll is activated by a processed small extracytoplasmic protein ligand, spätzle, that has a cysteine-knot structure. Binding of spätzle to Toll may lead to the activation of a cytoplasmic protein tube, which in turn activates the serine/threonine kinase, pelle, by recruiting it to the plasma membrane. Pelle then phosphorylates the inhibitory protein cactus that is complexed with the transcription factor dorsal. Phosphorylation of cactus leads to its own degradation, and this permits dorsal to relocate to the nucleus and activate genes controlling ventralization. Dorsal is a member of the rel/NF-κB family of proteins (102). Another Drosophila protein highly related to dorsal is Dif (dorsal related immunity factor), which is involved in activating the defense response in fat bodies (72). Mutation studies suggest that Toll may also play a role in the nuclear localization of Dif (79). The human interleukin-1/interleukin-1 receptor protein system is involved in both the inflammatory and immune responses (70, 85). IL-1R activates the transcription factor NF-κB by releasing it from a cytoplasmically localized complex with the inhibitor protein (IκB) and requires the protein kinase (IRAK), which has high homology to pelle (30.5% amino acid identity) and Pto (34% identity) (Figure 3B; 102). Considering both the sequence homology and related functions of these vertebrate, insect, and plant proteins, the N, L6, and RPP5 proteins could function in an analogous manner. This view is reinforced because one of the fastest recognized components of the N-mediated defense response is the generation of reactive oxygen species, ROS (15). During the mammalian innate immune response in macrophages, activation of NF-κB is also redox regulated (63). In both situations, once defense is activated rapid amplification of the initial response could be achieved by a positive feedback loop involving ROS. The induced assembly of a multisubunit, membrane-localized, NADPH-oxidase complex, is required for generation of ROS (42). In plants, a similar complex appears to be needed for rapid ROS generation during defense (27). The sequence alignments of the TIR domains (Figure 3A) indicate that the R proteins are more closely related to Toll than IL-1R, although distributed throughout are either R protein–specific or Toll/IL-1R–specific amino acids (indicated with r, d, and * in Figure 3A). Overall 25% of the amino acid sequence compared falls in this category. In addition, the homology at the C-terminal end of the Toll/IL-1R superfamily consensus (66), a region required for IL-1R function (29), is absent in the three R proteins. These findings suggest the TIR domain of R proteins may provide a novel function.

The serine/threonine kinase capacity possessed by Pto and Xa21 could clearly facilitate downstream signaling. Because Pto is highly homologous to the Drosophila protein pelle required for Toll mediated signaling (described above), Pto kinase may serve a similar function, i.e. transcription factor activation. When Pto was used as the "bait" in a yeast two hybrid system, a number of interacting gene products were identified (110). These included sequences with homology to transcription factors as well as another protein kinase called Pti1 (*Pto*-interacting gene*1*), but not Prf. Pto may simultaneously activate several distinct signaling pathways. Because the Pti1 gene product can be phosphorylated by Pto and is capable of autophosphorylation but cannot phosphorylate Pto, a protein kinase cascade initiated by Pti phosphorylation by Pto may be one of these downstream signaling pathways. The kinase domain of the rice Xa21 gene product is most homologous to that of the Arabidopsis protein RLK5. When RLK5 was used in an interaction cloning system, a type 2C phosphatase was identified (90). Moreover, for many gene–mediated resistances, the addition of either kinase or phosphatase inhibitors significantly blocked the induction of rapid defense responses (16, 56). It appears likely that both kinases and phosphatases are involved in downstream R protein–mediated signaling events.

For the tomato Cf proteins, in the absence of any obvious signaling domains, the molecular identity of the signaling partners remains enigmatic. Possibly the short cytoplasmic domains might interact with a protein kinase. For the membrane-anchored CD14 receptor of T-cells, activation of the downstream kinase $p56^{Lck}$ requires only a short cytoplasmic domain (103). Alternatively, Cf proteins may already exist in or become associated with a membrane receptor complex involving transmembrane LRR-kinase proteins analogous to TMK1, RLK5, or Xa21 or some of the other five classes of transmembrane kinase (3, 7). Such a mechanism would recruit a kinase domain for intracellular signaling. The carboxyl terminal 10 LRRs are the most likely to interact with either a common or conserved signaling partner(s) because of their high sequence conservation (Figure 2). The binding of the Avr ligands, either directly or in association with another protein(s), may lead to a conformational change to the Cf protein and cause these domains to activate signaling partner(s).

The alternate splicing products expressed from the wild-type *N* and *L6* genes lead to the synthesis of both full-length R protein and a shorter protein that lacks most of the LRR domain. Several mammalian cell-surface receptors, e.g. growth-factor and cytokine receptors, also exist as soluble truncated proteins. These truncated proteins, although incapable of signal activation themselves, compete with the full-length forms for ligand binding and thereby tightly control the concentrations of ligand available for signaling (70). Alternatively, these truncated forms have been shown to bind to the intact receptor and modulate its function (82, 107). A defined role for the truncated forms of L6 and N in resistance has yet to be established.

Other Shared Motifs of Unknown Function

Two short sequence motifs are found in the majority of NBS/LRR R proteins and Prf (Figures 2 and 3C) between the NBS and LRR domains. One, designated conserved domain 2 (23), encodes a hydrophobic domain (HD) with the consensus GLPL(A/T)ax(V/S)aaG(S/G)aa, where a is an aliphatic amino acid. The other, designated conserved domain 3 (23), is situated 50–70 amino acids carboxyl to the first and has the consensus L(R/K)xCFLY(C/I)(A/S)xF. There are also two other slightly less well conserved short domains in this region (Figure 3C). Computer data-base searches have revealed these domains to be unique to R proteins and Prf, with the exception of Genbank accession No. U19616 and several ESTs, which probably represent orphan R genes because of their high overall similarity to *RPM1* and *RPS2*. The function of these motifs is not known. The high amino acid sequence conservation and fixed location of domains 2 and 3 has permitted oligonucleotide primers to be designed that can specifically amplify R gene–related sequences in many plant species (54).

A MODEL FOR R PROTEIN ACTION

In Figure 4 a model for R protein function incorporating the current knowledge and predictions about R and Avr gene products is depicted. We anticipate that R proteins will activate multiple signaling pathways simultaneously. For numerous mammalian receptor proteins, this scenario is increasingly evident (70). Such a pivotal role for plant R proteins in resistance would ensure that a wide repertoire of potential defense responses are rapidly and coordinately induced in various cellular compartments. Immediate downstream signaling components will include kinase and phosphatase cascades, transcription factors, and reactive oxygen species. It is also likely that the plant's defense strategy relies on the induction of responses that are directed against pathogen attack in general and are not pathogen specific. This is suggested because R protein structure cannot be predicted from pathogen types (Table 1) and a similar array of defense responses are induced by unrelated organisms. For example, callose and lignin deposition, antimicrobial phytoalexin and salicylic acid synthesis, pathogenesis-related gene induction, and the hypersensitive cell death responses are frequent components of the local resistance response acti-

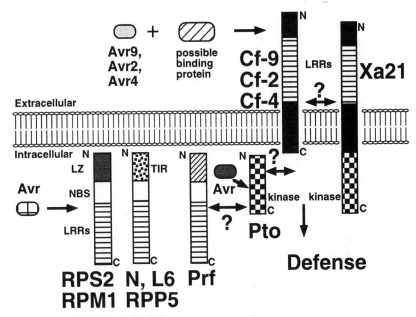

Figure 4 Representation of predicted R gene product structures and a model coupling the recognition of microbial Avr-dependent ligand and activation of plant defense. Pto can directly bind AvrPto (83, 92). The other R proteins probably bind the corresponding Avr gene products, either directly or in association with a binding protein. Both Pto and Xa21 have a protein kinase domain. It is likely that RPM1, RPS2, N, L6, and RPP5 and the Cf proteins also activate defense through a protein kinase, but the mechanism for this is not known. For example, the Cf proteins could interact with either an Xa21-like protein or a Pto-like protein to activate a protein kinase cascade. Prf is required for Pto-mediated resistance (80), but it is not understood why. Speculative interactions are indicated with a question mark. Abbreviations: LZ, putative leucine zipper region; TIR, region with homology to the cytoplasmic domain of the Drosophila Toll and human interleukin-1 receptors (see Figure 3A); LRR, leucine-rich repeat motifs; N, amino terminus; C, carboxyl terminus.

vated by many different avirulent microbes (27). Because the predicted R proteins have many similar features, it is likely that several of the downstream signaling cascades activated by distinct subclasses of R proteins will rapidly converge to execute this common defense response. The identification of the Arabidopsis *Ndr* mutation that compromises *RPM1, RPS2,* and some *RPP* gene-mediated resistances to *P. parasitica* establishes that resistances involving different R protein classes can have common component(s) (9). In addition, because the slow Arabidopsis HR phenotype mediated by the gene combination *avrRpm1/RPM1* can interfere with the fast resistance mediated by *avrRpt2/RPS2* genes (74, 76), some initial steps in the signal transduction pathway may be triggered by both R proteins.

The phenotype of Prf mutants demonstrates that Prf is required for Pto function. However, because avrPto and Pto directly interact in yeast, this leaves the precise function of Prf unclear. The LRR motif in Prf is unlikely to be directly involved in avr ligand recognition. Possibly Prf acts as an anchor protein that localizes the Pto kinase, or it might participate in a protein complex engaged by avrPto-Pto to stimulate additional signaling pathways. Alternatively, F Katagiri (personal communication) has proposed that Prf could recognize the phosphorylated form of a plant-encoded protein that is an avrPto-dependent substrate of Pto. Whatever the mechanism, the identification of Prf as a participating protein in Pto-mediated resistance provides a link between the NBS/LRR R proteins and protein kinases. This raises the possibility that all cytoplasmic-located R proteins with NBS/LRR domains will interact with a kinase(s) to activate downstream signaling events, and vice versa.

ADDITIONAL FEATURES OF *R* GENES

R Gene Function in Heterologous Plant Species

Several of the isolated *R* genes have been introduced, via *Agrobacterium*-mediated transformation, into other plant species, and have been demonstrated to retain biological activity. The tomato *Pto* gene functions in *N. tabacum* and *N. benthamiana,* the tobacco *N* gene is active in tomato where the N resistance response retains temperature sensitivity, and the tomato *Cf-9* gene mediates recognition of Avr9 peptide and necrosis formation in tobacco and potato (40, 78, 95, 106). Thus Avr-dependent R protein–triggered signaling cascades are conserved between plant species. However, when *Pto* or Fen is expressed at high levels in *Nicotiana clevelandii* by using a potato virus X vector system, necrotic symptoms develop in the absence of pathogen challenge (77; K Swords & B Staskawicz, unpublished data). These data highlight how finely tuned the relationship is between R proteins and signaling partners in their native plant species. Because sequences homologous to the introduced *R* genes can be detected in these other plant species, it is tempting to speculate that these may also function as *R* genes. We envision that *R* gene evolution is constrained not only by selection for pathogen recognition but also by selection against recognition of endogenous plant proteins, and that some other transgenic *R* gene transfer experiments may therefore lead to necrosis.

R Gene Expression

For plants to respond rapidly to microbial attack it was anticipated that R proteins should be present in healthy plants throughout life. RNA gel blot analyzes using *RPS2, RPM1, Pto, Cf-9,* and *Cf-2* as gene probes have revealed the presence of low abundance transcripts in unchallenged plants, indicating that at least the *RPM1* gene and some members of multigenic *R* families are expressed in the absence of the corresponding *Avr*-expressing pathogen (14, 23, 38, 59, 65). It is not yet known whether the levels of *R* gene expression increase at the site of microbial infection. However, this may not be a prerequisite because for the human immune response to be fully activated by the interleukin-1 receptor only 10 receptor molecules are required per cell to initiate the full response to ligand challenge (70).

R GENE ORIGIN AND EVOLUTION

From What Did R Genes Evolve?

The most likely ancestors of *R* genes probably coded for proteins involved in endogenous recognition/signaling systems required for the plant's normal growth or development, because a significant number of the mammalian, yeast, and insect proteins related to plant R proteins control endogenous signaling, development, and/or cell-to-cell adhesion. Two plant proteins similar to the extracytoplasmic LRR R protein, Xa21, are encoded by the Arabidopsis *erecta* and *clavata* genes that determine floral organ shape and size (96) (S Clark & E Meyerowitz, personal communication). Both the Erecta and Clavata proteins are thought to be involved in cell-to-cell communication events utilizing an extracellular ligand. Plant pathogens may have enhanced their own pathogenic potential by evolving a signaling capacity that could modify endogenous plant signaling systems. An alternative explanation for *R* gene evolution is that genes involved in multicellularity evolved from progenitor "*R*" genes involved in pathogen recognition by their unicellular ancestors; this seems less likely. The considerable structural homology between the NBS/LRR class of R proteins and the human major histocompatibility complex (MHC) class II transcription activator (CIITA), and between the exLRR class of R proteins and the mouse RP105 protein involved in B cell proliferation and protection against programmed cell death (see 37), suggests plant *R* genes and genes involved in mammalian immunity may have a common evolutionary origin.

Resistance to the same bacterial *Avr* genes has been observed in taxonomically distinct plant species (47, 104). This suggests that either there has been preservation of an ancient specificity or the same recognitional specificity evolved multiple times to a prevalent pathogen ligand. As additional plant *R* genes are isolated and related gene families recognized, it may be possible to determine which of these evolutionary scenarios is the more likely.

Generation of Evolutionary Novelty at R Gene Loci

Many plant pathogens exhibit a high mutation rate from avirulence to virulence that renders obsolete the effectiveness of individual *R* genes (10, 13, 44, 73). Because natural selection would favor the multiplication of these virulent races, plants must evolve novel R protein variants that can detect either the modified Avr determinant or another component of the pathogen. The most informative clue to the evolution of *R* gene diversity is their genomic organization. Different *R* loci can exist in one of four arrangements: They can consist of a single gene with an array of distinct alleles, each providing a different recognition specificity. The flax *L* locus is organized in this manner, so that only one of the 13 or more *L* specificities to *M. lini* are present within a pure breeding line (73). The *R* gene may exist as a single copy gene that is present in resistant lines but absent from susceptible lines. The Arabidopsis *RPM1* gene is of this type. For many *R* genes, the *R* locus is comprised of tandem arrays of closely linked *R* gene homologues with differing specificities. Examples of these "complex loci" include the *M* locus in flax, with 7 specificities to *M. lini* (73). Of the *R* genes isolated, *Cf-9, Cf-4, Cf-2, Cf-5, N, Pto,* and *Xa21* have been revealed by DNA gel blot analyses to reside within a linked cluster of related gene sequences (14, 38, 41, 59, 86, 105). Based on genetic analyses the *Rp1* locus of maize with 14 specificities to the rust *Puccinia sorghi* will likely also be of this type (34). Finally, in particular genomic regions, *R* genes to viral, bacterial, and fungal pathogens are loosely clustered, i.e. 1–2 cm apart (14, 51, 64). In Arabidopsis, five of these *R* gene–rich regions are now recognized and have been designated as "major resistance complexes" (MRCs) (31). It remains to be seen whether the *R* gene specificities identified in these "complexes" show enough relatedness to have had a common evolutionary origin. Tight clustering of *R* genes probably arose because of an initial duplication of the genomic segment that carried the ancestral gene. This was achieved by a rare crossing-over event, between homologous sequences at nonhomologous locations, possibly facilitated by the existence of linked repetitive elements (73). Because highly homologous gene sequences exist at the flax *L* and *M* loci, it is likely that repeated DNA structures surrounding the *M* locus but absent from the *L* region assisted the duplication of the intervening *M* genes (17).

At complex *R* loci, unequal cross-over via meiotic mispairing between different genes is currently thought to be the major way in which novel resistance specificities are generated and new combinations of parental *R* genes are created (Figure 5). Detailed genetic analysis of the highly unstable *Rp1* complex of maize, using DNA flanking markers to analyze the types of recombination events occurring, indicates that diversity has arisen via cross-overs and also to a lesser extent by gene conversion (34, 75). One example of an unequal cross-over at a cloned R gene locus has been molecularly characterized. The tomato Cf-2 and Cf-5 genes are essentially allelic. A Cf-2/Cf-5 trans-heterozygote was testcrossed and progeny were screened for individuals that carried neither Cf-2 nor Cf-5; one was recovered in 12,000. In this disease-susceptible individual, Cf-2 and Cf-5 homologue copy number was reduced from 3 (in Cf-2) or 4 (in Cf-5) to 2 (14). Further analysis has revealed this recombination event took place within a Cf-2 reading frame (M Dixon, personal communication). Overall, plants may not have devised a specialized mechanism to promote rapid R gene evolution (Figure 5). This contrasts strongly with the mechanism in mammals required for the recognition of nonself, where somatic events generate antibody diversity. An interesting feature of the plant mechanism to generate R gene diversity is that not all copies within the tandemly arranged R gene homologues would need to be expressed or functional.

At "simple" resistance loci, cross-over and gene conversion events probably play a similar role in generating R gene diversity. However, the number of distinct R gene sequences that can pair and recombine will be more limited, particularly in diploid self-fertile plant species. It is therefore likely that unequal intragenic recombination or slipped alignment during replication will play a significant role in modifying R gene product function. The sequence of the flax $L6$ gene has revealed how this could occur. The C-terminal half of the LRR domain comprises two direct repeats of 480 bp with 85% identity (53), and comparison of the $L6$, $L2$, and $L10$ alleles reveals variation in both LRR number and sequence in this region (17). In addition, at the flax M locus, loss of gene function is associated with the loss of one of these repeat units (P Anderson, J Ellis & G Lawrence, personal communication). Amplification or reduction in the number of LRR blocks within a single R protein may modify recognition specificity (87). However, none of the three mutant flax M alleles that have lost approximately six LRRs appears to confer novel resistance specificity (P Anderson, J Ellis & G Lawrence, personal communication). How the single copy Arabidopsis $RPM1$ gene evolved to provide dual specificity of recognition of avrB and avrRpm1 is currently unclear. In soybean the recognition mediated by these two sequence-unrelated avr genes involves either two different alleles of the $RPG1$ resistance gene or a second closely linked R gene (2).

For the N, $L6$, and $RPP5$ genes the recognized structural domains of these proteins are located within different exons: The TIR homology domain resides within exon 1, the NBS and HD in exon 2, and the LRR domain predominantly in exon 4 and subsequent exons (Figure 2). Because these intron locations are conserved in other members of each gene family, exon shuffling resulting in protein domain replacement is another potential mechanism that could facilitate the creation of novel R protein variants, at both simple and complex *R* loci. Such a mechanism has been proposed for other genes with this type of structural organization (61).

Several novel *R* gene variants have arisen at the meiotically unstable maize *Rp1* locus that have lost *R-Avr* specificity. These *R* alleles confer resistance to

A. *R* gene reassortment

B. Creating a novel variant *R* gene from a heterozygote

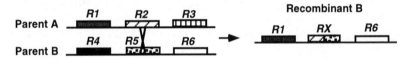

C. Creating a novel variant *R* gene from a homozygote by unequal crossover

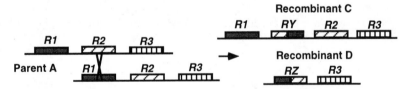

D. Resistance specificity

	Pathogen race								
	A1	A2	A3	A4	A5	A6	AX	AY	AZ
Parent A	I	I	I	C	C	C	C	C	C
Parent B	C	C	C	I	I	I	C	C	C
Recombinant A	I	C	C	C	I	I	C	C	C
Recombinant B	I	?	C	C	?	I	I	C	C
Recombinant C	I	?	I	C	C	C	C	I	C
Recombinant D	?	?	I	C	C	C	C	C	I

all *Puccinia sorghi* pathotypes tested and even other rust species, e.g. *P. polysora* (33, 34). The resistance is often weak and is invariably associated with a visible necrotic reaction. In the absence of pathogen challenge, plants with some of these variant *R* alleles still develop necrotic spotting (34). Other so-called "disease lesion mimics" are well known in maize, barley, and tomato (12, 36). A characteristic feature of many but not all, including the *Rp1* variant alleles, is that the necrotic spotting does not occur when the plants are grown under sterile conditions. Therefore a biotic stimulus (e.g. saprophytic microbe) is often required for phenotype expression. Possibly these variant R proteins are affected in the ligand binding/recognition domain, so that they are able to recognize a broad array of microbe/interaction-derived products. Alternatively, modifications to the signaling domain might constitutively activate defense signaling cascades, or provide a "hair trigger" that even weak Avr recognition events activate. Clearly, these novel *R* alleles are interesting to disease control because of the nonpathogen-specific nature of the resistance response. Their isolation is keenly awaited.

CONCLUDING REMARKS AND FUTURE PROSPECTS

With the isolation of the first few plant *R* genes, immense opportunities now unfold for protein biochemists, biologists, physiologists, and geneticists alike to elucidate how these gene products function and the gene families evolve.

─────────────────────────────────

Figure 5   Models for the creation of novel variant *R* genes. The lines and boxes represent contiguous regions of paired chromosomes during meiosis of either an F₁ hybrid between two different parental lines A and B that each carry three closely linked different *R* genes (*boxes*) (*panels A and B*) or after self-fertilization of the parent line A (*panel C*). Three different cross-over events and some of the different genetic outcomes are illustrated. (*A*) Crossover (X) in the intergenic region of correctly aligned genes creates Recombinant A, which carries a novel combination of *R* genes. The other recombinant chromosome not drawn would be *R4, R2, R3*. (*B*) Intragenic cross-over between two different *R* genes creates a novel *R* gene (*RX*) in Recombinant B capable of recognizing a different pathogen avirulence gene (*AX*). The other recombinant chromosome not drawn would be *R4, RW, R3*. (*C*) Misalignment during chromosome pairing caused by the high sequence relatedness of the clustered *R* genes permits unequal cross-over to create the novel genes, *RY* and *RZ* in Recombinant C and D, respectively, capable of recognizing additional pathogen avirulence genes *AY* and *AZ*, respectively. After an unequal cross-over event within a complex *R* locus, the number of homologues inherited will be either expanded or contracted. (*D*) Responses of the two parental lines and four recombinants to nine different pathogen races. I, incompatible interaction; C, compatible interaction. Recombinant A is resistant to the same assortment of the pathogen races as each parental line, but to no additional races. Recombinant B is resistant to one pathogen race that carries the *AX* gene, to which both parents were susceptible. Recombinants C and D are resistant to two different pathogen races that carry the *AY* or *AZ* gene, respectively, to which the parent plant was susceptible. The ? indicates that it cannot be predicted whether the recombinant gene will retain the recognition specificity of the original *R* genes as well as confer a novel recognition function.

Key unresolved questions include: What are the domain(s) or residues within each R protein that confer the specificity of microbial recognition? Do the Avr gene products always interact directly with R proteins? What are the immediate downstream signaling components and how do these activate multiple defenses? Which induced defense responses are crucial for conferring resistance against each microbial type? What roles do the gene products of other loci already identified by mutation analysis as required for *R-Avr* gene-mediated disease resistance, the *rdr* loci (9, 27, 51), play in defense? How are novel variant *R* genes naturally generated? Solving these questions is also likely to reveal new insight into the processes underlying normal plant growth and development and plant genome organization.

The identification of so few distinct classes of *R* genes is most intriguing. This suggests that plants have evolved only a limited number of mechanisms to defend themselves against microbial attack. Therefore, will nonrace-specific *R* genes, like the barley powdery mildew resistance gene *mlo,* code for a distinct R protein class? Likewise will the receptors for nonspecific microbial elicitors, for example chitin fragments, heptaglucosides, and the PEP-13 ligand from the fungus *Phytophthora sojae* (6, 24), which each confer defense response activation, be encoded by genes with homology to *R* genes? When resistance is inherited polygenically, will the quantitative trait loci (*QTLs*) that do not map to *R* gene clusters (64) identify other *R* gene classes, or will these *QTL* genes code for signaling or defense components? Also, for numerous pathogens, where completion of their life cycles involves two alternate hosts (1), for example the fungal rusts that require both wheat plants and barberry bushes, will the resistance manifested by these distinct plant species be mediated by related *R* genes or not?

The initial interest by plant breeders in resistant plant germplasm arose because it provided the possibility of a cheap solution to disease control in crops. Unfortunately, expectations of success for the new elite resistant cultivars were rarely achieved and the "Boom and Bust" cycle of disease control has prevailed in many crops for almost 50 years (1). Cloned *R* genes now provide novel tools for plant breeders to improve the efficiency of plant breeding strategies, via marker assisted breeding, and by using transformation for accelerating the introgression of useful *R* genes from related species (10, 64). It will also be possible to isolate and transfer homologous *R* genes between different plant species, e.g. wheat and barley, and therefore determine whether the nonhost status of individual plant species to *formae speciales* of a pathogen like *Erisyphe graminis* is caused by particular homologues of known *R* genes. Plant biotechnologists can attempt to manipulate both *Avr* and *R* gene sequences to provide broad-spectrum and durable disease control (26, 87).

Hopefully, a combination of strategies will reduce the requirement for agro-chemicals to control crop diseases and will accelerate effective retrieval and deployment of the natural variation in *R* genes of wild plant species.

ACKNOWLEDGMENTS

We thank numerous colleagues for sharing unpublished or prepublication material, and John Walker, Jeff Ellis, Jeff Dangl, Pam Ronald, Guido Van den Ackerveken, Mark Dixon, Mark Coleman, Martin Parniske, and Tom Tai for their helpful comments and critical reading of the manuscript. The research at the Sainsbury Laboratory is supported by grants from the Gatsby Charitable Foundation, the Biotechnology and Biological Sciences Research Council (BBSRC), and the European Community.

Visit the *Annual Reviews home page* at http://www.annurev.org.

Literature Cited

1. Agrios GN. 1988. *Plant Pathology.* London: Academic. 3rd ed.
2. Ashfield T, Keen NT, Buzzell RI, Innes RW. 1995. Soybean resistance genes specific for different *Pseudomonas syringae* avirulence genes are allelic, or closely linked, at the *RPG1* locus. *Genetics* 141:1597–604
3. Becraft PW, Stinard PS, McCarty DR. 1996. CRINKLY4: a TNFR-like receptor kinase involved in maize epidermal differentiation. *Science* 273:1406–9
4. Bent A. 1996. Function meets structure in the study of plant disease resistance genes. *Plant Cell* 8:1757–71
5. Bent AF, Kunkel BN, Dahlbeck D, Brown KL, Schmidt R, et al. 1994. *RPS2* of *Arabidopsis thaliana:* a leucine-rich repeat class of plant disease resistance genes. *Science* 265:1856–60
6. Boller T. 1995. Chemoperception of microbial signals in plant cells. *Annu. Rev. Plant Physiol. Plant Mol. Biol.* 46:189–214
7. Braun DM, Walker JC. 1996. Plant transmembrane receptors: new pieces in the signaling puzzle. *Trends Biol. Sci.* 21:70–73
8. Cao ZO, Henzel WJ, Gao XO. 1996. IRAK: A kinase associated with the interleukin-1 receptor. *Science* 271:1128–31
9. Century KS, Holub EB, Staskawicz BJ. 1995. *NDR1,* a locus of *Arabidopsis thaliana* that is required for disease resistance to both a bacterial and a fungal patho-

gen. *Proc. Natl. Acad. Sci. USA* 92: 6597–601
9a. Crute IR, Burdon JJ, Holub EB, eds. 1997. *The Gene-for-Gene Relationship in Host-Parasite Interactions.* Wallingford, UK: CAB Int.
10. Crute IR, Pink DAC. 1996. The genetics and utilization of pathogen resistance in plants. *Plant Cell* 8:1747–55
11. Dangl JL. 1995. The enigmatic avirulence gene of phytopathogenic bacteria. In *Bacterial Pathogenesis of Plants and Animals, Molecular and Cellular Mechanisms,* ed. JL Dangl, pp. 91–118. Berlin: Springer-Verlag
12. Dangl JL, Dietrich RA, Richberg MH. 1996. Death don't have no mercy: cell death programs in plant-microbe interactions. *Plant Cell* 8:1793–1807
13. Day PR. 1974. *Genetics of Host-Parasite Interactions.* New York: Freeman
14. Dixon MS, Jones DA, Keddie JS, Thomas CM, Harrison K, Jones JDG. 1996. The tomato *Cf-2* disease resistance locus comprises two functional genes encoding leucine-rich repeat proteins. *Cell* 84:451–59
15. Doke N, Ohashi Y. 1988. Involvement of an O_2^- generating system in the induction of necrotic lesions on tobacco leaves infected with tobacco mosaic virus. *Physiol. Mol. Plant Pathol.* 32:163–75
16. Dunigan DD, Madlener JC. 1995. Serine/

threonine protein phosphatase is required for tobacco mosaic virus-mediated programmed cell death. *Virology* 207:460–66

17. Ellis JG, Lawrence GJ, Finnegan EJ, Anderson PA. 1995. Contrasting complexity of two rust resistance loci in flax. *Proc. Natl. Acad. Sci. USA* 92:4185–88

18. Emsley P, Charles IG, Fairweather NF, Isaacs NW. 1996. Structure of *Bordetella pertussis* virulence factor P. 69 pertactin. *Nature* 381:90–92

19. Flor HH. 1971. Current status of the gene-for-gene concept. *Annu. Rev. Phytopathol.* 9:275–96

20. Gabriel DW, Rolfe BG. 1990. Working models of specific recognition in plant-microbe interactions. *Annu. Rev. Phytopathol.* 28:365–91

21. Galindo RL, Edwards DN, Gillespie SKH, Wasserman SA. 1995. Interaction of the pelle kinase with the membrane-associated protein tube is required for transduction of the dorsoventral signal in *Drosophila* embryos. *Development* 121:2209–18

22. Gopalan S, Bauer DW, Alfano JR, Loniello AO, He SY, Collmer A. 1996. Expression of the *Pseudomonas syringae* avirulence protein AvrB in plant cells alleviates its dependence on the hypersensitive response and pathogenicity (Hrp) secretion system in eliciting genotype-specific hypersensitive cell death. *Plant Cell* 8:1095–105

23. Grant MR, Godiard L, Straube E, Ashfield T, Lewald J, et al. 1995. Structure of the *Arabidopsis RPM1* gene enabling dual specificity disease resistance. *Science* 269:843–46

24. Hahlbrock K, Scheel D, Logemann E, Nürnberger T, Parniske M, Reinold S, et al. 1995. Oligopeptide elicitor-mediated defense gene activation in cultured parsley cells. *Proc. Natl. Acad. Sci. USA* 92:4150–57

25. Hamm HE, Gilchrist A. 1996. Heterotrimeric G proteins. *Curr. Opin. Cell Biol.* 8:189–96

26. Hammond-Kosack KE, Jones DA, Jones JDG. 1996. Ensnaring microbes: the components of plant disease resistance. *New Phytol.* 133:11–24

27. Hammond-Kosack KE, Jones JDG. 1996. Disease resistance gene-dependent plant defense mechanisms. *Plant Cell* 8:1773–91

28. Hashimoto C, Hudson KL, Anderson KV. 1988. The *Toll* gene of *Drosophila*, required for dorsal-ventral embryonic polarity, appears to encode a transmembrane protein. *Cell* 52:269–79

29. Heguy A, Baldari CT, Macchia G, Telford JL, Melli M. 1992. Amino acid conserved in interleukin-1 receptors (IL-1Rs) and the *Drosophila* Toll protein are essential for IL-1R signal transduction. *J. Biol. Chem.* 267:2605–9

30. Herbers K, Conrads-Strauch J, Bonas U. 1992. Race-specificity of plant resistance to bacterial spot disease determined by repetitive motifs in a bacterial avirulence protein. *Nature* 356:172–74

31. Holub EB. 1997. Organisation of resistance genes in *Arabidopsis*. See Ref. 9a, pp. 5–26

32. Hopkins CM, White FF, Choi S-H, Guo A, Leach JE. 1996. Identification of a family of avirulence genes from *Xanthomonas oryzae* pv. *oryzae*. *Mol. Plant-Microbe Interact.* 5:451–59

33. Hu G, Richter TE, Hulbert SH, Pryor T. 1996. Disease lesion mimicry caused by mutations in the rust resistance gene *rp1*. *Plant Cell* 8:1367–76

34. Hulbert S, Pryor T, Hu G, Richter T, Drake J. 1996. Genetic fine structure of resistance loci. See Ref. 9a, pp. 27–43

35. Johal GS, Briggs SP. 1992. Reductase activity encoded by the *HM1* disease resistance gene in maize. *Science* 258:985–87

36. Johal GS, Hulbert SH, Briggs SP. 1995. Disease lesion mimics of maize: a model for cell death in plants. *BioEssays* 17:685–92

37. Jones DA, Jones JDG. 1996. The roles of leucine-rich repeat proteins in plant defences. *Adv. Bot. Res. Inc. Adv. Plant Pathol.* 24:89–167

38. Jones DA, Thomas CM, Hammond-Kosack KE, Balint-Kurti PJ, Jones JDG. 1994. Isolation of the tomato *Cf-9* gene for resistance to *Cladosporium fulvum* by transposon tagging. *Science* 266:789–93

39. Jones JDG, Daniels M, Parker J, Szabo V, Coleman M. 1996. Plant pathogen resistance genes and uses thereof. *UK Patent* PCT/GB96/00849

40. Jones JDG, Hammond-Kosack KE, Jones DA. 1995. Method of introducing pathogen resistance in plants. *Patent* WO95/31564

41. Jones JDG, Thomas CM, Balint-Kurti PJ, Jones DA. 1995. Plant pathogen resistance gene and uses thereof. *Patent* WO96/35790

42. Jones OTG. 1994. The regulation of superoxide production by the NADPH oxidase of neutrophils and other mammalian cells. *BioEssays* 16:919–23

43. Joosten MHAJ, Cozijnsen TJ, de Wit PJGM. 1994. Host resistance to a fungal tomato pathogen lost by a single base-pair change in an avirulence gene. *Nature* 367:384–86

44. Keen NT. 1990. Gene-for-gene comple-

mentarity in plant-pathogen interactions. *Annu. Rev. Genet.* 24:447–63

45. Keen NT, Tamaki S, Kobayashi D, Gerhold D, Stayton M, Shen H, et al. 1990. Bacteria expressing avirulence gene *D* produce a specific elicitor of the soybean hypersensitive reaction. *Mol. Plant-Microbe Interact.* 3:112–21

46. Knogge W. 1996. Fungal infection of plants. *Plant Cell* 8:1711–22

47. Kobayashi DY, Tamaki SJ, Keen NT. 1989. Cloned avirulence genes from the tomato pathogen *Pseudomonas syringae* pv tomato confer cultivar specificity on soybean. *Proc. Natl. Acad. Sci. USA* 86:157–61

48. Kobe B, Deisenhofer J. 1994. The leucine-rich repeat: a versatile binding motif. *Trends Biochem. Sci.* 19:415–21

49. Kobe B, Deisenhofer J. 1995. Proteins with leucine-rich repeats. *Curr. Opin. Struct. Biol.* 5:409–16

50. Kobe B, Deisenhofer J. 1995. A structural basis of the interactions between leucine-rich repeats and protein ligands. *Nature* 374:183–86

51. Kunkel BN. 1996. A useful weed put to work: genetic analysis of disease resistance in *Arabidopsis thaliana. Trends Genet.* 12: 63–69

52. Landschulz WH, Johnson PF, McKnight SL. 1988. The leucine zipper: a hypothetical structure common to a new class of DNA binding proteins. *Science* 240: 1759–62

53. Lawrence GJ, Finnegan EJ, Ayliffe MA, Ellis JG. 1995. The *L6* gene for flax rust resistance is related to the *Arabidopsis* bacterial resistance gene *RPS2* and the tobacco viral resistance gene *N. Plant Cell* 7: 1195–206

54. Leister D, Ballvora A, Salamini F, Gebhardt C. 1996. A PCR based approach for isolating pathogen resistance genes from potato with potential for wide application in plants. *Nat./Genet.* 14:421–29

54a. Leister RT, Ausubel FM, Katagiri F. 1996. Molecular recognition of pathogen attack occurs inside of plant cells in plant disease resistance specified by the *Arabidopsis* genes *RPS2* and *RPM1. Proc. Natl. Acad. Sci. USA* 93:15497–502

55. Leppik EE. 1970. Gene centers of plants as sources of disease resistance. *Annu. Rev. Phytopathol.* 8:323–44

56. Levine A, Tenhaken R, Dixon R, Lamb C. 1994. H_2O_2 from the oxidative burst orchestrates the plant hypersensitive disease resistance response. *Cell* 79:583–93

57. Loh Y-T, Martin GB. 1995. The *Pto* bacte-

rial resistance gene and the *Fen* insecticide sensitivity gene encode functional protein kinases with serine/threonine specificity. *Plant Physiol.* 108:1735–39

58. Long SR, Staskawicz BJ. 1993. Prokaryotic plant parasites. *Cell* 73:921–35

59. Martin GB, Brommonschenkel SH, Chunwongse J, Frary A, Ganal MW, et al. 1993. Map-based cloning of a protein kinase gene conferring disease resistance in tomato. *Science* 262:1432–36

60. Martin GB, Frary A, Wu TY, Brommonschenkel S, Chunwongse J, et al. 1994. A member of the tomato *Pto* gene family confers sensitivity to fenthion resulting in rapid cell death. *Plant Cell* 6:1543–52

61. McKeown M. 1992. Alternative mRNA splicing. *Annu. Rev. Cell Biol.* 8:133–55

62. Meshi T, Motoyoshi F, Maeda T, Yoshiwoka S, Watanabe Y, Okada Y. 1989. Mutations in the tobacco mosaic virus 30-kD protein gene overcome *Tm-2* resistance in tomato. *Plant Cell* 1:515–22

63. Meyer M, Schreck R, Baeuerle PA. 1993. H_2O_2 and antioxidants have opposite effects on activation of NF-κB and AP-1 in intact cells: AP1 as secondary antioxidant-responsive factor. *EMBO J.* 12: 2005–15

64. Michelmore R. 1995. Molecular approaches to manipulation of disease resistance genes. *Annu. Rev. Phytopathol.* 15: 393–427

65. Mindrinos M, Katagiri F, Yu G-L, Ausubel FM. 1994. The *A. thaliana* disease resistance gene *RPS2* encodes a protein containing a nucleotide-binding site and leucine-rich repeats. *Cell* 78:1089–99

66. Mitcham JL, Parnet P, Bonnert TP, Garka KE, Gerhart MJ, et al. 1996. T1/ST2 signaling establishes it as a member of an expanding interleukin-1 receptor family. *J. Biol. Chem.* 271:5777–83

67. Miyata T, Miyazawa S, Yasonaga T. 1979. Two types of amino acid substitution in protein evolution. *J. Mol. Evol.* 12: 219–36

68. Morisato D, Anderson KV. 1995. Signaling pathways that establish the dorsal-ventral pattern of the *Drosophila* embryo. *Annu. Rev. Genet.* 29:371–99

69. Moyle WR, Campbell RK, Rao SNV, Ayad NG, Bernard MP, et al. 1995. Model of human chorionic gonadotropin and lutropin receptor interaction that explains signal transduction of the glycoprotein hormones. *J. Biol. Chem.* 270:20020–31

70. O'Neill LAJ. 1995. Interleukin-1 signal transduction. *Int. J. Clin. Lab. Res.* 25: 169–77

71. Padgett HS, Beachy RN. 1993. Analysis of a tobacco mosaic virus strain capable of

overcoming *N* gene-mediated resistance. *Plant Cell* 5:577–86

72. Petersen U-M, Björklund G, Ip YT, Engström Y. 1995. The *dorsal*-related immunity factor, Dif, is a sequence-specific *trans*-activator of *Drosophila Cecropin* gene expression. *EMBO J.* 14:3146–58

73. Pryor T, Ellis J. 1993. The genetic complexity of fungal resistance genes in plants. *Adv. Plant Pathol.* 10:281–305

74. Reuber TL, Ausubel FM. 1996. Isolation of *Arabidopsis* genes that differentiate between resistance responses mediated by the *RPS2* and *RPM1* disease resistance genes. *Plant Cell* 8:241–49

75. Richter TE, Pryor TJ, Bennetzen JL, Hulbert SH. 1995. New rust resistance specificities associated with recombination in the *Rp1* complex in maize. *Genetics* 141:373–81

76. Ritter C, Dangl JL. 1996. Interference between two specific pathogen recognition events mediated by distinct plant disease resistance genes. *Plant Cell* 8:251–57

77. Rommens CMT, Salmeron JM, Baulcombe DC, Staskawicz BJ. 1995. Use of a gene expression system based on potato virus X to rapidly identify and characterize a tomato *pto* homolog that controls fenthion sensitivity. *Plant Cell* 7:249–57

78. Rommens CMT, Salmeron JM, Oldroyd GED, Staskawicz BJ. 1995. Intergenic transfer and functional expression of the tomato disease resistance gene *Pto. Plant Cell* 7:1537–44

79. Rosetto M, Engström Y, Baldari CT, Telford JL, Hultmark D. 1995. Signals from the IL-1 receptor homolog, Toll, can activate an immune response in a *Drosophila* hemocyte cell line. *Biochem. Biophys. Res. Commun.* 209:111–16

80. Salmeron JM, Barker SJ, Carland FM, Mehta AY, Staskawicz BJ. 1994. Tomato mutants altered in bacterial disease resistance provide evidence for a new locus controlling pathogen recognition. *Plant Cell* 6:511–20

81. Salmeron JM, Oldroyd GED, Rommens CMT, Scofield SR, Kim H-S, et al. 1996. Tomato *Prf* is a member of the leucine-rich repeat class of plant disease resistance genes and lies embedded within the *Pto* kinase gene cluster. *Cell* 86:123–33

82. Schneider DS, Hudson KL, Lin T-Y, Anderson KV. 1991. Dominant and recessive mutations define functional domains of *Toll,* a transmembrane protein required for dorsal-ventral polarity in the *Drosophila* embryo. *Genes Dev.* 5:797–807

83. Scofield SR, Tobias CM, Rathjen JP, Chang

JH, Lavelle DT, et al. 1996. Molecular basis of gene-for-gene specificity in bacterial speck disease of tomato. *Science* 274:2063–65

84. Shelton CA, Wasserman SA. 1993. *pelle* encodes a protein kinase required to establish dorsovental polarity in the *Drosophila* embryo. *Cell* 72:515–25

85. Sims JE, Acres RB, Grubin CE, McMahan CJ, Wignall JM, et al. 1989. Cloning the interleukin 1 receptor from human T cells. *Proc. Natl. Acad. Sci. USA* 86:8946–50

86. Song W-Y, Wang G-L, Chen L-L, Kim H-S, Pi L-Y, et al. 1995. A receptor kinase-like protein encoded by the rice disease resistance gene, *Xa21. Science* 270:1804–6

87. Staskawicz BJ, Ausubel FM, Baker BJ, Ellis JG, Jones JDG. 1995. Molecular genetics of plant disease resistance. *Science* 268:661–67

88. Stein JC, Dixit R, Nasrallah ME, Nasrallah JB. 1996. SRK, the stigma-specific S locus receptor kinase of Brassica, is targeted to the plasma membrane in transgenic tobacco. *Plant Cell* 8:429–45

89. Steinbacher S, Seckler R, Miller S, Steipe B, Huber R, Reinemer P. 1994. Crystal structure of P22 tailspike protein: interdigitated subunits in a thermostable trimer. *Science* 265:383–86

90. Stone JM, Collinge MA, Smith RD, Horn MA, Walker JC. 1994. Interaction of a protein phosphatase with an *Arabidopsis* serine-threonine receptor kinase. *Science* 266:793–95

91. Suzuki N, Choe H-R, Nishida Y, Yamawaki-Kataoka Y, Ohnishi S, et al. 1990. Leucine-rich repeats and carboxyl terminus are required for interaction of yeast adenylate cyclase with RAS proteins. *Proc. Natl. Acad. Sci. USA* 87:8711–15

92. Tang X, Frederick RD, Zhou J, Halterman DA, Jia Y, Martin GB. 1996. Initiation of plant disease resistance by physical interaction of avrPto and Pto kinase. *Science* 274:2060–63

93. Taraporewala ZF, Culver JN. 1996. Identification of an elicitor active site within the three-dimensional structure of the tobacco mosaic tobamovirus coat protein. *Plant Cell* 8:169–78

94. Taylor SS, Knighton DR, Zheng J, Sowadski JM, Gibbs CS, Zoller MJ. 1993. A template for the protein kinase family. *Trends Biol. Sci.* 18:84–89

95. Thilmony RL, Chen Z, Bressan RA, Martin GB. 1995. Expression of the tomato *Pto* gene in tobacco enhances resistance to *Pseudomonas syringae* pv *tabaci* expressing *avrPto. Plant Cell* 7:1529–36

96. Torii KU, Mitsukawa N, Oosumi T, Matsuura Y, Yokoyama R, et al. 1996. The *Arabidopsis ERECTA* gene encodes a putative receptor protein kinase with extracellular leucine-rich repeats. *Plant Cell* 8: 735–46

97. Townsley FM, Pelham HRB. 1994. The KKXX signal mediates retrieval of membrane proteins from the Golgi to the ER in yeast. *Eur. J. Cell Biol.* 64:211–16

98. Traut TW. 1994. The functions and consensus motifs of nine types of peptide segments that form different types of nucleotide-binding sites. *Eur. J. Biochem.* 229: 9–19

99. Van den Ackerveken GFJM, Van Kan JAL, de Wit PJGM. 1992. Molecular analysis of the avirulence gene *avr9* of the fungal tomato pathogen *Cladosporium fulvum* fully supports the gene-for-gene hypothesis. *Plant J.* 2:359–66

99a. Van den Ackerveken G, Marios E, Bonas U. 1996. Recognition of the bacterial avirulence protein AvrBs3 occurs inside the host plant cell. *Cell* 87:1307–16

100. Walbot V. 1985. On the life strategies of plants and animals. *Trends Genet.* 1:165–69

101. Walton JD. 1996. Host-selective toxins: agents of compatibility. *Plant Cell* 8: 1723–33

102. Wasserman SA. 1993. A conserved signal transduction pathway regulating the activity of the *rel*-like proteins dorsal and NF-κB. *Mol. Biol. Cell* 4:767–71

103. Weinstein SL, June CH, Defranco AL. 1993. Lipopolysaccharide induced protein-tyrosine phosphorylation in human macrophages is mediated by CD14. *J. Immunol.* 151:3829–38

104. Whalen MC, Stall RE, Staskawicz BJ. 1988. Characterization of a gene from a tomato pathogen determining hypersensitive resistance in nonhost species and genetic analysis of this resistance in bean. *Proc. Natl. Acad. Sci. USA* 85:6743–47

105. Whitham S, Dinesh-Kumar SP, Choi D, Hehl R, Corr C, Baker B. 1994. The product of the tobacco mosaic virus resistance gene *N*: Similarity to Toll and the interleukin-1 receptor. *Cell* 78:1011–15

106. Whitham S, McCormick S, Baker B. 1996. The *N* gene of tobacco confers resistance to tobacco mosaic virus in transgenic tomato. *Proc. Natl. Acad. Sci. USA* 93:8776–81

107. Winans KA, Hashimoto C. 1995. Ventralization of the *Drosophila* embryo by deletion of extracellular leucine-rich repeats in the Toll protein. *Mol. Biol. Cell* 6:587–96

108. Yang Y, Gabriel DW. 1995. *Xanthomonas* avirulence/pathogenicity gene family encodes functional plant nuclear targeting signals. *Mol. Plant-Microbe Interact.* 8: 627–31

109. Zhang R, Cai H, Fatima N, Buczko E, Dufau ML. 1995. Functional glycosylation sites of the rat luteinizing hormone receptor required for ligand binding. *J. Biol. Chem.* 270:21722–28

110. Zhou J, Loh Y-T, Bressan RA, Martin GB. 1995. The tomato gene *Pti1* encodes a serine/threonine kinase that is phosphorylated by Pto and is involved in the hypersensitive response. *Cell* 83:925–35

Annu. Rev. Plant Physiol. Plant Mol. Biol. 1997. 48:609–39

MORE EFFICIENT PLANTS: A Consequence of Rising Atmospheric CO2?

Bert G. Drake and Miquel A. Gonzàlez-Meler

Smithsonian Environmental Research Center, P.O. Box 28, Edgewater, Maryland 21037

Steve P. Long

John Tabor Laboratories, The Department of Biological and Chemical Sciences, The University of Essex, Colchester, CO4 3SQ, United Kingdom

KEY WORDS: CO_2 and plants, CO_2 and photosynthesis, CO_2 and stomata, CO_2 and respiration, plants and climate change

ABSTRACT

The primary effect of the response of plants to rising atmospheric CO_2 (C_a) is to increase resource use efficiency. Elevated C_a reduces stomatal conductance and transpiration and improves water use efficiency, and at the same time it stimulates higher rates of photosynthesis and increases light-use efficiency. Acclimation of photosynthesis during long-term exposure to elevated C_a reduces key enzymes of the photosynthetic carbon reduction cycle, and this increases nutrient use efficiency. Improved soil–water balance, increased carbon uptake in the shade, greater carbon to nitrogen ratio, and reduced nutrient quality for insect and animal grazers are all possibilities that have been observed in field studies of the effects of elevated C_a. These effects have major consequences for agriculture and native ecosystems in a world of rising atmospheric C_a and climate change.

CONTENTS

609

INTRODUCTION[1]

Several lines of evidence suggest that terrestrial ecosystems are responding to rising atmospheric carbon dioxide (C_a) (39, 80, 116). The terrestrial biosphere responds to this increase solely through the response of plants. Photosynthesis (133) and transpiration (95) have long been known to be sensitive to increase in C_a, and it is now evident that respiration will also be affected (85). These three processes appear to be the only mechanisms by which plants and ecosystems can sense and respond directly to rising C_a. Understanding how these processes are affected by increase in C_a is therefore fundamental to any sound prediction of future response of both natural and agricultural systems to human beings' influence on the composition of the atmosphere and on the climate system.

Many detailed and thorough reviews identify the long list of changes at the whole plant level to rising C_a (e.g. 21, 26, 72, 81), but few focus on these initial steps in perceiving rising C_a. Influential ecological discussions appear sometimes to have ignored a physiological understanding. A common view is that the impact of rising C_a through stimulation of photosynthesis will be short-lived because other factors, particularly nitrogen, will be limiting in most ecosystems (21, 146, 197). Yet this view may ignore evidence from physiology that elevated C_a allows increased efficiency of nitrogen use. Thus the key effect is not removal of a limitation but increase in efficiency. An analysis of the available evidence shows that relative stimulations of plants grown with low N averaged across several studies appear just as large as those for plants grown with high N (130).

In this review, current understanding of the effects of C_a on transpiration, photosynthesis, and respiration are examined to help explain why rising C_a

---

1

Abbreviations: A, photosynthetic CO_2 assimilation; A_{sat}, light-saturated CO_2 assimilation; C_a, atmospheric CO_2 concentration; C_i, intercellular CO_2 concentration; Cyt, cytochrome pathway; Cytox, cytochrome-c-oxidase; ET, evapotranspiration; FACE, free-air carbon enrichment; g_s, stomatal conductance; HXK, hexokinase; KCN, potassium cyanide; LAI, leaf area index; LCP, light compensation point; LhcB, light-harvesting subunit; LUE, light-use efficiency; N, nitrogen; NEP, net ecosystem production; NUE, nitrogen use efficiency; Pa, pascal; PCO, photosynthetic carbon oxidation pathway; *RbcS,* Rubisco subunit gene; RH, relative humidity; RubP, Ribulose-1,5-bisphosphate; Rubisco, Ribulose-1,5-bisphosphate carboxylase/oxygenase; SDH succinate dehydrogenase; SHAM, salicylhydroxamic acid; S_r, Rubisco specificity; T, transpiration; T_{opt}, temperature optimum; TNC, total nonstructural carbohydrate; WUE, water use efficiency; $\emptyset$, photosynthetic light-use efficiency.

will increase resource-use efficiency and the implications of this increased efficiency. Each topic is introduced with a description of the mechanism by which elevated C$_a$ has its effect, followed by a discussion of acclimation of the process to elevated C$_a$. Acclimation is defined as those physiological changes that occur when plants are grown in elevated C$_a$. We have summarized the most relevant literature to indicate the intensity of the responses for key aspects of each of the three processes we discuss. Current C$_a$ is approximately 36 Pascals (Pa), although in many studies in our survey of the literature C$_a$ was lower than this by as much as 1.5 Pa. Elevated C$_a$ of the studies we reviewed varied considerably, from 55 Pa in the case of the Free Air Carbon Enrichment (FACE) studies to upward of 100 Pa in a few controlled environment studies. In most studies, however, the elevated C$_a$ was approximately 70 Pa, a concentration that is expected sometime during the twenty-first century.

STOMATA

In contrast with the effects of C$_a$ on photosynthetic CO$_2$ assimilation (A) and respiration, which are mediated by specific molecular targets, the mechanism by which stomata respond to C$_a$ remains unclear (152), although it appears most probable that it is linked to malate synthesis, which is known to regulate anion channels in the guard cell plasma membrane (96). Stomata of most species close with increase in C$_a$ as well as decrease in A and relative humidity (RH). For 41 observations covering 28 species, the average reduction of stomatal conductance was 20% (Table 1; see also 74). A recent analysis of responses in tree seedlings shows that the responses are highly variable, and in some species there is no response to elevated C$_a$ (46). It is not clear, however, whether failure to respond to elevated C$_a$ is due to a genetic trait or to acclimation of stomata to high humidity. For example, stomata of *Xanthium strumarium* grown in a greenhouse in high humidity failed to respond to elevated C$_a$ until given a cycle of chilling stress (62). Reduction of stomatal aperture and conductance (g$_s$) explains the reduction in leaf transpiration observed in plants grown in elevated C$_a$ (151). But does reduced g$_s$ in elevated C$_a$ limit photosynthesis in plants adapted to high C$_a$?

Stomatal Limitation of Photosynthesis

Two approaches to making a determination of the limitation of photosynthesis by g$_s$ have been applied (193), and both are based on analysis of the dependence of A on the intercellular CO$_2$ concentration (C$_i$), the A/C$_i$ curve. In plants grown in the present atmosphere, C$_i$ is generally maintained at 0.7 C$_a$, even when C_a is varied. In many plants, the value of A at the operational C_i is commonly about 90% of what it would be without the epidermis as a barrier to water loss and CO_2 diffusion into the intercellular spaces (i.e. A at C_i is about 0.9A at C_a). Here we use C_i/C_a as an index of the limitation of photosynthesis. If C_i/C_a in elevated C_a is less than C_i/C_a in normal ambient C_a, then the g_s would have decreased to be more of a limitation to A in elevated than in normal ambient C_a. In the literature we examined, mean and range of C_i/C_a were nearly identical for both normal ambient and elevated C_a grown plants in 26 species and 33 observations (Table 1). In six field studies, C_i/C_a was also very close to 0.7 for both treatments (0.73, 0.74 for normal ambient and elevated C_a). Thus, although the stomatal conductance is reduced in elevated C_a, this by itself does not limit photosynthesis. Similarly, reduced g_s at the leaf level does not necessarily mean that stand transpiration will be lower because there could be a compensatory increase in leaf area index (LAI). But does failure to limit photosynthesis mean that stomata do not acclimate to elevated C_a?

Acclimation of g_s to Elevated C_a.

Because stomatal conductance is mediated by changes in photosynthesis, lower conductance in plants having a reduced photosynthetic capacity is to be expected. There is some evidence that growth in high C_a alters the gain in the feedback loop for regulation of stomatal conductance (195). However, apart

Table 1 The effect of growth in elevated C_a on acclimation of stomatal conductance (g_s), transpiration (T), the ratio of intercellular to ambient CO_2 concentration (C_i/C_a), and leaf area index (LAI) (field grown species only); and the numbers of species (Sp) and studies (n).[a]

Attribute	R	Sp, n	References
g_s	$0.80^{b,c}$	28, 41	38, 43, 48, 66, 88, 89, 92, 107, 111, 140, 149, 160, 162, 181, 191, 192, 210, 217, 232, 241, 243
T	0.72^c	35, 80	2, 3, 13, 35, 43, 55, 67, 76, 92, 105, 112, 113, 128, 160, 186, 188, 189, 191, 192, 203, 216, 234, 235
C_i/C_a	0.99	26, 33	15, 19, 20, 43, 48, 73, 88, 89, 107, 108, 140, 149, 160, 162, 170, 181, 191, 192, 210, 241, 243, 248, 249
LAI	1.03	8, 12	13, 92, 142, 165, 173, 187

[a] R is the mean of n observations in various species (Sp) of the ratio of the attribute in plants grown in elevated to that for plants grown in current ambient C_a.

[b] Means statistically different from 1.0 ($p < 0.01$) by Student's t test.

[c] Means statistically different from 1.0 ($p < 0.01$) by Mann-Whitney rank sum test for data that failed normality test.

from a single paper (195), there is little evidence that stomata acclimate to elevated C$_a$ independently of acclimation of photosynthesis (65, 133, 193).

ACCLIMATION OF STOMATAL NUMBERS TO ELEVATED C$_a$ An acclimatory decrease in stomatal numbers appears a common but not universal response to growth at elevated C$_a$. In the absence of variation in stomatal dimensions, stomatal density will determine the maximum g_s that a unit area of leaf could attain. One expectation at increased C$_a$ is that fewer stomata are required because the rate of CO$_2$ diffusion into the leaf will be a decreasing limitation to photosynthesis as C$_a$ rises. Reported changes in stomatal density with growth at elevated C$_a$ include increases, decreases, and no change (90, 133). Long-term studies drawing on herbarium material and paleoecological evidence appear more conclusive, showing an inverse relation between variation in C$_a$ and variation in stomatal numbers (22, 23, 239). However, in a detailed study of variation in stomatal density within leaves from a single tree, Poole et al (175) showed that variation within a single tree is of the order found in herbarium specimens covering a 200-year period and previously attributed to the change in C$_a$. The authors further demonstrate that uncertainties in the environment from which palaeobotanical specimens have been sampled could explain the variation attributed to past variation in C$_a$.

RISING C$_a$ AND EVAPOTRANSPIRATION   Will reduced leaf transpiration by elevated C$_a$ also lead to reduced stand evapotranspiration (ET)? Whether elevated C$_a$ reduces ET depends on the effects of elevated C$_a$ on leaf area index (LAI) as well as on g_s. No savings in water can be expected in canopies where elevated C$_a$ stimulates increase in LAI relatively more than it decreases g_s. However, our survey shows that LAI did not increase in any of the long-term field studies of the effects of elevated C$_a$, on crops or native species (Table 1). This survey included studies of wheat (*Triticum aestivum*) and cotton in Arizona where FACE was used to expose the plants to 55 Pa (173) as well as open top chamber studies of native species. Elevated C$_a$ (>68 Pa) reduced ET compared with normal ambient in all the native species including the Maryland wetland (13), Kansas prairie (92), and the California grassland ecosystem (74). In the wetland ecosystem, ET was evaluated for a C$_3$-dominated and a C$_4$-dominated plant community (13). In these two communities, instantaneous values of ET averaged 5.5–6.5 for the C$_3$ and 7.5–8.7 mmol H$_2$0 m^{-2}s^{-1} for the C$_4$ communities at present ambient C$_a$ but at elevated C$_a$ (68 Pa), ET was reduced 17–22% in the C$_3$ and 28–29% in the C$_4$ community, indicating the relatively greater effect of elevated C$_a$ on g_s in the C$_4$ species. In the prairie ecosystem, cumulative ET over a 34-day period in midsummer was 180 kg m^{-2} at present ambient C$_a$, whereas it was 20% less at elevated C_a. In the grassland ecosystem, elevated C_a reduced ET sufficiently that the availability of soil water was increased (74). A four-year study of the responses of native Australian grass to elevated C_a in a phytotron reported higher water content of soils (138).

STOMATAL CONDUCTANCE AND THE ENERGY BUDGET   Reduced transpiration alters partitioning of energy between latent heat loss and convective exchange, potentially increasing leaf temperature (63). Elevating C_a to 55 Pa consistently decreased g_s and increased canopy temperature of cotton about 1°C (173).

SUMMARY   Reduced stomatal conductance is expected to be a feature of plants exposed to ever increasing C_a. Stomata do not appear to limit photosynthesis with elevation of C_a any more than they do at normal ambient C_a, even though g_s is usually decreased. A pattern of decreased g_s coupled with maintenance of a constant C_i/C_a will mean that water use efficiency will rise substantially, and there is evidence that this means increased yield for crops with no additional penalty in water consumption. Elevated C_a does not stimulate increased leaf area index in field studies with both crops and native species. Thus, reduced g_s leads to reduced ET and increased soil water content. However, reduced ET also causes increased warming of the plant canopy and surrounding air. Evidence for acclimation of stomatal development to elevated C_a is conflicting, though there is good evidence for a response of g_s to the acclimation of photosynthesis. The following section examines this acclimation.

PHOTOSYNTHESIS

The evidence that elevated C_a stimulates increased photosynthesis is overwhelming. In our survey of 60 experiments, growth in elevated C_a increased photosynthesis 58% compared with the rate for plants grown in normal ambient C_a (Table 2). Acclimation of photosynthesis to elevated C_a clearly reduces photosynthetic capacity but rarely enough to completely compensate for the stimulation of the rate by high C_a. This section of the paper reviews the mechanism for the fundamental effects of C_a on photosynthesis and what is known about acclimation to rising C_a.

Direct Effects of Rising C_a on Photosynthesis

Carbon dioxide has the potential to regulate at a number of points within the photosynthetic apparatus, including binding of Mn on the donor side of photosystem II (119), the quinone binding site on the acceptor side of photosystem II (86), and the activation of Ribulose-1,5-bisphosphate carboxylase/oxygenase (Rubisco) (178). While all these processes show a high affinity for

Table 2   Acclimation of photosynthesis to elevated C$_a$ determined as the ratio (R) of the value of the attribute for plants grown in elevated to that in normal ambient C$_a$ (R)[a]

Attribute	R	Sp,n	References
A at growth C$_a$			
Large rv	1.58[b]	45, 59	15, 34, 37, 55, 89, 92, 107, 108, 141, 144, 170, 180, 187, 203, 207, 213–215, 219, 224, 241, 247, 248
Small rv	1.28[c]	28, 103	12, 19, 20, 29, 31, 38, 41, 43, 48, 53, 54, 56, 64, 66, 68, 71, 77, 79, 88, 98, 101, 110, 111, 115, 125, 140, 141, 149, 159, 162, 168, 181, 191, 192, 194, 196, 202, 205, 210, 221, 226, 234, 235, 238, 240, 243
High N supply	1.57[b]	8, 10	12, 49, 55, 111, 160, 214, 234, 235
Low N supply	1.23[b]	8,10	12, 49, 55, 111, 160, 214, 234, 235
A, C$_a$ ≤ 35	0.93	28, 33	15, 34, 37, 49, 108, 180, 203, 207, 213, 224, 248, 249
Large rv			
Small rv	0.80[c]	18, 53	19, 20, 29, 38, 43, 48, 54, 56, 68, 71, 79, 88, 97, 110, 111, 115, 125, 140, 149, 162, 191, 192, 194, 202, 210, 221, 226, 234, 235, 238
High N supply	0.80[c]	4, 6	49, 111, 234, 235
Low N supply	0.61[b]	4, 6	49, 111, 234, 235
Starch	2.62[c]	21, 77	1, 7, 20, 24, 36, 49, 50, 54, 68, 93, 94, 98, 103, 108, 115, 139–141, 144, 155, 156, 170, 171, 177, 196, 203, 210, 215, 221, 228, 235, 243, 244
Sucrose	1.60[b]	9, 38	1, 7, 24, 50, 68, 82, 93, 94, 98, 103, 104, 139, 141, 144, 156, 169, 177, 203, 221, 229, 243
Protein	0.86[b]	11,15	7, 34, 37, 56, 93, 94, 108, 200, 202–204, 220, 229
[Rubisco]	0.85[b]	11,8	4–6, 34, 56, 108, 187, 194, 200, 202, 215, 220
Rubisco activity	0.76[c]	11,13	4–7, 34, 37, 56, 93, 94, 97, 106, 108, 124, 125, 140, 144, 169, 187, 194, 200, 202, 203, 214, 215, 226, 228, 234, 237, 243
Leaf [N]			
High N supply	0.85[b]	8, 10	12, 42, 49, 58, 138, 143, 170, 172, 234
Low N supply	0.81[b]	22, 39	12, 40, 42, 48, 49, 58, 99, 138, 140, 143, 149, 150, 160, 170, 172, 182, 187, 190–192, 215, 224, 233, 234

[a] Rooting volume (rv) is either large (>10 L) or small (<10 L). Other details as in Table 1.
[b,c] See table 1.

HCO$_3^-$ or CO$_2$ and are saturated at the current C$_a$, Rubisco has a low affinity for CO$_2$ on carboxylation, and this reaction is not saturated at the current C$_a$. Therefore, the carboxylation of Rubisco will respond to rising C$_a$.

The kinetic properties of Rubisco appear to explain the short- and many of the long-term responses of photosynthesis to this change in the atmosphere. Rising C_a increases the net rate of CO_2 uptake for two reasons. First, Rubisco is not CO_2-saturated at the current C_a. Second, Rubisco catalyzes the oxygenation of Ribulose-1,5-bisphosphate (RubP), a reaction that is competitively inhibited by CO_2 (18). Oxygenation of RubP is the first step of the photosynthetic carbon oxidation or photorespiratory pathway (PCO), which decreases the net efficiency of photosynthesis by 20–50%, depending on temperature (245), by utilizing light energy and by releasing recently assimilated carbon as CO_2. CO_2 is a competitive inhibitor of the oxygenation reaction, such that a doubling of concentration at Rubisco will roughly halve the rate of oxygenation (131). This second effect on the PCO may be of greater importance, because an increase in net photosynthesis will result regardless of whether photosynthesis is Rubisco- or RubP-limited and regardless of where metabolic control lies. The increase in uptake resulting from suppression of the PCO requires no additional light, water, or nitrogen, making the leaf more efficient with respect to each.

RUBISCO SPECIFICITY   Rubisco specificity (S_r) is the ratio of carboxylation to oxygenation activity when the concentrations of CO_2 and O_2 at Rubisco are equal. It determines directly the increase in efficiency of photosynthesis with rising C_a. This value is therefore of fundamental importance in predicting the direct responses of plants to rising C_a. S_r has been suggested to vary from 88–131 across a range of C_3 plants, with an average of about 100 (18). Terrestrial C_3 plants show both the highest and a fairly constant S_r in contrast with other photosynthetic groups such as C_4 plants and cyanophyta (26, 52, 225).

ELEVATED C_a AND TEMPERATURE   As temperature increases, S_r declines dramatically for two reasons: decreased solubility of CO_2 relative to O_2 and decreased affinity of Rubisco for CO_2 relative to O_2 (133). About 68% of the decline in S_r is calculated to result from the binding affinity of the protein for CO_2 (27, 131). The effect of this decline in S_r with temperature is to produce a progressive increase in the stimulation of photosynthesis by elevated C_a with temperature. The minimum stimulation of RuBP-limited photosynthesis by increasing C_a from 35 to 70 Pa rises from 4% at 10°C to 35% at 30°C. It also follows from this interaction that the temperature optimum (T_{opt}) of light-saturated CO_2 assimilation (A_{sat}) must increase with C_a by 2, 5, and 6°C with increase in C_a to 45, 55, and 65 Pa, respectively (137). The upper temperature at which a positive A_{sat} may be maintained is similarly increased. The change in these characteristic temperatures underlies the importance of considering rise in C_a not just as a factor that increases photosynthetic rate, but also as one that strongly modifies the response to temperature. Because increasing C_a is predicted to increase leaf temperature, both directly by decreasing latent heat loss and indirectly through radiative forcing of the atmosphere, this interactive effect of CO_2 and temperature has profound importance to future photosynthesis. It also suggests a much greater stimulation of photosynthesis in hot versus cold climates (118, 135, 136).

Acclimation of Photosynthesis to Elevated C_a

There is abundant evidence that in the long term, photosynthesis acclimates to elevated C_a, i.e. the photosynthetic properties of leaves developed at elevated C_a differ from those developed at the current C_a (46, 90, 133, 230). The vast majority of studies in our and others' surveys show a decrease in A of plants grown in elevated C_a, relative to controls grown at normal ambient, when both are measured at the current ambient C_a (Table 2; see also 90, 136, 193). Acclimation of photosynthesis is accompanied by higher carbohydrate concentration, lower concentration of soluble proteins and Rubisco, and inhibition of photosynthetic capacity. When there is no rooting-volume limitation, as for example in our survey when the rooting volume exceeded 10L, significant reduction in A caused by growth in elevated C_a is the exception rather than the rule (Table 2, A at $C_a < 35$ Pa) while, exceptionally, an increase in photosynthetic capacity is observed (15, 91).

Two reasons for this acclimation are apparent. First, the plant may be unable to use all the additional carbohydrate that photosynthesis in elevated C_a can provide; therefore a decrease in source activity must result. Second, less Rubisco is required at elevated C_a. Our survey shows an average reduction in the amount of Rubisco of 15% in eight studies including 11 species and a reduction in Rubisco activity of about 24% (Table 2). As a protein that can constitute 25% of leaf N, these reductions are a major component of the lower tissue N observed in foliage (15–19%) (Table 2).

SOURCE/SINK BALANCE   Arp (14) drew attention to the strong correlation between rooting volume and acclimation of photosynthesis of plants in elevated C_a. In small pots (i.e. <10 L), A of plants in elevated C_a was less than A of plants in normal ambient C_a. In Table 2 we separate the effects of elevated C_a on photosynthesis into the effects of small and large rooting volumes. In our survey of 163 studies, the stimulation of A was about 50% for large rooting volumes and field experiments but reduced by about half of this when the rooting volume is limited (Table 2). When there is no restriction of rooting volume, A_{sat} remains the same for plants grown in both elevated and ambient C_a. Similar conclusions are reported for tree seedlings (46). The effect of rooting volume on acclimation is probably confounded with effects of nutrient availability on photosynthesis.

NITROGEN-LIMITATION   Other factors, such as available nutrients, also reduce the sink strength. In a small number of studies, reducing the available N had an effect on A that was the same as the effect of limiting the rooting volume: At high N, the stimulation of A by elevated C_a was about 50%, but this stimulation dropped to about 25% when available N was low. Acclimation of photosynthesis to elevated C_a has frequently been suggested to be more marked when N supply is limiting (26, 46). Rubisco and large subunit Rubisco RNA (*RbcS*) expression in *Pisum sativum* and *Triticum aestivum* were unaffected by growth in elevated C_a when N supply was abundant but showed marked decreases in response to elevated C_a when N was deficient (158, 185).

For plants such as wheat and pea, which are able to rapidly form additional sinks during early vegetative growth, sink limitation is unlikely, whereas other requirements are not limiting. However, growth of additional sinks would be limited if N supply is limiting. Because less Rubisco is required under elevated C_a, this redistribution of N would greatly increase the efficiency of N use.

Although acclimation in many early experiments was exaggerated by the artifact of rooting restriction, there is also clear evidence that acclimation can occur in the absence of any rooting restriction (46). In the Maryland wetland ecosystem where open top chambers have been used to study the effects of elevated C_a (68 Pa), Rubisco was reduced 30–58%, and photosynthetic capacity, measured at normal ambient C_a, was reduced 45–53% in the sedge (*Scirpus olneyi*) after seven years of exposure (108). Wheat grown with an adequate supply of N and water showed no acclimation of photosynthesis to C_a elevated to 55 Pa in FACE until completion of flag leaf development when there was a significant loss of Rubisco followed by other photosynthetic proteins, relative to controls (157).

HOW MUCH RUBISCO IS REQUIRED IN HIGH C_a?   Rubisco can constitute 25% of leaf [N] in a C_3 leaf (18). Large quantities of this enzyme appear necessary to support light-saturated photosynthesis in present C_a (140). Calculations suggest that 35% of the Rubisco could be lost from the leaf before Rubisco will co-limit photosynthesis when C_a is increased to double the current concentration (133). *Nicotiana tabaccum* transformed with antisense *RbcS* to produce 13–18% less Rubisco showed lower rates of carbon gain and growth at the current C_a by comparison with the wild type from which they were derived. There was no difference in C gain or growth when both were grown at 80 Pa C_a (140), providing clear evidence of a decreased requirement for Rubisco at elevated C_a.

Woodrow (238) computed the amount of Rubisco required to maintain constant A as C_a increased from the present level to 100 Pa. At 25°C, the amount of Rubisco needed drops to 59% of present amount at 70 Pa, to 50% at 100 Pa (Figure 1). Because of the strong temperature dependence of S_r, the amount of Rubisco required will also decline strongly with increasing temperature. At 70 Pa and a leaf temperature of 35°C, only 42% of the Rubisco activity required at 35 Pa would be needed to maintain the same rate of photosynthesis. There would be a large need for Rubisco at low temperature, and this requirement changes very little as C_a rises (Figure 1). At 5°C, the requirement for Rubisco to maintain the same rate of photosynthesis at elevated C_a is 89% of that needed at normal ambient.

A wide range of studies have reported decreases in Rubisco content and activity with growth in elevated C_a. In our survey of 18 studies of 12 species, Rubisco was reduced 15% (Table 2). Growth in elevated C_a commonly results

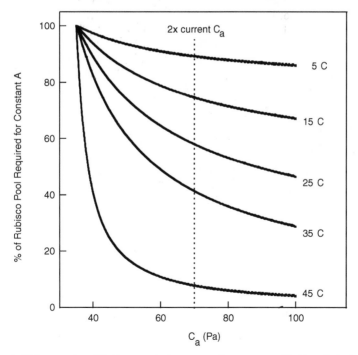

Figure 1   The proportion of Rubisco required to support the same rate of Rubisco-limited photosynthesis at 35 Pa as C_a increases at different leaf temperatures. (After 238.)

in decreased photosynthesis relative to controls when measured at the current atmospheric C_a, even though photosynthesis of the elevated grown leaves remains higher when they are measured at their elevated growth C_a. This could be explained by decreased in vivo Rubisco activity. In our survey of 13 studies and 11 species (Table 2), we indicate a reduction of Rubisco activity of 24%. Studies of *Phaseolus vulgaris* (194), *Pinus taeda* (215), and wheat (134) have shown A/C_i responses that indicate a selective loss of Rubisco activity in vivo without significant loss of capacity for regeneration of RuBP with growth at elevated C_a. A similar conclusion can be drawn from control analysis applied to *Helianthus annuus* (237).

THE MOLECULAR MECHANISM OF ACCLIMATION   Decrease in Rubisco is commonly correlated with an increase in leaf nonstructural carbohydrates. In our survey we found that sucrose and starch increased 60 and 160% in elevated C_a (Table 2). Regulation of the expression of photosynthetic genes, via increased soluble carbohydrate concentration, may underlie acclimation to growth in elevated C_a (Figure 2; 199, 206, 230). Decreased expression of several photosynthetic genes has occurred when sugar concentrations have been increased by directly feeding mature leaves through the transpiration stream (121, 123, 222), by expression of yeast-derived invertase in transgenic tobacco plants that directs the gene product to the cell wall to interrupt export from source leaves (227), and by cooling the petiole to decrease the rate of phloem transport in intact tobacco plants (122). Using chimeric genes created by fusing maize photosynthetic gene promoters with reporter genes, seven promoters including those for the light harvesting subunit *(LhcB)* and *RbcS* were repressed by soluble carbohydrates. The low concentration of glucose required for this repression suggests that other sugars, in particular sucrose and fructose, may be effective via metabolism in the cell to glucose. How might glucose suppress gene expression in the nucleus? Based on glucose signaling in yeasts, a hypothetical scheme whereby hexokinase (HXK) associated with a glucose channel or transporter in the plasmalemma or tonoplast would release an effector in response to glucose has been proposed (Figure 2; 121, 199). The effector would then interact with the promoters of nuclear genes coding for chloroplast components. This system would allow sensing of both an accumulation of sucrose in the vacuole and in the leaf vascular tissue, indicating an imbalance in sink capacity relative to source activity. Repression is blocked by the HXK inhibitor mannoheptulose, providing evidence of the role of HXK in this signal transduction pathway (109). Where carbohydrate repression has been demonstrated it appears to involve both *RbcS,* coding for Rubisco, and genes that will affect capacity for RubP regeneration. Optimum use of resources would require a system that would allow

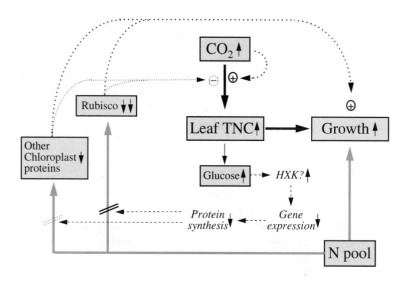

Figure 2   How rising C_a may support more growth when N is limiting. Elevated C_a ($CO_2\uparrow$) will stimulate photosynthesis and leaf total nonstructural carbohydrate (TNC) concentration, which in turn could support more growth of sink tissues. When growth is limited by N, TNC accumulation in the source leaves will be accentuated by elevated CO_2 concentration. Glucose, as a possible monitor of leaf TNC, represses expression of specific genes and in particular the *RbcS* gene coding for the small subunit of Rubisco. Glucose repression of nuclear gene expression is thought to occur via a hexokinase (*HXK*) signal transduction pathway. Decreased synthesis of Rubisco and to a lesser extent other chloroplast proteins will release a significant portion of the limiting supply of N.

decrease in Rubisco, without loss of capacity for RubP regeneration. Nie et al (156) showed in wheat that elevated C_a can result in decreased expression of *RbcS* but not other Calvin cycle of chloroplast membrane genes. This is consistent with Figure 2 because several different promoters are involved that could have different sensitivities to carbohydrate concentrations (199). Is carbohydrate repression consistent with observations of plants grown in elevated C_a? Although as a general rule Rubisco decreases with growth in elevated C_a and soluble carbohydrates rise, there are important exceptions (156). This suggests that other possible regulatory elements need to be identified before the mechanisms of acclimation can be fully understood.

ACCLIMATION AND CANOPY PHOTOSYNTHESIS   Our analysis of photosynthesis has only concerned the increase in leaf photosynthetic rates that result from growth in elevated C_a. If we consider a crop or natural canopy, carbon gain will only increase with increased leaf photosynthetic rates under elevated C_a in the absence of compensatory decreases in canopy size and architecture. If there is a compensatory decrease in canopy size, then gain at the leaf level might be offset by decrease at the canopy level. In Table 1 we show that for studies carried out in the field, canopy leaf area is not significantly increased or decreased by long-term growth in elevated C_a.

Considerable evidence supports the prediction that increase in CO_2 uptake will be greater in warm climates (131, 133, 145). Among the long-term experiments in which plants have grown under elevated CO_2 for successive seasons, most obvious is that in arctic tundra no sustained increase in net carbon gain was observed (163), whereas in warm temperate climates, e.g. the Maryland wetland ecosystem, stimulation of CO_2 uptake was observed for eight successive seasons (60). In two successive FACE experiments on the same site at Maricopa, Arizona, total daily canopy photosynthesis of *Gossypium hirsutum* in the middle of summer was increased by ca 40% in the canopy growing in 55 Pa. In wheat growing on the same site in the cooler temperatures of spring, canopy photosynthesis was increased by ca 10% (173). Relative stimulation of A by a doubled C_a in the evergreen *Pinus taeda* in the field was strongly correlated with seasonal variation in temperature (129).

Photosynthesis in the Shade

Photosynthesis is light limited for all leaves for part of the day, and for some leaves, those of the lower canopy, for all of the day. For a crop canopy, light-limited photosynthesis can account for half of total carbon gain, whereas photosynthesis of forest floor species might always be light limited. The initial slope of the response of photosynthesis to light defines the maximum quantum yield or photosynthetic light-use efficiency ($\varnothing$) of a leaf and determines the rate of CO_2 uptake under strictly light-limiting conditions.

At a given C_a, $\varnothing$ has been shown to be remarkably constant in C_3 terrestrial plants regardless of their taxonomic and ecological origins (158). This may reflect the constancy of the photosynthetic mechanism across C_3 species. Even under light-limited conditions net photosynthesis is reduced by the PCO, which consumes absorbed light energy and releases CO_2. Inhibition of the PCO by elevated C_a will therefore increase light-limited photosynthesis. This increase may be closely predicted from the kinetic properties of Rubisco (133). Forest floor vegetation commonly exists close to the light compensation point (LCP) of photosynthesis. Any increase in $\varnothing$ could therefore result in large increases in net photosynthesis. These predictions are consistent with recent observations of more than two- to fourfold increases in net carbon gain by leaves of both forest floor herbs (CP Osborne, BG Drake & SP Long, unpublished data) and tree seedlings (126) grown in elevated C_a in situ. Calculated from the kinetic constants of Rubisco, the maximum quantum yield of photosynthesis at 24°C will increase by 24% when C_a is doubled. The LCP should decline reciprocally by 20% if mitochondrial respiration remains unchanged. In *S. olneyi* grown and measured in 70 Pa C_a, $\varnothing$ was 20% greater than that of plants grown and measured at 36 Pa, close to theoretical expectation (132). LCP, however, was decreased by 42%, almost double the theoretical expectation. A similar increase in maximum quantum yield was observed in the forest floor herb *Duchesnea indica,* but here LCP decreased by 60% (CP Osborne, BG Drake & SP Long, unpublished data). These greater-than-predicted decreases in LCP could only be explained by a decrease in leaf mitochondrial respiration rate. The next section considers the mechanisms and evidence for such a decrease in respiration rate.

SUMMARY   Theory and experiments show that in rising C_a, photosynthesis will be stimulated in both light-limited and -saturated conditions and that the stimulation rises with temperature. Optimization theory suggests that substantial decreases in leaf Rubisco content could be sustained under elevated C_a while maintaining an increased rate of leaf photosynthesis, particularly at higher temperatures. Acclimation decreases Rubisco in response to elevated soluble carbohydrate levels. Higher quantum yield at elevated C_a reduces the light compensation point. Because of the temperature interaction between Rubisco activity and elevated C_a, we would expect higher rates of photosynthesis in tropical and subtropical species as well as shifts in the C:N for foliage.

MITOCHONDRIAL RESPIRATION   The earliest reported findings of a direct inhibition of dark respiration by elevated C_a are those of Mangin from 1896 (quoted in 153), although the 5% level employed far exceeds the doubling of current ambient C_a. It has now been established that the specific rate of dark respiration, measured either by CO_2 efflux or by O_2 uptake, decreases about 20% when the current ambient C_a is doubled (Table 3, Direct effect; 8, 17, 30, 85, 87, 242). Two different effects of elevated C_a have been suggested (28): an effect that occurs because of the growth or acclimation of the plant in high C_a (e.g. 17) and a readily reversible effect (e.g. 9, 28). These two effects are now referred to as the indirect and direct effects of elevated C_a on respiration. Although the mechanism for the indirect effect is not yet clear, the direct effect appears to be caused by inhibition of the activity of two key enzymes of the mitochondrial electron transport chain, cytochrome *c* oxidase (Cytox) and succinate dehydrogenase (85). We restrict our comments here to this emerging new direction in CO_2 effects research. For information on other aspects of the interaction of elevated C_a and respiration, we refer the reader to the numerous excellent reviews that have recently appeared (8, 16, 30, 70, 153, 176, 242).

Direct Effect of Elevated C_a on Dark Respiration

There are many reports of a decrease in respiration within minutes of increase in C_a (9, 28, 69, 87, 114, 166, 179, 183, 201). Respiration is reduced about 20% for a doubling of the atmospheric C_a (Table 3). This effect has been reported for many different kinds of tissues including leaves, roots, stems, and even soil bacteria, suggesting that whatever the basic mechanism, it involves a fundamental aspect of respiration.

MECHANISM OF DIRECT EFFECT OF C_a ON DARK RESPIRATION   A plausible mechanism underlying the direct effect is the inhibition of enzymes of the mitochondrial electron transport system. Experiments with enzymes in vitro showed that elevated C_a reduces the activity of both Cytox and succinate dehydrogenase (85, 166, 184a). Under experimental conditions in which Cytox controlled the overall rate of respiration in isolated mitochondria (148), O_2 uptake was inhibited by about 15% (85). Experiments with the enzymes in vitro indicated a direct inhibition by elevated C_a on their activity of about 20% for a doubling of the current ambient C_a (85; Figure 3). Measurements of O_2 consumption on isolated soybean mitochondria that were fully activated (State 3 conditions, i.e. sufficient ADP) and in which the respiration inhibitor salicylhydroxamic acid (SHAM) was used to inhibit the alternative pathway showed that doubling C_a inhibited the cytochrome (Cyt) pathway approximately 10–22% (85). By blocking the Cyt pathway with potassium cyanide (KCN) and using

Table 3   Respiration of shoots and leaves in elevated C_a[a]

Respiration	R	Sp, n	References
Direct Effect	0.82[b]	23, 53	9, 28, 32, 33, 45, 51, 57, 75, 78, 102, 114, 147, 191, 192, 208, 209, 212, 221, 223, 246
Indirect Effect	0.95	17, 37	17, 28, 32, 57, 103, 117, 120, 147, 154, 191, 192, 203, 221, 231, 246

[a] The direct effect refers to the ratio (R) of rates of dark respiration in the same samples when C_a is increased from the current ambient to the elevated level. The indirect effect refers to the ratio of rate of dark respiration of plants grown in elevated to the rate of plants grown in current ambient C_a when measured at the same background of C_a. Other details as in Table 1.

[b] Significantly different from 1.0 ($p < 0.05$) by Students' t test.

either succinate or NADH as electron donors, it was shown that the succinate dehydrogenase (SDH) in vivo was also inhibited by doubling C_a (85). The activity of the alternative pathway has been shown to be unaffected directly by changing the level of C_a (85, 184a). What is the specific mechanism for inhibition of Cytox by elevated C_a? Because the effect is time dependent (85; Figure 3) and appears to be dependent on CO_2 and not HCO_3 (166), one possibility is a carbamylation reaction. The structure of Cytox contains lysine residues (218), necessary for the proposed carbamylation.

Another proposed mechanism for the apparent inhibition of respiration is that elevated C_a stimulates dark CO_2 fixation (8). Measurements of the respiratory quotient (consumption of O_2/emission of CO_2) show that this is not a viable possibility because reduced CO_2 evolution is balanced by an equal reduction of O_2 uptake in elevated C_a (184).

The possibility that CO_2 inhibition of these enzymes mediates the direct effect of C_a on respiration in plants is supported by measurements on different types of plant organelles and tissues. Doubling present atmospheric C_a reduced

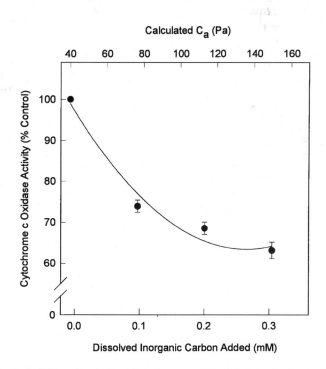

Figure 3   The inhibition of bovine heart Cytochrome *c* oxidase activity in vitro by elevated C_a (85).

in vivo O_2 uptake by soybean mitochondria, and by extracts from excised shoots of the sedge *S. olneyi* (84, 85). Experiments in which CO_2 efflux was used to measure dark respiration showed that doubling C_a reduced respiration in excised shoots removed from the field to the lab and from intact stands in which respiration was determined in the field on the C_3 sedge, *S. olneyi* (58). The importance of this effect for carbon balance of plants and ecosystems is that it apparently occurs at the most fundamental level of organization of the mitochondrial electron transport. Thus all respiring tissues are subject to this effect.

Acclimation of Respiration to Elevated C_a

Over days to months, the rate of dark respiration of foliage declines. This occurs in parallel with tissue declines in N concentration or protein content that is energetically costly (25), indicating that this decline reflects decreased demand for energy to sustain growth and/or maintenance. Plants grown in elevated C_a typically have lower protein and nitrogen concentrations (Table 2). Several reviews indicate the considerable potential for rising C_a to reduce respiration through effects on tissue composition (8, 46, 242). We reviewed data on measurements of respiration on leaves of 17 species grown in current ambient and elevated C_a. Acclimation of dark respiration was determined by comparison of the rate of CO_2 efflux or O_2 consumption measured on samples of tissue grown in current ambient or elevated C_a at a common background C_a (Table 3, Indirect effect). In our survey of the literature we found no overall difference between the specific rates of respiration of shoots and leaves grown in elevated or ambient C_a (Table 3).

However, some C_3—but not C_4—species do show the effects of acclimation to high C_a. Acclimation of the rate of respiration in the C_3 plants, *S. olneyi, Lindera benzoin,* and wheat, was due to reduction in activity of enzymatic complexes of the mitochondrial electron transport chain (Cytox and Complex III), which resulted in diminished capacity of tissue respiration (11, 17). Reduction of the activity of these enzymes was not found in the C_4 species *Spartina patens* (11).

SUMMARY Exposure of plants to elevated C_a usually results in a lower rate of dark respiration. Efflux of CO_2 from stands in the field; from excised leaves, roots, and stems; and from O_2 consumption of isolated mitochondria, suspensions of cells, and pieces of tissues are reduced about 20% for a doubling of current C_a. This effect appears to be caused mainly by the direct inhibition of the activity of the respiratory enzymes, cytox, and succinate dehydrogenase by

or for plant/insect/animal interactions. Reduced stomatal conductance results in greater WUE and reduced ET, and it may increase soil water content. However, reduced transpiration also alters canopy energy balance and shifts some heat loss from transpiration to convective heat loss. This effect has important consequences for climate. Incorporating a model of stomatal response to elevated C$_a$ into a coupled biosphere-atmosphere model (SiB2-GCM) showed that decreased g$_s$ and latent heat transfer will cause a warming of the order of 1–2°C over the continents (198) in addition to warming from the CO$_2$ greenhouse effect. Implicit in this development is that any loss of photosynthetic capacity, through acclimation, would lead to further decreased g$_s$ (198). These studies emphasize the need for an improved mechanistic understanding of stomatal response to atmospheric change.

Whereas the effects of CO$_2$ on these separate physiological processes occur via independent mechanisms, there are interactions among all three of them. Acclimation of photosynthesis reduces tissue [N], which may reduce the demand for energy generated by respiration. Reduction of g$_s$ improves water balance, which delays the onset of midday water stress and extends the period of most active photosynthesis; reduced ET increases soil water content and leads to increased N mineralization.

There are problems in moving across scales in the interpretation of processes on a global scale based upon effects at the molecular level. Yet the reduction of stomatal conductance, the improvement in the efficiency of photosynthesis, and the inhibition of the activity of respiratory enzymes are primary mechanisms by which terrestrial ecosystems will respond to rising atmospheric carbon dioxide.

ACKNOWLEDGMENTS

We acknowledge the work of Melanie Muehe in compiling the data from the literature for the tables and for valuable assistance in the preparation of the figures and text. I (BGD) also thank the students in my lab, Roser Matamala and Courtney Brown, for editorial assistance. We thank our colleagues, Damian Barrett, Paul Dijkstra, Roger Gifford, Bruce Hungate, and Hugo Rogers for generously reading and contributing valuable comments on early versions of the manuscript. This work was supported in part by the Smithsonian Institution and the Department of Energy.

Visit the *Annual Reviews home page* at http://www.annurev.org

Literature Cited

1. Ackerson RC, Havelka UD, Boyle MG. 1984. CO_2-enrichment effects on soybean physiology. II. Effects of stage-specific CO_2 exposure. *Crop Sci.* 24:1150–54
2. Akita S, Moss DN. 1972. Differential stomatal response between C_3 and C_4 species to atmospheric CO_2 concentration and light. *Crop Sci.* 12:789–93
3. Akita S, Tanaka I. 1973. Studies on the mechanism of differences in photosynthesis among species. *Proc. Crop Sci. Soc. Jpn.* 42(3):288–95
4. Allen LH, Baker JT, Boote KJ, Rowland-Bamford AJ, Jones JW, et al. 1988. *Effects of air temperature and atmospheric CO_2-plant growth relationships.* Washington, DC: US Dep. Energy
5. Allen LH, Bisbal EC, Campbell WJ, Boote KJ. 1990. Soybean leaf gas exchange responses to CO_2 enrichment. *Soil Crop Sci. Soc. Fla. Proc.* 49:124–31
6. Allen LH, Boote KJ, Jones PH, Rowland-Bamford AJ, Bowes G, et al. 1990. *Temperature and CO_2 Effects on Rice: 1988.* Washington, DC: US Dep. Energy, Off. Energy Res., Carbon Dioxide Res. Div.
7. Allen LH, Vu JCV, Valle RR, Boote KJ, Jones PH. 1988. Nonstructural carbohydrates and nitrogen of soybean grown under carbon dioxide enrichment. *Crop Sci.* 28:84–94
8. Amthor JS. 1997. Plant respiratory responses to elevated CO_2 partial pressure. In *Advances in CO_2 Effects Research,* ed. LH Allen, MH Kirkham, DM Olszyk, CE Whitman, LH Allen, et al. Madison, WI: Am. Soc. Agron.
9. Amthor JS, Koch GW, Boom AJ. 1992. CO_2 inhibits respiration in leaves of *Rumex crispus* L. *Plant Physiol.* 98:1–4
10. Andrews JT, Lorimer GH. 1987. Rubisco: structure, mechanisms, and prospects for improvement. In *The Biochemistry of Plants: A Comprehensive Treatise,* Vol. 10, *Photosynthesis,* ed. PK Stumpf, EF Conn, pp. 131–218. San Diego: Academic
11. Aranda X, Gonzàlez-Meler MA, Azcón-Bieto J. 1995. Cytochrome oxidase activity and oxygen uptake in photosynthetic organs of *Triticum aestivion* and *Scirpus olneyi* plants grown at ambient and doubled CO_2. *Plant Physiol.* 108:62 (Abstr. No. 262).
12. Arnone A III, Gordon JC. 1990. Effect of nodulation, nitrogen fixation and carbon dioxide enrichment on the physiology, growth and dry mass allocation of seedlings of *Alnus rubra* Bong. *New Phytol.* 116:55–66
13. Arp WJ. 1991. *Vegetation of a North American Salt Marsh and Elevated Atmospheric Carbon Dioxide.* Amsterdam: Vrije Univ.
14. Arp WJ. 1991. Effects of source sink relations on photosynthetic acclimation to elevated carbon dioxide. *Plant Cell Environ.* 14(8):869–76
15. Arp WJ, Drake BG. 1991. Increased photosynthetic capacity of *Scirpus olneyi* after 4 years of exposure to elevated CO_2. *Plant Cell Environ.* 14(9):1003–6
16. Azcón-Bieto J. 1992. Relationships between photosynthesis and respiration in the dark in plants. In *Trends in Photosynthesis Research,* ed. J Barber, J Barber, MG Guerrero, H Medranos, pp. 241–53. Andover, UK: Intercept
17. Azcón-Bieto J, Gonzàlez-Meler MA, Doherty W, Drake BG. 1994. Acclimation of respiratory O_2 uptake in green tissues of field grown native species after long-term exposure to elevated atmospheric CO_2. *Plant Physiol.* 106(3):1163–68
18. Bainbridge G, Madgwick P, Parmar S, Mitchell R, Paul M, et al. 1995. Engineering rubisco to change its catalytic properties. *J. Exp. Bot.* 46:1269–76
19. Barrett DJ, Gifford RM. 1995. Photosynthetic acclimation to elevated CO_2 in relation to biomass allocation in cotton. *J. Biogeogr.* 22:331–39
20. Barrett DJ, Gifford RM. 1995. Acclimation of photosynthesis and growth by cotton to elevated CO_2 interactions with severe phosphate deficiency and restricted rooting volume. *Aust. J. Plant Physiol.* 22:955–63
21. Bazzaz FA. 1990. The response of natural ecosystems to the rising global CO_2 levels. *Annu. Rev. Ecol. Syst.* 21:167–96
22. Beerling DJ, Chaloner WG. 1993. Evolutionary responses of stomatal density to global carbon dioxide change. *Biol. J. Linn. Soc.* 48(4):343–53
23. Beerling DJ, Chaloner WG, Huntley B, Pearson JA, Tooley MJ. 1993. Stomatal density responds to the glacial cycle of environmental change. *Proc. R. Soc. London Ser. B* 251:133–38
24. Betsche T, Morin F, Cote F, Gaugain F, Andre M. 1990. Gas exchanges, chlorophyll *a* fluorescence, and metabolite levels in leaves of *Trifolium subterraneum* during long-term exposure to elevated CO_2. In *Current Research in Photosynthesis,* ed. M

Baltcheffsky, pp. 409–12. Dordrecht: Kluwer

25. Bouma TJ, De Visser R, Janssen JHJA, De Kock MJ, Van Leeuwen PH, Lambers H. 1994. Respiratory energy requirements and rate of protein turnover in vivo determined by the use of an inhibitor of protein synthesis and a probe to assess its effect. *Physiol. Plant.* 92:585–94

26. Bowes G. 1993. Facing the inevitable: plants and increasing atmospheric CO_2. *Annu. Rev. Plant Physiol. Plant Mol. Biol.* 44:309–32

27. Brooks A, Farquhar GD. 1985. Effect of temperature on the CO_2/O_2 specificity of ribulose-1,5-bisphosphate carboxylase/oxygenase and the rate of respiration in the light. *Planta* 165:397–406

28. Bunce JA. 1990. Short-and-long term inhibition of respiratory carbon dioxide efflux by elevated carbon dioxide. *Ann. Bot.* 65(6):637–42

29. Bunce JA. 1993. Effects of doubled atmospheric carbon dioxide concentration on the responses of assimilation. *Plant Cell Environ.* 16(2):189–97

30. Bunce JA. 1994. Responses of respiration to increasing atmospheric carbon dioxide concentrations. *Physiol. Plant.* 90(2): 427–30

31. Bunce JA. 1995. Effects of elevated carbon dioxide concentration in the dark on the growth of soybean. *Ann. Bot.* 75(4):365–68

32. Bunce JA. 1995. The effect of carbon dioxide concentration on respiration of growing and mature soybean. *Plant Cell Environ.* 18(5):575–81

33. Byrd GT. 1992. *Dark Respiration in C₃ and C₄ Species*. Athens: Univ. Ga.

34. Campbell WJ, Allen LH, Bowes G. 1988. Effects of CO_2 concentration on rubisco activity, amount, and photosynthesis in soybean leaves. *Plant Physiol.* 88:1310–16

35. Carlson RW, Bazzaz FA. 1980. The effects of elevated CO_2 concentrations on growth, photosynthesis, transpiration, and water use efficiency of plants. In *Environmental and Climatic Impact of Coal Utilization*, ed. JJ Singh, A Deepak, pp. 609–23. New York: Academic

36. Cave G, Tolley LC, Strain BR. 1981. Effect of carbon dioxide enrichment on chlorophyll content, starch content and starch grain structure in *Trifolium subterraneum* leaves. *Physiol. Plant.* 51:171–74

37. Chen JJ, Sung JM. 1990. Crop physiology and metabolism. Gas exchange rate and yield responses of Virginia-type peanut to carbon dioxide enrichment. *Crop Sci.* 30: 1085–89

38. Chen XM, Begonia GB, Hesketh JD. 1995. Soybean stomatal acclimation to long term exposure to CO_2 enriched atmospheres. *Photosynthetica* 31(1):51–57

39. Ciais P, Tans PP, Trolier M, White JWC, Francey RJ. 1995. A large northern hemisphere terrestrial CO_2 sink indicated by the C^{13}/C^{12} ratio of atmospheric CO_2. *Science* 269: 1098–102

40. Coleman JS, Bazzaz FA. 1992. Effects of carbon dioxide and temperature on growth and resource use of co-occurring C_3 and C_4 annuals. *Ecology* 73(4):1244–59

41. Combe L, Kobilinsky A. 1985. Effet de la fumure carbonée sur la photosynthèse de Radis (*Raphanus sativus*) en terre en hiver. *Photosynthetica* 19(4):550–60

42. Conroy J, Hocking P. 1993. Nitrogen nutrition of C_3 plants at elevated atmospheric CO_2 concentrations. *Physiol. Plant.* 89(3): 570–76

43. Conroy JP, Kuppers M, Virgona J, Barlow EWR. 1988. The influence of CO_2 enrichment, phosphorus deficiency and water stress on the growth, conductance and water use of *Pinus radiata* D. Don. *Plant Cell Environ.* 11:91–98

44. Conroy JP, Milham PJ, Barlow EWR. 1992. Effect of nitrogen and phosphorus availability on the growth response of *Eucalyptus grandis* to high CO_2. *Plant Cell Environ.* 15(7):843–47

45. Cornic G, Jarvis JG. 1972. Effects of oxygen on CO_2 exchange and stomatal resistance in Sitka spruce and maize at low irradiances. *Photosynthetica* 6:225–39

46. Curtis PS. 1996. A meta-analysis of leaf gas exchange and nitrogen in trees grown under elevated CO_2 in situ. *Plant Cell Environ.* 19:127–37

47. Curtis PS, Drake BG, Whigham DF. 1989. Nitrogen and carbon dynamics in C_3 and C_4 estuarine marsh plants grown under elevated CO_2 in situ. *Oecologia* 78:297–301

48. Curtis PS, Teeri JA. 1992. Seasonal responses of leaf gas exchange to elevated carbon dioxide in *Populus grandidentata*. *Can. J. For. Res.* 22(9):1320–25

49. Curtis PS, Vogel CS, Pregitzer KS, Zak DR, Teeri JA. 1995. Interacting effects of soil fertility and atmospheric CO_2 on leaf area growth and carbon gain physiology in *Populus x euramericana* (Dode) Guinier. *New Phytol.* 129(2):253–63

50. Davis TD, Potter JR. 1989. Relations between carbohydrate, water status and adventitious root formation in leafy pea cuttings rooted under various levels of atmospheric CO_2 and relative humidity. *Physiol. Plant.* 77:185–90

51. Decker JP, Wien JD. 1958. Carbon dioxide surges in green leaves. *J. Sol. Energy Sci. Eng.* 2(1):39–41

52. Delgado E, Medrano H, Keys AJ, Parry MAJ. 1995. Species variation in rubisco specificity factor. *J. Exp. Bot.* 46(292): 1775–77

53. DeLucia EH, Callaway RM, Schlesinger WH. 1994. Offsetting changes in biomass allocation and photosynthesis in ponderosa pine (*Pinus ponderosa*) in response to climate change. *Tree Physiol.* 14(7–9): 669–77

54. DeLucia EH, Sasek TW, Strain BR. 1985. Photosynthetic inhibition after long-term exposure to elevated levels of atmospheric carbon dioxide. *Photosynth. Res.* 7:175–84

55. Diemer MW. 1994. Mid-season gas exchange of an alpine grassland under elevated CO_2. *Oecologia* 98(3–4):429–35

56. Downton WJS, Bjorkman O, Pike CS. 1980. Consequences of increased atmospheric concentrations of carbon dioxide for growth and photosynthesis of higher plants. In *Carbon Dioxide and Climate*, pp. 143–51. Canberra: Aust. Acad. Sci.

57. Downton WJS, Grant WJR. 1994. Photosynthetic and growth responses of variegated ornamental species to elevated CO_2. *Aust. J. Plant Physiol.* 21(3):273–79

58. Drake BG. 1992. A field study of the effects of elevated CO_2 on ecosystem processes in a Chesapeake Bay wetland. *Aust. J. Bot.* 40:579–95

59. Drake BG, Leadley PW. 1991. Canopy photosynthesis of crops and native plant communities exposed to long term elevated CO_2. *Plant Cell Environ.* 14(8):853–60

60. Drake BG, Muehe M, Peresta G, Gonzàlez-Meler MA, Matamala R. 1997. Acclimation of photosynthesis, respiration and ecosystem carbon flux of a Chesapeake Bay wetland after eight years exposure to elevated CO_2. *Plant Soil.* In press

61. Drake BG, Peresta G, Beugeling E, Matamala R. 1996. Long term elevated CO_2 exposure in a Chesapeake Bay wetland: ecosystem gas exchange, primary production, and tissue nitrogen. See Ref. 120a, pp. 197–214

62. Drake BG, Raschke K. 1974. Prechilling of *Xanthium strumarium* L. reduces net photosynthesis and, independently, stomatal conductance, while sensitizing the stomata to CO_2. *Plant Physiol.* 53:808–12

63. Drake BG, Raschke K, Salisbury FB. 1970. Temperature and transpiration resistances of *Xanthium* leaves as affected by air temperature, humidity, and wind speed. *Plant Physiol.* 46:324–30

64. duCloux HC, André M, Daguenet A, Massimino J. 1987. Wheat responses to CO_2 enrichment: growth and CO_2 exchanges at two plant densities. *J. Exp. Bot.* 38(194): 1421–31

65. Eamus D. 1991. The interaction of rising carbon dioxide and temperatures with water use efficiency. *Plant Cell Environ.* 14(8):843–52

66. Eamus D, Berryman CA, Duff GA. 1993. Assimilation, stomatal conductance, specific leaf area and chlorophyll responses to elevated CO_2 of *Maranthes corymbosa*, a tropical monsoon rain forest species. *Aust. J. Plant Physiol.* 20:741–55

67. Egli DB, Pendleton JW, Peters DB. 1970. Photosynthetic rate of three soybean communities as related to carbon dioxide levels and solar radiation. *Agron. J.* 62:411–14

68. Ehret DL, Jolliffe PA. 1985. Photosynthetic carbon dioxide exchange of bean plants grown at elevated carbon dioxide concentrations. *Can. J. Bot.* 63:2026–30

69. El Kohen A, Pontailler JY, Mousseau M. 1991. Effet d'un doublement du CO_2 atmospherique sur la respiration à l'obscurite des parties aeriennes des jeunes chataigniers (*Castanea sativa* Mill). *C. R. Acad. Sci. Paris III* 312:477–81

70. Farrar JF, Williams ML. 1991. The effects of increased atmospheric carbon dioxide and temperature on carbon partitioning. *Plant Cell Environ.* 14(8):819–30

71. Fetcher N, Jaeger CH, Strain BR, Sionit N. 1988. Long-term elevation of atmospheric CO_2 concentration and the carbon exchange rates of saplings of *Pinus taeda* L. and *Liquidambar styraciflua* L. *Tree Physiol.* 33:317–45

72. Field CB, Chapin FS III, Matson PA, Mooney HA. 1992. Responses of terrestrial ecosystems to the changing atmosphere: a resource-based approach. *Annu. Rev. Ecol. Syst.* 23:201–35

73. Field CB, Chapin FS III, Chiariello NR, Holland EA, Mooney HA. 1996. The Jasper Ridge CO_2 experiment: design and motivation. See Ref. 120a, pp. 121–45

74. Field CB, Jackson RB, Mooney HA. 1995. Stomatal responses to increased CO_2: implications from the plant to the global scale. *Plant Cell Environ.* 18:1214–25

75. Forrester ML, Krotkov K, Nelson CD. 1966. Effect of oxygen on photosynthesis, photorespiration and respiration in detached leaves. I. Soybean. *Plant Physiol.* 41:422–27

76. Fredeen AL, Field CB. 1995. Contrasting

leaf and 'ecosystem' CO$_2$ and H$_2$O exchange in *Avena fatua* monoculture. *Photosynth. Res.* 43(3):263–71

77. Frydych J. 1976. Photosynthetic characteristics of cucumber seedlings grown under two levels of carbon dioxide. *Photosynthetica* 10:335–38

78. Gale J. 1982. Evidence for essential maintenance respiration of leaves of *Xanthium strumarium* at high temperature. *J. Exp. Bot.* 33:471–76

79. Gay AP, Hauck B. 1994. Acclimation of *Lolium temulentum* to enhanced carbon dioxide concentration. *J. Exp. Bot.* 45(277): 1133–41

80. Gifford RM. 1980. Carbon storage by the biosphere. In *Carbon Dioxide and Climate: Australian Research*, ed. GI Pearman, pp. 167–81. Canberra: Aust. Acad. Sci.

81. Gifford RM. 1994. The global carbon cycle: a view point on the missing sink. *Aust. J. Plant Physiol.* 21:1–15

82. Gifford RM, Lambers H, Morison JIL. 1985. Respiration of crop species under CO$_2$ enrichment. *Physiol. Plant.* 63: 351–56

83. Gifford RM, Morison JIL. 1993. Crop responses to the global increase in atmospheric carbon dioxide concentration. In *International Crop Science*, pp. 325–31. Madison, WI: Crop Sci. Soc. Am.

84. Gonzàlez-Meler M. 1995. *Effects of increasing atmospheric concentration of carbon dioxide on plant respiration.* Barcelona: Univ. Barc.

85. Gonzàlez-Meler MA, Ribas-Carbó M, Siedow JN, Drake BG. 1997. The direct inhibition of plant mitochondrial respiration by elevated CO$_2$. *Plant Physiol.* 112:1349–55

86. Govindjee. 1993. Bicarbonate-reversible inhibition of plastoquinone reductase in photosystem-II. *Z. Nat.forsch. Teil C* 48(3–4):251–58

87. Griffin KL, Ball TJ, Strain BR. 1996. Direct and indirect effect of elevated CO$_2$ on whole-shoot respiration in ponderosa pine seedlings. *Tree Physiol.* 16:33–41

88. Grulke NE, Hom JL, Roberts SW. 1993. Physiological adjustment of two full sib families of ponderosa pine to elevated carbon dioxide. *Tree Physiol.* 12(4):391–401

89. Gunderson CA, Norby RJ, Wullschleger SD. 1993. Foliar gas exchange responses of two deciduous hardwoods during 3 years of growth at elevated CO$_2$: no loss of photosynthetic enhancement. *Plant Cell Environ.* 16(7):797–807

90. Gunderson CA, Wullschleger SD. 1994. Photosynthetic acclimation in trees to rising atmospheric CO$_2$: a broader perspective. *Photosynth. Res.* 39(3):369–88

91. Habash DZ, Paul MJ, Parry MAJ, Keys AJ, Lawlor DW. 1995. Increased capacity for photosynthesis in wheat grown at elevated CO$_2$: the relationship between electron-transport and carbon metabolism. *Planta* 197(3):482–89

92. Ham JM, Owensby CE, Coyne PI, Bremer DJ. 1995. Fluxes of CO$_2$ and water vapor from a prairie ecosystem exposed to ambient and elevated CO$_2$. *Agric. For. Meteorol.* 77:73–93

93. Havelka UD, Ackerson RC, Boyle MG, Wittenbach VA. 1984. CO$_2$-enrichment effects on soybean physiology. I. Effects of long-term CO$_2$ exposure. *Crop Sci.* 24: 157–69

94. Havelka UD, Wittenbach VA, Boyle MG. 1984. CO$_2$ enrichment effects on wheat yield and physiology. *Crop Sci.* 24: 1163–68

95. Heath OVS. 1948. Control of stomatal movement by a reduction in the normal carbon dioxide content of the air. *Nature* 161:179–81

96. Hedrich R, Marten I. 1993. Malate induced feedback regulation of plasma membrane anion channels could provide a CO$_2$ sensor to guard cells. *EMBO J.* 12(3):897–901

97. Hicklenton PR, Joliffe PA. 1980. Alterations in the physiology of CO$_2$ exchange in tomato plants grown in CO$_2$ enriched atmospheres. *Can. J. Bot.* 58:2181–89

98. Ho LC. 1977. Effects of CO$_2$ enrichment on the rates of photosynthesis and translocation of tomato leaves. *Ann. Appl. Biol.* 87:191–200

99. Hocking PJ, Meyer CP. 1991. Carbon dioxide enrichment decreases critical nitrate and nitrogen concentrations in wheat. *J. Plant Nutr.* 14(6):571–84

100. Hocking PJ, Meyer CP. 1991. Effects of CO$_2$ enrichment and nitrogen stress on growth, and partitioning of dry matter and nitrogen in wheat and maize. *Aust. J. Plant Physiol.* 18:339–56

101. Hollinger DY. 1987. Gas exchange and dry matter allocation responses to elevation of atmospheric CO$_2$ concentration in seedlings of three tree species. *Tree Physiol.* 3:193–202

102. Holmgren P, Jarvis PG. 1967. Carbon dioxide efflux from leaves in light and darkness. *Physiol. Plant.* 20:1045–51

103. Hrubec TC, Robinson JM, Donaldson RP. 1985. Effects of CO$_2$ enrichment and carbohydrate content on the dark respiration of soybeans. *Plant Physiol.* 79:684–89

104. Huber SC, Rogers HH, Israel DW. 1984.

Effects of CO_2 enrichment on photosynthesis and photosynthate partitioning in soybean leaves. *Physiol. Plant.* 62:95–101

105. Imai K, Murata Y. 1978. Effect of carbon dioxide concentration on growth and dry matter production of crop plants. *Jpn. J. Crop Sci.* 47:587–95

106. Israel AA, Nobel PS. 1994. Activities of carboxylating enzymes in the CAM species *Opuntia ficus-indica* grown under current and elevated CO_2 concentrations. *Photosynth. Res.* 40(3):223–29

107. Jackson RB, Sala OE, Field CB, Mooney HA. 1994. CO_2 alters water use, carbon gain, and yield for the dominant species in a natural grassland. *Oecologia* 98:257–62

108. Jacob J, Greitner C, Drake BG. 1995. Acclimation of photosynthesis in relation to rubisco and nonstructural carbohydrate contents. *Plant Cell Environ.* 18:875–84

109. Jang JC, Sheen J. 1994. Sugar sensing in higher-plants. *Plant Cell* 6(11):1665–79

110. Jarvis PG. 1989. Atmospheric carbon dioxide and forests. *Philos. Trans. R. Soc. London Ser. B* 324:369–92

111. Johnsen KH. 1993. Growth and ecophysiological responses of black spruce seedlings to elevated carbon dioxide. *Can. J. For. Res.* 23(6):1033–42

112. Jones P, Allen JLH, Jones JW, Boote KJ, Campbell WJ. 1984. Soybean canopy growth, photosynthesis, and transpiration responses to whole-season carbon dioxide enrichment. *Agron. J.* 76:633–37

113. Jones P, Allen JLH, Jones JW, Valle R. 1985. Photosynthesis and transpiration responses of soybean canopies to short-and long-term CO_2 treatments. *Agron. J.* 77: 119–26

114. Kaplan A, Gale J, Poljakoff-Mayber A. 1977. Effect of oxygen and carbon dioxide concentrations on gross dark CO_2 fixation and dark respiration in *Bryophyllum daigremontianum*. *Aust. J. Plant Physiol.* 4: 745–52

115. Kaushal P, Guehl JM, Aussenac G. 1989. Differential growth response to atmospheric carbon dioxide enrichment in seedlings of *Cedrus atlantica* and *Pinus nigra* ssp. *Laricia* var. *Corsicana*. *Can. J. Bot.* 19:1351–58

116. Keeling CD, Chine JFS, Whorf TP. 1996. Increased activity of northern vegetation inferred from atmospheric CO_2 measurements. *Nature* 382:146–49

117. Kendall AC, Turner JC, Thomas SM, Keys AJ. 1985. Effects of CO_2 enrichment at different irradiances on growth and yield of wheat. II. Effects on Kleiber spring wheat treated from anthesis in controlled environ-ments in relation to effects on photosynthesis and photorespiration. *J. Exp. Bot.* 36: 261–73

118. Kirschbaum MUF. 1994. The sensitivity of C_3 photosynthesis to increasing CO_2 concentration: a theoretical analysis of its dependence on temperature and background CO_2 concentration. *Plant Cell Environ.* 17(6):747–54

119. Klimov VV, Allakhverdiev SI, Feyziev YM, Baranov SV. 1995. Bicarbonate requirement for the donor side of photosystem II. *FEBS Lett.* 363(3):251–55

120. Knapp AK, Hamerlynck EP, Owensby CE. 1993. Photosynthetic and water relations responses to elevated CO_2 in the C_4 grass *Andropogon*. *Int. J. Plant Sci.* 154(4): 459–66

120a. Koch GW, Mooney HA, eds. 1996. *Carbon Dioxide and Terrestrial Ecosystems.* San Diego: Academic

121. Koch KE. 1996. Carbohydrate-modulated gene expression in plants. *Annu. Rev. Plant Physiol. Mol. Biol.* 47:509–40

122. Krapp A, Hofmann B, Schafer C, Stitt M. 1993. Regulation of the expression of rbcs and other photosynthetic genes by carbohydrates: a mechanism for the sink regulation of photosynthesis. *Plant J.* 3(6):817–28

123. Krapp A, Quick WP, Stitt M. 1991. Ribulose-1,5-bisphosphate carboxylase-oxygenase, other Calvin Cycle enzymes, and chlorophyll decrease when glucose is supplied to mature spinach leaves via the transpiration stream. *Planta* 186(1):58–69

124. Kriedemann PE, Sward RJ, Downton WJS. 1976. Vine response to carbon dioxide enrichent during heat stress. *Aust. J. Plant Physiol.* 3:605–18

125. Kriedemann PE, Wong SC. 1984. Growth response and photosynthetic acclimation to CO_2: comparative behaviour in two C_3 crop species. *Acta Hortic.* 162:113–20

126. Kubiske ME, Pregitzer KS. 1996. Effects of elevated CO_2 and light availability on the photosynthetic light response of trees of contrasting shade tolerance. *Tree Physiol.* 16(3):351–58

127. Lawler IR, Foley WJ, Woodrow IE, Cork SJ. 1997. The effects of elevated CO_2 atmospheres on the nutritional quality of *Eucalyptus* foliage and its interaction with soil nutrient and light availability. *Oecologia.* In press

128. Lenssen GM, Rozema R. 1990. *The Greenhouse Effect and Primary Productivity in European Agroecosystems.* Wageningen: Pudoc

129. Lewis JD, Tissue DT, Strain BR. 1996. Seasonal response of photosynthesis to ele-

vated CO_2 in loblolly pine (*Pinus-Taeda* L) over 2 growing seasons. *Global Change Biol.* 2(2):103–14

130. Lloyd J, Farquhar GD. 1996. The CO_2 dependence of photosynthesis, plant-growth responses to elevated atmospheric CO_2 concentrations and their interaction with soil nutrient status. I. General-principles and forest ecosystems. *Funct. Ecol.* 10(1): 4–32

131. Long SP. 1991. Modification of the response of photosynthetic productivity to rising temperature by atmospheric CO_2 concentrations: Has its importance been underestimated? *Plant Cell Environ.* 14(8): 729–39

132. Long SP, Drake BG. 1991. Effect of the long-term elevation of CO_2 concentration in the field on the quantum yield of photosynthesis of the C_3 sedge, *Scirpus olneyi*. *Plant Physiol.* 96:221–26

133. Long SP, Drake BG. 1992. Photosynthetic CO_2 assimilation and rising atmospheric CO_2 concentrations. In *Topics in Photosynthesis. Crop Photosynthesis: Spatial and Temporal Determinants*, ed. NR Baker, H Thomas, 2:69–107. Amsterdam: Elsevier Sci.

134. Long SP, Farage PK, Nie GY, Osborne CP. 1995. Photosynthesis and rising CO_2 concentration. In *Photosynthesis: From Light to Biosphere*, ed. P Mathis, 5:729–36. Dordrecht: Kluwer

135. Long SP, Hutchin PR. 1991. Primary production in grasslands and coniferous forests with climate change: an overview. *Ecol. Appl.* 1(2):139–56

136. Long SP, Nie GY, Baker NR, Drake BG, Farage PK, et al. 1992. The implications of concurrent increases in temperature, CO_2 and tropospheric O_3 for terrestrial C_3 photosynthesis. *Photosynth. Res.* 34(1):108

137. Long SP, Osborne CP, Humphries SW. 1997. Photosynthesis, rising atmospheric CO_2 concentration and climate change. In *Scope 56: Global Change*, ed. A Bremeyer, DO Hall, J Melillo. Chichester, UK: Wiley

138. Lutze JL. 1996. *Carbon and Nitrogen Relationships in Swards of* Danthonia richardsonii *in Response to Carbon Dioxide Enrichment and Nitrogen Supply.* Canberra: Aust. Natl. Univ.

139. Madsen E. 1968. Effect of CO_2-concentration on accumulation of starch and sugar in tomato leaves. *Physiol. Plant.* 21:168–75

140. Masle J, Hudson GS, Badger MR. 1993. Effects of ambient CO_2 concentration on growth and nitrogen in tobacco (*Nicotiana tabacum*) plants transformed with an antisense gene to the small subunit of ribulose-1,5-bisphosphate carboxylase/oxygenase. *Plant Physiol.* 103(4):1075–88

141. Mauney JR, Guinn G, Fry KE, Hesketh JD. 1979. Correlation of photosynthetic carbon dioxide uptake and carbohydrate accumulation in cotton, soybean, sunflower and sorghum. *Photosynthetica* 13:260–66

142. Mauney JR, Kimball BA, Pinter PJJ, LaMorte RL, Lewin KF, et al. 1994. Growth and yield of cotton in response to a free-air carbon dioxide enrichment (FACE) environment. *Agric. For. Meteorol.* 70: 49–67

143. McKee IF, Woodward FI. 1994. CO_2 enrichment responses of wheat: interactions with temperature, nitrate and phosphate. *New Phytol.* 127(3):447–53

144. McKee IF, Woodward FI. 1994. The effect of growth at elevated CO_2 concentrations on photosynthesis in wheat. *Plant Cell Environ.* 17(7):853–59

145. McMurtrie RE, Wang YP. 1993. Mathematical models of the photosynthetic response of tree stands to rising CO_2 concentrations and temperatures. *Plant Cell Environ.* 16(1):1–13

146. Melillo J, Callaghan TV, Woodward FI, Salati E, Sinha SK. 1990. Effects on ecosystems. In *Climate Change: The Ipcc Scientific Assessment*, ed. JT Houghton, GJ Jenkins, JJ Ephraums, pp. 283–310. Cambridge: Cambridge Univ. Press

147. Mitchell RJ, Runion GB, Prior SA, Rogers HH, Amthor JS, Henning FP. 1995. Effects of nitrogen on *Pinus palustris* foliar respiratory responses to elevated atmospheric CO_2. *J. Exp. Bot.* 46(291):1561–67

148. Moore AL. 1992. Factors affecting the regulation of mitochondrial respiratory activity. In *Molecular, Biochemical and Physiological Aspects of Plant Respiration,* ed. H Lambers, LHW van der Plas, H Lambers, pp. 9–18. The Hague: SPB Acad.

149. Morgan JA, Hung HW, Monz CA, Lecain DR. 1994. Consequences of growth at two carbon dioxide concentrations. *Plant Cell Environ.* 17(9):1023–33

150. Morgan JA, Knight WG, Dudley LM, Hunt HW. 1994. Enhanced root system C-sink activity, water relations and aspects of nutrient acquisition in mycotrophic *Bouteloua gracilis* subjected to CO_2 enrichment. *Plant Soil* 165(1):139–46

151. Morison JIL. 1987. Intercellular CO_2 concentration and stomatal response to CO_2. In *Stomatal Function,* ed. E Zeiger, GD Farquhar, IR Cowan, pp. 229–51. Stanford, CA: Stanford Univ. Press

152. Mott KA. 1990. Sensing of atmospheric CO_2 by plants. *Plant Cell Environ.* 13: 731–37

153. Murray DR. 1995. Plant responses to carbon dioxide. *Am. J. Bot.* 82(5):690–97

154. Musgrave ME, Strain BR, Siedow JN. 1986. Response of two pea hybrids to CO_2 enrichment: a test of the energy overflow hypothesis for alternative respiration. *Proc. Natl. Acad. Sci. USA* 83:8157–61

155. Nafziger ED, Koller HR. 1976. Influence of leaf starch concentration on CO_2 assimilation in soybean. *Plant. Physiol.* 57:560–63

156. Nie GY, Hendrix DL, Long SP, Webber AN. 1995. The effect of elevated CO_2 concentration throughout the growth of a wheat crop in the field on the expression of photosynthetic genes in relation to carbohydrate accumulation. *Plant Physiol.* 108(2): 92 (Suppl.)

157. Nie GY, Long SP, Garcia RL, Kimball BA, Lamorte RL, et al. 1995. Effects of Free-Air CO_2 Enrichment on the development of the photosynthetic apparatus in wheat, as indicated by changes in leaf proteins. *Plant Cell Environ.* 18(8):855–64

158. Nie GY, Long SP, Webber A. 1993. The effect of nitrogen supply on down-regulation of photosynthesis in spring wheat grown in an elevated CO_2 concentration. *Plant Physiol.* 102(1):138 (Suppl.)

159. Nilsen S, Hovland K, Dons C, Sletten SP. 1983. Effect of CO_2 enrichment on photosynthesis, growth and yield of tomato. *Sci. Hortic.* 20:1–14

160. Norby RJ, O'Neill EG. 1991. Leaf area compensation and nutrient interactions in CO_2 enriched seedlings of yellow poplar. *New Phytol.* 117:515–28

161. Norby RJ, Pastor J, Melillo JM. 1986. Carbon-nitrogen interactions in CO_2-enriched white oak: physiological and long-term perspectives. *Tree Physiol.* 2:233–41

162. Oberbauer SG, Strain BR, Fetcher N. 1985. Effect of CO_2-enrichment on seedling physiology and growth of two tropical tree species. *Physiol. Plant.* 65:352–56

163. Oechel WC, Cowles S, Grulke N, Hastings SJ, Lawrence B, et al. 1994. Transient nature of CO_2 fertilization in Arctic tundra. *Nature* 371(6497):500–3

164. O'Neill EG, Luxmoore RJ, Norby RJ. 1987. Elevated atmospheric CO_2 effects on seedling growth, nutrient uptake, and rhizosphere bacterial populations of *Liriodendron tulipifera* L. *Plant Soil* 104:3–11

165. Overdieck D, Reining F. 1986. Effect of atmospheric CO_2 enrichment on perennial ryegrass (*Lolium perenne* L.) and white clover (*Trifolium repens* L.) competing in managed model-ecosystems. I. Phytomass

production. *Acta Oecol./Oecol. Plant* 7: 357–66

166. Palet A, Ribas-Carbo M, Argiles JM, Azcón-Bieto J. 1991. Short-term effects of carbon dioxide on carnation callus cell respiration. *Plant Physiol.* 96:467–72

167. Pearson M, Davies WJ, Mansfield TA. 1995. Asymmetric responses of adaxial and abaxial stomata to elevated CO_2: impacts on the control of gas-exchange by leaves. *Plant Cell Environ.* 18(8):837–43

168. Peet MM. 1984. CO_2 enrichment of soybeans: effects of leaf/pod ratio. *Physiol. Plant.* 60:38–42

169. Peet MM, Huber SC, Patterson DT. 1986. Acclimation to high CO_2 in monoecious cucumbers. II. Carbon exchange rates, enzyme activities, and starch and nutrient concentrations. *Plant Physiol.* 80:63–67

170. Pettersson R, McDonald AJS. 1992. Effects of elevated carbon dioxide concentration on photosynthesis and growth of small birch plants (*Betula pendula* Roth.) at optimal nutrition. *Plant Cell Environ.* 15(8): 911–19

171. Pettersson R, McDonald AJS. 1994. Effects of nitrogen supply on the acclimation of photosynthesis to elevated CO_2. *Photosynth. Res.* 39(3):389–400

172. Pettersson R, McDonald AJS, Stadenberg I. 1993. Response of small birch plants (*Betula pendula* Roth.) to elevated CO_2 and nitrogen. *Plant Cell Environ.* 16(9): 1115–21

173. Pinter PJJ, Kimball BA, Garcia RL, Wall GW, Hunsaker DJ, LaMorte RL. 1996. Free-air CO_2 enrichment: responses of cotton and wheat crops. See Ref. 120a, pp. 215–49

174. Polley HW, Johnson HB, Marino BD, Mayeux HS. 1993. Increase in C_3 plant water-use efficiency and biomass over glacial to present CO_2 concentrations. *Nature* 361:61–64

175. Poole I, Weyers JDB, Lawson T, Raven JA. 1996. Variations in stomatal density and index: implications for palaeoclimatic reconstructions. *Plant Cell Environ.* 19: 705–12

176. Poorter H, Gifford RM, Kriedemann PE, Wong SC. 1992. A quantitative analysis of dark respiration and carbon content as factors in the growth response of plants to elevated CO_2. *Aust. J. Bot.* 40:501–13

177. Poorter H, Pot S, Lambers H. 1988. The effect of an elevated atmospheric CO_2 concentration on growth, photosynthesis and respiration of *Plantago major.* *Physiol. Plant.* 73:553–59

178. Portis AR. 1995. The regulation of Rubisco

by Rubisco activase. *J. Exp. Bot.* 46: 1285–91

179. Qi J, Marshall JD, Mattson KG. 1994. High soil carbon dioxide concentrations inhibit root respiration of Douglas-fir. *New Phytol.* 128:435–42

180. Radin JW, Kimball BA, Hendrix DL, Mauney JR. 1987. Photosynthesis of cotton plants exposed to elevated levels of carbon dioxide in the field. *Photosynth. Res.* 12: 191–203

181. Reekie EG, Bazzaz FA. 1989. Competition and patterns of resource use among seedlings of five tropical trees grown at ambient and elevated CO$_2$. *Oecologia* 79:212–22

182. Reeves DW, Rogers HH, Prior SA, Wood CW, Runion GB. 1994. Elevated atmospheric carbon dioxide effects on sorghum and soybean nutrient status. *J. Plant Nutr.* 17(11):1939–54

183. Reuveni J, Gale J. 1985. The effect of high levels of carbon dioxide on dark respiration and growth of plants. *Plant Cell Environ.* 8:623–28

184. Reuveni J, Gale J, Mayer AM. 1993. Reduction of respiration by high ambient CO$_2$ and the resulting error in measurements of respiration made with O$_2$ electrodes. *Ann. Bot.* 72(2):129–31

184a. Reuveni J, Gale J, Mayer AM. 1995. High ambient carbon-dioxide does not affect respiration by suppressing the alternative, cyanide-resistant respiration. *Ann. Bot.* 76: 291–95

185. Riviererolland H, Contard P, Betsche T. 1996. Adaptation of pea to elevated atmospheric CO2: Rubisco, phosphoenolpyruvate carboxylase and chloroplast phosphate translocator at different levels of nitrogen and phosphorus-nutrition. *Plant Cell Environ.* 19(1):109–17

186. Rogers HH, Sionit N, Cure JD, Smith JM, Bingham GE. 1984. Influence of elevated carbon dioxide on water relations of soybeans. *Plant Physiol.* 74:233–38

187. Rowland-Bamford AJ, Baker JT, Allen LH, Bowes G. 1991. Acclimation of rice to changing atmospheric carbon dioxide concentration. *Plant Cell Environ.* 14:577–83

188. Rozema J, Lenssen GM, Arp WJ, van de Staaij JWM. 1991. Global change, the impact of the greenhouse effect (atmospheric CO$_2$ enrichment) and the increased UV-B radiation on terrestrial plants. In *Ecological Responses to Environmental Stresses*, ed. J Rocema, JAC Verkleij, pp. 220–31. The Netherlands: Kluwer

189. Rozema J, Lenssen GM, Broekman RA, Arp WJ. 1990. Effects of atmospheric carbon dioxide enrichent on salt marsh plants.

In *Expected Effects of Climatic Change on Marine Coastal Ecosystems*, ed. JJ Beukema, WJ Wolff, JJWM Brouns, pp. 49–54. Dordrecht: Kluwer

190. Ryle GJA, Powell CE. 1992. The influence of elevated carbon dioxide and temperature on biomass production of continuously defoliated white clover. *Plant Cell Environ.* 15(5):593–99

191. Ryle GJA, Powell CE, Tewson V. 1992. Effect of elevated CO$_2$ on the photosynthesis, respiration and growth of perennial ryegrass. *J. Exp. Bot.* 43(251):811–18

192. Ryle GJA, Woledge J, Tewson V, Powell CE. 1992. Influence of elevated carbon dioxide and temperature on the photosynthesis and respiration of white clover dependent on N$_2$ fixation. *Ann. Bot.* 70(3): 213–20

193. Sage RF. 1994. Acclimation of photosynthesis to increasing atmospheric CO$_2$: the gas exchange perspective. *Photosynth. Res.* 39:39351–68

194. Sage RF, Harkey TD, Seeman JR. 1989. Acclimation of photosynthesis to elevated carbon dioxide in five C$_3$ species. *Plant Physiol.* 89:590–96

195. Šantrucek J, Sage RF. 1996. Acclimation of stomatal conductance to a CO$_2$-enriched atmosphere and elevated temperature in *Chenopodium album*. *Aust. J. Plant Physiol.* 23:467–78

196. Sasek TW, DeLucia EH, Strain BR. 1985. Reversibility of photosynthetic inhibition in cotton after long-term exposure to elevated CO$_2$ concentrations. *Plant Physiol.* 78:619–22

197. Schimel D. 1990. Biogeochemical feedbacks in the Earth system. In *Global Warming. The Greenpeace Report*, ed. J Leggett, pp. 68–82. Oxford: Oxford Univ. Press

198. Sellers PJ, Bounoua L, Collatz GJ, Randall DA, Dazlich DA, et al. 1996. Comparison of radiative and physiological-effects of doubled atmospheric CO$_2$ on climate. *Science* 271(5254):1402–6

199. Sheen J. 1994. Feedback-control of gene-expression. *Photosynth. Res.* 39:427–38

200. Sicher RC, Kremer DF. 1994. Responses of *Nicotiana tabacum* to CO$_2$ enrichment at low photon flux density. *Physiol. Plant.* 92(3):383–88

201. Silsbury JH, Stephens R. 1984. Growth efficiency of *Trifolium subterraneum* at high [CO$_2$]. In *Advances in Photosynthesis Research*, ed. C Sybesma, 4: 133–36. The Hague: Junk/Nijhoff

202. Socias FX, Medrano H, Sharkey TD. 1993. Feedback limitation of photosynthesis of *Phaseolus vulgaris* L. grown in elevated

carbon dioxide. *Plant Cell Environ.* 16(1): 81–86

203. Spencer W, Bowes G. 1986. Photosynthesis and growth of water hyacinth under CO_2 enrichment. *Plant Physiol.* 82:528–33

204. St. Omer L, Horvath SM. 1984. Developmental changes in anatomy, morphology and biochemistry of *Layia platyglossa* exposed to elevated carbon dioxide. *Am. J. Bot.* 72:693–99

205. Stewart JD, Hoddinott J. 1993. Photosynthetic acclimation to elevated atmospheric carbon dioxide and UV irradiation. *Physiol. Plant.* 88(3):493–500

206. Stitt M. 1991. Rising CO_2 levels and their potential significance for carbon flow in photosynthetic cells. *Plant Cell Environ.* 14:741–62

207. Teramura AH, Sullivan JH, Ziska LH. 1990. Interaction of elevated ultraviolet-B radiation and CO_2 on productivity and photosynthetic characteristics in wheat, rice, and soybean. *Plant Physiol.* 94:470–75

208. Teskey RO. 1995. A field study of the effects of elevated CO_2 on carbon assimilation, stomatal conductance and branch growth of *Pinus taeda* trees. *Plant Cell Environ.* 18(5):565–73

209. Thomas RB, Griffin KL. 1994. Direct and indirect effects of atmospheric carbon dioxide enrichment on leaf respiration of *Glycine max* (L.) Merr. *Plant Physiol.* 104: 355–61

210. Thomas RB, Strain BR. 1991. Root restriction as a factor in photosynthetic acclimation of cotton seedlings grown in elevated carbon dioxide. *Plant Physiol.* 96:627–34

211. Thompson GB, Drake BG. 1994. Insects and fungi on a C_3 sedge and a C_4 grass exposed to elevated atmospheric CO_2 concentrations in open-top chambers in the field. *Plant Cell Environ.* 17:1161–67

212. Thorpe N, Milthorpe FL. 1977. Stomatal metabolism: CO_2 fixation and respiration. *Aust. J. Plant Physiol.* 4:611–21

213. Tissue DT, Oechel WC. 1987. Response of *Eriophorum vaginatum* to elevated CO_2 and temperature in the Alaskan tussock tundra. *Ecology* 68:401–10

214. Tissue DT, Thomas RB, Strain BR. 1993. Long term effects of elevated CO_2 and nutrients on photosynthesis and rubisco in loblolly pine seedlings. *Plant Cell Environ.* 16(7):859–65

215. Tissue DT, Thomas RB, Strain BR. 1996. Growth and photosynthesis of loblolly pine (*Pinus taeda*) after exposure to elevated CO_2 for 19 months in the field. *Tree Physiol.* 16(1–2):49–59

216. Tolley LC, Strain BR. 1985. Effects of CO_2 enrichment and water stress on gas exchange of *Liquidambar styraciflua* and *Pinus taeda* seedlings grown under different irradiance levels. *Oecologia* 65:166–72

217. Tschaplinski TJ, Stewart DB, Hanson PJ, Norby RJ. 1995. Interactions between drought and elevated CO_2 on growth and gas exchange of seedlings of three deciduous tree species. *New Phytol.* 129:63–71

218. Tsukihara T, Aoyama H, Yamashita E, Tomizaki T, Yamaguchi K, et al. 1996. The whole structure of the 13-subunit oxidized cytochrome c oxidase at 2.8 Å. *Science* 272:1136–44

219. Valle R, Mishoe JW, Jones JW, Allen JLH. 1985. Photosynthetic responses of "brag" soybean leaves adapted to different CO_2 environments. *Crop Sci.* 25:333–39

220. Van Oosten JJ, Afif D, Dizengremel P. 1992. Long term effects of a CO_2 enriched atmosphere on enzymes of the primary carbon metabolism of spruce trees. *Plant Physiol. Biochem.* 30(5):541–47

221. Van Oosten JJ, Besford RT. 1995. Some relationships between the gas exchange, biochemistry and molecular biology of photosynthesis during leaf development of tomato plants after transfer to different carbon dioxide concentrations. *Plant Cell Environ.* 18(11):1253–66

222. Van Oosten JJ, Wilkins D, Besford RT. 1994. Regulation of the expression of photosynthetic nuclear genes by CO_2 is mimicked by regulation by carbohydrates: a mechanism for the acclimation of photosynthesis to high CO_2. *Plant Cell Environ.* 17(8):913–23

223. Villar R, Held AA, Merino J. 1994. Comparison of methods to estimate dark respiration in the light in leaves of two woody species. *Plant Physiol.* 105:167–72

224. Vogel CS, Curtis PS. 1995. Leaf gas exchange and nitrogen dynamics of N_2-fixing, field-grown *Alnus glutinosa* under elevated atmospheric CO_2. *Global Change Biol.* 1:55–61

225. von Caemmerer S, Evans JR, Hudson GS, Andrews TJ. 1994. The kinetics of Ribulose-1,5-bisphosphate carboxylase/oxygenase in vivo inferred from measurements of photosynthesis in leaves of transgenic tobacco. *Planta* 195(1):88–97

226. von Caemmerer S, Farquhar IR. 1984. Effects of partial defoliation, changes in irradiance during growth, short-term water stress and growth at enhanced $p(CO_2)$ on photosynthetic capacity of leaves of *Phaseolus vulgaris*. *Planta* 160:320–29

227. Vonschaewen A, Stitt M, Schmidt R, Sonnewald U, Willmitzer L. 1990. Expression

of a yeast-derived invertase in the cell-wall of tobacco and *Arabidopsis* plants leads to accumulation of carbohydrate and inhibition of photosynthesis and strongly influences growth and phenotype of transgenic tobacco plants. *EMBO J.* 9(10):3033–44

228. Vu CV, Allen LH, Bowes G. 1983. Effects of light and elevated atmospheric CO$_2$ on the ribulose bisphosphate carboxylase activity and ribulose bisphosphate level of soybean leaves. *Plant Physiol.* 73:729–34

229. Vu JCV, Allen LH, Bowes G. 1989. Leaf ultrastructure, carbohydrates and protein of soybeans grown under CO$_2$ enrichment. *Environ. Exp. Bot.* 29:141–47

230. Webber AN, Nie GY, Long SP. 1994. Acclimation of photosynthetic proteins to rising atmospheric CO$_2$. *Photosynth. Res.* 39(3):413–25

231. Williams ML, Jones DG, Baxter R, Farrar JF. 1992. The effect of enhanced concentrations of atmospheric CO$_2$ on leaf respiration. In *Molecular, Biochemical and Physiological Aspects of Plant Respiration*, ed. H Lambers, LHW van der Plas, pp. 547–51. The Hague: SPB Academic

232. Williams WE, Garbutt K, Bazzaz FA, Vitousek PM. 1986. The response of plants to elevated CO$_2$. IV. Two deciduous-forest tree communities. *Oecologia* 69:454–59

233. Wilsey BJ, McNaughton SJ, Coleman JS. 1994. Will increases in atmospheric CO$_2$ affect regrowth following grazing in C$_4$ grasses from tropical grasslands? A test with *Sporobolus kentrophyllus*. *Oecologia* 99(1–2):141–44

234. Wong SC. 1979. Elevated atmospheric partial pressure of CO$_2$ and plant growth. I. Interactions of nitrogen nutrition and photosynthetic capacity in C$_3$ and C$_4$ plants. *Oecologia* 44:68–74

235. Wong SC. 1980. Effects of elevated partial pressure of CO$_2$ on rate of CO$_2$ assimilation and water use efficiency in plants. In *Carbon Dioxide and Climate: Australian Research*, ed. GI Pearman, pp. 159–66. Canberra: Aust. Acad. Sci.

236. Wong SC. 1990. Elevated atmospheric partial pressure of CO$_2$ and plant growth. *Photosynth. Res.* 23:171–80

237. Woodrow IE. 1994. Control of steady-state photosynthesis in sunflowers growing in enhanced CO$_2$. *Plant Cell Environ.* 17:277–86

238. Woodrow IE. 1994. Optimal acclimation of the C$_3$ photosynthetic system under enhanced CO$_2$. *Photosynth. Res.* 39:401–12

239. Woodward FI. 1987. Stomatal numbers are sensitive to increases in CO$_2$ from preindustrial levels. *Nature* 327:617–18

240. Wulff RD, Strain BR. 1982. Effects of CO$_2$ enrichment on growth and photosynthesis of *Desmodium paniculatum*. *Can. J. Bot.* 60:1084–89

241. Wullschleger SD, Norby RJ, Hendrix DL. 1992. Carbon exchange rates, chlorophyll content, and carbohydrate status of two forest tree species exposed to carbon dioxide enrichment. *Tree Physiol.* 10:21–31

242. Wullschleger SD, Ziska LH, Bunce JA. 1994. Respiratory responses of higher plants to atmospheric CO$_2$ enrichment. *Physiol. Plant.* 90(1):221–29

243. Xu DQ, Gifford RM, Chow WS. 1994. Photosynthetic acclimation in pea and soybean to high atmospheric CO$_2$ partial pressure. *Plant Physiol.* 106(2):661–71

244. Yelle S, Beeson RCJ, Trudel MJ, Gosselin A. 1989. Acclimation of two tomato species to high atmospheric CO$_2$. I. Sugar and starch concentrations. *Plant Physiol.* 90:1465–72

245. Zelitch I. 1973. Plant productivity and the control of photorespiration. *Proc. Natl. Acad. Sci. USA* 70:579–84

246. Ziska LH, Bunce JA. 1994. Direct and indirect inhibition of single leaf respiration by elevated CO$_2$ concentrations: interaction with temperature. *Physiol. Plant.* 90(1):130–38

247. Ziska LH, Drake BG, Chamberlain S. 1990. Long term photosynthetic response in single leaves of a C$_3$ and C$_4$ salt marsh species grown at elevated atmospheric CO$_2$ in situ. *Oecologia* 83:469–72

248. Ziska LH, Hogan KP, Smith AP, Drake BG. 1991. Growth and photosynthetic response of nine tropical species with long term exposure to elevated carbon dioxide. *Oecologia* 86:383–89

249. Ziska LH, Sicher RC, Kremer DF. 1995. Reversibility of photosynthetic acclimation of swiss chard and sugarbeet grown at elevated concentrations of CO$_2$. *Physiol. Plant.* 95:355–64

Annu. Rev. Plant Physiol. Plant Mol. Biol. 1997. 48:641–71
Copyright © 1997 by Annual Reviews Inc. All rights reserved

STRUCTURE AND MEMBRANE ORGANIZATION OF PHOTOSYSTEM II IN GREEN PLANTS

Ben Hankamer and James Barber

Wolfson Laboratories, Department of Biochemistry, Imperial College of Science, Technology and Medicine, London SW7 2AY, United Kingdom

Egbert J. Boekema

Biophysical Chemistry, Groningen Biomolecular Sciences and Biotechnology Institute, University of Groningen, Nijenborgh 4, NL-9747 AG Groningen, The Netherlands

KEY WORDS: light harvesting, photosynthesis, photosystem II, structure, thylakoid membrane

ABSTRACT

Photosystem II (PSII) is the pigment protein complex embedded in the thylakoid membrane of higher plants, algae, and cyanobacteria that uses solar energy to drive the photosynthetic water-splitting reaction. This chapter reviews the primary, secondary, tertiary, and quaternary structures of PSII as well as the function of its constituent subunits. The understanding of in vivo organization of PSII is based in part on freeze-etched and freeze-fracture images of thylakoid membranes. These images show a resolution of about 40–50 Å and so provide information mainly on the localization, heterogeneity, dimensions, and shapes of membrane-embedded PSII complexes. Higher resolution of about 15–40 Å has been obtained from single particle images of isolated PSII complexes of defined and differing subunit composition and from electron crystallography of 2-D crystals. Observations are discussed in terms of the oligomeric state and subunit organization of PSII and its antenna components.

CONTENTS

1040-2519/97/0601-0641$08.00

INTRODUCTION[1]

Photosystem II (PSII) is a multisubunit complex embedded in the thylakoid membranes of higher plants, algae, and cyanobacteria. It uses light energy to catalyze a series of electron transfer reactions resulting in the splitting of water into molecular oxygen, protons, and electrons. These reactions occur on an enormous scale and are responsible for the production of atmospheric oxygen and indirectly for almost all the biomass on the planet. Despite its importance, the catalytic properties of PSII have never been reproduced artificially. Understanding PSII's unique chemistry is important and could have implications for agriculture as PSII is a main site of damage during environmental stress.

The chapter reviews our current knowledge of the three-dimensional structure of PSII in higher plants, an area of research that has developed rapidly over recent years. We first outline the photochemical reactions in this photosystem and summarize in a table the main structural features of individual subunits, focusing on their likely transmembrane helical content and cofactor binding properties. Electron microscopy of PSII is then reviewed to relate the subunit and cofactor composition of PSII to its three-dimensional structure. Low-resolution structural data on PSII, obtained from freeze-etch and freeze-fracture studies of thylakoid membranes, is reviewed initially. Such studies have provided information on the location, heterogeneity, and overall size and shape of PSII and its antenna system in the thylakoid membrane at resolutions of 40–50 Å. To obtain higher-resolution information (~15–40 Å), two other approaches have been used: single particle image averaging of detergent-solubilized PSII complexes and analysis of two-dimensional crystals. The former

---

[1] Abbreviations: β-Car, β-carotene; BChl, bacteriochlorophyll; Chl, chlorophyll; CD, circular dichroism; CP, chlorophyll protein; cyt b559, cytochrome b559; LHCII, light-harvesting complex of PSII; LH2, light-harvesting complex of purple photosynthetic bacteria; PSI, photosystem one; PSII, photosystem two; P680, primary electron donor of PSII; Pheo, pheophytin; PQ, plastoquinone; Q_A and Q_B, primary and secondary electron acceptors of PSII; RC, reaction center; STEM, scanning transmission electron microscopy

has yielded considerable information on the oligomeric state and subunit organization of PSII and its antenna system, whereas the latter offers the potential of a structure at atomic resolution. Results obtained from both approaches are discussed in terms of the question of whether PSII exists as a monomer or dimer in vivo. Finally, the conclusions emerging from these studies are compared with biochemical and cross-linking data.

THE PHOTOSYSTEM II SUBUNITS: STRUCTURE AND FUNCTION

Well over 20 subunits (PsbA–PsbX, Lhcb1–6) are associated with PSII of higher plants and green algae (Figure 1) and have been named after the genes

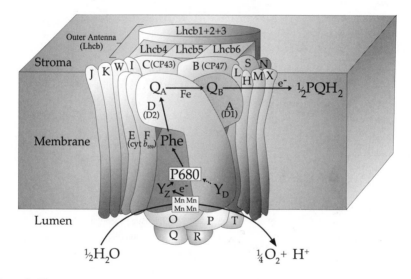

Figure 1   Photosystem II: an overview of subunit composition and electron transport. The PSII complex and its antenna system consists of more than 20 subunits that are either embedded in the thylakoid membrane or associated with its lumenal surface. Light energy is trapped predominantly in the outer antenna, consisting of the proteins Lhcb1–6. The excitation energy is transferred to the photochemically active reaction center (D1 and D2 proteins) via CP47 and CP43, where it is used to drive the water-splitting reaction. The electrons extracted from water are passed from the lumenally located four-atom Mn cluster to D1-Tyr 161 (Y_Z), P680$^+$, Phe, Q_A and on to Q_B, via a nonheme iron group. This electron transport pathway is marked with arrows. The protons and molecular oxygen produced during the water-splitting reaction are released into the lumen. The plastoquinone (PQ) bound at the Q_B site accepts two electrons derived from water via the electron transport chain and two protons from the stroma before being released into the thylakoid membrane in the form of PQH2. The letter notation used for the subunits of the core complex reflects gene origin (e.g. A = product of *psbA* gene).

Table 1 The PSII subunits: primary and secondary structure, cofactor content, and function[a]

Gene	Subunit	Mass (kDa)	No. of trans-membrane α-helices	Function
psbA	D1	38.021(S)	5	YZ & binds P680, Pheo, Q_B
psbD	D2	39.418(S)	5	YD & binds P680, Q_A
psbE	α-cyt b559	9.255(S)	1	Binds heme, photoprotection
psbF	β-cyt b559	4.409(S)	1	Binds heme, photoprotection
psbI	I protein	4.195(S)	1	?
psbB	CP47	56.278(S)	6	Excitation energy transfer, binds 33 kDa
psbC	CP43	50.066(S)	6	Excitation energy transfer, binds 33 kDa
psbH	H protein	7.697(S)	1	Photoprotection
psbK	K protein	4.283(S)	1	PSII assembly, PSII stability
psbL	L protein	4.366(S)	1	Involved in Q_A function
psbM	M protein	3.755(P)	1	?
psbN	N protein	4.722(T)	1	?
psbO	33-kDa ext. protein	26.539(S)	0	Stabilizes Mn cluster, Ca^{2+} & Cl^- binding
psbP	23-kDa ext. protein	20.210(S)	0	Ca^{2+} and Cl^- binding
psbQ	16-kDa ext. protein	16.523(S)	0	Ca^{2+} and Cl^- binding
psbR	R protein	10.236(S)	0	Donor and acceptor side functions
psbS	S protein	21.705(S)	4	Chl chaperonin/antenna component
psbT	T protein	3.283(S)	0	?
psbV	V protein*	15.121(Sy)	0	Donor side stability
psbW	W protein	5.928(S)	1	?
psbX	X protein	4.225(S)	1	Q_A function
lhcb4	Lhcb4 (CP29)	29	3	Excitation energy transfer & dissipation
lhcb5	Lhcb5 (CP26)	26	3	Excitation energy transfer & dissipation
lhcb6	Lhcb6 (CP24)	24	3	Excitation energy transfer & dissipation
psbJ	J protein	4.116(P)	1	PSII assembly
psbU	U protein*	10(Cy)	?	?
lhcb1	Lhcb1	25	3	Light harvesting
lhcb2	Lhcb2	25	3	Light harvesting
lhcb3	Lhcb3	25	3	Light harvesting

[a]Twenty-three putative PSII proteins are encoded by the psbA–psbX genes, whereas at least six outer antenna components are encoded by Lhcb1–L–6. The subunits encoded by these genes are listed according to the complexes with which they copurify (e.g. reaction centers), and those marked (*) have only been detected in cyanobacteria. The molecular masses of the mature PsbA–PsbX proteins, except PsbU, are calculated from the protein sequences reported in the SWISSPROT data base using the MacBioSpec program (Sciex Corp., Thornhill, Ontario, Canada). The abbreviations given in brackets after the molecular masses denote the organism for which the subunit mass is given, as

Chla	Chlb	β-Car	Pheo	Lut	Neo	Viol	
6	0	2	2	0	0	0	**PSII** reaction center
10–25	0	3	0	?	0	0	
9–25	0	5	0	?	0	0	
—	—	—	—	—	—	—	
—	—	—	—	—	—	—	
—	—	—	—	—	—	—	**Full PSII** core
—	—	—	—	—	—	—	
—	—	—	—	—	—	—	
0	0	0	0	0	0	0	
0	0	0	0	0	0	0	
0	0	0	0	0	0	0	
—	—	—	—	—	—	—	
5?	—	—	—	—	—	—	
—	—	—	—	—	—	—	
—	—	—	—	—	—	—	
—	—	—	—	—	—	—	
—	—	—	—	—	—	—	
9–10	3–4	—	0	1–2	1	1–2	**PSII core &**
7–9	4–5	—	0	2	0.5–1	0.5–1	**minor CAB**
6	7	—	—	2	1	1	proteins
—	—	—	—	—	—	—	
8	6	0	0	2	0.5–1	0.5	**Full PSII &** antenna
8	6	0	0	2	0.5–1	0.5	complex
8	6	—	—	—	—	—	

follows: S, spinach; Sy, *Synechococcus* sp.; P, pea; T, tobacco. For the Lhcb proteins and PsbU, only the apparent molecular masses are provided. The next column lists the number of predicted transmembrane helices of each subunit. The putative functions of each subunit are given in the next column. The cofactors associated with the individual subunits are also listed as follows: Chla, chlorophyll a; Chlb, chlorophyll b; β-Car, β-carotene; Pheo, Pheophytin; Lut, lutein; Neo, neoxanthin; Viol, violoxanthin. The references reporting the information summarized in this table are given in the text.

encoding them (38, 46–49, 58, 65). The structure (22, 28, 33, 53, 65, 69, 79, 89, 97, 105, 111, 123, 129), cofactor organization (16, 22, 25, 38, 46–49, 53, 65, 69, 79, 89, 97, 108, 111, 123, 129), and function (32, 38, 46–49, 53, 62, 65, 69, 89, 97, 111) of these subunits have been studied in detail and are summarized in Table 1.

Figure 1 shows the subunit composition of PSII and its antenna system and indicates the photochemical processes that this photosystem catalyzes. The overall reaction driven by PSII is given in Equation 1, and all the cofactors involved are either bound within the reaction center proteins D1 and D2 or are closely associated with them.

$$2H_2O + 2PQ \xrightarrow[\text{PSII}]{\text{LIGHT}} O_2 + 2PQH_2 \qquad \text{1.}$$

The light energy used to drive the water-plastoquinone (PQ) oxidoreductase reactions is captured predominantly by many molecules of chlorophyll (Chl) a and b [averaging about 250 per reaction center (70)] and carotenoid [β-carotene, lutein, neoxanthin, and violoxanthin (20, 69)] associated with light-harvesting antenna proteins, Lhcb1–6 (18, 21, 26, 46–49, 52, 65, 69, 97). The derived excitation energy is passed from the Lhcb proteins along an excitonically linked network of Chl molecules associated with CP47 (PsbB) and CP43 (PsbC) (18, 21, 22, 26, 65, 99, 100) to the PSII reaction center (RC). The RC consists of the D1 (PsbA) and D2 (PsbD) proteins (17, 91, 126), which are highly conserved between higher plant species and have a significant degree of local homology with the primary sequence of the L and M subunits of purple bacteria (16, 78, 110, 129). Ultimately the excitation energy derived from light is used to convert the primary oxidant, P680, to P680$^+$ (17, 91, 110, 111, 126). P680 is thought to consist of two Chl molecules ligated to the D1 and D2 proteins, though the excitonic coupling is much weaker than in the "special pair" of the purple bacterial reaction center (36, 111, 132). The excited state, P680*, donates a single high-energy unpaired electron to a molecule of pheophytin (Pheo), thereby forming the radical pair, P680$^+$Pheo$^-$ (17, 91, 126, 110, 111) (Figure 1). Each time P680$^+$ is formed, it accepts an electron from a specific amino acid residue (D1-Tyr161) and therefore is reduced to P680 (13, 25, 60, 98, 127). Illumination of PSII allows the P680, P680*, P680$^+$ cycle to be repeated and enables the sequential extraction of electrons from D1-Tyr161 (Figure 1). As D1-Tyr161 donates an electron to P680$^+$, it accepts another from water via a four-atom manganese (Mn) cluster, associated with the lumenal surface of PSII. Joliot et al (66) and Kok et al (68) used short light flashes to induce single turnovers of P680 and showed that the Mn cluster passed through a series of oxidation states referred to as the S_0-S_4 cycle. The electrons, which are accepted by P680$^+$, are passed along the electron transport chain (Figure 1). In this way, Pheo accepts electrons from P680 and passes them on to a plastoquinone molecule (Q_A), tightly bound to the D2 protein (34, 71, 78, 107, 129). Q_A^- passes its electron on to a second plastoquinone molecule, associated with the Q_B site on the D1 protein (35, 59, 71, 78, 79, 96, 103, 105, 113). This electron transfer is aided by the presence of a nonheme iron located between Q_A and Q_B (16, 34, 79, 110). Each plastoquinone associated with the Q_B site can accept two electrons derived from water and two protons from the stroma before being released into the lipid matrix in the form of reduced plastoquinone (PQH_2).

The lumenal surfaces of membrane embedded CP47, CP43, and reaction center proteins of higher plants and green algae are in close contact with a number of extrinsic proteins. These include the products of the *psbO* (33-kDa subunit), *psbP* (23-kDa subunit), and *psbQ* (16-kDa subunits) genes, which together form the oxygen-evolving complex (OEC). The OEC also includes the Mn cluster (9, 25, 44, 89) and the three extrinsic proteins are involved in the optimization of its function in water splitting. The 33-kDa subunit stabilizes the Mn cluster (89) while the 23-kDa subunit allows PSII to evolve oxygen under both Ca^{2+}- (43, 89) and Cl^--limiting conditions (86, 87). This has led to the suggestion that the 23-kDa subunit acts as a concentrator of these ions (89). The 16-kDa polypeptide aids PSII to evolve oxygen efficiently under severely Cl^--limiting (<3 mM) conditions (2).

Except for subunit PsbS, all other subunits associated with PSII (Figure 1, Table 1) have a mass under 10 kDa (38). The exact number of the small subunits associated with PSII in vivo is not known, primarily because some of these proteins stain poorly and/or are difficult to resolve even in high-resolution gels (73). The reader is referred to the detailed review by Erickson & Rochaix (38) and the SWISSPROT data base for further information on the low-molecular-weight subunits of PSII, because their properties and those of the other PSII subunits are only summarized here (Table 1). Table 1 lists the subunits, in a way that reflects the PSII particle type that they are associated with, the smallest being the isolated reaction center and the largest being PSII with its complete antenna. The cofactors bound by each subunit are given, together with subunit function, common name, and the number of α-helices that the subunit is predicted to contain.

We now relate the information on the subunit structure summarized in Table 1 to structural information available on PSII and its antenna system.

ARRANGEMENT OF PSII IN THE THYLAKOID MEMBRANE

Much of our understanding of the in vivo organization of PSII is based on freeze-etched and freeze-fracture images of thylakoid membranes. Such electron microscopy studies provide structural detail at a resolution of about 40–50 Å and have provided information on the localization, heterogeneity, dimensions, and shapes of PSII and its antenna system.

By thin sectioning of fixed chloroplasts, electron microscopy has revealed the overall architecture of the thylakoid membrane of higher plants. The thylakoid membrane consists of two main compartments, the grana and the stroma lamellae. The stroma membranes form unstacked (nonappressed) regions, whereas the grana membranes are mostly present as stacked (appressed) membranes. There is a marked heterogeneity in lateral distribution of the major complexes of this membrane. It is generally accepted that PSII, photosystem one (PSI), and ATP synthase are mainly laterally segregated, with PSI and ATP synthase excluded from the appressed grana membranes and PSII abundantly present in the stacked parts of the thylakoid membrane (10, 15). The total picture, however, is more complex, because there is also heterogeneity among both PSII and PSI in subunit composition and lateral distribution (3).

In freeze-etching studies, thylakoids are flash frozen before the evaporation of surface water under vacuum (typically at −100°C). This process exposes the membrane surface and so allows the visualization of extrinsic components in their near-to-native state. The organization of the membrane embedded parts of PSII and the antenna proteins have been studied using the freeze-fracture technique, which involves the cleavage of the lipid bilayer along its internal hydrophobic plane, before image analysis. The terminology (ESs, PSs, ESu, PSu) used by Staehelin (120, 121) has generally been adopted to describe the endoplasmic (E) and protoplasmic (P) surfaces (S) of stacked (s) and unstacked (u) freeze-etched thylakoid membranes. The corresponding freeze-fracture (F) planes are referred to as EFs, PFs, EFu, and PFu.

The ultrastructure of thylakoid membranes of barley (55, 80), spinach (85, 119, 120, 122), maize (84), lettuce (56), soybean (67), *Alocasia* (7), and pea (11, 101) have all been analyzed by freeze-etch and freeze-fracture electron microscopy and show marked similarities (see 80, 121). The protein complexes detected in these analyses were named according to the surfaces with which they were associated (e.g. ESs or PFu particles). They differed in size and shape, and the constituent components of many were identified by the analysis of mutant membranes. For example, the analysis of PSI-deficient mutant thylakoid membranes showed that this photosystem formed part of the large Pfu particles (117). Localization of complexes in the thylakoid membranes was further facilitated by antibody labeling (95).

Localization of PSII and LHCII

Comparative freeze-etch and freeze-fracture studies of wild-type and PSII-deficient mutants of tobacco showed that the thylakoid membranes of the latter were depleted of ESs and EFs particles (81). From these results, it was concluded that the ESs and EFs particles corresponded to the extrinsic and internal parts of PSII, respectively. Support for this conclusion came from parallel studies of PSII-deficient barley mutants (*xantha*-b12, *viridis*-c12, *viridis* e-64, and *viridis* zd69), which showed that their granal membranes were also greatly depleted of EFs particles (e.g. 115, 121). The antenna proteins were located using Lhcb protein–deficient mutants (e.g. barley mutants *xantha*-l35 and *viridis*-k23 and *chlorina* f2) and by comparing thylakoid membranes from light- and dark-grown plants, the latter being depleted in these antenna proteins (12, 83, 116, 118). These studies showed that membranes depleted of Lhcb proteins lacked the PFs particles found in wild-type membranes. Freeze-etch images of membranes depleted of Lhcb protein also showed the ESs particles to be smaller. Together, these results suggested that the PFs particles contained the antenna proteins and that they were closely associated with PSII (EFs and ESs) particles.

The close association of PSII and the Lhcb proteins was characterized in more detail by the analysis of 2-D crystalline arrays of ESs complexes. Images of such ordered arrays showing a section of their freeze-etch surface (ESs surface) and a part of the protoplasmic fracture face (PFs surface) were presented by Miller (80) and Simpson (116). They showed that the Lhcb proteins (PFs particles in the PFs fracture face) fitted in register into the grooves between the PSII complexes (ESs particles in the freeze-etched ESs surface).

Heterogeneity of PSII In Vivo

The ESs and EFs particles that contain PSII are not evenly distributed within the thylakoid membrane. Freeze-etch and freeze-fracture studies have shown that ESs and EFs particles, respectively, can be organized into 2-D arrays or be more randomly dispersed within the grana. PSII-like particles are also observed in the stroma lamellae. It has been calculated that 85% of PSII particles are located in the granal stacks. They are called PSIIα. The remaining 15% of PSII complexes are located in the stroma and are called PSIIβ (3, 10). PSIIα complexes have approximately twice the antenna size of the PSIIβ complexes (8) and are more efficient at reducing ferricyanide and duroquinone (57).

Because PSIIα complexes are more abundant and active than PSIIβ and because all the available structural data on PSII have been collected from such types of complexes, we focus on them in the sections below.

The PSIIα population in the grana can itself be divided into subpopulations of differing antenna size (4, 5). Interestingly, freeze-etch studies have shown that some ESs particles within the grana form 2-D arrays, whereas others are more randomly dispersed and so form at least two subpopulations (112, 116).

Several publications provide the dimensions of arrayed ESs particles. Wild-type ESs arrays in barley have unit cell dimensions of 175 × 247, 180 × 225, and 180 × 240 Å (82, 115). ESs arrays in spinach had an almost rectangular lattice with spacings of 175 × 204 Å (112), while those in the grana of two fatty acid desaturase mutants of *Arabidopsis thaliana* had unit cell dimensions of 190 × 230 Å and 180 × 230 Å (131). Closer analysis of the arrayed ESs particles showed them to have a tetramer-like appearance on their lumenal (ESs) surface and a height of 82 Å perpendicular to the membrane plane (112, 116). Sequential removal of the 16, 23, and 33 kDa proteins of the OEC resulted in a height reduction to 78, 74, and 61 Å, respectively (112). The removal of all three polypeptides exposed the membrane-embedded portion of the PSII-LHCII supercomplex and revealed it to have a dimer-like structure. This led to the conclusion that the ESs particle is a dimeric PSII-LHCII supercomplex and that the tetramer-like appearance of its lumenal surface reflects the organization of two copies of the OEC. Dimeric PSII-LHCII supercomplexes (Figure 2D) isolated by solubilization in detergent have been biochemically and structurally characterized (21, 52, 93). The structure of the isolated complex can be computationally aligned to form an array with a spacing of 190 × 210 Å and a similar lumenal surface topology to ESs particles (21, 52, 112). It is therefore possible that ESs-type arrays consist of dimeric PSII-LHCII supercomplexes of the type reported by Boekema et al (21).

One argument that could be raised against the proposal that the ESs particles consist of dimeric PSII-LHCII supercomplexes of this type is that, on average, PSII reaction centers in the grana are associated with about 250 Chl molecules (70). Thus a dimeric complex of PSII would be expected to bind about 500 Chl as opposed to the 200 Chl per RC found for the isolated PSII-LHCII supercomplex (21, 93). There is insufficient space within the unit cell of the constructed ESs lattice to accommodate the extra Lhcb subunits that would be required to bind an additional 300 Chl molecules. However, it should be stressed that the value of 250 Chl:RC obtained for granal membranes is an average value and that the randomly organized ESs particles within the grana are spaced sufficiently widely to allow them to be associated with much larger antenna systems (>250 Chl:RC). This hypothesis fits both the finding that the

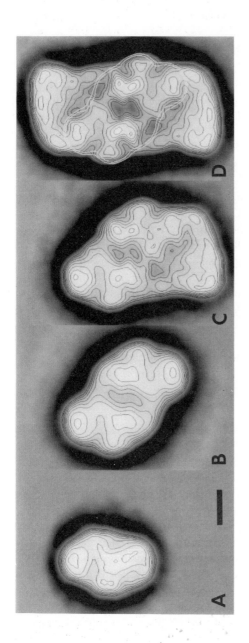

Figure 2   Top view projection maps of spinach PSII complexes obtained from nonperiodic (single particle) averaging of electron microscopy images obtained after negative staining. (*A*) A monomeric PSII core complex (240 kDa) consisting of CP47, CP43, the D1 and D2 proteins, cyt *b559*, the 33-kDa extrinsic protein, and associated low-molecular-weight polypeptides; (*B*) a dimeric PSII core complex (450 kDa) of the same subunit composition as the PSII core monomer; (*C*) a PSII-LHCII supercomplex consisting of a PSII core dimer and one set of Lhcb proteins (Lhcb1, 2, 4, and 5); (*D*) a PSII-LHCII supercomplex consisting of a PSII core dimer and two sets of Lhcb proteins (Lhcb1, 2, 4, and 5). The outline of the PSII core dimer shown in (*B*) is superimposed upon the PSII-LHCII supercomplex (*D*) to show the central location of the PSII core dimer. All the images are similar to those published by Boekema et al (21), but they differ in that they are the averages of larger data sets and so have an improved resolution (approximately 20 Å). Note: a twofold rotational symmetry has been imposed on images *B* and *D*. The scale bar measures 5 nm.

grana contain 250 Chl:RC on average and that the PSIIα population consists of subpopulations differing in antenna size (4, 5).

STRUCTURE DETERMINATION BY ELECTRON MICROSCOPY

X-ray crystallography is a powerful technique for elucidating the structure of proteins and has been particularly successful with photosynthetic membrane proteins. Atomic structures exist for the reaction center (6, 27–29) and the light-harvesting complex (LH2) (77) of photosynthetic purple bacteria. A map of near-atomic resolution also exists for PSI (42). Recently, the structure of the mitochondrial cytochrome b-c complex has been solved at atomic resolution, thus providing structural information relevant to the related photosynthetic cytochrome b_6f complex (137). Despite these striking successes, the X-ray approach has not revealed the structure of PSII. Adir et al (1) and Fotinou et al (40) have grown 3-D crystals of PSII cores, but they were too small and insufficiently ordered for high-resolution analyses.

In the absence of highly ordered 3-D crystals suitable for X-ray diffraction analyses, electron microscopy offers an alternative approach. Two techniques are available: single particle image averaging and electron crystallography using 2-D crystals. Both have been applied to elucidate the structure of PSII and in principle could yield maps at atomic resolution (54). To date, both techniques have yielded information on the oligomeric state (monomers vs dimers) of PSII in vivo and on its subunit organization.

Single Particle Image Averaging

Single particle analysis has been used to study the structure of a wide range of biological molecules, either imaged in negative stain or in vitreous ice (cryo-electron microscopy). Data obtained from biological molecules in vitreous ice correspond most clearly with the native protein structure and typically have a resolution up to 15 Å. Conventionally negatively stained samples (air dried and imaged at room temperature) usually give lower resolution (approximately 20 Å). The images obtained are classified according to the orientation of the particle (e.g. top and side views), and members of each class are then rotationally and translationally (i.e. shifted in the X-Y plane) aligned. The procedure is more useful for large particles (>250 kDa), which can be more accurately aligned.

The single particle image averaging approach has been used to analyze a number of detergent-solubilized PSII complexes of known subunit composition and oligomeric state (i.e. monomers and dimers) under negative stain conditions. The smallest higher plant PSII complex analyzed using this approach is a monomeric PSII core (approximately 240 kDa) consisting of CP47, CP43, D1, D2, cyt b559, the 33-kDa subunit, and associated low-molecular-weight polypeptides, together with about 36 Chla molecules (21, 51, 52). This complex is depicted in its top view orientation in Figure 2A. A dimeric complex of identical subunit composition and having an apparent molecular mass of 450 kDa is shown in its top view orientation in Figure 2B. By comparing Figures 2A and 2B, it can be seen that the projection map of the monomeric complex corresponds well with each half of the PSII core dimer. The dimensions of the latter were calculated to be 206×131 Å in the presence of the detergent (dodecyl maltoside) shell. When a detergent layer of 17 Å is deducted (30), the corrected dimensions of the isolated dimer are calculated to be 172×97 Å (21). Single particle analysis has also identified a PSII core dimer of similar size isolated from the cyanobacterium *Synechococcus elongatus* (21).

Although the dimeric organization of PSII has been readily accepted for cyanobacteria, there is currently a debate about whether the PSII particles in the grana are monomeric or dimeric complexes. It has been argued that dimeric PSII complexes, such as the one shown in Figure 2B, that were isolated from granal membranes, are the product of aggregation of PSII monomers rather than an in vivo form of PSII (39, 61, 92). A considerable volume of evidence speaks against this hypothesis. Freeze-etch images of thylakoid membranes showed that PSII complexes (ESs particles) located in the grana were most probably dimers (112). Holzenburg et al (61) argued that at the relatively low resolution of such freeze-etch images and of projection maps of crystallized dimer-like complexes (19), a pseudosymmetry relating to the similarity in structure of the D1 vs D2 proteins and CP47 vs CP43 proteins could be mistaken for a real twofold symmetry indicative of a dimeric complex. Consequently, it was suggested that PSII was actually a monomeric pseudosymmetric complex. To address this point, PSII complexes derived from gently solubilized granal and stromal membranes were analyzed under nondenaturing electrophoresis conditions. The granal PSII fraction consisted predominantly of dimeric PSII complexes associated with Lhcb proteins, while their stromal counterparts were monomeric (109). This result also showed that the nondenaturing electrophoresis conditions used did not induce aggregation of monomeric PSII (109). Other nondenaturing gel electrophoresis studies of this type came to the similar conclusion that PSII in the grana is dimeric (99). Furthermore, our own data (51, 52) has shown that isolated PSII core dimers (Figure 2B) were more active in terms of oxygen evolution, are associated with higher levels of Chla (39 Chla:2 Pheo), and have much lower levels of D1 and D2 breakdown fragments than their monomeric counterparts (Figure 2A). Based on this information it is difficult to see how two damaged monomers could aggregate to form an intact dimer of the type shown in Figure 2B (51, 52). Taken together, these and other studies (see section on "Heterogeneity of PSII In Vivo") strongly suggest that PSII is a dimer in the grana.

Recently, grana membranes were solubilized with a single gentle solubilization step, to obtain a more native PSII complex. The solubilized mixture was then resolved by sucrose density gradient centrifugation. The largest PSII complex (approximately 700 kDa) consisted of the PSII core dimer subunits (Figure 2B) plus Lhcb1, 2, 4, and 5 proteins and contained about 200 Chl molecules (156 Chla, 4 Chlb) (21, 93). Like the PSII monomer (Figure 2A) and dimer (Figure 2B) cores, this PSII-LHCII supercomplex was active in oxygen evolution. Single particle analysis showed an unsymmetrized form of this PSII-LHCII supercomplex (Figure 2D) to have a clear twofold rotational symmetry axis around the center of the complex, consistent with a dimeric structure (21). Superimposed upon the projection map of the PSII-LHCII supercomplex (Figure 2D) is the outline of the PSII core dimer depicted in Figure 2B. This alignment of Figures 2B and 2D shows that the PSII core dimer forms the central part of the PSII-LHCII supercomplex (Figure 2D) and that the two regions flanking it must contain the Lhcb proteins. Figure 2C shows a PSII-LHCII complex from which one of the Lhcb protein sets has dissociated. This form of PSII-LHCII complex (Figure 2C) lacks the twofold symmetry of the PSII-LHCII supercomplex (Figure 2D) due to the loss of this single Lhcb set. Nevertheless the PSII core region is clearly twofold symmetric and closely resembles the averaged PSII core dimer top view (Figure 2B). Furthermore the structural features of the remaining Lhcb set (Figure 2C) correspond closely with that of the symmetric PSII-LHCII supercomplex. These findings confirm the true twofold symmetry and hence the dimeric nature of the PSII core.

After correction for the detergent layer, the PSII-LHCII supercomplex shown in Figure 2D is calculated to have dimensions of 270×125 Å. In its side view orientation, the particle has a thickness of about 60 Å in the Lhcb region, consistent with the height of LHCII (69, 133). Its maximal thickness, close to the center of the complex, is about 90 Å. This dimension is similar to that of the arrayed ESs particles observed in freeze-etch and freeze-fracture images of thylakoid membranes of a number of higher plant species (see section on "Heterogeneity of PSII In Vivo"). This suggests further that the ESs (EFs) particles are probably dimeric PSII-LHCII supercomplexes, as discussed above.

Electron Crystallography

The successful application of the other type of high-resolution electron micros-copy (electron crystallography) in the determination of the atomic structure of trimeric LHCII demonstrates the power of this approach (69). Figure 3 shows that within each Lhcb protein monomer of LHCII, the two longest transmem-brane helices are surrounded by Chla and Chlb molecules, and two molecules of lutein. The chlorophylls are arranged in two layers, toward the lumenal and stromal membrane surfaces, and the main ligands are Glu, Asn, Glu, His, and Gly. The close association of the luteins with the excitonically linked chloro-phylls prevents singlet oxygen formation and the damage induced by it (69). The luteins also play a structural role in that they form a cross-brace in the center of the complex, providing a direct and strong link between the peptide

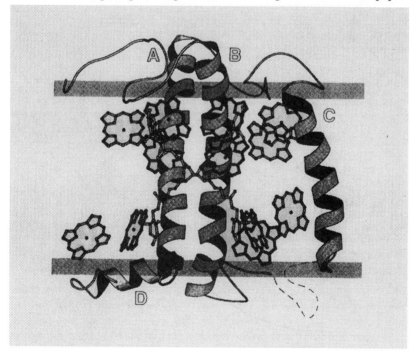

Figure 3   The structure of LHCII. This figure is taken from Kühlbrandt (69) and shows a folding model of a monomeric LHCII subunit. The LHCII monomer has three transmembrane α-helices (A–C) and a shorter helix (D) toward the lumenal surface of the membrane. The chlorophylls are arranged in two layers close to the stromal and lumenal surfaces of the membrane. Two Lut residues (L) form a cross-brace between helices A and B and are excitonically linked to the chlorophyll (Chl) network. Sequence homology between the subunits Lhcb1–6 suggests that all their structures will be similar.

loops at both surfaces (69). Based on the high degree of sequence homology, it is likely that the atomic structures of the other Lhcb proteins will also be similar to that of the LHCII monomer.

Isolated PSII complexes have also been crystallized when reconstituted with detergent solubilized lipid (31, 51, 90, 130). Another 2-D crystallization approach involves a detergent-induced delipidation step of granal membranes enriched in PSII and Lhcb proteins. This forces the complexes in the granal membranes into close contact and has yielded small 2-D crystals of PSII (19, 39, 61, 74–76, 109).

To compare the projection maps obtained using single particle image averaging (Figure 2) and electron crystallography (Figure 4), the isolated PSII core dimer shown in Figure 2B (51) was reconstituted with thylakoid lipid and crystallized, though the 33-kDa extrinsic protein dissociated during the process. The top view projection map, obtained from this D1-D2-cyt b559-CP47-CP43 complex, under negative stain conditions, is shown in Figure 4A. The first important point to note is that the unit cell of the crystallized PSII core dimer (a = 117 Å × b = 173 Å, γ = 110°) compares well with the size of the single particle averaged image of this complex (97 × 172 Å). The second is that the structural features of the two images are almost identical. The most prominent features are two densities (marked with * in Figure 4A), which are positioned on either side of a central hole. Although the projection map shown in Figure 4A is unsymmetrized, it clearly has a twofold rotational symmetry around its center, confirming its dimeric structure. Image sections taken through the 3-D image of this complex show it to have twofold symmetry throughout (E Morris, B Hankamer, D Zheleva & J Barber, unpublished data).

The top view projection map of another PSII crystal, imaged under cryo conditions, is shown in Figure 4B. The crystal was produced by delipidating granal membranes (75). A comparison of Figures 4A and 4B shows that the two projection maps have very similar features, including a central hole and two predominant regions of density, on either side of it (marked with a *). The two crystal forms also have very similar unit cell dimensions to that of the complex shown in Figure 4B (a = 114 Å, b = 173 Å, γ = 106.6°). Despite the similarity of the two projection maps and their unit cell dimensions, the crystals of Marr et al (75) were reported to contain Lhcb4, 5, and 6 and PsbS, in addition to the other PSII subunits of the complex shown in Figure 4A. Marr et al (75) came to this conclusion based on direct immunolabeling of their crystals. This apparent discrepancy remains to be resolved because the complex shown in Figure 4A does not contain these subunits.

Holzenburg et al (61) and Ford et al (39) also reported projection maps of crystallized PSII complexes with and without the extrinsic subunits of the

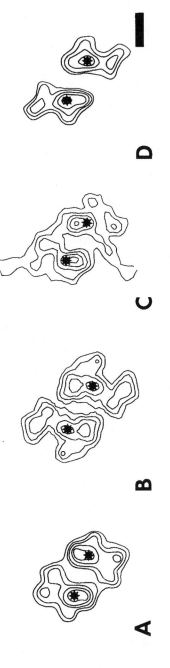

Figure 4   Projection structures of spinach PSII complexes obtained by crystallographic averaging of electron microscopy images of 2-D crystals. (*A*) A nonsymmetrized projection map of a crystallized PSII core dimer of the type shown in Figure 3A, consisting of CP47, CP43, the D1 and D2 proteins, cyt *b*559, and associated low-molecular-weight polypeptides but lacking the 33-kDa protein (E Morris, B Hankamer, D Zheleva & J Barber, unpublished results). (*B*) A filtered 2-D crystal projection map with twofold rotational symmetry imposed (redrawn from 75). (*C*) Nonsymmetrized section from the 3-D model of Ford et al (39), redrawn. (*D*) A nonsymmetrized filtered projection map of a 2-D crystal from Tsiotis et al (130), redrawn. In each of these four images, two prominent densities marked (*) are located in equivalent positions on either side of a central hole. The similarity in size and shape of these contoured maps suggests that all these complexes are dimeric. The scale bar measures 5 nm.

OEC. As in the case of Marr et al (75), these crystals were formed by partially delipidating granal membranes. To aid comparison with Figures 4A and 4B, the images of the crystal form lacking the extrinsic proteins are shown in Figure 4C (39). It is reported to have a unit cell (a = 177 Å, b = 201Å, γ = 91°) that is slightly larger than those of the dimeric PSII complexes shown in Figure 4A and 4B. On the basis of SDS-PAGE analysis, the authors concluded that their crystals contained both PSII core and Lhcb subunits. One of the difficulties in determining the subunit composition of a complex crystallized by the partial delipidation of granal membranes is to confirm the presence of given subunits in the crystalline fraction, when both crystalline and noncrystalline material is contained in the sample. Without direct immunolabeling of the crystals, their composition must remain unconfirmed. However, on the basis of the assumption that Lhcb proteins are associated with the crystallized complex, it was concluded that the PSII core complex must be monomeric because there was insufficient volume to accommodate a PSII core dimer and a large complement of Lhcb proteins (61). However, a comparison of the unsymmetrized images of Figures 4A, 4B, and 4C show them to be very similar. Once again, the same two prominent regions of density (*), also seen in Figures 4A and 4B, are apparent on either side of a central hole in Figure 4C. This suggests that this core complex may actually be a core dimer rather than a monomeric PSII complex and that its symmetry was mistaken as pseudosymmetry because of the relatively low resolution of the images.

Tsiotis et al (130) recently reported the projection maps of a PSII crystal (CP47, CP43, D1, D2, cyt b559, the 33-kDa extrinsic subunit, and associated low-molecular-weight polypeptides) obtained using detergent solubilized PSII core complexes (Figure 4D). This crystal was reported to have a unit cell of (a = 162 Å, b = 137Å, (γ = 142.°). However, if a choice of unit cell parameters is made to include all the densities shown in Figure 4D, the dimensions are actually very close to the ones reported by Hankamer et al (51) (Figure 4A) and Marr et al (75) (see Figure 4B). Furthermore, a central hole (or cavity) is visible in these complexes and two densities (*) are once again seen on either side of it, separated by about the same distance. All these points would suggest that this complex could also be a dimeric PSII core complex, although the authors concluded that the complex was a monomer, partly because of scanning transmission electron microscopy (STEM) measurements that suggested the complex had a mass of 318 kDa (130).

Recently, Nakazato et al (90) reported the crystallization of a D1-D2-cyt b559-CP47 complex, which yielded a projection map of 20-Å resolution. The projection map of this complex can be superimposed upon a PSII core monomer (Figure 3A) and comfortably fits into it (see Figure 5 and associated discussion). This finding confirms that the PSII complexes depicted in Figures 4*A* and 4*B*, and probably in 4*C* and 4*D*, are dimeric.

SUBUNIT ORGANIZATION OF PSII

This section reviews the subunit organization of PSII and its antenna system using information that has been obtained by electron crystallography, single particle image averaging, and crosslinking and other biochemical techniques. The combined data is summarized in the form of two currently favored subunit organization models (Figures 5*a* and 5*b*) because the information available is still insufficient to confirm which of these is correct.

Top (Figures 5*a* and 5*b*) and side view (Figure 5*C*) projection maps of the largest PSII-LHCII supercomplex structurally characterized to date are used as the framework for the two possible models of subunit organization. These contoured projection maps are very similar to those presented by Boekema et al (21) but improved in that they are the sums of larger data sets (1925 vs 500 top views, 2213 vs 80 side views) and have a higher resolution (approximately 20 Å). They also differ in that they contain the densities of two 23-kDa subunits in addition to those of the other core (CP47, CP43, D1, D2, the 33-kDa subunit, cyt *b*559) and antenna (Lhcb1, 2, 4, and 5) (21, 93) proteins. Both models are identical in terms of their depiction of the extrinsic and Lhcb protein components. They differ only in the attributed locations of the PSII core components, CP47, D1, D2, and cyt *b*559.

Localization of D1-D2-Cyt b559-CP47 Complex and CP43

Regions within each of the projection maps are shaded in dark, mid, and light gray. The dark-gray regions represent aligned monomeric D1-D2-cyt *b*559-CP47 complexes (and associated low-molecular-weight subunits) of the type reported by Dekker et al (31) and Nakazato et al (90). Together, the two dark- and two mid-gray regions (Figures 5*a* and 5*b)* depict the shape of the PSII core dimer (Figure 2*b*) consisting of the integral membrane protein components D1, D2, cyt *b*559, CP47, and CP43 and associated low-molecular-weight subunits. By elimination, it follows that the two mid-gray regions each contain CP43 (106).

Localization of the 33-kDa Extrinsic Subunits

Top and side view projection maps of PSII complexes (±33-kDa extrinsic subunit) were used to produce subunit difference maps (21, 30). In Figure 5 each monomeric portion of the dimer is associated with a region of density attributed to the 33-kDa subunit, and these densities overlap to form a single

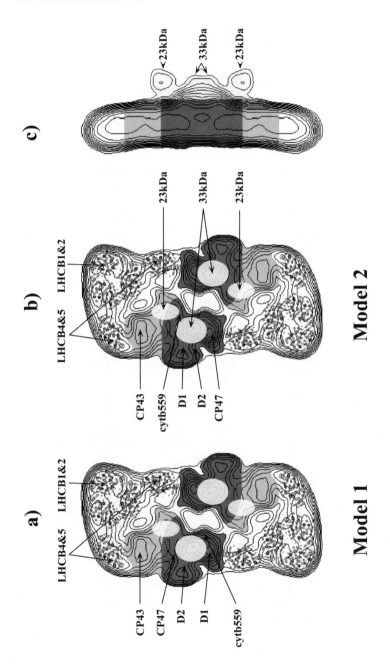

Figure 5   Subunit organization models of photosystem II. This figure shows top (Figures 5*a* and 5*b*) and side view (Figure 5*c*) projection maps of the largest PSII-LHCII supercomplex structurally characterized to date. It provides the framework for two possible models of subunit organization (Model 1 and Model 2), based predominantly on crosslinking studies and averaged images of PSII complexes of known and differing subunit composition. Both models are identical in terms of their depiction of the extrinsic and Lhcb protein components. They differ only in the attributed locations of the PSII core components, CP47, D1, D2, and cyt *b*559. In Model 1, CP47 is placed between CP43 and the reaction center components. In Model 2, the RC components are placed between CP47 and CP43. The three contoured projection maps (*a*, *b*, and *c*) are very similar to those presented in Boekema et al (21) but are improved in that they are the sums of larger data sets (1925 vs 500 top views, 2213 vs 80 side views) and have a higher resolution (approximately 20 Å). They also differ in that they contain the densities of two 23-kDa subunits in addition to those of the other core (CP47, CP43, D1, D2, the 33-kDa subunit, cyt *b*559) and the antenna (Lhcb1, Lhcb2, Lhcb4, and Lhcb5) (21, 93) proteins. Superimposed upon the central regions of each of these images are areas shaded in dark, mid, and light gray. Each of the two dark-gray regions corresponds to a monomeric CP47-RC complex similar to those reported by Dekker et al (31) and Nakazato et al (90). Together the two dark- and two mid-gray regions depict the centrally located CP47-CP43-RC core dimer (21). The four densities shaded in light gray in the top view projections (*a* and *b*) correspond to two 33-kDa and two 23-kDa subunits (21; see text). In the side view orientation (*c*) the two lumenally exposed 33-kDa subunits partially overlap. Positioned on either side of them are two densities attributed to the 23-kDa extrinsic proteins. The electron density map of an LHCII trimer (133) has been superimposed upon the two tips of the PSII-LHCII complex (*a* and *b*). It is suggested that these trimers consist of Lhcb1 and Lhcb2 and that they are linked to the centrally located PSII core via two monomeric Lhcb proteins (Lhcb4 and Lhcb5). The Lhcb4 and Lhcb5 proteins are depicted using the projection maps of monomeric LHCII components (133).

central protrusion in the side view (Figure 5*c*). The model implies that the 33-kDa subunit:PSII core monomer stoichiometry is 1:1. Some reports indicate the presence of two copies of the 33-kDa subunit per reaction center (72, 135, 136). However, the dimensions of the 33-kDa subunit protrusions shown in Figure 5 are consistent with single copies of the protein monomer (64).

Localization of the 23-kDa Extrinsic Subunits

When isolated in the presence of glycine betaine, PSII-LHCII supercomplexes additionally bind the 23-kDa subunit (EJ Boekema, J Nield, B Hankamer & J Barber, unpublished data). Preliminary difference mapping experiments suggested that the two lumenal protrusions in the side view projection map, located on either side of the 33-kDa components (Figure 5*c*), each contain a 23-kDa subunit. Similarly, positions in the top views, also identified by difference mapping, are shown in Figures 5*a* and 5*b*. These proposed positions are consistent with analysis of crystalline PSII arrays containing the 23-kDa subunit (76). Freeze-etching studies of the lumenal surface of the grana regions of thylakoid membranes (ESs surfaces) also show four lumenal protrusions (112) that could correspond to the four densities shown in Models 1 and 2 (Figures 5*a* and 5*b;* see section "Heterogeneity of PSII In Vivo"). If the PSII complexes studied, before and after the removal of extrinsic proteins by Ford et al (39), are interpreted as dimers, the positions attributed to the extrinsic subunits would be consistent with models shown in Figure 5.

Organization of the Membrane-Embedded PSII Core Components

Topological information is available on the organization of the subunits within the PSII reaction center complex. In its isolated form, the PSII reaction center consists of the α and β subunits of cyt $b559$ and PsbI (17, 91, 134). Crosslinking studies have shown that the α and β subunits of cyt $b559$ are closely associated with the D1 and D2 proteins (14, 88), whereas the stromally exposed N-terminus of PsbI is suggested to be in close contact with the D2 protein (128).

Sequence homology between the D1 and D2 subunits and the L and M subunits of purple bacteria suggest that the D1/D2 and L/M heterodimers are likely to have similar dimensions (approximately 70×30 Å) in the membrane plane (29). Cryo-electron crystallography studies have shown that the monomeric PSII RC-CP47 complex (dark-gray region) has dimensions of 81×75 Å (90), indicating that the estimated PSII RC size is reasonable.

Isolated RC-CP47 complexes (D1-D2-cyt $b559$-CP47) have also been subjected to crosslinking experiments, and the results indicate that CP47 is more closely associated with D2 than with the other reaction center components (88). It is for this reason that CP47 is placed close to the D2 protein in both subunit organization models (Figures 5a and 5b).

There are currently no crosslinking data that help to position CP43 with respect to the D1-D2-cyt b559-CP47 complex. Consequently, at least two models can be proposed. In the first, CP47 is positioned between the reaction center components and CP43 (Model 1, Figure 5a). In the second, the reaction center components are located between CP43 and CP47 (Model 2, Figure 5b). Recently, PSII core dimer crystals were labeled with Fab antibody fragments, specific to the C-termini of the D1 protein and cyt b559 (75). These studies suggested that the D1 and cyt b559 subunits were located farthest from CP43 but within the dark-gray region as shown in Model 1 (Figure 5a). However, crosslinking results of Seidler (114) detected interactions between the 23-kDa subunit and cyt b559, favoring Model 2 depicted in Figure 5b. The positions of cyt b559 as shown in Models 1 and 2 might allow the formation of an α-cyt b559 homodimer in crosslinking experiments carried out using D1-D2-cyt b559-CP47 complexes, assuming these preparations contained dimeric complexes (88).

Crosslinking and other biochemical studies have shown that CP43 (63) and cyt b559 (124, 125) are also closely associated with the 33-kDa subunit. Both cyt b559 and PsbI are estimated to be within 11 Å of the 33-kDa extrinsic polypeptide (37). Model 2 (Figure 5b) is more consistent with this combined data because it is difficult to see how CP43 and cyt b559 could interact simultaneously with the 33-kDa subunit in Model 1 (Figure 5a). Perhaps the most compelling evidence for the relative positioning of CP43 and CP47 on either side of the D1/D2 subunits (Model 2) comes from the comparison with the reaction center structure of PSI (42). This complex is composed of a heterodimer of PsaA and PsaB proteins each having 11 transmembrane helices. In each case, 5 of these helices are similarly arranged to the 5 helices of the L and M subunits of the purple bacterial reaction center (and presumably with the D1 and D2 proteins). The remaining 6 helices in each PSI subunit are positioned farther from the pseudo twofold axis that relates the two subunits in the heterodimer. Furthermore, the two sets of peripheral helices show sequence homology with those of CP47 and CP43. By analogy, therefore, it would seem reasonable to place CP43 and CP47 on either side of the D1 and D2 proteins.

Recently 3-D maps obtained from dimeric RC-CP47-CP43 complexes showed that the binding region for the 33-kDa subunit protrudes into the lumen (E Morris, B Hankamer, D Zheleva & J Barber, unpublished data). The D1 and D2 proteins, the α and β subunits of cyt b559 and PsbI, are all very hydrophobic proteins with small lumenally exposed loops. In contrast, the lumenal loops E of CP47 and CP43 are large (22, 41, 45), suggesting that they could form part of this extrinsic region. This conclusion agrees well with the finding that loop E of CP47 is in close contact with the 33-kDa extrinsic polypeptide (23, 24, 41, 50, 94, 102). However, this information does not provide a sufficiently clear distinction between Models 1 and 2. This is because in Model 1, the 33-kDa subunit is positioned directly above CP47, allowing direct contact with loop E of the latter. However, in Model 2, CP47 is adjacent rather than directly under the 33-kDa subunit, and it is quite conceivable that loop E could fold over the lumenal surface of the centrally located D1/D2 heterodimer and so come into contact with the 33-kDa subunit.

Localization of the Lhcb Proteins

Western blot analysis of isolated PSII-LHCII supercomplexes showed them to be enriched in Lhcb1, Lhcb2, Lhcb4 (CP29), and Lhcb5 (CP26), but depleted of Lhcb3 and Lhcb6 (CP24) (93). Crosslinking studies have shown that CP47 and CP43 are both in close contact with CP29 (Lhcb4) and that CP43 is also in close contact with CP26 (Lhcb5) (26, 104). By superimposing a high-resolution electron density map of the Lhcb1/Lhcb2 heterotrimer (133) upon the top-view projection maps shown in Figure 5a and 5b, the LHCII complex can be seen to fit snugly into the two tips of the PSII-LHCII supercomplex. Furthermore, because Lhcb4 and 5 are similar in mass and have strong sequence homology with the LHCII proteins, two monomeric electron density maps based on the data of Wang & Kühlbrandt (133) were positioned in the PSII-LHCII supercomplex (in both Models 1 and 2) between the PSII core and the LHCII heterotrimer. There appears to be insufficient space to fit any additional monomeric Lhcb proteins. This conclusion is in agreement with pigment analyses of the isolated PSII-LHCII supercomplex, which showed it to be associated with 156 Chla and 46 Chlb molecules (see Figure 2d and 51, 52, 93). It was suggested that Lhcb6 (CP24) might also be present in these isolated PSII-LHCII supercomplexes (21), but this was a precautionary conclusion based on the detection of low levels of the protein present in early and less pure preparations. This conclusion has been amended in the light of the Western blot data presented by Hankamer et al (52) and Nield et al (93). The positions of Lhcb4 and 5 between the LHCII trimer and the PSII core is in agreement not only with the available crosslinking data (26, 104) but with the hypothesis that Lhcb4, 5, and 6 are involved in the regulation of the rate of excitation energy transfer from the outer Lhcb (Lhcb1, 2, and 3) proteins to the reaction center via CP47 and CP43 (see 65). However, the depletion of Lhcb3 and 6 in these complexes with respect to granal membranes must be explained.

One explanation is that the isolated PSII-LHCII supercomplex characterized originates from the population of ESs particles equivalent to those observed as arrays in freeze-etching studies of grana regions, and that Lhcb3 and Lhcb6 are associated predominantly with the nonarrayed ESs particles in the grana (see section on "Heterogeneity of PSII In Vivo"). This hypothesis is consistent with the isolation of a Lhcb6-Lhcb4-LHCII supramolecular complex (18). It is also consistent with the finding that PSIIα complexes in the grana consist of more than one subpopulation differing in antenna size (4, 5). Furthermore, the large distances between the nonarrayed ESs particles means that they may have larger antenna systems than their arrayed counterparts and could additionally bind Lhcb3 and Lhcb6 (see section on "Heterogeneity of PSII In Vivo").

The low-resolution subunit organizations presented in Figure 5, though incomplete, give two models that it is hoped will aid more detailed determinations of subunit organization and PSII structure in the future.

CONCLUDING REMARKS

The PSII-LHCII supercomplex is in projection possibly one of the largest integral membrane protein complexes in nature. Despite its large size, this discrete unit has escaped attention until recently. One of the reasons may be its thickness, which is only 60 Å at the outer ends. In fact, only the central parts, which are 90 Å in height, substantially protrude from the membrane and contribute to the appearance of the approximately 200 × 200 Å "particles," already seen long ago by freeze-fracture/freeze-etching techniques. No doubt, the PSII-LHCII supercomplex will provide a further basis for structural work, although its stability after isolation will possibly prevent the growing of highly ordered crystals. It is, however, highly suited for further single particle analysis using cryoelectron microscopy to produce a more detailed 3-D map suitable for the insertion of high-resolution data. Such data will almost certainly be obtained by crystallography (2-D and 3-D) of PSII complexes with smaller numbers of subunits or of individual subunits. In this way, the complete structure of PSII at atomic resolution will be built. Whether the high-resolution data will be obtained from X-ray or electron crystallography is uncertain, but with the successful outcome of the LHCII structure, the electron microscopy approach has yielded the most information to date for PSII. The instability of this photosystem, particularly when illuminated, presents a major hurdle that is unique to this complex. The authors are confident, however, that this hurdle will be surmounted in the not-too-distant future and that PSII, like the other major complexes of the photosynthetic electron transport chain, will have its structure determined at atomic resolution.

ACKNOWLEDGMENTS

In particular, we thank Lyn Barber for all her hard work and care in preparing this review and Jon Nield for his help in the figure preparation. We are also very grateful to Alain Brisson, Ed Morris, Jon Nield, Peter Nixon, Jyoti Sharma, Alison Telfer, and Daniella Zheleva for their critical proofreading of sections of the review and their constructive comments on them.

> **Visit the *Annual Reviews home page* at http://www.annurev.org.**

Literature Cited

1. Adir N, Okamura MY, Feher G. 1992. Crystallization of the PSII-reaction center. See Ref. 88a, 2:195–98
2. Akabori K, Imaoka A, Toyoshima Y. 1984. The role of lipids and 17-kDa protein in enhancing the recovery of O_2 evolution in cholate-treated thylakoid membranes. *FEBS Lett.* 173:36–40
3. Albertsson PA. 1995. The structure and function of the chloroplast photosynthetic membrane: a model for the domain organization. *Photosynth. Res.* 46:141–49
4. Albertsson PA, Yu SG. 1988. Heterogeneity among photosystem II. Isolation of thylakoid membrane vesicles with different functional antennae size of photosystem II. *Biochim. Biophys. Acta* 936:215–21
5. Albertsson PA, Yu SG, Larsson UK. 1990. Heterogeneity in Photosystem II. Evidence from fluorescence and gel electrophoresis experiments. *Biochim. Biophys. Acta* 1016: 137–40
6. Allen JF, Feher G, Yeates TO, Komiya H, Rees DC. 1987. Structure of the reaction center from *Rhodobacter sphaeroides* R-26: the co-factors. *Proc. Natl. Acad. Sci. USA* 84:5730–34
7. Anderson JM, Goodchild DM, Boardman NK. 1973. Composition of the photosystems and chloroplast structure in extreme shade plants. *Biochim. Biophys. Acta* 325: 573–85
8. Anderson JM, Melis A. 1983. Localization of different photosystems in separate regions of chloroplast membranes. *Proc. Natl. Acad. Sci. USA* 80:745–49
9. Andersson B, Åkerlund HE. 1987. Proteins of the oxygen-evolving complex. In *Topics in Photosynthesis, The Light Reactions,* ed. J Barber, 8:379–420. Amsterdam: Elsevier
10. Andersson B, Anderson JM. 1980. Lateral heterogeneity in the distribution of chloro-
phyll-protein complexes of the thylakoid membranes of spinach chloroplasts. *Biochim. Biophys. Acta* 593:427–40
11. Armond PA, Arntzen CJ. 1977. Localization and characterization of PSII in grana and stroma lamellae. *Plant Physiol.* 59: 398–404
12. Armond PA, Staehelin LA, Arntzen CJ. 1977. Spatial relationship of photosystem I, Photosystem II and the light harvesting complex in chloroplast membranes. *J. Cell. Biol.* 73:400–18
13. Babcock GT. 1987. The photosynthetic oxygen-evolving process. In *Photosynthesis,* ed. J Amesz, 15:125–58. Amsterdam: Elsevier
14. Barbato R, Friso G, Ponticos M, Barber J. 1995. Characterization of the light-induced cross-linking of the α-subunit of cytochrome b559 and the D1 protein in isolated Photosystem II reaction centers. *J. Biol. Chem.* 270:24032–37
15. Barber J. 1982. Influence of surface charges on thylakoid structure and function. *Annu. Rev. Plant Physiol.* 33:261–95
16. Barber J. 1987. Photosynthetic reaction centres: a common link. *Trends Biochem. Sci.* 12:323–26
16a. Barber J, ed. 1992. *The Photosystems: Structure, Function and Molecular Biology.* Amsterdam: Elsevier
17. Barber J, Chapman DJ, Telfer A. 1987. Characterisation of a photosystem II reaction centre isolated from chloroplasts of *Pisum sativum. FEBS Lett.* 220:67–73
18. Bassi R, Dainese P. 1992. A supramolecular light harvesting complex from chloroplast photosystem-II membranes. *Eur. J. Biochem.* 204:317–26
19. Bassi R, Ghiretti Magaldi A, Tognon G, Giacometti GM, Miller KR. 1989. Two-dimensional crystals of the Photosystem II

reaction center complex from higher plants. *Eur. J. Cell Biol.* 50:84–93

20. Bassi R, Pineau B, Dainese P, Marquardt J. 1993. Carotenoid-binding proteins of photosystem-II. *Eur. J. Biochem.* 212:297–303

21. Boekema EJ, Hankamer B, Bald D, Kruip J, Nield J, et al. 1995. Supramolecular structure of the photosystem II complex from green plants and cyanobacteria. *Proc. Natl. Acad. Sci. USA* 92:175–79

22. Bricker TM. 1990. The structure and function of Cpa-1 and Cpa-2 in photosystem II. *Photosynth. Res.* 24:1–13

23. Bricker TM, Frankel LK. 1987. Use of monoclonal antibody in structural investigations of the 49 kDa polypeptide of photosystem II. *Arch. Biochem. Biophys.* 256:295–301

24. Bricker TM, Odom WR, Queirolo CB. 1988. Close association of the 33-kDa extrinsic protein with the apoprotein of CPA1 in photosystem-II. *FEBS Lett* 231:111–17

25. Britt RD. 1996. Oxygen evolution. See Ref. 95a, pp. 137–59

26. Dainese P, Santini C, Ghiretti-Magaldi A, Marquardt J, Tidu V, et al. 1992. The organisation of pigment proteins within photosystem II. See Ref. 88a, 2:13–20

27. Deisenhofer J, Epp O, Miki K, Huber R, Michel H. 1984. X-ray structure analysis of a membrane protein: electron density map at 3Å resolution and a model of the photosynthetic reaction center from *Rhodopseudomonas viridis. J. Mol. Biol.* 180:385–98

28. Deisenhofer J, Epp O, Miki K, Huber R, Michel H. 1985. Structure of the protein subunits in the photosynthetic reaction centre of *Rhodopseudomonas viridis* at 3Å resolution. *Nature* 318:618–24

29. Deisenhofer J, Michel H. 1989. The photosynthetic reaction center from the purple bacterium *Rhodopseudomonas viridis. EMBO J.* 8:2149–70

30. Dekker JP, Boekema EJ, Witt HT, Rögner M. 1988. Refined purification and further characterization of oxygen-evolving and Tris-treated photosystem-II particles from the thermophilic cyanobacterium *Synechococcus* sp. *Biochim. Biophys. Acta* 936:307–18

31. Dekker JP, Bowlby NR, Yocum CF, Boekema EJ. 1990. Characterization by electron microscopy of isolated particles and two-dimensional crystals of the CP47-D1-D2-cytochrome b–559 complex of Photosystem II. *Biochemistry* 29:3220–25

32. Demmig-Adams B. 1990. Carotenoids and photoprotection in plants: a role for the xanthophyll zeaxanthin. *Biochim. Biophys. Acta* 1020:1–24

33. de Vitry C, Olive J, Drapier D, Recouvreur M, Wollman FA. 1989. Post-translational events leading to the assembly of photosystem II protein complex: a study using photosynthesis mutants from *Chlamydomonas reinhardtii. J. Cell Biol.* 109:991–1006

34. Diner BA, Petrouleas V, Wendoloski JJ. 1991. The iron-quinone electron-acceptor complex of photosystem II. *Physiol. Plant.* 81:423–36

35. Dostatni R, Meyer HE, Oettmeier W. 1988. Mapping of two tyrosine residues involved in the quinone (Q_B) binding site of the D1 reaction center polypeptide of photosystem II. *FEBS Lett.* 239:207–10

36. Durrant JR, Klug DR, Kwa SLS, van Grondell R, Porter G, Dekker JP. 1995. A multimer model for P680, the primary electron donor of PSII. *Proc. Natl. Acad. Sci. USA* 92:4798–802

37. Enami I, Ohta S, Mitsuhashi S, Takahashi S, Ikeuchi M, Katoh S. 1992. Evidence from crosslinking for a close association of the extrinsic 33 kDa protein with the 9.4 kDa subunit of cytochrome b559 and the 4.8 kDa product of the psbI gene in oxygen evolving photosystem II complexes from spinach. *Plant Cell Physiol.* 33:291–97

38. Erickson JM, Rochaix JD. 1992. The molecular biology of photosystem II. See Ref. 16a, 2:101–77

39. Ford RC, Rosenberg MF, Shepherd FH, McPhie P, Holzenburg A. 1995. Photosystem II 3-D structure and the role of the extrinsic subunits in photosynthetic oxygen evolution. *Micron* 26:133–40

40. Fotinou C, Kokkinidis M, Fritzsch G, Haase W, Michel H, Ghanotakis DF. 1993. Characterization of a photosystem-II core and its 3-D crystals. *Photosynth. Res.* 37:41–48

41. Frankel LK, Bricker TM. 1992. Interaction of CPa-1 with the manganese-stabilizing protein of photosystem II: identification of domains on CPa-1 which are shielded from N-hydroxysuccinimide biotinylation by the manganese-stabilizing protein. *Biochemistry* 31:11059–64

42. Fromme P, Witt HT, Schubert WD, Klukas O, Saenger W, Krauss N. 1996. Structure of photosystem-I at 4.5-Å resolution: a short review including evolutionary aspects. *Biochim. Biophys. Acta* 1275:76–83

43. Ghanotakis DF, Topper JN, Babcock GT, Yocum CF. 1984. Water-soluble 17-kDa and 23-kDa polypeptides restore oxygen evolution activity by creating a high-affin-

ity binding-site for Ca^{-2} and on the oxidizing side of photosystem-II. *FEBS Lett.* 170: 169–73

44. Ghanotakis DF, Yocum CF. 1985. Polypeptides of photosystem II and their role in oxygen evolution. *Photosynth. Res.* 7: 97–114

45. Gleiter HM, Haag E, Shen JR, Eaton-Rye JJ, Inoue Y, et al. 1994. Functional characterization of mutant strains of the cyanobacterium *Synechocystis* sp. PCC 6803 lacking short domains within the large, lumen-exposed loop of the chlorophyll protein CP47 in photosystem II. *Biochemistry* 33:12063–71

46. Green BR, Durnford DG. 1996. The chlorophyll-carotenoid proteins of oxygenic photosynthesis. *Annu. Rev. Plant Physiol. Plant Mol. Biol.* 47:685–714

47. Green BR, Durnford DG, Aebersold R, Pichersky E. 1992. Evaluation of structure and function in the Chl a/b and Chl a/c antenna protein family. See Ref. 88a, 1: 195–202

48. Green BR, Pichersky E. 1994. Hypothesis for the evolution of three-helix Chl a/b and Chl a/c light-harvesting antenna proteins from two-helix and four-helix ancestors. *Photosynth. Res.* 39:149–62

49. Green BR, Pichersky E, Kloppstech K. 1991. Chlorophyll a/b binding proteins: an extended family. *Trends Biochem. Sci.* 16: 181–86

50. Haag E, Eaton-Rye JJ, Renger G, Vermaas WFJ. 1993. Functionally important domains of the large hydrophilic loop of CP47 as probed by oligonucleotide directed mutagenesis in *Synechocystis* sp. PCC6803. *Biochemistry* 32:4444–54

51. Hankamer B, Morris E, Zheleva D, Barber J. 1995. Biochemical characterisation and structural analysis of monomeric and dimeric photosystem II core preparations. See Ref. 76a, 3:365–68

52. Hankamer B, Nield J, Zheleva D, Boekema E, Jansson S, Barber J. 1996. Isolation and biochemical characterisation of monomeric and dimeric PSII complexes from spinach and their relevance to the organisation of photosystem II *in vivo. Eur. J. Biochem.* 243:422–29

53. Hansson Ö, Wydrzynski T. 1990. Current perceptions of photosystem II. *Photosynth. Res.* 23:131–62

54. Henderson R. 1995. The potential and limitations of neutrons, electrons and X-rays for atomic resolution microscopy of unstained biological molecules. *Q. Rev. Biophys.* 28:171–93

55. Henriques F, Park RB. 1975. Further

chemical and morphological characterization of chloroplast membranes from a chlorophyll b-less mutant of *Hordeum vulgare. Plant Cell Physiol.* 55:763–67

56. Henriques F, Park RB. 1976. Development of the photosynthetic unit in lettuce. *Proc. Natl. Acad. Sci. USA* 73:4560–64

57. Henrysson T, Sundby C. 1990. Characterisation of photosystem II in stroma thylakoid membranes. *Photosynth. Res.* 25: 107–17

58. Hiratsuka J, Shimada H, Whittie R, Ishibashi T, Sakamoto M, et al. 1989. The complete sequence of the rice (*Oryza sativa*) chloroplast genome: intermolecular recombination between distinct transfer-RNA genes accounts for a major plastid DNA inversion during the evolution of the cereals. *Mol. Gen. Genet.* 217:185–94

59. Hirschberg J, McIntosh L. 1983. Molecular basis of herbicide resistance in *Amaranthus hybridus. Science* 222:1346–49

60. Hoganson CW, Lydakis-Simantiris N, Tang XS, Tommos C, Warncke K, et al. 1995. A hydrogen-atom abstraction model for the function of Y_Z in photosynthetic oxygen-evolution. *Photosynth. Res.* 46: 177–84

61. Holzenburg A, Bewly MC, Wilson FH, Nicholson WV, Ford RC. 1993. Three-dimensional structure of photosystem II. *Nature* 363:470–72

62. Horton P, Ruban AV, Walters RG. 1996. Regulation of light-harvesting in green plants. *Annu. Rev. Plant Physiol. Plant Mol. Biol.* 47:655–84

63. Isogai Y, Yamamoto Y, Nishimura M. 1985. Association of the 33-kDa polypeptide with the 43-kDa component in photosystem II particles. *FEBS Lett.* 187:240–44

64. Jansen T, Rother C, Steppuhn J, Reinke H, Beyreuther K, et al. 1987. Nucleotide sequence of cDNA clones encoding the complete 23-kDa and 16-kDa precursor proteins associated with the photosynthetic oxygen evolving complex from spinach. *FEBS Lett.* 216:234–40

65. Jansson S. 1994. The light-harvesting chlorophyll a/b-binding proteins. *Biochim. Biophys. Acta* 1184:1–19

66. Joliot P, Barbier IG, Chabaud R. 1969. Un nouveau modúle des centres photochimique du systeme II. *Photochem. Photobiol.* 10:309–29

67. Keck RW, Dilley RA, Allen CF, Biggs S. 1970. Chloroplast composition and structural differences in a soybean mutant. *Plant Physiol.* 46:692–98

68. Kok B, Forbush B, McGloin M. 1970. Cooperation of charges in photosynthetic evo-

lution. I. A linear four step mechanism. *Photochem. Photobiol.* 11:457–75

69. Kühlbrandt W, Wang DN, Fujiyoshi Y. 1994. Atomic model of plant light-harvesting complex by electron crystallography. *Nature* 367:614–21

70. Lam E, Baltimore B, Ortiz W, Chollar S, Melis A, Malkin R. 1983. Characterization of a resolved oxygen-evolving photosystem-II preparation from spinach thylakoids. *Biochim. Biophys. Acta* 724:201–11

71. Lancaster CRD, Michel H, Honig B, Gunner MR. 1996. Calculated coupling of electron and proton-transfer in the photosynthetic reaction-center of *Rhodopseudomonas viridis. Biophys. J.* 70:2469–92

72. Leuschner C, Bricker TM. 1996. Interaction of the 33 kDa protein with photosystem II: rebinding of the 33 kDa extrinsic protein to photosystem II membranes which contain four, two, zero manganese per photosystem II reaction center. *Biochemistry* 35: 4551–57

73. Lorkovic ZJ, Schröder WP, Pakrasi HB, Irrgang KD, Herrmann RG, Oelmuller R. 1995. Molecular characterization of PsbW, a nuclear-encoded component of the photosystem-II reaction-center complex in spinach. *Proc. Natl. Acad. Sci. USA* 92: 8930–34

74. Lyon MK, Marr KM, Furcinitti PS. 1993. Formation and characterization of two-dimensional crystals of photosystem II. *J. Struct. Biol.* 110:133–40

75. Marr KM, Mastronarde DM, Lyon MK. 1996. Two-dimensional crystals of photosystem II: biochemical characterization, cryoelectron microscopy and localization of the D1 and cytochrome b559 polypeptides. *J. Cell Biol.* 132:823–33

76. Marr KM, McFeeters RL, Lyon MK. 1996. Isolation and structural analysis of two-dimensional crystals of photosystem II from *H. vulgare viridis* zb63. *J. Struct. Biol.* 117: 86–98

76a. Mathis P, ed. 1995. *Photosynthesis: From Light to Biosphere.* The Netherlands: Kluwer

77. McDermott G, Prince SM, Freer AA, Hawthornthwaite-Lawless AM, Papiz MZ, et al. 1995. Crystal structure of an integral membrane light-harvesting complex from photosynthetic bacteria. *Nature* 374:517–21

78. McPherson PH, Schönfeld M, Paddock ML, Okamura MY, Feher G. 1994. Protonation and free energy changes associated with formation of $Q_B H_2$ in native and Glu-L212→Gln mutant reaction centers from *Rhodobacter sphaeroides. Biochemistry* 33:1181–93

79. Michel H, Deisenhofer J. 1988. Relevance of the photosynthetic reaction center from purple bacteria to the structure of photosystem II. *Biochemistry* 27:1–7

80. Miller KR. 1981. Freeze-etching studies of photosynthetic membranes. In *Electron Microscopy in Biology,* ed. JD Griffith, 1:1–30. New York: Wiley-Interscience

81. Miller KR, Cushman RA. 1978. A chloroplast membrane lacking photosystem II. *Biochim. Biophys. Acta* 546:481–99

82. Miller KR, Jacob JS. 1991. Surface structure of the photosystem II complex. In *Proc. EMSA Meet., 49th,* ed. GW Bailey. San Francisco: San Francisco Press

83. Miller KR, Miller GJ, McIntyre KR. 1976. The light-harvesting chlorophyll-protein complex of photosystem II. *J. Cell Biol.* 71:624–38

84. Miller KR, Miller GJ, McIntyre KR. 1977. Organization of the photosynthetic membrane in maize mesophyll and bundle sheath chloroplasts. *Biochim. Biophys. Acta* 459:145–56

85. Miller KR, Staehelin LA. 1976. Analysis of the thylakoid outer surface. Coupling factor is limited to unstacked membrane regions. *J. Cell Biol.* 68:30–47

86. Miyao M, Murata N. 1984. Role of the 33kDa polypeptide in preserving Mn in the photosynthetic oxygen-evolving system and its replacement by chloride ions. *FEBS Lett.* 170:350–54

87. Miyao M, Murata N. 1984. Calcium ions can be substituted for the 24-kDa polypeptide in photosynthetic oxygen evolution. *FEBS Lett.* 168:118–20

88. Moskalenko AA, Barbato R, Giacometti GM. 1992. Investigation of the neighbour relationships between PSII polypeptides in the two types of isolated reaction centres (D1/D2/cytb559 and CP47/D1/D2/cytb559 complexes). *FEBS Lett.* 314:271–74

88a. Murata N, ed. 1992. *Research in Photosynthesis.* Dordrecht: Kluwer

89. Murata N, Miyao M. 1985. Extrinsic membrane proteins in the photosynthetic oxygen-evolving complex. *Trends Biochem. Sci.* 10:122–24

90. Nakazato K, Toyoshima C, Enami I, Inoue Y. 1996. Two-dimensional crystallization and cryo-electron microscopy of photosystem II. *J. Mol. Biol.* 257:225–32

91. Nanba O, Satoh K. 1987. Isolation of a photosystem II reaction center consisting of D-1 and D-2 polypeptides and cytochrome b-559. *Proc. Natl. Acad. Sci. USA* 84:109–12

92. Nicholson WV, Shepherd FH, Rosenberg MF, Ford RC, Holzenburg A. 1996. Struc-

ture of photosystem II in spinach thylakoid membranes: comparison of detergent-solubilized and native complexes by electron microscopy. *Biochem. J.* 315:543–47

93. Nield J, Hankamer B, Zheleva D, Hodges ML, Boekema EJ, Barber J. 1995. Biochemical characterisation of PSII-LHCII complexes associated with and lacking the 33 kDa subunit. See Ref. 76a, 3:361–64

94. Odom WR, Bricker TM. 1992. Interaction of CPa-1 with the Manganese-Stabilizing protein of photosystem II: identification of domains cross-linked by 1-ethyl-3-[3-(dimethylamino)-propyl] carbimide. *Biochemistry* 31:5616–20

95. Olive J, Vallon O. 1991. Structural organization of the thylakoid membrane: freeze-fracture and immuno-cytochemical analysis. *J. Electron Microsc. Tech.* 18:360–74

95a. Ort DR, Yocum CF, eds. 1996. *Oxygenic Photosynthesis: The Light Reactions.* Dordrecht: Kluwer

96. Paddock ML, Feher G, Okamura MY. 1995. Pathway of proton transfer in bacterial reaction centers: further investigations on the role of Ser-L223 studied by site-directed mutagenesis. *Biochemistry* 34: 15742–50

97. Paulsen H. 1995. Chlorophyll a/b binding proteins. *Photochem. Photobiol.* 62:367–82

98. Pecoraro VL, Gelasco A, Baldwin M. 1995. Modeling the chemistry and properties of multinuclear manganese enzymes. In *Bioinorganic Chemistry,* ed. DP Kessissoglu, pp. 287–98. Dordrecht: Kluwer

99. Peter GF, Thornber JP. 1991. Biochemical evidence that the higher plant photosystem II core complex is organized as a dimer. *Plant Cell Physiol.* 32:1237–50

100. Peter GF, Thornber JP. 1991. Biochemical composition and organisation of higher plant photosystem II light harvesting pigment-proteins. *J. Biol. Chem.* 266: 16745–54

101. Popov VI, Tageeva SV, Kaurov BS, Gavrilov AG, Rubin AB, Rubin LB. 1977. Effect of ruby laser illumination on the ultrastructure of pea chloroplast membranes. *Photosynthetica* 11:76–80

102. Putnam-Evans C, Bricker TM. 1992. Site-directed mutagenesis of the CP-1 protein of photosystem II: alteration of the basic pair 384, 385R to 484, 385G leads to a defect association with the O_2 evolving complex. *Biochemistry* 31:11482–88

103. Renger G, Rutherford AW, Völker M. 1985. Evidence for resistance of the micro environment of the primary plastoquinone ac-

ceptor ($Q_A^-Fe^{2+}$) to mild trypsination in PSII particles. *FEBS Lett.* 185:243–47

104. Rigoni F, Barbato R, Friso G, Giacometti GM. 1992. Evidence for direct interaction between the chlorophyll-proteins CP29 and CP47 in Photosystem II. *Biochem. Biophys. Res. Commun.* 184:1094–100

105. Rochaix JD, Erickson J. 1988. Function and assembly of photosystem II-genetic and molecular analysis. *Trends Biochem. Sci.* 13:56–59

106. Rögner M, Boekema EJ, Barber J. 1996. How does photosystem 2 split water? The structural basis of efficient energy conversion. *Trends Biochem. Sci.* 21:44–49

107. Ruffle SV, Donelly D, Blundell TL, Nugent JHA. 1992. A three dimensional model of the photosystem II reaction centre of *Pisum sativum. Photosynth. Res.* 34:287–300

108. Rutherford AW, Zimmermann JL, Boussac A. 1992. Oxygen evolution. See Ref. 16a, 2:179–229

109. Santini C, Tidu V, Tognon G, Ghiretti Magaldi A, Bassi R. 1994. Three-dimensional structure of higher plant photosystem II reaction centre and evidence for its dimeric organization. *Eur. J. Biochem.* 221:307–15

110. Satoh K. 1993. Isolation and properties of the photosytem II reaction center. In *The Photosynthetic Reaction Center,* ed. J Deisenhofer, JR Norris, pp. 289–318. New York: Academic

111. Satoh K. 1996. Introduction to the Photosystem II reaction center: isolation and biochemical and biophysical characterisation. See Ref. 95a, pp. 193–207

112. Seibert M, DeWit M, Staehelin LA. 1987. Structural localization of the O_2-evolving apparatus to multimeric (tetrameric) particles on the lumenal surface of freeze-etched photosynthetic membranes. *J. Cell Biol.* 105:2257–65

113. Shinkarev VP, Takahashi E, Wraight CA. 1993. Flash-induced electric potential generation in wild type and L212EQ mutant chromatophores *of Rhodobacter sphae roides:* Q_BH_2 is not released from L212EQ mutant reaction centers. *Biochim. Biophys. Acta* 1142:214–16

114. Seidler A, 1996. The extrinsic polypeptides of photosystem II. *Biochim. Biophys. Acta* 1277:35–60

115. Simpson DJ. 1978. Freeze-fracture studies on barley plastid membranes II. Wild-type chloroplast. *Carlsberg Res. Commun.* 43: 365–89

116. Simpson DJ. 1979. Freeze-fracture studies on barley plastid membranes. III. Location of the light harvesting chlorophyll-protein. *Carlsberg Res. Commun.* 44:305–36

117. Simpson DJ. 1982. Freeze-fracture studies on barley plastid membranes *V. viridis-n34,* a photosystem I mutant. *Carlsberg Res. Commun.* 47:215–25

118. Simpson DJ, Lindberg Møller B, Høyer-Hansen G. 1978. Freeze-fracture structure and polypeptide composition of thylakoids of wild type and mutant barley plastids. In *Chloroplast Development,* ed. G Akoyunoglou, pp. 507–12. Amsterdam: Elsevier/North-Holland Biomed.

119. Staehelin LA. 1975. Chloroplast membrane structure. Intramembranous particles of different sizes make contact in stacked membrane regions. *Biochim. Biophys. Acta* 408:1–11

120. Staehelin LA. 1976. Reversible particle movements associated with unstacking and restacking of chloroplast membranes *in vitro. J. Cell Biol.* 71:136–58

121. Staehelin LA. 1986. Chloroplast structure and supramolecular organization of photosynthetic membranes. In *Photosynthesis III, Photosynthetic Membranes and Light Harvesting Systems,* ed. LA Staehelin, CJ Arntzen, pp. 1–84. Berlin: Springer

122. Staehelin LA, Armond PA, Miller KR. 1976. Chloroplast membrane organisation at the supramolecular level and its functional implications. *Brookhaven Symp. Biol.* 28:278–315

123. Svensson B, Vass I, Cedergren E, Styring S. 1990. Structure of donor side components in photosystem II predicted by computer modelling. *EMBO J.* 9:2051–59

124. Tae GS, Black MT, Cramer WA, Vallon O, Bogorad L. 1988. Thylakoid membrane protein topography: Transmembrane orientation of the chloroplast cytochrome b559 *psbE* gene product. *Biochemistry* 27:9075–80

125. Takahashi Y, Asada K. 1991. Determination of the molecular size of the binding site for the manganese-stabilizing 33 kDa protein in photosystem II membranes. *Biochim. Biophys. Acta* 1059:36–64

126. Tang XS, Fushimi K, Satoh K. 1990. D1-D2 complex of the photosystem II reaction center from spinach: isolation and partial characterization. *FEBS Lett.* 273:257–60

127. Tommos C, Tang XS, Warncke K, Hoganson CW, Styring S, et al. 1995. Spin-density distribution, conformation and hydrogen bonding of the redox-active tyrosine Y_Z in

PSII from multiple electron magnetic-resonance spectroscopies: implications for photosynthetic oxygen evolution. *J. Am. Chem. Soc.* 117:10325–35

128. Tomo T, Enami I, Satoh K. 1993. Orientation and nearest neighbor analysis of psbI gene product in the photosystem II reaction center complex using bifunctional cross-linkers. *FEBS Lett.* 323:15–18

129. Trebst A. 1986. The topology of plastoquinone and herbicide binding peptides of photosystem II in the thylakoid membrane. *Z. Naturforsch. Teil C* 41:240–45

130. Tsiotis G, Walz T, Spyridaky A, Lustig A, Engel E, Ghanotakis D. 1996. Tubular crystals of a photosystem II core complex. *J. Mol. Biol.* 259:241–48

131. Tsvetkova NM, Apostolova EL, Brain APR, Williams WP, Quinn PJ. 1995. Factors influencing PSII particle array formation in *Arabidopsis thaliana* chloroplasts and the relationship of such arrays to the thermostability of PSII. *Biochim. Biophys. Acta* 1228:201–10

132. van Mieghem FJE, Satoh K, Rutherford AW. 1991. A chlorophyll tilted 30° relative to the membrane in the PSII reaction centre. *Biochim. Biophys. Acta* 1058:375–78

133. Wang DN, Kühlbrandt W. 1991. High-resolution electron crystallography of light-harvesting chlorophyll-a/b-protein complex in 3 different media. *J. Mol. Biol.* 217:691–99

134. Webber AN, Packman LC, Chapman DJ, Barber J, Gray JC. 1989. A 5th chloroplast-encoded polypeptide is present in the photosystem II reaction center complex. *FEBS Lett.* 242:259–62

135. Xu Q, Bricker TM. 1992. Structural organization of proteins on the oxidizing side of Photosystem II. *J. Biol. Chem.* 267:25816–21

136. Yamamoto Y, Tabata K, Isogai Y, Nishimura M, Okayama S, et al. 1984. Quantitative analysis of membrane components in a highly active O_2-evolving photosystem II preparation from spinach chloroplasts. *Biochim. Biophys. Acta* 767:493–500

137. Yu C-A, Xia J-Z, Kachurin AM, Yu L, Xia D, et al. 1996. Crystallization of preliminary structure of beef heart mitochondrial cytochrome-bc1 complex. *Biochim. Biophys. Acta* 1275:47–53

Annu. Rev. Plant Physiol. Plant Mol. Biol. 1997. 48:673–701

GENETICS OF ANGIOSPERM SHOOT APICAL MERISTEM DEVELOPMENT

Matthew M. S. Evans and M. Kathryn Barton

Department of Genetics, University of Wisconsin-Madison, Madison, Wisconsin 53706

KEY WORDS: plant developmental genetics, leaf development, stem cell, branching, embryogenesis

ABSTRACT

Recent progress has been made in the genetic dissection of angiosperm shoot apical meristem (SAM) structure and function. Genes required for proper SAM development have been identified in a variety of species through the isolation of mutants. In addition, genes with expression patterns indicating they play a role in SAM function have been identified molecularly. The processes of SAM formation, self-renewal, and pattern formation within the SAM are examined with an emphasis on the contributions of recent classical and molecular genetic experiments to our understanding of this basic problem in plant developmental biology.

CONTENTS

1040-2519/97/0601-0673$08.00

INTRODUCTION

Angiosperm shoot and root apical meristems consist of a small number of morphologically undifferentiated, dividing cells located at the tips of shoots and roots, respectively. The word meristem itself derives from the Greek *merizein,* meaning "to divide." Both shoot and root apical meristems arise during embryogenesis and act throughout the lifetime of the plant as sources of new cells and as sites of organ formation. It is at the shoot apical meristem (SAM) that leaf, stem, and floral primordia are formed. It is not surprising that much research in plant developmental biology has concentrated on how apical meristems work (reviewed in 91). Recently a number of mutants affecting the SAM have been described. This review focuses on how such mutants have contributed to our understanding of development within the SAM.

Although SAMs appear as small regions of morphologically undifferentiated, dividing cells, a group of morphologically undifferentiated, dividing cells does not necessarily constitute a SAM. Rather, increasing genetic and molecular evidence indicates that SAMs are highly organized, or patterned, regions of the plant in which many important events in early organogenesis occur. Thus, the term "SAM" is used to denote a highly organized structure and site of pattern formation, while the more generic term "meristematic" is used to denote mitotically dividing, densely cytoplasmic cells found at many locations in the plant, e.g. in young leaf primordia.

The small size of the SAM and the presence of the plant cell wall pose technical barriers to performing the types of surgical and biochemical experiments traditionally used in developmental biology. Genetics provides a powerful way to circumvent the technical difficulties involved in working with rare molecules, which developmental regulators are likely to be. Genes that encode rare products are, generally speaking, just as susceptible to mutation as those encoding abundant products. Once genes important for a particular process have been identified, molecular cloning and genetics can provide the tools necessary to initiate a biochemical approach to the problem. Further, a combination of molecular and classical genetics can be used to generate genetic mosaics, the study of which can answer questions otherwise approached by transplantation and surgery.

Issues to be discussed in this review concern SAM formation and function. SAMs are formed at distinct times and locations—in the embryo, in the axils of leaves, and in modified form as floral meristems on the inflorescence axis. The pattern of meristem formation can be a major determinant of plant architecture. Once formed, SAMs have two main functions. First, they maintain a self-renewing population of initial cells. (These are analogous to stem cells in animals. The term "stem cell" has, of course, another use in plant biology.) Second, they produce the two fundamental body parts of the shoot system: leaves and stem. These two functions of the SAM must be spatially coordinated; leaves develop on the flanks of the SAM in a defined arrangement, whereas initial cells reside in the center of the SAM. It has long been recognized that cell fate determination within the SAM depends on positional cues, but the nature and organization of this positional information is largely a mystery. Mutants defective in aspects of SAM formation and/or function have been found in several plant species, and careful study of these has yielded important advances in our understanding of the inner workings of the angiosperm SAM.

STRUCTURE OF THE SHOOT APICAL MERISTEM

The typical SAM is about 100 to 200 microns in diameter, although species with SAMs as small as 50 microns and as large as 900 microns have been reported (25). The morphology of the SAM varies from flat [e.g. sunflower 90)] through mound-shaped [e.g. the Arabidopsis vegetative SAM (59; Figure 1)] to finger-shaped [e.g. maize (92)]. The size and shape of the SAM vary within a plant during the plastochron (the period between the formation of successive leaves). Toward the end of the plastochron, just before the emer-

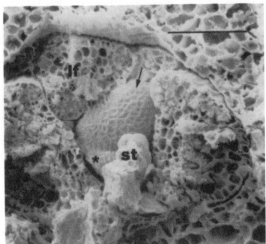

Figure 1   Scanning electron micrograph of a vegetative SAM of *Arabidopsis*. Leaves (lf) have been dissected so that the SAM can be viewed. st, stipule; *, leaf primordium; *arrow* indicates position at which the next leaf primordium will form. Scale bar equals 50 microns. (Photograph courtesy of J McConnell.)

gence of a morphologically distinct leaf primordium, the SAM is at its largest, whereas at the beginning of a plastochron, just after a leaf primordium has been produced, the SAM is at its smallest. Presumably, near the end of the plastochron the SAM includes a leaf primordium that is not yet morphologically distinct from the SAM.

One of the most obvious features of the angiosperm SAM is its layered arrangement (Figure 2). The outermost region of the meristem consists of one or more sheets of cells that together comprise the tunica. This sheet-like organization is a consequence of the regular orientation of cell divisions within the tunica. The tunica overlies a mass of cells called the corpus. The irregular arrangement of cells within the corpus reflects the variable orientation of cell division in this region. However, in the basal/central region of the SAM, the rib meristem, composed of files of cells parallel to the long axis of the plant, can often be distinguished (Figure 2). Derivatives of the rib meristem give rise to internal layers of the stem.

Cell layers within the SAM are designated L1, L2, L3, etc, where L1 refers to the most superficial layer, L2 to the next subjacent layer, and so on. Thus, the Arabidopsis SAM consists of three layers: L1 and L2 comprise the tunica while L3 is the corpus (35, 45, 59). In maize, there are two layers in the SAM, the single-layered tunica (L1) and the corpus (L2) (92). Despite the nearly universal occurrence of a stratified SAM among the angiosperms(25), the function of this layering, if any, in SAM development or function is unknown.

Although highly regular patterns of cell division are found in the angiosperm SAM, the study of chimeric plants shows that the position of a cell

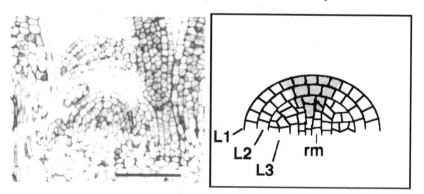

Figure 2. (*A*) Longitudinal section through Arabidopsis SAM. (*B*) Schematic drawing of SAM organization. The L1 and L2 together constitute the tunica. The L3 makes up the corpus. Stippling indicates the central zone. The peripheral zone surrounds the central zone and the rib zone lies just below it. rm, rib meristem. Scale bar equals 50 microns. (Photograph courtesy of J McConnell.)

rather than its lineage determines its fate [for a recent review, see Szymkowiak & Sussex (96)]. Traditionally, three positions or zones—central, peripheral, and rib—have been recognized within the SAM (91; Figure 2). Cells in the central zone stain less intensely with standard histological stains. However, a central zone is not histologically distinguishable in some species, and in others detection depends on the stage of SAM development. Cells in the central zone also have a slower rate of cell division than surrounding cells. The peripheral zone surrounds the central zone and is the site of leaf formation. Both the peripheral and central zones overlap the tunica and the corpus. The rib zone is subjacent to the central zone and does not overlap the tunica layers.

SHOOT APICAL MERISTEM FORMATION

Shoot Apical Meristem Formation During Embryogenesis

The SAM develops in different positions within dicot and monocot embryos. In dicots, the SAM develops between the two cotyledon primordia, in the central and apical portion of the embryo (Figure 3). In monocots, the SAM develops laterally on the embryo, at the base of the single scutellum, a leaf-like structure comparable in its embryonic origin to the cotyledons of the dicot embryo (72; Figure 4). Once formed, how much the SAM is active during embryogenesis varies. In maize the first five leaves are produced during embryogenesis (72), whereas only two small leaf primordia are detected by the end of Arabidopsis embryogenesis (59).

In Arabidopsis, most of the cotyledons and the entire SAM develop from the apical half of the globular embryo (Figure 3). It is clear both from histological studies in species with regular patterns of cell division during embryogenesis, such as Arabidopsis (2), and from clonal analysis studies of species with irregular patterns of cell division, such as cotton (10), that the cotyledons and SAM develop from distinct regions of the globular embryo. Three models describe the development of the apical region of the globular embryo. The cotyledons and SAM may be specified independently of one another (Figure 5, Model A). In this model, cotyledons and SAM respond to positional cues independently to establish their respective cell fates. Alternatively, the specification of one may be dependent on the prior specification of the other (Figure 5, Models B, C). In Model B, the cotyledons are required for SAM formation, whereas in Model C the SAM is required for cotyledon formation. The existence of mutations that "delete" the SAM but not the cotyledons (e.g. mutations in the *STM* gene described below) suggest that the specification of the cotyledons is not dependent on the specification of the SAM, weakening the case for

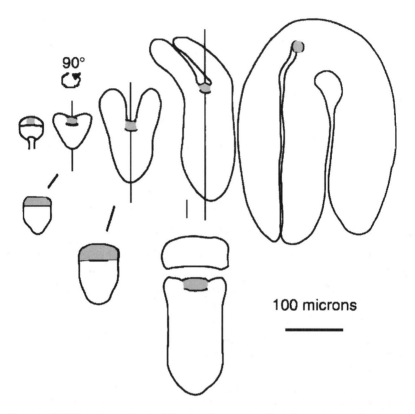

Figure 3 SAM formation during Arabidopsis embryogenesis. Shaded areas indicate the pattern of *STM* expression (49). This pattern coincides with those cells predicted to give rise to the SAM based on histological observations (2). Upper five diagrams represent embryo development from the globular through the heart, torpedo, walking stick, and mature stages of embryogenesis. Lower three diagrams represent sagittal sections through the heart, torpedo, and walking stick stage embryos and show loss of *STM* expression on the flanks of the embryo late in embryogenesis.

Model C. The existence of mutations that "delete" both the cotyledons and the SAM (8; J Long & K Barton, unpublished observations) are consistent with both Models A and B. Such mutations could prevent the specification of the cotyledons and thereby also prevent the SAM from forming (Model B), or they could affect gene products used in the transduction of positional information to both presumptive cotyledon and SAM primordia (Model A).

Mutations exist in several species that result in a failure of the SAM to form properly during embryogenesis. Some were discovered during systematic screens for mutants affecting SAM formation (2, 42, 57) or embryo develop-

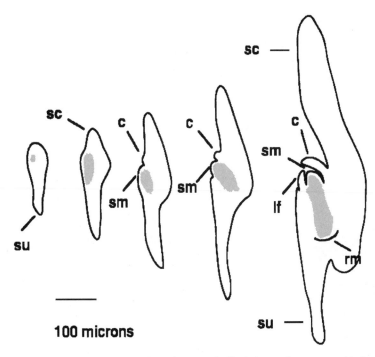

sc —

sc

c

c

c

sm

sm

sm

sm

lf

su

rm

100 microns

su —

Figure 4   SAM formation during maize embryogenesis. Shaded areas show pattern of *kn1* expression (85). The maize SAM forms on the side of the embryo. The coleoptile is a leaf-like organ that sheaths and protects the developing shoot as it grows through the soil. Note that in contrast with the pattern of *STM* expression in Arabidopsis, *kn1* expression extends to the root pole of the embryo. sm, shoot apical meristem; c, coleoptile; sc, scutellum; lf, first leaf; su, suspensor; rm, root apical meristem. Stages of embryogenesis are after Randolph (72).

ment (56, 61), whereas others were discovered fortuitously by observant researchers in more general screens (8). Recessive mutations in the *GURKE* (*GK*) gene cause the cotyledons and SAM to be reduced or absent (56). *gurke* mutants also have reduced roots and malformed hypocotyls (100). *gk* mutations are allelic with *emb22* mutations (D Meinke, personal communication), which were discovered by Meinke & Sussex (61) in a broad screen for embryo lethal mutants. *emb22* mutants described by Franzmann et al (20) do not have the apical deletion phenotype reported for *gk* mutants but are more generally defective in morphogenesis and have been described as "green blimps." *emb22* mutants are also defective in cellular differentiation, as determined by the failure of protein bodies to accumulate during embryogenesis (67). When cultured, *emb22* tissue forms abnormal, thick, tube-shaped leaves (20). The

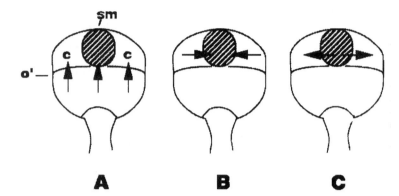

Figure 5. Three simple models for cotyledon and SAM specification during embryogenesis. In Model A, cotyledon and SAM fates are specified independently of one another, in response to positional cues from other parts of the embryo. For the sake of illustration only, the positional cues that determine SAM and cotyledon fate are shown as emanating from the lower half of the embryo in Model A. Other sources of positional information are, of course, possible. In Model B, the specification of SAM fate is dependent on input from the presumptive cotyledon. In Model C, the specification of cotyledon fate is dependent on input from the SAM. sm, region from which shoot apical meristem derives; c, region from which cotyledons derive; o', o' boundary that separates top and bottom halves of embryo. The cell walls that form this boundary are laid down at the eight-cell stage of embryogenesis.

less severe phenotype of *gurke* mutants relative to *emb22* mutants suggests that *gk* alleles are weak alleles of *emb22*. Given the pleiotropic nature of the *gk/emb22* phenotype, it is unlikely that wild-type *GK/EMB22* gene function is required specifically for the development of apical pattern elements as was originally hypothesized (56). While pleiotropy does not rule out an important role for *GK/EMB22* in apical development, additional studies are required to determine what that role might be.

A recessive mutation in the Arabidopsis *TOPLESS* (*TPL*) gene also causes a failure in cotyledon and SAM formation during embryogenesis (J Long & K Barton, unpublished observations). Only one mutant allele of this locus exists thus far, and the phenotype conferred by it is highly variable. Severely affected *tpl* mutants consist of only root and hypocotyl with no evidence of cotyledons or SAM. Less severely affected seedlings are radially symmetric with a collar, or tube, of cotyledon tissue and no SAM. The least severely affected seedlings make a single nearly normal cotyledon and a SAM. These SAMs form shoots that exhibit only minor floral defects (failure of carpels to fuse and the production of chimeric organs, such as petaloid sepals). The observation that shoots, once formed, are nearly normal, has several possible explanations. The *TPL* gene may be required only for some aspect of embryonic pattern formation, but not for postembryonic SAM function. Alternatively, less *TPL* activity may be required postembryonically than during embryogenesis. Because the *TPL* gene is represented by only one mutant allele, it is possible that stronger alleles would result in more extreme embryonic and/or postembryonic phenotypes.

In contrast with the recessive mutations in the *TPL* and *GK/EMB22* genes, the *Lanceolate* (*La*) mutation of tomato is incompletely dominant (8). Wild-type tomatoes have compound leaves, whereas *La* heterozygotes have simple, lanceolate leaves. Homozygotes are more profoundly affected and can be grouped into three classes: those with extremely lanceolate leaves, those that produce one cotyledon and sometimes a SAM, and those that entirely lack both cotyledons and a SAM. The latter class is similar to the most severe class of *TPL* mutants. Because it is dominant, *La* may be a gain-of-function mutation, making it difficult to deduce the function of the wild-type LA^+ gene in development. If the *La* mutation causes LA^+ function to be over- or ectopically expressed, then LA^+ would be required to repress cotyledon and SAM development in positions, or at times, at which apical structures do not normally develop. However, if the lesion in the LA^+ gene is a dominant negative mutation, then LA^+ would be required for apical structures to form.

The Arabidopsis *SHOOTMERISTEMLESS* (*STM*) gene is required for embryonic SAM formation (2). Seedlings homozygous for recessive loss-of-function alleles germinate with roots, hypocotyl, and cotyledons, but no SAM is formed. The *STM* gene encodes a *Knotted1* (*Kn1*) type of homeodomain protein (49). Homeodomain proteins regulate transcription in many species (24) and have been shown to play a role in the regulation of translation as well (16). The maize KN protein was the first plant protein shown to contain a homeodomain (103). Many *KN1*-like genes have been found since then and have been organized into categories on the basis of their sequence similarities (40). In both maize (40), and Arabidopsis (48), *KN1*-like genes make up a multigene family. *STM* belongs to the class of *KN1*-like genes that is most closely related to *kn1* (49).

During Arabidopsis embryogenesis, the SAM forms as a small bump consisting of three layers of cells between the cotyledons. By the end of embryogenesis, cell division patterns have been established in these layers that are characteristic of the tunica (L1 and L2)/corpus (L3) organization of the vegetative SAM (2). In *stm* mutant embryos, cells that normally divide to form this bump fail to do so (2). In wild-type embryos, the *STM* transcript accumulates in cells that give rise to the embryonic SAM (49). *STM* transcript is found earliest in one or two cells of the late globular stage embryo and subsequently spreads to form a stripe between the presumptive cotyledon primordia (Figure

3). Thus, cotyledons derive from a region of the embryo that never expresses *STM*, and they differ from true leaves in this regard.

The precise localization of the *STM* transcript within the embryo indicates that *STM* expression is sensitive to positional information present in the globular stage embryo. Moreover, the expression of *STM* as a stripe of expression along the top half of the globular embryo indicates that the embryo possesses bilateral symmetry at this early stage. The mechanism by which this positional information is transduced is unknown, but the *TPL* gene may play a role because *STM* transcript fails to accumulate in *tpl* mutant embryos (J Long & K Barton, unpublished observations). Is bilateral symmetry required for a SAM to develop? Evidence in favor of such a requirement is the correlation between loss of bilateral symmetry and failure to form a SAM in *tpl* and *La* mutants (and in the *cuc* mutant described below): mutant seedlings that develop with radial symmetry fail to form a SAM, whereas those that form with bilateral symmetry frequently manage to form a SAM (8; J Long & K Barton, unpublished observations; M Aida & M Tasaka, personal communication).

Expression of *STM* is dynamic during embryogenesis. Cells on the flanks, or periphery, of the stripe of STM expression initially express *STM* at levels similar to the central regions of the stripe but cease expression by the end of embryogenesis (J Long & K Barton, unpublished observations; Figure 3). These lateral regions of the stripe are positions that form the "valleys" between the cotyledons. *STM* may function in this region to cause separation of the cotyledons because in strong *STM* alleles the petioles of the cotyledons frequently are fused (12) (J Long & K Barton, unpublished observations). Thus, loss of *STM* function appears to have different consequences for the center of the embryonic SAM (too few cell divisions resulting in failure of the SAM "bump" to form) and the flanks of the embryonic SAM (too many cell divisions resulting in cotyledon fusion).

Genes homologous to *STM* may play similar roles in SAM formation in other species. The *kn1* gene of maize and the *OSH1* gene of rice both are expressed at the site of the presumptive SAM and are not expressed in the scutellum primordium (73, 85; Figure 4). In later stages of embryogenesis the expression of *kn1* differs from that of *STM* in that it extends into the root pole of the embryo (85). Because no loss-of-function alleles of the *kn1* or *OSH1* genes have been reported, the role of these genes in monocot SAM development is unknown. However, the similarity in expression pattern and in amino acid sequence suggest that this class of genes may be required generally for SAM formation in the angiosperm embryo.

As a homeodomain-containing protein, *STM* may promote SAM formation by transcriptionally or translationally regulating other genes. A candidate target for *STM* is the *UNUSUAL FLORAL ORGANS (UFO)* gene. *UFO* is hypothesized to create floral organ boundaries in the floral meristem (44, 106). *UFO* is expressed in both the vegetative and the embryonic SAM (I Lee & D Weigel, personal communication). In the embryo, *UFO* is initially expressed in a domain similar to that of *STM*. This expression is dependent on *STM* function, because *stm* mutant embryos fail to express *UFO* (J Long & K Barton, unpublished results). Later *UFO* expression encircles the embryonic SAM, suggesting that it plays a role in creating a boundary between the SAM and the cotyledons. Because *ufo* mutants do not exhibit a phenotype during embryonic or vegetative development, the importance of the *UFO* gene product in SAM development is unclear, it may play no role in SAM development, or it may be redundant to the function of another, as yet unidentified, gene.

Similar to the *STM* gene, the *CUPSHAPED COTYLEDON 2 (CUC2)* locus of Arabidopsis (M Aida & M Tasaka, personal communication) and the *no apical meristem (nam)* (88) locus of Petunia are required for both SAM formation and cotyledon separation. The predicted amino acid sequences of their products and their expression patterns suggest that these genes do so by a different mechanism than *STM*. *CUC2* and *nam* encode similar proteins of unknown biochemical function and confer similar phenotypes when mutant. Arabidopsis *cuc1;cuc2* double mutant seedlings and Petunia *nam* mutant seedlings frequently lack a SAM and exhibit variable degrees of cotyledon fusion, with cotyledon tissue sometimes forming a collar or cup around the entire apex of the seedling. In Petunia, only the *nam* gene need be mutant for a SAM-less phenotype to be observed. In contrast, Arabidopsis seedlings must be homozygous for *CUC2* as well as for a mutation at the unlinked *CUC1* locus for a SAM-less phenotype to be apparent. The only phenotype seen in *cuc1* or *cuc2* single mutants is fusion of some floral organs. The NAM and CUC2 proteins are members of multigene families in both Arabidopsis and Petunia. Given its apparent functional redundancy to *CUC2* in the embryo, *CUC1* may encode another member of this gene family. In contrast to *STM*, *nam* gene expression is first detected at the late heart stage of embryogenesis, where, as in the later phase of *UFO* expression, it appears as a ring around the developing SAM. The failure of *nam* expression to coincide with the SAM suggests that *NAM* acts in a non-cell-autonomous manner to allow SAM formation. However, the presence of low levels of *nam* mRNA within the SAM cannot be ruled out. Postembryonically, *nam* RNA is found in rings around organ and meristem primordia in vegetative and floral tissues. Because *nam* mutants exhibit defects in only a subset of these tissues (*nam* mutants have increased numbers of petals and variably abnormal gynoecia), it has been hypothesized that other members of the *nam* gene family may compensate for lack of *nam* function in regions that are phenotypically unaffected in *nam* mutants (88).

Another gene required for embryonic SAM formation is the Arabidopsis *PINHEAD* (*PNH*) gene (57). All *pnh* alleles isolated thus far exhibit highly variable phenotypes: the most severely affected *pnh* mutants form a small, flat group of cells at the site normally occupied by the SAM. More commonly, the SAM terminates in a single, determinate organ, examples of which are trumpet-shaped leaves, single leaves, or slender pin-like organs. The production of any one of these structures at the apex in *pnh* mutants requires wild-type *STM* activity because *stm* is epistatic to *pnh* (57). Consistent with this, early expression of the *STM* gene in *pnh* mutants appears to be normal (J Long & K Barton, unpublished observations). However, later *STM* expression in *pnh* mutants extends further laterally than in the wild type. The apical placement of leaf primordia within the SAM and the abnormal localization of *STM* expression in *pnh* mutants suggest that *PNH* is required for some aspect of positional information in the Arabidopsis embryo. The radialization of leaf primordia may reflect a direct role for *PNH* in specifying positional information within the leaf primordium, or an indirect effect due to the development of the leaf primordium in the wrong position on the meristem.

In *pnh* mutants in which the SAM terminates early, new SAMs form in the axils of the cotyledons. These SAMs go on to produce leaves and flowering shoots (57). The positions of leaves and flowers, and the number and positions of floral organs in these shoots, are frequently abnormal (K Barton, unpublished observations). Nonetheless, such *pnh* mutant shoots do not terminate prematurely, indicating that SAMs, once established, are less sensitive to loss of *PNH* function than newly organizing SAMs are.

Finally, mutations in the *ZWILLE* (*ZLL*) gene also block the development of the SAM at an early stage (17a, 37). These mutants, similar to *pnh* mutants, form SAMs that produce flowering shoots in the axils of the cotyledons. *ZLL* function seems to be required only for embryonic SAM formation because no postembryonic defect is seen in *zll* mutants.

Mutations in the *MONOPTEROS* (*MP*) gene differ from other mutants in SAM formation described thus far (with the exception of the *gk*/*emb22* mutants) in that they affect both root and shoot apical meristem formation during embryogenesis. *mp* mutants were first characterized as basal deletion mutants that lack root and hypocotyl (56). However, *rootless* mutants [*rootless* mutations are allelic to *mp* mutations (K Barton, unpublished observations)] frequently fail to form a SAM (2) and often exhibit a single cotyledon or two fused cotyledons (5). It is unclear whether the failure of *mp* mutants to form a

SAM is an indirect effect of the defect in hypocotyl and root formation or whether the *MP* gene plays a more direct role in embryonic SAM formation.

Shoot Apical Meristem Formation on the Axis of the Plant: Making Branches and Flowers

SAMs form throughout the lifetime of most higher plants in lateral positions along the main shoot axis. Three types of SAMs form postembryonically: vegetative lateral SAMs, inflorescence lateral SAMs, and floral meristems. Molecular and classical genetic evidence demonstrates that lateral SAMs share many common genetic requirements with the primary SAM formed in the embryo. Gene expression data support a role for *kn1* in maize (36), *STM* in Arabidopsis (49), and other *KN*-like genes (48, 77) in lateral SAM function. The phenotypes of the Arabidopsis *stm, clavata1* (*clv1*), *clv3, pnh,* and *wuschel* mutants demonstrate that these genes act in lateral SAMs as well as in the primary SAM (13, 14, 19, 42, 57).

Axillary SAMs, particularly those that only become evident several nodes from the apex, have been proposed to originate either through de novo SAM formation or through a detached meristem (91). The important distinction is whether SAM cell identity is only acquired once during embryogenesis and is perpetuated throughout the life of the plant, or whether groups of cells acquire SAM identity at different times during plant development. Genetic regulation of lateral SAM formation is only one step in the control of plant architecture. Many mutants that affect the number of visible branches do not affect lateral SAM formation. Plants can change the number of visible branches by regulating bud outgrowth or by replacing one type of lateral SAM with another. One factor that appears to be important for both postembryonic SAM formation and bud outgrowth is cytokinin, which can cause SAM formation in tissue culture and in normally empty axils of *Stellaria media* (80, 98).

To understand the effect of different genes on lateral SAM formation, it is important to be familiar with branching patterns in a few model species (Figure 6). These descriptions also serve to illustrate the influence that postembryonic SAM formation can have in determining plant architecture. In Arabidopsis, vegetative branches (rosette coflorescences) are produced by lateral SAMs initiated in the axils of rosette leaves several nodes below the shoot apex. Inflorescence branches (cauline coflorescences) are initiated in a basipetal direction upon floral induction starting in the axils closest to the apex (31). Rarely, a second (i.e. accessory) branch will arise in the axil of a leaf in Arabidopsis (97).

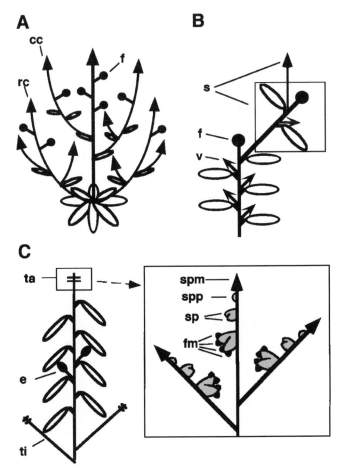

Figure 6 Branching patterns in Arabidopsis, Petunia, and maize. (*A*) Branching in Arabidopsis. Arabidopsis makes rosette coflorescences (*rc*) during the vegetative phase of development and makes cauline coflorescences (*cc*) and flowers (*f*) during the reproductive phase. (*B*) Sympodial branching in Petunia. Vegetative lateral SAMs (*v*) are initiated in the axils of the leaves until plants undergo the transition to flowering. Then the shoot apex terminates in a flower (*f*), and the sympodial meristem (*s*) is initiated in the axil of the last leaf. After making two leaves, the sympodial meristem also terminates in a flower and a new sympodial meristem (*s*) is initiated in the axil of the last leaf repeating the process. (*C*) Branching in maize. The vegetative lateral SAMs of maize can form tillers (*ti*) or ear shoots (*e*), although many vegetative lateral SAMs remain dormant. In the staminate inflorescence, the tassel (*ta*) and spike meristems (*spm*), which produce indeterminate lateral branches, form from the primary inflorescence SAM at the base of the inflorescence. Spikelet pair primordia (*spp*) form from the spike and primary inflorescence SAMs. Each spikelet pair primordia forms two spikelet primordia (*sp*), and each spikelet forms two floral meristems (*fm*). Inflorescence branching is the same in the ear except that spike meristems are not formed.

Maize has a more complex branching pattern. Vegetative lateral SAMs, which can form either tillers or ear shoots, develop in the axils of leaves when the axil is several nodes from the apex (43). The inflorescences of maize develop in terminal positions: the tassel from the primary SAM and the ears from the uppermost lateral SAMs. Within the inflorescences, the branching pattern is even more complex, with three orders of branching before floral meristem formation.

Tomato and Petunia exhibit a branching pattern termed sympodial growth (30, 70, 74). Vegetative lateral SAMs are formed in the axils of leaves several nodes from the apex, as in Arabidopsis and maize. However, upon transition to flowering the SAM terminates in either a floral meristem (Petunia) or a determinate inflorescence (tomato). Further indeterminate growth is taken over by the sympodial meristem, which forms in the axil of the last leaf directly from the primary SAM (70). The sympodial meristem then repeats the process.

Lateral SAMs frequently develop in the axils of subtending lateral organs. In species with obvious bracts (e.g. snapdragon) this relationship is clear. However, in many species mature flowers and inflorescence branches lack subtending organs. Nevertheless, during ontogeny these SAMs may develop in the axil of an organ that remains rudimentary or degenerates, as in the case of the leaf rudiments that subtend tassel branches in maize (23). When a lateral SAM develops in the axil of a leaf, it may develop from the adaxial surface of the leafbase, as in Arabidopsis (97), or from the "stem side" of the axil, as in maize. This difference in lateral SAM development is also reflected in cell lineage; the lateral SAM is usually clonally related to the leaf below it in Arabidopsis (22, 35) and to the leaf above it in maize (58). It is unclear whether the appropriate positional cues for maize lateral SAM development are abaxial signals from the stem or, as seems likely in Arabidopsis, the medial adaxial signals from the leaf below.

Some mutations increase the frequency of formation of lateral SAMs. In maize, the dominant *Fascicled ear1* mutant has highly branched inflorescences because the primary inflorescence and spike SAMs bifurcate (28). The *auxin resistant1* (*axr1*) mutant of Arabidopsis (47) and the *decreased apical dominance1* (*dad1*) mutant of Petunia (63) both produce accessory branches in leaf axils, and mutant *ap1* plants produce flowers in the axils of the sepals (34). The production of additional lateral SAMs in recessive mutants suggests the presence of factors that actively repress the formation of lateral SAMs in certain contexts.

Many more mutations decrease the frequency of lateral SAM formation. Some are specific for one type of lateral SAM, while others affect multiple types of SAMs. In Arabidopsis, *pnh* mutants have fewer rosette and cauline inflorescence SAMs and are defective in floral meristem formation (57; K Barton, unpublished observations). Mutations in *revoluta* (*rev*) similarly affect lateral SAM formation but more severely affect floral meristems, reducing them all to filaments (97). Strikingly, mutations in the *UFO* gene reduce lateral SAM formation but also increase coflorescence production by switching meristem identity (44, 106). The Arabidopsis *pinformed* (*pin*) and *pinoid* (*pid*) mutations disrupt the formation of floral meristems and cauline coflorescences but not rosette coflorescence SAMs (4, 65). *pin* and *pid* mutants make ridges of tissue instead of floral meristems, suggesting that floral meristems initiate and then arrest. The *MP* gene also is required for lateral SAM formation. Like *pin* and *pid*, mutant *mp* seedlings that survive until flowering bolt but fail to make flowers (71).

In tomato the *lateral suppressor* (*ls*) and *torosa-2* (*to-2*) mutations reduce formation of lateral SAMs (51, 52). *ls* specifically affects the initiation of vegetative lateral SAMs but not of floral or sympodial meristems. *ls* has pleiotropic effects; flowers of *ls* mutants fail to initiate petals. The *ls* mutation has no affect when present in the L1, whereas chimeras with the mutation in the L2 and L3 have the *ls* phenotype, which suggests that internal cell layers are important for regulating lateral SAM formation (95). In *Nicotiana*, the genotype of the L3 is capable of influencing the frequency of accessory lateral SAM formation, which provides further evidence for the importance of internal cell layers in lateral SAM regulation (99). *to-2* has a less severe effect on lateral SAM formation than *ls* and no effect on floral morphology (52, 53). Both *ls* and *to-2* mutants have reduced cytokinin levels, which suggests that this is a possible cause of the lack of axillary SAMs (53, 87, 101). Increasing the levels of cytokinin in both *to-2* and *ls* plants failed to increase lateral SAM formation, although dormancy of existing buds was removed in *to-2* (27, 52).

In maize, many mutations have been identified that reduce lateral branch formation. *barren stalk1* (*ba1*) mutants lack vegetative branches (i.e. they have no ear shoots or tillers), and mutant tassels have many fewer branches and spikelets (33, 66). Recessive mutations in the *ba2, ba3,* and *barren stalk fastigiate* (*BAF*) genes also block the formation of ears and tillers, but they have little (*ba2* and *baf*) or no effect (*ba3*) on tassel branch and spikelet formation (33, 66, 102). The exact point at which branch formation is blocked in *ba1, ba2, ba3,* and *baf* is unknown.

Several mutations specifically reduce inflorescence branches. *Suppressor of sessile spikelet1* (*Sos1;* dominant) blocks initiation of the sessile spikelet meristem in the ear and less frequently in the tassel (15). *Sos1* also reduces the number of tassel branches and the number of rows of spikelet pair meristems. *Barren inflorescence1* (*Bif1;* dominant), *bif2* (recessive), and *barren sterile* (*bs;* recessive) reduce formation of tassel branches and spikelet pairs in both ears and tassels (7, 64, 84, 102). A recessive mutation that is a potential allele of *bif1* has a similar phenotype to the dominant *Bif1* mutation (84). *bif2* also reduces the formation of sessile spikelets (7). Another mutation with a similar phenotype, *tasselless,* may be an allele of *bs* (1). The inflorescences of *bif* mutants, at least superficially, resemble those of the Arabidopsis *pin, pid,* and *mp* mutants in the production of bare stem without lateral shoots.

The existence of genes that are required for lateral SAM formation but not for primary SAM formation demonstrates that there are different (or additional) genetic requirements for lateral SAM formation. In addition, because they are affected by different mutations, the different classes of lateral SAMs may be regulated by partially independent pathways. Alternatively, the different development contexts of the different classes of lateral SAMs may make them sensitive to different genetic lesions.

Adventitious Shoot Apical Meristem Formation

Plants can also be induced to form adventitious shoots in positions on the plant that normally do not produce them. SAMs produced in the normally empty axils of *S. media* upon cytokinin treatment (98) and in the axils of the cotyledons in Arabidopsis *pnh* mutants (57) can be called adventitial. However, these SAMs have similar positional contexts, the axils of lateral organs, as normal lateral SAMs. Here we focus on adventitious SAM development in tissue culture and from determinate organs in plants.

SAMs can form in culture in the apparent absence of positional information normally required to promote SAM formation. Adventitious SAM formation in vitro requires multiple steps (reviewed in 32). In tobacco, at least some of these SAMs form from multiple cells rather than a single cell, and cells from each layer of the leaf have this ability (54). In addition, cells incapable of responding to the shoot-inducing conditions used can be incorporated into newly formed SAMs if cells competent to respond are also incorporated (54).

Early work demonstrated the requirement for cytokinin to induce shoot formation in vitro (80). However, little has been done to elucidate the genetic requirements for SAM formation in vitro. Callus from *stm* mutant roots is unable to form shoots in vitro, demonstrating that a functional *STM* gene is required for shoot formation in culture as well as in the embryo (2). Unlike *STM, PNH* is not required for in vitro shoot formation (57).

Adventitious SAMs have also been produced in transgenic plants. Ectopic expression of two classes of genes, cytokinin biosynthetic genes and *KN*-like genes, promotes adventitious SAM formation. The ability of cytokinin to induce adventitious shoots has been tested in planta by overexpressing the *isopentenyl transferase* (*ipt*) gene (reviewed in 6). Ectopic expression of the *ipt* gene from the CaMV-35S promoter leads to adventitious shoot formation at the site of *Agrobacterium* infection, as well as a shooty phenotype of the transformed cells in culture (82). In these studies, whole plants are difficult to establish because of the inhibitory effect of cytokinin on root development, and a precise analysis of the timing and mode of cytokinin action on determined tissues is impossible because of the constitutive expression. Attempts to use a heat shock promoter gave mixed results (60, 76, 81). Transformants showed characteristic cytokinin phenotypes under noninductive conditions and in some experiments failed to exhibit phenotypic alterations upon heat shock even though cytokinin levels rose, suggesting that increased basal levels of cytokinin desensitized the plants to further increases. Greater success in demonstrating the ability of cytokinin to induce adventitious shoot formation was seen by limiting the tissues expressing the *ipt* gene either by driving its expression from the *SAUR* promoter or by somatically activating an *ipt* gene in clones of cells (18, 46). Both lines produced adventitious buds over veins on the adaxial surfaces of leaves. In both cases adventitious buds were associated with a localized increase of cytokinin rather than the dramatic increase seen with constitutive expressors. Some of the buds in CaMV-35S-*ipt* plants lack *ipt* expression, suggesting that the cells making the adventitious SAM need not produce cytokinin themselves.

Tobacco plants constitutively expressing maize *kn1* and Arabidopsis plants constitutively expressing *KNAT1,* an Arabidopsis *KN*-like gene, from CaMV35S promoters make adventitious shoots on leaves in the most severe transformants (11, 79). In other species ectopic expression of *KN*-like genes has not led to adventitious shoot production. Tomato plants carrying a CaMV35S-*kn1* construct have supercompound leaves but lack adventitious shoots (29). As a possible explanation, a tomato *KN*-like gene, in apparent contrast with the situation in maize and Arabidopsis, is normally expressed in leaf primordia where it may act to regulate subdivision of the compound tomato leaf (29). In the barley *Hooded* mutant, caused by a dominant mutation in a *KN*-like gene, floral meristems initiate on the lemmas (sepal-like organs) (62). Rice with increased gene copy number of *OSH1* and maize mutants with ectopic expression of *kn1* or *roughsheath* (*rs1*), a maize *KN*-like gene, have altered leaf morphology and ectopic outgrowths on leaf surfaces but make no adventitious shoots (3, 21, 55). Whether the differences in the shoot-inducing ability of *KN*-like genes in these different species are a function of differences in protein levels, fundamental differences between species in the competence of leaves to form SAMs, or a combination of both remains to be determined.

Adventitious shoots in Arabidopsis and tobacco plants that overexpress *kn1* and *KNAT1* all develop on the adaxial surfaces of leaves near veins, the same pattern as seen in cytokinin overexpressing plants. The normal developmental cues that cause the lateral SAM to develop on the adaxial leafbase may also predispose the adaxial portion of the dicot leaf to adventitious SAM formation.

SELF-RENEWAL WITHIN THE SHOOT APICAL MERISTEM

A key attribute of the SAM is its capacity for self-renewal. The self-renewing initial cell population resides within the central zone of the SAM. Clonal analysis has yielded insight into the dynamics of cell proliferation within the SAM (reviewed in 91). A small number of slowly dividing initial cells (typically 2 to 4 per layer) act as a self-replenishing population, whereas some of their descendants, pushed out onto the flanks of the SAM, differentiate into leaves. Other descendants, displaced below the SAM, differentiate into stem. The immediate descendants of the initial cells divide further, amplifying the cell population before being incorporated into leaf or stem primordia.

Arabidopsis plants homozygous for weak *stm* alleles form vegetative, inflorescence, and floral meristems that terminate prematurely, indicating that *STM* is required for self-renewal within all three types of SAMs (12, 19, 20). [The *waldmeister* mutant described in (19) is an allele of *STM* (P-C Morris, personal communication).] The *STM* transcript (and the maize KN protein) is found throughout the SAM except in presumptive leaf primordia (49, 83; Figure 7). *STM* and *kn1* expression domains include the central zone, where the initial cells reside, but also extend into the peripheral zone between organ primordia, where these genes may act to prevent organ fusion. This is supported by the observation that mutants carrying weak *stm* alleles exhibit fusion of floral organs (19).

wuschel (*wus*) mutants are among several mutants proposed to act specifically in the development and/or function of the central zone (42). In *wus* mutants, the first 1–4 leaves are made with normal adaxial/abaxial polarity and at normal positions, though their appearance is delayed relative to the wild type. Subsequently, the SAM terminates in a flat, enlarged apex in *wus* mutants. Leaves and new SAMs develop on these enlarged flat apices. Most of the new SAMs also terminate, but some go on to produce wild-type-appearing rosettes. The *wus* defect is apparent in mature embryos where the region normally occupied by the SAM harbors fewer cells than in the wild type but more cells than in *stm* mutants, consistent with the observed epistasis of *stm* relative to *wus*. In some cases, leaves form in the central regions of the SAM in *wus* mutants. This latter observation combined with the premature termination of the SAM in *wus* mutants led to the hypothesis that the WUS gene product is required for central zone identity (42).

Mutations in the *CLV1* and *CLV3* genes have been proposed to specifically increase the size of the central zone, either through a failure to regulate cell division in the central zone or through failure to negatively regulate the spatial extent of the central zone (13, 14, 45). The SAM in Arabidopsis *clv1* and *clv3* mutants is larger than in wild-type embryos, and its size increases over time with inflorescence SAMs becoming as large as 1 mm across and frequently exhibiting fasciation. The *CLV1* gene is predicted to encode a membrane-bound receptor kinase expressed in the L3 and possibly the L2 layer of the SAM (S Clark, R Williams & E Meyerowitz, personal communication). Receptor kinases in other species mediate intercellular signaling, suggesting that *CLV1* may act to regulate meristem size through an intercellular signaling pathway, and providing a first glimpse at the molecular mechanisms of signaling within the SAM. The predicted existence of a *CLV1*-mediated signaling pathway within the L3 layer that is responsible for restricting SAM size is intriguing in light of the finding that in tomato periclinal chimeras carrying

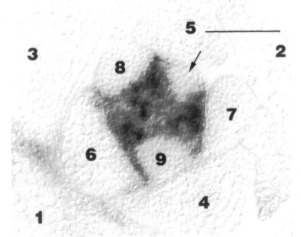

Figure 7. Pattern of expression of the *STM* gene, as detected by in situ hybridization to *STM* mRNA, in the vegetative Arabidopsis SAM. Leaves are numbered from oldest to youngest in a counterclockwise spiral. New leaves develop at a 137° angle from the previously formed leaf. *STM* expression is absent from the region where the next leaf is expected to form (*arrow*). Note that *STM* expression is found throughout the SAM, even in the peripheral zone where it is found at the boundaries between leaves. Scale bar equals 100 microns. (Photograph courtesy of J Long.)

combinations of wild-type and *fasciated* mutant tissue (the tomato *fasciated* mutant has a similar phenotype to that of *clv1* mutants), the genotype of the L3 could strongly influence SAM size (94). *CLV1* mRNA is also found in leaves (S Clark, personal communication). Consistent with this, *clv1* mutant leaves have a different shape than wild-type leaves (45). The widespread distribution of *CLV1* mRNA suggests that it may play a more general role in the plant, perhaps by regulating cell division in growing plant tissues.

In plants doubly mutant for *clv1* or *clv3,* and *stm,* the *stm* phenotype is epistatic, with regard to embryonic SAM formation (12). This is consistent with two models of gene action. In one, *CLV1* and *CLV3* act as negative regulators of *STM.* In the absence of *stm,* the presence or absence of *CLV* function is immaterial, and *stm* is epistatic. In the other model, *STM* is required to initiate SAM development, and *CLV* activity is expressed as a part of this SAM program where it acts to regulate meristem size.

Postembryonically, a different interaction between *clv* and *stm* mutations is observed. They mutually suppress each other (12). In *stm-1* single mutants, leaves develop from the cotyledon bases. Additional leaves form from the petioles of these leaves, and the plants can become very bushy. It is not known whether an organized meristem forms in these "escapes." In *clv stm* double mutants, the proportion of seedlings that escape is higher, and some form both rosettes and inflorescences with flowers, indicating that organized SAMs form. Conversely, floral meristem size is reduced in the double mutants relative to *clv* single mutants. This mutual suppression suggests that *STM* and *CLV* regulate SAM development through opposing effects on a third process (e.g. cell division) or regulator (as yet unidentified). In this model, postembryonic *CLV* function is not dependent on *STM* function. However, because it is unknown whether the alleles used in this study are null alleles, other models for the postembryonic action of *CLV* and *STM* cannot be ruled out.

Genetically, *CLV3* has been interpreted as operating in the same pathway as *CLV1* because double mutants homozygous for both *clv1* and *clv3* are similar in phenotype to either single mutant (14). Thus, the *CLV3* gene behaves as would be predicted for a gene encoding a ligand for the CLV1 receptor; whether this is true awaits the molecular characterization of the *CLV3* gene.

Finally, the Arabidopsis *ALTERED MERISTEM PROGRAM* (*AMP*) mutant also has an enlarged SAM (S Poethig, personal communication). These mutants also have increased levels of cytokinins (9). Which of these phenomena is cause and which is effect is unknown.

POSITIONAL INFORMATION IN THE SAM AND ITS ROLE IN LEAF DEVELOPMENT

Positional Information in the SAM and Its Role in Positioning Leaves

Experiments in which leaf primordia were isolated from the SAM by an incision changed the positions at which new primordia developed (86). This led to the hypothesis that existing leaf primordia inhibit the formation of new leaf primordia around them such that new leaf primordia develop at the position in the SAM where inhibition is least (105). The nature of this proposed inhibition is unknown. In addition to inhibition from young leaf primordia, this model also postulates inhibition from the center of the SAM so that leaves do not develop in the central zone.

Mutants that show altered phyllotaxis need not be defective in this hypothetical inhibitory process. As discussed in Leyser & Furner (45), the correct functioning of leaf spacing signals in a larger SAM will result in leaves emanating at smaller divergence angles relative to one another. Several Arabidopsis mutants that affect SAM size also affect phyllotaxis: *clv1*, *clv3*, and *amp1* mutants all make leaves with abnormal spacing patterns (9, 12, 14, 45). Similarly, in maize, the *abphyll* mutant has a larger SAM with an associated change in the spacing of leaves (26). Instead of the normal distichous arrangement, decussate (i.e. two opposite leaves at a node) or spiral leaf arrangement is observed in *abphyll* plants. Although both *clv* and *amp* mutants make leaves more rapidly than the wild type, the leaves in *amp* mutants are organized in a spiral with divergence angles of about 120° (9), whereas *clv1* mutants exhibit a more irregular spacing of leaves with frequent "reversals" of the phyllotactic spiral (45). Thus, *amp* seems to make the meristem larger in a fundamentally different way than *clv* mutations. *amp* mutations also differ from *clv* mutations in their effect on cotyledon spacing and number: *amp* mutants frequently have three or more cotyledons (9).

Regions lacking *kn1* mRNA and protein within the maize SAM provided the first molecular indications of early events in leaf formation (36, 83). Such clear regions arise before any morphological indications of leaf initiation. In maize, the number of cells that exhibit this clearing is consistent with the number of cells predicted from clonal analysis studies to give rise to the leaf (69). The regions free of *kn1* or *STM* gene products encompass a substantial group of cells, suggesting that the clearing process is rapid and may involve the active degradation of *kn1* or *STM* message and protein.

Positional Information in the SAM and Its Role in Patterning Leaves

While the specification of leaf position occurs within the SAM, it is unknown to what extent the basic pattern of a leaf is established while the leaf founder cells are still within the SAM. This issue has been addressed recently in the characterization of the maize *narrowsheath* mutant (75). The *narrowsheath* mutant phenotype is caused by two unlinked recessive mutations that together cause the maize leaf to be narrower than the wild type. Scanlon et al (75) have presented evidence that these leaves lack a specific marginal domain of the leaf and that associated with this phenotype the domain of "cleared" *KN*-like staining in the maize SAM is less extensive than in the wild type, suggesting that marginal leaf fate is specified when the leaf primordium is still morphologically continuous with the SAM.

Adaxial/abaxial leaf polarity is a fundamental asymmetry of the leaf that may also develop within the SAM. The top of the leaf (adaxial side) is actually oriented toward the axis of the plant (or center of the SAM) in the primordial stages of leaf development, whereas the bottom of the leaf (abaxial side) is initially oriented away from the axis. Experiments performed on potato in which incipient leaf primordia were isolated from the meristem by an incision indicated that the SAM may be the source of a signal required for correct adaxial/abaxial development of the leaf (93). The snapdragon *phantastica* (*phan*) gene has been proposed to be required for the development of adaxial cell fates within the leaf; *phan* mutant plants have leaves that exhibit abaxial characteristics around their circumference (104). The *phan* gene encodes a myb-like protein that is expressed throughout the primordium while it is still within the SAM (A Hudson, personal communication). The symmetric distribution of *phan* indicates that some other component of a hypothetical adaxial/abaxial signaling system must be asymmetrically localized, perhaps the SAM-derived adaxializing signal originally hypothesized by Sussex (93).

Molecular Patterns Within the Meristem

Several genes of unknown function are expressed in patterns in the SAM that suggest they are important for SAM development. Among these are the maize *knox3, knox8,* and *roughsheath* (*rs1*) genes, and the Arabidopsis *KNAT1* gene; all are members of the *KN* family of genes, and all are expressed within the SAM and excluded from incipient leaf primordia (36, 48, 77). The *knox8* gene is expressed at a lower level but in a pattern similar to *kn1* (36). *rs* and *knox3* are expressed in a basal, abaxial region of the maize SAM predicted to contain progenitors of the internode and axillary bud (36, 77). Likewise, *KNAT1* is expressed in a basal, peripheral region of the Arabidopsis vegetative SAM (48).

Some genes show layer-restricted expression within the SAM (39). This includes the maize *kn1* gene, transcripts of which are detected in the corpus (L2) but not the tunica (L1) (36). KNOTTED1 protein is detected in both layers (83), an observation that led to the hypothesis that KNOTTED protein may be transported between cells. The non-cell-autonomous behavior of dominant *knotted1* mutations in genetic mosaics (78) and the ability of *knotted1* transcripts and KNOTTED1 protein to move between cells in a tobacco mesophyll system support this notion (50). More direct evidence for movement of transcription factors between plant cells comes from studies of the localization of the snapdragon GLOBOSA and DEFICIENS proteins, both transcription factors, in genetically mosaic SAMs (68). Thus, movement of transcription factors from cell to cell may be a generalized phenomenon in plants. If so, regulating their movement will be an additional step in regulating their pattern of expression.

The tobacco *NFL* gene, a homolog of the snapdragon *FLORICAULA* and Arabidopsis *LEAFY* genes, and the Arabidopsis *AINTEGUMENTA1* (*ANT1*) genes are both expressed in young leaf primordia (17, 38, 41). In addition, *NFL* is expressed in a region that may correspond to the peripheral zone of the SAM (38). Because loss-of-function mutations for *NFL* do not exist and because single mutants in the *ANT1* gene show wild-type embryonic and vegetative development—*ant1* mutants were identified based on their ovule defects—the role of these genes in the SAM is unknown. However, double mutants between *ant1* and *apetala2*, a homolog of *ANT1*, fail to make all floral organs except the carpel, indicating a possible role for *ANT1* in organ development (17).

CONCLUDING REMARKS

The classical and molecular genetic study of genes required for SAM development and/or function is rapidly advancing our understanding of this basic area in plant developmental biology. Nevertheless, much work remains to be done. Many genes have yet to be molecularly cloned. Other genes remain unidentified, either because mutations in them result in embryo or gametophyte lethality or because they encode redundant functions. Yet others have been identified molecularly, but the corresponding loss-of-function mutant does not exist. As the genetic sophistication of the system increases, so will our ability to identify these genes and/or their function. In addition, with the increasing availability of molecular markers for distinct regions within the SAM, we will become better at accurately interpreting the phenotypes of SAM-defective mutants and at separating those deficient in developmental functions within the SAM from those defective in more basic aspects of cellular metabolism. There is much to look forward to learning in the area of SAM development.

ACKNOWLEDGMENTS

We would like to acknowledge our colleagues Steve Clark, Andrew Hudson, Robert Martiensson, members of the Barton lab, and especially Scott Poethig for challenging and illuminating discussions on the shoot apical meristem and how it works. We also thank Megan Maguire for excellent help with figures.

Visit the *Annual Reviews home page* at http://www.annurev.org.

Literature Cited

1. Albertsen MC, Trimnell MR, Fox TW. 1993. Description and mapping of the tassel-less (*tls1*) mutation. *Maize Genet. Coop. Newsl.* 67:51–52

2. Barton MK, Poethig S. 1993. Formation of the shoot apical meristem in *Arabidopsis thaliana:* an analysis of development in the wild type and in the *shoot meristemless* mutant. *Development* 119:823–31

3. Becraft P, Freeling M. 1994. Genetic analysis of *Rough sheath1* developmental mutants of maize. *Genetics* 136:295–311

4. Bennett SRM, Alvarez J, Bosslinger G, Smyth DR. 1995. Morphogenesis in *pinoid* mutants of *Arabidopsis thaliana. Plant J.* 8:505–20

5. Berleth T, Jürgens G. 1993. The role of the *monopteros* gene in organising the basal body region of the *Arabidopsis* embryo. *Development* 118:575–87

6. Binns AN. 1994. Cytokinin accumulation and action: biochemical, genetic, and molecular approaches. *Annu. Rev. Plant Physiol. Plant Mol. Biol.* 45:173–96

7. Briggs S, Johal G. 1992. A recessive barren-inflorescence mutation. *Maize Genet. Coop. Newsl.* 66:51

8. Caruso JL. 1968. Morphogenetic aspects of a leafless mutant in tomato. I. General patterns in development. *Am. J. Bot.* 55: 1169–76

9. Chaudhury AM, Letham S, Craig S, Dennis ES. 1993. *amp1:* a mutant with high cytokinin levels and altered embryonic pattern,

faster vegetative growth, constitutive photomorphogenesis and precocious flowering. *Plant J.* 4:907–16

10. Christianson ML. 1986. Fate map of the organizing shoot apex in *Gossypium. Am. J. Bot.* 73:947–58

11. Chuck G, Lincoln C, Hake S. 1996. *KNAT1* induces lobed leaves with ectopic meristems when overexpressed in *Arabidopsis. Plant Cell* 8:1277–89

12. Clark SE, Jacobsen SE, Levin JZ, Meyerowitz EM. 1996. The *CLAVATA* and *SHOOTMERISTEMLESS* loci competitively regulate meristem activity in *Arabidopsis. Development* 122:1567–75

13. Clark SE, Running MP, Meyerowitz EM. 1993. *CLAVATA1,* a regulator of meristem and flower development in *Arabidopsis. Development* 119:397–418

14. Clark SE, Running MP, Meyerowitz E. 1995. *CLAVATA3* is a specific regulator of shoot and floral meristem development affecting the same processes as *CLAVATA1. Development* 121:2057–67

15. Doebley J, Stec A, Kent B. 1995. *Suppressor of sessile spikelets1 (Sos1):* a dominant mutant affecting inflorescence development in maize. *Am. J. Bot.* 82:571–77

16. Dubnau J, Struhl G. 1996. RNA recognition and translational regulation by a homeodomain protein. *Nature* 379:694–99

17. Elliott RC, Betzner AS, Huttner E, Oakes MP, Tucker WQJ, et al. 1996. *AINTEGUMENTA,* an *APETALA2*-like gene of Arabi-

dopsis with pleiotropic roles in ovule development and floral organ growth. *Plant Cell*
8:155–68

17a. Endrizzi K, Moussian B, Haecker A, Levin
JZ, Laux T. 1996. The *SHOOTMER-
ISTEMLESS* gene is required for maintenance of undifferentiated cells in Arabidopsis shoot and floral meristems and acts
at a different regulatory level than the meristem genes *WUSCHEL* and *ZWILLE*. *Plant
J.* 10:967–79

18. Estruch JJ, Prinsen E, Onckelen HV, Schell
J, Spena A. 1991. Viviparous leaves produced by somatic activation of an inactive
cytokinin-synthesizing gene. *Science* 254:
1364–67

19. Felix G, Altmann T, Uwer U, Jessop A,
Wollmitzer L, Morris P-C. 1996. Characterization of *waldmeister*, a novel developmental mutant in *Arabidopsis thaliana*. *J.
Exp. Bot.* 47:1007–17

20. Franzmann L, Patton DA, Meinke DW.
1989. In vitro morphogenesis of arrested
embryos from lethal mutants of *Arabidopsis thaliana*. *Theor. Appl. Genet.* 77:
609–16

21. Freeling M, Hake S. 1985. Developmental
genetics of mutants that specify knotted
leaves in maize. *Genetics* 111:617–34

22. Furner IJ, Pumfrey JE. 1992. Cell fate in the
shoot apical meristem of *Arabidopsis
thaliana*. *Development* 115:755–64

23. Galinat WC. 1959. The phytomer in relation to floral homologies in the American
Maydea. *Bot. Mus. Leaflets, Harvard Univ.*
19:1–32

24. Gehring WJ, Affolter M, Bürglin T. 1994.
Homeodomain proteins. *Annu. Rev. Biochem.* 63:487–526

25. Gifford EM. 1954. The shoot apex in angiosperms. *Bot. Rev.* 20:447–29

26. Greyson RI, Walden DB, Hume JA, Erickson RO. 1978. The ABPHYL syndrome in
Zea mays. II. Patterns of leaf initiation and
the shape of the shoot meristem. *Can. J.
Bot.* 56:1545–50

27. Groot SPC, Bouwer R, Busscher M, Lind-
hout P, Dons HJ. 1995. Increase of endogenous zeatin riboside by introduction of
the *ipt* gene in wild type and the *lateral
suppressor* mutant of tomato. *Plant Growth
Regul.* 16:27–36

28. Haas G, Orr A. 1994. Organogenesis of the
maize mutant *Fascicled ear (Fas)*. *Maize
Genet. Coop. Newsl.* 68:18–19

29. Hareven D, Gutfinger T, Parnis A, Eshed Y,
Lifschitz E. 1996. The making of a compound leaf: genetic manipulation of leaf
architecture in tomato. *Cell* 84:735–44

30. Hareven D, Gutfinger T, Pnueli L, Bauch

L, Cohen O, Lifschitz E. 1994. The floral
system of tomato. *Euphytica* 79:235–43

31. Hempel FD, Feldman LJ. 1994. Bi-direc-
tional inflorescence development in *Arabidopsis thaliana:* acropetal initiation of
flowers and basipetal initiation of
paraclades. *Planta* 192:276–88

32. Hicks GS. 1994. Shoot induction and organogenesis in vitro: a developmental perspective. *In Vitro Cell. Dev. Biol.* 30:10–15

33. Hofmeyr JDJ. 1930. *The inheritance and
linkage relationships of barren-stalk-1 and
barren-stalk-2, two mature plant characters of maize*. PhD thesis. Cornell Univ.,
Ithaca, NY

34. Irish VF, Sussex IM. 1990. Function of the
apetala-1 gene during *Arabidopsis* floral
development. *Plant Cell* 2:741–53

35. Irish VF, Sussex IM. 1992. A fate map of
the *Arabidopsis* embryonic shoot apical
meristem. *Development* 115:745–53

36. Jackson D, Veit B, Hake S. 1994. Expression of maize *KNOTTED1* related homeobox genes in the shoot apical meristem
predicts patterns of morphogenesis in the
vegetative shoot. *Development* 120:
405–13

37. Jürgens G, Ruiz RAT, Laux T, Mayer U,
Berleth T. 1994. Early events in apical-ba-
sal pattern formation in *Arabidopsis*. In
*Plant Molecular Biology: Molecular-Ge-
netic Analysis of Plant Development and
Metabolism,* ed. G Coruzzi, P Puigdomenech, pp. 95–103. Berlin: Springer-Verlag

38. Kelly AJ, Bonnlander MB, Meeks-Wagner
DR. 1995. *NFL,* the tobacco homolog of
FLORICAULA and *LEAFY* transcriptionally expressed in both vegetative and floral
meristems. *Plant Cell* 7:225–34

39. Kelly AJ, Meeks-Wagner DR. 1995. Characterization of a gene transcribed in the L2
and L3 layers of the tobacco shoot apical
meristem. *Plant J.* 8:147–53

40. Kerstetter R, Vollbrecht E, Lowe B, Veit B,
Yamaguchi J, Hake S. 1994. Sequence
analysis and expression patterns divide the
maize *Knotted1*-like homeobox genes into
two classes. *Plant Cell* 6:1877–87

41. Klucher K, Chow H, Reiser L, Fischer RL.
1996. The *AINTEGUMENTA* gene of
Arabidopsis required for ovule and female
gametophyte development is related to the
floral homeotic gene *APETALA2*. *Plant
Cell* 8:137–53

42. Laux T, Mayer KFX, Berger J, Jürgens G.
1996. The *WUSCHEL* gene is required for
shoot and floral meristem integrity in
Arabidopsis. *Development* 122:87–96

43. Lejeune P, Bernier G. 1996. Effect of environment on the early steps of ear initiation

in maize (*Zea mays* L.). *Plant Cell Environ.* 19:217–24

44. Levin JZ, Meyerowitz EM. 1995. *UFO:* an Arabidopsis gene involved in both floral meristem and floral organ development. *Plant Cell* 7:529–48

45. Leyser HMO, Furner IJ. 1992. Characterization of three apical meristem mutants of *Arabidopsis thaliana. Development* 116:397– 403

46. Li Y, Hagen G, Guilfoyle TJ. 1992. Altered morphology in transgenic tobacco plants that overproduce cytokinins in specific tissues and organs. *Dev. Biol.* 153:386–95

47. Lincoln C, Britton JH, Estelle M. 1990. Growth and development of the *axr1* mutants of *Arabidopsis. Plant Cell* 2:1071–80

48. Lincoln C, Long J, Yamaguchi J, Serikawa K, Hake S. 1994. A *Knotted1*-like homeobox gene in Arabidopsis is expressed in the vegetative meristem and dramatically alters leaf morphology when overexpressed in transgenic plants. *Plant Cell* 6:1869–76

49. Long JA, Moan EI, Medford JI, Barton MK. 1996. A member of the KNOTTED class of homeodomain proteins encoded by the *STM* gene of *Arabidopsis. Nature* 379: 66–69

50. Lucas WJ, Bouché-Pillon S, Jackson DP, Nguyen L, Baker L, et al. 1995. Selective trafficking of KNOTTED1 homeodomain protein and its mRNA through plasmodesmata. *Science* 270:1980–83

51. Malayer JC, Guard AT. 1964. A comparative developmental study of the mutant sideshootless and normal tomato plants. *Am. J. Bot.* 51:140–43

52. Mapelli S, Kinet JM. 1992. Plant growth regulator and graft control of axillary bud formation and development in the *TO-2* mutant tomato. *Plant Growth Regul.* 11: 385–90

53. Mapelli S, Lombardi L. 1982. A comparative auxin and cytokinin study in normal and *to-2* mutant tomato plants. *Plant Cell Physiol.* 23:751–57

54. Marcotrigiano M. 1986. Origin of adventitious shoots regenerated from cultured tobacco leaf tissue. *Am. J. Bot.* 73:1541–47

55. Matsuoka M, Ichikawa H, Saito A, Tada Y, Fujimura T, Kano-Murakami Y. 1993. Expression of a rice homeobox gene causes altered morphology of transgenic plants. *Plant Cell* 5:1039–48

56. Mayer U, Ruiz RAT, Berleth T, Misera S, Jürgens G. 1991. Mutations affecting body organization in the *Arabidopsis* embryo. *Nature* 353:402–7

57. McConnell J, Barton MK. 1995. Effect of

mutations in the *PINHEAD* gene of *Arabidopsis* on the formation of shoot apical meristems. *Dev. Genet.* 16:358–66

58. McDaniel CN, Poethig RS. 1988. Cell-lineage patterns in the shoot apical meristem of the germinating maize embryo. *Planta* 175: 13–22

59. Medford JI, Behringer FJ, Callos JD, Feldmann KA. 1992. Normal and abnormal development in the Arabidopsis vegetative shoot apex. *Plant Cell* 4:631–43

60. Medford JI, Horgan R, El-Sawi Z, Klee HJ. 1989. Alterations of endogenous cytokinins in transgenic plants using a chimeric isopentenyl transferase gene. *Plant Cell* 1: 403–13

61. Meinke DW, Sussex IM. 1979. Embryo-lethal mutants of *Arabidopsis thaliana:* a model system for genetic analysis of plant embryo development. *Dev. Biol.* 72:50–61

62. Muller KJ, Romano N, Gerstner O, Garcia-Maroto F, Pozzi C, et al. 1995. The barley *Hooded* mutation caused by a duplication in a homeobox gene intron. *Nature* 374: 727–30

63. Napoli C. 1996. Highly branched phenotype of the Petunia *dad1-1* mutant is reversed by grafting. *Plant Physiol.* 111: 27–37

64. Neuffer MG, Sheridan KA. 1977. Dominant mutants from EMS-treated pollen. *Maize Genet. Coop. Newsl.* 51:59–60

65. Okada K, Ueda J, Komaki MK, Bell CJ, Shimura Y. 1991. Requirement of the auxin polar transport system in early stages of *Arabidopsis* floral bud formation. *Plant Cell* 3:677–84

66. Pan YB, Peterson PA. 1992. *ba3:* a new barrenstalk mutant in *Zea mays* L. *J. Genet. Breed.* 46:291–94

67. Patton DA, Meinke DW. 1990. Ultrastructure of arrested embryos from lethal mutants of *Arabidopsis thaliana. Am. J. Bot.* 77:653–61

68. Perbal M-C, Haughn G, Saedler H, Schwarz-Sommer Z. 1996. Non-cell-autonomous function of the *Antirrhinum* floral homeotic proteins *DEFICIENS* and *GLOBOSA* is exerted by their polar cell-to-cell trafficking. *Development* 122:3433–41

69. Poethig S, Szymkowiak G. 1995. Clonal analysis of leaf development in maize. *Maydica* 40:67–76

70. Prior PV. 1957. Alterations in the shoot apex of *Petunia hybrida* Vilm. at flowering. *Proc. Iowa Acad. Sci.* 64:104–9

71. Przemeck GKH, Mattsson J, Hardtke CS, Sung ZR, Berleth T. 1996. Studies on the

role of the *Arabidopsis* gene *MONOP-TEROS* in vascular development and plant cell axialization. *Planta* 200:229–37

72. Randolph LF. 1936. The developmental morphology of the caryopsis in maize. *J. Agric. Res.* 53:881–916

73. Sato Y, Hong SK, Tagiri A, Kitano H, Yamamoto N, et al. 1996. A rice homeobox gene, *OSH1*, is expressed before embryo differentiation in a specific region during early embryogenesis. *Proc. Natl. Acad. Sci. USA* 93:8117–22

74. Sawhney VK, Greyson RI. 1972. On the initiation of the inflorescence and floral organs in tomato (*Lycopersicon esculentum*). *Can. J. Bot.* 50:1493–95

75. Scanlon M, Schneeberger RG, Freeling M. 1996. The maize mutant *NARROW SHEATH* fails to establish leaf margin identity in a meristematic domain. *Development* 122:1683–91

76. Schmulling T, Beinsberger S, Greef JD, Schell J, Onckelen HV, Spena A. 1989. Construction of a heat-inducible chimaeric gene to increase the cytokinin content in transgenic plant tissue. *FEBS Let.* 249: 401–6

77. Schneeberger RG, Becraft PW, Hake S, Freeling M. 1995. Ectopic expression of the *Knox* homeobox gene *rough sheath1* alters cell fate in the maize leaf. *Genes Dev.* 9:2292–304

78. Sinha N, Hake S. 1990. Mutant characters of *Knotted* maize leaves are determined in the innermost tissue layers. *Dev. Biol.* 141: 203–10

79. Sinha NR, Williams RE, Hake S. 1993. Overexpression of the maize homeobox gene, *Knotted-1*, causes a switch from determinate to indeterminate cell fates. *Genes Dev.* 7:787–95

80. Skoog F, Miller CO. 1957. Chemical regulation of growth and organ formation in plant tissues cultured *in vitro*. *Symp. Soc. Exp. Biol.* 11:118–31

81. Smigocki AC. 1991. Cytokinin content and tissue distribution in plants transformed by a reconstructed isopentenyl transferase gene. *Plant Mol. Biol.* 16:105–15

82. Smigocki AC, Owens LD. 1988. Cytokinin gene fused with a strong promoter enhances shoot organogenesis and zeatin levels in transformed plant cells. *Proc. Natl. Acad. Sci. USA* 85:5131–35

83. Smith LG, Greene B, Veit B, Hake S. 1992. A dominant mutation in the maize homeobox gene, *Knotted-1*, causes its ectopic expression in leaf cells with altered fates. *Development* 116:21–30

84. Smith LG, Hake S. 1993. A new mutation

affecting tassel and ear morphology. *Maize Genet. Coop. Newsl.* 67:2–3

85. Smith LG, Jackson D, Hake S. 1995. Expression of *knotted1* marks shoot meristem formation during maize embryogenesis. *Dev. Genet.* 16:344–48

86. Snow M, Snow R. 1931. Experiments on phyllotaxis. *Phil. Trans. R. Soc. London Ser. B* 221:1–43

87. Sossountzov L, Maldiney R, Sotta B, Sabbagh I, Habricot Y, et al. 1988. Immunocytochemical localization of cytokinins in Craigella tomato and a sideshootless mutant. *Planta* 175:291–304

88. Souer E, Vanhouwelingen A, Kloos D, Mol J, Koes R. 1996. The *NO APICAL MERISTEM* gene of petunia is required for pattern formation in embryos and flowers and is expressed at meristem and primordia boundaries. *Cell* 85:159–70

89. Deleted in proof

90. Steeves TA, Hicks MA, Naylor JM, Rennie P. 1969. Analytical studies on the shoot apex of *Helianthus annuus*. *Can. J. Bot.* 47:1367–75

91. Steeves TA, Sussex IM. 1989. *Patterns in Plant Development*. Cambridge: Cambridge Univ. Press

92. Steffensen DM. 1968. A reconstruction of cell development in the shoot apex of maize. *Am. J. Bot.* 55:354–69

93. Sussex IM. 1955. Morphogenesis in *Solanum tuberosum* L: Experimental investigation of leaf dorsiventrality and orientation in the juvenile shoot. *Phytomorphology* 5:286–300

94. Szymkowiak EJ, Sussex IM. 1992. The internal meristem layer (l3) determines floral meristem size and carpel number in tomato periclinal chimeras. *Plant Cell* 4: 1089–100

95. Szymkowiak EJ, Sussex IM. 1993. Effect of *lateral suppressor* on petal initiation in tomato. *Plant J.* 4:1–7

96. Szymkowiak EJ, Sussex IM. 1996. What chimeras can tell us about plant development. *Annu. Rev. Plant Physiol Plant Mol. Biol.* 47:351–76

97. Talbert P, Adler H-T, Parks DW, Comai L. 1995. The *REVOLUTA* gene is necessary for apical meristem development and for limiting cell divisions in the leaves and stems of *Arabidopsis thaliana*. *Development* 121:2723–35

98. Tepper HB. 1992. Benzyladenine promotes shoot initiation in empty leaf axils of *Stellaria media* L. *J. Plant Physiol.* 140: 241–43

99. Tian H-C, Marcotrigiano M. 1994. Cell-layer interactions influence the number and

position of lateral shoot meristems in *Nicotiana. Dev. Biol.* 162:579–89

100. Torres-Ruiz RA, Fisher T, Haberer G. 1996. Genes involved in the elaboration of apical pattern and form in *Arabidopsis thaliana:* genetic and molecular analysis. In *Embryogenesis, The Generation of a Plant,* ed. TL Wang, A Cuming, pp. 15–34. Oxford: Bios Sci.

101. Tucker DJ. 1976. Endogenous growth regulators in relation to side shoot development in the tomato. *New Phytol.* 77:561–68

102. Veit B, Schmidt RJ, Hake S, Yanofsky MF. 1993. Maize floral development: new genes and old mutants. *Plant Cell* 5: 1205–15

103. Vollbrecht E, Veit B, Sinha N, Hake S. 1991. The developmental gene *Knotted1* is a member of a maize homeobox gene family. *Nature* 350:241–43

104. Waites R, Hudson A. 1995. *phantastica:* a gene required for dorsiventrality of leaves in *Antirrhinum majus. Development* 121: 2143–54

105. Wardlaw CW. 1949. Experiments on organogenesis in ferns. *Growth* 13:93–31 (Suppl.)

106. Wilkinson MD, Haughn GW. 1995. *UNUSUAL FLORAL ORGANS* controls meristem identity and organ primordia fate in Arabidopsis. *Plant Cell* 7:1485–99

Annu. Rev. Plant Physiol. Plant Mol. Biol. 1997. 48:703–34
Copyright © 1997 by Annual Reviews Inc. All rights reserved

ALTERNATIVE OXIDASE: From Gene to Function

Greg C. Vanlerberghe

Department of Botany and Division of Life Science, University of Toronto at Scarborough, 1265 Military Trail, Scarborough, Ontario M1C 1A4, Canada

Lee McIntosh

Department of Energy Plant Research Laboratory and Biochemistry Department, Michigan State University, East Lansing, Michigan 48824

KEY WORDS: alternative pathway, plant mitochondria, covalent modification, pyruvate, transgenic, promoter

ABSTRACT

Plants, some fungi, and protists contain a cyanide-resistant, alternative mitochondrial respiratory pathway. This pathway branches at the ubiquinone pool and consists of an alternative oxidase encoded by the nuclear gene *Aox1*. Alternative pathway respiration is only linked to proton translocation at Complex 1 (NADH dehydrogenase). Alternative oxidase expression is influenced by stress stimuli—cold, oxidative stress, pathogen attack—and by factors constricting electron flow through the cytochrome pathway of respiration. Control is exerted at the levels of gene expression and in response to the availability of carbon and reducing potential. Posttranslational control involves reversible covalent modification of the alternative oxidase and activation by specific carbon metabolites. This dynamic system of coarse and fine control may function to balance upstream respiratory carbon metabolism and downstream electron transport when these coupled processes become imbalanced as a result of changes in the supply of, or demand for, carbon, reducing power, and ATP.

CONTENTS

1040-2519/97/0601-0703$08.00

INTRODUCTION

The alternative terminal oxidase in higher plant mitochondria was first a thermogenic curiosity observed during anthesis (reviewed in 83; 137) and was more recently recognized as part of a plant's ability to regulate its energy/carbon balance responses to a changing environment (81, 147). The alternative pathway of mitochondrial respiration branches from the cytochrome pathway in the inner mitochondrial membrane at the ubiquinone pool and passes electrons to a single terminal oxidase, the alternative oxidase (Figure 1). The alternative oxidase apparently reduces molecular oxygen to water in a single four-electron transfer step (reviewed in 20, 24, 25, 94, 133). The alternative pathway is nonphosphorylating. The oxidase is resistant to cyanide and inhibited by substituted hydroxamic acids such as salicylhydroxamic acid (SHAM) and n–propyl gallate (128, 132). All angiosperms, many algae, and some fungi contain the genetic capacity to express this pathway (43, 81, 103).

   The significance of this pathway may be that its electron transport from ubiquinone to water does not contribute to a transmembrane potential and thus wastes two of the three energy coupling sites that are part of the cytochrome pathway. The phosphorylating potential from site I (NADH dehydrogenase) is retained, thus allowing some energy production (Figure 1; 149, 153, 159). In plants and fungi the alternative pathway of mitochondrial respiration requires a single nuclear gene (existing as a small gene family in some plants) *Aox1*, encoding the alternative oxidase (73, 111, 124). All plant species tested possess the genetic capacity to express the alternative pathway under any number of developmental and environmental conditions (70, 81–83). The question then arises, What is the function of this alternative pathway? The physiology

---→

Figure 1 The plant mitochondrial complexes and electron transport are shown in schematic fashion: Complex I, NADH dehydrogenase; Complex II, succinate dehydrogenase; Complex III, cytochrome bc_1; Complex IV, cytochrome oxidase (COX); AOX, alternative oxidase; UQ, ubiquinone; Ext., an external plant NADH dehydrogenase; ?, another proposed NADH-dehydrogenase for plants; AA, antimycin A; SHAM, salicylhydroxamic acid; TCA, tricarboxylic acid.

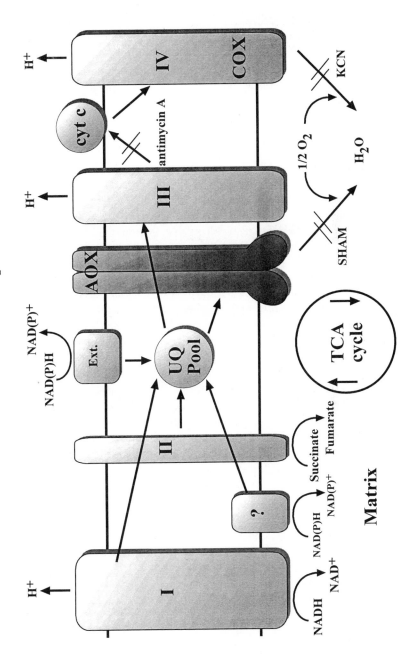

and regulation of the alternative oxidase including biophysical measurements, adenylate control, biochemical control of alternative oxidase activity, and methods to measure electron transport and partitioning have been reviewed with regularity (24, 25, 94, 133).

In this review, we summarize the genetic characterization of the alternative oxidase, its transcriptional and posttranscriptional regulation, major factors controlling the partitioning of electrons to this pathway, and the more recent use of molecular/genetic approaches to explore the in vivo function of alternative pathway respiration in higher plants.

BACKGROUND AND HISTORY

Many organisms possess more than one terminal oxidase. *Escherichia coli* hosts two terminal oxidases, the cytochrome o and cytochrome d complexes (5, 141). Besides an aa_3-type cytochrome c oxidase, the prokaryote *Paracoccus denitrificans* can express several alternative terminal oxidases (36). When cultured under microaerobic or anaerobic conditions, the level of cytochrome aa_3 drops, and an alternative oxidase, cytochrome co, arises (13, 36). A protist, *Trypanosoma brucei brucei*, is a bovine extracellular parasite and cycles between a mammalian host and the digestive tract of the tsetse fly (*Glossina* spp.). The bloodstream form of *T. brucei* lacks functional cytochromes, limiting most of its energy metabolism to glycolysis while passing the reducing equivalents produced to water via a plant–like alternative oxidase (15, 107).

Many fungi (43), including *Candida lipolytica* (42), *Neurospora crassa* (9), and *Hansenula anomala* (169), contain a plant-like cyanide-resistant alternative oxidase. For *H. anomala* and *N. crassa*, the alternative oxidase is induced when inhibitors limiting cytochrome pathway respiration [e.g. chloramphenicol inhibiting mitochondrial-encoded cytochrome pathway components (9, 67)] are added. Antimycin A inhibits electron transfer between cytochrome b and cytochrome c (120), suppressing the cytochrome pathway and inducing expression of the alternative pathway in both fungi (9, 169) and plants (150). Reduced forms of sulfur, including the amino acids cysteine and methionine, also induce expression of alternative pathway respiration in *H. anomala* (89).

Several hypotheses explain cyanide-resistant alternative pathway respiration in the fungi. First, the "plant hypothesis" proposes that it acts as an energy overflow, releasing the tricarboxylic acid cycle from adenylate regulation and allowing synthesis of carbon skeletons (65). A second hypothesis is that the alternative pathway replaces, in part, the phosphorylating cytochrome pathway when it is restricted by inhibitors produced by bacteria, fungi, and plants (69, 138). Approximately 800 plant species produce cyanide upon wounding or when attacked by pathogens (78). A fungal pathogen *Stemphylium loti,* of the cyanogenic plant birdsfoot trefoil (*Lotus corniculatus* L.), possesses a cyanide–resistant alternative pathway (119). The alternative pathway is thought to provide energy for a cyanide-detoxifying enzyme formamide hydro-lyase (FHL). This enzyme would decrease the levels of plant-produced cyanide, allowing the cytochrome pathway to function (119). The alternative oxidase has been identified in the yeast *Pichia stipitis* (54, 55), where it was proposed to act as a redox sink to prevent xylitol formation in *P. stipitis* during oxygen–limited xylose fermentation (54).

The only confirmed function for alternative pathway respiration is the thermogenic respiration in arums and other species where heat produced during anthesis volatilizes aromatic compounds to attract pollinators (83, 84). Thermogenesis in *Arum* was first described by Lemark where he commented that "le gouet d'Italie...acquiert à une certaine époque de la floraison, une chaleur considérable....Cette chaleur s'élève à 21, 8 degrès, l'air ambiant étant à 14.9 degrès" (63). Okunaki (cited in 83) first proposed two possible respiratory pathways after observing *Lilium auratum* pollen. Van Herk (44) postulated a respiratory pathway containing a noncytochrome, autoxidizable flavoprotein in *Sauromatum guttatum.* Low phosphorylative capacity was documented in tissues possessing high relative alternative pathway respiration, and all the data were consistent with Complex I (NADH–dehydrogenase), the first coupling site present. The two cytochrome coupling sites were not components of this alternative pathway (140). Van Herk (44) and Meeuse (83) concentrated on identifying the thermogenic-inducing principle termed "calorigen" in *S. guttatum,* recently identified as salicylic acid (109, 146).

Initial efforts to purify the alternative oxidase enzyme began its identification as a quinol oxidase (51, 118). In 1986, partial purification of the alternative oxidase from *Arum maculatum* (12) and *S. guttatum* (29) was achieved. Rhoads & McIntosh (111) isolated the first cDNA clone for the alternative oxidase gene *Aox1* using polyclonal and monoclonal antibodies raised against the partially purified *S. guttatum* alternative oxidase (30, 31). Following this work, *Aox1* cDNA for the yeast *Hansenula anomala* was isolated with the plant antibodies and *S. guttatum* cDNA sequence (124). Berthold & Siedow (8) further purified the alternative oxidase in skunk cabbage (*Symplocarpus foetidus*) by means of monoclonal antibodies specific to the *S. guttatum* oxidase.

GENETICS OF THE ALTERNATIVE OXIDASE

The genetic basis for alternative pathway activity and its regulation was first postulated with *N. crassa.* A number of respiratory mutants, including *poky*

708 VANLERBERGHE & McINTOSH

(*mi–l*), a mutant lacking a- and b-type cytochromes, were shown to have a cyanide- and antimycin A–resistant respiratory pathway (67). The other pathway was grouped with plant alternative oxidase when it was also shown to be resistant to SHAM, the alternative oxidase–specific inhibitor. Bertrand et al (9) characterized a number of respiratory mutants of *N. crassa* that, in the presence of antimycin A, a potent inducer of alternative pathway activity, were unable to express cyanide-resistant respiration. Two nuclear complementation groups were described: *aodl* for the structural gene and *aod2* for a proposed regulatory gene.

The *N. crassa* mutants were further characterized using monoclonal antibodies developed to the alternative oxidase of *S. guttatum* (68). In uninduced *N. crassa* cells, there was little or no detectable alternative oxidase protein, whereas in cells grown in the presence of chloramphenicol, or in the *poky* mutant, two polypeptide bands of around 36 kDa reacted with the *S. guttatum* antibody. A multiplicity of bands are found in plants reacting with the monoclonal or polyclonal antibodies (31).

Alternative oxidase activity has been noted for algae, possibly first for *Chlorella* (160) and more recently for *Chlamydomonas reinhardtii* (32, 161) and wild-type *Selenastrum minutum*. *S. minutum* apparently lacks alternative oxidase (162) but is capable of azide-stimulated oxygen consumption (77). Alternative pathway respiration in *Chlamydomonas* is stimulated by cyanide (38). Mendelian mutations affecting respiration in *Chlamydomonas* have been isolated as obligate photoautotrophs [*dark dier* (dk)] growing only with light and supplied CO_2 and dying when transferred to heterotrophic growth conditions, dark plus acetate (168). Although these strains had much-reduced cytochrome pathway activity, cyanide-resistant respiration was present and apparently supporting, in part, TCA cycle activity (38, 168). Studies with Angiosperms have yielded no verified mutants in alternative oxidase or its regulation. Early reports that Progress No. 9 peas lacked the alternative oxidase (97–99) were not supported on reevaluation (37, 81, 102).

ALTERNATIVE OXIDASE GENES AND GENE EXPRESSION

The first alternative oxidase gene cloned was from *S. guttatum*, the voodoo lily (111). Antibodies raised to the alternative oxidase (31) were used to isolate a cDNA clone, *Aox1*, encoding a 42-kDa polypeptide. An *Aox1* clone was also isolated from *Hansenula anomala*. In this case, the gene was selected by its increase in transcript abundance after addition of KCN or antimycin A (124), both of which induce alternative oxidase activity in *H. anomala*.

Do the *S. guttatum* and *H. anomala* cDNA clones encode the entire alternative oxidase enzyme? This question was answered, at least partly, when a project to clone heme biosynthesis genes "backed into" the *Arabidopsis thaliana* alternative oxidase gene. Kumar & Soll (62) used an *E. coli* strain (SASX41B) low in glutamyl–tRNA reductase, and thus hemeA-deficient and low in cytochromes, to screen for complementation among Arabidopsis cDNAs. Cytochromes are essential cofactors for many redox enzymes such as cytochrome oxidase. A clone complementing SASX41B and supporting aerobic growth encoded the alternative oxidase. AOA monoclonal antibody raised to the plant alternative oxidase also recognized the product of this clone, albeit at a lower molecular weight than is found for alternative oxidase from many plants (31). These data indicate that the holoenzyme activity of alternative oxidase in higher plants is encoded by a single gene. This clone yields very low activity of the enzyme in *E. coli*. Regulatory genes and/or posttranscriptional modifications of the alternative oxidase may occur in plants and other eukaryotes.

Following isolation of the *S. guttatum Aox*1 cDNA and identification of the Arabidopsis clone, *Aox*1 DNA clones (Figure 2) have been isolated from a number of plants including *Nicotiana tabacum* (150, 166), soybean (164), and *Mangifera indica L.*, mango (18). Recently, the *Neurospora crassa* gene has been isolated and characterized (73), and the putative alternative oxidase from *Trypanosoma brucei brucei* demonstrates great sequence similarity to *Aox*1 (Genback Accession #1399589).

All alternative oxidase genes sequenced to date encode a highly similar protein. Conserved features include (*a*) two possible alpha–helical membrane–spanning regions, more or less in the center of the oxidase, (*b*) a possible "surface-exposed" alpha-helix, and (*c*) N- and C-terminal hydrophilic regions (Figure 2). Experiments employing mitoplasts (right-side-out isolated mitochondria lacking an outer membrane) and submitochondrial particles exposed to trypsin and/or proteinase K indicate that the large N- and C-terminal regions are exposed to the matrix (110, 134). This would place the "surface helix" in intermembrane space. The conserved cysteine residues in the N-terminal region (Figure 2) are postulated to form a labile disulfide-bridge(s) in the alternative oxidase. The oxidation or reduction of this intramolecular bridge affects the oxidase's activity (see below; reviewed in 135). On the basis of published sequence data, it has been suggested that four potential small, C-terminal alpha-helices, with two of the four each containing one of two totally conserved Glu-X-X-His motifs (Figure 2), may form a binuclear iron center (Figure 2; 135, 136). If this is true, then the long-sought metal cofactor for the alternative oxidase, if there is a metal, may be iron. All these speculations on biochemical regulation remain to be directly demonstrated.

```
         1                                                      50
TobBY    MMTRGATRMT RTVLGHMGPR YFSTAIFRND AGTGVMSGAA VFMHGVPANP SEKAVVTWVR HFPVMGSRSA MSMALNDK.. .....QHDKK
TobSR1   .......... .......... .......... .......... .......... ......MWVR HFPVMGPRSA STVALNDK.. .....QHDKK
potato   ......MMSS RFAGTAL.RQ LGPVLFAS.A PGARAAAEPA YALLAGAPAA APTRAAVWLV ..RFPLSRAA STMSAPAA.. ....PEG.ET
AtAOX    .......... .......... .......... .......... .......... ......MDTR APTIGGMRFA STITLGEKTP MKEEDANQKK
Soybean  .......... .......... ........MM MMMSRSGANR VANTAMFVAK GLSGEVGGLR ALYGGGVRSE STLALSEK.. ....EKIEKK
mango    .......... .......... .......... .......... ...MTVMRGL LNGGRYGNRY IWTAISLRHP EVMEGNGLES AVMQWRRMLS
SgAOX    ......MISS RLAGTALCRQ LSHVPVPQYL PALRPTADTA SSLLHRCSAA APAQRAGLWP PSWFSPPRHA STLSARAQ.. ....DGGKEK

NcAOX    .......... .......... .......... .......... .........M NTPKVNILHA PGQAAQLSRA LISTCHTRPL LLA...GSRV
HaAOX    .......... .......... .......... .......... .......... .......... .......... MIKTYQYRSI LNSRNVGIRF
TAO      .......... .......... .......... .......... .......... .......... ....MFRNHA SRITAAAAPW VLRTACRQKS

                   100                                    *i1              150
TobBY    AENGSAAA.T GGGDGGDEKS VVSYWGVQPS KVTKEDGTEW KWNCFRPWET Y.KADLSIDL TKHHAPTTFL DKFAYWTVKS LRYPTDI...
TobSR1   VENGGAAA.S GGGDGGDEKS VVSYWGVPPS KVTKEDGTEW KWNCFRPWET Y.KADLSIDL TKHHAPTTFL DKFAYWTVKA LRYPTDI...
potato   AAKGDVDVTK KAEGDTEQKA VVSYWGVPPS RVTKEDGSPW RWACFRPWEA Y.ESDMSIDL KKHHAPTTFL DKMAFWTVKS LRWPTDI...
AtAOX    TENESTGGDA AGGNNKGDKG IASYWGVEPN KITKEDGSEW KWNCFRPWET Y.KADITIDL KKHHVPTTFL DRIAYWTVKS LRWPTDL...
Soybean  V......GLS SAGGNKEEKV IVSYWGIQPS KITKKDGTEW KWNCFSPWGT Y.KADLSIDL EKHMPPTTFL DKMAFWTVKV LRYPTDV...
mango    NAGGAEAQVK EQKEEKKDAM VSNYWGISRP KITREDGSEW PWNCFMPWET Y.RSDLSIDL KKHHVPRTFM DKFAYRTVKI LRVPTDI...
SgAOX    AAGTAGKVPP GEDGGAEKEA VVSYWAVPPS KVSKEDGSEW RWTCFRPWET Y.QADLSIDL HKHHVPTTIL DKLALRTVKA LRWPTDI...

                                                                                           *i1
NcAOX    ATSLHPTQTN LSSPSPRNFS TTSVTRLKDF FPAKETAYIR QTPPAWPHHG WTEEEMTSVV PEHRKPETVG DWLAWKLVRI CRWATDIATG
HaAOX    LKTLSPSPHS KDPNSKSIFD IG..TKLIVN PPPQMADNQY VTHPLFPHPK YSDEDCEAVH FVHREPKTIG DKIADRGVKF CRASFDFVTG
TAO      DAKTPV.... WGHTQLNRLS FLETVPVVPL RVSDESSEDR P.......T WSLPDIENVA ITHKKPNGLV DTLAYRSVRT CRWLFDTFSL

                              200  i2   *     M1                                  250      S
TobBY    .......... .......... .FFQRRYGCR AMMLETVAAV PGMVGGMLLH CKSLRRFEQS GGWIKTLLDE AENERMHLMT FMEVAKPNWY
TobSR1   .......... .......... .FFQRRYGCR AMMLETVAAV PGMVGGMLLH CKSLRRFEQS GGWIKALLEE AENERMHLMT FMEVAKPNWY
potato   .......... .......... .FFQRRYGCR AMMLETVAAV PGMVGGLLLH LKSLRRFEHS GGWIKALLEE AENERMHLMT FMEVSQPRWY
AtAOX    .......... .......... .FFQRRYGCR AMMLETVAAV PGMVGGMLLH CKSLRRFEQS GGWIKALLEE AENERMHLMT FMEVAKPKWY
Soybean  .......... .......... .FFQRRYGCR AMMLETVAAV PGMVAGMLLH CKSLRRFEHS GGWFKALLEE AENERMHLMT FMEVAKPKWY
mango    .......... .......... .FFQRRYGCR AMMLETVAAV PGMVGGMLLH LKSLRKLEQS GGWIKALLEE AENERMHLMT MVELVQPKWY
SgAOX    .......... .......... .FFQRRYACR AMMLETVAAV PGMVGGVLLH LKSLRRFEHS GGWIRALLEE AENERMHLMT FMEVAQPRWY
                                    i2
NcAOX    IRPEQQVDKH HPTTATSADK PLTEAQWLVR FIFLESIAGV PGMVAGMLRH LHSLRRLKRD NGWIETLLEE SYNERMHLLT FMKMCEPGLL
HaAOX    YKKPKDVNGM LKSWE.GTRY EMTEEKWLTR CIFLESVAGV PGMVAAFIRH LHSLRLLKRD KAWIETLLDE AYNERMHLLT FIKIGNPSWF
TAO      YRFG...... .......... SITESKVISR CLFLETVAGV PGMVGGMLRH LSSLRYMTRD KGWINTLLVE AENERMHLMT FIELRQPGLP
```

```
                   M2                     300       H1         H2          H3        350       H4
TobBY    ERALVFAVQG VFFNAYFVTY LLSPKLAHRI VGYLEEEAIH SYTEFLKELD KGNI.....E NVPAPAIAID YCRLPKD.ST LLDVVLVVRA
TobSR1   ERALVFAVQG VFINAYFVTY LLSPKLAXRI VGYLEEEAIH SYTEFLKELD KGNI.....E NVPAPAIAID YWRLPKD.ST LRDVVLVVRA
potato   ERALVLAVQG VFFNAYFLGY LLSPKFAHR. WGYLEEEAIH SYTEFLKEID KGTI.....D NVPAPAIALD YWRLPPG.ST LRDWM.VVRA
AtAOX    ERALVITVQG VFFNAYFLGY LISPKFAHRM VGYLEEEAIH SYTEFLKELD KGNI.....E NVPAPAIAID YWRLPAD.AT LRDVVMVVRA
Soybean  ERALVITVQG VFFNAYFLGY LLSPKFAHRM FGYLEEEAIH SYTEFLKELD KGNI.....E NVPAPAIAID YWQLPPG.ST LRDVVMVVRA
mango    ERLLVLAVQG VFFNSFFVLY VLSPKLAHRI VGYLEEEAIH SYTEYLKDID SGAI.....K NIPAPAIAID YWRLPKD.AT LKDVITVVRA
SgAOX    ERALVLAVQG VFFNAYFLGY LLSPKFAHRV VGYLEEEAIH SYTEFLKDID NGAI.....Q DCPAPAIALD YWRLPQG.ST LRDVVTVVRA

NcAOX    MKTLILGAQG VFFNAMFLSY LISPKITHRF VGYLEEEAVH TYTRCIREIE EGHLPKWSDE KFEIPEMAVR YWRMPEGKRT MKDLIHYIRA
HaAOX    TRFIIYMGQG VFANLFFLVY LIKPRYCHRF VGYLEEEAVS TYTHLIKDID SKRLPKFDD. .VNLPEISWL YWTDLNEKST FRDLIQRIRA
TAO      LRVSIIITQA IMYLFLLVAY VISPRFVHRF VGYLEEEAVI TYTGVMRAID EGSCPYQKCM FPKWLACTGT SAKMPHSATS ....STCIRA

           H4                i3                   400                                            441
TobBY    DEAHHRDVNH F........A SDIHYQGQQL KDSPAPIGYH .......... .......... .......... .......... .
TobSR1   DEAHHRDVNH F........A PDIHYQGQQL KDSPAPIGYH .......... .......... .......... .......... .
potato   DEAHHRDVNH F........A SDVHYQGHQL KEAPAPLGYH .......... .......... .......... .......... .
AtAOX    DEAHHRDVNH F........A SDIHYQGREL KEAPAPIGYH .......... .......... .......... .......... .
Soybean  DEAHHRDVNH F........A SDIHYQGREL REAAAPIGYH .......... .......... .......... .......... .
mango    DEAHHRDVNH F........A SDVQVQGKEL RDAPAPVGYH .......... .......... .......... .......... .
SgAOX    DEAHHRDVNH F........A SDVHYQDLEL KTTPAPLGYH .......... .......... .......... .......... .

NcAOX    DEAVHRGVNH T........L SNL.....DQ KEDPNPFVSD YKEGEGGRRP VNPALK...P TGFERAEVIG .......... .
HaAOX    DESKHREVNH T........L ANL.....EQ KKDRNPFALK VEDVPKEQQP DEYSLKTPHP EGWNREQMRL .......... .
TAO      DEAEHRVVNH TFVTCTKNAC KTVSTPSLYL KKNPEEMYSN QPSGKTRTDF ASEGAKTASN VNKQRVRAGI TFDGVPPFRV I
```

Figure 2 Compilation of deduced alternative oxidase sequences from various plants (top seven lines), two fungi, and a protist (bottom three sequences): TobBY, *N. tabacum* Bright Yellow (150); TobSR1, *N. tabacum* SR1 (166); potato (45); AtAOX, *A. thaliana* (62); soybean (164); mango, *Mangifera indica* L. (18); SgAOX, *S. guttatum* (111, 113); NcAOX, *N. crassa* (73); HaAOX, *H. anomala* (124); TAO, *Trypanosoma brucei brucei* (Accession #U52964). The first amino acid residues, alanines, of the mature alternative oxidase for potato and *S. guttatum* have been determined and are boxed (45, 111). Highly conserved cysteines in both plant and nonplant sequences are boxed with an *. The position of the three introns determined for the *S. guttatum* *Aox*1 gene (113) are designated i1, i2, and i3. In the nonplant sequences, the position of the two *N. crassa* introns are designated i1 and i2. Two predicted membrane-spanning helices are designated as M1 and M2, with a possible surface-exposed helix noted as S (81). H1, H2, H3, and H4 are highly conserved regions in all organisms. These sequences have also been proposed to be involved in a possible di-iron active site for the alternative oxidase (95, 136).

Early genomic hybridization experiments with the *Aox*1 cDNAs of *S. gut-tatum* and *A. thaliana* indicated that the alternative oxidase might be a low- or single-copy gene (62, 111). More recently, polymerase chain reaction (PCR) techniques have revealed at least three alternative oxidase genes in soybean and two or more copies in *N. tabacum* (165). Preliminary experiments employing reverse transcription–PCR (RT-PCR) indicated that soybean *Aox*1 is expressed primarily in cotyledons, whereas *Aox*3 is expressed in both cotyledons and leaves. No expression was found for *Aox*2 in leaves or cotyledons; however, *Aox*2 was isolated as a cDNA clone, indicating expression at some level (165). The significance of these gene families is unknown. It is possible that the multiple protein isoforms detected immunologically are products of different genes.

The *S. guttatum Aox*1 gene contains three introns and four exons (113). The third exon contains all the alpha helices proposed to both cross the membrane and form the di-iron center, whereas the conserved cysteines lie in exons one and three. A genomic clone of *N. crassa* strain NCN53 has two introns, with the second intron situated near the insertion of the second *S. guttatum* intron, and with exon three containing the most conserved, and membrane-spanning, regions of the protein (73). The transcription initiation site for *S. guttatum* Aox1 starts 79 bp before the start methionine, and there is a putative TATA box (TATAAA) located at −32 to −27 (113). There are two putative CAAT boxes at −83 to −79 (CCCAT) and −67 to −62 (CAAAAT), respective to the transcript initiation base (113), although there are a number of putative regulatory motifs in the promoter region. From region −524 to −568 there is a segment containing sequence similar to the salicylic acid–responsive elements described for the glycine rich protein GRP8 and the pathogenesis-related (PR) protein gene *PR1a* (113). Salicylic acid participates in thermogenesis in aroids and triggers increased expression of alternative oxidase in *S. guttatum* (109, 112). In vitro nuclear transcription assays indicated that this accumulation of *Aox*1 mRNA was not the result of a simple increase in *Aox*1 transcription. This contrasts with earlier experiments demonstrating that addition of actinomycin D, a transcriptional inhibitor, to immature floral tissue of *S. guttatum* blocked the cyanide–resistant respiration and the thermogenic response (80).

Transcription of the *N. crassa Aox1* gene begins 55 bp upstream from the initiation codon, with putative TATA and CAAT boxes upstream from the C nucleotide determined to be the initiation site (73). A perfectly conserved cAMP–responsive element, (TGACGTCA), was found at −746 to −739. The CRE elements bind transcription factors elicited by numerous signal transduction pathways (40). Although we do not know what this possible signal transduction pathway in *Neurospora* might be, it is interesting to note that alternative oxidase was induced in cultures grown in the presence of citrate or pyruvate as the sole carbon source (73). It is possible that the alternative pathway of *Neurospora* may be under some type of metabolite regulation, as has been proposed for tobacco (151).

ALTERNATIVE OXIDASE EXPRESSION IN TRANSGENIC PLANTS

The relatively recent isolation of *Aox*1 is one reason there are comparatively few studies of its expression in higher plants. In *S. guttatum*, the mRNA pools of *Aox*1 increased before thermogenesis. Similar increases occurred after application of salicylic acid to tissue sections of the immature appendix, the sterile upper portion of the aroid inflorescence (112). In further studies with tobacco suspension cell cultures, antimycin A induced *Aox*1 mRNA, while having little or no influence on the accumulation of mRNA for nuclear-encoded genes for the apoprotein of cytochrome *c* and no effect on expression of the mitochondrial gene encoding cytochrome oxidase subunit I (150). Alternative pathway activity increases in potato tuber slices upon aging, as does accumulation of the oxidase protein (46). Alternative oxidase mRNA is extremely low in both aged and fresh tuber slices (as detected by PCR quantification), and even though the activity increases as slices age, there was little detectable difference in mRNA amounts (47). We concluded that regulation of activity might depend on posttranscriptional and/or posttranslational modifications of the oxidase. Alternative pathway activity increases in many fruits during ripening (70). In mango fruits, alternative pathway activity and amounts of protein and mRNA all increase during the ripening process (18). From the few studies to date, it appears that increases in alternative oxidase protein generally parallel increased alternative pathway activity; however, it is too early to say whether this increase generally results from increased transcription.

Alternative oxidase gene expression has been specifically altered through genetic transformation for both tobacco (152) and potato (45). The native *Aox*1 genes, under CaMV 35S control, have been introduced into tobacco and potato in both "sense" and "antisense" orientations. In potato, only sense transgenic plants with Aox1 overexpressed were isolated; antisense plants with drastically lowered alternative oxidase were not found. Overexpression of potato alternative oxidase increased mitochondrial capacity for alternative oxidase activity in tubers and leaves, alternative oxidase protein levels, and *Aox*1 mRNA abundance (45). In all seven overexpressing lines studied in detail, an additional, slightly higher molecular mass, alternative oxidase–antibody reactive band appeared, in addition to the normal AOX bands present in potato.

This band has an authentic N-terminal sequence, therefore its alteration in mass could indicate a secondary modification. Another possibility is that in plants vastly overexpressing the alternative oxidase protein, a secondary processing or modification enzyme is unable to accommodate such a large pool of alternative oxidase, and the unmodified alternative oxidase accumulates in the membrane. Such an event may demonstrate a posttranslational step in alternative oxidase regulation that needs to be investigated further. In tobacco, both sense- and antisense-transformed lines were isolated. Overexpressing lines displayed greater alternative oxidase capacity and increased mRNA abundance, but it was difficult to ascertain whether modified alternative oxidase protein was present as found in the potato lines (152).

One antisense line of transgenic tobacco, AS8, has been isolated. It contains very low levels of alternative oxidase activity and protein (152). The mRNA pool in this line was much reduced compared with wild type and was of a larger molecular size. Suspension cell lines of AS8 grew at near normal levels unless the cytochrome pathway was inhibited with antimycin A, when the cells ceased growth. Wild-type cells grew at about one third of their normal rate of cells with addition of antimycin A. These data indicate alternative pathway respiration can serve to maintain cell growth, albeit at a lower rate, when the cytochrome pathway is inhibited. Tobacco lines overexpressing alternative oxidase were also isolated; they showed dramatic increases in alternative oxidase activity, but only when the cytochrome pathway was inhibited, for example, by KCN.

BIOCHEMISTRY OF ALTERNATIVE OXIDASE: REGULATION OF THE PLANT ENZYME

The common substrate of the cytochrome pathway and alternative oxidase is reduced ubiquinone. Moore et al (92) reported a voltametric method whereby oxygen consumption and the level of reduction of ubiquinone were monitored simultaneously and continuously in isolated plant mitochondria. Early experiments with this "Q electrode" method showed that partitioning of electrons to the cytochrome pathway varied linearly with the degree of ubiquinone reduction. Alternative oxidase was not active until the level of reduced ubiquinone reached a high threshold value (28, 92). This nonlinear response of the alternative pathway confirmed numerous studies on both isolated mitochondria and whole tissue, which suggested that alternative oxidase would function as an "energy overflow," only becoming active in O_2 consumption when the cytochrome pathway was saturated with electrons (64–66, 105). Such conditions were envisioned to arise when respiratory substrate was plentiful or when cellular adenylate energy charge was high, thus restricting oxidative phosphorylation (6, 22). The Q electrode technique has also allowed the development of kinetic models to describe electron flux to the plant alternative oxidase (53, 115, 133).

Two recently discovered mechanisms of biochemical control of the alternative oxidase now indicate that the regulation of electron partitioning to the alternative pathway is considerably more sophisticated than that of a simple overflow of the cytochrome pathway. First, it has been shown that certain alpha-keto acids (particularly pyruvate) stimulate alternative oxidase activity in isolated soybean mitochondria (23, 87). The stimulation is independent of pyruvate metabolism, suggesting that its effect is direct; that is, pyruvate interacts with the alternative oxidase protein. When care is taken to remove pyruvate from isolated mitochondria, alternative oxidase activity is nearly completely dependent on its addition. It has also been shown that the reported ability of succinate and malate to stimulate alternative oxidase activity directly in isolated mitochondria (74, 156, 158) may actually occur because these organic acids are converted to pyruvate by mitochondrial malic enzyme (86).

The second mechanism of biochemical control of the alternative oxidase enzyme was reported by Umbach & Siedow (143). They used chemical cross-linkers to show that soybean alternative oxidase exists in the inner mitochondrial membrane as either a covalently or noncovalently linked dimer. The relative amounts of the two forms could be visualized using nonreducing SDS-PAGE in which the covalently bound alternative oxidase has an apparent molecular mass approximately twice that of noncovalently bound alternative oxidase. The dimer, when covalently linked by putative intermolecular disulfide bond(s), was a less active form of the enzyme. Reduction of the disulfide bond(s) in isolated mitochondria by dithiothreitol produced the more active form; oxidation of the thiol(s) with diamide produced the less active form (143). This represents a reversible covalent modification of alternative oxidase activity. It has also been shown that only the more active reduced form is susceptible to pyruvate activation (144). All plant alternative oxidase protein sequences possess two completely conserved cysteine residues. These are potential candidates for the residue(s) involved in the disulfide linkage of the alternative oxidase proteins (Figure 2).

Using the Q electrode technique, it has now been shown that the fully reduced and activated enzyme has a higher affinity for reduced ubiquinone than was seen in earlier experiments (144). Considering this higher affinity for reduced ubiquinone, alternative oxidase may compete with the cytochrome pathway for electrons, as had been demonstrated by Wilson (167). By simultaneously monitoring cytochrome pathway activity spectrophotometrically (in the presence of ferricyanide and the cytochrome oxidase inhibitor CN) and alternative oxidase activity with an oxygen electrode, Hoefnagel et al (48) were able to show that the fully activated alternative oxidase could compete with the cytochrome pathway for electrons and did not simply act as an overflow pathway.

Transgenic tobacco containing high levels of mitochondrial alternative oxidase protein (152) facilitated further characterization of these biochemical mechanisms (147). Tobacco alternative oxidase is subject to covalent modification and pyruvate activation similar to that found in soybean, suggesting that these are common mechanisms. Furthermore, it was shown that reduction of tobacco alternative oxidase to its more active form was mediated by the oxidation of specific TCA cycle substrates, notably isocitrate and malate (147). Other substrates (succinate, glycine, 2-oxoglutarate, pyruvate) were relatively ineffective. The most likely explanation is that intramitochondrial reducing power generated by the activity of isocitrate dehydrogenase (when mitochondria are supplied with citrate or isocitrate) or malate dehydrogenase (when mitochondria are supplied with malate) supports the reduction of alternative oxidase. The results further suggest that the generation of intramitochondrial NADPH may be required for alternative oxidase reduction, because only isocitrate and malate metabolism are coupled to reduction of NADP in plant mitochondria (110). These findings are consistent with alternative oxidase reduction mediated by thioredoxin or glutathione, both of which require NADPH (50). Components of each of these systems have been identified in plant mitochondria (11, 79, 110), but a role for either in alternative oxidase reduction has not yet been reported. Extramitochondrial NADH or NADPH was ineffective in promoting alternative oxidase reduction, indicating that this is an intramitochondrial process (147). The biochemical controls of tobacco alternative oxidase activity are outlined in Figure 3.

---→

Figure 3 A high rate of AOX activity is dependent upon *both* a TCA cycle–mediated covalent modification of the AOX protein to its reduced (more active) form as well as an activation of the reduced form by pyruvate (pyr). In this experiment, leaf mitochondria were isolated from transgenic tobacco containing high levels of AOX protein, and the O_2 electrode analysis and Western analysis shown above were done in parallel. With external NADH and ADP as substrate, the mitochondria showed no O_2 uptake when myxothiazol (myxo) was added to inhibit the cyt pathway. Subsequent addition of pyruvate stimulated O_2 uptake only slightly. However, in the presence of pyruvate, citrate (Cit) addition caused a dramatic stimulation of O_2 uptake within a few minutes, all of which was sensitive to the AOX inhibitor n-propyl gallate (n-PG). The ability of citrate to increase AOX activity was associated with loss of the less active, covalently bound AOX dimer and an increase in the more active, noncovalently associated AOX dimer, as shown by nonreducing SDS-PAGE and Western analysis. Similar effects on AOX activity and on AOX protein forms were seen when isocitrate or

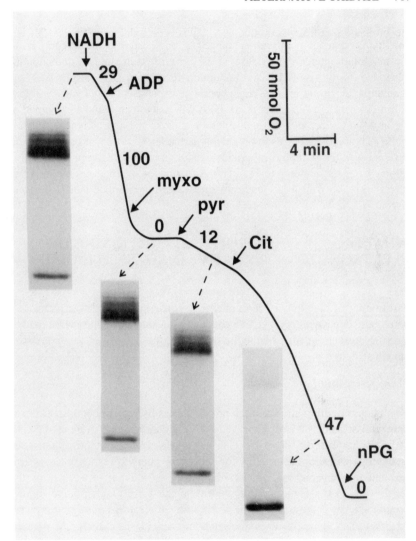

malate substituted for citrate. If the same experiment is done in the absence of pyruvate, citrate addition still results in the reduction of AOX to its more active form, but high rates of AOX activity are not seen until pyruvate is added. Additions were made at the following final concentrations: 2 mM NADH, 1 mM ADP, 16 μM myxo, 1 mM pyr, 10 mM Cit, and 100 μM n-PG. Rates are expressed as nmol O_2 mg^{-1} protein min^{-1}. Mitochondrial proteins (25 μg) were separated by nonreducing SDS-PAGE, transferred to nitrocellulose, and probed with a monoclonal antibody to AOX. Nonreducing SDS-PAGE allows visualization of less active covalently bound AOX dimer (~70 kDa, upper band) and more active noncovalently associated AOX dimer (~35 kDa, lower band). See text and Reference 147 for further details.

Before the discovery of the biochemical control of alternative oxidase activity by covalent modification and activation, it had often been reported that the maximum alternative oxidase activity measured in isolated mitochondria depended on the substrate used to support respiration (20, 156). It was suggested that electron transport chain components were not equally accessible to alternative oxidase (20). These substrate differences can now be explained by the ability of substrates to reduce and/or activate the alternative oxidase enzyme (23, 49, 147). For example, the apparent lesser ability of electrons from external NADH to support alternative oxidase activity (as seen in Figure 3) results from its inability to facilitate reduction of alternative oxidase to its active form, as well as its inability to generate intramitochondrial pyruvate (147).

WHAT FACTORS CONTROL THE PARTITIONING OF ELECTRONS TO ALTERNATIVE OXIDASE IN HIGHER PLANTS?

Partitioning of electrons to the nonenergy-conserving alternative pathway is potentially dependent upon coarse control of the amount of alternative oxidase enzyme present and fine metabolic control of the activity of the preexisting enzyme (Figure 4).

Coarse Control

The amount of alternative oxidase protein varies by growth conditions, developmental state, tissue-specific differences, and chemical treatments (16, 18, 46, 56, 58, 59, 96, 101, 109, 112, 114, 139, 148, 151, 155). Transgenic technology can alter the capacity for alternative pathway respiration, confirming that this capacity is primarily dependent on alternative oxidase protein levels (45, 152). Hence, coarse control of the level of alternative oxidase protein is a key determinant in the partitioning of electrons in respiration.

Given that alternative oxidase may compete with the cytochrome pathway for electrons and thus affect respiratory efficiency, it seems logical that plants regulate these pathways in a coordinated fashion and in response to metabolic requirements (149). Chemical inhibition of the cytochrome pathway can induce the synthesis of alternative oxidase in many organisms (7, 9, 124, 149, 150, 159). In tobacco suspension cells, inhibition of the cytochrome pathway by antimycin A induced a large increase in alternative oxidase mRNA and alternative oxidase protein, enabling the cells to maintain high respiration rates using this oxidase (149, 150). Hence, a mechanism exists whereby *Aox*1

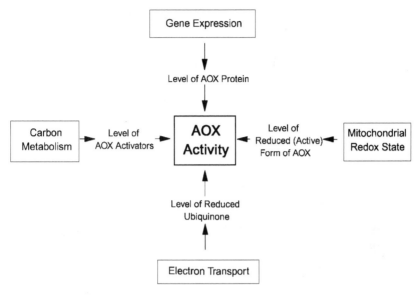

Figure 4   Major factors that may control the partitioning of respiratory electrons to the nonenergy-conserving alternative pathway in higher plant mitochondria. Partitioning of electrons to AOX may be dependent upon coarse control of the amount of mitochondrial AOX enzyme present as well as fine metabolic control of the activity of preexisting AOX enzyme. See text for details.

expression can respond to changes in cytochrome pathway activity. How the status of mitochondrial electron transport is perceived and then transmitted to the nucleus to activate *Aox*1 expression is unknown. Such a mechanism will almost certainly involve both physiological signals and the products of other genes.

In one current model for the regulation of genes encoding components of energy transduction, the existence of a "two component redox regulatory system" is proposed. The redox state of a particular electron carrier would modulate the expression of genes encoding components of electron transport via a signal transduction pathway involving protein phosphorylation/dephosphorylation cascades and DNA-binding proteins (2). Such models are based largely on regulatory examples from bacterial energy transduction, but they have also been extended to the photosyntheses in plastids (3, 4, 9). For example, transcription of the nuclear *cab* gene (encoding a chloroplast protein involved in light harvesting) appears to be regulated by the redox state of plastoquinone, an electron carrier in the photosynthetic electron transport chain (33). Evidence is presented for the existence of a redox-sensing kinase and a specific

DNA-binding protein. The model predicts that any environmental parameter (such as light quantity) altering the level of reduced plastoquinone will affect expression of the *cab* gene in such a way that the amount of this component of the light–harvesting apparatus can be adjusted to optimize photosynthesis under the new conditions (33). Related to this is the finding that translation of a photosynthetic electron transport chain protein in the unicellular green alga *Chlamydomonas reinhardtii* is light-regulated, apparently via changes in the chloroplast redox state altering an RNA binding protein (19).

Mitochondrial biogenesis may also depend on redox regulation of nuclear gene expression. The level of reduced ubiquinone may act as a signal regulating expression of alternative oxidase. In this way, an imbalance between upstream carbon metabolism and downstream electron transport resulting in changes in the level of reduced ubiquinone could be corrected by new gene expression. As a first step toward identifying potential components of a redox–regulated signal transduction pathway from mitochondrion to nucleus, Hakansson & Allen (41) showed that the phosphorylation of three proteins in pea mitochondria reflected the redox state of the reaction medium.

*Aox*1 gene expression may also respond to a particular active oxygen species such as superoxide, H_2O_2, or a hydroxyl radical. The generation of harmful active oxygen species is a consequence of metabolism in an aerobic environment; active oxygen moieties are signals regulating the expression of a diverse range of genes (71, 129, 130). Recent work in *Saccharomyces cerevisiae* indicates that mitochondrial respiration is an important source of active oxygen species in vivo (75), confirming in vitro studies showing that progressive reduction of electron transport chain components (such as the ubiquinone pool) leads to increased generation of active oxygen (14). Active oxygen species may therefore act as sensitive indicators to signal the nucleus about changes in mitochondrial redox state and thus serve to modulate expression of alternative oxidase. Some recent studies support this hypothesis. First, H_2O_2 treatment of plant cells induces the alternative oxidase pathway (151, 155). Addition of 5 mM H_2O_2 to tobacco suspension cells increased the level of *Aox*1 mRNA within 2 h (151). Second, inhibition of the cytochrome pathway by antimycin A or other inhibitors of respiratory Complex III results in the generation of active oxygen species in mitochondria (88, 90). In *Hansenula anomala* and in the fungus *Pyricularia grisea,* induction of alternative oxidase by these cytochrome pathway inhibitors is suppressed by low oxygen conditions or by the addition of scavengers of active oxygen such as plant flavonoids (88, 90). Third, the generation of active oxygen species in plants is pronounced during periods of biotic and abiotic stress (4, 76, 125), and the level of alternative oxidase protein appears to be sensitive to diverse stress treatments (81).

Another possibility is that alternative oxidase gene expression responds to a particular metabolite whose level reflects some key parameter of mitochondrial status. In this regard, it was reported that an increase in the cellular level of the TCA cycle intermediate citrate (either after its exogenous supply to cells, after inhibition of aconitase by monofluoroacetate, or after H_2O_2 treatment of cells) was rapidly followed by increased levels of $Aox1$ mRNA (151). Citrate is the first organic acid of the TCA cycle and its accumulation (e.g. because of slowed carbon flow through the TCA cycle) could be an important physiological signal linking mitochondrial metabolism and nuclear gene expression. The importance of metabolite-regulated expression of plant genes has already been recognized in the control of photosynthetic metabolism (60, 131).

Whelan et al (165) reported multiple Aox genes in soybean, and it is possible that different genes are expressed in particular tissues or at particular developmental stages (16, 18, 56, 59, 112). In this regard, Kearns et al (59) observed that the number of alternative oxidase protein bands seen on Western blots differs in soybean shoots versus roots. The existence of multiple alternative oxidase protein bands has been seen in a number of species, but in no case to date has it been conclusively determined that isoforms are distinct gene products. In addition, the relative activities of these different protein bands is not known. It is possible that import of the alternative oxidase precursor protein into the mitochondrion as well as subsequent processing or modification of alternative oxidase represent other points for coarse control of alternative oxidase activity (45, 163).

Fine Control

Although the level of alternative oxidase protein present in a tissue will determine the maximum possible partitioning of electrons to the alternative pathway, the actual flux appears to be subject to fine metabolic controls (Figure 4).

As explained above, the redox poise of the ubiquinone pool is an important factor regulating alternative oxidase activity. This redox poise will be directly dependent upon the relative activities of the dehydrogenases supplying electrons for ubiquinone reduction and the terminal oxidases driving ubiquinone oxidation (145). To date, most Q electrode experiments have focused on how alternative oxidase activity responds to the ratio of reduced to oxidized ubiquinone. However, it has been pointed out that it is the absolute level of reduced ubiquinone that will determine alternative oxidase activity (117). This activity is dependent not only on the relative activities of the dehydrogenases and terminal oxidases but on the absolute level of total ubiquinone in a tissue (117). Hence, each of these factors must be considered as potentially influencing the partitioning of electrons to alternative oxidase. For example, it has been suggested that the low amount of total ubiquinone in soybean roots may limit alternative oxidase activity in this tissue (117).

Another important fine control of alternative oxidase activity may be the status of the proposed alternative oxidase intermolecular disulfide bond. As explained above, the status of this disulfide bond appears to depend on the redox state of the mitochondrial pyridine nucleotide pool (147). Reduction of this pool via the oxidation of specific TCA cycle substrates facilitates reduction of alternative oxidase to its active form. It is expected that turnover of the TCA cycle will rapidly reduce the intramitochondrial pyridine nucleotide pool if the reoxidation of this pool by electron transport becomes limiting. Hence, the regulation of alternative oxidase activity by the redox state of the pyridine nucleotide pool represents a potential mechanism to couple TCA cycle metabolism and electron transport. For example, limitation of TCA cycle turnover by electron transport will favor reduction of alternative oxidase to its more active form. This will effectively increase the capacity of electron transport, favoring oxidation of the pyridine nucleotide pool and allowing increased turnover of the TCA cycle. In chloroplasts, light-dependent redox modulation of enzyme activities via the redox state of regulatory cysteine(s) is a major mechanism regulating photosynthetic metabolism (17, 52, 127). However, such covalent redox modulation of plant mitochondrial enzyme activity has not, to our knowledge, been previously reported. The in vivo significance of this regulatory mechanism could be tested by generating transgenic plants expressing large amounts of a recombinant alternative oxidase protein in which the potential regulatory cysteine(s) (Figure 2) is (are) replaced by alanines. Such an experiment would directly test the significance of the proposed covalent modification, identify the regulatory cysteine(s) involved, and may generate transgenic plants with a constitutively high in vivo partitioning of electrons to alternative oxidase.

Another major mechanism of fine control of alternative oxidase activity may be the intramitochondrial concentration of activators of the reduced enzyme, particularly pyruvate (23, 144). Regulation of alternative oxidase activity by pyruvate level may be well suited to couple respiratory carbon metabolism and electron transport, because pyruvate level can integrate several aspects of respiratory status. First, pyruvate synthesis from phospho*enol*pyruvate by the enzyme pyruvate kinase represents a key regulatory step in plant glycolysis (106, 142). The reaction is far removed from equilibrium, such that increases in glycolytic flux result in decreases in the level of phospho*enol*pyruvate and increases in pyruvate. Pyruvate can be a sensitive indicator of the rate of glycolytic flux. Second, pyruvate kinase requires ADP as a substrate, and the synthesis of pyruvate may depend on the degree to which glycolytic flux is restricted by adenylate control. Hence, a strict limitation of glycolytic flux by ADP limitation of pyruvate kinase may lower pyruvate level, leading to inactivation of alternative oxidase under conditions when substrate supply to the mitochondrion may be limiting. Note that this contrasts with adenylate restriction of oxidative phosphorylation in the mitochondrion, in which case alternative oxidase activity could be favored by an accumulation of pyruvate and/or increased reduction of the mitochondrial pyridine nucleotide pool and ubiquinone pool. Third, pyruvate is at an important branch point in metabolism, a substrate for the TCA cycle, fermentative pathways, and amino acid biosynthesis. Thus, its level may influence the direction of respiratory carbon flow. At present, neither a detailed in vivo analysis of the concentration of potential activators of alternative oxidase (particularly pyruvate) and of the activity of the alternative pathway nor a metabolic control model has been completed. Such studies would test the in vivo significance of this regulatory mechanism.

Finally, it should be emphasized that conditions hypothesized to favor partitioning of electrons to alternative oxidase under the "energy overflow" model (high substrate supply, high adenylate energy charge) also generate the requirements now thought to favor alternative oxidase activity (high pyruvate, reduced mitochondrial pyridine nucleotide pool, reduced ubiquinone pool). As we learn more about the coarse and fine controls of alternative oxidase activity, it becomes more apparent that the partitioning of electrons to alternative oxidase is not a static switch (a simple overflow of the cytochrome pathway), but rather a dynamic system able to respond to the availability of carbon and reducing power in the mitochondrion.

WHAT IS THE FUNCTION OF ALTERNATIVE OXIDASE IN HIGHER PLANTS?

From increased appreciation of the coarse and fine controls of alternative oxidase, greater insight should lead to testable hypotheses about alternative oxidase function. Given the biochemical controls of the alternative oxidase enzyme, any metabolic condition that leads to accumulation of either reduced ubiquinone, mitochondrial NADPH, or pyruvate has the potential to increase electron flow to the alternative pathway. Such conditions may arise when there is an imbalance between upstream respiratory carbon metabolism and downstream electron transport (Figure 5). Such imbalances may result from changes in the supply of, or demand for, carbon, reducing power, and ATP. The most general function of alternative oxidase may therefore be to balance carbon metabolism and electron transport. Rapid changes in either of these two coupled processes could be counteracted by rapid fine biochemical control of alternative oxidase activity. Likewise, longer-term changes in carbon metabolism or electron transport resulting in imbalances could be offset by coarse control of the amount of alternative oxidase protein present. Given this general function, alternative oxidase may serve specific roles, some of which we discuss.

The ability of mitochondrial electron transport to adjust rapidly in capacity via activation of alternative oxidase may be a mechanism to prevent over-reduction of respiratory chain components that might otherwise result in the formation of harmful active oxygen species (108, 157). Photosynthetic electron transport, another important source of active oxygen species, is thought to be rigorously protected against over-reduction by several mechanisms (34). In plant mitochondria, alternative oxidase may act to shunt electrons from reduced ubiquinone (thus preventing the generation of active oxygen species) when the cytochrome pathway is saturated with electrons or restricted by the availability of ADP.

Given that pyruvate is at such a central position in respiratory carbon metabolism (Figure 5), it is possible that alternative oxidase activation by pyruvate is a sensitive mechanism to regulate pyruvate level. Pyruvate accumulation could activate alternative oxidase, thus increasing rates of electron

———————————————————————————————→

Figure 5 A model of alternative oxidase function and regulation in respiratory metabolism of higher plants. Partitioning of electrons (e⁻) from reduced ubiquinone (Q) to alternative oxidase is regulated in response to the availability of carbon and reducing power in the mitochondrion. This is accomplished by a reversible covalent modification of the alternative oxidase protein responsive to the redox state of the mitochondrial pyridine nucleotide pool (NADPH) and by an activation of the alternative oxidase protein by specific metabolites such as pyruvate. Metabolic conditions resulting in the accumulation of pyruvate, intramitochondrial NADPH or reduced ubiquinone have the potential to increase electron transport to alternative oxidase. Control of the partitioning of electrons to alternative oxidase by this dynamic system may function to balance upstream respiratory carbon metabolism and downstream electron transport when these coupled processes become imbalanced because of changes in the supply of, or demand for, carbon, reducing power, and ATP. This may serve to prevent over-reduction of electron transport chain components (which can result in the generation of harmful active oxygen species), prevent aerobic fermentation from accumulated pyruvate, or adjust respiratory metabolism in response to photosynthesis or biotic and abiotic stress. For simplicity, we have only included some metabolic intermediates and have not maintained correct stoichiometry between intermediates. Numbers refer to particular enzymatic reactions mentioned in the text: 1, pyruvate kinase; 2, NAD(P)-isocitrate dehydrogenase; 3, NAD(P)-malate dehydrogenase; 4, malic enzyme. Roman numerals refer to the mitochondrial electron transport complexes I through IV.

transport and drawing down the pyruvate level. In this regard, Vanlerberghe et al (147) analyzed transgenic plants lacking alternative oxidase protein. When the cytochrome pathway was inhibited by antimycin A and was limiting respiration, the lack of alternative oxidase resulted in the production of large quantities of ethanol. Similar experiments in transgenic suspension cells showed that this aerobic fermentation is the result of a large accumulation of pyruvate

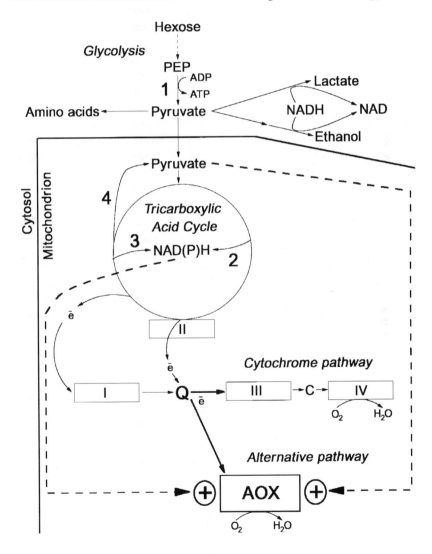

(153). These results imply that feed-forward activation of alternative oxidase by upstream carbon metabolism can prevent pyruvate accumulation and aerobic fermentation when there is an imbalance between carbon metabolism and electron transport. This hypothesis needs to be further tested because, to date, it has only been demonstrated using antimycin A to artificially generate the imbalance.

Alternative oxidase may play a role in photosynthetic metabolism because photosynthesis will dramatically influence the supply of carbon, reducing equivalents, and ATP available to cellular metabolism. It has been suggested that mitochondrial electron transport in the light may be a mechanism to oxidize excess photosynthetic reducing power, but to date the contribution, if any, of alternative oxidase is unknown (61). Photorespiration is a major source of reducing power generated in the mitochondrion in the light. It is interesting that two photorespiratory intermediates—hydroxypyruvate and glyoxylate—also activate reduced alternative oxidase similar to pyruvate (86). There are also indications that alternative oxidase participates in the oxidation of reducing equivalents generated during either the rapid malate decarboxylation in the light that occurs in CAM species (121, 123) or the bundle-sheath localized malate decarboxylation that occurs in C_4 species (1, 35). Detailed study will be required to understand the potential contribution of alternative oxidase to photosynthetic metabolism.

Finally, imbalances between carbon metabolism and electron transport may be brought about or magnified by the changing environmental conditions that plants regularly face or by biotic and abiotic stresses. A wide range of such conditions appears to influence the capacity of the alternative pathway (10, 57, 104, 108, 139, 148, 161). Hence, alternative oxidase may be one of a battery of metabolic adjustments that plants will use to overcome such adversities.

FUTURE DIRECTIONS

One promising area for future research is molecular genetic alteration of mitochondrial electron transport chain components. Our ability to silence in large part and to overexpress alternative oxidase in transgenic plants (45, 152) indicates the feasibility of this strategy. Transgenic plants with an altered capacity for alternative pathway respiration should allow a critical analysis of alternative oxidase function in plant metabolism.

$Aox1$ is one of the few examples of the isolation and characterization of a plant nuclear gene encoding a mitochondrial protein involved in electron transport. It therefore represents an excellent opportunity to study the physiological signals and molecular mechanisms that coordinate mitochondrial function and nuclear gene expression. The signals and mechanisms that regulate mitochondrial biogenesis are beginning to be elucidated in yeast and animals (26, 100, 126, 154).

Another area for future research will be the biochemical analysis of purified and active alternative oxidase enzyme. Such work has previously met with limited success because the enzyme rapidly loses activity after solubilization. Recently, purification of active enzyme was improved by adding pyruvate to the isolation medium (170). This stabilizing influence is further evidence of a direct interaction of pyruvate with the alternative oxidase protein.

Further research that should provide exciting insights into alternative oxidase function will be direct measurements of the in vivo partitioning of electrons to alternative oxidase in various plant tissues and under various environmental or metabolic conditions. Most previous estimates of in vivo partitioning have been based on the use of inhibitors of the cytochrome pathway and alternative oxidase (for a review of methodology, see 91). The use of inhibitors has always been problematic: lack of specificity of the inhibitors and problems with penetration into tissues, among other problems. A more fundamental problem with inhibitors is now evident: Their usefulness depends on the assumption that alternative oxidase cannot compete with the cytochrome pathway for electrons and is active only when the cytochrome pathway is saturated. On the basis of recent work (see above), this assumption is no longer considered valid. The implications of this are twofold. First, the partitioning of electrons to alternative oxidase (when determined using inhibitors) may have been considerably underestimated in the past; second, a noninvasive method is required (21, 85). The basis for such a noninvasive method was reported by Guy et al (39), who showed that alternative oxidase discriminated against heavy labeled O_2 ($^{18}O^{16}O$) substantially more than did cytochrome oxidase. Discrimination was determined by observing progressive changes in the isotopic composition of O_2 substrate over time within a closed system (i.e. a buildup of $^{18}O^{16}O$ relative to $^{16}O^{16}O$ as the O_2 was consumed by respiration). To determine the degree of partitioning to each pathway, the discrimination by each separate pathway must first be known. Discrimination by the cytochrome pathway was determined in the presence of an inhibitor of alternative oxidase, whereas discrimination by alternative oxidase was determined in the presence of an inhibitor of the cytochrome pathway. Having established these endpoints, discrimination in the absence of inhibitors was measured and the steady state partitioning of electrons between the two pathways was calculated (39). Recently, systems for both gas-phase and aqueous-phase measurements of plant respiration using an on-line gas chromatograph–mass spectrometer have been developed (116, 122). The gas-phase system is ideal for measuring discrimination on whole plant tissues (122). The aqueous-phase system is useful for isolated mitochondria and/or cell suspensions (116). Such experiments will help us to quantify the contribution of alternative oxidase to total mitochondrial electron flux and to identify factors that influence this flux.

ACKNOWLEDGMENTS

We acknowledge research support from the Department of Energy grant DE-FG02-91ER20021 (LM), the National Science Foundation grant IPB 9407979 (LM), and US Department of Agriculture–CSRS grant 94-37304-0952 (LM).

Visit the *Annual Reviews home page* at http://www.annurev.org.

Literature Cited

1. Agostino A, Heldt HW, Hatch MD. 1996. Mitochondrial respiration in relation to photosynthetic C_4 acid decarboxylation in C_4 species. *Aust. J. Plant Physiol.* 23:1–7

2. Allen JF. 1993. Redox control of transcription: sensors, response regulators, activators and repressors. *FEBS Lett.* 332:203–7

3. Allen JF, Alexciev K, Hakansson G. 1995. Regulation by redox signalling. *Curr. Biol.* 5:869–72

4. Allen RD. 1995. Dissection of oxidative stress tolerance using transgenic plants. *Plant Physiol.* 107:1049–54

5. Anraku Y, Gennis RB. 1987. The aerobic respiratory chain of *Escherichia coli*. *Trends Biochem.* 12:262–66

6. Azcon-Bieto J, Lambers H, Day DA. 1983. Effect of photosynthesis and carbohydrate status on respiratory rates and the involvement of the alternative pathway in leaf respiration. *Plant Physiol.* 72:598–603

7. Benichou P, Calvayrac R, Claisse M. 1988. Induction by antimycin A of cyanide-resistant respiration in heterotrophic *Euglena gracilis:* effects on growth, respiration and protein biosynthesis. *Planta* 175:23–32

8. Berthold DA, Siedow JN. 1993. Partial purification of the cyanide-resistant alternative oxidase of skunk cabbage (*Symplocarpus foetidus*) mitochondria. *Plant Physiol.* 101:113–19

9. Bertrand H, Argan CA, Szakacs NA. 1983. Genetic control of the biogenesis of cyanide insensitive respiration in *Neurospora crassa*. In *Mitochondria*, ed. RJ Schweyen, K Wolf, F Kaudewitz, pp. 495–507. Berlin: de Guyter

10. Bingham IJ, Farrar JF. 1989. Activity and capacity of respiratory pathways in barley roots deprived of inorganic nutrients. *Plant Physiol. Biochem.* 27:847–54

11. Bodenstein-Lang J, Buch A, Follmann H. 1989. Animal and plant mitochondria contain specific thioredoxins. *FEBS Lett.* 258: 22–26

12. Bonner WD, Clarke SD, Rich PR. 1986. Partial purification and characterization of the quinol oxidase activity of *Arum maculatum* mitochondria. *Plant Physiol.* 80: 838–42

13. Bosma G, Braster M, Stouthamer AH, van Verseveld HW. 1987. Isolation and characterization of ubiquinol oxidase complexes from *Paracoccus denitrificans* cells cultured under various limiting growth conditions in the chemostat. *Eur. J. Biochem.* 165:657

14. Boveris A, Cadenas E. 1982. Production of superoxide radicals and hydrogen peroxide in mitochondria. In *Superoxide Dismutase*, ed. LW Oberley, 2:15–30. Boca Raton, FL: CRC Press

15. Chaudhuri M, Ajoyi W, Hill GC. 1995. Identification and partial purification of a stage-specific 33 kDa mitochondrial protein as the alternative oxidase of the *Trypanosoma brucei brucei* blood-stream trypomastigotes. *J. Euk. Microbiol.* 42: 467–72

16. Conley CA, Hanson MR. 1994. Tissue-specific protein expression in plant mitochondria. *Plant Cell* 6:85–91

17. Crawford NA, Droux M, Kosower NS, Buchanan BB. 1989. Evidence for function

of the ferredoxin/thioredoxin system in the reductive activation of target enzymes of isolated intact chloroplasts. *Arch. Biochem. Biophys.* 271:223–39

18. Cruz-Hernandez A, Gomez-Lim MA. 1995. Alternative oxidase from mango (*Mangifera indica* L.) is differentially regulated during fruit ripening. *Planta* 197: 569–76

19. Danon A, Mayfield SP. 1994. Light-regulated translation of chloroplast messenger RNAs through redox potential. *Science* 266:1717–19

20. Day DA, Dry IB, Soole KL, Wiskich JT, Moore AL. 1991. Regulation of alternative pathway activity in plant mitochondria: deviations from Q-pool behavior during oxidation of NADH and quinols. *Plant Physiol.* 95:948–53

21. Day DA, Krab K, Lambers H, Moore AL, Siedow JN, et al. 1996. The cyanide-resistant oxidase: to inhibit or not to inhibit, that is the question. *Plant Physiol.* 110:1–2

22. Day DA, Lambers H. 1983. The regulation of glycolysis and electron transport in roots. *Physiol. Plant.* 58:155–60

23. Day DA, Millar AH, Wiskich JT, Whelan J. 1994. Regulation of alternative oxidase activity by pyruvate in soybean mitochondria. *Plant Physiol.* 106:1421–27

24. Day DA, Whelan J, Millar AH, Siedow JN, Wiskich JT. 1995. Regulation of alternative oxidase in plants and fungi. *Aust. J. Plant Physiol.* 22:497–509

25. Day DA, Wiskich JT. 1995. Regulation of alternative oxidase activity in higher plants. *J. Bioenerg. Biomembr.* 27:379–85

26. de Winde JH, Grivell LA. 1993. Global regulation of mitochondrial biogenesis in *Saccharomyces cerevisiae. Prog. Nucleic Acid Res. Mol. Biol.* 46:51–91

27. Douce R, Neuburger M. 1989. The uniqueness of plant mitochondria. *Annu. Rev. Plant Physiol. Plant Mol. Biol.* 40:371–414

28. Dry IB, Moore AL, Day DA, Wiskich JT. 1989. Regulation of alternative pathway activity in plant mitochondria: nonlinear relationship between electron flux and the redox poise of the quinone pool. *Arch. Biochem. Biophys.* 273:148–57

28a. Ducet G, Lance C, eds. 1978. *Plant Mitochondria.* Amsterdam: Elsevier/North Holland Biomed. Press

29. Elthon TE, McIntosh L. 1986. Characterization and solubilization of the alternative oxidase of *Sauromatum guttatum* mitochondria. *Plant Physiol.* 82:1–6

30. Elthon TE, McIntosh L. 1987. Identification of the alternative terminal oxidase of higher plant mitochondria. *Proc. Natl. Acad. Sci. USA* 84:8399–403

31. Elthon TE, Nickels RL, McIntosh L. 1989. Monoclonal antibodies to the alternative oxidase of higher plant mitochondria. *Plant Physiol.* 89:1311–17

32. Eriksson M, Gardestrom P, Samuelsson G. 1995. Isolation, purification, and characterization of mitochondria from *Chlamydomonas reinhardtii. Plant Physiol.* 107: 479–83

33. Escoubas J-M, Lomas M, LaRoche J, Falkowski PG. 1995. Light intensity regulation of *cab* gene transcription is signaled by the redox state of the plastoquinone pool. *Proc. Natl. Acad. Sci. USA* 92:10237–41

34. Foyer CH, Lelandais M, Kunert KJ. 1994. Photooxidative stress in plants. *Physiol. Plant.* 92:696–717

35. Gardestrom P, Edwards GE. 1983. Isolation of mitochondria from leaf tissue of *Panicum miliaceum,* a NAD-malic enzyme type C_4 plant. *Plant Physiol.* 71:24–29

36. Gier JWL de, Lubben M, Reijnders WN, Tipker CA, Slotboom DJ, et al. 1994. The terminal oxidases of *Paracoccus denitrificans. Mol. Microbiol.* 13:183–96

37. Goyal A, Hiser C, Tolbert NE, McIntosh L. 1991. Progress No. 9 cultivar of pea has alternative respiratory capacity and normal glycine metabolism. *Plant Cell Physiol.* 32: 247–51

38. Goyal A, Tolbert NE. 1989. Variations in the alternative oxidase in *Chlamydomonas* grown in air or high CO_2. *Plant Physiol.* 89:958–62

39. Guy RD, Berry JA, Fogel ML, Hoering TC. 1989. Differential fractionation of oxygen isotopes by cyanide-resistant and cyanide-sensitive respiration in plants. *Planta* 177: 483–91

40. Habner JF. 1990. Cyclic AMP response element binding proteins: a cornucopia of transcription factors. *Mol. Endocrinol.* 4: 1087–94

41. Hakansson G, Allen JF. 1995. Histidine and tyrosine phosphorylation in pea mitochondria: evidence for protein phosphorylation in respiratory redox signalling. *FEBS Lett.* 372:238–42

42. Henry MF, Hamaide-Deplus MC, Nyns EJ. 1974. Cyanide-insensitive respiration of *Candida lipolytica. Antonie van Leeuwenhoek* 40:79–91

43. Henry MF, Nyns EJ. 1975. Cyanide-insensitive respiration: an alternative mitochondrial pathway. *Sub-Cell. Biochem.* 4:1–65

44. Herk AWH van. 1937. Die chemischen Vorgange im Sauromatum Lokben. I. *Mitt. Rec. Trav. Bot. Neerl.* 34:69–156

45. Hiser C, Kapranov P, McIntosh L. 1996. Genetic modification of respiratory capacity in potato. *Plant Physiol.* 110:277–86
46. Hiser C, McIntosh L. 1990. Alternative oxidase of potato is an integral membrane protein synthesized *de novo* during aging of tuber slices. *Plant Physiol.* 93:312–18
47. Hiser CH, McIntosh L. 1994. Potato alternative oxidase: detection of mRNA by PCR and tissue-specific differences in the protein level. In *The Molecular and Cellular Biology of the Potato*, ed. WR Belknap, ME Vayda, WD Park, pp. 143–50. Wallingford, UK: CAB Int.
48. Hoefnagel MHN, Millar AH, Wiskich JT, Day DA. 1995. Cytochrome and alternative respiratory pathways compete for electrons in the presence of pyruvate in soybean mitochondria. *Arch. Biochem. Biophys.* 318: 394–400
49. Hoefnagel MHN, Wiskich JT. 1996. Alternative oxidase activity and the ubiquinone redox level in soybean cotyledon and *Arum* spadix mitochondria during NADH and succinate oxidation. *Plant Physiol.* 110: 1329–35
50. Holmgren A. 1989. Thioredoxin and glutaredoxin systems. *J. Biol. Chem.* 264: 13963–66
51. Huq S, Palmer JM. 1978. Isolation of a cyanide-resistant duroquinol oxidase from *Arum maculatum* mitochondria. *FEBS Lett.* 95:217–20
52. Issakidis E, Saarinen M, Decottignies P, Jacquot JP, Cretin C, et al. 1994. Identification and characterization of the second regulatory disulfide bridge of recombinant sorghum leaf NADP-malate dehydrogenase. *J. Biol. Chem.* 269:3511–17
53. James AT, Venables WN, Dry IB, Wiskich JT. 1994. Random effects and variances as a synthesis of nonlinear regression analyses of mitochondrial electron transport. *Biometrika* 81:219–35
54. Jeppsson H. 1996. *Pentose utilization in yeasts: physiology and biochemistry.* PhD thesis. Dep. Appl. Microbiol., Lund Univ., Sweden
55. Jeppsson H, Alexander NJ, Hahn-Hagerdal B. 1995. Existence of cyanide-insensitive respiration in the yeast *Pichia stipitis* and its possible influence on product formation during xylose utilization. *Appl. Environ. Microbiol.* 61:2596–600
56. Johns C, Nickels R, McIntosh L, MacKenzie S. 1993. The expression of alternative oxidase and alternative respiratory capacity in cytoplasmic male sterile common bean. *Sex. Plant. Reprod.* 6:257–65
57. Jolivet Y, Pireaux JC, Dizengremel P. 1990.

Changes in properties of barley leaf mitochondria isolated from NaCl-treated plants. *Plant Physiol.* 94:641–46
58. Kapulnik Y, Yalpani N, Raskin I. 1992. Salicylic acid induces cyanide resistant respiration in tobacco cell suspension cultures. *Plant Physiol.* 100:1921–26
59. Kearns A, Whelan J, Young S, Elthon TE, Day DA. 1992. Tissue-specific expression of the alternative oxidase in soybean and siratro. *Plant Physiol.* 99:712–17
60. Koch KE. 1996. Carbohydrate-modulated gene expression in plants. *Annu. Rev. Plant Physiol. Plant Mol. Biol.* 47:509–40
61. Kromer S. 1995. Respiration during photosynthesis. *Annu. Rev. Plant Physiol. Plant Mol. Biol.* 46:45–70
62. Kumar AM, Soll D. 1992. *Arabidopsis* alternative oxidase sustains *Escherichia coli* respiration. *Proc. Natl. Acad. Sci. USA.* 89: 10842–46
63. Lamarck JB de. 1815. *Flore Fr.* 3: 151
64. Lambers H. 1980. The physiological significance of cyanide-resistant respiration in higher plants. *Plant Cell Environ.* 3:293–302
65. Lambers H. 1982. Cyanide-resistant respiration: a nonphosphorylating electron transport pathway acting as an energy overflow. *Physiol. Plant.* 55:478–85
66. Lambers H. 1985. Respiration in intact plants and tissues: its regulation and dependence on environmental factors, metabolism and invaded organisms. In *Encyclopedia of Plant Physiology. Higher Plant Cell Respiration,* ed. R Douce, DA Day, 18:418–73. New York: Springer-Verlag
66a. Lambers H, van der Plas LHW, eds. 1992. *Molecular, Biochemical and Physiological Aspects of Plant Respiration.* Amsterdam: Academic
67. Lambowitz AM, Sabourin JR, Bertrand H, Nickels R, McIntosh L. 1989. Immunological identification of the alternative oxidase of *Neurospora crassa* mitochondria. *Mol. Cell. Biol.* 9:1362–64
68. Lambowitz AM, Slayman CW. 1971. Cyanide-resistant respiration in *Neurospora crassa. J. Bacteriol.* 108:1087–96
69. Lambowitz AM, Zannoni D. 1978. Cyanide-insensitive respiration in *Neurospora* genetic and biophysical approaches. See Ref. 28a, pp. 283–91
70. Laties GG. 1992. The cyanide-resistant, alternative pathway in higher plant respiration. *Annu. Rev. Plant Physiol.* 33:519–55
71. Levine A, Tenhaken R, Dixon R, Lamb C. 1994. H_2O_2 from the oxidative burst orchestrates the plant hypersensitive disease resistance response. *Cell* 79:583–93

72. Deleted in proof
73. Li Q, Ritzel RG, McLean LLT, McIntosh L, Ko T, et al. 1996. Cloning and analysis of the alternative oxidase gene of *Neurospora crassa. Genetics* 142:129–40
74. Liden AC, Akerlund HE. 1993. Induction and activation of the alternative oxidase of potato tuber mitochondria. *Physiol. Plant.* 87:134–41
75. Longo VD, Gralla EB, Valentine JS. 1996. Superoxide dismutase activity is essential for stationary phase survival in *Saccharomyces cerevisiae:* mitochondrial production of toxic oxygen species *in vivo. J. Biol. Chem.* 271:12275–80
76. Low PS, Merida JR. 1996. The oxidative burst in plant defense: function and signal transduction. *Physiol. Plant.* 96:533–42
77. Lynnes JA, Weger HG. 1996. Azide-stimulated oxygen consumption by the green alga *Selenastrum minutum. Physiol. Plant.* 97:132–38
78. Mansfield JW. 1983. Antimicrobial compounds. In *Biochemical Plant Pathology,* ed. JA Callow, pp. 237–65. London: Wiley
79. Marcus F, Chamberlain SH, Chu C, Masiarz FR, Shin S, et al. 1991. Plant thioredoxin h: an animal-like thioredoxin occurring in multiple cell compartments. *Arch. Biochem. Biophys.* 287:195–98
80. McIntosh L. 1977. *A developmental analysis of anthesis in* Sauromatum guttatum *Schott.* PhD thesis. Univ. Wash., Seattle
81. McIntosh L. 1994. Molecular biology of the alternative oxidase. *Plant Physiol.* 105: 781–86
82. McIntosh L, Meeuse BJD. 1978. Control of the development of cyanide resistant respiration in *Sauromatum guttatum* (Araceae). See Ref. 28a, pp. 339–45
83. Meeuse BJD. 1975. Thermogenic respiration in aroids. *Annu. Rev. Plant Physiol.* 26:117–26
84. Meeuse BJD, Buggeln RG. 1969. Time, space, light and darkness in the metabolic flare-up of the *Sauromatum* appendix. *Acta Bot. Neerl.* 18:159–72
85. Deleted in proof
86. Millar AH, Hoefnagel MHN, Day DA, Wiskich JT. 1996. Specificity of the organic acid activation of alternative oxidase in plant mitochondria. *Plant Physiol.* 111: 613–18
87. Millar AH, Wiskich JT, Whelan J, Day DA. 1993. Organic acid activation of the alternative oxidase of plant mitochondria. *FEBS Lett.* 329:259–62
88. Minagawa N, Koga S, Nakano M, Sakajo S, Yoshimoto A. 1992. Possible involvement of superoxide anion in the induction of cyanide-resistant respiration in *Hansenula anomala. Agric. Biol. Chem.* 55: 1573–78
89. Minagawa N, Sakajo S, Yoshimoto A. 1991. Sulfur compounds induce alternative oxidase in *Hansenula anomala. Argric. Biol. Chem.* 55:1573–78
90. Mizutani A, Miki N, Yukioka H, Tamura H, Masuko M. 1996. A possible mechanism of control of rice blast disease by a novel alkoxyiminoacetamide fungicide, SSF126. *Phytopathology* 86:295–300
91. Moller IM, Berczi A, van der Plas LHW, Lambers H. 1988. Measurement of the activity and capacity of the alternative pathway in intact plant tissues: identification of problems and possible solutions. *Physiol. Plant.* 72:642–49
92. Moore AL, Dry IB, Wiskich JT. 1988. Measurement of the redox state of the ubiquinone pool in plant mitochondria. *FEBS Lett.* 235:76–80
93. Moore AL, Leach G, Whitehouse DG, van den Bergen CWM, Wagner AM, Krab K. 1994. Control of oxidative phosphorylation in plant mitochondria: the role of nonphosphorylating pathways. *Biochim. Biophys. Acta* 1187:145–51
94. Moore AL, Siedow JN. 1991. The regulation and nature of the cyanide-resistant alternative oxidase of plant mitochondria. *Biochim. Biophys. Acta* 1059:121–40
95. Moore AL, Umbach AL, Siedow JN. 1995. Structure-function relationships of the alternative oxidase of plant mitochondria: a model of the active site. *J. Bioenerg. Biomembr.* 27:367–77
96. Moynihan MR, Ordentlich A, Raskin I. 1995. Chilling-induced heat evolution in plants. *Plant Physiol.* 108:995–99
97. Musgrave ME, Murfet IC, Siedow JN. 1986. Inheritance of cyanide resistant respiration in two cultivars of pea (*Pisum sativum* L.). *Plant Cell Environ.* 9:153–56
98. Musgrave ME, Siedow JN. 1985. A relationship between plant responses to cytokinins and cyanide-resistant respiration. *Physiol. Plant.* 64:161–66
99. Musgrave ME, Strain BR, Siedow JN. 1986. Response of two pea hybrids to CO_2 enrichment: a test of the energy overflow hypothesis for alternative respiration. *Proc. Natl. Acad. Sci. USA* 83:8157–61
100. Nunnari J, Walter P. 1996. Regulation of organelle biogenesis. *Cell* 84:389–94
101. Obenland D, Diethelm R, Shibles R, Stewart C. 1990. Relationship of alternative respiratory capacity and alternative oxidase amount during soybean seedling growth. *Plant Cell Physiol.* 31:897–901

102. Obenland D, Hiser C, McIntosh L, Shibles R, Stewart CR. 1988. Occurrence of alternative respiratory capacity in soybean and pea. *Plant Physiol.* 88:528–31

103. Ordentlich A, Linzer RA, Raskin I. 1991. Alternative respiration and heat evolution in plants. *Plant Physiol.* 97:1545–50

104. Palet A, Ribas-Carbo M, Argiles JM, Azcon-Bieto J. 1991. Short-term effects of carbon dioxide on carnation callus cell respiration. *Plant Physiol.* 96:467–72

105. Palmer JM. 1976. The organization and regulation of electron transport in plant mitochondria. *Annu. Rev. Plant Physiol.* 27: 133–57

106. Plaxton WC. 1996. The organization and regulation of plant glycolysis. *Annu. Rev. Plant Physiol. Plant Mol. Biol.* 47:185–214

107. Priest JW, Hajduk SL. 1994. Developmental regulation of mitochondrial biogenesis in *Trypanosoma brucei*. *J. Bioenerg. Biomembr.* 26:179–91

108. Purvis AC, Shewfelt RL. 1993. Does the alternative pathway ameliorate chilling injury in sensitive plant tissues? *Physiol. Plant.* 88:712–18

109. Raskin I, Turner IM, Melander WR. 1989. Regulation of heat production in the inflorescences of an *Arum* lily by endogenous salicylic acid. *Proc. Natl. Acad. Sci. USA* 86:2214–18

110. Rasmusson AG, Moller IM. 1990. NADP-utilizing enzymes in the matrix of plant mitochondria. *Plant Physiol.* 94:1012–18

111. Rhoads DM, McIntosh L. 1991. Isolation and characterization of a cDNA clone encoding an alternative oxidase protein of *Sauromatum guttatum* (Schott). *Proc. Natl. Acad. Sci. USA,* 88:2122–26

112. Rhoads DM, McIntosh L. 1992. Salicylic acid regulation of respiration in higher plants: alternative oxidase expression. *Plant Cell* 4:1131–39

113. Rhoads DM, McIntosh L. 1993. The salicylic acid-inducible alternative oxidase gene *aox1* and genes encoding pathogenesis-related proteins share regions of sequence similarity in their promoters. *Plant Mol. Biol.* 21:615–24

114. Rhoads DM, McIntosh L. 1993. Cytochrome and alternative pathway respiration in tobacco. Effects of salicylic acid. *Plant Physiol.* 103:877–83

115. Ribas-Carbo M, Berry JA, Azcon-Bieto J, Siedow JN. 1994. The reaction of the plant mitochondrial cyanide-resistant alternative oxidase with oxygen. *Biochim. Biophys. Acta* 1188:205–12

116. Ribas-Carbo M, Berry JA, Yakir D, Giles L, Robinson SA, et al. 1995. Electron partitioning between the cytochrome and alternative pathways in plant mitochondria. *Plant Physiol.* 109:829–37

117. Ribas-Carbo M, Wiskich JT, Berry JA, Siedow JN. 1995. Ubiquinone redox behavior in plant mitochondria during electron transport. *Arch. Biochem. Biophys.* 317:156–60

118. Rich P. 1978. Quinol oxidation in *Arum maculatum* mitochondria and its application to the assay, solubilization and partial purification of the alternative oxidase. *FEBS Lett.* 96:252–56

119. Rissler JF, Millar RL. 1977. Contribution of a cyanide-insensitive alternate respiratory system to increases in formamide hydro-lyase activity and to growth in *Stemphylium loti in vitro. Plant Physiol.* 60: 857–61

120. Roberts H, Smith SC, Marzuki S, Linnane AW. 1980. Evidence that cytochrome b is the antimycin-binding component of the yeast mitochondrial cytochrome bc_1 complex. *Arch. Biochem. Biophys.* 200:387–95

121. Robinson SA, Ribas-Carbo M, Yakir D, Giles L, Reuveni Y, Berry JA. 1995. Beyond SHAM and cyanide: opportunities for studying the alternative oxidase in plant respiration using oxygen isotope discrimination. *Aust. J. Plant Physiol.* 22:487–96

122. Robinson SA, Yakir D, Ribas-Carbo M, Giles L, Osmond B, et al. 1992. Measurements of the engagement of cyanide-resistant respiration in the crassulacean acid metabolism plant *Kalanchoe daigremontiana* with the use of on-line oxygen isotope discrimination. *Plant Physiol.* 100: 1087–91

123. Rustin P, Queiroz-Claret C. 1985. Changes in oxidative properties of *Kalanchoe blossfeldiana* leaf mitochondria during development of crassulacean acid metabolism. *Planta* 164:415–22

124. Sakajo S, Minagawa N, Komiyama T, Yoshimoto A. 1991. Molecular cloning of cDNA for antimycin A-inducible mRNA and its role in cyanide-resistant respiration in *Hansenula anomala. Biochim. Biophys. Acta* 1090:102–8

125. Scandalios JG. 1990. Response of plant antioxidant defence genes to environmental stress. *Adv. Genet.* 28:1–41

126. Scarpulla RC. 1996. Nuclear respiratory factors and the pathways of nuclear-mitochondrial interaction. *Trends Cardiovasc. Med.* 6:39–45

127. Scheibe R. 1991. Redox-modulation of chloroplast enzymes. A common principle for individual control. *Plant Physiol.* 96: 1–3

128. Schonbaum GR, Bonner W, Storey BT,

Bahr JT. 1971. Specific inhibition of cyanide-insensitive respiratory pathway in plant mitochondria by hydroxamic acids. *Plant Physiol.* 47:124–28

129. Schreck R, Rieber P, Baeuerle PA. 1991. Reactive oxygen intermediates as apparently widely used messengers in the activation of the NF-B transcription factor and HIV-1. *EMBO J.* 10:2247–58

130. Sen CK, Packer L. 1996. Antioxidant and redox regulation of gene-transcription. *FASEB J.* 10:709–20

131. Sheen J. 1994. Feedback control of gene expression. *Photosynth. Res.* 39:427–38

132. Siedow JN, Bickett DM. 1981. Structural features required for inhibition of cyanide-insensitive electron transfer by propyl gallate. *Arch. Biochem. Biophys.* 207:32–39

133. Siedow JN, Moore AL. 1993. A kinetic model for the regulation of electron transfer through the cyanide-resistant pathway in plant mitochondria. *Biochim. Biophys. Acta* 1142:165–74

134. Siedow JN, Whelan J, Kearns A, Wiskich JT, Day DA. 1992. Topology of the alternative oxidase in soybean mitochondria. See Ref. 66a, pp. 19–27

135. Siedow JN, Umbach AL. 1995. Plant mitochondrial electron transfer and molecular biology. *Plant Cell* 7:821–31

136. Siedow JN, Umbach AL, Moore AL. 1995. The active site of the cyanide resistant oxidase from plant mitochondria contains a binuclear iron center. *FEBS Lett.* 362:10–14

137. Skubatz H, Nelson TA, Meeuse BJD, Bendich AJ. 1991. Heat production in the voodoo lily (*Sauromatum guttatum*) as monitored by infrared thermography. *Plant Physiol.* 95:1084–88

138. Slayman CW. 1977. The functin of an alternative terminal oxidase in *Neurospora*. In *Functions of Alternative Terminal Oxidase*, ed. H Degn, D Lloyd, GC Hill, pp. 159–68. Oxford: Pergamon

139. Stewart CR, Martin BA, Reding L, Cerwick S. 1990. Seedling growth, mitochondrial characteristics, and alternative respiratory capacity of corn genotypes differing in cold tolerance. *Plant Physiol.* 92: 761–66

140. Storey BT, Bahr JT. 1969. The respiratory chain of plant mitochondria. II. Oxidative phosphorylation in skunk cabbage mitochondria. *Plant Physiol.* 44:126–34

141. Trumpower BL, Gennis RB. 1994. Energy transduction by cytochrome complexes in mitochondrial and bacterial respiration: the enzymology of coupling electron transfer reactions to transmembrane proton translocation. *Annu. Rev. Biochem.* 63:675–716

142. Turner JF, Turner DH. 1980. The regulation of glycolysis and the pentose phosphate pathway. In *The Biochemistry of Plants, A Comprehensive Treatise. Metabolism and Respiration*, ed. DD Davies, 2:279–315. New York: Academic

143. Umbach AL, Siedow JN. 1993. Covalent and noncovalent dimers of the cyanide-resistant alternative oxidase protein in higher plant mitochondria and their relationship to enzyme activity. *Plant Physiol.* 103:845–54

144. Umbach AL, Wiskich JT, Siedow JN. 1994. Regulation of alternative oxidase kinetics by pyruvate and intermolecular disulfide bond redox status in soybean seedling mitochondria. *FEBS Lett.* 348:181–84

145. van den Bergen CWM, Wagner AM, Krab K, Moore AL. 1994. The relationship between electron flux and the redox poise of the quinone pool in plant mitochondria: interplay between quinol-oxidizing and quinone-reducing pathways. *Eur. J. Biochem.* 226:1071–78

146. Van Der Straeten D, Chaerle L, Sharkov G, Lambers H, Van Montagu M. 1995. Salicylic acid enhances the activity of the alternative pathway of respiration in tobacco leaves and induces thermogenicity. *Planta* 196:412–19

147. Vanlerberghe GC, Day DA, Wiskich JT, Vanlerberghe AE, McIntosh L. 1995. Alternative oxidase activity in tobacco leaf mitochondria: dependence on tricarboxylic acid cycle-mediated redox regulation and pyruvate activation. *Plant Physiol.* 109: 353–61

148. Vanlerberghe GC, McIntosh L. 1992. Lower growth temperature increases alternative pathway capacity and alternative oxidase protein in tobacco. *Plant Physiol.* 100:115–19

149. Vanlerberghe GC, McIntosh L. 1992. Coordinate regulation of cytochrome and alternative pathway respiration in tobacco. *Plant Physiol.* 100:1846–51

150. Vanlerberghe GC, McIntosh L. 1994. Mitochondrial electron transport regulation of nuclear gene expression: studies with the alternative oxidase gene of tobacco. *Plant Physiol.* 105:867–74

151. Vanlerberghe GC, McIntosh L. 1996. Signals regulating the expression of the nuclear gene encoding alternative oxidase of plant mitochondria. *Plant Physiol.* 111: 589–95

152. Vanlerberghe GC, Vanlerberghe AE, McIntosh L. 1994. Molecular genetic alteration of plant respiration: silencing and overexpression of alternative oxidase in transgenic tobacco. *Plant Physiol.* 106:1503–10

153. Vanlerberghe GC, Vanlerberghe AE, McIntosh L. 1996. Molecular genetic evidence of the ability of alternative oxidase to support respiratory carbon metabolism. *Plant Physiol.*

154. Virbasius JV, Scarpulla RC. 1994. Activation of the human mitochondrial transcription factor A gene by nuclear respiratory factors: a potential link between nuclear and mitochondrial gene expression in organelle biogenesis. *Proc. Natl. Acad. Sci. USA* 91:1309–13

155. Wagner AM. 1995. A role for active oxygen species as second messengers in the induction of alternative oxidase gene expression in *Petunia hybrida* cells. *FEBS Lett.* 368:339–42

156. Wagner AM, Kraak MHS, van Emmerik WAM, van der Plas LHW. 1989. Respiration of plant mitochondria with various substrates: alternative pathway with NADH and TCA cycle derived substrates. *Plant Physiol. Biochem.* 27:837–45

157. Wagner AM, Krab K. 1995. The alternative respiration pathway in plants: role and regulation. *Physiol. Plant.* 95:318–25

158. Wagner AM, van den Bergen CWM, Wincencjusz H. 1995. Stimulation of the alternative pathway by succinate and malate. *Plant Physiol.* 108:1035–42

159. Wagner AM, Van Emmerik WAM, Zwiers JH, Kaagman HMCM. 1992. Energy metabolism of *Petunia hybrida* cell suspensions growing in the presence of antimycin A. See Ref. 66a, pp. 609–14

160. Warburg O. 1919. Uber die Geschwindigkeit der photochemischen Kohlensaeurezersetzung in lebenden Zellen. *Biochem. Z.* 100:230–70

161. Weger HG, Dasgupta R. 1993. Regulation of alternative pathway respiration in

Chlamydomonas reinhardtii (CHLOROPHYCEAE). *J. Phycol.* 29:300–8

162. Weger HG, Guy RD, Turpin DH. 1990. Cytochrome and alternative pathway respiration in green algae. *Plant Physiol.* 93:356–60

163. Whelan J, Hugosson M, Glaser E, Day DA. 1995. Studies on the import and processing of the alternative oxidase precursor by isolated soybean mitochondria. *Plant Mol. Biol.* 27:769–78

164. Whelan J, McIntosh L, Day DA. 1993. Sequencing of a soybean alternative oxidase cDNA clone. *Plant Physiol.* 102:1481

165. Whelan J, Millar AH, Day DA. 1996. The alternative oxidase is encoded in a multigene family in soybean. *Planta* 198:197–201

166. Whelan J, Smith MK, Meijer M, Yu JW, Badger MR, et al. 1995. Cloning of an additional cDNA for the alternative oxidase in tobacco. *Plant Physiol.* 107:1469–70

167. Wilson SB. 1988. The switching of electron flux from the cyanide-insensitive oxidase to the cytochrome pathway in mung-bean (*Phaseolus aureus* L.) mitochondria. *Biochem. J.* 249:301–3

168. Wiseman A, Gillham NW, Boynton JE. 1977. Nuclear mutations affecting mitochondrial structure and function in *Chlamydomonas. J. Cell Biol.* 73:56–77

169. Yoshimoto A, Sakajo S, Minagawa N, Komiyama T. 1989. Possible role of a 36 kDa protein induced by respiratory inhibitors in cyanide-resistant respiration in *Hansenula anomala. J. Biochem.* 105:864–66

170. Zhang Q, Hoefnagel MHN, Wiskich JT. 1996. Alternative oxidase from *Arum* and soybean: its stabilization during purification. *Physiol. Plant.* 96:551–58

AUTHOR INDEX

SUBJECT INDEX

A

A23187 calcium ionophore
oxidative burst in disease resistance and, 264
AAL-toxin
oxidative burst in disease resistance and, 264
programmed cell death and,
528–29
Abaxial surfaces
trichome development in
Arabidopsis and, 141
Abcission
ethylene response pathway in
Arabidopsis and, 278
pollination regulation and,
547
Abiotic stress
jasmonates and, 369–71
Abscisic acid
ethylene response pathway in
Arabidopsis and, 280
fluorescent microscopy of living plant cells and, 169–70,
184
ACC oxidase
gibberellin biosynthesis and,
445
pollination regulation and,
564–65
ACC synthase
pollination regulation and,
562–63
ACE1 activator
chemical control of gene induction and, 101
Acetosyringone
plant transformation and,
309
Acetyl-CoA
auxin biosynthesis and, 60
Acetyl-CoA carboxylase
fatty acid synthesis regulation and, 109, 112–21, 124,
128, 130–31
Acidosis
cytoplasmic
and oxygen deficiency and
root metabolism, 223
Acl1.2 gene
fatty acid synthesis regulation and, 128
Acorus calamus
and oxygen deficiency and
root metabolism, 226, 230,
238
Acrylodan
fluorescent microscopy of living plant cells and, 179

act1 gene
fatty acid synthesis regulation and, 112
Actin
pollen germination and, 461,
469–71
and transport of proteins and
nucleic acids through plasmodesmata, 29, 34
Actinomycin D
oxidative burst in disease resistance and, 260
Activation
chemical control of gene induction and, 101–4
Acyl chains
fatty acid synthesis regulation and, 110–11
Adaptation
cyanobacterial circadian
rhythms and, 347–48
Adenosine triphosphate (ATP)
alternative oxidase and, 703,
724–25
auxin biosynthesis and, 60
and oxygen deficiency and
root metabolism, 223, 225,
227–29, 233, 235, 239–41
pollen germination and, 470
symbiotic membranes from
legume nodules and,
509–11
and transport of proteins and
nucleic acids through plasmodesmata, 29, 31, 40
Adh1 gene
and oxygen deficiency and root
metabolism, 228, 237–38,
240
ADP-glucose
starch granule synthesis and,
67–71, 82–83
ADP-glucose pyrophosphorylase
phloem unloading and, 207
AE gene
starch granule synthesis and, 78
Aequoria victoria
fluorescent microscopy of living plant cells and, 180–82
Aequorin
fluorescent microscopy of living plant cells and, 165,
179–81, 185
Aerenchyma
and oxygen deficiency and
root metabolism, 223,
241–43
Aerobic fermentation
alternative oxidase and, 725

African green monkey kidney
cells
programmed cell death and,
528
AGPase
starch granule synthesis and,
68–72, 83
AGP genes
starch granule synthesis and,
69, 71
Agriculture
and CO_2 and more efficient
plants, 609–10, 614, 628
phloem unloading and, 194
Agrobacterium rhizogenes
chemical control of gene induction and, 99
Agrobacterium sp.
angiosperm shoot apical
meristem development and,
690
chemical control of gene induction and, 98
plant transformation and,
299, 302, 304–6, 308–9,
311–13, 317–18
and transport of proteins and
nucleic acids through plasmodesmata, 41
Agrobacterium tumefaciens
auxin biosynthesis and, 57
Agrostemma githago
gibberellin biosynthesis and,
450
Alanine
alternative oxidase and, 711,
722
aquaporins and, 408
auxin biosynthesis and,
61–62
and transport of proteins and
nucleic acids through plasmodesmata, 41
Aldehydes
cyanobacterial circadian
rhythms and, 335–36
Ald gene
and oxygen deficiency and
root metabolism, 238
Aleurone
fluorescent microscopy of
living plant cells and, 169,
175–76, 185
alf3 mutant
auxin biosynthesis and, 60
Alfalfa
chemical control of gene induction and, 94
oxidative burst in disease resistance and, 266

CUMULATIVE INDEXES

CONTRIBUTING AUTHORS, VOLUMES 38–48

CHAPTER TITLES, VOLUMES 38–48

ACCLIMATION AND ADAPTATION

Annual Reviews
THE INTELLIGENT SYNTHESIS OF SCIENTIFIC LITERATURE

ANNUAL REVIEW OF:

	Individuals U.S.	Individuals Other countries	Institutions U.S.	Institutions Other countries
ANTHROPOLOGY				
• Vol. 26 (avail. Oct. 1997)	$55	$60	$110	$120
• Vol. 25 (1996)	$49	$54	$49	$54
ASTRONOMY & ASTROPHYSICS				
• Vol. 35 (avail. Sept. 1997)	$70	$75	$140	$150
• Vol. 34 (1996)	$65	$70	$65	$70
BIOCHEMISTRY				
• Vol. 66 (avail. July 1997)	$68	$74	$136	$148
• Vol. 65 (1996)	$59	$65	$59	$65
BIOPHYSICS & BIOMOLECULAR STRUCTURE				
• Vol. 26 (avail. June 1997)	$70	$75	$140	$150
• Vol. 25 (1996)	$67	$72	$67	$72
CELL & DEVELOPMENTAL BIOLOGY				
• Vol. 13, 1997 (avail. Nov. 1997)	$64	$69	$128	$138
• Vol. 12 (1996)	$56	$61	$56	$61
COMPUTER SCIENCE				
• Vols. 3-4 (1988-1989/90) (suspended)	$47	$52	$47	$52
• Vols. 1-2 (1986-1987)	$41	$46	$41	$46
• Vols. 1-4 Price for all four, ordered together.	$100	$115	$100	$115
EARTH & PLANETARY SCIENCES				
• Vol. 25 (avail. May 1997)	$70	$75	$140	$150
• Vol. 24 (1996)	$67	$72	$67	$72
ECOLOGY & SYSTEMATICS				
• Vol. 28 (avail. Nov. 1997)	$60	$65	$120	$130
• Vol. 27 (1996)	$52	$57	$52	$57
ENERGY & THE ENVIRONMENT				
• Vol. 22 (avail. Oct. 1997)	$76	$81	$152	$162
• Vol. 21 (1996)	$76	$81	$76	$81
ENTOMOLOGY				
• Vol. 42 (avail. Jan. 1997)	$60	$65	$120	$130
• Vol. 41 (1996)	$52	$57	$52	$57

ANNUAL REVIEW OF:

	Individuals U.S.	Individuals Other countries	Institutions U.S.	Institutions Other countries
FLUID MECHANICS				
• Vol. 29 (avail. Jan. 1997)	$60	$65	$120	$130
• Vol. 28 (1996)	$52	$57	$52	$57
GENETICS				
• Vol. 31 (avail. Dec. 1997)	$60	$65	$120	$130
• Vol. 30 (1996)	$52	$57	$52	$57
IMMUNOLOGY				
• Vol. 15 (avail. April 1997)	$64	$69	$128	$138
• Vol. 14 (1996)	$56	$61	$56	$61
MATERIALS SCIENCE				
• Vol. 27 (avail. Aug. 1997)	$80	$85	$160	$170
• Vol. 26 (1996)	$80	$85	$80	$85
MEDICINE				
• Vol. 48 (avail. Feb. 1997)	$60	$65	$120	$130
• Vol. 47 (1996)	$52	$57	$52	$57
MICROBIOLOGY				
• Vol. 51 (avail. Oct. 1997)	$60	$65	$120	$130
• Vol. 50 (1996)	$53	$58	$53	$58
NEUROSCIENCE				
• Vol. 20 (avail. March 1997)	$60	$65	$120	$130
• Vol. 19 (1996)	$52	$57	$52	$57
NUCLEAR & PARTICLE SCIENCE				
• Vol. 47 (avail. Dec. 1997)	$70	$75	$140	$150
• Vol. 46 (1996)	$67	$72	$67	$72
NUTRITION				
• Vol. 17 (avail. July 1997)	$60	$65	$120	$130
• Vol. 16 (1996)	$53	$58	$53	$58
PHARMACOLOGY & TOXICOLOGY				
• Vol. 37 (avail. April 1997)	$60	$65	$120	$130
• Vol. 36 (1996)	$52	$57	$52	$57
PHYSICAL CHEMISTRY				
• Vol. 48 (avail. Oct. 1997)	$64	$69	$128	$138
• Vol. 47 (1996)	$56	$61	$56	$61

ANNUAL REVIEW OF:

	Individuals U.S.	Individuals Other countries	Institutions U.S.	Institutions Other countries
PHYSIOLOGY				
• Vol. 59 (avail. March 1997)	$62	$67	$124	$134
• Vol. 58 (1996)	$54	$59	$54	$59
PHYTOPATHOLOGY				
• Vol. 35 (avail. Sept. 1997)	$62	$67	$124	$134
• Vol. 34 (1996)	$54	$59	$54	$59
• Vol. 33 (1995) and 10 Year CD-ROM Archive (volumes 24-33)	$49	$54	$49	$54
• 10 Year CD-ROM Archive only	$40	$45	$40	$45
PLANT PHYSIOLOGY & PLANT MOLECULAR BIOLOGY				
• Vol. 48 (avail. June 1997)	$60	$65	$120	$130
• Vol. 47 (1996)	$52	$57	$52	$57
PSYCHOLOGY				
• Vol. 48 (avail. Feb. 1997)	$55	$60	$110	$120
• Vol. 47 (1996)	$48	$53	$48	$53
PUBLIC HEALTH				
• Vol. 18 (avail. May 1997)	$64	$69	$128	$138
• Vol. 17 (1996)	$57	$62	$57	$62
SOCIOLOGY				
• Vol. 23 (avail. Aug. 1997)	$60	$65	$120	$130
• Vol. 22 (1996)	$54	$59	$54	$59

BACK VOLUMES ARE AVAILABLE
Visit www.annurev.org for a list and prices

The Excitement & Fascination Of Science

	Individuals U.S.	Individuals Other countries	Institutions U.S.	Institutions Other countries
• Vol. 4, 1995	$50	$55	$50	$55
• Vol. 3 (1990) 2-part set, sold as a set only	$90	$95	$90	$95
• Vol. 2 (1978)	$25	$29	$25	$29
• Vol. 1 (1965)	$25	$29	$25	$29
Intelligence And Affectivity by Jean Piaget (1981)	$8	$9	$8	$9

ANNUAL REVIEWS INDEX on Diskette (updated quarterly) DOS format only. Prices are the same to all locations.	single copy $15	1 yr. (4 els) $50	single copy $15	1 yr. (4 els) $50

Annual Reviews

BB97

A nonprofit scientific publisher • P.O. Box 10139
4139 El Camino Way •
Palo Alto, CA 94303-0139 USA

STEP 1 : ENTER YOUR NAME & ADDRESS

NAME

ADDRESS

CITY   STATE/PROVINCE   COUNTRY   POSTAL CODE

TODAY'S DATE   DAYTIME PHONE

E-MAIL ADDRESS   FAX NUMBER

Phone 800-523-8635 *(U.S. or Canada)*
Orders 415-493-4400 ext. 1 *(worldwide)*
8 a.m. to 4 p.m. Pacific Time, Monday–Friday

Mention priority code **BB97** when placing phone orders

FAX 415-424-0910
Orders 24 hours a day

STEP 4 : CHOOSE YOUR PAYMENT METHOD

☐ Check or Money Order Enclosed (US funds, made payable to "Annual Reviews")
☐ Bill Credit Card ☐ AmEx ☐ MasterCard ☐ VISA

Account No. _____

Signature _____

Exp. Date MO/YR Name _____
                   (print name exactly as it appears on credit card)

STEP 2 : ENTER YOUR ORDER

QTY	ANNUAL REVIEW OF:	VOL.	Place on Standing Order? SAVE 10% NOW WITH PAYMENT	PRICE	TOTAL
		#	☐ Yes, save 10% ☐ No	$	$
		#	☐ Yes, save 10% ☐ No	$	$
		#	☐ Yes, save 10% ☐ No	$	$
		#	☐ Yes, save 10% ☐ No	$	$
		#	☐ Yes, save 10% ☐ No	$	$

30% STUDENT/RECENT GRADUATE DISCOUNT (past 3 years) *Not for standing orders. Include proof of status.*

CALIFORNIA CUSTOMERS: Add applicable California sales tax for your location. $

CANADIAN CUSTOMERS: Add 7% GST (Registration # 121449029 RT). $

STEP 3 : CALCULATE YOUR SHIPPING & HANDLING

HANDLING CHARGE (Add $3 per volume, up to $9 max.). Applies to all orders. $

SHIPPING OPTIONS:
(No UPS to P.O. boxes)

U.S. Mail 4th Class Book Rate (surface). Standard option. FREE.
UPS Ground Service ($3/ volume. 48 contiguous U.S. states.)

Please note expedited shipping preference:
☐ UPS Next Day Air ☐ UPS Second Day Air ☐ US Airmail
☐ UPS Worldwide Express ☐ UPS Worldwide Expedited

Note option at left. We will calculate amount and add to your total

Abstracts and content lists available on the World Wide Web at www.annurev.org. **E-mail orders welcome: service@annurev.org**

TOTAL $ _____

Orders may also be placed through booksellers or subscription agents or through our Authorized Stockists.
From Europe, the UK, the Middle East and Africa, contact: **Gazelle Book Service Ltd.** Fax (0) 1524-63232.
From India, Pakistan, Bangladesh or Sri Lanka, contact: **SARAS Books,** Fax 91-11-941111.